Analysis and design of geotechnical structures

Analysis and design of geotechnical structures

Manuel Matos Fernandes

CRC Press
Taylor & Francis Group
Boca Raton London New York

CRC Press is an imprint of the
Taylor & Francis Group, an **informa** business

First edition published 2020
by CRC Press
2 Park Square, Milton Park, Abingdon, Oxon, OX14 4RN

and by CRC Press
6000 Broken Sound Parkway NW, Suite 300, Boca Raton, FL 33487-2742

© 2021 Taylor & Francis Group LLC

CRC Press is an imprint of Taylor & Francis Group, LLC

No claim to original U.S. Government works

Printed on acid-free paper

ISBN: 978-0-367-02662-2 (hbk)
ISBN: 978-0-367-02663-9 (pbk)
ISBN: 978-0-429-39845-2 (ebk)

Library of Congress Cataloging-in-Publication Data

Names: Fernandes, Manuel Matos, author.
Title: Analysis and design of geotechnical structures / Manuel Matos Fernandes.
Description: First edition. | Abingdon, Oxon ; Boca Raton, FL : CRC Press, 2020. | Includes bibliographical references and index.
Identifiers: LCCN 2020009402 (print) | LCCN 2020009403 (ebook) | ISBN 9780367026639 (pbk) | ISBN 9780367026622 (hbk) | ISBN 9780429398452 (ebk) | ISBN 9780429676161 (adobe pdf) | ISBN 9780429676147 (mobi) | ISBN 9780429676154 (epub)
Subjects: LCSH: Geotechnical engineering--Textbooks.
Classification: LCC TA705 .F428 2020 (print) | LCC TA705 (ebook) | DDC 624.1/51--dc23
LC record available at https://lccn.loc.gov/2020009402
LC ebook record available at https://lccn.loc.gov/2020009403

Typeset in Sabon
by Deanta Global Publishing Services, Chennai, India

Contents

3 Basis of geotechnical design   123

The leaning bell-tower of Pisa 287

6 Shallow foundations 295

7 Pile foundations

375

8 Earth-retaining structures   435

10 Introduction to earthworks: Soil compaction   551

Preface

This book deals with the theories and methodologies used by engineers for the analysis and design of the main geotechnical structures.

The word *structures* is used here with a clear intention: indeed, the book adopts a structural outlook, inspired by the approach of the Structural Eurocodes and similar documents based on the Limit State Design philosophy.

The assumption is made that the reader is familiar with the basic principles of soil mechanics. Out of respect for this admirable discipline, the book does not attempt a summary of these issues, as it is considered that a meaningful and didactic approach to them would be incompatible with such an option.

In writing this book – capitalizing, to the best of my ability, on the experience of thirty five years as a university teacher and geotechnical consultant – I had in mind a two-pronged objective: to combine, in a single endeavor, a textbook that helps students understand the behavior of geotechnical works, with a reference text that has its place "on the working desk" of geotechnical professionals, for regular perusal.

As regards the pedagogical component, the reader will find innovative approaches to various issues, resulting from many years of teaching and discussing their main difficulties and subtleties with generations of enthusiastic students.

As far as the professional component is concerned, the reader will find an in-depth and up-to-date approach to the issues covered in each chapter, supplemented by a plethora of informative material used daily by geotechnical designers.

For the benefit of both students and professionals, the book includes exercises, at the end of most chapters, either of a conceptual nature or exploring practical problems of geotechnical design.

When writing this book I benefited from the inestimable help of many colleagues to whom I would like to pay tribute, namely: Cristiana Ferreira, Mafalda Laranjo, Sara Rios, Teresa Bodas Freitas, Alexandre Pinto, Eduardo Fortunato, Jorge Almeida e Sousa, José Couto Marques, José Freitas, José Grazina, José Menezes, Nuno Guerra, and Paulo da Venda Oliveira.

My special thanks go to Paulo Pinto and Pedro Alves Costa, who co-authored Chapter 7, on Pile Foundations.

I am greatly indebted also to the expertise and endless patience of Manuel Carvalho who prepared the figures for the book.

I thank all my colleagues who, throughout the years, were part of the geotechnical staff of my university, for contributing to a stimulating work environment and for their enthusiasm in both teaching and research. The majority of the application exercises included in the book were developed with their help.

Finally, I would like to thank Tony Moore, of Taylor & Francis, for challenging me to write this book.

Porto, December 2019

xix

Author

Manuel Matos Fernandes is Professor of Soil Mechanics and Geotechnical Engineering at the University of Porto, Portugal. He also serves as president of the Portuguese Geotechnical Society, and, for fifteen years, he was a member of the committee responsible for Eurocode 7 – Geotechnical Design.

Chapter 1

Geotechnical characterization

In situ testing

1.1 INTRODUCTION

Geotechnical structures are the constructions whose conception and design are essentially controlled by the mechanical and hydraulic behavior of the soil or rock masses present at the site. Examples of geotechnical structures include foundations, earth-retaining structures, embankments, cuts and slopes, and underground works, to name but a few.

This chapter is dedicated to the operations carried out at a construction site for its geotechnical characterization. This usually involves:

(i) the identification, in geological and geotechnical terms, of the sequence of layers or strata that may affect the behavior of the structure;
(ii) the physical, mechanical, and hydraulic characterization of the soils that form these layers or strata;
(iii) the characterization of water conditions in the ground.

The physical, mechanical, and hydraulic characterization of soils requires laboratory and *in situ* tests. Laboratory tests have been studied in soil mechanics, such as triaxial tests, direct and simple shear tests, confined and isotropic compression tests, and other more sophisticated tests, namely the hollow cylinder and the resonant column tests.

The importance assigned to laboratory tests in theoretical soil mechanics is understandable, since these address well-defined and controllable stress, strain, and drainage conditions. These conditions enable the interpretation of test results and, using the principles of soil mechanics, the derivation of soil parameters.

This practice has been of critical importance for the development of soil mechanics and it continues to be invaluable for the design of geotechnical structures. However, when dealing with conventional types of structures and foundations, with no exceptional soil or loading conditions, the design is essentially based on field tests.[1] This explains the importance assigned to *in situ* tests in this introductory chapter of the book, after some brief considerations on the preliminary operations for site characterization.

The final part of the chapter is devoted to soil stiffness characterization. This is one of the most challenging topics in geotechnical engineering, one that has seen outstanding progress in recent decades, and treatment of which requires a special combination of laboratory and *in situ* tests.

Readers particularly interested in studying the theme of this chapter are advised to study the "Manual on Subsurface Investigations" (Mayne et al., 2001) published by the National

---

[1] A discussion on the advantages and limitations of laboratory and *in situ* tests is presented in Section 1.4.2.

Highway Institute of the United States of America, available online (http://geosystems.ce .gatech.edu/Faculty/Mayne/papers/NHI%202002%20Subsurface%20Investigations.pdf).

1.2 GEOTECHNICAL INVESTIGATION

1.2.1 Preliminary surface survey

The geotechnical investigation is preceded by the so-called preliminary geological-geotechnical survey. This consists of an on-site superficial survey, sometimes involving small investigation activities, such as the opening of shallow pits and shafts. Usually, the field visit is preceded by a desk study comprising the collection of existing written and drawn information about the site, in particular, topographic and geological charts and, if available, geotechnical charts. Currently, the use of two- and three-dimensional aerial images obtained from satellite and free web navigation tools is also very useful, particularly for large works outside urban areas, such as roads, dams, and for the stabilization of natural slopes. When dealing with densely populated areas, geotechnical characterization reports for nearby constructions can generally be found, which represent important sources to be collected and checked.

Data analysis of all this information, as well as the preliminary site survey data, is compiled in a report. This will form the basis of the preliminary design or viability studies stage of the construction works. This report will also allow definition of the most appropriate geotechnical investigation program for the project. In general, the geotechnical investigation comprises geophysical and geomechanical investigations.

1.2.2 Geophysical investigation

1.2.2.1 Introduction

In the past few decades, the application of geophysical methods to geotechnical engineering has expanded considerably. This has resulted from a combination of several factors:

(i) the advent of "new" methods, such as the surface seismic wave method;
(ii) the combination of conventional geomechanical *in situ* tests with geophysical methods (such as the seismic cone penetration test and the seismic flat dilatometer test);
(iii) the comparative application of geophysical methods in the field and in the laboratory on samples under well-defined test conditions;
(iv) the strong advances in the acquisition, treatment, and interpretation of results, a critical aspect in this type of investigation.

Table 1.1 presents a tentative classification of geophysical methods applied in geotechnical engineering. The non-intrusive methods are covered in this section. The intrusive field methods will be presented together with the *in situ* tests. Finally, the laboratory methods are mentioned in the last part of the chapter where soil stiffness characterization is discussed.

Traditionally, geophysical investigation is applied to large-scale projects, such as roads and railway works, dams, and other extensive facilities, using non-intrusive methods, and typically preceding the geomechanical investigation. In general, the results of geophysical investigation influence the final program of the geomechanical investigation, by drawing attention to the more delicate and doubtful issues, ground zones, and strata.

The non-intrusive methods listed in Table 1.1 may be considered those most frequently employed in geophysical investigations. As is often the case with technologies used for a particular purpose, the advantageous aspects of these methods are closely related to their

Table 1.1 Main geophysical methods employed in geotechnical engineering.

Location	Type	Designation	Physical phenomenon	Measured parameters	Issues under evaluation
Field	Non-intrusive methods	Electrical resistivity	Electrical field	ρ (electrical resistivity)	See Table 1.2
		Seismic refraction	Propagation of P waves	V_p	
		Spectral analysis of surface waves (SASW)	Propagation of R (Rayleigh) surface waves and S waves	V_s	
		Ground-penetrating radar	Propagation of EM waves	V	
	Intrusive methods – *In situ* tests	Cross-hole seismic test	Propagation of P and S waves	V_p, V_s	Soil stiffness Poisson's ratio
		Down-hole seismic test	Propagation of P and S waves	V_p, V_s	Porosity CRR (cyclic liquefaction resistance)
		CPTu seismic test	Propagation of P and S waves	V_p, V_s	
		DMT seismic test	Propagation of P and S waves	V_p, V_s	
Laboratory	Non-intrusive methods	Bender elements Ultrasonic transducers	Propagation of P and S waves	V_p, V_s	Soil stiffness Sample quality Process monitoring

limitations. On the one hand, they make it possible to study, relatively quickly and economically, large volumes of the ground, without being intrusive. On the other hand, since they involve no sampling or borehole drilling, interpretation of their results under certain conditions may prove difficult or even inconclusive. For this reason, the execution of some borings in parallel, even in the context of a preliminary investigation, is frequently undertaken.

Examples of purposes of the application of non-intrusive geophysical methods include:

(i) the evaluation of the depth of the bedrock;
(ii) the evaluation of the depth of the water level;
(iii) the assessment of the excavatability and rippability of rock masses to predict the excavation cost, for example, in a road project;
(iv) the study of borrow pits or quarry sites for dams (earth, rockfill, and concrete), ports, etc.;
(v) the identification of underground cavities when building on karstic formations.

1.2.2.2 Electrical resistivity method

The resistivity method is one of the most widely used among the non-intrusive geophysical methods (González de Vallejo and Ferrer, 2011). The technique determines the apparent electrical resistivity of the ground. This property in a layered ground is variable in depth with the lithology and microstructure and, particularly, is very sensitive to the water content.

Figure 1.1 shows a conventional arrangement (the so-called Schlumberger configuration) with a pair (AB) of current electrodes and a pair (MN) of potential electrodes.

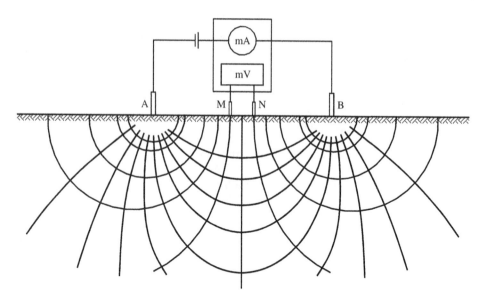

Figure 1.1 The electrical resistivity method.

By establishing an electrical field from AB, the measured potential in MN allows calculation of the resistivity of a given soil section encompassed by the electrical field.

There are two main ways of performing the investigations. In the first mode, which can be designated a *geoelectric sounding*, the electrodes are maintained on the same line but the current electrodes are progressively expanded with reference to the (fixed) central point. This induces an electrical field deeper and deeper in the ground, with a consequent variation of the measured resistivity. The result is interpreted as a vertical 1-D sounding below the central position.

In the second mode, which can be designated a *geoelectric profile*, the relative distance between the electrodes is fixed but the array is laterally moved. This enables the covering of a given rectangular area in plan, the subsurface of which is characterized in terms of resistivity to a given depth, thus providing 2-D and 3-D models of the investigated ground.

Traditionally, the interpretation of the results of the geoelectric soundings was achieved by matching with theoretical resistivity charts. At present, the interpretation of the results of any kind of geoelectric investigation is done by applying specific software (inversion programs).

1.2.2.3 Seismic refraction method

The seismic refraction method, shown in Figure 1.2a, is a well-established and traditional non-intrusive method (González de Vallejo and Ferrer, 2011). This method consists of the generation of elastic waves in a given point of the surface and the detection of these waves by receivers (vertically-sensitive geophones) placed along a linear array at the surface. The source of the waves may be a small explosion or a mechanical vertical impact of a hammer on a steel plate on the ground surface.

If soil 1, closer to the surface, is underlain by soil 2, with higher wave propagation velocity, some waves are refracted at the interface. It can be demonstrated that the so-called *critical angle of refraction*, i_c, which corresponds to refracted waves travelling along the interface with the velocity of the lower layer, is related to the ratio of the wave velocities by the following equation:

$$\frac{V_1}{V_2} = \sin i_c \tag{1.1}$$

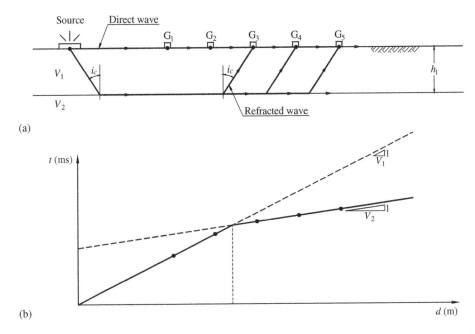

(a)

(b)

Figure 1.2 The seismic refraction method: a) basic layout; b) typical record and interpretation of results.

Figure 1.2b depicts a typical record of results, relating the distance, d, of each receiver from the source with the time interval between the generation of the elastic waves and the detection of the first wave arrival. (The first wave is a compression wave, the velocity of which is greater than of a shear wave). It can be seen that, beyond a given distance from the source, the refracted waves are received prior to direct waves, which travel at the surface with lower velocity. For direct waves, the arrival time is:

$$t_1 = \frac{d}{V_1} \tag{1.2}$$

Taking h_1 as the thickness of soil 1, the arrival time for a refracted wave is:

$$t_2 = \frac{2h_1}{\cos i_c V_1} + \frac{d}{V_2} - \frac{2h_1 \tan i_c}{V_2} \tag{1.3}$$

For the point of intersection of the two straight lines:

$$t_1 = t_2 \tag{1.4}$$

Then, combining Equations 1.2 and 1.3, and taking from the experimental graph (Figure 1.2b) the values of V_1, V_2 and d (the abscissa of the intersection point of the two lines), the thickness of the upper layer, h_1, is obtained.

Interpretation methods for more complex cases, such as the existence of more than two layers as well as non-horizontal interfaces, are available in the literature. The linear distance covered by the receivers at the surface should be no less than four times the depth of interest.

1.2.2.4 Spectral-analysis-of-surface-waves (SASW) method

The SASW method involves the generation of Rayleigh waves at a given point on the surface, and the record of the induced vertical motion at various distances by receivers placed along a linear array, as shown by Figure 1.3a (Stokoe and Santamarina, 2000; Stokoe et al., 2004).

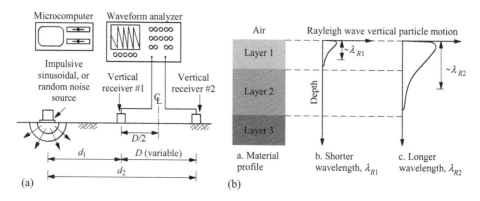

Figure 1.3 The SASW method (after Stokoe and Santamarina, 2000).

Rayleigh waves propagate in a layered ground with varying wavelengths and phase velocities, due to the variation of the shear stiffness from layer to layer. The greater the wavelength, the deeper the layer involved by the wave dispersion, as shown by Figure 1.3b.

The interpretation of these records, through numerical codes specially developed for this purpose, leads to a profile with the V_s distribution in depth. Therefore, this method can be considered a non-intrusive alternative to seismic field (intrusive) tests such as the cross-hole or down-hole tests.

1.2.2.5 Ground-penetrating radar (GPR)

The application of GPR is based on the fact that electromagnetic (EM) waves propagate in the ground at the velocity of light divided by the square root of the dielectric constant of the material:[2]

$$V = \frac{c}{k^{1/2}} \tag{1.5}$$

where c is the velocity of light and k is the dielectric constant. Different soil layers will have different values for this constant.

Basically, GPR at the surface transmits EM waves into the ground. When the waves meet an interface between materials, part of the energy is reflected towards the ground surface. The greater the difference between the values of the dielectric constant of the two materials, the greater the reflected energy.

In general, the transmitter and the receiver are associated in the same apparatus. By moving this apparatus at the surface along a straight line, and maintaining a high emission rate, a continuous record, as represented in Figure 1.4, is obtained. The interpretation of these results is performed with the help of specialized software.

Table 1.2 summarizes the main advantages and limitations of the non-intrusive methods described and their applicability.

Readers interested in learning more on this subject are advised to study the textbook *Environmental and Engineering Geophysics* (Sharma, 1997) and the manual *Seeing into the Earth* (NRC, 2000).

---

[2] The dielectric constant is a measure of the ability of a material to store electrical charge, like a battery (NCR, 2000).

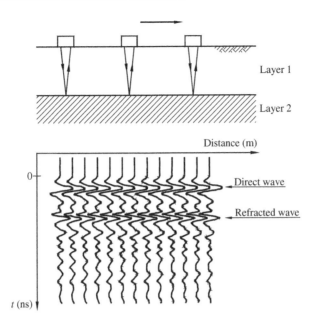

Figure 1.4 The ground-penetrating radar.

Table 1.2 Advantages and limitations of the non-intrusive methods listed in Table 1.1.

Method	Advantages	Limitations	Objectives and capabilities
Electrical resistivity	Inexpensive technique. Fairly straightforward interpretation.	Multiple measurements using different source–receiver layouts are required. Non-uniqueness of results.	Geological mapping Hydrogeology Water table depth Top of bedrock Cavity detection
Seismic refraction	Inexpensive technique. Fairly straightforward interpretation.	Relatively large source–receiver offset. Only works if V_p increases with depth.	Geological mapping Hydrogeology Water table depth Top of bedrock
SASW	Good characterization of shallow material. Accurate profile of V_s with depth.	Multiple measurements using different source–receiver layouts are required. Interpretation requires high-level expertise.	Geological mapping Profile of V_s with depth
GPR (ground-penetrating radar)	Portable equipment. Easy for non-expert to visualize information.	Very limited penetration in clay-rich environments. Dispersion and scattering higher than in other seismic tests.	Geological mapping Hydrogeology Cavity detection

1.2.3 Geomechanical investigation

1.2.3.1 Introduction

In relatively concentrated works, such as buildings, bridges or other structures, geotechnical investigation is mostly limited to mechanical ground investigation. In soil masses, this comprises:

    i) trial pits and shafts;

ii) borehole drilling; and
iii) rotary coring.

The following paragraphs include brief references to these types of mechanical investigations in order to cover the core of the chapter, related to *in situ* tests, many of which are carried out in combination or simultaneously with the aforementioned mechanical investigation operations.

1.2.3.2 Trial pits and shafts

Trial pits and shafts are investigation methods, which have considerable advantages:

i) the possibility of visual inspection of the face of the pit in depth;
ii) ease of collection of disturbed samples in large quantities, irrespective of soil type;
iii) the onset of ideal conditions for the collection of high-quality undisturbed samples, in particular large block samples, for subsequent trimming at the laboratory for preparation of testing specimens.

In addition, typical scenarios favorable to the use of these *in situ* examination operations are:

i) the study of the stability of natural slopes, as these allow observation of the sliding surface;
ii) the evaluation of the depth of the bedrock and its observation, beneath a soft and relatively thin soil layer.

Trial pits and shafts can be excavated rather quickly by machine excavation, if accessible, depending on the digging depth and machine width of the backhoe or excavator.

An important limiting aspect of these operations is the difficulty of working below the ground water level in soft soils. On the other hand, construction site safety codes require supported trenches in pits from relatively low depths, unless the slope faces are adequately inclined, depending on the soil strength. Under these circumstances, the cost of these investigation operations becomes relatively high, and it is more convenient to use boreholes.

1.2.3.3 Borehole drilling

Figure 1.5 illustrates the execution of a percussion drill and the equipment involved. Percussion drilling is used in all types of soils and on soft rocks. Drilling techniques are based on the technique of fatigue disintegration. The disintegrating effort is achieved by repeated impacts of a trephine or chisel (Figure 1.5b), with considerable mass, which, in successive maneuvers is first positioned above the ground surface and then lowered in free fall to the bottom of the hole. The material is removed through a tube, with a beveled shoe to facilitate penetration into the ground, called cutter or bailer (Figure 1.5c). When the cutter or bailer is lowered into the hole and inserted into the ground by means of its self-weight, a hinged lid (valve), located in the interior just above the shoe, rotates to a nearly vertical position, allowing the entrance of the disaggregated soil. When the cutter or bailer is removed, the valve resumes to the horizontal position, thus sealing the tube. In the case of soft soils, drilling can be carried out directly with the cutter or bailer without the need for pre-drilling.

The down and up maneuvers of the trephine and the cutter or bailer are generally performed with a steel cable, as in the case of Figure 1.5a, but hollow or solid steel rods can also be used. In order to facilitate soil extraction and to prevent hydraulic failure at the bottom

(b)

(a)

(c)

Figure 1.5 Percussion drilling: a) overview of the operation; b) trephines or chisels; c) cutter or bailer (photos: Nuno Cruz).

of the hole, in sandy soils below the ground water table, the hole is filled with water (or, in some cases, with bentonite mud) up to the surface of the ground. With the exception of stiff clayey soils and soft rocks, the hole is continuously cased with steel tubing, immediately following each drilling stage.

At any drilling stage, the operation may be interrupted for sampling or *in situ* testing. For this purpose, the bottom of the hole is cleaned, the sampler or testing probe is lowered with the help of the string of rods to the bottom of the hole and then sampling or testing takes place. As the name implies, the string of rods is composed of a set of several steel rods, each usually 1 m in length, which are successively screwed together, and it is employed to drive the sampler or testing probe through the borehole to depth.

The main advantages associated with percussion drilling are their relative simplicity and adaptability to current situations. In turn, the main drawbacks are the greater need for human resources and the time required for its execution. The diameters of the drilling holes range between 100 and 200 mm.

1.2.3.4 Rotary coring

Figure 1.6 shows equipment for performing rotary coring. Traditionally, this type of drilling was only performed in strong ground, involving very hard soils and rocks, in which percussion drilling is ineffective. However, with the improvement of the technology, in particular with the use of triple samplers, described below, such operations have also become relatively common in soft soils.

(a)

(b)

(c)

(d)

(e)

Figure 1.6 Rotary coring: a) overview of the operation; b) coring in progress; c) tungsten carbide bit; d) surface set diamond bit; e) diamond impregnated bit (photos: Nuno Cruz).

In addition to the drilling advance mode, discussed below, there is a very important aspect that distinguishes percussion drilling from rotary coring. In the former, the drilling operation is detached from the sampling operation, which is discontinuous, carried out at specific points (in depth) of the soil profile. The opposite happens in rotary coring.

Rotary coring is based on the friction caused by an abrasive tool called core bit, installed at the lower end of a sampler tube or corebarrel, which abrades away the surface to be drilled by rotation friction. As the tube penetrates the ground, because of the rotation of the corebit, a cylindrical sample (the core) is collected inside the tube.

The rotation is transmitted to the tube through the string of rods. Coring is performed by injection from the surface of a flush fluid (usually water), which flows down inside the rods and returns to the surface between the sampler and the walls of the borehole. Circulating water is employed to cool down the corebit and to evacuate coring debris to the surface.

The bits used for coring can be of three types, as shown in Figure 1.6:

(i)   hard metal bits, usually tungsten, suitable for materials with a hardness of less than 4 on the Mohs scale, or with uniaxial compression strength below 10 MPa;

(ii) surface set diamond bits, suitable for materials with a hardness of 3 to 6 on the Mohs scale, or with uniaxial compressive strength between 10 and 60 MPa;

(iii) diamond-impregnated bits, suitable for materials with a hardness of 4 to 8 on the Mohs scale, or with uniaxial compressive strength between 20 and 200 MPa.

Inside the bit, there is a catcher spring, which prevents the sample from dropping during the removal maneuvers of the equipment. The most common coring diameters are 76, 86, 101, and 116 mm.

Regardless of the corebit used, the design of rotary coring equipment consists of the following three types:

a) single tube, in which the rotation of the sampler tube is integral with the rotational movement of the bit; this system naturally causes friction between the collected sample and the inner wall of the sampler tube; in addition, the flush water is always in contact with the sample, which enhances its disturbance;

b) double tube, in which the inner tube holding the sample is disconnected from the rotational movement of the outer tube responsible for coring progress; this system allows a substantial reduction in the sample disturbance by canceling the friction with the sampler and by reducing the contact of the flush water with the sample;

c) triple tube, similar to the previous type but with a third inner tube (liner) where the sample is collected; in this system the contact of flush water with the sample is completely prevented, which can be of fundamental importance in soils and even in certain soluble or very fractured rock masses.

It should be noted that the sample length obtained is not necessarily equal to the drilling length. This length, as well as the quality of the sample, depends fundamentally on the quality of the ground and the type of sampler tube, although it also depends on aspects such as rotation speed, bit pressure, and flow rate of the flush fluid. In fact, in a soft soil mass or in a highly weathered and fractured rock mass, the use of a single pipe can cause disturbance to the sample by friction and by the contact with the flush water, leading to a core length much shorter than the drilling length. Significant material losses occur even with double tube samplers in very fractured and/or weathered rock masses.

In cases where double or triple tube barrels, with a coring diameter of 76 mm (3 inches) or greater, are used, it is common to calculate the so-called core recovery, generally expressed as a percentage, as the ratio of the recovered core length with the potential core length. Percentages close to 100% are typical of very good quality rock masses. In highly disaggregated core samples, the calculation of core recovery may involve significant subjectivity, hence, the preference for the so-called *RQD* (rock quality designation index), which corresponds to the core recovery but only takes into account pieces of core more than 10 cm long.

1.2.4 Note on undisturbed soil sampling

1.2.4.1 Introduction

In order to ensure that mechanical and hydraulic characterization, based on laboratory tests, provides results that lead to reliable predictions of soil behavior, these tests must be performed on undisturbed samples. Ideally, an undisturbed sample will differ from the soil mass from which it was collected only in terms of stress state.

In the strict sense of the term, it is not possible to obtain undisturbed samples, since some disturbance is always inevitable, either in the removal from the soil mass or in the whole sequence of operations that precedes the laboratory test itself. As will be discussed later, samples can be objectively classified with regard to their quality, by comparing the shear wave velocity, V_s, in the samples with those measured *in situ* on the soil mass from which these were collected (Ferreira et al., 2011). In-depth discussion on undisturbed soil sampling exceeds the scope of this book. However, some considerations are deemed necessary and appropriate in the context of this chapter.

1.2.4.2 Direct sampling

As shown in Figure 1.7a, when there is direct access to the soil to be sampled, by means of trial pits or trenches, it is possible to cut soil blocks of relatively large dimensions, which are then carefully wrapped and protected in order to avoid vibrations, shocks, and loss of moisture and taken to the laboratory. The specimens are then extracted from the blocks by trimming in the laboratory using particularly careful techniques (Figure 1.7b).

This technique enables to obtain very high-quality samples in a wide range of materials, such as residual soils of granite and overconsolidated clays. However, this sampling process is limited to relatively small depths.

1.2.4.3 Indirect sampling

In the vast majority of cases, sampling is performed through a borehole, as shown schematically in Figure 1.8, requiring the use of samplers.

Baligh et al. (1987) listed the main causes of soil disturbance during tube sampling, namely: changes in soil conditions ahead of the advancing borehole during drilling operations; penetration of the sampling tube and sample retrieval to the ground surface; water content redistribution in the tube; extrusion of the sample from the tube; drying and/or changes in water pressures; and trimming and other processes required to prepare specimens for laboratory testing.

Clayton et al. (1998) examined the influence of the following issues of sampling tube geometry, defined in Figure 1.9:

- the area ratio (AR), which relates the volume of displaced soil with the volume of the sample:

(a)   (b)

Figure 1.7 Undisturbed sampling of granite residual soils: a) block sampling in a pit; b) trimming cylindrical samples for triaxial testing from block samples (Ferreira, 2009).

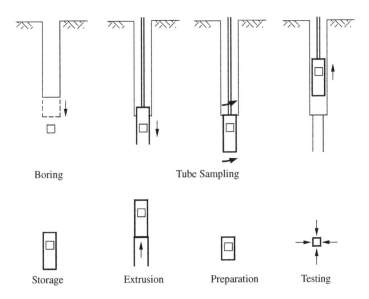

Boring   Tube Sampling

Storage   Extrusion   Preparation   Testing

Figure 1.8 Stages in tube sampling and soil specimen preparation for laboratory testing (Hight, 2000).

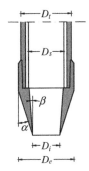

Figure 1.9 Definition of the geometric parameters of a tube sampler.

$$AR = \left[\frac{D_e^2 - D_i^2}{D_i^2}\right] \times 100 \qquad (1.6)$$

- the inside clearance ratio (*ICR*), which controls internal friction:

$$ICR = \left[\frac{D_s - D_i}{D_i}\right] \times 100 \qquad (1.7)$$

- the outside cutting edge angle (*OCA*), α;
- the inside cutting edge angle (*ICA*), β.

These authors identified the major factors controlling the magnitude of tube sampling strains as the increasing *AR* or increasing *OCA*, either of which contributes significantly to the peak axial strain in compression, and the increasing *ICR*, which has a strong influence

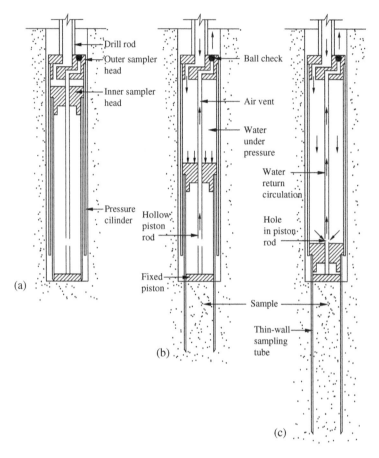

Figure 1.10 Osterberg composite hydraulic fixed piston sampler (Osterberg, 1973): a) position for being dropped in the borehole; b) static driving; c) end of driving.

on the peak axial strain in extension. Their results emphasize the sensitivity of imposed strain levels to details of sampling tube geometry.

It should be noted that the use of higher values for those geometric parameters derives from the need to obtain robust samplers in order to support the loads associated with their function, which naturally increase with the increase in soil resistance. As an example, Figure 1.10 shows a sampler traditionally used in soft clays with excellent results, where the geometric parameters mentioned above are very low (Osterberg, 1973).

1.2.4.4 Sampling of sands

The above-mentioned processes are not feasible for sampling sands or other granular soils, especially for clean sands and below the water table.

Soil freezing

An undisturbed sampling process of saturated sands, developed by Japanese researchers, involves soil freezing (Yoshimi et al., 1994). Essentially, it consists of inducing freezing of the water present in a given volume of soil prior to sampling, in order to then collect the

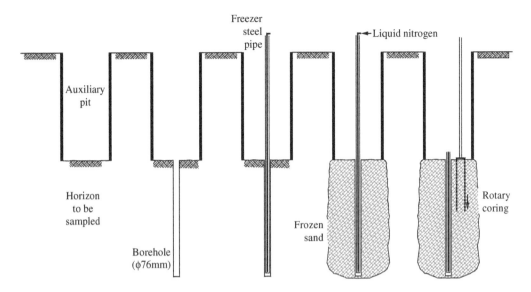

Figure 1.11 Sampling procedure by freezing.

sample with a rotary advancing sampler, such as those used for rock masses. Figure 1.11 shows a schematic of the process.

From the description in Figure 1.11, it is clear that this sampling process is highly complex and involves high cost and execution time, which are justifiable in only exceptionally important case studies.

Gel-Push sampling

An alternative to the use of soil freezing is the so-called Gel-Push sampling technique, also developed in Japan, and whose first application occurred there in 2004 (Lee et al., 2012). This sampler follows the main concept of fixed-piston sampling, which is the basis of the Osterberg sampler represented in Figure 1.10.

This sampler consists of a triple core barrel and uses a viscous polymer gel as its drilling fluid, hence the name Gel-Push (GP). The gel lubricates and reduces friction between the sample and the sampling tube, during both sampling and laboratory extrusion. Such innovation is a key factor in preserving the natural structure and fabric of the soil (Figure 1.12).

There are four different variations of GP: GP-Rotary, GP-Drilling, GP-Triple, and GP-Static. In the context of this book, only the GP-Static (GP-S) variant will be discussed, as it is particularly suited for sampling sandy soils.

Unlike a conventional hydraulic activated fixed-piston, GP-S has a triple core barrel that includes three pistons: the stationary piston, the sampling tube-advancing piston, and the core-catcher activating piston. The first piston is fixed and the other two are traveling pistons. The outer tube secures the borehole and keeps the penetration rod and piston fixed in alignment during penetration. The advancing piston contains the gel, ensures the downward movement of the system, and activates the catcher while it is inserted into the soil. The core-catcher piston captures the sample inside a metallic liner tube. GP-S allows retrieving samples with an approximate diameter of 71 mm and a length of 1 m (Viana da Fonseca et al., 2019).

The sampling methodology of the GP-S comprises three operation phases. The first phase includes the assembly of the sampler and the preparation of the gel. The gel is prepared, generally at a concentration ratio of 1% (v/v) viscous polymer in clean water and is immediately

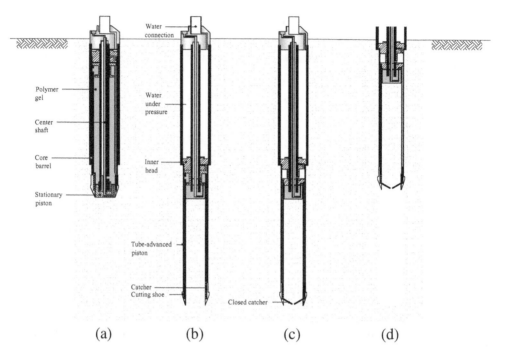

Figure 1.12 GP-S sampler operation stages: a) lowered down cased hole; b) pushed into virgin soil; c) closure of core catcher; d) removal to the surface (Viana da Fonseca et al., 2020).

poured into the tube sampler. The second phase refers to the positioning of the sampler at the desired depth of sampling inside the borehole. For sampling, a penetration rate of 1 m/min is usually adopted. During this stage, the core barrel starts to advance and the cutting shoe penetrates the soil. Simultaneously, the hydraulic piston closes a bypass valve and the fixed piston squeezes gel into the core catcher, lubricating the end of the collected soil. The third phase corresponds to the downward advance of the core barrel into the soil until the maximum liner length (1 m) is reached. The remnant gel flows through the liner, allowing a smooth sliding of the soil, while the catcher blades close, holding the soil inside the sampler.

1.3 *IN SITU* TESTING

1.3.1 Introduction

In the following sections, the most relevant and frequently used *in situ* tests will be discussed.
   In view of the nature of this book, it has been decided to describe each test in only its essence:

(i) the general concept of the equipment and its interaction with the ground under investigation;
(ii) the quantities measured in the test, how they are interpreted, and how soil parameters or other results of interest can be derived, including practical examples.

There is abundant information, as well as numerous details related to these tests, which are essential for their proper implementation, which are not included in this text. For each

Table 1.3 Relevant standards for geotechnical investigation and in situ testing.

Operation/Test[a]	ASTM	EN ISO
Geotechnical investigation and sampling	D 420, D 1452 / D 1452M - 16, D 4700	–
Identification and description	D 2488	ISO 14688-1
SPT	D 1586 / D 1586M - 18	ISO 22476-3
CPTu	D 3441 - 16 D 5778 - 12	ISO 22476-1
DP	–	ISO 22476-2
PLT	D 1195 / D 1195M - 09 D 1196 / D 1196M - 12	
CHT	D 4428 / D 4428M - 14	
VST	D 2573 / D 2573M - 18	ISO 22476-9
SBPT		ISO 22476-6
PMT		ISO 22476-4
DMT	D 6635 - 15	ISO 22476-11
Permeability tests	D 4050	ISO 22282-1 ISO 22282-2
Pumping tests	D 4050	ISO 22282-4

[a] The complete designation of each test is included in the subsequent sections.

in situ test, the reader is directed to the extensive normative documents and standards, such as those summarised in Table 1.3.

1.3.2 Standard penetration test (SPT)

1.3.2.1 Essential aspects of the equipment and test procedure

The approach to geotechnical in situ testing necessarily starts with the SPT, by far the most used field test worldwide. It was introduced to the United States by the Raymond Pile Company in 1902, but its widespread use only began in the 1940s with the publication of the book Soil Mechanics in Engineering Practice (Terzaghi and Peck, 1948).

The test consists of driving a standardized sampler to the bottom of a borehole, shown in Figure 1.13a, by tapping a 63.5 kgf (140 lb) weight hammer falling from a height of 76 cm (30 inches). The sampler is a steel split barrel (with outer and inner diameters of 51 mm and 35 mm, respectively) with a length of about 80 cm and an approximate weight of 6.8 kgf. At the lower end of the sampler, a beveled driving shoe is attached, which facilitates penetration into the ground. At the opposite end, a part is threaded for the connection with the string of rods, and provided with a non-return ball check valve and side vents for purging air and water during penetration of the sampler.

In order to carry out the test, the drilling operation is interrupted, followed by cleaning of the bottom of the borehole (removal of the material disaggregated by drilling and contact with the water used to fill the hole) and descent of the sampler carried by the string of rods. After positioning the sampler in contact with the ground at the bottom of the borehole, the top rod (i.e., at the surface) is adjusted with an anvil that will receive the hammer blows (Figure 1.13b).

The test is performed in two successive phases: in the 1st phase, with a penetration of the sampler of 15 cm; and (sequentially) in the 2nd phase, with 15 cm plus 15 cm, counting the respective number of hammer blows. The number referring to the 1st phase is taken as merely informative, since it is essentially intended to go through the most disturbed ground

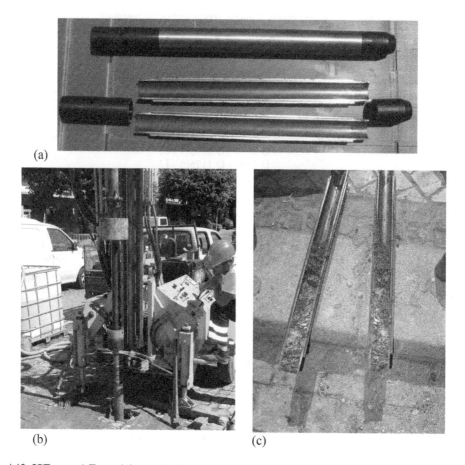

Figure 1.13 SPT test: a) Terzaghi's standard sampler; b) view of the test just before the free-fall of the hammer; c) open split barrel sampler after testing (photos: Carlos Rodrigues).

immediately below the bottom of the borehole. The total number of blows in the 2nd phase (i.e., the sum of the two 15-cm sub-phases), termed N, "N-value" or "standard penetration resistance", is considered the test result.

If, in the 1st phase, the number of blows reaches 50 without a penetration of 15 cm, or the same number (50) is reached in the 2nd phase without a penetration of 30 cm, the test is interrupted and the penetration length reached is registered. In this case, the N value for a penetration of 30 cm is usually obtained by a simple extrapolation.

Before further detailing of some aspects of the test itself, it is important to emphasize that the SPT is, first and foremost, a process for collecting disturbed samples. For this purpose, sampler driving is usually carried out with a spacing of 1.5 m or less. When brought to the surface, the sampler is opened in two barrels, allowing examination of the soil inside (Figure 1.13c). A fraction of the last 30 cm of sampled soil is then stored in a small, properly identified, watertight box, which is later examined for the definition of the stratigraphy and lithology of the ground.

The geological–geotechnical profiles, containing the succession of soil layers or strata used as the basis for the geotechnical studies of a construction site, are usually established based on the disturbed samples collected by driving the SPT sampler. The idea of associating sampler driving with a geotechnical test should indeed be acknowledged, as it provides a basic descriptive parameter of the (mechanical) quality of the ground that can be used in current practice by engineers.

1.3.2.2 SPT corrections

The generalized use of SPT for decades in the most diverse geotechnical conditions has allowed an empirical association between the number of blows, N, in a given soil type with certain characteristics, for example, relative density (in sands) and consistency (in clays). It can also be associated with soil behavior, such as liquefaction susceptibility of sandy deposits, and with geomechanical parameters, such as the angle of shearing resistance of granular soils. Many of these correlations exhibit relatively limited reliability because, despite its name, the SPT is not truly standardized!

In fact, several aspects of the test, which are likely to have a substantial effect on the results, vary from country to country and even from company to company. Among those aspects are, for example, the operation mode and fall of the hammer, the weight of the anvil, the type of rods and connections, the hole diameter, etc. (De Mello, 1971). It was only in 1988 that a document was published, establishing internationally accepted procedures for the SPT (ISSMFE, 1988).

Experimental studies have led to the conclusion that, for this purpose, it was of utmost importance to establish comparisons with the level of energy actually transmitted to the string of rods in each blow of the hammer. Thus, if in a given soil, at a given depth, SPT tests are performed with two equipments, I and II, differing in terms of the hammer maneuver system and the anvil weight, with corresponding energies transmitted to the string of rods, E_I and E_{II}, respectively, the obtained values of N with such systems are related by the following equation:

$$N_I E_I = N_{II} E_{II} \tag{1.8}$$

where N is inversely proportional to the energy transmitted to the rods.

This is an extremely relevant aspect, since the same studies have demonstrated that the energy E of each blow may be substantially lower than the potential energy of the hammer prior to the fall, E_p ($E_p = 63.5 \times 0.76 = 48.26$ kgm or $E_p = 474$ J). In effect, in some equipment, the operating procedure of the hammer does not allow a completely free fall, which obviously reduces the kinetic energy of the impact on the anvil. On the other hand, some energy is also dissipated in the anvil, and such dissipation increases with its weight.

Considering E_R, the ratio of the energy transmitted to the string of rods, as:

$$E_R = \frac{E}{E_p} \times 100 \tag{1.9}$$

a standard value of 60% has been established.

As such, any SPT result, N, conducted with equipment, the characteristics of which correspond to a specific energy ratio, E_R, should be converted to the result that would be obtained using the standard energy ratio. Such a result is denoted as N_{60} and is obtained from the following expression:

$$N_{60} = C_E N \tag{1.10}$$

in which C_E, the energy corrective coefficient, is

$$C_E = \frac{E_R}{60} \tag{1.11}$$

Table 1.4 Correction factor, C_R, for the rod length (Skempton, 1986).

Rod length	Correction factor C_R
>10 m	1.0
6–10 m	0.95
4–6 m	0.85
3–4 m	0.75

In addition to energy-related correction of the results, other corrections are advised in the literature. Firstly, dynamic studies have shown that the energy applied at each blow of the hammer is in fact only completely absorbed by the string of rods when the total rod length is such that the weight of the rods is greater than or equal to that of the hammer. Therefore, for rod lengths below a given value, N must be corrected by means of a minorative correction factor, C_R. Table 1.4 includes a proposal for this factor.

Another important aspect is the diameter of the borehole. It is recognized that the wider the borehole, the greater the depth of the disturbed ground below the bottom of the hole, due to the change (reduction) in the mean effective stress. Although the test was designed for holes up to 100 mm in diameter, drilling diameters up to 200 mm (see Section 1.2.3) are common practice. The effect of the drilling diameter, although qualitatively easy to understand, is not well quantified by means of experimental results, and the values in Table 1.5 for the respective corrective coefficient, C_D, may be regarded as prudent.

A final correction is necessary to account for the effect of the effective stress at the testing depth in the case of sandy soils. It should be noted that, by fixing all aspects of the test as well as the soil itself, N will naturally increase with mean effective stress, that is, with depth. For this reason, the concept of normalized value, N_1, has been established, corresponding to the result of a given system in a given soil for an at-rest vertical effective stress equal to 1 atmosphere (approximately 1 bar or 100 kPa). The introduction of an overburden correction factor, C_N, is therefore indispensable, so that:

$$\left(N_1\right)_{60} = C_N\, N_{60} \tag{1.12}$$

in which C_N is calculated as follows:

$$C_N = \left(\frac{p_a}{\sigma'_{v0}}\right)^{0.5} \tag{1.13}$$

where p_a represents the atmospheric pressure and σ'_{v0} is the vertical effective stress at the depth in which N_{60} was obtained. This empirically derived expression is based on calibration chamber test data, where SPT was carried out on reconstituted granular soils under a wide range of mean effective stresses (Liao and Whitman, 1986). In general, it is not recommended to use C_N values lower than 0.5 nor greater than 2.0.

Table 1.5 Correction factor, C_D, for the borehole diameter (Skempton, 1986).

Borehole diameter	Correction factor, C_D
65–115 mm	1.0
150 mm	1.05
200 mm	1.15

Thus, taking N as the raw test result, the normalized and corrected test result is obtained from the equation:

$$(N_1)_{60} = C_E C_R C_D C_N N \tag{1.14}$$

It should be noted that, in contrast to the correction factors for energy and overburden stress (C_E and C_N), the application of correction factors for rod length and borehole diameter (C_R and C_D) is not consensual. For certain authors, $(N_1)_{60}$ is computed from Equation 1.14, assuming that these two latter factors are equal to unity.

1.3.2.3 Equipment calibration

In view of the above, it is clear that it is extremely important, in each investigation campaign, to know precisely the energy ratio, E_R, of the equipment used in the SPT tests. For this purpose, proof of recent calibration of the equipment should be requested.

Calibration methods are outside the scope of this book. In any case, it is anticipated that the energy transferred to the rods in each blow is equal to the integral of the applied force multiplied by the velocity, defined from the moment of impact of the hammer on the rods until the moment the integral reaches the maximum value, according to the equation:

$$E = \int_0^t F(t)v(t)dt \tag{1.15}$$

in which $F(t)$ and $v(t)$ correspond to the force and velocity values, respectively, as a function of time.

The evaluation of the energy involved in each blow, therefore, requires the instrumentation of one or more rods by means of strain gauges (for obtaining the force, F) and of accelerometers (which allow determination of the propagation velocity of the shock wave, v). These should be linked to a data acquisition system, and its interpretation can be done using appropriate commercial software. The time corresponding to the greatest transferred energy determines the integration interval.

Likewise, it is also advisable to check the hammer drop height, sampler dimensions and sampler and hammer weights: since the SPT requires very simple equipment, it can be produced in non-specialized workshops.

1.3.2.4 Correlations of $(N_1)_{60}$ with soil properties and parameters

Table 1.6 shows the correlation between $(N_1)_{60}$ and the relative density or density index for normally consolidated sands, as proposed by Skempton (1986).

Table 1.6 Relation between $(N_1)_{60}$ and density index for sands (Skempton, 1986).

$(N_1)_{60}$	0–3	3–8	8–25	25–42	>42
I_D (%)	0–15	15–35	35–65	65–85	85–100
Relative density	Very loose	Loose	Medium dense	Dense	Very dense

Notes:

1. For $I_D \geq 0.35$ $(N_1)_{60} / I_D^2 \cong 60$.

2. For coarse sands, N should be multiplied by 55/60.

3. For fine sands, N should be multiplied by 65/60.

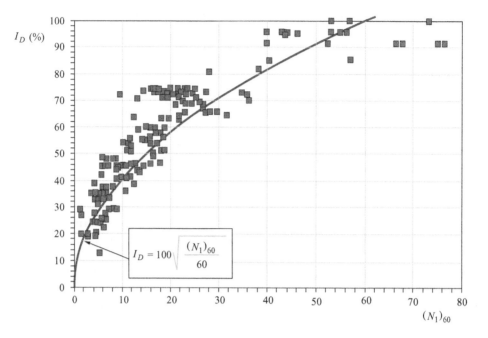

Figure 1.14 Correlation between $(N_1)_{60}$ and density index for clean sands (Mayne et al., 2001).

Figure 1.14 (Mayne et al., 2001) essentially presents the same correlation together with a large number of experimental determinations obtained by several authors.

Figure 1.15 shows two correlations between $(N_1)_{60}$ and the angle of shearing resistance (peak values) proposed by Décourt (1989) and Hatanaka and Uchida (1996). It can be seen that the proposals are in reasonable agreement with each other.

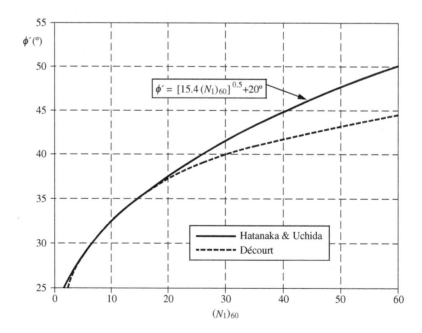

Figure 1.15 Correlations between $(N_1)_{60}$ and the angle of shearing resistance for sands (Décourt, 1989; Hatanaka and Uchida, 1996).

Table 1.7 Correlation between the density index and the angle of shearing resistance for quartz sands (US Army Corps of Engineers, 1993).

I_D (%)	Fine sands		Medium sands		Coarse sands	
	Uniform	Well-graded	Uniform	Well-graded	Uniform	Well-graded
40	34	36	36	38	38	41
60	36	38	38	41	41	43
80	39	41	41	43	43	44
100	42	43	43	44	44	46

Table 1.8 Relation between N_{60} and the consistency of clays (Clayton et al., 1995).

N_{60}	0–4	4–8	8–15	15–30	30–60	>60
Consistency	very soft	soft	medium	stiff	very stiff	hard

Table 1.7 shows a correlation between the density index and the angle of shearing resistance for quartz sands (US Army Corps of Engineers, 1993).

The correlations between SPT results and liquefaction susceptibility of sandy soils will be discussed in Chapter 6 (see 6.5).

Table 1.8 includes a soil classification proposed by Clayton et al. (1995), based on the SPT for clayey soils, in terms of consistency.

Before concluding this presentation of the SPT, it should be pointed out that the estimate of mechanical parameters of the soil, based on correlations, such as those previously mentioned, does not deplete its usefulness for geotechnical design. In fact, numerous empirical methods of the design of shallow and deep foundations (piles), directly based on SPT results (N_{60} or $(N_1)_{60}$), can be found in the literature.

1.3.3 Cone penetrometer test (CPT/CPTu)

1.3.3.1 Essential aspects of the equipment and test procedure: measured parameters

The cone penetrometer test was first developed in the Netherlands in the 1930s and is one of the most commonly used field tests. It has the obvious advantage over SPT that it is completely automated, so that its results are fully reproducible, that is, independent of the operator. However, it does not allow for sample collection. Therefore, the CPT is strictly an *in situ* test, and it is often complemented by borehole drilling for visual identification of the soil stratigraphy. However, in cases where stratigraphy is already clearly established in the context of previous investigation campaigns in the vicinity, the CPT is sometimes used on its own (Lunne et al., 1997).

The test consists of the continuous driving of a penetrometer into the soil by means of a hydraulic system at a rate of 20 mm/s, as shown in Figure 1.16. The penetrometer comprises a conical tip (tip angle of 60° and base of the cone area equal to 10 cm²) and a friction sleeve (length of 134 mm, area of 150 cm²).

Since the 1980s, the version of the device known as piezocone or CPTu has been generalized, allowing the measurement of pore water pressure near the tip during pushing. As shown in Figure 1.16a a ring filter (consisting of a porous metal) is located immediately

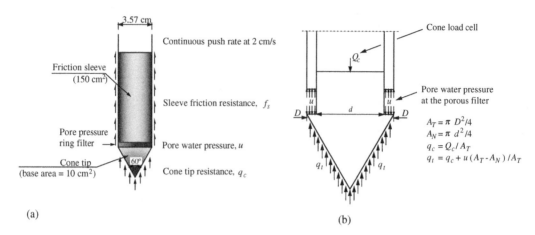

Figure 1.16 CPTu tip: a) details and measured parameters; b) tip resistance correction for the device with the filter element above the cone.

above the cone, which allows the transmission of pore water pressure to a pressure transducer housed inside the nozzle tip.[3]

The measured parameters are:

(i) the cone tip resistance, q_c, the ratio of the vertical reaction force of the ground (measured in a load cell housed at the conical tip) to the area of the base of the cone;

(ii) the sleeve resistance, f_s, the ratio of the friction force developed along the sleeve (measured by another load cell next to the sleeve) to its surface area;

(iii) the pore water pressure (measured on the inner transducer in the filter element).

Besides the three measured parameters, the equipment software allows for the calculation of the friction ratio, R_f, by the following equation:

$$R_f = \frac{f_s}{q_c} \times 100 \tag{1.16}$$

Figure 1.17 shows an image of the system. In the foreground, the set of rods (each 1 m in length) is visible, rods being sequentially added as penetration advances; inside the rods runs a cable, with the set of electric wires connecting the tip transducers to the power source and to the acquisition box, both located at the surface; the latter is connected to a computer. In the background, the structure of the pushing equipment can be seen, anchored to the ground by a set of augers, in order to achieve the necessary reaction for the hydraulic jacks to drive the cone and the rods through the soil. An alternative reaction system can be achieved using concrete blocks (heavy weight).

It is desirable that the reaction structure allows the application of a driving force of at least 10 tonnes. Otherwise, the tip will not be able to cross relatively resistant layers, and the test may be limited to shallow depths in many situations. In any case, this test is not appropriate to characterize very hard soils nor those with large particles, such as medium and coarse gravels.

---

[3] In certain devices, the point of pore pressure measurement is located elsewhere in the tip, namely in the face of the cone.

Figure 1.17 Image of the pushing equipment (anchored drill-rig) of the piezocone, with the string of rods in the foreground (photo: Carlos Rodrigues).

The use of a piezocone requires a correction of the values taken for the cone tip resistance, and the parameter q_t is used instead of q_c. As shown in Figure 1.16b, since the ring-shaped filter is located immediately above the cone, pore water pressures will be exerted downwardly on the top of the conical tip, more precisely on a circular crown of inner diameter d and outer diameter D. Thus, the total resistance of the soil to the penetration of the conical tip, q_t, is equal to q_c added to the ratio between the resultant of the water pressures on that circular crown and the area of the base of the cone.

This correction is only relevant in soft fine-grained soils, in which very low values of q_c, combined with high values of pore water pressure (due to the excess pore pressure induced by pushing the cone), are observed. For stiff clays and sandy soils, this correction is negligible, and q_c and q_t are practically identical. In any case, the latter parameter will be considered in the following sections.

1.3.3.2 Interpretation of results: soil classification charts

As has already been mentioned, the CPT does not allow sampling, so that the soil type must be determined, either directly, by means of boreholes carried out in parallel, or indirectly, by associating certain trends of the measured quantities with soil characteristics, naturally based on the experience gained in cases where the first-mentioned option was adopted.

When the original version of the apparatus was used (without measuring pore water pressure), charts were developed as shown in Figure 1.18. As can be seen, typically fine soils

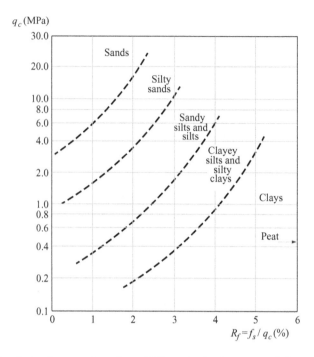

Figure 1.18 Simplified soil classification chart from CPT results (Robertson and Campanella, 1983).

tend to exhibit lower values of q_c and higher values of the friction ratio. The limitation of this type of chart arises from the fact that it is based on test results that have not reached very great depths, typically less than 30 m. In a very deep hole, for example, in a normally consolidated clay, q_c (or q_t) will tend to reach relatively high values, for which the application of this chart would be misleading, resulting in an incorrect classification of the soil.

The continuous measurement of the pore pressures during driving, which was made possible by the CPTu, represented, in this regard, remarkable progress.[4] In fact:

(i) crossing layers of soft or medium clayey soils occurs in undrained conditions, i.e., generating excess pore pressures (positive, in the present case), so that values of pore water pressure generally greater than u_0 (the hydrostatic at-rest pore pressure) are measured;

(ii) on the contrary, crossing sandy layers (with less than 10% fines) typically occurs under drained conditions, with the piezocone recording pore pressure values typically very close to u_0;

(iii) in some overconsolidated and fissured clayey soils or very dense sandy soils, i.e., dilatant soils, pore pressures lower than u_0 may be recorded (Jamiolkowski et al., 1985; Mayne et al., 2001).

Figures 1.19 to 1.21 illustrate examples of CPTu results in recent (Holocene) alluvial deposits, involving different soils. As can be seen, the equipment provides nearly continuous logging in depth (generally, at 2-cm intervals) of the four quantities defined above. In each of the figures, the stratigraphic profile of the site was added to the right side, deduced from conventional boreholes.

---

[4] For this purpose, it is crucial that the filter element is fully pre-saturated so that the pore pressures can actually be transmitted to the pore pressure cell.

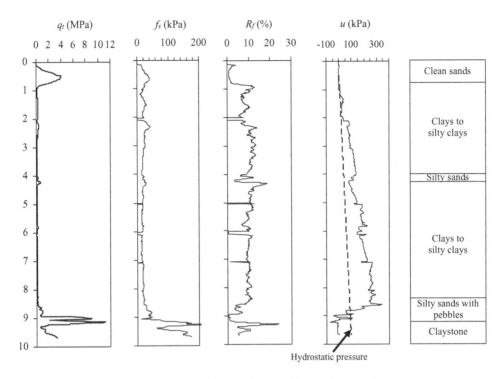

Figure 1.19 CPTu results in the University of Aveiro Campus, Portugal – soft clays scenario.

Figure 1.19 corresponds to a scenario of about 9 m of soft clayey soils over dense sandy soils and overconsolidated clayey soils. The CPTu results corroborate the above-mentioned trends, in which the very low cone resistance in the soft clays is particularly evident.

On the other hand, Figure 1.20 represents a scenario of essentially sandy soils over 30 m thick. As would be expected, the cone resistance is clearly greater than that shown in the previous figure for the soft clays and tends to increase with depth; below a depth of 20 m, the occurrence of a denser layer is clear. In this sandy deposit, soft clay (mud) intercalations occur, particularly between approximately 11 m and 15 m in depth. As can be seen, these interspersed clayey layers imply: i) low values of cone resistance; (ii) positive excess pore pressures; and iii) increases in the friction ratio.

Finally, Figure 1.21 shows a scenario essentially characterized by soft clays, extending from 5 m to about 25 m depth. What is essentially different in this test is the appearance of several (fine) interbedded sands in this thick layer, namely at about 13 m and at 23 m depth. These intercalations are characterized by: i) higher values of cone resistance; (ii) pore pressures close to hydrostatic pressures; and iii) decreases in the friction ratio.

From the previous examples, it can be recognized that the CPTu, by its ability to identify very thin layers within other, larger layers, gains exceptional utility in the context of embankment works on soft, clayey soils, in the correct definition of the hydraulic boundary conditions, fundamental to a realistic approximate prediction of the consolidation time.

The experience with the CPTu allowed the development of much more reliable and complete charts for identification of the soil type and soil behavior than the one in Figure 1.18. Among those, the chart proposed by Robertson (1990) shown in Figure 1.22, which is already considered to be a classic example, is based on the comparison of the normalized

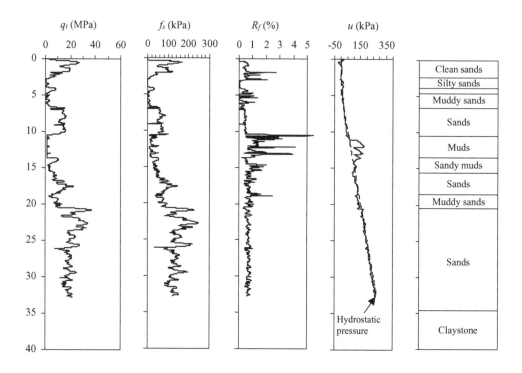

Figure 1.20 CPTu results in Aveiro Harbour, Portugal – loose sandy soils scenario, with interbedded clays and muds.

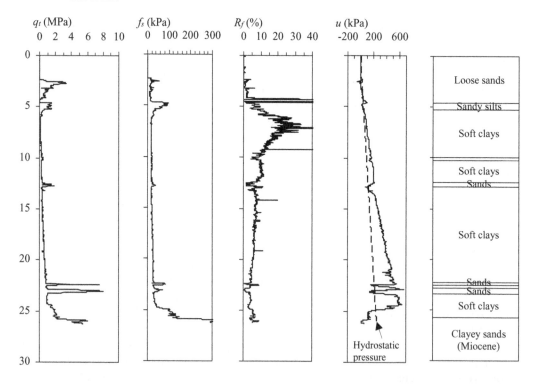

Figure 1.21 CPTu results in the River Tagus Valley, Portugal – soft clayey soils scenario, with interbedded sandy layers.

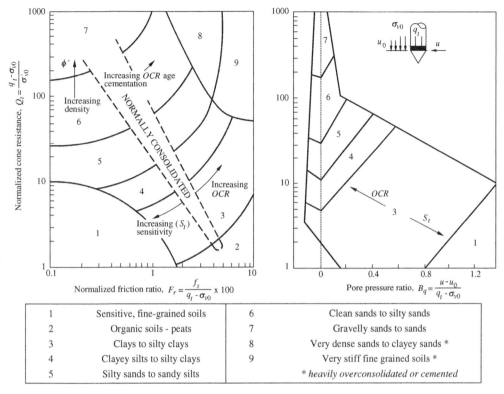

Note: *OCR* corresponds to the overconsolidation ratio of the soil, the ratio of the pre-consolidation effective stress to σ'_{v0}

Figure 1.22 Normalised CPT or CPTu soil behavior type chart (Robertson, 1990).

cone penetration resistance, Q_t, the normalized friction ratio, F_R, and the pore pressure ratio, B_q, which are expressed as follows, respectively:

$$Q_t = \frac{q_t - \sigma_{v0}}{\sigma'_{v0}} \tag{1.17}$$

$$F_R = \frac{f_s}{q_t - \sigma_{v0}} \times 100 \tag{1.18}$$

$$B_q = \frac{u - u_0}{q_t - \sigma_{v0}} = \frac{\Delta u}{q_t - \sigma_{v0}} \tag{1.19}$$

where *u* is the instantaneous pore pressure measured by the piezocone.

The left-hand side of the chart represents CPT results while the right-hand side represents CPTu results. As can be seen, on the diagonal area of the left-hand figure are the normally consolidated soils, with the coarser soils (intersected by the cone in drained conditions) above and the finer soils (in undrained conditions) below. Progressive upward displacement of the diagonal corresponds to older and overconsolidated soils, while downward displacement corresponds to increasingly sensitive soils.[5]

---

[5] The sensitivity of a clay, S_t, is the ratio between the undrained shear strength of the undisturbed soil and that of the reconstituted or remolded soil (see equation 1.30).

Since then, the right-hand chart has been scarcely used, whereas, on the contrary, the left-hand chart has been extensively used, motivating the introduction of a new parameter. In fact, Jefferies and Davies (1993) proposed a curious complement to Robertson's Q_t-F_r chart. They noticed that the boundaries between the classes could be fitted by concentric circles, the radius of which was designated the *soil behavior type index*, I_c.

Robertson and Wride (1998) modified the definition of I_c, as defined by:

$$I_c = \left[\left(3.47 - \log Q_t\right)^2 + \left(\log F_r + 1.22\right)^2\right]^{0.5} \tag{1.20}$$

The boundaries based on I_c provide a good approximation to the soil behavior type descriptions, especially in the center of the chart, where normally to lightly overconsolidated soils are located, as shown by Figure 1.23.

Robertson (2009) pointed out that $I_c = 2.6$ may be considered a good proposal for a boundary between sandy ($I_c < 2.6$) and clayey soils ($I_c > 2.6$), based on cyclic liquefaction case histories that essentially involved normally consolidated soils.

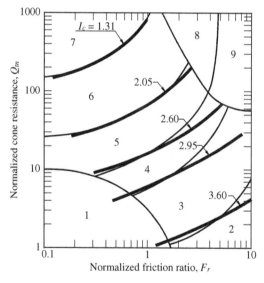

Zone	Soil behaviour type	I_c
1	Sensitive, fine grained	N/A
2	Organic soils – clay	> 3.6
3	Clays – silty clay to clay	2.95 – 3.6
4	Silt mixtures - clayey silt to silty clay	2.60 – 2.95
5	Sand mixtures – silty sand to sandy silt	2.05 – 2.6
6	Sands – clean sand to silty sand	1.31 – 2.05
7	Gravelly sand to dense sand	< 1.31
8	Very stiff sand to clayey sand*	N/A
9	Very stiff, fine grained*	N/A

Figure 1.23 Contours of soil behavior type index, I_c (thick lines), on soil behavior type Q_t-F_r Robertson (1990) chart (Robertson, 2009)

1.3.3.3 Correlations with soil characteristics and parameters

1.3.3.3.1 Sands

Although cone penetration into a homogeneous and elasto-plastic half-space is susceptible to a theoretical approach, attempts to interpret CPT results in order to obtain soil strength and stiffness parameters have not yet been sufficiently developed. Thus, empirical correlations, based on experience, are used to estimate those parameters, as for the SPT. These correlations are considered to be much more reliable than those applied for the SPT, mainly because the CPT does not suffer from the diversity of aspects and test procedures previously discussed with regard to the SPT, among other reasons.

When shallow depths are involved, with low values of σ'_{v0}, as in the case of shallow foundations, Table 1.9 can be applied, which relates q_c with the angle of shearing resistance and the deformation modulus of quartz or feldspar sands (Bergdahl et al., 1993).

In order to increase the reliability of the correlations, it is preferable to use the value of cone penetration resistance, properly corrected to take into account the effective stresses at depth, q_{c1} or q_{t1}; as previously mentioned, for granular soils, q_{c1} and q_{t1} are practically identical. Analogous to the SPT correlations (see Equations 1.12 and 1.13), the normalized cone resistance can be obtained by means of the following equation:

$$q_{c1} \approx q_{t1} = C_N \, q_t = \left(\frac{p_a}{\sigma'_{v0}} \right)^{0.5} q_t \tag{1.21}$$

For example, the correlation between q_{c1} and the density index of sands (Mayne et al., 2001) is shown in Figure 1.24.

1.3.3.3.2 Clays

In clayey soils, CPTu naturally implies an undrained loading. The relationship between the total cone penetration resistance, q_t, and the undrained shear strength, c_u, at a given depth can be expressed as follows:

$$q_t = N_{kt} \, c_u + \sigma_{v0} \tag{1.22}$$

where N_{kt} is a dimensionless parameter.[6]

Table 1.9 Relation between q_c and the angle of shearing resistance and deformation modulus of quartz or feldspar sands (proposed for shallow foundations, low values of σ'_{v0}).

Relative density	q_c (MPa)	ϕ' (°)	E' (MPa)
very low	0.0–2.5	29–32	< 10
low	2.5–5.0	32–35	10–20
medium	5.0–10.0	35–37	20–30
high	10.0–20.0	37–40	30–60
very high	>20.0	40–42	60–90

Notes:

1. ϕ' values are valid for sands; for silty soils, a reduction of 3° should be considered; for gravels, an increase of 2° should be adopted.
2. E' values correspond to secant moduli for the estimate of settlements; it is possible that, for silty soils, the most appropriate values are about 50% lower than those provided, whereas, for gravels, these can be about 50% higher; in overconsolidated soils, this modulus may be considerably higher.

---

[6] Equation 1.22 started being used prior to the introduction of the total resistance concept, using q_c in the first term; at the time, the adopted symbol for the dimensionless parameter affecting c_u was N_k instead of N_{kt}

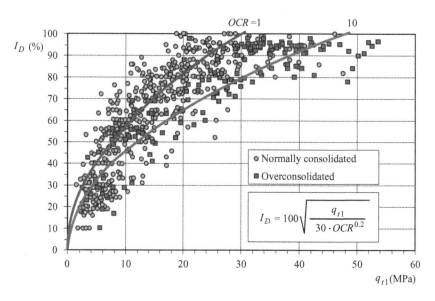

Figure 1.24 Correlation between q_t and the density index of sands (677 results from sands at 26 different locations) (adapted from Mayne et al., 2001).

This is a theoretical equation, applicable to the bearing capacity to vertical loading of footings and piles in undrained conditions. The theoretical value of the dimensionless factor multiplying the undrained shear strength of the soil, for situations in which the foundation has a circular cross section and a diameter much less than the depth – a situation that is similar to the CPT case – is about 9.0 (Skempton, 1951).[7] Numerical simulations of the test, using several highly refined finite element models, showed that the cone penetration resistance in clays depends on several factors beside the undrained shear strength itself, such as the magnitude of the horizontal total stresses, the soil stiffness, and the roughness of the tip surface (Yu, 2004). For most current conditions, these studies provided values of N_{kt} between 9 and 12.

However, N_{kt} values deduced from field investigations and available in the literature – from CPT results and c_u evaluations based on other *in situ* tests, such as the vane test (VST), or laboratory tests, such as the triaxial test – exhibit considerable dispersion, ranging from 8 to 24. This is clearly in disagreement with what would be expected from the theoretical or numerical simulation studies cited above (Salgado, 2008).

This dispersion derives partly from the fact that the values for N_{kt} were deduced based on undrained shear strength assessed by means of diverse laboratory or field tests. Soft clays often show undrained strength anisotropy, which is reflected in the fact that this parameter is typically maximal in triaxial compression tests, minimal in triaxial extension tests, and intermediate in the direct simple shear tests, all in the laboratory, and also intermediate in the vane shear test (see Section 1.3.5).

Given the type of soil failure ahead of the CPT cone, it seems reasonable to say that, as for an embankment or a vertically loaded foundation, the undrained shear strength may correspond to an average value of the soil strength for the three mentioned stress paths. Therefore, it would be convenient, when discussing the N_{kt} factor, to take c_u values obtained from the vane shear test as a preferred reference. If this criterion is followed, the dispersion

---

[7] As will be discussed in Chapter 6, in foundation design this parameter is designated N_c.

of values is substantially reduced, and a considerable part of the experimental results reasonably corroborates the values of the mentioned theoretical and numerical studies. Thus, in the current state of the art, it seems reasonable to adopt N_{kt} factors equal to 12, to obtain c_u values corresponding to vane shear test estimates.

One of the most useful applications of CPTu in clays is the evaluation of the horizontal consolidation coefficient of the soil, c_h, which is necessary for the design of vertical drainage systems for consolidation acceleration (see Section 4.6.3). As previously mentioned, pushing the cone tip in clayey soils occurs under undrained conditions, generating a positive excess pore pressure in soft soils. If the pushing operation is suspended at a certain depth and the tip remains in place, it is possible to follow the evolution of the excess dissipation over time, using the piezocone. Such dissipation occurs essentially in the horizontal direction, being controlled by the horizontal or radial consolidation coefficient, c_h.[8]

The problem of the dissipation over time of the excess pore pressure, generated by the introduction of the CPT tip, can be theoretically described. Two distinct formulations leading to very similar results are known, the older one approximating the process to the expansion of a spherical or cylindrical cavity in the soil (Torstensson, 1977), with the more recent one using the strain path method (Teh and Houlsby, 1991). The presentation of these theories is outside the scope of this book but it is considered useful here to present some of the solutions of practical interest.

Figure 1.25 shows the curve that relates the evolution of normalized excess pore pressure, $\Delta u(t)/\Delta u(0)$, with T^*, the so-called modified time factor of the second-mentioned theory, which relates to the remaining parameters, according to the equation:

$$T^* = \frac{c_h t}{R^2 I_R^{0.5}} \qquad (1.23)$$

in which t is time, R is the radius of the cone, and I_R is the rigidity index (see below).

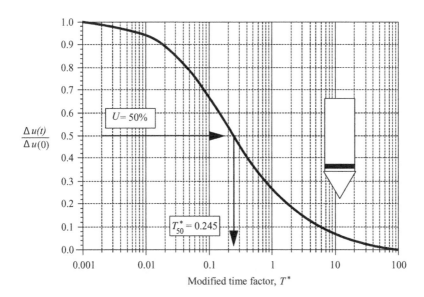

Figure 1.25 Radial consolidation around the CPT tip during a dissipation test: relation $\Delta u(t)/\Delta u(0)$ with modified theoretical time factor T^* (Teh and Houlsby, 1991).

---

[8] Usually, these dissipation tests are carried out when driving is suspended to add another rod.

Sensitivity studies show that the application of Equation 1.23 to a 50% consolidation ratio, that is, entering the second term with t_{50}, the time corresponding to 50% dissipation of the excess pore pressure in the test, minimizes the associated errors with the initial pore pressure measurement, as well as with the estimates of equilibrium pore pressure (Baligh and Levadoux, 1986). As shown in Figure 1.25, the theoretical value of T^* for a consolidation ratio of 50% is 0.245. The rigidity index, I_R, in the previous equation has the following expression:

$$I_R = \frac{G}{c_u} \tag{1.24}$$

where G is the shear modulus of the soil. Mayne et al. (2001) recommends the application of Equation 1.23 together with the chart in Figure 1.26 (Keaveny and Mitchell, 1986).

Robertson et al. (1992) evaluated a large number of dissipation test results. They concluded that the estimated c_h values by these tests are generally higher than those estimated in the laboratory. As both values typically underestimate real values, observed in practice, it seems valid to conclude that the estimates provided by the CPTu are closer to reality. This aspect reinforces the great convenience of using CPTu in projects involving soft soils, already emphasized in the comments made in Section 1.3.3.2.

1.3.3.3.3 Correlations between CPT and SPT results

Since SPT and CPT are the most widely used field tests in most countries, their execution at the same site in countless geotechnical investigation campaigns soon encouraged attempts to correlate the parameters N and q_c.

Figure 1.27 illustrates some relationships between the cone penetration resistance and the SPT blow count as a function of the soil behavior type index, I_c. Atmospheric pressure, p_a, is

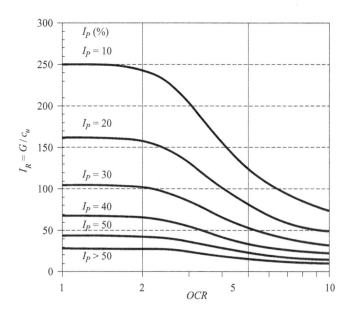

Figure 1.26 Chart for the assessment of the rigidity index, I_R, to be used with Equation 1.23 (Keaveny and Mitchell, 1986).

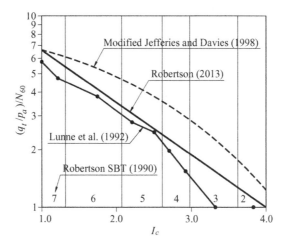

Figure 1.27 Correlations between q_t and N_{60} as a function of the soil behavior type index, I_c (adapted from Robertson, 2013).

involved in order to make the correlations independent on the units used for q_t. The proposal by Robertson (2013) may be expressed by the following approximate equation:

$$\frac{q_t}{p_a N_{60}} = 10^{(1.1268 - 0.2817\, I_c)}$$

(1.25)

The author notes that, in fine-grained soils with high sensitivity, this equation may overestimate the equivalent N_{60}.

It can be observed that the $(q_t/p_a)/N_{60}$ ratio is greater for coarser soils (lower values of the soil behavior type index, I_c). A qualitative explanation for this trend can be associated with the fact that the penetration resistance in SPT is highly dependent on the shear stresses mobilized at the external surface of the sampler, where the phenomena involved are clearly similar to those occurring at the CPT sleeve. Furthermore, the results of this test show that for fine-grained soils, the sleeve resistance is typically high when compared with cone resistance (see, for example, Figure 1.18).

The usefulness of this type of chart is due to the fact that relationships between CPT results and geotechnical parameters are numerous and considered to be reasonably accurate. On the other hand, there are a number of design methods based on the results of this test. Therefore, when, for a given site, only SPT results are available, it may be convenient, by using charts such as those of Figure 1.27, to "estimate" q_t values in order to apply the correlations and methods based on this parameter.

1.3.4 Dynamic probing test (DP)

1.3.4.1 Essential aspects of the equipment and test procedure

Dynamic probing tests are probably similar to the earliest means by which the characteristics of the subsoil have been investigated by our predecessors.

The test consists of determining the number of blows (N) of a hammer of mass (M) in freefall from a height (H) onto the assembly composed of, from top to bottom, an anvil, a string of rods and a conical tip (the base of which has an area A), so that a certain length of penetration (L) occurs. The diameter of the rods is smaller than the tip cone base, so,

Table 1.10 Type of dynamic penetrometers.

Type	Abbreviation	Mass of the hammer (kg)
Light	DPL	$M \leq 10$
Medium	DPM	$10 < M < 40$
Heavy	DPH	$40 \leq M \leq 60$
Super-heavy	DPSH	$60 < M$

Table 1.11 Referential characteristics for penetrometers (ISSMFE, 1989).

Characteristics	DPL	DPM	DPH	DPSH
Hammer mass, M (kg)	10	30	50	63.5
Height of fall, H (m)	0.5	0.5	0.5	0.75
Mass of the anvil and guide rod, B (kg)	6	18	18	30
Rod length (m)	1	1–2	1–2	1–2
Maximum mass of the rods, V (kg)	3	6	6	8
Outer diameter of the rods (mm)	22	32	32	32
Inner diameter of the rods (mm)	6	9	9	–
Cone apex angle (°)	90	90	90	90
Base area of the cone, S (cm²)	10	10	15	20
Penetration length, L (cm)	10	10	10	20
Test output	N_{10}	N_{10}	N_{10}	N_{20}
Range of number of blows	3–50	3–50	3–50	5–100
Specific energy per blow, $E_{s,DP}$ (kJ/m²)	50	150	167	238

theoretically, the penetration resistance results only from ground reaction forces on the conical surface of the tip.

Table 1.10 includes a classification of the penetrometers in terms of light, medium, heavy and super-heavy, depending on the mass of the hammer. In turn, Table 1.11 includes some referential characteristics for the four types of penetrometers while Figure 1.28 shows the rods and tips used in some of the dynamic penetrometers.

In order to compare the results of different penetrometers, it is common to adopt the so-called specific energy per blow, $E_{s,DP}$, which represents the kinetic energy of the hammer per unit area of the tip section, expressed by:

$$E_{s,DP} = \frac{MgH}{S} \tag{1.26}$$

Using two systems, I and II, with specific energy values $E_{s,DP}^{I}$ and $E_{s,DP}^{II}$, the respective results, N_I and N_{II}, corresponding to probing of rod lengths L_I and L_{II} in a given soil, will be related as follows:

$$E_{s,DP}^{I} \frac{N_I}{L_I} = E_{s,DP}^{II} \frac{N_{II}}{L_{II}} \tag{1.27}$$

This means that the number of blows required to obtain a unitary penetration length is inversely proportional to the specific energy per stroke. This type of comparison involves some reservations, since, as previously discussed, the energy transmitted to the ground also depends on other parameters not included in the previous equation, such as the weight of the anvil, the weight of the rods, etc.

Figure 1.29 illustrates tests using the light and super-heavy dynamic penetrometers.

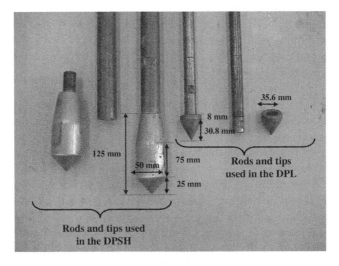

Figure 1.28 Rods and tips used in some of the dynamic penetrometers (photo: Carlos Rodrigues).

(a)   (b)

Figure 1.29 Execution of dynamic probing tests (a) DPL and (b) DPSH (photos: Carlos Rodrigues).

1.3.4.2 Comment regarding the use of dynamic probing tests: interpretation of results

There are no known methods by which to theoretically interpret test results with dynamic penetrometers in order to estimate mechanical parameters of the ground. On the other hand, the reliability of the available empirical correlations is not comparable to those associated with the SPT or especially the CPT. It should be reckoned that the diversity of equipment and the respective probing energies are clearly related to that lack of reliability.

Nevertheless, dynamic probing tests may be, in many situations, quite convenient because the systems are simple and economical, and the lighter versions are easily transportable, even to locations without access to motor vehicles.

It is, however, essential to take into account that dynamic penetrometers need to be combined with other *in situ* tests and borehole drilling. It may be a good option, for example in the geotechnical characterization of large areas, to carry out, in the immediate vicinity, conventional boreholes with SPT and penetration probes with a dynamic penetrometer. The sequence of the layers would be identified by the SPT tests and, for each layer – with the results of both types of tests – correlations between N(SPT) and N(DP) would be established, exclusively valid for the layer in question and for the penetrometer used. This would allow a wider grid of conventional boreholes (more expensive and time-consuming) to be combined with a tighter grid of dynamic probes (faster and cheaper), thus achieving a more complete knowledge of the ground. This option is particularly appropriate when defining the position of a stronger layer, the bedrock, between the (furthest) points of the main grid of boreholes.

In this context, it is important that the penetrometer used should match the energy appropriate to the type of ground and the maximum probing depth.

Another context in which the use of dynamic penetrometers, in particular DPL, has proved very useful is in the compaction control of road and railway earthfill works, in the latter case in conjunction with plate load tests (see Section 10.3.8).

1.3.5 Vane shear test (VST)

1.3.5.1 Essential aspects of the equipment and test procedure

The vane shear test is particularly suitable for estimating the undrained shear strength of soft clayey soils (from very soft to soft and medium clays). The test essentially consists of inserting into the ground a rod, to which a set of four rectangular blades, arranged in a cruciform pattern, are welded, as shown in Figure 1.30a. Once the device is positioned at the desired depth, a torque is applied to the rod, forcing it to rotate. This rotation movement will induce shear failure in a cylindrical shape, as the soil is detached from the surrounding ground (Figure 1.30b). The torsional moment is transmitted to the rod by a mechanical device placed on the ground surface, which imposes a constant rotation speed on the vane and allows determination of the applied moment versus angle of rotation diagram, typically similar to that of Figure 1.30c.

The vane can be introduced into the ground on a pre-drilled borehole to the depth of the layer to be characterized or pushed from the ground surface. In this case, it is essential that the vane crosses the ground, namely coarser and more resistant layers, surrounded by a protective device, as shown in Figure 1.31a. Once the layer to be characterized is reached, the vane is then detached from the protective casing and is positioned at the testing point (Figure 1.31b). At this last stage, the insertion of the apparatus has to be carried out with particular care, avoiding any rotation that may disturb the ground. In general, the test should be performed five minutes after insertion into place. Table 1.12 summarizes the main aspects of this test.

1.3.5.2 Interpretation of test results for deriving c_u

Considering an isotropic soil mass and assuming that the undrained shear strength of the soil is fully mobilized in the outer surface of the cylinder (lateral surface plus the two bases), such strength, c_{fv}, can be obtained from the following equation:

$$c_{fv} = \frac{2 \cdot M_{tf}}{\pi D^2 \left(H + \dfrac{D}{3} \right)} \tag{1.28}$$

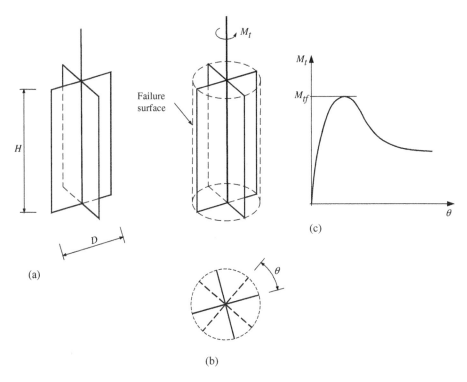

Figure 1.30 Vane shear test: a) vane; b) failure plans; c) torque versus angle of rotation diagram.

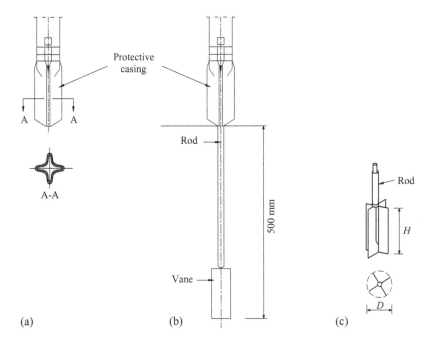

Figure 1.31 Advancement equipment for pushing the vane test: a) protective casing and vane shoe; b) vane ready for testing; c) general view of the vane.

Table 1.12 Main aspects regarding the vane shear test (ENV 1997-3: 1999).

Parameter	Value
H/D relation	2
Vane dimensions	Maximum $D \times H = 100$ mm $\times 200$ mm Minimum $D \times H = 40$ mm $\times 80$ mm
Blade thickness	0.8 to 3 mm
Rod diameter near the vane	≤ 16 mm
Extension rod diameter	≥ 20 mm
Testing procedure	
Rate of penetration in the ground near the testing point	≤ 20 mm/s
Depth of penetration (push method)	$\geq 5D$
Depth of insertion (pre-drilling method)	$\geq 5 \times$ borehole diameter
Minimum vertical distance between tests	0.5 m
Time interval between insertion and testing	2 to 5 min
Vane rotation rate	6°/min to 12°/min

This equation results from a simple equilibrium equation of moments when the soil resistance is mobilized: the maximum applied torque, or torsional moment, M_{tf}, and the integral of the undrained shear strength moment in each unit area of the outer cylindrical surface of the soil, in relation to the vertical axis of the apparatus.

The results of the application of Equation 1.28 are usually corrected by a dimensionless parameter, μ, in order to obtain a more accurate estimate of undrained shear strength (Bjerrum, 1972):

$$c_u = \mu \, c_{fv} \tag{1.29}$$

In Figure 1.32 are included two proposals for the parameter, μ, as a function of the plasticity index of the clay.

These corrections resulted from back-analyses of well-documented failure cases of embankments on soft clays. Both proposals were based on the same case histories. The most recent proposal by Azzouz et al. (1983) provides a lower corrective factor because the authors analyzed these failure cases, taking three-dimensional effects into account, which had been neglected in Bjerrum's original work, where a plane strain approach had been adopted.

Note that μ is a reducing coefficient for most soils, which means that the undrained shear strength directly derived from the test (Equation 1.28) corresponds to an overestimation of the observed resistance. According to Bjerrum (1972), this deviation is essentially linked to the very high rate of application of the shear stresses in the test. The vane test values of c_u after correction are generally close to those determined in the laboratory by direct simple shear tests.

The vane test is the most widely used means of characterizing the so-called sensitivity of clays, which is defined as the ratio of undrained shear strength of the undisturbed soil, c_u, and that of the remolded soil, c_{ur}:

$$S_t = \frac{c_u}{c_{ur}} \tag{1.30}$$

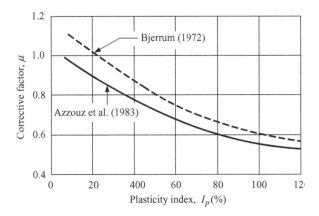

Figure 1.32 Corrections to the vane shear test results (Bjerrum, 1972; Azzouz et al., 1983).

Once the maximum torque has been determined, the vane is rapidly rotated ten times to remold the soil in the failure surface. Then, a new test is performed, similar to the initial one, both in the performance aspects and in the interpretation, leading to the value of the remolded strength, to be introduced into the denominator of Equation 1.30. This sensitivity parameter is useful for assessing soil susceptibility to loss of strength with strain.

1.3.6 Plate load test (PLT)

1.3.6.1 Essential aspects of the equipment and test procedure

The plate load test consists of the incremental loading of a circular steel plate, placed at the surface of the ground to be tested, by measuring the resulting settlement. It can thus be considered to be a simulation, on a reduced scale, of a shallow foundation. Notwithstanding this fact, as will be seen below, its purpose is not necessarily related to the design of such foundations.

Figure 1.33a shows a simplified schematic of the plate load test. The photo in Figure 1.33b illustrates a test setup in which the hydraulic jack acts against the rear axle of a loaded

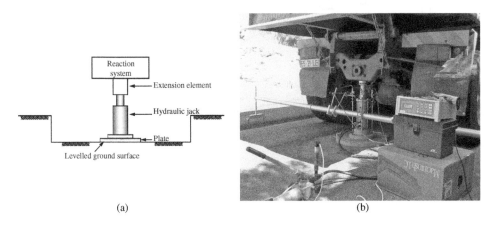

(a)   (b)

Figure 1.33 Plate load test: a) simplified schematic; b) test setup where the hydraulic jack acts against the rear axle of a truck, also showing the settlement measuring system (photo: Carlos Rodrigues).

truck (with cement bags, concrete blocks or other material) in order to increase the available weight, hence the maximum reaction. In the figure, the system for settlement measurement is also visible: deflectometers (in this case, three) are installed at the jack and connected to a steel beam positioned sufficiently far from the plate so that it can be considered fixed.

It is known that the depth of the soil mass beneath the loaded area, that influences the measured deformation response, depends on the dimensions of such area, namely the diameter of the plate. Hence, the diameter of the plate should be as large as possible because: i) for the prediction of the behavior of shallow foundations, given their dimensions, a substantial thickness of the soil mass will be involved; and ii) the quality of the estimates improves with the increase in the tested volume of soil, especially in heterogeneous soil masses.[9]

However, it must be noted that the use of large diameter plates for application of stresses similar to those transmitted by the structure under study will normally require very high loads. The implementation of a system capable of providing a reaction to the jack under such circumstances generally entails very high costs and extended deadlines. In view of that, it is common to adopt stresses equal to or greater than those expected for the structure under study and, for these, to select the appropriate plate diameter, considering the maximum available reaction. The most common diameters generally range between 0.30 m and 0.80 m, and the respective thickness is large enough for the plate to behave rigidly.

1.3.6.2 Interpretation of results

1.3.6.2.1 Theoretical approach to obtaining mechanical soil parameters

Figure 1.34 shows a typical applied pressure–settlement diagram. As can be seen, two loading and unloading cycles are usually performed. The first, lower-amplitude cycle is intended to ensure close contact between the plate and the surface of the ground, since this, despite being previously leveled, may not be completely flat.

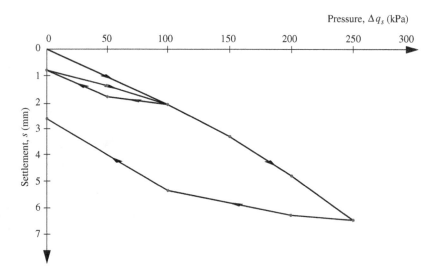

Figure 1.34 Typical pressure-settlement diagram of a plate load test.

---

[9] In particular, at a depth twice the plate diameter, the incremental stresses in the ground correspond to a minimal fraction (about 10%) of the applied pressure at the surface (see Figure A6.2.3).

If the test is conducted until failure of the foundation soil, Equations 6.16 and 6.15 in Chapter 6 may be applied for the evaluation of the undrained shear strength, c_u, or the effective strength parameters, c' or ϕ', according to the type of loading in undrained or drained conditions, respectively.[10]

This is not, however, the most usual way of interpreting the results of plate load tests. Indeed, under current conditions, the test is conducted involving much lower stresses than those corresponding to soil failure, allowing the estimation of the deformation modulus, by means of the interpretation of results in the light of the theory of elasticity. As will be seen in Chapter 6, the settlement, s, of a rigid shallow foundation, of circular area of diameter B, for an applied loading pressure Δq_s, on a linear and homogeneous elastic half-space, with a Young's modulus E and Poisson's ratio ν, corresponds to (according to Section 6.3.2):

$$s = \frac{0.79 \Delta q_s B \left(1 - \nu^2\right)}{E}$$

(1.31)

This equation allows to evaluate the stiffness modulus of the soil, taking the coordinates s and Δq_s from a point on the test diagram and adopting a value for ν. In the case of clayey soils, the loading of the test occurs in undrained conditions, for which a Poisson's ratio equal to 0.5 should be adopted. For granular soils, the stiffness modulus is estimated under drained conditions, and it is reasonable to adopt values of 0.2 to 0.3 for Poisson's ratio. For these soils, intended for loading under drained conditions, there is a need to wait for the settlements to stabilize at each loading step. In certain cases, when the soil has a significant fines content, this may require some time, leading to a rather time-consuming test.

There is no single criterion for choosing the point in the diagram at which to proceed with the calculation presented above. If the objective is to predict stiffness for the loaded soil at a given value of Δq_s, it seems reasonable to select the point where the ordinate corresponds to that value.[11]

When these tests are performed to characterize foundation soils or embankments for road or railway works, it is common to designate the stiffness moduli obtained for the 1st and 2nd loading cycles by EV_1 and EV_2, respectively. In view of what was previously mentioned regarding the two loading cycles, it should be understood that the EV_2 modulus is considered the more representative, although the EV_2/EV_1 ratio should also be taken into account (AFNOR, 2000).

It should also be noted that Equation 1.31 is valid when the test is performed on the ground surface or at the base of a preliminary excavation (which should be at least five times wider than the plate diameter). Otherwise, the ground stresses above the test platform will tend to influence the results, reducing the settlement, and thus leading to an overestimation of the modulus. In these cases, the use of a correction factor is recommended (see Section 6.3.3.1).

---

[10] In this case, with the plate at the ground surface or at the base of an excavation (much larger than the plate diameter): (i) in a sandy soil, the vertical bearing capacity, q_{ult}, is reduced to the third term of the second member of Equation 6.15, whereby the test allows determination of the value of the coefficient N_γ and then, by means of Equations 6.13 or 6.14, the angle of shearing resistance; or ii) in a soil that also exhibits shear strength due to cohesion, the vertical bearing capacity, q_{ult}, results from the first and third terms of the second member of Equation 6.15, for which the evaluation of c' and ϕ' will require the determination of q_{ult} by means of two tests with different plate diameters, thus obtaining those two effective shear strength parameters from two independent equations.

[11] If the second load cycle is taken into account, the value of the ordinate, s, should naturally ignore the settlement corresponding to the end of the first unloading cycle.

1.3.6.2.2 Direct use of test results

When the test is conducted to predict settlements of a shallow foundation, the previously exposed approach may not be the most advantageous because the deformation modulus obtained is representative of a very superficial ground horizon, the thickness of which is of the same order of magnitude as the plate diameter. That modulus may be considerably lower than that affecting the behavior of the actual foundation because, even in a homogeneous soil mass, stiffness typically increases with depth, due to the increase of the effective stresses.

Considering, as shown in Figure 1.35a, a footing and a plate with the same geometric shape and under the same applied pressure, the ratio of the respective settlements, s_f/s_p, would increase linearly with the ratio of the respective diameters, B_f/B_p, in cases where the modulus is constant in depth (see Equation 1.31). As will be discussed in detail in Chapter 6, the increase of the modulus with depth reduces the influence of the size of the foundation on the settlement (see Section 6.3.2.5). This is reflected in Figure 1.35b, which illustrates a proposal for the evaluation of the footing settlement, based on the plate settlement, for the same pressure, in sands (Bergdahl et al., 1993). This proposal is valid if the sand layer has a thickness of at least twice the diameter of the foundation.

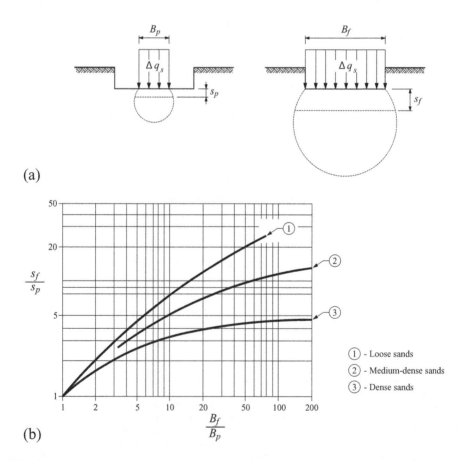

Figure 1.35 Example of direct use of plate load test results: a) plate and footing of the same geometry, under the same pressure, on sandy soil; b) the ratio of the footing and plate settlements versus the ratio of the respective diameters (Bergdahl et al., 1993).

1.3.6.3 Final remark

As will be seen in Chapter 10, plate load tests are of great importance in characterizing the quality of compacted fills for road and railway works.

In addition, when the test is performed with the largest plates of the aforementioned range, allowing for a much larger volume of soil than most other field tests, it is especially suited for the characterization of the stiffness of natural soil masses or embankments composed of coarse gravels or other large particles. One aspect that severely limits a broader application of the test is that it requires direct access to the layer to be characterized, so that its implementation becomes difficult – and therefore even more costly and time consuming – at significant depths.

1.3.7 Cross-hole seismic test (CHT)

1.3.7.1 Essential aspects of the equipment and test procedure

The cross-hole seismic test consists of inducing the generation of shear waves (S-waves) in the ground at a certain depth by means of an impact inside a borehole and recording the wave arrival at a receiver at the same depth in an adjacent borehole. Since the distance between the two boreholes is known and the time interval between the impact and the wave arrival has been measured, the propagation velocity of the S-wave, V_s, can be directly determined by the distance-over-time ratio. This test is commonly performed at consecutive depths with a spacing of about 2 m.

Figure 1.36 represents the setup for a cross-hole seismic test.

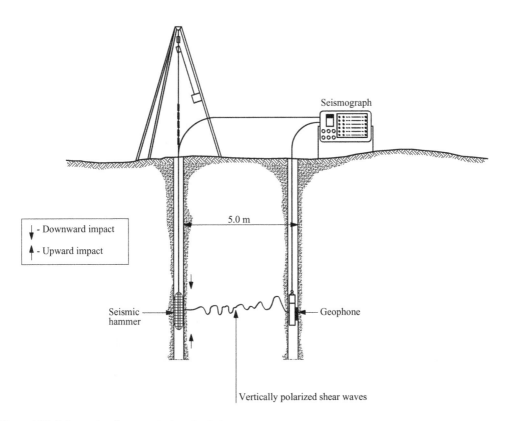

Figure 1.36 Schematic of a cross-hole seismic test.

The distance between the parallel boreholes is usually 4 m to 5 m (distances outside the range of 3 m to 6 m are not adopted). Usually, at least one of the boreholes is used for definition of the layer sequence, sample collection, execution of other tests, namely SPT, etc. The boreholes are lined with metal casing tubes sealed to the borehole walls with cement grout, to ensure coupling for adequate wave transmission. It is necessary to wait for the grout to harden before performing the seismic tests.

The basic testing equipment consists of the following devices:

a) a seismograph for observing, recording, and storing the test results for later analysis and interpretation;
b) a seismic hammer, triggered from the surface, which is responsible for the downward or upward impact on the anvil, which is attached to the walls of the borehole casing at the testing depth, by a hydraulic or pneumatic system;
c) a receiver – usually a triaxial geophone – placed in the other borehole at the same depth.

Since the drilling methods currently used may allow for significant deviations from the vertical in relatively deep boreholes, it is advisable to carry out inclinometer measurements on the two boreholes in order to accurately determine the wave travel length.

1.3.7.2 Interpretation of results

Assuming a linear elastic behavior for the soil, the shear wave velocity is related to the elastic shear modulus, G_0, by the following theoretical relationship:

$$G_0 = \rho\, V_s^2 = \frac{\gamma}{g}\, V_s^2 \tag{1.32}$$

where ρ is the mass density of the soil at the testing depth and g is the acceleration due to gravity.

It is also common to use the G_{max} symbol (as an alternative to G_0) for the elastic shear modulus calculated from V_s, since the strain levels involved in the cross-hole seismic tests are within a very small range, typically of the order of 10^{-6}, and fully reversible. Hence, it is completely legitimate to interpret this test in light of the theory of elasticity and to designate the derived modulus as elastic.

This seismic test also allows measurement of the compression wave (P-wave) velocity, V_p, which is higher than V_s, so that it is not difficult to distinguish its arrival in the received signal. Based on V_p and V_s, it is possible to determine Poisson's ratio by means of the theoretical expression:

$$\nu_{dyn} = \frac{1}{2}\, \frac{\left(\dfrac{V_p}{V_s}\right)^2 - 2}{\left(\dfrac{V_p}{V_s}\right)^2 - 1} \tag{1.33}$$

The index *dyn* used in the previous expression to compose the symbol ν_{dyn} aims to emphasize that the values of the Poisson's ratio obtained are normally used in dynamic analyses. In these analyses, taking $G_{dyn} = G_0$, the corresponding elastic modulus, E_0 (or E_{max}), can be obtained from the equation:

$$E_0 = 2(1+\nu)G_0 \tag{1.34}$$

where it is recommended, for small-strain levels, to take $\nu = 0.2$ for drained static monotonic loading and $\nu = 0.5$ for undrained loading conditions (Mayne et al., 2001).

From V_p and V_s, it is also possible to estimate the porosity of saturated soils, by means of the following theoretical equation (Foti et al., 2002):

$$n = \frac{\rho^S - \sqrt{\left(\rho^S\right)^2 - \dfrac{4\left(\rho^S - \rho^F\right)K^F}{V_p^2 - 2\left(\dfrac{1-\nu^{SK}}{1-2\nu^{SK}}\right)V_s^2}}}{2\left(\rho^S - \rho^F\right)} \tag{1.35}$$

in which ρ^S represents the mass density of the solid particles, ρ^F and K^F represent the mass density and the bulk modulus of the pore fluid (usually water), respectively, and ν^{SK} represents the Poisson's ratio of the soil skeleton. It is known that, in most soils, ρ^S is about 2.65–2.70 t/m³, ρ^F corresponds to 1 t/m³ and K^F can be taken to be equal to 2.25×10^6 kPa. Foti et al. (2002) show that the results are rather insensitive to the value of ν^{SK}, suggesting the use of 0.25 for this parameter.

Equation 1.35 results from the simplification of another exact equation, under the hypothesis that the volumetric compressibility of the solid particles is negligible. This simplification leads to a slight overestimation of the porosity (generally of about 10%).

Based on porosity, the void ratio, e, and the soil unit weight, γ, can be calculated. The usefulness of this equation can be understood by bearing in mind that, when direct access to the soil is not feasible, the determination of those parameters would require the collection of undisturbed samples.

1.3.7.3 Final remark

In recent decades, cross-hole seismic tests have become one of the most important field tests in soil mechanics, by allowing the evaluation of a fundamental reference parameter, the small-strain shear modulus, G_0, as well as the corresponding Young's modulus, E_0, which, in this case, can appropriately be designated the *elasticity modulus*.

As will be seen below, the strain levels associated with the vast majority of engineering works are substantially higher than those induced by the cross-hole seismic test. Since soil behavior is markedly non-linear, stiffness decreases with the strain level. Therefore, the stiffness moduli obtained from this test cannot be directly used in deformation analyses, requiring correction. This will be discussed later in this chapter (see Section 1.5).

Another important aspect related to the cross-hole seismic test is that there is no depth limitation, since it is performed between drilled boreholes, and it is applicable to all types of ground, from the softest to the most resistant, and from the finest to the coarsest soils. None of the other tests introduced in this chapter exhibit such an aspect, which alone would attest to the relevance and applicability of this test. In certain situations, it is in fact the only applicable *in situ* test, as in the case of gravelly soil layers, which often compose the base strata of the Holocene formations in alluvial valleys, where pebbles over 20 cm in diameter may appear.

The results of the cross-hole seismic tests are also fundamental for dynamic analyses of the soil masses. For example, the classification of soil types for foundation design in Eurocode 8 (EN 1998-1:2004), for the calculation of seismic actions, is based on shear wave velocity values, V_s (see Table 3.12).

1.3.8 Down-hole seismic tests

Down-hole seismic methods include the down-hole seismic test, DHT, the seismic cone penetration test, SCPT, and the seismic flat dilatometer test, SDMT.

An economical alternative to the cross-hole seismic test is the so-called down-hole test, which makes use of only one borehole, into which the receiver is positioned, while the S-waves are generated at the surface near the top of the borehole, as schematically shown in Figure 1.37a.

In these methods, the most common system for generating S-waves makes use of a wooden or steel beam laid on the ground surface, on which the wheel of a loaded vehicle or truck is positioned. The waves are generated by means of the percussion of a hammer at one end of the beam. As in the cross-hole seismic test, the geophone is successively placed at different depths in the borehole. Another economical aspect in the present case is that any vertical deviation of the borehole has very little repercussion on the wave travel length, and therefore inclinometer measurements are dispensable.

One particularly attractive down-hole test is the so-called seismic cone, SCPT (or seismic piezocone, SCPTu), which consists of the test with the cone (or piezocone), the tip of which is provided with a receiver, as shown in Figure 1.37b (Robertson et al., 1986). In this case, it is common to measure V_s at every 1 m depth, when the CPTu driving is interrupted to add another rod to the rod string.

A conceptually similar setup has been devised using the flat dilatometer test, designated the seismic dilatometer test, SDMT, as detailed in Mayne et al. (1999) and Foti et al. (2006). The flat dilatometer test is treated below (see Section 1.3.11).

The distance between the point at the surface where the S-waves are generated and the geophone, divided by the travel time, provides a weighted average value of V_s in the several crossed layers in each record. However, it is possible to calculate V_s for each depth (Salgado, 2008). Consider two consecutive records, with the geophone at depths z and $z+\delta z$, travel times of t_z and $t_{z+\delta z}$, respectively, and x as the horizontal distance from the wave source to the top of the borehole. The shear wave velocity V_s between the two depths is the ratio of the difference of the distances traveled by the difference between the respective travel times:

$$V_s = \frac{\left[(z+\delta z)^2 + x^2\right]^{0.5} - \left(z^2 + x^2\right)^{0.5}}{\left(t_{z+\delta z} - t_z\right)}$$

(1.36)

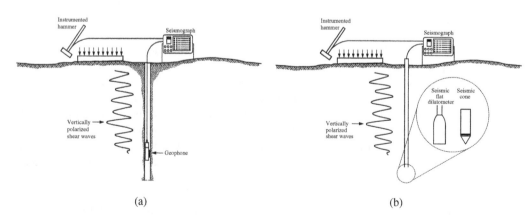

(a) (b)

Figure 1.37 Schematic of the down-hole test: a) using a borehole; b) using the seismic cone or the seismic flat dilatometer.

1.3.9 Self-boring pressuremeter test (SBPT)

1.3.9.1 Essential aspects of the equipment and test procedure

The self-boring pressuremeter (SBPT) was developed at the University of Cambridge and it is marketed under the name "Camkometer" (Wroth and Hughes, 1973). Figure 1.38 shows an image of the apparatus prior to its insertion into the ground.

It is a costly device, with a complex and time-consuming operation, requiring highly specialized staff for its operation. On the other hand, it is the only *in situ* test capable, at least potentially, of providing estimates of the stress state, stiffness, and strength parameters of the soil mass, and the respective results can be interpreted theoretically with acceptable approximations. Among the potentialities of this test, the possibility of estimating the at-rest horizontal stresses of the soil should be noted, thus resulting in perhaps the only reliable means of directly determining the coefficient of earth pressure at-rest, K_0. Although, in current testing practice, difficulties may appear that compromise the quality of its determinations, there is no doubt that the possibilities offered by the SBPT are extraordinary.

Figure 1.38 Cambridge self-boring pressometer, SBPT (photo: António Sousa Coutinho).

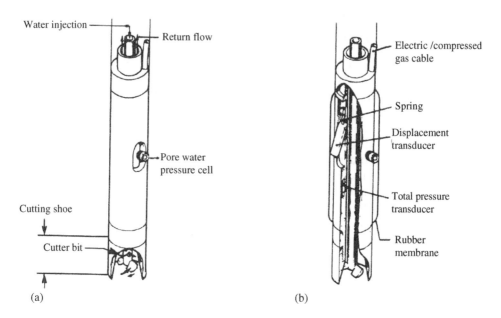

Figure 1.39 Schematic of the SBPT: a) prior to the application of the internal pressure; b) under the application of the internal pressure (adapted from Jamiolkowski et al., 1985).

Figure 1.39 shows a schematic of the apparatus (Windle and Wroth, 1977; Jamiolkowski et al., 1985). In simplified terms, the pressuremeter consists of a cylindrical cell measuring 1.0 m in height and 8 cm in diameter, where the central area of the cylindrical surface, more precisely the intermediate 64 cm, is covered by a rubber membrane. In Figure 1.38, the surface of the cell is surrounded by a protective metal jacket – the so-called Chinese lantern – in order to avoid direct contact of the solid grains with the membrane.

The pressuremeter is self-boring, that is, it is capable of opening a borehole in the ground by its own means with removal of the soil as it penetrates the soil mass. At the desired depth, by means of internal compressed air pressure, the membrane is forced to expand, as shown in Figure 1.39b. The radial deformation is measured by three transducers arranged at mid-height of the membrane at circumferentially spaced points of 120°. In addition to the internal pressure and radial deformation, the pore water pressure is also measured by two diametrically opposed transducers located in the central zone of the membrane.

The cell body is a rigid hollow cylinder, so that, during insertion of the apparatus into the ground, before the pressure acts on the membrane, it remains cylindrical in shape with the same diameter as the upper and lower rigid parts. This detail is essential to avoid decompression of the surrounding soil.

For boring, the pressuremeter is provided with a cutting shoe at the bottom with the same outer diameter, which houses a rotary tool capable of demolishing the ground. Through an axial tube, water is injected to the front at low pressure, which returns to the surface with the drilling debris through the space between the tube and the cell body.

The self-insertion in the soil is of fundamental importance, not so much for eliminating the need for pre-drilling, but mainly because the apparatus is positioned in the testing point without introducing deformations in the soil, without changing its stress state, and without inducing disturbances in the soil that would significantly affect its response to the loading subsequently applied by the device.

Nevertheless, the pressuremeter test can naturally be carried out, as many other field tests already mentioned, together with boreholes. In order to do this, it is necessary to introduce the apparatus to the bottom of the borehole and then allow it to penetrate the ground by its own means to a depth considered sufficiently distant from the disturbed soil due to borehole drilling. It should be noted that this process is often followed exclusively to facilitate the pressuremeter tests, because its drilling process is very slow, even unfeasible in some coarser soils, and also because this is the most damaging operation for the apparatus, particularly for the membrane.

1.3.9.2 Interpretation of results

The pressuremeter loading on clayey soils can usually be considered undrained, in view of the relatively rapid rate at which the cell expands. For sandy soils, the loading conditions can be considered to be drained, which can also be confirmed during the test by the pore pressure measurement.

Figure 1.40a shows the results of a pressuremeter test in a soft clay of the Tagus River valley. The horizontal axis represents the radial strain (that is, the ratio of the radius variation of the cylindrical unit to the initial value of the radius), ε, while the ordinates represent the pressure applied to the membrane, Ψ.

The determination of the at-rest total horizontal stress (marked in the figure on the ordinate axis, 129 kPa) is carried out by analyzing the initial part of the stress–strain experimental curve, as shown in Figure 1.40b. In fact, assuming that the penetration by the pressuremeter did not alter the at-rest stress state, the membrane expansion should only start when the applied pressure to the ground exceeds σ_{h0}. In fact, there is a small deformation of the system up to a certain pressure value, the so-called *lift-off*, from which an abrupt increase in the radial strains occurs as a result of the expansion of the cylindrical cavity by deformation of the surrounding soil.

The at-rest horizontal stress, σ_{h0}, is considered to be equal to the value of the lift-off pressure. Since the at-rest total stress and the equilibrium pore pressure are known, the coefficient of earth pressure at-rest, K_0, can be computed.

The interpretation of the test, with respect to the stiffness parameters and shear strength, is carried out in the light of the theory of expansion of a cylindrical cavity of infinite height in an infinite, homogeneous, and elastic–perfectly plastic space.

In this case, the cavity expansion implies that the material is loaded in pure shear (the normal incremental octahedral stress is null), so that the measured stiffness parameter is the shear modulus, G.

By means of this theory, it is possible to prove that, in the small-strain domain:

$$G = \frac{1}{2}\frac{d\psi}{d\varepsilon}$$ (1.37)

In order to ensure that the estimates of G are not affected by any potential errors in the determination of the lift-off pressure, it is common to determine G from the slope of the unloading-reloading cycles, as shown in Figure 1.40a.

By means of the same theory, assuming undrained loading conditions, the undrained shear strength is given by the expression:

$$c_u = \frac{d\psi}{d\left(\dfrac{\Delta V}{V}\right)}$$ (1.38)

where $\Delta V/V$ is the volumetric strain of the cavity (Figure 1.40c).

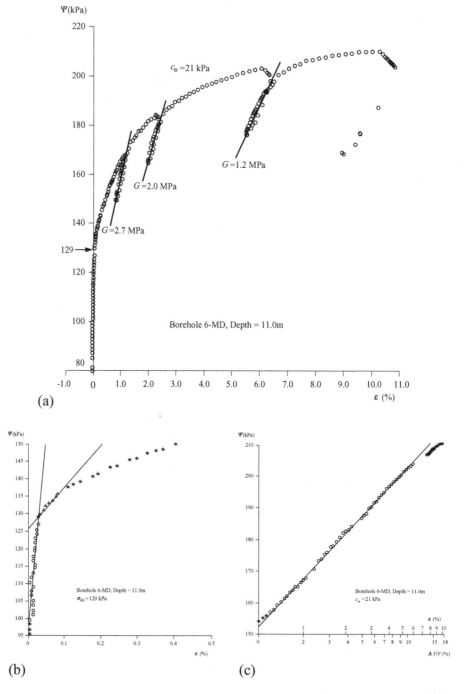

Figure 1.40 Results of a Camkometer test in the Tagus River muds: a) stress-radial strain diagram; b) determination of σ_{h0} (*lift-off* pressure); c) determination of c_u (courtesy of Estradas de Portugal).

For drained loading conditions, the angle of shearing resistance can be determined from the following equation:

$$\frac{d\ln(\psi - u_0)}{d\ln\varepsilon} = \frac{2\sin\phi'}{(1 + \sin\phi') + K(1 - \sin\phi')} \tag{1.39}$$

where u_0 is the equilibrium pore pressure at the testing depth and K can be computed as:

$$K = \frac{1 + \sin\phi'_{cv}}{1 - \sin\phi'_{cv}} \tag{1.40}$$

in which ϕ'_{cv} is the critical state or constant volume friction angle.

1.3.10 Ménard pressuremeter test (PMT)

1.3.10.1 Essential aspects of the equipment and test procedure

The pressuremeter test developed by Ménard (1956) preceded the Cambridge self-boring pressuremeter test by about two decades and it is distinguished by the fact that a borehole must be executed prior to the test itself and that the cell in contact with the ground is not equipped with instrumentation. Figure 1.41 shows a simplified scheme of the apparatus. The pressuremeter consists of three separate cylindrical rubber cells, in which the upper and lower cells are called *guard cells* and the central cell is the *measuring cell*.

After the pressuremeter is introduced into the borehole, equal internal pressure is installed in all three cells by means of water injection into the measuring cell and of gas into the guard cells. The pressure is increased by increments, normally of 1 min in duration, and, for each increment, the following data are recorded: i) the volume injected into the cell after 1 min; and ii) the injected volume change between 30 s and 1 min. From these records, the graphs outlined in Figure 1.42 are plotted. The number of increments is usually around ten.

The guard cells are intended to restrict the deformation of the measuring cell in the vertical direction. The soil deformation process, in the horizontal plane at the depth of the

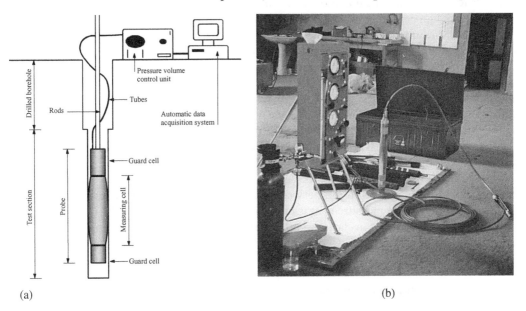

(a)   (b)

Figure 1.41 Ménard pressuremeter test: a) schematic of the test; b) image of the apparatus (photo: Carlos Rodrigues).

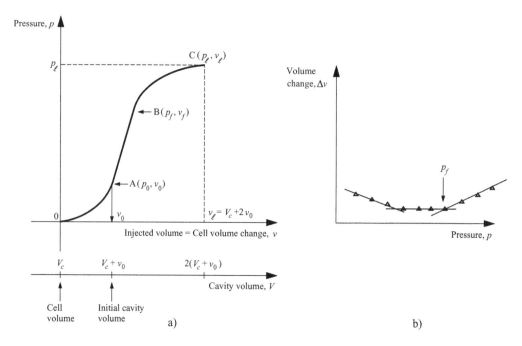

Figure 1.42 Typical PMT diagrams: a) injected volume versus applied pressure; b) applied pressure versus injected volume variation of the last 30 s for each increment.

measuring cell, corresponds to the expansion of a cylindrical cavity of infinite vertical dimension, therefore, to a plane strain state.

There are at least six different diameters for the pressuremeter, between 35 mm and 73 mm, varying the ratio of the measuring cell height to the diameter between four and six. For example, the pressuremeter with 58 mm in diameter and 210 mm in height is frequently used. The borehole for introducing the pressuremeter is usually performed using a manual or mechanical auger, with a diameter about 10% greater than that of the pressuremeter. This means that, in the case of PMT, unlike other tests such as SPT, VST or CHT, conventional boreholes are not used, rather that specific boreholes are drilled for the execution of PMT.

The drilling operation seriously affects the quality of the test results. Boreholes executed with insufficient care, inducing significant disturbance in the surrounding soil, or with equipment unsuitable for the type of soil, may compromise the interpretation of the results, giving rise to relationships that are different from those shown in Figure 1.42. This is considered one of the major disadvantages of the test.

As stated above, the pressuremeter cell is not instrumented and the measurements, namely the injected volume and pressure, are recorded in the control unit. This requires the introduction of corrections: for example, the cell pressure is equal to the pressure measured in the control unit plus the water column between the surface and the testing depth.

If pore water pressure measurement is not available, it is assumed that the test is conducted under undrained conditions in clays and under drained conditions in clean or low-fines-content sands. In intermediate graded soils, such as residual soils from granite, it may be recommended to carry out longer increments to ensure that the test is conducted under drained conditions.

1.3.10.2 Interpretation of results

Figure 1.42a illustrates the fundamental relationship between pressure and the water volume injected into the cell, while the relationship in Figure 1.42b represents the so-called creep relationship, relating the (final) pressure with the volume change at the last 30 s of each increment. Note that the test is initiated with the cell filled with water, under a pressure corresponding to the water column up to the surface. Therefore, the injected volume coincides with the volume variation of the cell. For these quantities, the symbol v will be used, whereas, for the cavity volume, the symbol V will be used.

The injected volume versus applied pressure relationship first has a curved section (OA), concave upward and nearly horizontal tangent at the origin, because, at the beginning, there is a gap between the pressuremeter and the walls of the borehole. This is followed by a linear section (AB), interpreted as the phase where the soil surrounding the cell deforms in an "elastic" regime. The final part of the diagram (BC), interpreted as the deformation of the soil in a "plastic" regime, is curved and concave downward, tending towards a horizontal asymptote, which defines the so-called limit pressure, p_ℓ.

It should be noted that, strictly speaking, the initial volume of the cavity is not *a priori* known. Considering that the cell is perfectly adjusted to the ground at the beginning of the linear central section, as shown in Figure 1.42a, the initial volume of the cavity, V_0, is then defined as:

$$V_0 = V_c + v_0 \tag{1.41}$$

where V_c is the initial (known) volume of the cell and v_0 is the abscissa of point A.

Sometimes, the horizontal asymptote is not reached, in which case the limit pressure is taken as the ordinate of the relationship, which corresponds to an injected volume that doubles the initial volume of the cavity.[12] Therefore, the respective abscissa, v_ℓ, is equal to (taking into account, in particular, Equation 1.41):

$$v_\ell = v_0 + V_0 = V_c + 2v_0 \tag{1.42}$$

Point B defines the end of the central part of the relationship and is identified on the basis that its ordinate, p_f, termed creep pressure, is defined according to the relationship in Figure 1.42b.

Thus, assuming that, between A and B, the soil deforms in pure shear in an elastic regime, the following can be written:

$$\Delta p = G \frac{\Delta V}{V} \tag{1.43}$$

where ΔV is the volume change in the cavity of volume V. In the test, $\Delta V = \Delta v$, hence the previous equation can be rewritten as follows:

$$G = V \frac{\Delta p}{\Delta v} \tag{1.44}$$

The ratio $\Delta p/\Delta v$ is constant between A and B, whereas V is not. In the conventional interpretation of PMT, this cavity volume can be taken as the average value between A and B, therefore:

$$V = V_{\text{ave}} = \frac{v_0 + v_f}{2} + V_c \tag{1.45}$$

---

[12] It should be noted that, if the initial diameter of the borehole is much larger than that of the cell, the volume v_0 required to adjust the cell to the ground can be quite high, with the risk of depleting the available volume of water in the control unit before the limit pressure is reached.

from which results:

$$G_{PMT} = V_{ave} \frac{\Delta p}{\Delta v} \tag{1.46}$$

and the so-called pressuremeter modulus (with the physical meaning of Young's modulus):

$$E_{PMT} = 2(1+v)\, G_{PMT} \tag{1.47}$$

Thus, it can be concluded that PMT results are the values of the pressuremeter modulus and the limit pressure, E_{PMT} and p_ℓ, respectively. Since the test is performed in a pre-drilled borehole, the disturbance and decompression of the surrounding soil lead to curves that are quite different from those obtained with the self-boring pressuremeter. This fact represents a serious limitation to the use of PMT results to estimate strength and stiffness parameters – and even more so the at-rest stress state – to be applied in the analysis of geotechnical structures by theoretical methods.

From early on, the French authors most closely involved with PMT developments understood this limitation. They proposed to overcome this limitation, on the basis of far-reaching studies, empirically seeking, through experimental evidence, to relate the settlements of shallow foundations and the bearing capacity of shallow and pile foundations on different types of ground, with PMT results (Frank, 2003). These well-documented design methods are established in official technical documents in France, where these design methods are most popular, but its use has also been reported internationally. Examples of such design methods for shallow and pile foundations are provided in Chapters 6 and 7, respectively.

PMT is particularly useful in the characterization of hard soils and soft rocks, formations where other tests, such as SPT and especially CPT, have limited or no application. Coincidently, it is in this type of soils that the execution of pre-drilled boreholes is most satisfactory.

1.3.11 Flat dilatometer test (DMT)

1.3.11.1 Essential aspects of the equipment and test procedure

The flat dilatometer, also referred to as the Marchetti dilatometer, since this author was responsible for the development of the device and its first characterization studies (Marchetti, 1975, 1980), consists of a narrow stainless steel cell (height 225 mm, width 95 mm and thickness 15 mm), with a beveled lower end (cutting angle of 14°), which is statically pushed through the ground at a rate of 20 mm/s (equal to the CPT). Figure 1.43 shows an outline of the apparatus. In the central part of one of the cell faces, there is a 60-mm diameter flexible circular steel membrane, where the center point of its inner face is connected to a displacement transducer.

When the dilatometer is positioned at the testing point, the membrane is inflated by injecting gas (nitrogen, carbon dioxide or air) under pressure into the cell, and two types of readings are taken: i) the A-pressure, recorded about 15 s after injection, which corresponds to the so-called lift-off pressure, that is, to the pressure which places the membrane, initially retracted into the cell, in the same plane as the surrounding soil face; and ii) the B-pressure, recorded 15 to 30 s after A, corresponding to the pressure required to move the center point of the membrane 1.1 mm towards the soil. After this reading, the pressure in the cell is removed, the membrane automatically retracts inwards, and further advancement is performed by pushing the blade to the next testing depth, typically 20 cm to 30 cm below the previous one.

The two pressure readings mentioned, after applying corrections related to membrane stiffness and the transducer, give rise to pressures p_0 and p_1, known as lift-off and expansion pressures, respectively. From these pressures, the following three parameters, designated

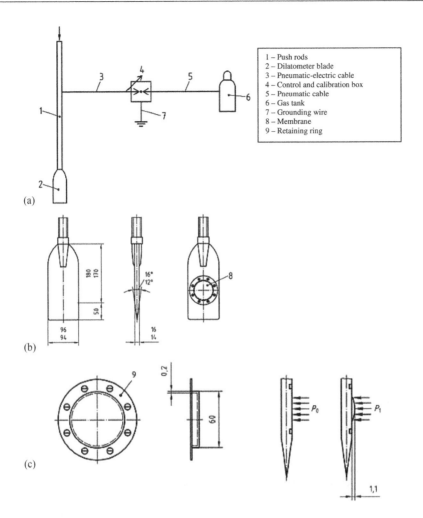

1 – Push rods
2 – Dilatometer blade
3 – Pneumatic-electric cable
4 – Control and calibration box
5 – Pneumatic cable
6 – Gas tank
7 – Grounding wire
8 – Membrane
9 – Retaining ring

Figure 1.43 Flat dilatometer test: a) general layout of the test; b) blade details; c) flexible membrane details (adapted from EN 1997-2, 2007).

material index, I_{DMT}, dilatometer modulus, E_{DMT}, and horizontal stress index, K_{DMT}, are calculated, as follows:

$$I_{DMT} = \frac{p_1 - p_0}{p_0 - u_0} \qquad (1.48)$$

$$E_{DMT} = 34.7\left(p_1 - p_0\right) \qquad (1.49)$$

$$K_{DMT} = \frac{p_0 - u_0}{\sigma'_{v0}} \qquad (1.50)$$

where u_0 is the value of the equilibrium pore pressure and σ'_{v0} is the overburden effective vertical stress at the testing depth.[13]

---

[13] For the quantities defined in Equations 1.48 to 1.50 the symbols I_D, E_D and K_D are most commonly used. These

1.3.11.2 Interpretation of results

Experience has shown that the material index, I_{DMT}, is very consistent with respect to soil classification, varying accordingly with grain size distribution, as summarized in Table 1.13 (Marchetti, 1980).

With regard to the meaning of the horizontal stress index, K_{DMT}, it seems reasonable to accept that it is related to the coefficient of earth pressure at-rest. Indeed, if the insertion of the blade did not cause any deformations to the surrounding ground, p_0 would correspond to σ_{h0}, hence K_{DMT} would coincide with K_0. Since that hypothesis is not true, the relationship between these two coefficients can be established only empirically.

The following empirical relationship between K_{DMT} and K_0 has been applied with satisfactory results for soils with I_{DMT} below 1.2 (Marchetti et al., 2001):

$$K_0 = \left(\frac{K_{DMT}}{1.5}\right)^{0.47} - 0.6 \tag{1.51}$$

Empirical relationships have also been established between K_{DMT} and soil strength parameters, namely the undrained shear strength of clays and the angle of shearing resistance of sands. For the latter, the following expression has been proposed (Marchetti et al., 2001):

$$\phi'\left(^{\circ}\right) = 28 + 14.6\log_{10} K_{DMT} - 2.1\left(\log_{10} K_{DMT}\right)^2 \tag{1.52}$$

valid for I_{DMT} greater than 1.8.

For the estimate of the undrained shear strength, the following equation is recommended (Marchetti et al., 2001):

$$c_u = 0.22\,\sigma'_{v0}\left(0.5 K_{DMT}\right)^{1.25} \tag{1.53}$$

valid for I_{DMT} less than 1.2.

The dilatometer modulus, E_{DMT}, derives from an interpretation of the test based on the theory of elasticity, assuming that the soil surrounding the cell corresponds to two elastic half-spaces, separated by the symmetry plane of the cell.

For estimating foundation settlement, the following empirical correlation was proposed by Marchetti et al. (2001) between the dilatometer modulus and the oedometer modulus (stiffness modulus for confined loading; Equation 4.18):

$$E_{oed} = R_M\, E_{DMT} \tag{1.54}$$

in which R_M is a dimensionless parameter, defined according to the values in Table 1.14.

Table 1.13 Soil classification based on the material index, I_{DMT}, from DMT (Marchetti, 1980).

Clays			Silts			Sands	
Sensitive	Normal	Silty	Clayey	Pure	Sandy	Silty	Clean
$I_{DMT}<0.1$	$0.1<I_{DMT}<0.35$	$0.35<I_D<0.6$	$0.6<I_{DMT}<0.9$	$0.9<I_{DMT}<1.2$	$1.2<I_{DMT}<1.8$	$1.8<I_{DMT}<3.3$	$3.3<I_{DMT}$

symbols may lead to confusion: for example, I_D is used as the density index for sands. For this reason, in this work as well as in others, it was preferred to use the subscript DMT. It should be recognized, however, that this option may also entail the risk of some confusion: for example, the E_{DMT} symbol is often used to designate the stiffness estimate obtained from the DMT test and not the "raw" value resulting from the tests, as in Equation 1.49.

Table 1.14 Values for the parameter R_M for use in Equation 1.54.

Values of I_{DMT}	Expression for R_M
$I_{DMT} \leq 0.6$	$R_M = 0.14 + 2.36 \log K_{DMT}$
$0.6 < I_{DMT} < 3.0$	$R_M = R_{M0} + (2.5 - R_{M0}) \log K_{DMT}$ where $R_{M0} = 0.14 + 0.15\ (I_{DMT} -0.6)$
$3.0 \leq I_{DMT} < 10$	$R_M = 0.5 + 2.0 \log K_{DMT}$
$10 < I_{DMT}$	$R_M = 0.32 + 2.18 \log K_{DMT}$

Note: If a value of $R_M < 0.85$ is obtained, use $R_M = 0.85$.

It is also possible to derive the unit weight of the soil from the following equation (Mayne et al., 2001):

$$\gamma = 1.12\,\gamma_w \left(\frac{E_{DMT}}{p_a}\right)^{0.1} \left(I_{DMT}\right)^{-0.05} \tag{1.55}$$

1.3.12 Tests for permeability characterization

1.3.12.1 Pumping tests

Pumping tests in wells for evaluating the coefficient of soil permeability are recommended for medium to high permeability soil masses, e.g., for sandy soils. Figure 1.44 summarizes the conditions of these tests and the expressions for the assessment of the permeability coefficient (Dupuit, 1863).

The tests consist of pumping water from a well at a constant flow rate and observing the effect of this pumping on the water level drop in piezometers or wells at specified distances. Figure 1.45 shows a layout for the observation devices suggested by Mayne et al. (2001). As can be seen, the observation piezometers (or wells) are arranged radially with respect to the pumping well. According to those authors: i) the nearest piezometers should be at a well distance of about 7.5 m; (ii) the furthest piezometers should be within the predicted limit of

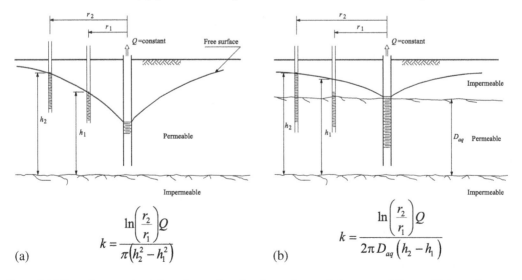

(a) $$k = \frac{\ln\left(\dfrac{r_2}{r_1}\right) Q}{\pi\left(h_2^2 - h_1^2\right)}$$

(b) $$k = \frac{\ln\left(\dfrac{r_2}{r_1}\right) Q}{2\pi D_{aq}\left(h_2 - h_1\right)}$$

Figure 1.44 Pumping tests in wells and equations for the estimate of the permeability coefficient: a) unconfined aquifer; b) confined aquifer.

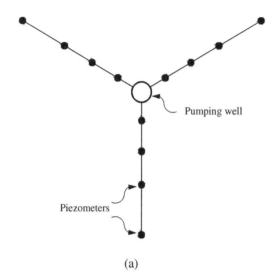

Time since the start of pumping	Intervals between readings
0 – 10 min	0.5 min
10 – 60 min	2.0 min
1 – 6 h	15.0 min
6 – 9 h	30.0 min
9 – 24 h	1.0 h
24 – 48 h	3.0 h
> 48 h	6.0 h

(a) (b)

Figure 1.45 Pumping tests in wells (Mayne et al., 2001): a) layout of the observation piezometers (or wells); b) time intervals between readings for observation of the water level.

the lowering effect; and (iii) the two intermediate piezometers shall be positioned so that the water level lowering can be approximately defined (to this end, the distance between adjacent piezometers should increase progressively with the distance to the well).

The pumped water must be discharged sufficiently far from the test site to prevent recharge of the aquifer under study during the test.

The determination of the stabilized water level position in the piezometers requires observations of the water position in each of them over time. The time intervals between readings are shown in the table alongside Figure 1.45. The cited authors recommend that the pumping test should continue for at least 4 h after achieving stabilization. They also recommend pumping at three increasing flow rates. Upon completion of pumping, the progressive recovery of the initial ground water level should be observed.

The advantages of pumping tests in wells are very relevant: these tests allow characterization of the permeability of large volumes of the ground, thus incorporating certain permeability variations between different layers or sublayers of the aquifer and are susceptible to theoretical interpretation. However, from the previous description, it is clear that these are expensive tests, due, in particular, to the requirement of installation of the observation piezometers (wells), which involves time-consuming operations. Readers interested in learning further about these tests should refer to the study by Jimenez Salas et al. (1976).

1.3.12.2 Borehole or Lefranc tests

Borehole tests are faster and cheaper to set up than those previously described. For these tests, drilling is interrupted, the borehole is carefully cleaned and the casing tube is maneuvered in order to leave an uncased section of a specified length near the bottom of the borehole. One of the conditions for carrying out these tests (see below) is that the uncased section is located below ground water level. There are basically three testing modes:

i) falling head test: the hole is filled with water and the change (decrease) in the water level in the hole over time is observed until the equilibrium position is reached (corresponding to the ground water table in the vicinity);

ii) rising head test: the water in the borehole is pumped to a level lower than the ground water table in the vicinity and the change (increase) in water level in the borehole over time is observed until the equilibrium position is reached;

iii) constant head test: the hole is filled with water and the (constant) flow rate required to maintain the water level in the hole is measured.

The first two types of tests are applicable to high- to medium-permeability materials. The third type is recommended for medium- to low-permeability soils. These tests are also known as Lefranc (1936) tests. However, the theoretical solutions available for the interpretation of these test results are due to the work of Hvorslev (1948, 1951).

In terms of the tests with a variable hydraulic head, the falling head test is generally considered the more convenient. In fact, since the rising head test implies an initial reduction of the hydraulic head at the bottom of the borehole in relation to the surrounding soil mass, hydraulic instability may occur in the vicinity of the hole, which would likely compromise the test results.

Figure 1.46a illustrates the conditions of the falling head test. Since this test is sometimes performed using piezometers, the diameter of the pipe, d, may differ from the diameter of the borehole section, D, where flow takes place. This is the situation illustrated in the figure.

For the interpretation of the test, the following hypotheses are formulated:

i) the soil does not experience volumetric variations during flow;

ii) the uncased borehole section is sufficiently below the ground water table, so that the shape of the flow lines is not significantly affected; similarly, with respect to any impermeable layer;

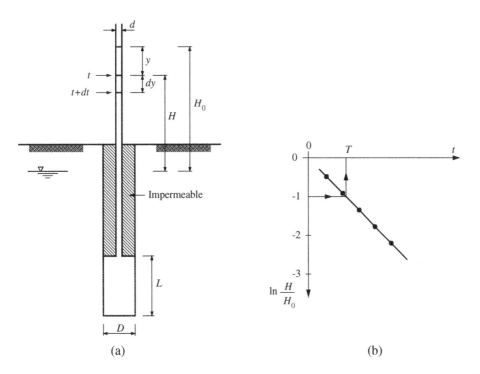

(a) (b)

Figure 1.46 Falling head permeability test: a) schematic of the test setup; b) diagram for the estimate of factor T (basic time lag).

iii) the drilling operation does not cause significant changes in the permeability of the soil in the walls of the hole, in relation to that of the surrounding ground.

By applying Darcy's law, at the instant t, the flow rate, Q, is equal to:

$$Q = FkH = Fk(H_0 - y) \tag{1.56}$$

where F is a factor dependent on the geometry of the section where the flow takes place.

For a short time interval, dt, the volume of water flowing to the ground is equal to the flow leaving the tube, thus:

$$Q\,dt = \frac{\pi\,d^2}{4}\,dy \tag{1.57}$$

Combining the two previous equations results in:

$$\frac{dy}{H_0 - y} = \frac{4Fk}{\pi\,d^2}\,dt \tag{1.58}$$

Hvorslev (1951) introduced the *basic time lag* concept, as the time interval required to reach equilibrium if the initial flow conditions were maintained throughout the test (in reality, since the difference in hydraulic heads between the tube and the ground reduces, the flow during the test tends to decrease). Hence, the *basic time lag* can be defined as:

$$T = \frac{V}{Q_{t=0}} = \frac{\pi\,d^2 H_0}{4FkH_0} = \frac{\pi\,d^2}{4Fk} \tag{1.59}$$

Combining Equations 1.58 and 1.59, the following equation is obtained:

$$\frac{dy}{H_0 - y} = \frac{dt}{T} \tag{1.60}$$

which, by integration:

$$\int_0^{H_0 - H} \frac{dy}{H_0 - y} = \int_0^t \frac{dt}{T} \tag{1.61}$$

results in the following relationship:

$$T = -\frac{t}{\ln\dfrac{H_0}{H}} \tag{1.62}$$

Therefore, as shown in Figure 1.46b, the factor T can be obtained by representing the change with time of the ratio H_0/H in a semi-logarithmic axis system; such a factor corresponds to the value of t when $\ln(H_0/H) = -1$, hence, when $H_0/H = 0.368$.

Consequently, once T is known, the permeability coefficient can be obtained, taking Equation 1.59 into account:

$$k = \frac{\pi\,d^2}{4FT} \tag{1.63}$$

Taking D and L as the diameter and length, respectively, of the borehole (or piezometer) section in contact with the ground, the shape factor F is given by the following equation:

$$F = \frac{2\pi L}{\ln\left[\dfrac{L}{D} + \sqrt{1 + \left(\dfrac{L}{D}\right)^2}\right]} \tag{1.64}$$

This equation was deduced by Hvorslev (1951) by assimilating the volume of the borehole section through which flow occurs to be an ellipsoid, with a minor axis D and focal length L.

For constant head tests, where H is the difference in hydraulic head between the tube and the ground, and Q is the flow rate required to maintain such head differences, by applying Darcy's law:

$$Q = FkH \tag{1.65}$$

where the shape factor F is given by Equation 1.64.

In clean sandy soils, it is particularly difficult to ensure the stability of the borehole walls without tube casing. For this purpose, the uncased borehole section may be filled with gravel, that is, with a granular material much more permeable than the surrounding soil. Such difficulty sometimes leads to the execution of the test with flow exclusively through the bottom of the hole, for which there is also a theoretical solution given by Equation 1.63, but with a different value of the shape factor ($F = 2.75D$). Jimenez Salas et al. (1976) advised against the use of these tests, indicating the following pertinent reasons: (i) the results are dependent on a possible irregularity or peculiarity of the ground (e.g., a potentially less permeable sublayer); and (ii) since this test is carried out in a borehole, regardless of how careful it is cleaned after boring, the presence of a small sediment layer at the bottom will inevitably affect the results. The same authors also included some recommendations about the water to be used in the tests: i) it has to be clean water, in order to avoid clogging of the ground by fine particles in suspension; and (ii) it must be at a temperature above that of the ground to prevent the release of dissolved air bubbles, which would lead to underestimation of permeability.

Lambe and Whitman (1979) present a compilation of Hvorslev's (1951) solutions for the conditions, other than those presented above, with regard to both the soil mass and the borehole geometry.

1.4 GLOBAL OVERVIEW ON SITE CHARACTERIZATION

1.4.1 Summary of *in situ* tests

For the reader who is only recently familiar with geotechnical engineering, some difficulties may persist in distinguishing the essentials from such diverse tests, such as those presented in the previous section. It could hardly be otherwise because clear ideas about each of the testing methods can only be established when these are applied to solve real characterization problems. The text of this chapter, given its more informative rather than formative nature, must therefore be understood, in essence, as a reference tool when used in professional practice.

In order to facilitate the readability of the text, Table 1.15 summarizes some of the essential aspects of the previously described tests (excluding those related to the evaluation of permeability), thus providing a quick guide to the less-experienced reader for the selection of the most convenient test(s) in each case scenario.

Table 1.15 Essential aspects regarding the most common in situ tests.

Test	Type of loading	Complexity	Cost	Applicable soils	Applicability to soft rocks	Access	Depth	Measured parameters	Estimated parameters	Interpretation	Reproducibility of results	Notes
SPT		Low	Low	Clays to fine gravels	Yes	Borehole	Any	N_{60}	I_D, ϕ'	Empirical	Weak to fair	It is the only test that involves disturbed sample collection. Several corrections are needed as it is not completely standardized.
CPTu		Medium to high	Medium to high	Clays to fine gravels	No	Its own means	Limited to the available reaction	$q_c (q_t), f_s, u$	I_D, ϕ', E, c_u, c_v	Empirical or theoretical	Very good	It is the only test that provides continuous recording of the results in depth, allowing the detection of thin layers of distinct soils interspersed with thicker strata.
DP		Low	Low	Clays to coarse sands	No	Its own means	Low to medium	N	ϕ', E	Empirical	Fair	Due to its simplicity, it can be used for embankment compaction control or for on-site checking of the position of the bedrock.
VST		Low	Low to medium	Clays to soft silts	No	Borehole or pre-driving	Any (in the case of borehole)	M_t (torque) - θ (rotation)	c_u, S_t	Theoretical	Fair	It is best suited to characterize the undrained shear strength of soft soils. The results need correction.

(Continued)

Table 1.15 (Continued) Essential aspects regarding the most common *in situ* tests.

Test	Type of loading	Complexity	Cost	Applicable soils	Applicability to soft rocks	Access	Depth	Measured parameters	Estimated parameters	Interpretation	Reproducibility of results	Notes
PLT		Medium	Medium to high	All	Yes	Direct	Low	Δq_r (pressure at the surface) – s (settlement)	E, c', ϕ', c_u	Theoretical	Fair	It is recommended for the characterization of the stiffness of compacted fills and of natural coarse soils near the surface.
CHT or DHT		Medium	Medium to high	All	Yes	Borehole	Any	V_s, V_p	G_0, ν	Theoretical	Very good	It is the only test applicable to all soils and rocks and can be performed to any depth. It provides the true elastic stiffness modulus of the ground.
SBPT		Very high	Very high	Clays to fine gravels	No	Its own means	Any (in the case of boreholes)	ε (radial strain) - ψ (pressure)	K_0, G, ϕ', c_u	Theoretical	Good	It is the only test that allows evaluation of strength and stiffness parameters and also the at-rest stress state, through theoretical interpretation.
PMT		Medium to high	High	Clays to medium gravels	Yes	Borehole	Any	v (volume) - p (pressure)	$E_{PMT}, p_l \rightarrow q_{ult}$ (strength) and s (settlement) of foundations	Essentially empirical	Fair	It is a good alternative to characterize hard soils and soft rocks. The results are very dependent on the quality of the pre-drilled borehole. There are well-calibrated empirical methods of direct application to test results.
DMT		Low	Medium	Clays to sands	No	Borehole or self-driving	Any (in the case of pre-drilled borehole)	p_0, p_1	$I_{DMT}, K_{DMT}, E_{DMT} \rightarrow K_0, E, \phi', c_u$	Essentially empirical	Very good	Through correlations, it permits the evaluation of the strength and stiffness parameters and also the coefficient of earth pressure at-rest.

1.4.2 *In situ versus* laboratory tests

1.4.2.1 Introduction

Having addressed the main *in situ* tests, it seems appropriate to compare them with the relevant laboratory tests to describe their relative advantages and limitations. This is not meant to say that the two groups of tests are mutually exclusive, although it should be recognized that, in the profession, preferences are often declared of one type over the other.

As will be seen from the following discussion, the limitations of one of these two groups of tests usually correspond to the potentialities of the other and *vice versa*, so the most satisfactory approach for the mechanical characterization of the soil usually results from the combined use of *in situ* and laboratory tests.

1.4.2.2 Advantages and limitations of laboratory tests

The fundamental advantage of the tests for the mechanical characterization of soils in the laboratory is that the stress state (in terms of total and effective stresses, and pore water pressures), the strain state and the drainage conditions are generally clearly defined. As a result, the interpretation of these test results can be based on various theories, allowing one to determine diverse defining parameters of mechanical behavior.

On the other hand, the determination of the main identification properties (grain size distribution and Atterberg limits) and physical indices (water content, void ratio, unit weight, etc.) on the same soil makes it possible to establish, in conjunction with mechanical parameters, a coherent depiction of the ground formations involved in the project.

Nevertheless, there are considerable limitations to laboratory testing. As previously pointed out, possibly the main limitation refers to the need for undisturbed samples of sandy soils, which requires sophisticated and costly techniques. However, even for other soil types, sample disturbance can seriously affect the reliability of laboratory results. In this respect, it should be noted that the stiffness characteristics of soils are much more sensitive to such disturbance than are the strength parameters. This issue will be discussed later (see Section 1.5).

On the other hand, a laboratory-based soil characterization, involving small-scale test elements, limited in number and collected from discrete points of the soil mass, may lead to a less accurate description of the ground, which may affect the prediction of soil behavior and consequently the geotechnical design.

Despite this reservation with regard to laboratory testing, the following also applies, in the author's opinion: a few well-conducted and well-interpreted tests on high-quality samples are worth more than many ill-conducted tests on poor samples. Laboratory testing is generally expensive and time-consuming. Therefore, a judicious selection of tests on high-quality samples and the application of high standards in laboratory practice and procedures are strongly recommended.

1.4.2.3 Advantages and limitations of in situ tests

As a first fundamental advantage of *in situ* tests, it can be pointed out that their execution does not exclude any type of soil.

With *in situ* tests, a large number of points of the soil mass are generally characterized; some tests can, in fact, provide a continuous recording of results with respect to depth (the CPTu, for example), which makes it possible to detect the presence of very thin layers and the exact position of the boundaries between the various layers.

Another relevant issue is that the soil is tested in its own environment, with no change in its stress state. It is not surprising, therefore, that the only tests that can reasonably estimate the at-rest stress state are *in situ* tests, namely the self-boring pressuremeter test.

Finally, since some field tests are carried out taking advantage of already-drilled boreholes, which are indispensable for reliably identifying the sequence of soil strata, the position of water levels, etc., they also generally involve lower costs and are much faster to perform than laboratory tests.

On the other hand, the most important limitation of *in situ* tests is that, in many of these tests, the stress and strain states and the drainage conditions are not clearly defined. Therefore, the rational interpretation of these test results is difficult, if not impossible. Consequently, the mechanical parameters of the soil mass must be, in such cases, obtained by empirical correlations derived from experience in similar works and soils, with the results from these tests. Linked with this limitation, it is clear that the changes in stress state and deformation conditions imposed on the ground by these tests can be very different from those caused by engineering works.

Finally, with the exception of the SPT test, *in situ* tests do not allow direct identification of the nature of the tested soil as they do not involve any soil sampling.

1.5 STIFFNESS CHARACTERIZATION BY MEANS OF *IN SITU* AND LABORATORY TESTS

1.5.1 General overview: definitions

Before concluding this chapter, it is worthwhile discussing the evaluation of soil stiffness or deformability parameters.

First of all, a comment on the terms *stiffness* and *deformability* should be made. Etymologically, these two concepts are opposites: an increase in stiffness means a reduction in deformability, and *vice versa*. The word deformability was more common in the past. The related mechanical parameter is usually the E modulus, with the physical meaning of Young's modulus or elasticity modulus. However, as soils are highly inelastic materials, it is common to designate this parameter as a deformability modulus or a deformation modulus. However, within this terminology, the larger the deformation modulus, the smaller the deformation of the material, which has, of course, drawbacks.

More recently, the literature has preferred the term stiffness. The advantage is that, the greater the stiffness, the greater are all other related parameters, which are essentially the same as those used with the alternative terminology. In the following considerations, the use of the term stiffness will be preferred, although deformability is still used.

Table 1.16 summarizes the definitions of different elastic parameters for various loading conditions. Figure 1.47 summarizes, for triaxial loading, the definitions of the diverse stiffness moduli, of which two require specific comments. The E_{max} modulus is the true elastic modulus (associated with purely reversible deformation), which is determined for very small strains of the order of 10^{-6}. As seen above, this parameter can be evaluated through cross-hole or down-hole tests (see Sections 1.3.7 and 1.3.8). The importance of this elastic parameter in the current approach to soil stiffness characterization will be discussed below. This modulus differs from the initial tangent modulus, E_i, in the figure because, in many cases, whether from *in situ* or laboratory stress–strain curves, the initial part of the curve, in the small-strain range, is poorly defined. Thus, as Figure 1.47 suggests, the initial tangent modulus does not have a consistent definition.

The discussion on stiffness characterization at the end of this chapter is justified by the following reasons: i) as a follow-up to the introduction of *in situ* tests (also used for that purpose) and their comparison with laboratory tests; ii) since the following chapters will

Table 1.16 Summary of the elastic parameters for different types of loading.

Type of loading	Definition of strains	Elastic moduli	Relations between moduli
Triaxial compression ($\Delta\sigma_a \neq 0$)	$\varepsilon_a = \dfrac{\delta_a}{H_0}$ $\varepsilon_r = \dfrac{\delta_r}{R_0}$ or $\varepsilon_r = \dfrac{\varepsilon_{vol} - \varepsilon_a}{2}$	$E = \dfrac{\Delta\sigma_a}{\varepsilon_a}$ $\nu = \dfrac{-\varepsilon_r}{\varepsilon_a}$	–
Triaxial extension ($\Delta\sigma_r \neq 0$)	$\varepsilon_a = \dfrac{\delta_a}{H_0}$ $\varepsilon_r = \dfrac{\delta_r}{R_0}$ or $\varepsilon_r = \dfrac{\varepsilon_{vol} - \varepsilon_a}{2}$	$E = \dfrac{-2\nu\,\Delta\sigma_r}{\varepsilon_a}$ $\nu = \dfrac{-\varepsilon_r}{2\varepsilon_a}$	–
Isotropic compression	$\varepsilon_{vol} = \varepsilon_a + 2\varepsilon_r = 3\varepsilon_a$	$E = \dfrac{3\Delta\sigma\,(1-2\nu)}{\varepsilon_{vol}}$	–
Confined compression	$\varepsilon_a = \dfrac{\delta_a}{H_0} = \varepsilon_{vol}$	$E_{oed} = \dfrac{\Delta\sigma_a}{\varepsilon_{vol}}$	$E_{oed} = \dfrac{E(1-\nu)}{(1+\nu)(1-2\nu)}$
Simple shear	$\gamma = \dfrac{\delta_t}{H_0}$	$G = \dfrac{\Delta\tau}{\gamma}$	$G = \dfrac{E}{2(1+\nu)}$

address theories and methods applicable to various geotechnical structures, where stiffness parameters are used; and iii) unlike shear strength, the framework and meaning of the stiffness concept have evolved significantly in recent decades. This last point is related to the generalization of the application of numerical methods, in particular the finite element method, to geotechnical engineering problems.

1.5.2 Small is beautiful

With regard to stiffness, it is of first and foremost important to take into account that the stress–strain relations in soils, for a given stress path and a given loading rate, are highly

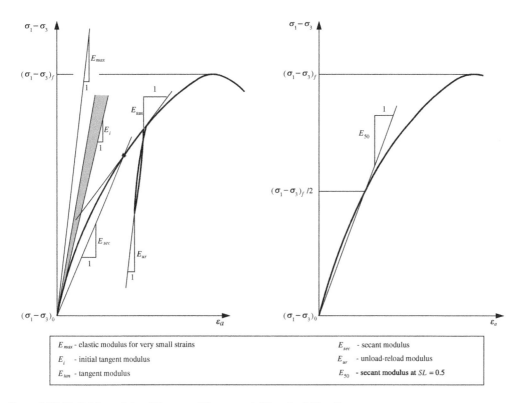

E_{max} - elastic modulus for very small strains	E_{sec} - secant modulus
E_i - initial tangent modulus	E_{ur} - unload-reload modulus
E_{tan} - tangent modulus	E_{50} - secant modulus at $SL = 0.5$

Figure 1.47 Definition of the different stiffness moduli in triaxial loading.

nonlinear. Figure 1.48 helps to explain this issue for a particularly simple situation, a drained triaxial compression test on a sample of residual soil from granite (Topa Gomes, 2008).

Figure 1.48a shows the specimen with the local instrumentation used for axial and radial strain measurements. Figure 1.48b shows the deviatoric stress *versus* axial strain curve. Figure 1.48c relates the ratio of the secant stiffness modulus, E_{sec}, at each loading stage to the value of the initial tangent modulus, E_i, with the so-called stress level, SL, which represents the ratio of the incremental deviatoric stress by its value at failure:

$$SL = \frac{(\sigma_1 - \sigma_3) - (\sigma_1 - \sigma_3)_0}{(\sigma_1 - \sigma_3)_f - (\sigma_1 - \sigma_3)_0} \tag{1.66}$$

Therefore, the stress level ranges from 0 to 1 between the start of the test and failure, and can be considered as the inverse of the safety factor, e.g. $F = 1/SL$; for example, when the stress level is 0.25, the safety factor to failure is 4.0. Figure 1.48d has the same ordinate axis as the previous figure but the strains are represented on the abscissa axis on a logarithmic scale. The initial tangent modulus was, in the present case, defined for the point of the relationship corresponding to a strain of 5×10^{-5}, since, for strains below this value, the measurement system does not have sufficient sensitivity.[14]

---

[14]The system shown in Figure 1.48a consists of local strain instrumentation of the soil specimen. This allows strain measurements with much greater accuracy than the conventional method, which, for axial strains, makes use of a deflectometer attached to the top of the triaxial cell.

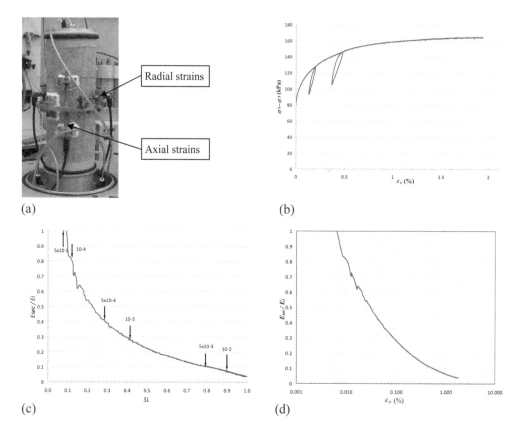

(a)   (b)

(c)   (d)

Figure 1.48 Triaxial compression test on a sample of residual soil from granite (Topa Gomes, 2008): a) specimen with local strain instrumentation; b) deviatoric stress *versus* axial strain; c) ratio of secant stiffness modulus to initial tangent modulus *versus* stress level; d) ratio of secant stiffness modulus to initial tangent modulus *versus* axial strain (on a logarithmic scale).

Based on these results, it can be said that: i) stiffness progressively decreases with the increase in stress level, approaching zero at failure, that is, when $SL = 1$; and ii) for very low values of SL, hence for relatively high safety factors, the strains are still relatively small.

From the results presented, an initial conclusion can be drawn: when dealing with a deformation problem, the parameter expressing stiffness must be representative of the curve relating stresses and strains for the stress level induced in the ground. Since, for a given problem, the stress level and the strain level are interrelated, the previous statement is naturally valid by replacing stress with strain.

A second, very important conclusion is as follows: in many soil-structure interaction problems under serviceability conditions, where the safety factor to failure is high, the strain levels are typically low. Characterization of soil stiffness for such conditions – note that those are indeed where deformation analyses are particularly necessary – requires testing methodologies where such strain levels are involved and appropriately measured.

Contrary to the former conclusion, the latter represents an idea that was consolidated in the last thirty years, with relevant contributions of researchers from Imperial College, London. Not accidentally, one of the reference publications on this subject has the significant title "Small is beautiful" (Burland, 1989)!

As a reference, Figure 1.49 includes the typical orders of magnitude of the strains involved in different types of *in situ* tests and engineering works (Ishihara, 1996; Mayne et al., 2001).

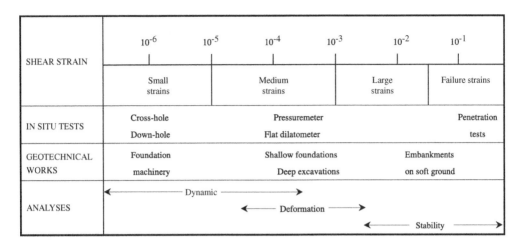

SHEAR STRAIN	10^{-6}		10^{-5}		10^{-4}		10^{-3}		10^{-2}		10^{-1}
	Small strains				Medium strains				Large strains		Failure strains
IN SITU TESTS	Cross-hole				Pressuremeter						Penetration
	Down-hole				Flat dilatometer						tests
GEOTECHNICAL WORKS	Foundation				Shallow foundations				Embankments		
	machinery				Deep excavations				on soft ground		
ANALYSES	←———— Dynamic ————→										
					←——— Deformation ———→						
								←——— Stability ———→			

Figure 1.49 Strain levels for some *in situ* tests and geotechnical works and range of geotechnical analyses.

1.5.3 Relation of the tests to the strain level in the soil

From previous considerations, a clear advantage of laboratory testing could be anticipated in this context, by allowing characterization of stress–strain relations between the at-rest state and failure under well-known and controlled stress, strain, and drainage conditions and with adequate strain monitoring. It is not exactly so, because sampling quality issues still prevail as limiting the reliability of laboratory stiffness estimates.

In addition to the aforementioned difficulty in collecting undisturbed samples, particularly in clean granular soils, test results on high-quality samples of cohesive soils, such as those trimmed from block samples, can still reveal considerable underestimation of stiffness. It is also the case of residual soils from granite, where the stress relief associated with sampling seems to induce breakage or damage of many cementitious bonds, reducing stiffness in relation to its natural state (Viana da Fonseca et al., 1997).

In relation to *in situ* tests, pressuremeter and other tests use the theory of elasticity for the interpretation of results to derive "one" stiffness modulus, E. However, as Figure

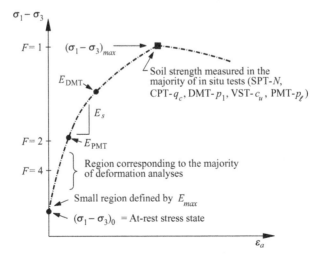

Figure 1.50 Idealized stress–strain curve and soil stiffness for small to large strains (adapted from Mayne et al., 2001).

1.50 suggests, since the tests involve different degrees of strain disturbance in the soil, these stiffness estimates cannot be consistent with each other. On the other hand, those tests that induce higher (or much higher) strains than those prevailing in most soil-structure interaction problems will tend to underestimate stiffness. As the figure also suggests, the most common tests, such as SPT and CPT, imply the mobilization of soil strength, and therefore very large strains. For this reason, it is clear that stiffness estimates for deformation analyses based on these tests can only be made through empirical correlations.

Tests with the Ménard pressuremeter (PMT) and the flat dilatometer (DMT) allow the characterization of the ground at a much lower strain level than penetration tests, but still probably higher than that prevailing in many soil-structure interaction problems. Out of these tests, the most reliable for stiffness characterization is clearly the self-boring pressuremeter (SBPT), which, as previously described, provides a complete stress–strain curve. However, even in this test, the stiffness for small deformations is often affected by the disturbance of the ground associated with the drilling of the hole to insert the apparatus. Careful execution of the unload–reload cycles when the stress level is still low is considered to be very convenient because the corresponding modulus, G_{ur} or E_{ur}, is a good approximation to the small-strain modulus.

In this complex context – still with regard to Figure 1.50 – seismic tests (cross-hole or down-hole) are particularly useful because they involve very low strains (of the order of 10^{-6}), thus providing the true elastic shear modulus, G_0. Indeed, for the aforementioned order of magnitude, or even one order of magnitude higher (10^{-5}), the strains are fully reversible because, in uncemented soils, they are related to the elastic deformation of the particles, especially at the contact points, without significant sliding phenomena. In turn, in cemented soils, the strains are associated with deformations at the bonds between the particles without breakage or damage.

In the past, the use of the shear modulus obtained from seismic tests was restricted to the cyclic loading for dynamic analysis, and the associated symbol was G_{dyn}. Experimental field and laboratory studies have shown that such a shear modulus, now designated G_0 (or G_{max}), does not depend on the stress path, nor on the (monotonic or cyclic) loading type or rate, so is also applicable to static loading (Tatsuoka and Shibuya, 1992). For this reason, it has become, in recent decades, a fundamental reference parameter for soil stiffness: elastic (in the strict sense of the term), maximum, experimentally measurable by relatively simple tests, and suitable for all types of soils and rocks.

1.5.4 Methodology for characterization of soil stiffness for all strain levels

It was previously pointed out that, when addressing a deformation problem, the parameter expressing stiffness must be appropriate to the stress level induced in the ground. However, full compliance with this requirement makes it highly convenient to characterize the relationship between stiffness and both stress and strain levels over a relatively wide range.

Contrary to what happens in a simple soil specimen in a triaxial test, the relationship between either the stress level or the safety factor of a structure interacting with the ground and the induced strain level cannot be established, even for simple problems, without a deformation analysis. The reason for this is that, in one specimen, for each stress level, the stress and strain states are uniform. On the contrary, in the field, even with a deformation problem as simple as the centered vertical loading of a footing, this is far from the case; different stress levels are mobilized in the various elements (points) of the ground, and consequently different stiffness values are also being mobilized. In addition, in soil masses, stiffness and strength generally also depend on the initial effective stress, that is, they typically increase with increasing depth.

In classical soil mechanics, deformation analyses are generally based on the theory of linear elasticity, where soil stiffness is represented by a single parameter, generally E. Thus, in more advanced analyses within this methodology, an iterative process is required: i) adopt a representative value of stiffness; ii) carry out the deformation analysis; iii) verify if the strain level obtained is compatible with the adopted stiffness; and (iv) if it is, accept the result as satisfactory; if not, correct the stiffness values and carry out a new deformation analysis. It should be added that this process is not exempt from subjective decisions because the strain level is variable in the soil mass. For example, the analysis of the settlements of shallow foundations can be improved by subdividing the ground into layers and attributing specific stiffness values to each layer.

With the application of modern numerical methods, especially those based on the finite element method, it is now possible to consider a complete stress–strain–strength constitutive law (equation) (for instance, as that in Figure 1.47). For this purpose, the computational model adjusts, for each soil mass element, the stiffness to the corresponding stress–strain level (and to the stress path) at each loading phase. These analytical potentialities reveal the great interest in characterizing stiffness for a wide range of strain levels.

In dynamic problems involving cyclic loading of the soil, the traditional way of representing stiffness degradation with strain level is through graphs such as the one in Figure 1.51, with the shear strains represented on the abscissa (on a logarithmic scale) and the ratio of G, the secant shear modulus for each strain level, to G_0 on the ordinate. Note that all curves start from a unit ordinate value for very small strains (hence G_0 is also called G_{max}). As shown in the figure, the stiffness degradation with strain is more pronounced in soils showing lower plasticity.[15]

These charts were developed by combining G_0 values provided from field seismic tests, with G values for higher strain levels obtained from laboratory tests in which samples were cyclically loaded (resonant column, triaxial, direct cyclic simple shear, and cyclic torsional tests).

The adoption of G_0 as the reference stiffness for static loading problems, as mentioned above, has suggested the use of analogous charts for these cases. Experience has shown that the stiffness degradation with increasing strain is, for the same type of soil, more pronounced in static loading than in dynamic (cyclic) loading.

The current methodology for characterizing the stiffness degradation laws for static loading also consists of combining the assessment of G_0 by means of seismic tests with other field and laboratory tests, covering a wide range of strain levels. The methodology is outlined in Figure 1.52 and can be described as follows:

1. assessment of G_0 from *in situ* seismic tests;
2. laboratory testing (under triaxial conditions or other), with adequate local strain instrumentation, in order to obtain the stress–strain curves from small to large strains, up to failure (as the case of the test in Figure 1.48);
3. complementation, by *in situ* testing, namely using pressuremeter tests, to obtain the stress–strain curve for different strain levels;

---

[15]Very plastic clay soils can exhibit virtually linear elastic behavior and very low damping up to relatively high strain levels when loaded in cyclic shear. This fact explains why the seismic behavior of these soils is highly unfavorable. In fact, where thick layers of high-plasticity clays occur, the seismic motions induce a near-linear elastic behavior, amplifying the movements coming from the base, thus substantially aggravating the seismic actions on the structures at the surface. This type of behavior was responsible for the failure of a large number of tall buildings in Mexico City during the 1985 earthquake.

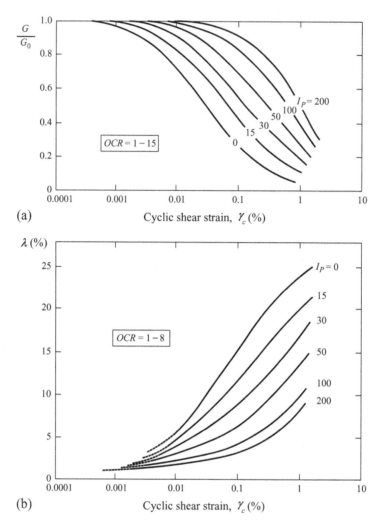

Figure 1.51 Stiffness modulus degradation curve (a) and damping ratio curve (b) with the strain level in cyclic shear loading (Vucetic and Dobry, 1991).

4. combining the types of characterization described in i) to iii), taking into account that the complete relationship between G and shear strain must be continuous and start at a (known) G_0 value for very small strains; in practice, this implies *scaling* the curves described in ii) and iii), in order to compensate or correct for the effects of soil distur-bance, induced by sampling in laboratory tests or by the *in situ* testing procedures.

Alternatively, E_0 can be used, which is related to G_0 through Equation 1.34, taking the val-ues for Poisson's ratio previously suggested in Section 1.3.7.2.

A particularly objective and simple procedure for scaling or adjusting the curves from laboratory tests is through the $V_{s,lab}/V_{s,field}$ ratio, in which the numerator refers to the shear wave velocity measured in the laboratory in soil specimens after consolidation at the over-burden effective stress state, before shearing. This can be done by means of *bender ele-ments*, already mentioned in Table 1.1 (Atkinson, 2000; Ferreira et al., 2014).

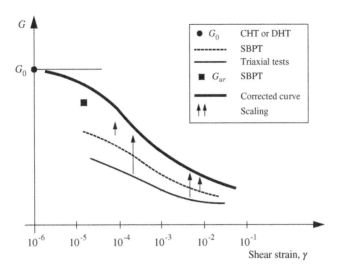

Figure 1.52 Methodology for stiffness characterization for a wide range of strain levels (adapted from Tatsuoka and Shibuya, 1992).

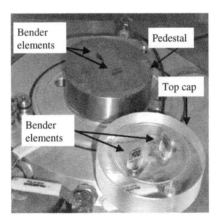

Figure 1.53 Arrangement of bender elements for a triaxial test (Ferreira, 2009).

Bender elements are small piezoelectric transducers, capable of transmitting and receiving shear waves, which can easily be installed in the various laboratory soil-testing devices. Figure 1.53 illustrates a triaxial cell equipped with bender elements in its platens. One transducer acts as the transmitter, being excited at one end of the specimen with a single pulse excitation, and the time required for this wave to arrive through the specimen to the receiver at the other end is measured by an oscilloscope, from which $V_{s,lab}$ is directly computed.

Typically, this velocity is lower (or even substantially lower) than $V_{s,field}$, due to the effects of sample disturbance.[16] It is worth noting that, when comparing *in situ* and laboratory

---

[16]There are proposals for sample quality classification based on the ratio $V_{s,lab}/V_{s,field}$. Values above 70%–80% are considered very good or excellent.

shear wave velocities, the wave propagation and polarization directions should be taken into account, particularly in the case of anisotropic soils (Atkinson, 2000).

It can be seen that this methodology, which has been consolidated in recent years, combines laboratory and *in situ* tests in a particularly interesting way towards the achievement of good and reliable estimates of stiffness. These estimates can be used in the classical "elastic" approaches or in non-linear deformation analyses with numerical models, such as those based on the finite element method.

Chapter 2

Overall stability of soil masses

2.1 INTRODUCTION

The laboratory characterization of a series of undisturbed samples, taken from a soil mass and/or the execution of *in situ* tests on that soil mass, may shed light on the stress states that cause *local failure* but not – at least directly – on those stress states that might lead to a *global failure* of that mass.

In fact, a soil mass behaves like a highly redundant or hyperstatic structure, i.e., *local failure* does not mean or imply *global failure*. The latter requires the existence of a ground mass completely enveloped by a *failure surface* or *sliding surface*, where the shearing resistance has been exhausted. Figure 2.1 shows an example of a global failure.

Therefore, the characterization of global failure from local failure data (i.e., from the soil strength parameters) demands the application of theories or methodologies that depend on the type of structure, on the type of soil, and on the nature of the loading. These theories are the base of the so-called *stability analyses*, which are central to geotechnical design. Such theories will be discussed in this book for a wide range of geotechnical structures.

This chapter will treat the overall stability of soil masses, essentially subjected to their self-weight and the action of water, without relevant surcharge loading and soil–structure interaction. Examples will encompass embankments over soft soil, unsupported excavations, natural slopes, and embankment dams.

As will be seen in this book, the stability analysis methods used in geotechnical engineering have been suggested (or inspired) by observation of the mechanisms associated with global failures in Nature. On the other hand, in many of the failures observed, it is possible to calculate, with some accuracy, the destabilizing force and, therefore, the shearing resistance mobilized along the sliding surface during collapse.[1] Real failure events are therefore valuable opportunities for comparing *local resistance*, estimated by laboratory or field tests, with the *resistance mobilized on a large sliding surface*.

This is important because the latter does not always simply coincide with the integral of the former. In fact, local resistance may be calculated through several distinct criteria, using peak, critical state, or residual strength parameters under both drained and undrained conditions! This complex matter is discussed later in this chapter and in Chapter 9 (see Section 9.5.2).

Loading a soil mass is a mechanical problem that essentially requires the establishment of:

   i) equilibrium equations;
   ii) strain compatibility equations;
   iii) a constitutive law, i.e., stress-strain-strength relations for the soil mass.

---

[1] This type of approach is generally termed back-analysis.

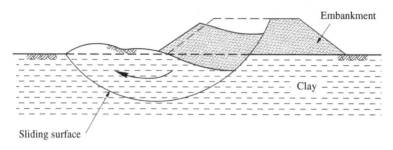

Figure 2.1 Example of a global failure.

In the stability analysis of soil masses, the constitutive law usually adopted is of the rigid-plastic type, whereby the strain developed before full-strength mobilization is neglected. Even with the adoption of this simplified constitutive law, the availability in geotechnical engineering of mathematically exact stability problem solutions, satisfying the equilibrium, compatibility, and boundary conditions, is scarce and applies only to highly simplified situations.

Nowadays, this limitation tends to be overcome mostly by recourse to numerical modelling based on the finite element method, which has been developed and refined in recent decades, coupled with fast-expanding computing power. The treatment of these numerical techniques is outside the scope of this book, but occasionally results will be presented or mentioned, as occurs later in this chapter.

The following section presents the main classical approaches used in geotechnical engineering for the solution of stability problems, which, as will be seen, involve simplifications in the formulation of the mechanical problem, as stated above.

2.2 INTRODUCTION TO LIMIT ANALYSIS THEORY

2.2.1 Formulation: upper and lower bound theorems

Within the methodologies that involve simplification of the mechanical problem, the most consistent is the limit analysis theory, essentially based on two theory of plasticity theorems, the *upper bound theorem,* and the *lower bound theorem.*

This is a topic of some complexity, both formal and analytical, so that a treatment essentially aiming at a physical (or mechanical) perception will be provided. The interested reader is directed to the works of Salençon (1974, 2001) and Atkinson (1993).

According to the lower bound theorem (LBT), considering a given structure under the action of a set of external loads, if a stress distribution can be found that equilibrates the loading without exceeding the material strength at any point, then the structure is stable.

This theorem contemplates, therefore, the consideration of the equilibrium conditions and the material properties (which determine the resistance) but does not involve strain compatibility.

Considering a given structure under the action of a set of external loads, the upper bound theorem (UBT) postulates that, if a compatible displacement field can be assigned to the structure, for which the work of the external loads equals that of the internal stresses, then structural collapse will occur.

This theorem contemplates, therefore, the consideration of the compatibility conditions and the material properties (which govern the work of the internal stress) but disregards the equilibrium conditions.

In summary: the lower bound theorem deals with *statically admissible stress states*, so it is also called the static theorem, while the upper bound theorem deals with *kinematically admissible collapse mechanisms*, so it is known as the kinematic theorem. The first theorem leads to solutions on the safe side or by default – *lower bound solutions* – while the second provides solutions which are unsafe or by excess – *upper bound solutions*.

The application of these theorems to some soil mechanics problems leads to a very narrow interval around the true (unknown) solution, with each theorem providing an *approximate solution*. In some simpler cases, the solutions provided by the two theorems coincide, and the problem can be said to have an *exact solution*.

2.2.2 Example of application – excavation with vertical face under undrained conditions

As these theorems are somewhat cryptic to those unfamiliar with the theory of plasticity, they will be now applied to a classic geotechnical problem illustrated in Figure 2.2a: the evaluation of the maximum or *critical depth*, h_{cr}, of an unsupported vertical cut, in a homogeneous soil layer of unit weight, γ, and undrained shear strength, c_u.

As shown in Figure 2.2b, the point with the largest shear stress is located on the face of the excavation, immediately above the base plane. If h_m is the excavation depth for which point P reaches shear failure, then, as shown in Figure 2.2c:

$$\gamma h_m = 2 c_u \tag{2.1}$$

and

$$h_m = \frac{2 c_u}{\gamma} \tag{2.2}$$

Recalling the theorems presented before, the reader will agree that this excavation depth value corresponds to a solution obtained with recourse to the lower bound theorem, thus on the safe side, i.e.:

$$h_m \leq h_{cr} \tag{2.3}$$

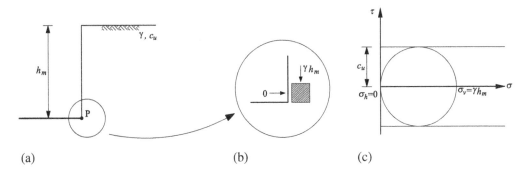

Figure 2.2 Total stress analysis of a vertical cut in a homogeneous soil under undrained conditions: a) layout; b) location of the largest shear stress; c) Mohr circle at point P when failure is reached due to increase of the excavation depth.

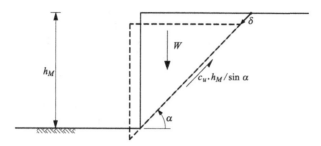

Figure 2.3 Kinematically admissible collapse mechanism for a vertical cut in a homogeneous soil formation under undrained conditions.

Consider now Figure 2.3 that shows a block of the same soil with weight W, delimited by a plane surface passing through the base of the cut face, at an angle α with the horizontal. The aim is to calculate the excavation depth value h_M at which sliding of the block will occur.

Imagining that the block slides down by δ along the inclined plane, an upper bound theorem solution can be obtained by equating the work of the external forces to the plastic deformation work of the resisting forces along the sliding plane.[2] Therefore:

$$W \delta \sin \alpha = \frac{c_u h_M \delta}{\sin \alpha} \tag{2.4}$$

which leads to

$$h_M = \frac{W \sin^2 \alpha}{c_u} \tag{2.5}$$

Taking into account that:

$$W = 0.5 \gamma h_M^2 \cot \alpha \tag{2.6}$$

then

$$h_M = \frac{2 c_u}{\gamma} \cdot \frac{1}{\sin \alpha \cos \alpha} \tag{2.7}$$

Deriving h_M with respect to α and equating to zero, one gets, for this angle, the value $\pi/4$, which then leads to the minimum value for h_M as:

$$h_M = \frac{4 c_u}{\gamma} \tag{2.8}$$

The critical depth is then bounded by the values obtained from the application of the two theorems:

$$\frac{2 c_u}{\gamma} \leq h_{cr} \leq \frac{4 c_u}{\gamma} \tag{2.9}$$

---

[2] In the present case, as undrained conditions are assumed , there is no volume change and so the dilatancy angle is zero. Therefore, the displacement δ must coincide with the inclined sliding plane, if the compatibility conditions are to be verified.

As can be seen, the two solutions obtained define a very wide interval (the upper limit is twice the lower one), which merely serves to illustrate the application of the two theorems. This problem will be revisited later, and closer upper and lower bounds will be discussed (see Section 2.5.4).

2.3 LIMIT EQUILIBRIUM METHODS

2.3.1 Introduction

The so-called *limit equilibrium methods* are more employed in geotechnical engineering practice than are those based on the upper and lower bound theorems. Although lacking the theoretical soundness of the theorems, such methods are tools of enormous practical importance. They have been developed, as would be expected, from the observation and interpretation made by engineers of real collapse case histories.

Basically, these methods comprise the following steps:

1. postulate a collapse mechanism involving a soil mass limited by a (curved, plane, or mixed) sliding surface;
2. compute for that mass the effect[3] of the destabilizing forces, S, on the sliding surface;
3. for the assumed collapse mechanism, calculate the *resisting* tangential forces, R, on the sliding surface;
4. make a comparison between the forces (or their effects) referred to in b) and c), which may be carried out in several ways, as shall be seen below;
5. repeat the steps 1 to 4 for other assumed mechanisms and soil masses, so as to identify the so-called *critical surface*, which will be the one delimiting the soil mass that leads to the least favorable comparison of S with R.

This methodology combines features derived from the two limit analysis theorems. In fact, the potential sliding surfaces naturally correspond to predetermined collapse mechanisms, as is usually assumed when applying the upper bound theorem, but they may fail to comply with the deformation compatibility requisites. On the other hand, although the global equilibrium conditions of the potentially unstable mass have to be satisfied, the local equilibrium conditions may be occasionally violated.

In spite of these limitations, long experience in the application of limit equilibrium methods attests that the results obtained approximate closely to real observed failures.

The traditional soil mechanics approach for comparing the forces (or their effects) mentioned in b) and c) is by computing the ratio:

$$F = \frac{R}{S} \tag{2.10}$$

where F represents what is conventionally designated as the *global safety factor*.

In Chapter 3, this question will be further explored, considering alternative ways of comparing the two effects. In the present chapter, only the above definition will be applied.

Next is presented the method of slices, the limit equilibrium method most-used for performing the global stability analysis of soil masses.

---

[3] The term *effect* has a global meaning: it may, for example, correspond to the moment of forces relative to the center of rotation, such as the center of the sliding surface, when it is a circular arc.

2.3.2 Method of slices: general formulation

In natural slopes or embankments lacking zones or layers possessing highly contrasting mechanical characteristics, failure very often occurs as a result of sliding surfaces, the transverse cross section of which is approximately circular in shape.

In such cases, the equilibrium analysis of the potentially sliding mass is conveniently carried out by dividing it into slices with vertical faces, as indicated in Figure 2.4a. The pore pressure distribution in the soil mass is known, from the flow net under hydrodynamic conditions, or simply from the water level location in the hydrostatic case. Figure 2.4b presents a generic slice and its applied forces: i) the self-weight, W_i; ii) the resultant of the effective stress normal to the slice base, N_i'; iii) the resultant of the pore pressure on the base of the slice, U_i; iv) the resultant of the tangential stresses mobilized on the slice base, T_i; and v) the normal and tangential components of the interaction forces on the left slice face, E_{li} and X_{li}, respectively, and on the right face, E_{ri} and X_{ri}, respectively. The resultants of the normal effective stresses on the lateral slice faces, E_{li}' and E_{ri}', are related to the forces E_{li} and E_{ri}, respectively, through the resultants of pore water pressure on those faces.

The safety factor is defined in terms of moments relative to the center of the circular arc:

$$F = \frac{M_R}{M_S} \tag{2.11}$$

with M_R being the moment of the resisting forces opposed to sliding along the arc, and M_S the moment of the forces that tend to cause such sliding, i.e., the weight of the soil mass above the circular surface.

So, taking into account the notation of Figure 2.4 and considering the general case in which the strength parameters vary along the sliding surface, with c_i' and ϕ_i' being their values at the base of the generic slice, the expression for the resisting tangential force, T_{fi}, at the base of the slice, may be written as:

$$T_{fi} = c_i' \Delta \ell_i + \tan \phi_i' N_i' \tag{2.12}$$

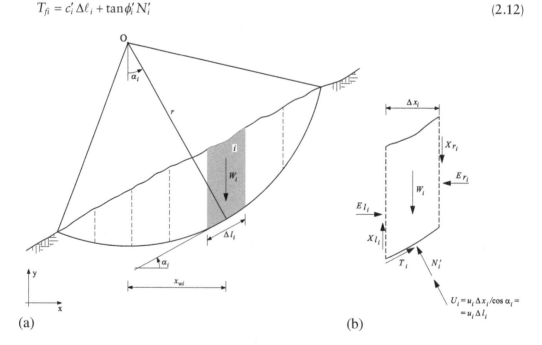

(a) (b)

Figure 2.4 Method of slices: a) soil mass considered; b) generic slice with the applied forces.

The resisting moment is then:

$$M_R = r \sum_{i=1}^{n} T_{fi} = r \sum_{i=1}^{n} \left(c_i' \Delta \ell_i + N_i' \tan \phi_i' \right) \tag{2.13}$$

and M_S is given by:

$$M_S = \sum_{i=1}^{n} W_i x_{wi} = r \sum_{i=1}^{n} W_i \sin \alpha_i \tag{2.14}$$

Using Equations 2.13 and 2.14 in Equation 2.11, one gets

$$F = \frac{\sum_{i=1}^{n} \left(c_i' \Delta \ell_i + N_i' \tan \phi_i' \right)}{\sum_{i=1}^{n} W_i \sin \alpha_i} \tag{2.15}$$

This may be rewritten as:

$$\sum_{i=1}^{n} \left(\frac{c_i'}{F} \Delta \ell_i + \frac{\tan \phi_i'}{F} N_i' \right) = \sum_{i=1}^{n} W_i \sin \alpha_i \tag{2.16}$$

The global safety factor can therefore be interpreted as being the value by which the strength parameters of the soil mass should be divided in order to achieve a limit equilibrium situation.

The only unknown quantity in Equation 2.15 is N_i', the effective normal stress resultant on the slice base, because the problem is statically indeterminate. In fact, as shown in Table 2.1, the number of available equations, $3n$, is lower than the number of unknowns, $4n - 2$. In the latter, the n unknown coordinates of the application point of N_i' are not included; if slices of a narrow width are considered, it is reasonable to admit that such a point coincides with the middle of the slice base arc, with no significant error.

The various approaches available within the general framework of the method of slices differ due to the hypotheses adopted to overcome this indeterminate nature. In general, such hypotheses are related to the slice interaction forces. Three of these methods are presented next.

2.3.3 Fellenius method

The first method developed within the described approach is the so-called *Swedish* or *Fellenius method* (Fellenius, 1927, 1936), which assumes that the interaction forces between each slice and those adjacent are parallel to the base of the slice in question.

Table 2.1 Equations and unknowns in the method of slices (adapted from Whitman and Bailey, 1967).

Equations		Unknowns	
Force equilibrium:		Related to force equilibrium:	
• 2 per slice:	$2n$	• forces N_i'	n
		• forces E_i and X_i	$2(n-1)$
		• value of F relating T_i with T_{fi}	1
Moment equilibrium:		Related to moment equilibrium:	
• 1 per slice:	n	• application points of force E_i	$n-1$
Total:	$3n$	Total:	$4n-2$

Therefore, these interaction forces give zero contribution to the force equilibrium equation in the direction normal to the slice base:

$$W_i \cos \alpha_i - N_i' - U_i = 0 \tag{2.17}$$

or:

$$N_i' = W_i \cos \alpha_i - u_i \Delta \ell_i \tag{2.18}$$

Using this result in Equation 2.15 leads to the Fellenius safety factor:

$$F = \frac{\sum_{i=1}^{n} c_i' \Delta \ell_i + (W_i \cos \alpha_i - u_i \Delta \ell_i) \tan \phi_i'}{\sum_{i=1}^{n} W_i \sin \alpha_i} \tag{2.19}$$

2.3.4 Simplified Bishop method

The simplified version of the method proposed by Bishop (1955), the so-called *simplified Bishop method*, assumes that the slice interaction forces are horizontal. This enables the calculation of the N_i' forces from a force projection equation in the vertical direction:

$$W_i - N_i' \cos \alpha_i - u_i \Delta x_i - T_i \sin \alpha_i = 0 \tag{2.20}$$

The force T_i in this equation represents the tangential force *mobilized*, i.e., necessary for equilibrium, at the slice base. Noting that the *resisting* tangential force, T_{fi}, is given by Equation 2.12, the force T_i can be obtained from the force T_{fi} means of the safety factor, so:

$$T_i = \frac{c_i'}{F} \Delta \ell_i + \frac{\tan \phi_i'}{F} N_i' \tag{2.21}$$

Using this result in Equation 2.20 leads to:

$$N_i' = \frac{W_i - u_i \Delta x_i - (1/F) c_i' \Delta x_i \tan \alpha_i}{\cos \alpha_i \left[1 + (1/F) \tan \phi_i' \tan \alpha_i \right]} \tag{2.22}$$

Combining Equations 2.22 and 2.15, the safety factor is then obtained:

$$F = \frac{\sum_{i=1}^{n} \left[c_i' \Delta x_i + (W_i - u_i \Delta x_i) \tan \phi_i' \right] \left[1/M_i(\alpha) \right]}{\sum_{i=1}^{n} W_i \sin \alpha_i} \tag{2.23}$$

with:

$$M_i(\alpha) = \cos \alpha_i \left(1 + \frac{\tan \alpha_i \tan \phi_i'}{F} \right) \tag{2.24}$$

The simplified Bishop method does not provide an explicit solution for the safety factor, given that F appears in the two members of Equation 2.23. It is, therefore, necessary to perform an iterative procedure for the resolution of the problem: i) select a starting value for F;

ii) compute F using Equations 2.24 and 2.23; iii) assess the difference between this F value and the starting one; iv) select a new value for F and repeat the process until the difference between consecutive values becomes sufficiently small. The convergence of this process is usually quite fast.

2.3.5 Spencer method

The two previous methods can be criticized in view of the lack of a global force equilibrium verification, given that the safety factor is defined exclusively as a ratio of moments. The method developed by Spencer (1967) overcomes this question by computing a safety factor simultaneously in terms of moments and forces.

Consider Figure 2.5, which represents the soil mass under analysis and one generic slice. If the slices are narrow, it is reasonable to assume that W_i, N_i' and U_i are applied at the center of the slice base. As those three forces and also T_i are concurrent, then equilibrium demands that the resultant, Q_i, of the interaction forces acting on the slice must also pass through the same point, as shown in Figure 2.5b, where ρ_i denotes the angle of force Q_i with the horizontal.

The force equilibrium equations along the normal and tangential directions to the slice base are:

$$N_i' + u_i\,\Delta\ell_i - W_i\cos\alpha_i - Q_i\sin(\alpha_i - \rho_i) = 0 \tag{2.25}$$

$$T_i - W_i\sin\alpha_i - Q_i\cos(\alpha_i - \rho_i) = 0 \tag{2.26}$$

From Equation 2.25, the force N_i' is obtained. Using this result in the equation relating the mobilized tangential force with the resistant tangential force leads to:

$$T_i = \frac{c_i'\,\Delta\ell_i + \left[W_i\cos\alpha_i + Q_i\sin(\alpha_i - \rho_i) - u_i\,\Delta\ell_i\right]\tan\phi_i'}{F} \tag{2.27}$$

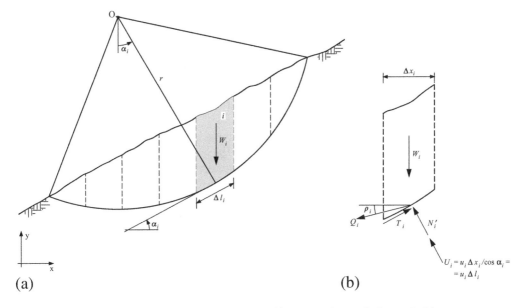

(a)   (b)

Figure 2.5 Spencer method: a) soil mass under analysis; b) generic slice with the applied forces.

Now, replacing T_i in Equation 2.26, the following expression is obtained for the resultant, Q_i, of the interaction forces acting on the slice:

$$Q_i = \frac{\dfrac{c_i' \Delta \ell_i}{F} + \dfrac{\left(W_i \cos \alpha_i - u_i \, \Delta \ell_i\right) \tan \phi_i'}{F} - W_i \sin \alpha_i}{\cos\left(\alpha_i - \rho_i\right)\left[1 - \dfrac{\tan \phi_i' \tan\left(\alpha_i - \rho_i\right)}{F}\right]} \tag{2.28}$$

As the sum of the moments relative to point O of the forces external to the soil mass is null, the same will happen to the sum of the moments of the forces Q_i:

$$\sum Q_i \, r \cos\left(\alpha_i - \rho_i\right) = 0 \tag{2.29}$$

or, since r is constant:

$$\sum Q_i \cos\left(\alpha_i - \rho_i\right) = 0 \tag{2.30}$$

Analogously, as the external forces are in equilibrium, the same has to occur in relation to the forces Q_i, which leads to the following equilibrium equations:

$$\sum Q_i \cos\left(\alpha_i - \rho_i\right) = 0 \tag{2.31}$$

$$\sum Q_i \sin\left(\alpha_i - \rho_i\right) = 0 \tag{2.32}$$

As the resultants of the interaction forces are parallel, ρ_i will be constant, and the two previous equations can be replaced by:

$$\sum Q_i = 0 \tag{2.33}$$

The procedure of the Spencer method is then as follows:

1. a starting value for ρ is selected, for example $\rho_1 = 0$, which is used in Equation 2.28 to obtain the forces Q_i, as a function of F;
2. introducing these forces in Equation 2.33, the value of F corresponding to ρ_1 is obtained;
3. the procedure described in the two previous items is repeated for successive values of ρ, in order to obtain the curve relating F to ρ; the corresponding values of F are designated as F_f, since they result from satisfying the force equilibrium equation;
4. the selection process of successive ρ values is repeated but the forces Q_i obtained from Equation 2.28 are introduced now into equation 2.30, so as to obtain the corresponding values of F, now designated as F_m, due to being obtained from the moment equilibrium equation;
5. the intersection of the two $F_f(\rho)$ and $F_m(\rho)$ curves corresponds to the values of ρ and F that satisfy the force and moment equilibrium, as shown in Figure 2.6.

The interesting point is that, with F and ρ now being known, it is possible to calculate the forces Q_i from Equation 2.28. Subsequently, and starting the computation from slice 1,

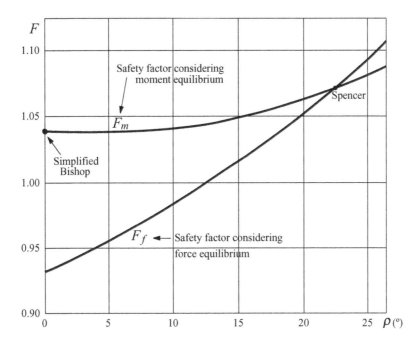

Figure 2.6 Determination of the safety factor by the Spencer method as the intersection of the curves obtained, considering force equilibrium and moment equilibrium (adapted from Spencer, 1967).

the X and E components of the interaction forces and the respective application point can be determined for the successive slices by means of the force and moment equilibrium equations.

2.3.6 Comment

It should be noted that, in both the Fellenius and the simplified Bishop methods, the specific hypotheses adopted concerning the direction of the $n-1$ slice interaction forces render the number of unknowns equal to $4n-2-(n-1)=3n-1$, and so lower than the number of equations, $3n$ (see Table 2.1). The indeterminate problem therefore becomes overdeterminate. The consequence of this fact is that local equilibrium, i.e., for each slice, is not necessarily satisfied by application of these two methods.

The same does not occur with the Spencer method. In fact, assuming that the inclination of the interaction forces is constant in the whole soil mass is equivalent to making $n-1$ assumptions; but the value of the angle, ρ, defining such inclination is being simultaneously introduced as a supplementary unknown. Thus, the problem becomes statically determinate, and a mathematically correct solution is achieved. The Spencer method therefore provides a force system in equilibrium for all slices.

Returning to Figure 2.6, it is also interesting to note that the value of F_m for ρ equal to zero corresponds to the safety factor of the simplified Bishop method. In fact, it merely defines the safety factor in terms of moments and postulates horizontal interaction forces, therefore normal to the interslice boundaries.

Because the trend of the safety factor curves is usually similar to that depicted in Figure 2.6, the simplified Bishop safety factor is lower than that for the Spencer method. The error of the former method is normally very small because, as the figure illustrates, the value of F_m

is not very sensitive to the inclination angle, ρ, assumed for the interaction forces between slices.

The low numerical sensitivity of F_m relative to ρ, has a simple physical explanation: the moment equilibrium is not significantly influenced by the tangential forces acting on the vertical slice faces because the rotation of the unstable mass does not require sliding between slices. On the contrary, the lateral movement of the same mass is not possible without substantial sliding between slices; consequently, the safety factor F_f is quite sensitive to the assumption adopted in relation to the tangential interslice forces (Krahn, 2003).

The very small error value associated with the simplified Bishop method (typically below 2%, as demonstrated by Spencer (1967) for a wide range of conditions) and the fact of it erring on the safe side, explain its generalized adoption in stability analyses when the sliding surfaces are circular.

The Fellenius method also provides results on the safe side but, in certain circumstances, the associated error may be considerable. Taking into consideration that all stability analyses are nowadays assisted by numerical tools, it does not make sense to use the Fellenius method in design. This method is mentioned here due to its historical relevance and, by leading to an explicit determination of the safety factor, it provides a simple introductory exercise for novice students.

In Chapter 9, still within the general approach of the method of slices, the method of Morgenstern and Price (1965) will be presented, which, as the Spencer method, provides mathematically rigorous solutions. That method has the additional advantage of dealing with sliding surfaces other than circular ones.

The safety factors, calculated with the various expressions already described, correspond to the soil mass delimited by the potential sliding surface that was taken into account. Therefore, for stability analysis of any given work – embankment on soft soil, natural slope, embankment dam, or any other – it is indispensable to perform a careful search for other potential sliding surfaces, applying to each the same calculation procedure, in order to identify the one with the lowest safety factor value. This search is most conveniently carried out by recourse to software for slope stability analysis (see, for example, Figure 2.8c).

2.4 STABILITY OF EMBANKMENTS ON SOFT CLAYEY SOIL

2.4.1 Introduction

The problems of stability of embankments on soft clayey soils are perhaps those in which the method of slices is most frequently applied. The observation of numerous failures shows that the sliding surfaces, when analyzed on a vertical plane transverse to the embankment development, are approximately circular arcs, as indicated in Figure 2.7a.

In this type of problem, the safety factor reaches its minimum value at the end of construction. In fact, as shown in Figure 2.7b, the application of load leads to an increase in the mean total stress and in the shear stress in the soil mass, which causes the development of positive excess pore pressure, the maximum value of which is reached at the end of construction.

The excess pore pressure generated at a point in the soil can be expressed by the equation (Skempton, 1954):

$$\Delta u = \Delta\sigma_3 + A\left(\Delta\sigma_1 - \Delta\sigma_3\right)$$

(2.34)

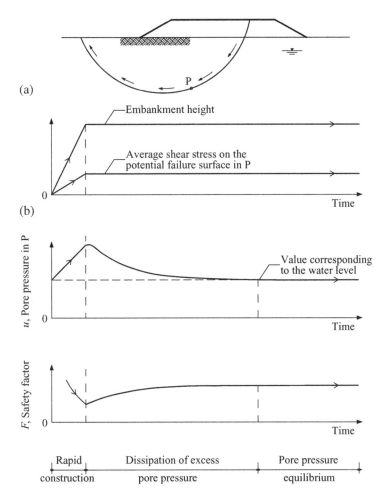

(a)

(b)

Figure 2.7 Embankment on soft clayey soil: a) layout and failure type; b) change of shear stress, pore pressure and safety factor during and after construction (adapted from Bishop and Bjerrum, 1960).

which may be rewritten in the following form:

$$\Delta u = \frac{\Delta\sigma_1 + \Delta\sigma_3}{2} + \frac{(2A-1)(\Delta\sigma_1 - \Delta\sigma_3)}{2} \tag{2.35}$$

This has the advantage of separating, in the two terms of the right-hand side, the increments
of the mean normal stress and of the maximum shear stress.

In the problems under consideration, the first term is positive. Taking into account the
typical range of values of the Skempton A pore pressure parameter, the second term will be,
in principle, positive in normally consolidated to lightly overconsolidated soils and negative
in highly overconsolidated soils.

In summary, and as a global result, the excess pore pressure generated will be positive, but
higher in normally consolidated soils than in overconsolidated ones.

After loading, during a period which may vary widely, depending on the prevailing drainage conditions and on the soil permeability, the excess pore pressure tends to dissipate, the
soil water content diminishes, the effective mean stress increases, as does the shear resistance, and, therefore, so does the safety factor.

This implies that the stability analyses for the present case will have to be carried out for the end-of-construction conditions. Since the time for construction is normally much shorter than the time required for the consolidation process to develop, it is usually admitted that the foundation soil is loaded under undrained conditions, an assumption on the safe side.

When such conditions prevail, the analyses are generally conducted in terms of total stress, due to its inherent simplicity. The soil mass is then considered as a continuous medium, without distinguishing between stress in the solid skeleton and in the pore water, with the shear resistance defined by the Tresca criterion, in which the strength is independent from the total normal stress. This is equivalent to taking $c=c_u$ and $\phi=0$. Some authors usually classify such an approach as a $\phi=0$ analysis.

2.4.2 Application of the method of slices in total stress analyses

Figure 2.8 illustrates an embankment over a clay mass, representing, on the right-hand side of the figure, the typical change with depth of the undrained shear strength. The figure also includes a generic sliding surface and displays the division into slices of the respective soil mass.

Neglecting the shear strength in the part of the sliding surface crossing the embankment – which is currently done because sliding failure is usually preceded by the development of tension cracks progressing from the base to the top of the embankment – the application of Equation 2.15 to this problem assumes the simpler form:

$$F = \frac{\sum_{i=1}^{n} c_{ui}\Delta \ell_i}{\sum_{i=1}^{n} W_i \sin \alpha_i} \tag{2.36}$$

with c_{ui} being taken from the relationship that expresses the change with depth of the undrained shear strength. In the present case, and as the shear strength at the base of the slices does not depend on their weight, the solutions provided by the Fellenius and simplified Bishop methods coincide.

The profile expressing the change of c_u with depth should take into account the considerations presented in the previous chapter on the evaluation of the undrained strength of clays through the vane test (see Section 1.3.5).

An aspect that should be discussed, due to its relevant influence on the safety factor value, has to deal with the strength value to be adopted for the superficial desiccated crust. As this crust is frequently fissured, its resistance may be overestimated by laboratory or *in situ* tests, which are less affected by (possible) fissures as they involve a small soil volume. Figure 2.8b presents two proposals for the strength to be assigned to that layer in order to account for this question (Tavenas and Leroueil, 1980; Lefebvre et al., 1987).

Figure 2.8c shows the results of a stability analysis of the type presented above, performed with specific software. The contour lines represented correspond to sliding surface centers with identical safety factor values. The minimum safety factor value is also represented, together with the corresponding surface.

Analyses of the type presented, taking into account the final embankment geometry, very often provide unacceptable (low) values for the global safety factor. Normally, minimum values in the range 1.3 to 1.5 are required, depending on a certain number of factors, such as experience with similar projects on the same soil, which is clearly the most important, the sensitivity of the clay, which may lead to progressive failure, and the consequences of the occurrence of sliding, among others. In case the safety factor obtained is considered unacceptable, various methods can be employed to achieve a solution for the embankment, some of which are now briefly presented.

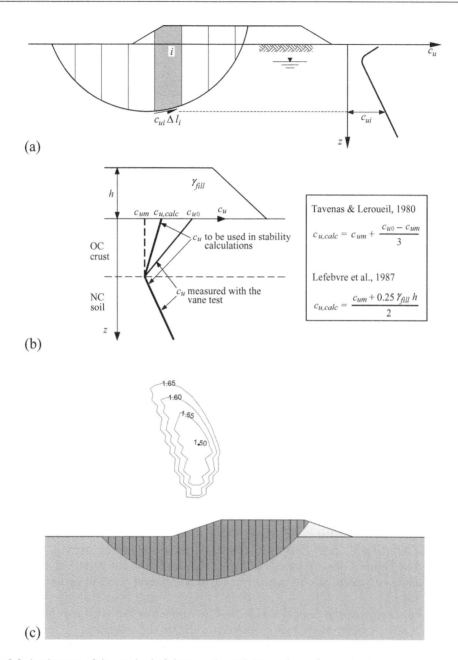

Figure 2.8 Application of the method of slices to the stability analysis of an embankment on soft clayey soil: a) general scheme and typical change of c_u with depth; b) proposals for the consideration of c_u in the superficial crust; c) graphical output of a real case analyzed with the SLOPE/W program.

2.4.3 Methods to improve stability conditions

2.4.3.1 Lateral stabilizing berms

The contribution of lateral berms to the increase in the safety factor can be easily understood by examining Figure 2.9. Imagine that, for the conditions of Figure 2.9a, the safety factor is insufficient and that the critical sliding surface is as represented. The increase in the

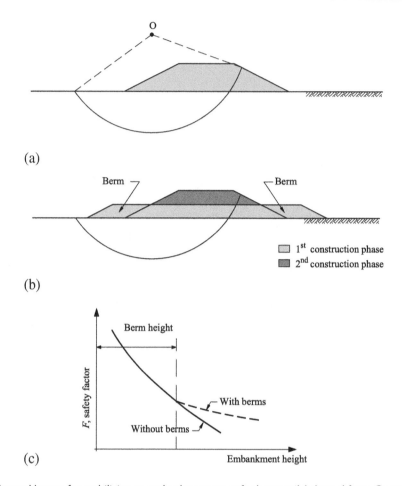

Figure 2.9 Lateral berms for stabilizing an embankment on soft clayey soil (adapted from Correia, 1982).

weight of soil to the left of the center of the circular arc containing the sliding surface will contribute to reducing M_S, and thus to increasing F. Such weight increase can be achieved, as illustrated in Figure 2.9b, by means of lateral berms of the embankment material, the width and height of which can be defined in various different ways in order to reach the target value for F.

It should be noted that the critical surface will no longer be the same for the new geometry, so that a new stability analysis is required to validate the solution.

The disadvantage of this solution is that it increases the area to be expropriated for the works, as well as increasing the negative landscape and environmental impacts associated with embankment construction. This explains why this method is much less frequently used than are those presented below.

It should be mentioned that reducing the inclination of the lateral surfaces of the embankment is similar in its effect to the inclusion of lateral berms.

2.4.3.2 Staged construction

One of the classic procedures for enabling embankment construction on soft clayey soil consists of phasing the load application to the foundation soil. After constructing the initial layer, each additional layer is placed only after dissipation of most of the excess pore pressure generated by the previous construction phase.

Figure 2.10a represents the construction of an embankment in two phases, while Figure 2.10b shows the total and effective stress paths (TSP and ESP, respectively) at point P on the embankment symmetry axis. After the 1st construction phase, the consolidation process is concluded before starting the 2nd phase and the total stresses remain practically unchanged during the consolidation process.

The TSP and ESP associated with the two loading phases are similar, as is the ESP, during the ensuing consolidation processes. Taking into account the ESP orientation during undrained loading, it is easy to verify that, in a situation where the embankment had been completed in a single phase, point P would have suffered shear failure before the total height had been reached.[4] As can be seen, the fact of waiting for the dissipation of excess pore pressure in P after the 1st loading phase allows the 2nd construction phase to proceed without failure having been reached at *P*.

The idea behind this methodology is quite simple: the undrained shear strength depends on the effective stress in the soil mass before undrained loading, increasing as the effective stress increases. When the soil is loaded in the 1st construction stage, the undrained shear strength available at point P is $c_u(0)$, as shown in Figure 2.10. Then, when the soil receives the 2nd loading increment, the available undrained shear strength, $c_u(1)$, also indicated, is now larger, as determined by the average effective stress achieved by the consolidation associated with the 1st construction stage.

In general, due to demanding work execution deadlines, construction phasing will only be viable when combined with the acceleration of the consolidation process, namely with vertical drains (see Section 4.6.3).

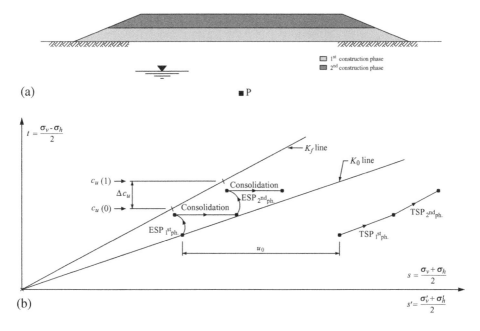

Figure 2.10 Embankment staged construction: a) layout; b) total and effective stress paths at a point under the embankment center, with indication of the undrained shear strength for both loading phases.

---

[4] In this event, the TSP marked in the figure would not take place, because, after shear failure in P, the total stress change with loading would be different, given that the ESP could not go beyond the K_f line.

Annex A2.1 includes some considerations about the stability analyses when construction phasing is adopted. They lead to the conclusion that significant difficulties are to be expected for the accurate estimation of the increase in undrained shear strength due to the preceding loading increments. It is therefore highly recommended to perform *in situ* tests, such as vane shear tests, for validation of these estimates before applying a new loading stage.

2.4.3.3 Foundation soil reinforcement with stone columns

Reinforcing soft clayey soil may also be a method for enabling the application of a given surface loading. The most common method for reinforcing clayey soils consists of the construction of *stone columns*, made of compacted gravel, which are extended down to the base of the soft layer.

As shown in Figure 2.11a, the resistance is increased along the potential sliding surfaces that intersect the stone columns. Their diameter varies, in most cases, between 0.8 m and 1.2 m, depending both on the execution process and the strength of the soil to be reinforced (being greater in weaker soils); the gravel diameter usually ranges between 20 mm and 80 mm.

The degree of reinforcement is usually expressed by the so-called *area replacement ratio*, a_s, which merely expresses the percentage of the total column cross section area relative to the total area (Mitchell, 1981):

$$a_s = \frac{A_{col}}{A_{col} + A_{soil}} \tag{2.37}$$

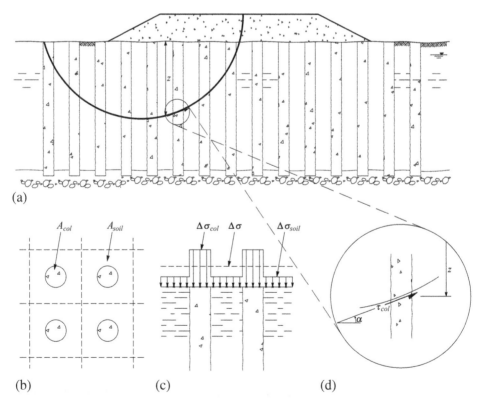

Figure 2.11 Stone columns for reinforcing the foundation of an embankment on soft clayey soil (adapted from Aboshi et al., 1979): a) general layout; b) typical layout in plan; c) column stress concentration; d) detail of the sliding surface intersection with one column.

Values of 0.1 to 0.3 are currently adopted for a_s. For example, in the case of stone columns with 1.0 m diameter in a square layout, as represented in Figure 2.11b, those values correspond to distances of 2.8 m and 1.6 m, respectively, between the axes of adjacent columns.

The contrast between column stiffness and that of the surrounding clayey soil – being naturally higher that of the columns – will induce a concentration of vertical stresses in the columns, associated with their reduction in the soft clay, as illustrated in Figure 2.11c. This phenomenon is very favorable with respect to stability, because it increases the effective vertical stress in the stone columns, and so their shear strength in the intersection points with the sliding surface.

For stability analysis, it is not strictly necessary that the division into slices of the potentially unstable mass explicitly involves the columns. What must be done is to account for the favourable effect of the shear strength of the stone columns intersected by the sliding surface, as shown in detail in Figure 2.11d. To this end, the shear strength at the base of the generic slice i may be obtained from the equation (Aboshi et al., 1979):

$$\tau_{fi} = \left(1 - a_s\right)c_{ui} + a_s\left[\gamma_{col}z_i + \mu_{col}\Delta\sigma_i - u_i\right]\cos^2\alpha_i\tan\phi'_{col} \qquad (2.38)$$

where, in addition to already defined variables:

- c_{ui} represents the undrained shear strength of the clay at the slice base (unreinforced, obviously);
- z_i is the depth of the midpoint of the slice base;
- γ_{col} and ϕ'_{col} are the unit weight and the angle of shearing resistance of the column material, respectively;
- μ_{col} is a factor (>1) to account for the stress concentration in the stone columns; this factor is given by Expression A2.2.4 (see Annex A2.2);
- $\Delta\sigma_i$ is the average vertical incremental stress on the slice area at the soil surface; for slices under the fill, $\Delta\sigma_i = \gamma_{fill}h_{fill,i}$, where γ_{fill} and $h_{fill,i}$ represent the unit weight of the fill material and the average fill height in the slice, respectively;
- u_i is the equilibrium pore pressure at the slice base.[5]

Before closing this topic, it should be highlighted that, besides ensuring an increment in stability, the stone columns provide two further relevant contributions with regard to consolidation settlements: i) they reduce the magnitude of the settlements, due to the vertical stress concentration on the columns, mentioned above; and ii) they also work as vertical drains, thus accelerating the consolidation process. This explains why stone columns are frequently used together with embankments on soft soils. The aspects concerning the effect of the stone columns on the consolidation settlements will be treated in Section 4.6.5.

2.4.3.4 Reinforcement of the base of the embankment with geosynthetics

In the framework of the methods for enabling fill placement on soft soil, a very interesting solution will now be presented, which consists in the reinforcement of the embankment base with a load-bearing geosynthetic, as shown in Figure 2.12 (Borges and Cardoso, 2002).

For each sliding surface intersecting the fill, and therefore the geosynthetic, one has to add to the resisting moment (see Equation 2.13) the moment, M_{Rg}, of the geosynthetic tensile force, T_g, relative to point O, at the center of the arc containing that surface in the plane of the figure:

$$M_{Rg} = T_g\,r\cos\alpha \qquad (2.39)$$

---

[5] Note that Equation 2.38 is based on the hypothesis that the slice interaction forces are parallel to the slice base, in accordance with the Fellenius assumption.

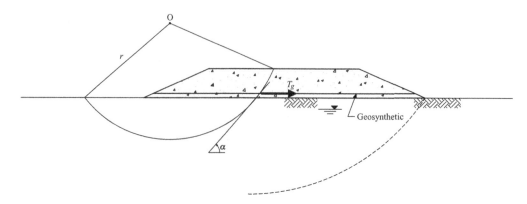

Figure 2.12 Reinforcement of the embankment base with geosynthetic.

A current stability analysis procedure consists of adopting a trial value for the geosynthetic tensile force, T_g, and then calculating the embankment safety factor. The force magnitude will be successively adjusted until a satisfactory value for F is reached. In this way, the required geosynthetic force, $T_{g,req}$, to ensure adequate stability conditions, is obtained, which will have to be lower than: i) the geosynthetic tensile strength; and ii) its pull-out resistance. This resistance is controlled by the tangential stresses between the fill and the geosynthetic, on one side, and between the geosynthetic and the foundation soil, on the other, along the shorter of the two stretches of reinforcement defined by the intersection with the critical sliding surface (in the case of Figure 2.12, the one to the left of the intersection point).

The sliding surfaces that intersect the embankment are therefore prevented by adequate selection of the geosynthetic reinforcement. On the other hand, the surfaces that encompass the embankment as a whole (see the dashed line) will reach deeper and more resistant zones of the clayey soil, or a firm layer below. Therefore, for them, the safety factor will, in principle, be much higher.

In Annex A2.3 is included an alternative approach for the stability analysis proposed by Leroueil and Rowe (2001).

2.4.3.5 Use of lightweight aggregates as fill material

A solution distinct from all the previous ones involves employing, as fill material, a lightweight aggregate, i.e., an artificial granular material of low unit weight. In most cases, this corresponds to an expanded clay.

Some types of clay, when heated to very high temperature (above 1000°C), suffer profound chemical and structural alterations, assuming a spherical shape with a volume many times higher than that of the original particles, with a strong skin and a highly porous honeycomb structure. These particles exhibit high strength, very low unit weight, and are practically inert, and therefore durable under the typical environmental conditions of embankment works.

Table 2.2 lists some typical properties of expanded clay.

As a granular material of appreciable mechanical strength, good durability, and a unit weight around 25% of that of conventional fill materials, the relevance of this alternative is readily understood. Figure 2.13 shows a transverse cross section of an embankment incorporating expanded clay. As can be seen, this material is enveloped by conventional fill material.

Table 2.2 Typical properties of expanded clay.

Property	Typical value range
Particle diameter	0–30 mm 10–20 mm (more frequently)
Unit weight	Loose state and factory moisture content: 2.5–3.0 kN/m³ Compacted and humid: 3.3–4.0 kN/m³ Compacted and saturated: 5.0 kN/m³
Friction angle	37–40°
Maximum allowable compression stress	100 kPa

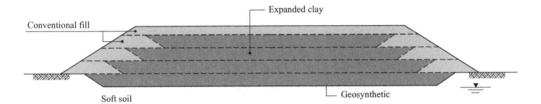

Figure 2.13 Cross section of a fill incorporating expanded clay.

2.4.4 Comment

The methods already described are not exhaustive, although they are probably the most currently utilized. Table 2.3 summarizes their main features in comparative terms, while also addressing settlement-related questions.

Monitoring the construction of an embankment over soft soil is of paramount importance in the control of safety against a rotational sliding during the construction phase. This issue will be treated in Chapter 4, together with monitoring of the subsequent process related to consolidation (see Section 4.7).

2.5 UNSUPPORTED CUTS IN CLAYEY SOIL

2.5.1 Introduction: basic features

Consider now a slope created by a cut on a clay formation, as shown in Figure 2.14a, and consider that it has been performed in a much shorter time than that required for dissipation of the excess pore pressure it induced in the soil mass.

The excavation is simultaneously responsible for a decrease in the mean total stress and an increase in the shear stress, namely in the soil close to the cut face.

Considering again the Skempton equation rewritten in the form:

$$\Delta u = \frac{\Delta\sigma_1 + \Delta\sigma_3}{2} + \frac{(2A-1)(\Delta\sigma_1 - \Delta\sigma_3)}{2} \tag{2.40}$$

it may be understood that the mean stress decrease will generate negative excess pore pressure, i.e., will reduce the pore water pressure. In normally consolidated or lightly overconsolidated soils (in which the A parameter is positive and relatively high), such a reduction is, to a certain extent, attenuated by the positive excess pore pressure associated with the increase in shear stress. In medium to highly overconsolidated soils (in which the A parameter is low

Table 2.3 Summary of methods for enabling embankment construction on soft soils.

Method	Increase of stability	Settlement magnitude	Consolidation time	Remarks
Lateral berms	Reduces M_S	Increases the total but reduces the differential	Increases	May be applied as a remedial measure; increases the expropriation area
Staged construction with acceleration of consolidation	Increases M_R (by increasing c_u)	Does not significantly affect the total but reduces the differential	Increases (1)	Difficulty for estimating c_u after the 1st phase
Stone columns	Increase M_R (due to the shear forces in the columns)	Reduces substantially	Reduces substantially	Also reduces immediate settlement
Base reinforcement with geosynthetics	Increases M_R (due to the tensile force in the geosynthetic)	Does not significantly affect the total but reduces the differential	Does not affect	–
Lightweight fill material	Reduces M_S	Reduces substantially	Reduces	Also reduces immediate settlement

Note:

1. In comparison with non-phased loading, with acceleration of consolidation

or even negative), the reverse tends to happen, so that, as illustrated in Figure 2.14b, the decrease in pore pressure is much more pronounced in such soils. Even more considerable negative excess pore pressures occur in the soil under the base of the excavation.

After completion of the excavation, and because of the negative excess pore pressure, the pore water pressure in the adjacent soil is smaller or much smaller than that in regions further away from the cut face. The excess pore pressure dissipation process, i.e., the consolidation process that follows, will therefore involve a water flow directed to the zone adjacent to the cut, where the water content, as well as the volume, will increase.

In comparison with the moment after the end of excavation, the pore water pressure will rise, the effective stress will decrease, and so will the shear strength and the safety factor. This factor will reach its minimum value at the end of consolidation, when an equilibrium situation in terms of pore pressure is re-established in the soil mass, as illustrated in Figure 2.14b.

It is interesting to underline that, in the present situation – as opposed to the previously analyzed case of soft clay soil loaded at the surface by an embankment – the long-term pore water equilibrium conditions are no longer hydrostatic, because the excavation has altered the hydraulic boundary conditions. If the cut is permanent, a steady state flow will develop, discharging at the toe of the slope, as suggested in Figure 2.14a.

This leads to the conclusion that stability evaluation in the long-term demands an analysis in terms of effective stresses, taking into consideration the pore pressure distribution at the end of consolidation, which, in turn, requires the respective steady-state flow net. The procedure for performing the stability analysis may naturally be the method of slices, as explained in Section 2.3.2.

Given the importance of understanding the basic features relevant to excavation works, as well as those related to surface loading works (embankment construction or others), Table 2.4 provides a brief summary.

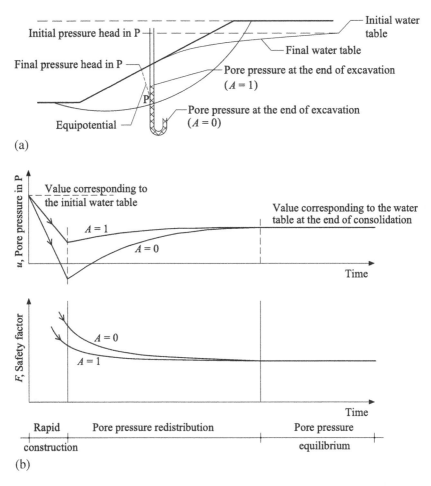

Figure 2.14 Unsupported excavation in clay: a) initial and final conditions of the water table and pore water pressure in a generic point P at the end of excavation; b) change over time of the pore water pressure in P and of the safety factor (adapted from Bishop and Bjerrum, 1960).

2.5.2 The question of safety and its evolution with time

The considerations presented lead to a conclusion of enormous practical importance: in an excavation in a fine-grained soil mass, safety diminishes with time! In other words, the fact of performing a cut in a fine-grained soil, which is stable at the moment of execution, is by no means a guarantee that stability will be maintained in the long-term.

The verification of stability in the long-term demands, as seen before, the performance of effective stress analyses, taking into consideration the groundwater conditions corresponding to a steady-state regime; in these analyses, the water table at a remote location (the hydraulic boundary condition) must be assumed at the most unfavorable elevation.

What has just been stated does not mean, however, that undrained analyses for the end-of-construction conditions are not useful, in certain well-defined situations, such as when the excavation is of a temporary type. In such situations, whenever the period in question is much smaller than that required for the dissipation of a significant fraction of the excess pore pressure, undrained analyses will be the more appropriate for stability assessment.

Table 2.4 Comparison of features associated with surface loading works and to cuts in clayey soil.

Item	Embankment works	Excavation works
Total mean stress	Increases	Decreases
Shear stress (under the embankment or behind the cut face)	Increases	Increases
Excess pore pressure, Δu	Positive	Negative
Soil type in which $|\Delta u|$ is maximum	Normally consolidated	Heavily overconsolidated
Evolution of u with time (after the end of construction/excavation)	Decreases	Increases
Flow during consolidation	Directed outwards	Directed towards the excavation
Evolution of w and e with time	Decrease	Increase
Evolution of volumetric deformation with time	Positive (compression)	Negative (expansion)
Ground surface displacement	Settlement	Heave
Change with time of the mean effective stress and of the shear strength (after the end of construction/excavation)	Increase	Decrease
Change with time of the safety factor	Increases	Decreases
Phase in which the stability analyses should be performed	End of construction (loading)	End of consolidation
Conditions for the stability analyses	Undrained	Drained
Approach in terms of stress for the stability analyses	Total stress	Effective stress
Strength parameters involved in the analyses	c_u, considering its evolution with depth	Effective strength parameters, c' and ϕ'
Other data items required for the analyses	Unit weight	Steady-state flow net; unit weight

A comment on these issues is nevertheless essential: the time corresponding to the consolidation process in these problems is extraordinarily variable! For instance, Skempton (1964) and Vaughan and Walbancke (1973) reported cut sliding in overconsolidated clays, close to railway lines in the United Kingdom, which took place decades after construction. In contrast, there are studies on stiff fissured clays in Scandinavia, in which the dissipation of excess pore pressure close to trenches developed in a matter of days (DiBiagio and Bjerrum, 1957).

In the present state of knowledge, it is not possible to estimate with accuracy the change in time of the excess pore pressure generated by an excavation in a clayey soil or in a soil with a significant fine fraction, such as a residual soil. The issue is due not to a lack of theoretical and numerical tools, but rather to the difficulty in estimating, in a reliable manner, the permeability of the soil mass, which controls the development of the consolidation rate. This implies that *it is not possible to evaluate in an approximate manner the decrease in soil resistance with time*.

Therefore, assessment of the stability of a temporary cut by means of an undrained analysis can only be made if it can be proved, based on solid *comparable experience*, that the time period during which it will remain open is much smaller than that required for consolidation.[6] Disregarding this common-sense basic rule has contributed to many accidents in trench construction, with loss of human life.

---

[6] The definition of comparable experience in Eurocode 7 (EN 1997-1, 2004) is as follows: documented or other clearly established information related to the ground being considered in the design, involving the same types

In temporary excavations, where stability cannot be ascertained under undrained conditions, the following rules must be followed:

1. the stability analyses shall be undertaken for the end-of-consolidation conditions; these analyses will be carried out in terms of effective stresses;
2. in cases where these analyses indicate that the excavation is not stable, a reformulation of the solution becomes indispensable;
3. one way of increasing the safety is by reducing the angle of the cut face with the horizontal;
4. in certain cases, as in frequently conducted works for the installation or repair of pipes and other buried infrastructures, it may be more convenient to resort to supported trenches (with vertical faces).

2.5.3 Note about the maximum depth of tension cracks on the ground surface

When a cut is performed in a clayey soil, surface tension cracks tend to develop in the vicinity. As the soil tensile strength is very low and can become practically nil in the short-term, vertical tension cracks tend to initiate from the surface, as suggested in Figure 2.15a.

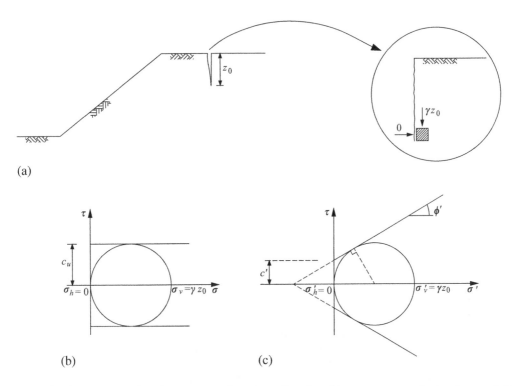

Figure 2.15 Excavation in clay: a) tension crack at the soil surface; b) stress state at the tension crack tip, in a homogeneous soil under undrained conditions and in a total stress analysis; c) stress state at the tension crack tip, in a soil with cohesion and friction angle in an effective stress analysis.

of soil and rock and for which similar geotechnical behavior is expected, and involving similar structures. Information gained locally is considered to be particularly relevant.

Consider a homogeneous soil with undrained shear strength c_u and let z_0 be the maximum depth reached by tension cracks in that soil.

At any point in the soil on the crack wall at a depth z lower than or equal to z_0, the stress on a vertical surface element is zero and the stress on a horizontal surface element represents the total soil weight, γz, above that point.

Therefore, as shown in Figure 2.15b, the crack can only progress to the maximum depth z_0, for which:

$$\gamma z_0 = 2c_u \tag{2.41}$$

hence:

$$z_0 = \frac{2c_u}{\gamma} \tag{2.42}$$

In the case of an effective stress analysis for a soil with cohesion and friction angle, as shown in Figure 2.15c, the crack can reach a depth for which:

$$\left(c' \cot an \phi' + \frac{\gamma z_0}{2} \right) \sin \phi' = \frac{\gamma z_0}{2} \tag{2.43}$$

so:

$$z_0 = \frac{2\, c' \cos \phi'}{\gamma \left(1 - \sin \phi' \right)} \tag{2.44}$$

or

$$z_0 = \frac{2c'}{\gamma \left(\dfrac{1 - \sin \phi'}{1 + \sin \phi'} \right)^{1/2}} \tag{2.45}$$

2.5.4 Cuts under undrained conditions

2.5.4.1 Vertical cuts

2.5.4.1.1 Approximate solutions without considering tension cracks

In Section 2.2.2, an analytical solution based on the UBT has been found for the critical depth, assuming a plane sliding surface. The consideration of a circular-shaped surface leads to lower values, so it is more accurate. Actually, the observation of real failures clearly suggests that this a more realistic type of sliding surface.

Figure 2.16 represents the problem under scrutiny (Cardoso et al., 2007): the soil wedge limited by the circular surface with the center at point O will undergo a rotation defined by the angle δ around that point. The surface, defined by the angles λ and α, passes at the base of the cut face. No restriction is applied to the location of point O.

Table 2.5 presents in its upper part the external forces, i.e., the weight of the two parts into which the soil wedge has been divided, the respective vertical displacements associated with the rotation δ around point O, and their products, which represent the work of the external forces. The same table includes, in its lower part, the computation of the plastic work of the resisting forces along the sliding surface for the same rotation, δ.

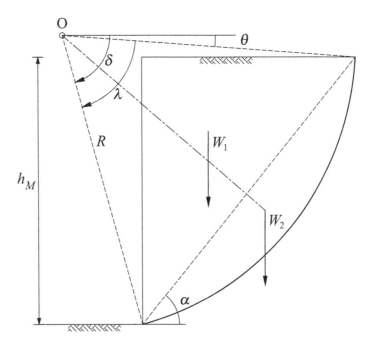

Figure 2.16 Conditions allowed in the upper bound solution for the critical excavation depth in a purely cohesive soil (Cardoso et al., 2007).

Equalizing the external force work and the plastic work, and applying the relationships listed in Table 2.5, after some transformations, one obtains

$$h_M = \frac{4c_u}{\gamma} \frac{\dfrac{\lambda}{\sin 2\alpha \ \sin \lambda}}{1 + \dfrac{1}{3} \tan\alpha \ \tan\dfrac{\lambda}{2}} \tag{2.46}$$

In this equation, the values of the angles α and λ that minimize h_M are, respectively, $\alpha^* = 47.55°$ and $\lambda^* = 30.02°$. Using these values in Equation 2.46 (expressed in radians) leads to:

$$h_M = 3.83 \frac{c_u}{\gamma} \tag{2.47}$$

This result coincides with that obtained by Taylor (1948), using the so-called friction circle method, an approach based on the philosophy of the upper bound theorem.

At present, the application of the two limit analysis theorems (upper and lower bound theorems) may be performed through very powerful computational models based on the finite element method. Some of these models have been applied to solve this classical problem of the vertical cut in a purely cohesive material. Pastor et al. (2000) published results corresponding to the upper and lower bound solutions, confining the "correct" solution within a very narrow interval. Taking into account these results, a very approximate solution for the critical depth is given by:

$$h_{cr} \cong 3.77 \frac{c_u}{\gamma} \tag{2.48}$$

Table 2.5 Calculation of the external force work and of the plastic work for the problem of Figure 2.16 (Cardoso et al., 2007).

Forces, F_i	Displacements, d_i	Work, $F_i\,d_i/\mu\,h_M\,\delta$
External force work		
$W_1 = \dfrac{\gamma h_M^2}{2\tan\alpha} = \dfrac{\mu}{\tan\alpha}$	$\left(\dfrac{R}{h_M}\cos\theta - \dfrac{2}{3\tan\alpha}\right)h_M\,\delta$	$\left(\dfrac{R}{h_M}\cos\theta - \dfrac{2}{3\tan\alpha}\right)\dfrac{1}{\tan\alpha} = \dfrac{1}{2\tan\alpha}\left(\dfrac{1}{\tan(\lambda/2)} - \dfrac{1}{3\tan\alpha}\right)$
$W_2 = \left(\dfrac{R}{h_M}\right)^2(\lambda - \sin\lambda)\,\mu$	$\dfrac{4}{3}\dfrac{R}{h_M}\dfrac{\left(\sin(\lambda/2)\right)^3}{\lambda - \sin\lambda}\sin\alpha\,h_M\,\delta$	$\dfrac{4}{3}\left(\dfrac{R}{h_M}\sin(\lambda/2)\right)^3\sin\alpha = \dfrac{1}{6(\sin\alpha)^2}$
Plastic work		
$\lambda R c_u = \lambda\dfrac{2}{N_S}\dfrac{R}{h_M}\mu$	$\dfrac{R}{h_M}h_M\,\delta$	$\lambda\dfrac{2}{N_S}\left(\dfrac{R}{h_M}\right)^2 = \dfrac{\lambda}{2N_S}\dfrac{1}{\left(\sin\alpha\,\sin(\lambda/2)\right)^2}$

With:

$$N_S = \frac{\gamma h_M}{c_u} \qquad R = \frac{h_M}{2\sin\alpha\sin(\lambda/2)}$$

$$\mu = \frac{1}{2}\gamma h_M^2 \qquad \frac{R}{h_M}\cos\theta = \frac{1}{2}\left(\frac{1}{\tan(\lambda/2)} + \frac{1}{\tan\alpha}\right)$$

Considering the work of Lyamin and Sloan (2002), who obtained 3.772 as a lower bound solution for the number that multiplies the ratio c_u/γ, it can be concluded that Equation 2.48 errs on the side of safety.

In a cut or slope context, it is current practice to designate the ratio of the total vertical stress, γh, at the base of the excavation, to the undrained shear strength, c_u, as the *stability number*, N_s:

$$N_s = \frac{\gamma h}{c_u} \tag{2.49}$$

When $h = h_{cr}$, that parameter becomes the *critical stability number*, and so, given Equation 2.48:

$$N_{s,cr} = \frac{\gamma h_{cr}}{c_u} \cong 3.77 \tag{2.50}$$

2.5.4.1.2 Zero tensile strength case

The solutions presented above assume that the soil has some tensile strength, since no tension crack has been considered in the analytical or numerical calculations.

When the tensile strength is zero, the failure mechanism changes from that represented in Figure 2.16, assuming instead the form depicted in Figure 2.17a, according to Drucker (1953). The block represented, with very low thickness, d, is delimited by a vertical tension crack and a plane inclined at 45°, passing through the base of the cut face.

As shown in Figure 2.17b, in the upper bound theorem approach, it is considered that the block has a displacement δ along the inclined base plane. With d being, admittedly, very small, the weight of the block can be given by:

$$W = \gamma \, d \, h_M \tag{2.51}$$

The work produced by the block weight, when the displacement occurs, will be equal to the external plastic work (which has zero contribution from the block vertical face that corresponds to the tension crack):

$$\gamma \, d \, h_M \, \delta \frac{\sqrt{2}}{2} = c_u \, d \sqrt{2} \, \delta \tag{2.52}$$

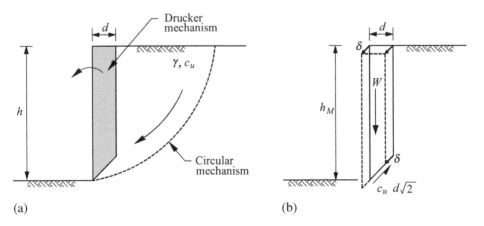

(a)   (b)

Figure 2.17 Vertical cut in a purely cohesive soil with zero tensile strength: a) Drucker mechanism; b) application of the UBT for calculating the critical excavation depth.

and so

$$h_M = \frac{2c_u}{\gamma} \tag{2.53}$$

Noting that the solution obtained for h_m in Section 2.2.2 is applicable to soils with zero tensile strength,[7] it can be concluded that, in this case, the two limit analysis theorems provide coincident solutions. Therefore, the expression:

$$h_{cr} = \frac{2c_u}{\gamma} \tag{2.54}$$

corresponds to an *exact solution* for the case of a vertical cut under undrained conditions with zero tensile strength.

2.5.4.1.3 Analysis of the failure mechanism for various tensile strength conditions

After what has been stated in the two previous points, and moving on to discuss, in practical terms, the vertical cut problem, it shall be concluded that the *real* value of the critical depth will be within the interval:

$$\frac{2c_u}{h} < h_{cr} < 3.77\frac{c_u}{\gamma} \tag{2.55}$$

being closer to the upper bound while the tensile strength endures – naturally in temporary terms, as is actually the case with undrained behavior – and getting closer to the lower limit, when that strength is practically null. In real soils, intermediate situations will certainly exist, i.e., between the maximum tensile strength ($2c_u$) and zero.

Figure 2.18 shows results obtained by Antão and Guerra (2009) with a finite element limit analysis model, which illustrate the development of the collapse of a vertical cut in a homogeneous soil under undrained conditions. Three scenarios were considered, concerning tensile strength: i) the classic Tresca yield surface; ii) tensile strength equal to half the undrained shear strength; iii) zero tensile strength (equivalent to a truncated Tresca envelope). Note that, in this problem, the classic Tresca yield surface is equivalent to a tensile strength of $2c_u$.

The numerical results, displaying the zones where the plastically dissipated energy concentrates, replicate very closely, both qualitatively and quantitatively, the analytical solutions which have already been discussed for cases 1 and 3. The critical depth for case 2 is intermediate in relation to the others, with the tensile crack being apparent near the surface.

2.5.4.2 Inclined face excavations

When sloped cuts, undrained conditions, and a homogeneous soil are considered, the critical depth will naturally increase with a decreasing slope inclination, for a given value of the undrained shear strength. The same will also happen to the critical stability number, given its direct relationship with depth.

---

[7] In fact, the stress field implied in Figure 2.2 does not comply with the existence of tensile stress at any point of the soil

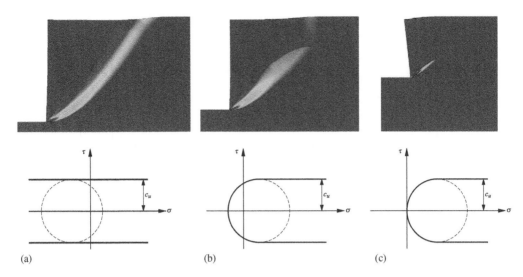

Figure 2.18 Failure mechanisms of a vertical cut in a purely cohesive soil: a) classic Tresca yield surface; b) truncated Tresca envelope, with tensile strength equal to $0.5c_u$; c) truncated Tresca envelope, with zero tensile strength (Antão and Guerra, 2009).

This is confirmed by the stability chart represented in Figure 2.19, developed by Janbu (1954). The chart presents the critical stability number, $N_{s,cr}$ as a function of the cut slope angle with the horizontal, β, for several values of the parameter d, which is the ratio between the distance, D, from the excavation base to the top of a firm stratum and the excavation depth, h. This is a relevant parameter in the analysis of the problem, because it affects the mode of failure for lower values of β.

Basically, as shown in the figure:

1. for β values above 53°, failure involves a toe circle, a surface passing through the toe of the slope face, as seen before in the case of the vertical cut (the situation corresponding to $\beta = 90°$);
2. for β values below 53°, failure may occur with an over-slope circle (the sliding surface intersects the slope face above its toe), a toe circle or an under-slope circle (the sliding surface intersects the base of the cut beyond the toe of the slope face), depending on the depth of the rigid boundary;
3. the curious aspect expressed by the figure is that, for large values of d, $N_{s,cr}$ does not increase with a reduction in the slope face angle below 53°; this may be explained because, above a certain value of N_s (around 5.5), the critical failure mode switches from sliding of the soil behind the excavation face to one involving failure as a result of insufficient bearing resistance of the soil below the bottom of the excavation.

It should also be noted that the figure contains, as a particular case, the solution discussed in Section 2.5.4.1.1 (Equation 2.47) for the vertical cut case.

The charts produced by Taylor (1937) about this problem are also well known. Charts of this type are useful for a preliminary analysis of the problem, which is nowadays very expediently carried out by recourse to commercial software, providing solutions with diverse methods, namely the method of slices, on which the presented chart has been based.

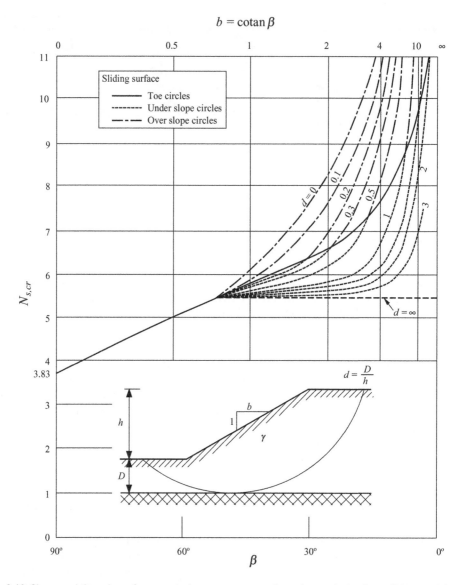

Figure 2.19 Slope stability chart for cuts in homogeneous soil, under undrained conditions, with a rigid boundary at a certain depth (adapted from Janbu, 1954, 1996; Leroueil et al., 2001).

2.5.5 Excavations in drained conditions: effective stress analyses: Hoek and Bray charts

As emphasized in Section 2.5.2, excavation safety is minimal at the end of consolidation. Therefore, in most cases, the stability analyses should be performed in terms of effective stress, considering the long-term soil water conditions, which will correspond to a steady-state seepage regime directed towards the excavation. Such analyses use the strength parameters in terms of effective stress, the effective cohesion, c', and the effective angle of shearing resistance, ϕ', the former being practically zero in normally consolidated clays.

It is well known that the maximum inclination of the surface of a cohesionless soil mass above the water table coincides with its angle of shearing resistance. In soils where the shearing resistance arises from a combination of c' and ϕ', the inclination of the surface, β, may be higher than this angle, if the slope height does not exceed a given value, also designated as critical (see Section 9.2).

However, in practical applications, the question is typically posed in a different manner: the excavation depth is prescribed, and it is the slope inclination that must be established. This inclination depends on: i) the soil strength parameters; ii) the existence of tension cracks at the surface, which have to be taken into account, in view of their unfavorable effect on stability; iii) the ground water conditions, given that seepage adjacent to the slope is highly detrimental; and d) the adopted value for the safety factor (F). Other issues may also influence stability, such as seismic action.

The literature provides several stability chart proposals for the conditions under discussion, relating some or all of the indicated issues. The most complete are probably those developed by Hoek and Bray (1981), based on a limit equilibrium method which considers circular sliding surfaces. These charts, extremely simple to use, take into account the existence of tension cracks at the surface, as well as the steady-state seepage pattern into the excavation.

Figure 2.20 illustrates the two groundwater flow types considered in the charts. In Figure 2.20a the water level prior to the excavation coincides with the ground surface and remains

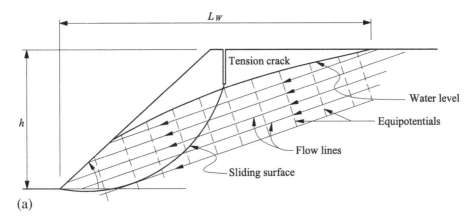

(a)

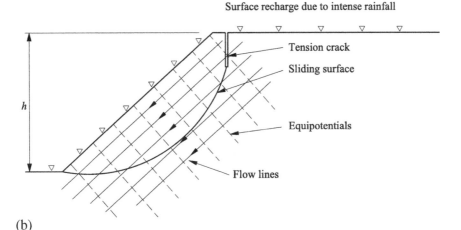

(b)

Figure 2.20 Soil water conditions considered in the Hoek and Bray (1981) stability charts: a) flow associated to the phreatic level drawdown imposed by the excavation, the water level coinciding with the soil surface at a distance L_w from the excavation base; b) flow affected by permanent surface recharge.

so beyond a distance L_w from the toe of the slope; the figure presents the steady state flow net, whose flow lines run parallel to the phreatic surface. In the complementary example, Figure 2.20b illustrates a situation in which, due to very intense rainfall, there is a permanent recharge zone extending to the whole ground surface, the water level coinciding with it in this case.

The stability charts are represented in Figures 2.21 to 2.25, being indicated the corresponding situation in hydraulic terms.

In order to obtain the slope angle, β, the use of the charts involves the following steps: i) select the appropriate chart for the prevailing hydraulic conditions; ii) calculate the abscissa, $c'/(\gamma h F)$, and the ordinate, $\tan \phi'/F$, and mark the respective point (which will generally be located between the two curves corresponding to the two values of β); and iii) determine, by interpolation, the value of β.

When β is already established and the aim is to obtain the respective safety factor, the use of the charts requires the following steps: i) select the appropriate chart for the prevailing

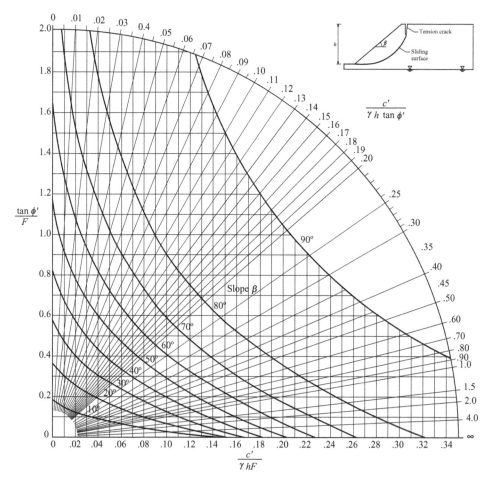

Figure 2.21 Hoek and Bray stability chart (1981) – deep phreatic surface.

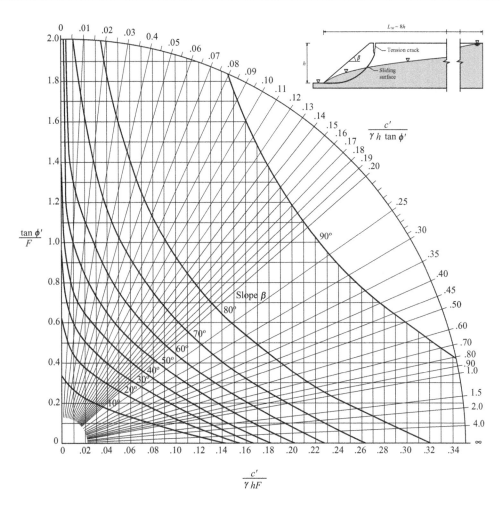

Figure 2.22 Hoek and Bray stability chart (1981) – phreatic surface with $L_w = 8h$.

hydraulic conditions; ii) calculate the ratio $c'/(\gamma h \tan \phi')$ and locate the corresponding value on the external circular scale of the chart; iii) follow the radial line from that point until it intersects the arc corresponding to the β value under consideration; iv) take the abscissa (or the ordinate) of that intersection point on the horizontal (or vertical) axis; and v) compute the safety factor, F, using the abscissa (or the ordinate) value.

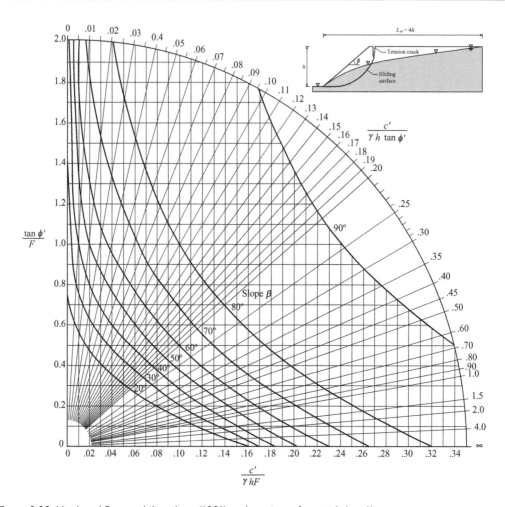

Figure 2.23 Hoek and Bray stability chart (1981) – phreatic surface with $L_w = 4h$.

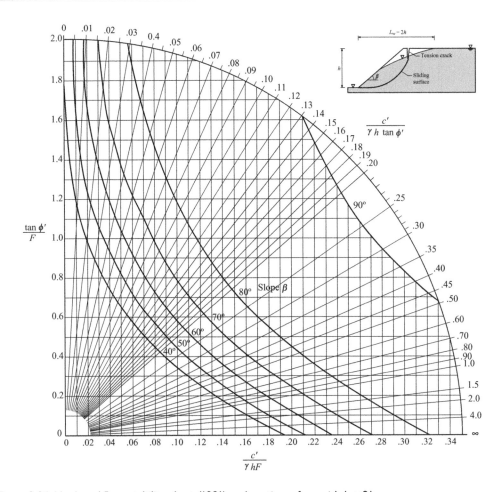

Figure 2.24 Hoek and Bray stability chart (1981) – phreatic surface with $L_w = 2h$.

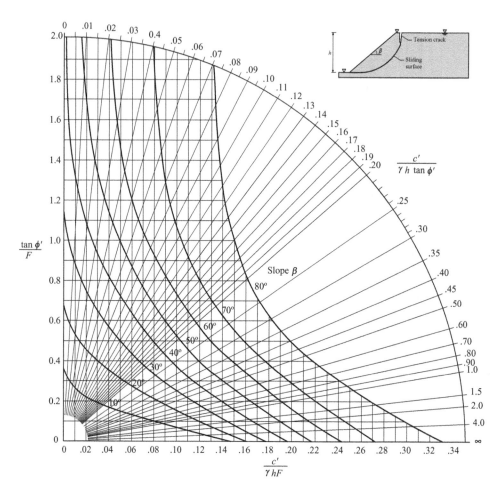

Figure 2.25 Hoek and Bray stability chart (1981) – phreatic surface at the soil surface.

ANNEX A2.1 STABILITY ANALYSIS OF EMBANKMENTS DURING STAGED CONSTRUCTION

There are several approaches in the literature concerning stability analysis during staged construction. The one which seems most consistent, in theoretical terms, is designated by *undrained strength analysis* (USA), as described by Ladd (1991): for each loading phase after the first, a stability analysis under undrained conditions is required, considering the current embankment geometry and the undrained shear strength values, after their update, as discussed below.

The difficulty resides in evaluating the undrained shear strength increments. The proposal of Ladd, if applied to normally consolidated or lightly overconsolidated clays, consists of the following simple steps:

1. consider, for the first loading stage, the relationship between the undrained shear strength and depth, as given by the tests performed under at-rest conditions; this will correspond to a relationship of the type illustrated in Figure 2.8a, in which, below the surface-desiccated crust, c_u/σ'_{v0} remains approximately constant;
2. assume, in each subsequent loading stage, that the increase in the undrained shear strength at each point, Δc_u, represents the same fraction of the effective vertical stress increment as in the first stage, i.e., $\Delta c_u/\Delta\sigma'_v = c_u/\sigma'_{v0} = constant$.

The first loading stage induces effective vertical stress increments, $\Delta\sigma'_v$, that may be represented by the stress bulb depicted in Figure A2.1.1. Multiplying this bulb by the c_u/σ'_{v0} ratio will allow the distribution of the strength increase, Δc_u, to be then considered.

Figure A2.1.1 shows that the more significant vertical stress increments are concentrated under the central zone of the embankment, for very small values of the z/B ratio. Curiously, the reduction in depth is even more rapid for the horizontal stress increments (see, incidentally, Figure 6.11). Therefore, the effective mean stress increments fade out with the increases in z/B and x/B even more markedly than what is illustrated in Figure A2.1.1 for the vertical stress increments.

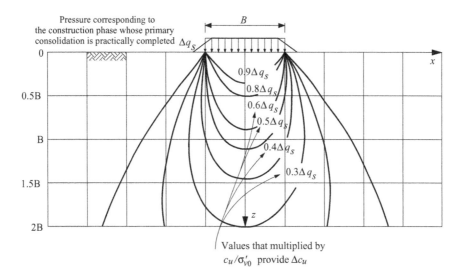

Figure A2.1.1 Simplified scheme for estimating the increase of c_u along construction phases with intermediate consolidation periods.

Therefore, this approach appears to be more applicable when the embankment width is large, in comparison with the thickness of the soft layer, a situation that favors a relatively shallow critical sliding surface, mostly located underneath the embankment. When such conditions do not apply, the critical sliding surface tends to be relatively deep and extending laterally to the embankment, i.e., involving zones where the increase in c_u has been small in the previous loading-consolidation phases.

The experimental results reported by Law (1985) relative to works in Canada and in Sweden clearly support the considerations presented above. Even for very wide embankments, when compared to the thickness of the clayey soil, this author found values for $\Delta c_u / \Delta \sigma'_v$ somewhat lower than c_u / σ'_{v0}. This result seems reasonable given that, even in normally consolidated soils, ageing by secondary consolidation induces an increase in soil strength, which naturally does not exist in a staged construction context.

It can be concluded that, in many situations, the adoption of staged construction may be unable to provide a substantial safety factor increase in embankment works on soft soil. Therefore, *in situ* confirmation of the increase in shear strength by vane tests (or other appropriate tests) is highly recommended.

ANNEX A2.2 STONE COLUMN REINFORCEMENT OF THE FOUNDATION OF EMBANKMENTS ON SOFT SOIL

Consider Figure 2.11. The stone column stress concentration factor, m may be defined as the ratio:

$$m = \frac{\Delta \sigma_{col}}{\Delta \sigma_{soil}} \tag{A2.2.1}$$

where $\Delta \sigma_{col}$ and $\Delta \sigma_{soil}$ represent the average total stress increments in the stone columns and in the soil, respectively.

Representing, with $\Delta \sigma$, the total stress increment applied at the ground surface, one may write:

$$\left(A_{col} + A_{soil} \right) \Delta \sigma = A_{col} \Delta \sigma_{col} + A_{soil} \Delta \sigma_{soil} \tag{A2.2.2}$$

Combining the two previous equations with Equation 2.37, leads to:

$$\Delta \sigma_{soil} = \frac{\Delta \sigma}{1 + \left(m - 1 \right) a_s} = \mu_{soil} \, \Delta \sigma \tag{A2.2.3}$$

and

$$\Delta \sigma_{col} = \frac{m \Delta \sigma}{1 + \left(m - 1 \right) a_s} = \mu_{col} \, \Delta \sigma \tag{A2.2.4}$$

Figure A2.2.1 presents a theoretical solution, developed by Priebe (1995), for the stress concentration factor, based on the following hypotheses: i) the stone columns are founded on a rigid stratum; ii) the column material is incompressible (null volumetric deformation); iii) the settlement of each column gives rise to an increase in the respective diameter, constant from top to base; iv) this increase mobilizes reactive pressure in the surrounding soil, in an elastic regime; and v) after column installation, the total stress state is isotropic in the surrounding clayey soil, i.e., $\sigma_v = \sigma_h$.

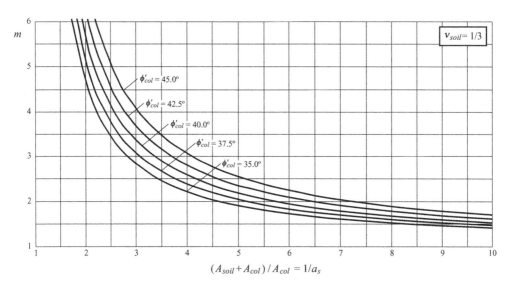

$$(A_{soil} + A_{col}) / A_{col} = 1/a_s$$

Figure A2.2.1 Stress concentration factor in stone columns (Priebe, 1995).

ANNEX A2.3 METHOD OF LEROUEIL AND ROWE (2001) FOR THE STABILITY ANALYSIS OF EMBANKMENTS ON SOFT SOIL WITH BASE REINFORCEMENT WITH GEOSYNTHETIC

The method comprises the following steps, taking Figure A2.3.1 as reference.

1. Establish the design values of c_u by application of a safety factor F_d considered adequate, $c_{u,d} = c_u/F_d$. Typically, the minimum values for F_d are between 1.3 and 1.5.

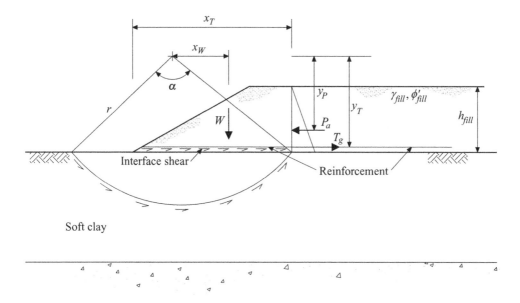

Figure A2.3.1 Stability analysis of an embankment with geosynthetic base reinforcement on soft clayey soil (adapted from Rowe and Mylleville, 1993).

2. For the fill material, select the maximum expected unit weight and the critical value for the angle of shearing resistance, ϕ'_{cv}, taking into account that, in embankments on very soft soil, it is often difficult to achieve a proper compaction and that shear deformation of the embankment may be high.

3. Consider a potential sliding surface of circular shape, as shown in the figure, so that the vertical line through its intersection with the base reinforcement crosses the upper horizontal boundary of the embankment; it should be noted that the sliding surface does not cross the embankment (see the following points).

4. Calculate the stabilizing moment supplied by the shear strength of the foundation soil as given by:

$$M_R = r^2 \int_0^\alpha c_{u,d}\, d\alpha \tag{A2.3.1}$$

5. The contribution of the embankment to the destabilizing moment is due to the combined effect of two forces: i) the weight, W, of the fill mass to the left of the vertical line mentioned in step 3; and ii) the horizontal force P_a, applied at one-third of the embankment height, measured from its base, the value of which is the Rankine active thrust (see Chapter 5):

$$P_a = \frac{1}{2} K_a \gamma_{fill}\, b^2_{fill} \tag{A2.3.2}$$

6. Calculate, for the adopted sliding surface, the force, T_g, in the geosynthetic necessary for moment equilibrium as:

$$T_g = \frac{Wx_w + P_a\, y_P - M_R}{y_T} \tag{A2.3.3}$$

7. Repeat steps 3 to 6 in order to find, for the so-called critical sliding surface, the force in the geosynthetic, $T_{g,req}$, required for stability.

8. The force obtained in step 7 will have to verify the condition:

$$T_{g,req} \leq T_{g,lim} \tag{A2.3.4}$$

where $T_{g,lim}$ is the geosynthetic limit tensile force defined as:

$$T_{g,lim} = \min\left(T_{g,f}; T_{g,p}\right) \tag{A2.3.5}$$

where $T_{g,f}$ is the design value for the geosynthetic tensile strength and $T_{g,p}$ is the geosynthetic pull-out resistance, obtained from:

$$T_{g,p} = P_a + P_s \tag{A2.3.6}$$

where P_a is given by Equation A2.3.2 and P_s is:

$$P_s = \frac{\chi\, c_{uc}\, x_T}{F_d} \tag{A2.3.7}$$

with c_{uc} being the shear strength of the foundation soil in contact with the geosynthetic, χ a factor (lower than or equal to one) to account for the possible reduction in the foundation soil-geosynthetic interface strength, and x_T the distance from the toe of the embankment to the intersection point between the critical sliding surface and the reinforcement.

ANNEX E2 EXERCISES (RESOLUTIONS ARE INCLUDED IN THE FINAL ANNEX)

E2.1 – Figure E2.1 shows a cross section of a cutting in an overconsolidated clay. The flow net represented in the figure corresponds to the steady-state seepage at the end of the consolidation process induced by the excavation.

Calculate the safety factor for the potential slip surface represented in the figure, applying the Fellenius method and the simplified Bishop method. Take $\gamma_w = 9.8$ kN/m³.

E2.2 – Figure E2.2 illustrates a cross section of a vertical cut executed in a homogeneous clay mass. Assuming undrained clay behavior and the further conditions of the figure:

a) calculate the stability number of the excavation and confirm that it is close to the estimates for the respective critical value;

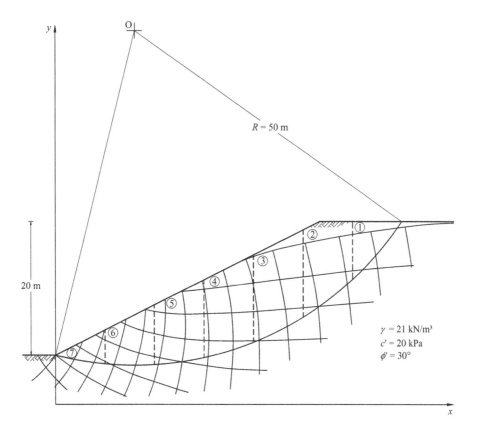

$R = 50$ m

$\gamma = 21$ kN/m³
$c' = 20$ kPa
$\phi' = 30°$

20 m

Figure E2.1

b) using software for stability analysis by the method of slices, confirm that the safety factor is close to unity.

E2.3 – Figure E2.3 illustrates a cross section in a homogeneous clay in which a temporary cut will be executed. Adopting a safety factor of 1.4 for the undrained shear strength:

a) calculate the inclination of the slope using the slope stability chart of Janbu (Figure 2.19);
b) check the result, using software for stability analysis by the method of slices.

E2.4 – Figure E2.4 illustrates a cross section of an embankment on a soft clay deposit underlain by a very stiff layer. Table E2.4 includes the geotechnical parameters for the stability analysis.

a) Calculate the safety factor against a rotational failure sliding of the embankment and the foundation soil, using software for stability analysis by the method of slices.
b) Using the same software, design a solution with lateral stabilizing berms that provides a safety factor close to 1.5.

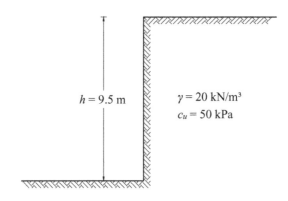

$h = 9.5$ m

$\gamma = 20$ kN/m³
$c_u = 50$ kPa

Figure E2.2

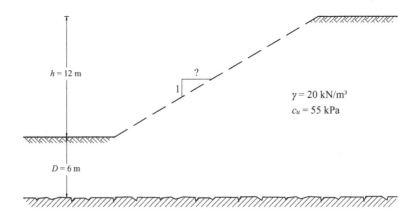

$h = 12$ m

?

1

$\gamma = 20$ kN/m³
$c_u = 55$ kPa

$D = 6$ m

Figure E2.3

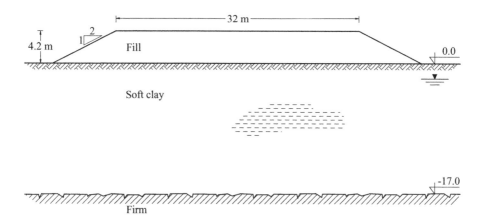

Figure E2.4

Table E2.4

Material	z, depth (m)	γ (kN/m³)	Strength parameters
Fill	–	22	$\phi' = 35°$
Clay – layer 1	z = 0.0 to −2.0	17	$c_u = 24$ kPa
Clay – layer 2	z = −2.0 to −4.0	17	$c_u = 22$ kPa
Clay – layer 3	z = −4.0 to −10.0	17	$c_u = 20$ kPa
Clay – layer 4	z = −10.0 to −17.0	17	$c_u = 28$ kPa

c) Using the same software, establish an inclination of the lateral sides of the embankment that provides a safety factor close to 1.5.

d) Using the same software and applying the method of Leroueil and Rowe (see Annex A2.3), design a solution with reinforcement of the embankment base with a geosynthetic that provides a safety factor close to 1.5.

Chapter 3

Basis of geotechnical design

3.1 INTRODUCTION

The constructions designed by civil engineers must satisfy a number of requisites or conditions. Traditionally, such requisites have been condensed into the following:

i) *stability*, i.e., safety against failure or structural collapse;
ii) *serviceability*, i.e., the capacity to fulfill the design purpose without significant constraints for the users;
iii) *durability*, in order to satisfy the two preceding requisites across the life span of the project within reasonable maintenance costs.

Naturally, these requisites should be accomplished with acceptable costs and within a reasonable execution time.

More recently, the *sustainability* requisite has emerged, i.e., all operations and materials involved in the design, construction, exploration, maintenance, and, finally, building demolition, should comply with the environmental sustainability paradigm.

The present chapter, dedicated to the basis of design of geotechnical works, will discuss the methodologies for fulfilling the first two requisites.

Civil engineering structures induce stress states in the soil masses with which they interact, which modify their at-rest conditions to a certain extent. The structural design issues related to the interaction with the (underlying, adjacent, or surrounding) soil are generally governed by two essential criteria:

i) the stress state in the ground resulting from that interaction should be sufficiently distant from the one that, for similar loading type, would lead to collapse or global failure;
ii) the deformation associated with that stress state should be acceptable for the serviceability of the structure itself and of others in the vicinity.

The objective of the present chapter is to present and discuss the philosophies behind geotechnical design, i.e., the methods employed to satisfy the two above-mentioned criteria. The considerations that follow have much in common with the design of other civil engineering structures and fit into the general framework of structural safety (Borges and Castanheta, 1968).

3.2 VARIABLES AND UNCERTAINTIES IN GEOTECHNICAL DESIGN

The process of safety evaluation in the general framework of structural design and, in particular, of geotechnical design, involves variables of diverse categories: i) independent or primary (measurable) variables that include actions, material strength properties, and geometrical parameters; and ii) dependent variables, namely the *effects of actions*, on the one hand, and the *resistances* on the other.

In this book, for the sake of simplicity, the term *demand*, denoted by the symbol S, will be generically employed to designate the effect of actions, due to its tradition of use in structural engineering.[1] Also, again for the sake of simplicity, the term *resistance* will be employed to designate the capacity to withstand actions without mechanical failure, denoted by the symbol R.

In general, the dependent variables are calculated from the primary ones by the so-called *calculation models*. Some of these calculation models, namely for steel structures and reinforced concrete, are probably already familiar to the reader. As to what concerns geotechnical works, one may recall, for example, the method of slices for the stability analysis of earth masses, presented in Chapter 2. In the following chapters, calculation models with other objectives will be presented.

Figure 3.1 summarizes the overall format of the safety evaluation procedure, specifying the diverse variables involved, as well as their respective relationships. As to what concerns these relationships, it is convenient to draw the reader's attention to the following fact, represented in the figure by the crossing arrows: in most geotechnical problems, the demand (*also*) depends on the ground resistance parameters and the resistance (*also*) depends on the actions. It should be noted that this interdependence does not in general occur in problems within the scope of structural engineering.

Figure 3.2 illustrates two examples of this interdependence. In the case of Figure 3.2a, which shows a cantilever retaining wall of reinforced concrete, the soil thrust, P_a, controlling the effect of actions on the retaining wall in any limit state is essentially dependent on the strength properties of the retained ground. On the other hand, the resistance to global sliding of the slope, represented in Figure 3.2b, depends on the weight (permanent action) of the potentially unstable mass, which controls the resistant shear stresses on the sliding surface through normal effective stresses.

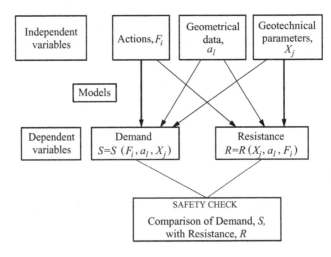

Figure 3.1 General scheme for the evaluation of structural safety.

---

[1] In other publications, E is currently used as the symbol for the effect of actions, as, for instance, in the Structural Eurocodes.

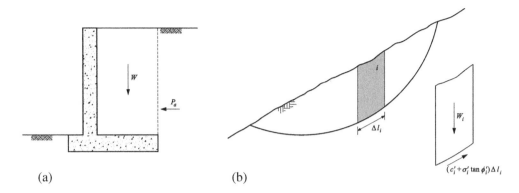

Figure 3.2 Examples of structures in which the effect of actions depends on the soil strength parameters and the resistance depends on the actions: a) reinforced concrete cantilever retaining wall; b) natural slope.

Due to this interdependence, it becomes quite complex to apply, in geotechnical design, the structural safety approach adopted in structural engineering. This subject will be further explored later.

Regardless of the philosophy on which the design is based, what the two criteria mentioned in Section 3.1 aim to ensure is that the *resistance* of the system or structure – bridge, building, dam, foundation, slope, etc. – be greater than the *demand*, for a level of safety considered acceptable. In mathematical terms this can be expressed in the following way:

$$\text{Resistance}(R) > \text{Demand}(S) \tag{3.1}$$

Taking Figure 3.3 as reference, the design process must provide a solution that is located in the "safe" region or, more rigorously, that satisfies the *design criterion* during the life span of the structure. As shown in the figure, design criteria have some levels or *margins of safety* relative to the boundary that separates safe solutions from unsafe solutions. As shall be

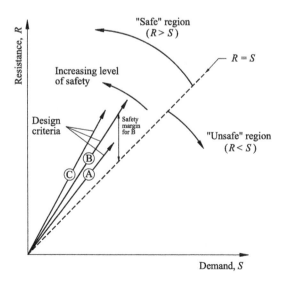

Figure 3.3 General criterion for structural design (adapted from Oliphant, 1993).

seen, these safety margins are related to the safety factors adopted in design, whether *global safety factors*, or *partial safety factors*.

Those safety margins and the respective safety factors are intended to deal with the *uncertainties* that affect the design process.

Such uncertainties are related to:

i)   action estimation, involving permanent actions (in general, the self-weight of the structure and of the ground), and variable actions related to service life (traffic, people, movable property, etc.), and to the environment (wind, earthquake, temperature variation, snow, etc.);
ii) spatial variability of ground properties;
iii) evaluation of ground mechanical parameters from laboratory and field tests (sample disturbance, test interpretation, empirical correlations, etc.);
iv) deviation of geometrical parameters relative to the values assumed in design (foundation dimensions, errors in foundation location relative to the structure, etc.);
v)   limitations of design methods in accurately modelling the mechanical phenomena that control the behavior of both the structure and the soil, as well as their interaction.

The uncertainties relative to the ground resistance properties are clearly the most important, and they naturally derive from the complex geological processes involved in the formation of soil and rock masses. On the other hand, in many geotechnical works, the variable actions (with moderate to high uncertainty) have little relevance, compared with the magnitude of the permanent actions (with relatively low uncertainty), the latter therefore generally controlling action variability in overall terms.

Incidentally, it is useful to add that the line which in Figure 3.3 separates "safe" from "unsafe" solutions, is not actually known, due to the uncertainties that affect the design process.

3.3 GLOBAL SAFETY FACTOR METHOD

3.3.1 Definition of global safety factor: typical values

Over millennia, it has been possible to support structural design by experience from similar structures on similar ground, with the building codes of practice being upgraded from generation to generation based on the (satisfactory or deficient) behavior of the constructions. This empirical method has proved to be appropriate in an era of essentially unchanging building materials (mostly stone and brick), and with a very slow evolution of structural solutions.

Over the past 150 years, the emergence of new structural materials, especially steel and concrete, and the rapid structural evolution encouraged by their use, have demanded more rational methods for the design of a wider range of structural systems.

The first rational approach that supported structural design consisted of comparing the resistance, R, with the demand, S, through their ratio:

$$F = \frac{R}{S} \tag{3.2}$$

where F represents what is conventionally known as the *global safety factor*.

Another way of expressing the previous equation is:

$$S = \frac{R}{F} \tag{3.3}$$

Table 3.1 Values of global factors of safety (Terzaghi and Peck, 1948, 1967).

Failure modes	Item	Factor of safety
Shearing	Earthworks	1.3–1.5
	Earth retaining structures, excavations, offshore foundations	1.5–2.0
	Foundations on land	2.0–3.0
Seepage	Uplift, heave	1.5–2.0
	Exit gradient, piping	2.0–3.0
Ultimate pile loads	Load tests	1.5–2.0
	Dynamic formulas	3.0

This alternative presentation shows clearly why this approach is known in the literature as *working stress design*, WSD. The main idea is that the applied or service stress, *S*, does not exceed, in each structural member or cross section, a given admissible stress that is equal to the ratio of the yield or ultimate stress, *R*, by a factor, *F*, greater than one.

In this methodology, abandoned in structural engineering after the 1960s and 1970s, but still applied in geotechnical engineering practically until the present, the uncertainties involved in design, whichever they may be, are globally taken into account by means of a single safety factor. The greater the value of *F* is above one, the greater will naturally be the safety margin for a given practical situation.

Table 3.1 presents a list of global safety factor values, the ranges of which are generally considered to be satisfactory. This means that these values, when combined with an appropriate ground characterization and the application of well-established design methods, lead to solutions that may be considered "safe" from a deterministic point of view. The upper values in each interval should be adopted for current loading conditions, also designated as *service conditions*, while the lower values are to be employed for maximum loads and/or more extreme actions and worst environmental conditions. The lower values are also acceptable in conjunction with real-scale tests, with the observational method or for temporary works (Meyerhof, 1995).

3.3.2 Limitations of the global safety factor method

As shown in Figure 3.4 (Becker, 1996a), in this approach both the demand and the resistance are considered to be deterministic variables, and are characterized, in each case, by a single

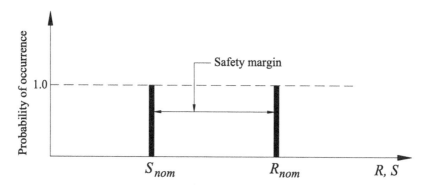

Figure 3.4 Definition of the global safety factor (adapted from Becker, 1996a).

value, called the *nominal value*. This designation is given to a value of a physical variable selected without considerations of a statistical nature. In general, when dealing with the nominal value, the symbol of the physical variable in question is represented without any auxiliary index (for example, simply ϕ' for the angle of shearing resistance). However, in the context of this chapter, for reasons of clarity, the subscript *nom* will be used to identify nominal values.

Due to the uncertainties previously mentioned, both resistance and demand are, in reality, random variables with a given statistical distribution, as expressed (merely for indicative purposes) in Figure 3.5a by means of probability density functions. Annex A.3.1 provides a brief summary of the probabilistic concepts employed in this chapter.

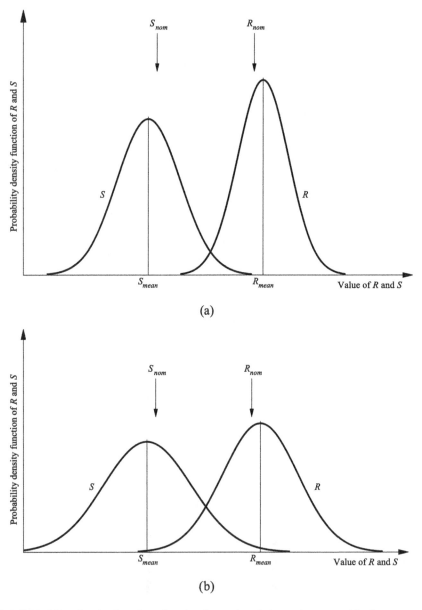

Figure 3.5 a) Probability density functions for the demand and the resistance, with the respective mean values and hypothetical nominal values; b) idem, for a case with broader distribution curves (adapted from Becker, 1996a).

It is important to recognize that the random character of both resistance and demand is, in fact, taken into account, though implicitly, in the selection process of the nominal values of these variables. In reality, as suggested in Figure 3.5a, a judicious designer will tend to adopt a nominal resistance lower than the mean value ($R_{nom} < R_{mean}$) and a nominal demand greater than the mean value ($S_{nom} > S_{mean}$), even without explicitly invoking statistical considerations.

A fundamental point in this context is to note that the intersection of the resistance and demand distribution curves represented in Figure 3.5a shows that, under certain conditions, the demand may exceed the resistance. This means that the probability of failure is not null.

Now consider Figure 3.5b, where, for the same mean and nominal values of resistance and demand, the probability density functions are broader due to greater uncertainty. The area under both curves is larger than in the previous figure, so that the probability of failure increases, even though the global safety factor remains the same. That area is not equal to the probability of failure, though it increases with it.

In summary: i) a global safety factor value greater than one does not necessarily mean safety, i.e., the probability of failure is not zero and may even be quite substantial; ii) on the other hand, failure will not necessarily occur if the global safety factor is less than one; iii) an identical safety factor value for different situations may correspond to different safety levels or probabilities of failure, a greater probability of failure being exhibited for the situation in which, due to greater uncertainty, the probability density functions for resistance and/or demand are broader.

From the preceding considerations, an essential limitation of the global safety factor method can be identified: by accounting, *via* a single factor, for all uncertainties inherent in the design process, it does not encourage the user to ponder about the distinct sources of uncertainty.

Moreover, the global safety factor method does not distinguish stability requisites (i.e., safety relative to failure) from serviceability requirements (i.e., safety relative to excessive deformation). In certain cases, as in both shallow and pile foundations, the relatively large global safety factor values derived from experience are implicitly meant to limit the settlements to tolerable values, which signifies that, on the whole, the requisites of stability and functionality are, in fact, satisfied.

Finally, taking into account the comments made about Figure 3.5, the true level of structural safety is not known, using the global safety factor method.

3.4 LIMIT STATE METHOD AND PARTIAL SAFETY FACTORS

3.4.1 General considerations

The structural codes of practice for concrete, steel, and other structural materials started to adopt *limit state design* (LSD) as the basic philosophy from the 1960s and 1970s. This method uses partial safety factors and is based on considerations of a probabilistic type.

More recently, especially since the 1990s, geotechnical design codes have been progressively adapted to the limit state design philosophy. This occurred in Europe, with the approval in 1994 of Eurocode 7 as a European Pre-norm (ENV 1997-1:1994), but also in the United States, Canada and Japan (Ovesen and Orr, 1991; Becker, 1996ab; Coduto, 2001; Honjo and Kusakabe, 2002).

This evolution of the codes applied in geotechnics results from the obvious advantage inherent in the adoption of the "same language" and coherent design concepts by both structural and geotechnical engineers in the design teams of civil engineering structures, preventing misunderstandings that may cause more or less serious errors. Moreover, a common background philosophy for the design of structures, whichever they may be, will also facilitate the teaching of civil engineering.

A limit state can be defined as a state beyond which the structure no longer satisfies, to some manner, the functions for which it was designed. Within the limit states, there are: i) *ultimate limit states*, associated with collapse or other similar forms of failure; and ii) *serviceability limit states*, which represent conditions beyond which the structure, or a structural element, ceases to satisfy some performance requisite. In terms of geotechnical works, the serviceability limit states are essentially related to excessive ground deformation.

The limit state method basically consists of the following steps:

1. identification of all potential limit states of the structure;
2. individualization of all variables conditioning safety, which can be sorted into actions, strength properties, and geometrical data;
3. establishment of deformation limits and of other parameters related to serviceability conditions (cracking width, vibrations, and the like);
4. evaluation for each limit state, by theoretical models, numerical calculations or other means, of the *demand*, on the one hand, and of the *resistance*, on the other; this step involves the application of partial safety factors in the evaluation of both the demand and the resistance; the way in which demand and resistance are factored depends on the safety methodology adopted and will be discussed next;
5. *verification of safety*, i.e., demonstration that the occurrence of the limit state is sufficiently improbable.

When dealing with ultimate limit states, the *verification of safety* takes the following form:

$$R_d, \text{Factored Resistance} > S_d, \text{Factored Load Effects} \tag{3.4}$$

For a serviceability limit state, the verification assumes the following form in geotechnical works:

$$\text{Deformation} \leq \text{Tolerable Deformation} \tag{3.5}$$

Two essential aspects distinguish, in a positive way, the present approach from the previous one. Firstly, there is a clear distinction between the verifications relative to ultimate limit states and to serviceability limit states. This distinction – that is the core of the LSD approach – is positive because, in practical terms, it prompts the performance of both stability and deformation analyses.

The second distinctive aspect is that both actions and material properties are treated as random variables. The partial safety factors assume values adjusted to the greater or lesser uncertainty affecting each variable. For instance, the partial safety factors for variable actions are typically greater than those for permanent actions, because there are substantially fewer uncertainties concerning the latter.

The values of the partial safety factors, for the evaluation of demand and resistance, that are included in the codes of practice, were established from two types of studies involving the so-called *calibration calculations*. One type of study determines which sets of partial safety factor values lead to design results close to those obtained by the traditional approach with global safety factors. The other type of study investigates, by means of probabilistic studies, accounting for the statistical dispersion of the distinct variables, those sets of partial safety factor values which ensure a given failure probability (Meyerhof, 1993, 1995; Becker, 1996b; Cardoso and Matos Fernandes, 2001).

From these considerations, it may be understood that the partial safety factor values that are involved in the evaluation of demand and resistance are inter-related, forming a set which corresponds to a given level of safety.[1] The values of the partial safety factors will be addressed later.

Despite the referred advantages of the limit state method over the global safety factor approach, they still have in common the drawback of not quantifying the true level of structural safety.

3.4.2 Distinct forms of application of the limit state method

As mentioned above, the limit state method has been applied in several distinct ways, which can be grouped under two major categories, designated by several authors as the *European concept* and the *American concept* (Ovesen and Orr, 1991; Becker, 1996a). Figure 3.6 illustrates, in a simplified manner, the difference between the two concepts, which is essentially based on the form of calculation of the resistance, R.

As shown in Figure 3.6a, in the so-called European concept, after selecting the characteristic values of the ground resistance properties, these are reduced by partial safety factors in order to obtain the corresponding design values, which are then introduced into the calculation models and lead to the design value of the resistance, R_d. The designation of this approach as European is justified because of its adoption by the Danish Geotechnical Code in 1965, following the pioneering work of Brinch Hansen (1953, 1956).

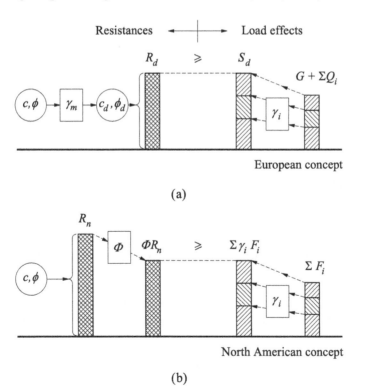

Figure 3.6 Comparison of limit state design approaches for ultimate limit states: a) European concept; b) North American concept (Ovesen and Orr, 1991).

---

[1] In order to preserve this level, it does not make sense to modify one of these factors without adjusting the others.

As shown in Figure 3.6b, in the American concept, the characteristic values of the ground strength properties are directly used in the calculation models that supply the characteristic or nominal value of the resistance, R_n, which is then reduced by a resistance factor, Φ, in order to obtain the value of the factored resistance, ΦR_n. The American concept is known as Load and Resistance Factor Design (LRFD).

There is considerable debate concerning the advantages and disadvantages of either methodology (Becker, 1996a; Simpson and Driscoll, 1998).

The European concept seems more elaborate, because it applies the partial safety factors to the material strength properties, which are the variables most affected by uncertainty. In so doing, it manages to *calibrate* the value of the partial safety factors according to the nature of the strength parameters (for instance, decreasing the undrained shear strength more than the angle of shearing resistance). The most pertinent limitation that is generally pointed out involves the fact that the failure mechanism considered in the calculation models depends on the values of the soil strength properties. Therefore, by reducing these properties, the mechanism can change and hence so can the computed resistance.

In the American concept, on the other hand, the resistance factor is meant to cover not only the uncertainties relative to the strength properties, but also those affecting the geometrical data and, additionally, the limitations of the methods or models used in the evaluation of the resistance.[1]

Another aspect to be considered in this context is that, in many cases, the resistance in geotechnical design is evaluated not by means of theoretical calculations (introducing the ground strength parameters), but with load tests (as in the case of pile foundations and ground anchors), or by empirical or semi-empirical correlations with *in situ* test results (for the design of shallow foundations from PMT results, for example). In these situations, the application of a resistance factor, in accordance with the American concept, does not raise difficulties, as would happen with the application of the European concept.

As will be seen later, the current version of Eurocode 7 (EN 1997-1: 2004) establishes three design approaches that follow not only the European concept but also the American concept (see Section 3.6). The latter is at the foundation of the American and Canadian design codes and will later be presented in more detail (see Section 3.7).

3.5 PROBABILISTIC METHODS

3.5.1 Safety margin, reliability index, and probability of failure

In spite of the advantages of the limit state method over the global safety factor method, both share the limitation of not providing a quantification of the real safety level of the structure.

The safety level of a structure is adequately expressed by the *probability of failure*. As can be understood, its evaluation demands the use of probabilistic methods. This is a vast and complex subject, addressed by the reliability theory, and one which exceeds the scope of this book. The following considerations are meant to provide a simple introduction to this topic.

Figure 3.7 displays the probability density functions of demand and resistance for a given limit state of a generic structure, following a normal distribution.

The so-called *margin of safety*, M, is defined (Whitman, 1984; Christian, 2004) as:

$$M = R - S \tag{3.6}$$

---

[1] In what concerns the limitations of the calculation models, it is always possible in either methodology (European or American) to include a model factor, both in resistance and demand, to ensure that the results of the design calculation model are either accurate or err on the safe side.

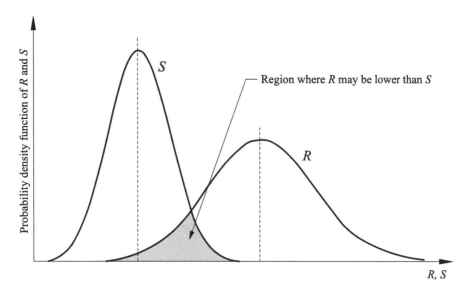

Figure 3.7 Probability density functions of demand and resistance for a given limit state of a generic structure, assuming a normal distribution.

R and S are random variables with normal distribution, as is M, with a probability density function of the same type, as depicted in Figure 3.8a. The theory of probability allows us to prove that the mean value and standard deviation (σ) of the three variables are related through:

$$M_{\text{mean}} = R_{\text{mean}} - S_{\text{mean}} \tag{3.7}$$

and

$$\sigma_M^2 = \sigma_R^2 + \sigma_S^2 - 2\rho_{RS}\,\sigma_R\,\sigma_S \tag{3.8}$$

where ρ_{RS} is the so-called correlation between R and S. If R and S are independent variables, equation 3.8 becomes:

$$\sigma_M^2 = \sigma_R^2 + \sigma_S^2 \tag{3.9}$$

The occurrence of the limit state corresponds to the equality $R = S$, i.e., to $M = 0$. Therefore, the *probability of failure*, P_f, as shown in Figure 3.8, corresponds to the probability of $M \leq 0$.

The so-called *reliability index*, β, is the number by which the standard deviation, σ_M, has to be multiplied to locate the zero of the probability density function, i.e., the failure region below the mean value, M_{mean}. So:

$$\beta = \frac{M_{\text{mean}}}{\sigma_M} \tag{3.10}$$

In cases where R and S obey a normal distribution and are independent, and taking into account Equations 3.7 and 3.9, β is given by:

$$\beta = \frac{R_{\text{mean}} - S_{\text{mean}}}{\sqrt{\sigma_R^2 + \sigma_S^2}} \tag{3.11}$$

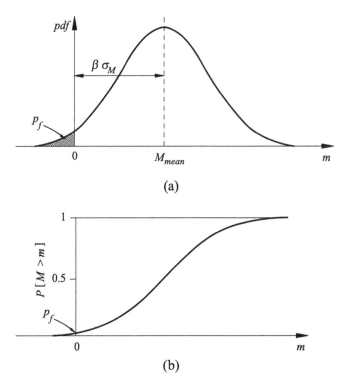

Figure 3.8 Margin of safety relative to a limit state of a generic structure: a) probability density function and definition of the reliability index, β; b) cumulative distribution function, with indication of the probability of failure.

or:

$$\beta = \frac{1 - \dfrac{S_{\mathrm{mean}}}{R_{\mathrm{mean}}}}{\sqrt{V_R^2 + \left(\dfrac{S_{\mathrm{mean}}}{R_{\mathrm{mean}}}\right)^2 V_S^2}} \tag{3.12}$$

where V_R and V_S are the coefficients of variation of R and S, respectively.

When the variables R, S, and M obey a lognormal distribution, β is given by:

$$\beta = \frac{\ln\left(\dfrac{S_{\mathrm{mean}}}{R_{\mathrm{mean}}}\sqrt{\dfrac{1 + V_S^2}{1 + V_R^2}}\right)}{\sqrt{\ln\left[\left(1 + V_R^2\right)\left(1 + V_S^2\right)\right]}} \tag{3.13}$$

Table 3.2 shows the relationship between the reliability index and the probability of failure. This relation can easily be obtained from Table A3.1.2 of Annex A3.1 of the present chapter.

The Structural Eurocodes are, in general, calibrated for reliability indices above 3.8 for a reference period of 50 years, which correspond to probabilities of failure below 10^{-4} (EN 1990: 2002), as shown in Table 3.2. The reference period affects the probability of failure through the values considered for variable actions, in particular, environmental actions.

Table 3.2 Relation between the reliability
index and the probability of failure.

Reliability index, β	Probability of failure, P_f
1.0	0.1587
1.645	0.05
2.0	0.0228
3.0	1.35×10^{-3}
3.5	1.1×10^{-4}
4.0	3.2×10^{-5}
4.5	3.3×10^{-6}

For a given probability of failure, structures with a greater reference period will have to be designed for greater values of those actions.

It should be mentioned, in this context, that the occurrence of failure in civil engineering structures depends significantly on human error, which naturally is not accounted for by (global or partial) safety factors. Therefore, the reliability index does not necessarily supply an approximate estimate of the frequency with which structural failures do, in fact, occur. The influence of gross human error may be minimized by means other than by increasing the safety factor value. The most common process consists of the so-called project revision by independent entities.

3.5.2 Application of probabilistic methods in geotechnical design

The application of probabilistic methods to geotechnical problems, as in the one presented above, involves some difficulties.

The first is the fact that, in general, the number of experimental determinations of ground strength parameters is relatively modest, preventing their conventional statistical treatment. This limitation may be partly overcome by assuming a normal distribution defined by the mean value, computed from the available data, in combination with a coefficient of variation selected from databases of soils of the same type (Alonso, 1976; Duncan, 2000; Sayão et al., 2012). For indicative purposes, Table 3.3 shows values for the coefficients of variation of soil properties collected from published sources.

As seen before, both demand and resistance depend, in general, on several independent variables, so the same will happen with the function, M, the margin of safety. The probability density function of M will therefore generally depend on the statistical variability of a significant number of those independent variables.

In simpler situations, M may be given by an equation, as in the case of the vertical loading of a shallow foundation. In these circumstances, analytical methods may be employed to account for the influence of the statistical variability of the independent variables on the variability of M.

When the margin of safety cannot be expressed by an equation, the Monte Carlo simulation method is usually employed. This method consists basically of the following procedures:

1. given the probability density functions of all independent variables, a random value is assigned to each of them; random generation is made by recourse to specialist computer codes;
2. from that set of values, one value is calculated for the dependent variable, M in this case;
3. repeating the process a large number of times (which naturally requires the use of computational procedures), the probability density function for M can be generated.

Table 3.3 Values of coefficients of variation of soil properties collected from published sources (adapted from Cardoso and Matos Fernandes, 2001).

Parameter, X_i	Mean value, $X_{mean,i}$	Coefficient of variation, V_{Xi} (mean value)
Undrained shear strength, c_u	<50 kPa	0.26–0.82
	50–150 kPa	0.19–0.66
	150–300 kPa	0.19–0.53
	>300 kPa	0.13–0.41
	All ranges	0.12–0.85 (0.34)
Angle of shearing resistance, ϕ'	<30°	0.03–0.15
	30°–40°	0.10–0.22
	All ranges	0.05–0.25 (0.13)
tan ϕ'		0.07–0.15
Unit weight, γ	All ranges	0.04–0.16 (0.07)

In the design of geotechnical works, the most serious difficulty in the briefly outlined application of probabilistic methods – designated in the literature as *incomplete probabilistic analyses* – results from demand and resistance rarely being independent variables, as already discussed in Section 3.2.

When R and S are not independent, a so-called *complete probabilistic analysis* can be performed, which is normally justified only in works of exceptional importance. Its treatment, even superficially, exceeds the scope of this book.

Before concluding this section, it should be noted that, given the substantial variability typical of ground resistance properties, the application potential of probabilistic methods in geotechnical engineering is, in fact, very large, probably even larger than in structural engineering. This fact explains the growing interest of many authors in these methodologies, in particular, the interest of many professionals involved in geotechnical design.

3.6 INTRODUCTION TO EUROCODE 7 – GEOTECHNICAL DESIGN

3.6.1 The Structural Eurocodes: generalities

The Structural Eurocodes were the result of a political decision of the European Commission, taken in 1975, with the aim of harmonizing the technical specifications for the design of civil engineering structures, which varied widely within the member countries of the European Economic Community (now the European Union). This initiative aimed to foster the creation of a European market for the construction and public works industry.

At the end of the following decade, the Eurocode elaboration process had been transferred to the European Committee for Standardization (Comité Européen de Normalization, CEN), an organization outside the scope of the European Union, in which most European countries are represented.

Table 3.4 presents the complete list of Structural Eurocodes currently applicable, in which is included Eurocode 7, dedicated to the design of geotechnical works. The first generation of these documents was published in the 1990s as European Pre-norms. A new generation of the Eurocodes has been published in the 21st century as European Norms (EN). Each

Table 3.4 List of Structural Eurocodes – European Norms (EN).

EN 1990	Eurocode 0: Basis of Structural Design
EN 1991	Eurocode 1: Actions on Structures
EN 1992	Eurocode 2: Design of Concrete Structures
EN 1993	Eurocode 3: Design of Steel Structures
EN 1994	Eurocode 4: Design of Composite Steel and Concrete Structures
EN 1995	Eurocode 5: Design of Timber Structures
EN 1996	Eurocode 6: Design of Masonry Structures
EN 1997	Eurocode 7: Geotechnical Design
EN 1998	Eurocode 8: Design of Structures for Earthquake Resistance
EN 1999	Eurocode 9: Design of Aluminium Structures

Eurocode, generally consisting of a number of parts, is complemented in each country with the respective National Annex.

The Structural Eurocodes process constituted a powerful incentive to speed up the adoption in Europe of the limit state method in the design of geotechnical works. Indeed, it was intended that this set of codes constituted a suite with a common design philosophy, established in Eurocode 0. At the starting date of the process, the generality of codes for concrete, steel, and other structures at country-level were already based on the limit state method. The need to extend this methodology to geotechnical design became very clear.

This explains why the development of Eurocode 7 has involved additional difficulties in relation to the other Eurocodes, in view of the alteration introduced in most European countries to the basic design philosophy of geotechnical works.[1] These difficulties were highlighted in a particularly clear manner by Becker (1996a), a member of the team that elaborated the current Canadian geotechnical design code, also based on the limit state design approach:

> Design using partial factors has worked well in structural engineering primarily because a sufficient quality control on the manufacturing process of structural materials exists and design calculations are based on a specified theory or approach. On the other hand, for geotechnical materials, the implementation of the concept runs into considerable difficulty. The reasons for this include the inherent variability of natural geological materials; the fact that many different methods exist for measuring soil strength parameters and different values will be obtained from different tests; that different theories are available to calculate the same type of resistance (...); and that much geotechnical design is based on empirical or semi-empirical methods. Furthermore, in a large country such as Canada, different geotechnical design methods are commonly preferred and used in different regions based on the specific conditions in a given region.

It is those difficulties that essentially explain why Eurocode 7 comprises three design approaches, each one corresponding to a given set of partial safety factor values for the verification of ultimate limit states. The option for the design approach, and hence the corresponding safety factors, to be used in each country is established in the corresponding national annex.

The following sections provide a brief presentation of Chapter 2 of Eurocode 7, which establishes the basis of geotechnical design. This presentation is meant to facilitate the

---

[1] As mentioned before, in 1965, the limit state method had already been adopted in the Danish geotechnical code.

understanding of the key aspects of the code. The study of the document itself is indispensable for the reader intending to apply it in a professional context. The work of Bond and Harris (2008) is also highly recommended for a proper application of Eurocode 7.

3.6.2 Design values of the actions, material properties, and geometric data

The design values of the actions, F_d, are obtained from:

$$F_d = \gamma_F\, F_{\text{rep}} \qquad (3.14)$$

where γ_F is the partial safety factor for the actions, taking into account the possibility of unfavorable deviations relative to representative values, F_{rep}, given by:

$$F_{\text{rep}} = \psi\, F_k \qquad (3.15)$$

where F_k represents the characteristic value of the actions and ψ, a coefficient converting characteristic values into representative values. Values of ψ are established in Eurocode 0 (EN 1990:2002).

The design values of the material strength parameters, X_d, are given by:

$$X_d = \frac{X_k}{\gamma_M} \qquad (3.16)$$

where X_k represents the characteristic value of the strength parameter and γ_M is the partial safety factor that accounts for possible unfavorable deviations from the characteristic value.

Finally, the design values, a_d, of the geometrical data are obtained from:

$$a_d = a_{\text{nom}} \pm \Delta a \qquad (3.17)$$

where a_{nom} represents the nominal value of the geometrical data and Δa accounts for possible unfavorable deviations of the geometrical data from the nominal value.

It should be noted that Eurocode 7 permits the direct assessment of the design values of the actions and ground properties, i.e., without the preliminary selection of their characteristic values and the application of partial safety factors. Anyway, the code establishes that, in such cases, the recommended value of the partial safety factors should be employed as a guide to obtain the required level of safety.

3.6.3 The question of the characteristic value of a geotechnical parameter

3.6.3.1 The concept of characteristic values of a structural and a geotechnical material parameter: Eurocode 0 versus Eurocode 7

The characteristic value of a material property is defined in Eurocode 0 as the "value of a material or product property having a prescribed probability of not being attained in a hypothetical unlimited test series. This value generally corresponds to a specified fractile of the assumed statistical distribution of the particular property of the material or product." It is known in structural engineering that the percentile is generally equal to 5% for materials like concrete and steel.

In the process of elaboration of Eurocode 7, the concept of a characteristic value of a ground property motivated a long debate (Schneider, 1997; Simpson and Driscoll, 1998; Cardoso and Matos Fernandes, 2001; Frank et al., 2004). As shall be seen, the definition inserted in the code is far less objective than that quoted above. This is a question of great relevance and will therefore be the focus of our attention.

Some of the principles of Eurocode 7 with relevance to this subject are listed here:

1. The selection of characteristic values for geotechnical parameters shall be based on results and derived values from laboratory and field tests, *complemented by well-established experience.*
2. The characteristic value of a geotechnical parameter shall be selected as a cautious estimate of the value affecting the occurrence of the *limit state.*
3. The selection of characteristic values for geotechnical parameters shall take account of the following:
   – geological and other background information, such as *data from previous projects;*
   – the variability of the measured property values and other relevant information, e.g., from *existing knowledge;*
   – the extent of the field and laboratory investigations;
   – the type and number of samples;
   – the *extent of the zone of ground governing the behavior of the geotechnical structure* at the *limit state being considered;*
   – the *ability of the geotechnical structure* to transfer loads from weak to strong zones in the ground.
4. If statistical methods are used, the characteristic value should be derived such that the calculated probability of a worse value governing the occurrence of the *limit state under consideration* is not greater than 5%.

As can be concluded from this citation, the selection in geotechnical works of the characteristic value of a ground property for use in the safety check, relative to a given limit state, is not exclusively determined by the variability of that property in the construction site. On the contrary, consideration must be given not only to *"data from previous projects"*, *"well-established experience"* and *"existing knowledge"* about that property on the soil type in question, but also to the very *"geotechnical structure"* and the *"limit state under consideration"*.

Therefore, the selection of the characteristic value in Eurocode 7 does not do without a careful weighting of diverse aspects, related to the ground, the structure, and the limit state being considered. If this procedure is assisted by probabilistic studies, so much the better. But that is not essential, because it is important to acknowledge that the incorporation of the previously mentioned factors makes the statistical treatment of the available data quite complex (Frank et al., 2004).

The contrast between the objectivity of the definition of characteristic value in Eurocode 0 (very directed to structural materials, as expected), and the approach dedicated by Eurocode 7 to the same concept, is masterfully explained by Simpson and Driscoll (1998) with the following words:

> The designer of a structure is concerned with the properties of materials which generally do not exist at the time of design, but which can be specified with fair precision. The range of uncertainty of their properties is reasonably well known, and, in many cases, may be better understood by the drafters of codes than by designers in practice. Hence, it is appropriate that codes give specific rules about the measurements of characteristic

values and that the possible range of uncertainty is entirely accommodated in factors prescribed by the code writers.

In geotechnical design, however, the designer is in possession of information not available to the code drafters. He knows where the site is located, what its geology is, and he has the test results, relevant publications, observations of nearby constructions, and so on. The designer is therefore in a much better position than the code drafter to make allowance for the range of uncertainty of the parameter values. It is this extra information which Eurocode 7 requires the designer to incorporate in his selection of characteristic values.

3.6.3.2 The dependence on the structure and on the limit state under consideration

The question of the dependence of the characteristic value of a given ground property, relative to the limit state being considered, may be explained with an example, inspired in the work of Simpson and Driscoll (1998). Consider the conditions of Figure 3.9, which represents a building with a low stiffness structure and shallow foundations on a reasonably competent residual soil derived from granite but with scattered occurrences of highly plastic softer pockets. The site is also close to a slope, as shown in the figure.

Given this geological-geotechnical scenario, the characteristic values of the ground strength parameters for the bearing capacity limit state of the footings should be affected by the (lower) resistance parameters of the pockets, due to the possibility that one or more footings would be implanted over zones of that softer soil.

Taking now into consideration the sliding limit state of the slope, it will be reasonable to adopt higher characteristic values for the strength parameters than in the previous limit state, since it will be unlikely that a sliding surface, such as the one represented, would be totally located in the softer soil.

From the preceding considerations, it is also possible to understand the dependence of the characteristic value relative to the structure itself. In fact, a stiffer structure will have a greater capacity for redistributing loads over the foundations. So, in case a footing is located over a softer soil and tends to settle more than the adjacent footings, its load will be reduced and partially redistributed to these other footings, without the occurrence of a failure limit state of the foundation. Therefore, an identical probability of failure will correspond to a higher characteristic value of the strength parameter for the stiffer structure than for the more flexible structure.

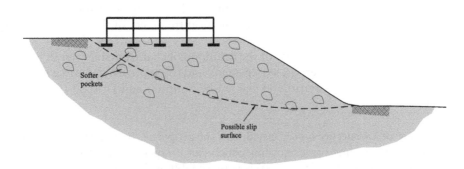

Figure 3.9 Example of the dependence of the characteristic value of soil strength properties relative to the structure and to the limit state under consideration.

3.6.3.3 The dependence on the size of the ground region that affects a given limit state

The questions discussed about Figure 3.9 may be examined from a different perspective, that of the dependence of the characteristic value relative to the dimension of the ground zone, affecting a given limit state. On this issue, EC 7 includes the following statement:

> The zone of ground governing the behavior of a geotechnical structure at a limit state is usually much larger than a test sample or the zone of ground affected in an *in situ* test. Consequently, the value of the governing parameter is often the mean of a range of values covering a large surface or volume of the ground. The characteristic value should be a cautious estimate of this mean value.

To illustrate this with an example, consider a given ground where several structures will be constructed with shallow foundations. The structures will be of two types, represented in Figure 3.10: a low-stiffness structure (Figure 3.10a), without a significant capacity for load redistribution between foundations in case of failure or excessive deformation, and a very stiff structure (Figure 3.10b), whose capacity for load redistribution is considerable.

The figures highlight, for mere indicative purposes, the ground zones likely to affect a foundation limit state. Given the greater capacity of the stiffer structure to transfer loads between foundations, it is easy to understand that a greater volume of soil will affect the occurrence of a possible limit state.

Suppose that several experimental determinations were performed in the area, which provided a large set of estimates of soil strength parameter(s) – the angle of shearing resistance, for instance. Let curve (1) of Figure 3.11 be the probability density function of the

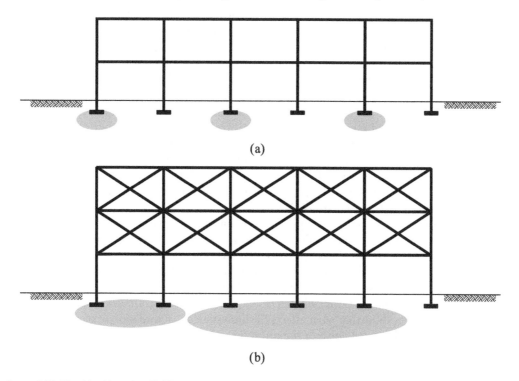

(a)

(b)

Figure 3.10 Flexible (a) and stiff (b) structures with shallow foundation on a given ground and soil zones, whose mean strength parameters control the failure limit state of the foundation.

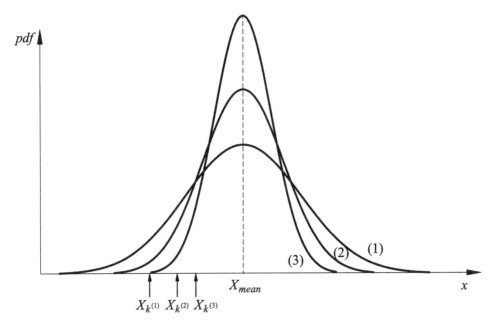

Figure 3.11 Dependence of the characteristic value of a ground resistance parameter on the volume affecting a given limit state.

experimental determinations and let $X_k(1)$ be the value that bounds the 5% least-favorable results.

Should $X_k(1)$ be taken as a characteristic value for a safety check relative to the limit state of foundation ground failure for the structures in Figure 3.10? According to Eurocode 7, the answer would be negative, because the soil zones likely to govern that limit state for both structures are larger than the volume involved in each experimental determination.

Let curves 2 and 3 (in the same figure) be the probability density functions of the mean value of the experimental determinations in volumes of the same order of magnitude of those that will control the foundation limit state of the structures of Figure 3.10a,b, respectively. The mean values of curves 2 and 3 will naturally coincide with that of curve 1, but the respective standard deviation decreases as the volume increases. Moreover, the value bounding the 5% least-favorable results for each distribution increases with the ground volume concerned. Therefore, for the structure of Figure 3.10a, in the context of Eurocode 7, the characteristic value should be $X_k(2)$, whereas, for the structure of Figure 3.10b, the appropriate characteristic value will be $X_k(3)$.

Let us conclude with another example, concerning the gravity-retaining wall with large linear development, represented in Figure 3.12. If the backfill strength parameters, at a given cross section(s), are much lower than average, the effect of the soil thrust may be greater than that of the wall weight. However, this does not necessarily imply the occurrence of an ultimate limit state: this *deficit* in local resistance may be easily compensated for by the *surplus* in adjacent sections, where the backfill strength parameters are higher.

These considerations show that what affects the limit state of a foundation or of a geotechnical structure is normally the *mean value of the soil resistance on a given volume* of ground, which may vary in size. *The characteristic value is therefore a cautious estimate of this mean value.*

Therefore, it seems reasonable to consider that, in the traditional practice of sound design by the global safety factor method, nominal values adopted for the ground parameters are, in the Eurocode 7 context, characteristic values.

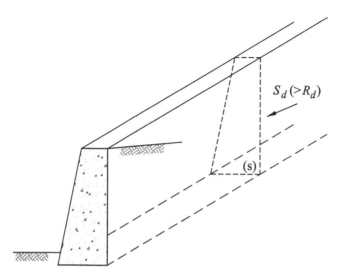

$S_d (> R_d)$

(s)

Figure 3.12 Gravity retaining wall with large linear development where, at a given section, the demand exceeds the resistance but, nevertheless, stability is ensured because the mean value of demand is lower than that of resistance.

Before closing this subject, an important caveat is in order. In structures of large linear development, the approximation of the characteristic value to the global mean value is valid only when a possible *deficit* of local resistance can be compensated for by a *surplus* in neighboring zones, as in the case of the retaining wall of Figure 3.12. The opposite, i.e., a detrimental effect of the large linear development of the structure, happens, for instance, in the case of embankments for flood protection. In this case, the occurrence at a given location of low resistance values may lead to a failure, first locally, but, as it evolves to dyke overtopping, ending up in global failure. In these cases, the large linear development of the work has a penalizing effect on the selection of the characteristic value (CUR, 1996).

3.6.4 Types of limit states

In Eurocode 7, in line with the set of Structural Eurocodes, five ultimate limit state types are provided, namely:

   i)   loss of equilibrium of the structure or the ground, considered to be a rigid body, in which the strengths of structural materials and the ground are insignificant in providing resistance (EQU);
   ii) internal failure or excessive deformation of the structure or structural elements, (including, for example, footings, piles, or basement walls), in which the strength of the structural materials is significant in providing resistance (STR);
   iii) failure or excessive deformation of the ground, in which the strength of soil or rock is significant in providing resistance (GEO);
   iv) loss of equilibrium of the structure or the ground due to uplift by water pressure (buoyancy) or other vertical actions (UPL);
   v)   hydraulic heave, internal erosion and piping in the ground caused by hydraulic gradients (HYD).

Annex A3.2 includes the partial safety factors for actions and material properties prescribed in Eurocode 7 for the EQU, UPL, and HYD limit states.

3.6.5 Verification of safety for STR and GEO limit states

In practical terms, the STR and GEO limit states are the most relevant because they are present in the great majority of civil engineering structures. In general, the former affect the resistance of structural elements (for example, the cross section and the area of steel of a reinforced concrete retaining wall) and the latter control the dimensions of the structural elements in foundations or retaining structures (for example, the dimensions in plan of a shallow foundation).

In this section, dedicated to Eurocode 7, the *effect of the actions* is represented by the symbol E, as used in that document. In the remainder of the book, the symbol S is preferentially used for that variable. For each of the STR or GEO limit states, it shall be verified that:

$$E_d \leq R_d \tag{3.18}$$

where E_d represents the design value of the effect of actions and R_d is the design value of the resistance.

The design value of the effect of actions can be expressed by:

$$E_d = E\{\gamma_F F_{\text{rep}}; X_k / \gamma_M; a_d\} \tag{3.19a}$$

and

$$E_d = \gamma_E E\{F_{\text{rep}}; X_k / \gamma_M; a_d\} \tag{3.19b}$$

where γ_E represents the partial safety factor for the effect of an action.

Equations 3.19a and 3.19b suggest two comments. The first draws the attention to an issue already highlighted: in the geotechnical limit states, the effect of actions depends in general on the ground strength parameters. The second comment explains the reason for having two equations to express E_d.

As can be seen, in certain cases, partial safety factors are applied to the action representative values, which are introduced in the calculation model to obtain the design value of the effect of actions (Equation 3.19a). In other cases, the effect of actions is calculated from the action representative values and then *factored by a safety factor*, γ_E (Equation 3.19b).

In linear analyses, which constitute the majority of analyses in structural engineering, this distinction is not relevant since both equations lead to the same result. However, in non-linear analyses and under certain conditions, which are relatively frequent in geotechnical problems, the distinction becomes relevant. Figure 3.13 presents two examples in which the application of each of the presented equations is justified.

Figure 3.13a concerns a given structure with shallow foundation which, in addition to the self-weight, is subjected to a variable action with a relevant horizontal component in comparison with that weight. Consider the ULS relative to the vertical loading of the foundation. This is a situation that clearly justifies the application of equation 3.19a. In fact, factoring the action representative values will lead to an increased E and to a reduced R, because: i) the eccentricity of the vertical foundation load is greater, so the effective area is lower; ii) the inclination of the foundation load is greater, so that the value of the i corrective factors for calculation of the vertical soil resistance are lower (see Section 6.2.5).

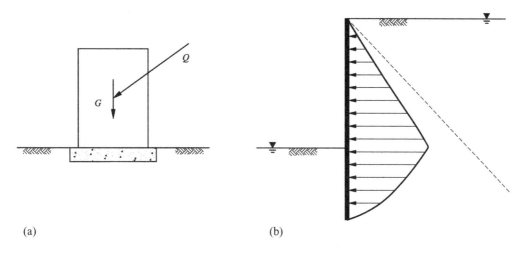

(a)   (b)

Figure 3.13 Examples of situations in which partial safety factors should be applied (a) to the representative values of the actions or (b) to the effect of the actions.

Consider now Figure 3.13b, which presents the retaining structure of an excavation where the water level is close to the surface in the supported ground, whereas, on the other side, it is maintained by pumping close to the base of the excavation. The figure shows the graph that represents the distribution of the net water pressure on the structure, the resultant of which naturally constitutes an unfavorable (variable) action. In this case, the direct application of a partial safety factor, γ_F, to the resultant of the represented diagram would lead to a result which is physically not reasonable, because it would correspond to a water level located above the ground surface on the right side of the retaining wall. In this case, it would be more appropriate to apply equation 3.19b, whereby the effect of the action is calculated from the representative value of the action of the water, and only then factored by γ_E.

Complementarily, the design value of the resistance may be calculated, in Eurocode 7, in the three following ways:

$$R_d = R\left\{\gamma_F F_{\text{rep}}; X_k/\gamma_M; a_d\right\}$$ (3.20a)

$$R_d = \frac{1}{\gamma_R} R\left\{\gamma_F F_{rep}; X_k; a_d\right\}$$ (3.20b)

and

$$R_d = \frac{1}{\gamma_R} R\left\{\gamma_F F_{rep}; X_k / \gamma_M; a_d\right\}$$ (3.20c)

where γ_R represents a safety factor applied to the characteristic value of the resistance. This factor is equivalent to the inverse of the resistance factor Φ used in the LRFD (see Section 3.7).

The reader should note that these expressions reflect the dependence of the resistance on the actions, as already emphasized earlier.

Tables 3.5 to 3.7 present the values established in the code for the partial safety factors γ_F or γ_E, γ_M, and γ_R.

Table 3.5 Partial factors on actions (γ_F) or the effects of actions (γ_E) (EN 1997-1:2004).

Action		Symbol	Set	
			A1	A2
Permanent	Unfavorable	γ_G	1.35	1.0
	Favorable		1.0	1.0
Variable	Unfavorable	γ_Q	1.5	1.3
	Favorable		0	0

Table 3.6 Partial factors for soil parameters (γ_M) (EN 1997-1:2004).

Soil parameter	Symbol	Set	
		M1	M2
Angle of shearing resistance[a]	$\gamma_{\phi'}$	1.0	1.25
Effective cohesion	$\gamma_{c'}$	1.0	1.25
Undrained shear strength	γ_{cu}	1.0	1.4
Unconfined strength	γ_{qu}	1.0	1.4
Weight density	γ_{γ}	1.0	1.0

[a] This factor is applied to $\tan \phi'$.

Table 3.7 Partial resistance factors (γ_R) (EN 1997-1:2004).

Limit state		Symbol	Set			
			R1	R2	R3	R4
Bearing resistance to vertical loading of spread foundations and retaining structures		$\gamma_{R;v}$	1.0	1.4	1.0	–
Sliding resistance of spread foundations and retaining structures		$\gamma_{R;h}$	1.0	1.1	1.0	–
Pile base resistance	Driven	γ_b	1.0	1.1	1.0	1.3
	Bored		1.25	1.1	1.0	1.6
	CFA		1.1	1.1	1.0	1.45
Pile shaft resistance	Compression	γ_s	1.0	1.1	1.0	1.3
	Tension	$\gamma_{s;t}$	1.25	1.15	1.1	1.6
Pile total resistance (compression)	Driven	γ_t	1.0	1.1	1.0	1.3
	Bored		1.15	1.1	1.0	1.5
	CFA		1.1	1.1	1.0	1.4
Anchors		γ_a	1.1	1.1	1.0	1.1
Earth passive resistance in front of retaining structures		$\gamma_{R;e}$	1.0	1.4	1.0	–
Earth passive resistance for slopes and overall stability		$\gamma_{R;e}$	1.0	1.1	1.0	–

3.6.6 The three design approaches

It is for the STR and GEO limit states that Eurocode 7 establishes the three different design approaches mentioned previously. The existence of these three distinct approaches contributed significantly to the complexity of the document. The differences between them has to deal with how the partial safety factors are assigned to actions (γ_F), to the effect of actions (γ_E), to material properties (γ_M) and to resistances (γ_R) in Equations 3.19 and 3.20. This

Table 3.8 Design approaches and the respective partial factors (see Tables 3.5 to 3.7).

Design approach	Actions or effects of actions (γ_F or γ_E)	Soil parameters (γ_M)	Resistances (γ_R)	Notes
1 - Combination 1	A1	M1	R1	1
1 - Combination 2	A2	M2	R1	1
		M1 or M2 (piles)	R4 (piles)	2
2	A1	M1	R2	3, 4
3	A1 or A2 (7) (8)	M2	R3	5, 6

Notes:

1. In Combinations 1 and 2, the partial factors are applied to actions and to ground strength parameters. For each limit state, the verification must be made for both combinations.

2. In Combination 2, set M1 is used for calculating resistances of piles or anchors and set M2 for calculating unfavorable actions on piles as a result of negative skin friction or transverse loading.

3. In this approach, the partial factors are applied to actions or to the effect of actions and to ground resistance.

4. If this approach is employed in slope stability or global stability analyses, the effect of actions on the failure surface is multiplied by γ_E and shear resistance along the failure surface is divided by $\gamma_{R;e}$.

5. In this approach, the partial factors are applied to actions or to the effect of actions from the structure, and to ground strength parameters.

6. In slope stability or global stability analyses, the actions on the ground (for example, structural actions or traffic loads) are treated as geotechnical actions, using set A2 of partial safety factor values.

7. In structural actions.

8. In geotechnical actions. In Eurocode 7, a geotechnical action is defined as an action transmitted to the structure by the ground, fill, standing water or groundwater.

reflects different perspectives as to the way uncertainties are taken into account in the modeling of the effects of actions and of the resistances. Annex A3.3 is recommended to the particularly interested reader.

Table 3.8 summarizes the partial safety factors for the three design approaches, the respective values having been presented in Tables 3.5 to 3.7.

In Design Approach 1, for each limit state, it is necessary to perform two safety checks through independent calculations, using the sets of safety factors of Combinations 1 and 2. However, if it is obvious that one of the two combinations governs the design, it is not necessary to perform calculations for the other combination. Nevertheless, different combinations may be critical to different aspects of the same design. Generally, the factors are applied to actions and not to the effect of actions. Generally, the factors are applied to ground strength parameters, but, in the design of piles and anchors, they are applied to resistances. This approach may be considered to fit into what has been designated as the European Concept of LSD application.

In Design Approaches 2 and 3, a single calculation is required for each limit state, and the way in which the factors are applied varies, according to the calculation considered.

In Design Approach 2, the factors are applied, on the one hand, to actions or to the effect of actions and, on the other hand, to resistances. One may say that this approach is close to what has been termed the American Concept of LSD application.

In Design Approach 3, the factors are applied, on the one hand, to actions or to the effect of actions from the structure and, on the other hand, to ground strength parameters. This approach may be considered to fit into the European Concept of LSD application.

Figure 3.14 shows an application example of the three design approaches to the safety check relative to an ULS as a result of insufficient resistance to vertical loading of the foundation soil of a retaining wall, which provides support to the deck slab of a road viaduct.

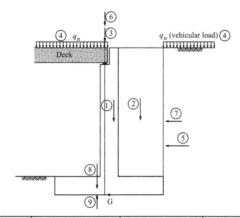

Force	Description	Type of action	DA 1.1	DA1.2	DA 2	DA3
1	Weight of the retaining wall	Permanent (favorable)	$G_d = 1.0\ G_k$	$G_d = 1.0\ G_k$	$G_d = 1.0\ G_k$	$G_d = 1.0\ G_k$
2	Weight of soil above the heel	Permanent (favorable)	$G_d = 1.0\ G_k$	$G_d = 1.0\ G_k$	$G_d = 1.0\ G_k$	$G_d = 1.0\ G_k$
3	Weight of the deck	Permanent (unfavorable)	$G_d = 1.35\ G_k$	$G_d = 1.0\ G_k$	$G_d = 1.35\ G_k$	$G_d = 1.35\ G_k$
4	Surcharge (vehicles)	Variable (unfavorable)	$q_d = 1.50\ q_k$	$q_d = 1.30\ q_k$	$q_d = 1.50\ q_k$	$q_d = 1.50\ q_k$
5	Soil thrust	Permanent (unfavorable)	$E_{G,d} = 1.35\ E_G\ (\phi'_k, \gamma_k)$	$E_{G,d} = 1.0\ E_G\ (\phi'_d, \gamma_k)$	$E_{G,d} = 1.35\ E_G\ (\phi'_k, \gamma_k)$	$E_{G,d} = 1.0\ E_G\ (\phi'_d, \gamma_k)$
6	Resultant of the surcharge on the deck	Variable (unfavorable)	$Q_d = 1.50\ Q_k$	$Q_d = 1.30\ Q_k$	$Q_d = 1.50\ Q_k$	$Q_d = 1.50\ Q_k$
7	Thrust due to the surcharge on the backfill	Variable (unfavorable)	$E_{Q,d} = E_Q\ (\phi'_k, q_d)$	$E_{Q,d} = E_Q\ (\phi'_d, q_d)$	$E_{Q,d} = E_Q\ (\phi'_k, q_d)$	$E_{Q,d} = E_Q\ (\phi'_d, q_d)$
8	Vertical load on the foundation	Effect of the actions	$V_d = \Sigma V_{G,d} + \Sigma V_{Q,d}$	$V_d = \Sigma V_{G,d} + \Sigma V_{Q,d}$	$V_d = \Sigma V_{G,d} + \Sigma V_{Q,d}$	$V_d = \Sigma V_{G,d} + \Sigma V_{Q,d}$
9	Bearing capacity	Resistance	$R_{V,d} = R_V\ (V_d, H_d, \phi'_k, c'_k, \gamma_k)$	$R_{V,d} = R_V\ (V_d, H_d, \phi'_d, c'_d, \gamma_k)$	$R_{V,d} = R_V\ (V_d, H_d, \phi'_k, c'_k, \gamma_k)\ /\ 1.4$	$R_{V,d} = R_V\ (V_d, H_d, \phi'_d, c'_d, \gamma_k)$

Notes (the note number coincides with that of the corresponding force in the table):

1. It is assumed that the retaining wall weight is favorable. This is equivalent to assume that its favorable effect (reducing the eccentricity and the inclination of the foundation load) is larger than its unfavorable effect (increasing the vertical force on the foundation soil). In a concrete problem, it is convenient to perform a verification with the opposite hypothesis.
2. It is assumed that the weight of the soil above the heel is favorable. See also note 1.
3. It is assumed that this weight is unfavorable. In a concrete problem, it is convenient to perform a verification with the opposite hypothesis.
4. The surcharge acting on the ground to the left of the virtual back of the wall has not be considered in coherence with the hypothesis that force 2 is favorable.
5. In DA1.2 and in DA3, the soil thrust is calculated with factored values of the angle of shearing resistance. In DA3, this thrust is considered a geotechnical action (see Table 3.8).
6. The surcharge has been considered to act on the viaduct in coherence with the hypothesis that force 3 is unfavorable.
7. In DA1.2 and in DA3, the soil thrust is calculated with factored values of the angle of shearing resistance. In DA3, this thrust is considered a geotechnical action (see Table 3.8).
8. In the general case, this summation includes the vertical components of the soil thrusts.
9. In DA1.2 and in DA3, this resistance is calculated with factored values of the angle of shearing resistance and of the effective cohesion of the foundation soil. In DA2, the factor $\gamma_R = 1.4$ is applied to the resistance calculated from the characteristic values of the angle of shearing resistance and of the effective cohesion of the foundation soil.

Figure 3.14 Example of application of the design approaches of Eurocode 7.

3.6.7 Comment about the justification for adopting a unit safety factor for the permanent actions

The following considerations are intended to supply a brief explanation to the fundaments of Design Approach 1 – Combination 2, which assigns a unit partial safety factor to permanent actions.

As mentioned in Section 3.2, in many geotechnical problems, the demand depends on the ground strength properties and the resistance depends on the actions. In addition, in the two examples of Figure 3.2, as well as in many geotechnical structures, the permanent actions, particularly the ground self-weight, are largely predominant. In the case of the retaining wall (Figure 3.2a), the soil thrust on the wall is naturally governed by the gravity action (the weight of the ground behind the wall), but, simultaneously, it is this weight that, by being applied on the base of the wall, ensures stability. With regard to the slope (Figure 3.2b), as a result of the above-mentioned interdependence, the weight of a given slice of soil may contribute unfavorably to the effect of the action but, on the other hand, contributes favorably to the resistance.

Therefore, it seems very difficult, for a given limit state, to distinguish *a priori* between the zone of the ground whose weight is favorable and the zone whose weight is unfavorable. So – and this is the reason for the Combination 2 – it might be inadequate to affect the permanent actions with a partial safety factor different from one due to the difficulty in controlling the "effect" of such a factor. In other words, given the diversity of practical situations that may arise, it would be difficult to guarantee that the application of a non-unit partial safety factor would be in fact (and always) *on the safe side*!

Moreover, the adoption of a unit partial safety factor for the permanent actions seems acceptable since the soil unit weight is known with much less uncertainty than are the soil strength parameters.

Combination 2 typically conditions the so-called *external sizing* in limit states essentially controlled by the ground resistance (of GEO type), leading to the general geometry of the structure or of the structural element in contact with the ground (for example, the dimensions in plan of a shallow foundation).

Although Combination 1 seldom governs the foundation external sizing, by imposing its use for all limit states (within the option for Design Approach 1), the code intends to guarantee, in any case, coherence in the design of the superstructure and the foundation. In fact, if this imposition did not exist, this could lead in some cases, though not very frequent, to the situation illustrated in Figure 3.15; the force obtained from the analysis of the superstructure (with partial safety factors for actions relative to Combination 1) might not intersect the foundation designed (only) according to Combination 2 (Simpson and Driscoll, 1998).

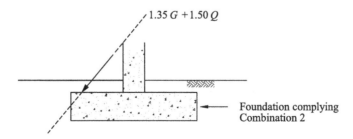

Figure 3.15 Situation in which Combination 1 of Design Approach 1 governs the external sizing of a shallow foundation.

3.6.8 Final comment and perspectives

It may be concluded that Eurocode 7 in its current version is not yet, in fact, a fully coherent code, as diverse design approaches coexist within it. Nevertheless, it constitutes a very relevant step towards the harmonization of the terminology and the design *praxis* of geotechnical works with those of other structures.

As a result of its generalized application, with its distinct Design Approaches, it is predictable that contributions may arise in the near future towards a more coherent code. In any case, it is indispensable to continue teaching geotechnical design based on global safety factors. This option has been assumed throughout this book.

3.7 APPLICATION OF LRFD: AASHTO CODE (2012)

The so-called LRFD (Load and Resistance Factor Design) constitutes the basis of the civil engineering design codes, either in the USA, or in Canada, and, as seen in Section 3.4, it is based on the philosophy of Limit State Design (Becker, 1996a; Coduto, 2001). Some of these codes include the design of foundations and other geotechnical structures, like the LRFD Bridge Design Specifications (AASHTO, 2012) and the Ontario Highway Bridge Design Code (MTO, 1991).

In accordance with the LRFD concept, the verification of safety must satisfy the following criterion:

$$\Sigma \gamma_i S_{ni} \leq \Phi R_n \tag{3.21}$$

where S_{ni} represents the nominal value of the actions (permanent and variable of various types), γ_i represents the load factors to be applied to S_{ni}, in order to account for the action-related uncertainty, $\Sigma \gamma_i S_{ni}$ is the summation of factored overall load effects, for a given load combination, R_n is the nominal resistance, and Φ represents the resistance factor.

The nominal resistance is determined basically in one of three ways: i) through calculations, based on nominal values of the ground strength parameters, and applying models of an essentially theoretical nature; ii) through calculations, based on field test results, and applying correlations or approaches of an essentially empirical or semi-empirical nature; or iii) through performance data, such as pile or anchor load tests.

The resistance factor, which will affect the nominal value of the resistance, is meant to account for: i) the variability of the geotechnical parameters; ii) the limitations of the calculation models; iii) the limitations of the correlations of empirical or semi-empirical nature; iv) the limitations of the methods of interpretation of the load test results; and v) in general, the uncertainties associated with the estimation of the nominal resistance.

Both the load factor values (typically greater than 1.0), and the resistance factor values (typically lower than 1.0), and also the combinations of actions, prescribed in the codes, have been established on the basis of probabilistic criteria for the safety check of the various limit states (Becker, 1996b).

As an example, Table 3.9 presents an extract of the AASHTO code with the combinations of actions for verification of ultimate limit states for bridge design, including the structural components whose behavior is controlled by the ground.

Table 3.10 displays the load factor values for permanent actions taken from the same code. The minimum values indicated should be applied when the permanent actions are favorable. Note that the table prescribes load factors for the ground's active thrust. The calculation of this thrust is based on the nominal value of the ground parameters.

Table 3.9 Load combinations and load factors (AASHTO, 2012).

Load combination limit state	DC DD DW EH EV ES EL PS CR SH	LL IM CE BR PL LS	WA	WS	WL	FR	TU	TG	SE	EQ
Strength I (unless noted)	γ_p	1.75	1.00	–	–	1.00	0.50/1.20	γ_{TG}	γ_{SE}	–
Strength II	γ_p	1.35	1.00	–	–	1.00	0.50/1.20	γ_{TG}	γ_{SE}	–
Strength III	γ_p	–	1.00	1.40	–	1.00	0.50/1.20	γ_{TG}	γ_{SE}	–
Strength IV	γ_p	–	1.00	–	–	1.00	0.50/1.20	–	–	–
Strength V	γ_p	1.35	1.00	0.40	1.0	1.00	0.50/1.20	γ_{TG}	γ_{SE}	–
Extreme Event I	γ_p	γ_{EQ}	1.00	–	–	1.00	–	–	–	1.00

Note: this is just part of the complete table of the cited document.

Legend:

1. Permanent Loads: DC = dead load of structural components and non-structural attachments; DD = down-drag force; DW = dead load of wearing surfaces and utilities; EH = horizontal earth pressure load; EV = vertical pressure from dead load of earth fill; ES = earth surcharge load; EL = miscellaneous locked-in force effects resulting from the construction process, including jacking apart of cantilevers in segmental construction; PS = secondary forces from post-tensioning; CR = force effects due to creep; SH = force effects due to shrinkage.

2. Transient Loads: LL = vehicular live load; IM = vehicular dynamic load allowance; CE = vehicular centrifugal force; BR = vehicular braking force; PL = pedestrian live load; LS = live load surcharge; WA = water load and stream pressure; WS = wind load on structure; WL = wind on live load; FR = friction load; TU = force effect due to uniform temperature; TG = force effect due to temperature gradient; SE = force effect due to settlement; EQ = earthquake load.

3. Limit States: Strength I — Basic load combination relating to the normal vehicular use of the bridge without wind; Strength II — Load combination relating to the use of the bridge by owner-specified special design vehicles, evaluation permit vehicles, or both without wind; Strength III — Load combination relating to the bridge exposed to wind velocity exceeding 55 mph; Strength IV — Load combination relating to very high dead load to live load force effect ratios; Strength V — Load combination relating to normal vehicular use of the bridge with wind of 55 mph velocity; Extreme Event I — Load combination including earthquake (load factor for live load γ_{EQ}, shall be determined on a project-specific basis).

Table 3.11 presents values of the resistance factor for shallow foundations taken from the same code. Values of the resistance factor for pile foundations and retaining walls, also from the AASHTO code, are included in Annex A3.4 of this chapter.

The examination of the tables for the resistance factor values reveals an option clearly distinct from that adopted in Eurocode 7, particularly for shallow foundations and for pile foundations. In fact, the AASHTO code establishes very different resistance factor values, for a given limit state, depending on the method applied. Therefore, these values have been *calibrated*, taking into account, in a very detailed way, the limitations of the (theoretical, semi-empirical, empirical or experimental) models used to estimate the resistance. It may be concluded that the resistance factor also performs, and in an efficient way, the role of *model factor*.[1]

Figure 3.16 illustrates the application of the AASHTO code to the verification of the safety of the abutment wall foundation of Figure 3.14 under vertical loading. The hypotheses

---

[1] In Eurocode 7, the concept of model factor is established, but its use is left to the discretion of those responsible for each national annex.

Table 3.10 Load factors (γ_p) for permanent loads (AASHTO, 2012).

Type of load, foundation type	Load factor Maximum	Load factor Minimum
Dead load of structural components and non-structural attachments (DC)	1.25	0.90
Dead load of wearing surfaces and utilities (DW)	1.50	0.65
Horizontal earth pressure (EH):		
• Active	1.50	0.90
• At-rest	1.35	0.90
• AEP for anchored structures	1.35	N/A
Vertical earth pressure (EV)		
• Overall stability	1.00	N/A
• Retaining walls and abutments	1.35	1.00
• Rigid buried structures	1.30	0.90
Earth surcharge (ES)	1.50	0.75

Note: this is just part of the complete table of the cited document.

Table 3.11 Resistance factors for geotechnical resistance of shallow foundations at the strength limit state (AASHTO, 2012).

		Method/Soil/Condition	Resistance factor
Bearing resistance	Φ_b	Theoretical method (Munfakh et al., 2001), in clay	0.50
		Theoretical method (Munfakh et al., 2001), in sand, using CPT	0.50
		Theoretical method (Munfakh et al., 2001), in sand, using SPT	0.45
		Semi-empirical methods (Meyerhof, 1957), all soils	0.45
		Footings on rock	0.45
		Plate load test	0.55
Sliding	Φ_τ	Precast concrete placed on sand	0.90
		Cast-in-place concrete on sand	0.80
		Cast-in-place or precast concrete on clay	0.85
		Soil on soil	0.90
	Φ_{ep}	Passive earth pressure component of sliding resistance	0.50

concerning the (favorable or unfavorable) actions were retained. The combination of actions and the load factor values corresponding to the case termed Strength I in Table 3.9 have been considered.

3.8 NOTE ABOUT EUROCODE 8: DEFINITION OF SEISMIC ACTION

Seismic action assumes particular importance for many geotechnical works. Its treatment is, therefore, indispensable in the context of the present chapter. In the Eurocodes context, structural design for earthquake resistance is treated in a set of specific documents that constitute Eurocode 8 (EN 1998-1: 2004). In these, Part 5 is especially dedicated to the geotechnical aspects (EN 1998-5: 2004).

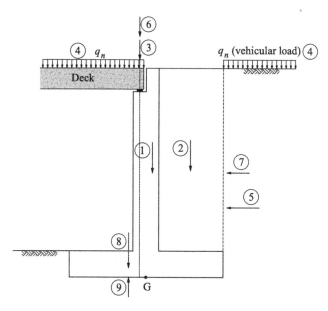

Force	Description (symbols are in agreement with Table 3.9.)	Type of action	Application of LRFD
1	Weight of the retaining wall (DC)	Permanent (favorable)	$\gamma_P DC_n = 0.90\,DC_n$
2	Weight of soil above the heel (DC)	Permanent (favorable)	$\gamma_P DC_n = 0.90\,DC_n$
3	Weight of the deck of the viaduct (DC)	Permanent (unfavorable)	$\gamma_P DC_n = 1.25\,DC_n$
4	Surcharge (vehicular live load) (LL)	Variable (unfavorable)	$\gamma_{LL}\,q_n = 1.75\,q_n$
5	Soil thrust (EH)	Permanent (unfavorable)	$\gamma_P EH_n = 1.50\,EH\,(\phi'_n, \gamma_n)$
6	Resultant of the surcharge on the deck	Variable (unfavorable)	$\gamma_{LL}\,LL_n = 1.75\,LL_n$
7	Thrust due to the surcharge on backfill (EH)	Variable (unfavorable)	$\gamma_{LL} EH_n = 1.75 EH\,(\phi'_n, q_n)$
8	Vertical load on the foundation	Effect of the actions	$\Sigma\,\gamma_i\,V_{ni}$
9	Bearing capacity (it is assumed that it is calculated by a theoretical method in clay or in sand using CPT, see Table 3.11)	Resistance	$\Phi R_n = 0.5(\gamma_i V_{ni},\ \gamma_i H_{ni},\ \phi'_n,\ c'_n,\ \gamma_n)$

Notes (the note number coincides with that of the corresponding force in the table):

1. It is assumed that the retaining wall weight is favorable (see also note 1 of Figure 3.14).

2. It is assumed that the weight of the soil above the heel is favorable (see also note 2 of Figure 3.14). This force is considered as a "nonstructural attachment-DC" and not a "vertical earth pressure-EV", since in the latter case it would be multiplied by a factor (1.0) greater than the one used for the weight of the retaining wall (see Table 3.10).

3. It is assumed that this weight is unfavorable   (see note 3 of Figure 3.14).

4. The surcharge acting on the ground to the left of the virtual back of the wall has not be considered in coherence with the hypothesis that force 2 is favorable.

5. The soil thrust has been calculated with the nominal value of the angle of shearing resistance.

6. The surcharge has been considered to act on the viaduct in coherence with the hypothesis that force 3 is unfavorable.

7. The soil thrust is calculated with the nominal value of the angle of shearing resistance.

8. In the general case, this summation includes the vertical components of the soil thrusts.

9. This resistance is calculated with factored values of the vertical and horizontal actions and with nominal values of the angle of shearing resistance and of the effective cohesion of the foundation soil.

Figure 3.16 Application of the AASHTO code (2012) to the example of Figure 3.14, considering the load combination and load factors for the case "Strength I" of Table 3.9.

In this chapter, only the questions exclusively related to the definition of the seismic action will be discussed. The aspects related to each type of work (namely retaining walls, foundations, and slopes) will be dealt with in the respective chapters. As an example, the conditions of the city of Lisbon, which, in 1755, suffered one of the most severe historical earthquakes (Chester, 2001), will be considered.

Table 3.12 describes the various types of ground for considering the influence of local conditions in the seismic action (EN 1998-1: 2004).

Table 3.12 Ground types for the consideration of the influence of local conditions on the seismic action (EN 1998-1: 2004).

Ground type	Description of the stratigraphic profile	Parameters		
		$V_{s,30}$ (m/s) (1)	N_{SPT}	c_u (kPa)
A	Rock or other rock-like geological formation, including at most 5 m of weaker material at the surface.	>800	–	–
B	Deposits of very dense sand, gravel, or very stiff clay, at least several tens of meters in thickness, characterized by a gradual increase of mechanical properties with depth.	360–800	>50	>250
C	Deep deposits of dense or medium-dense sand, gravel or stiff clay, from several tens to many hundreds of meters in thickness.	180–360	15–50	70–250
D	Deposits of loose-to-medium cohesionless soil (with or without some soft cohesive layers), or of predominantly soft-to-firm cohesive soil.	<180	<15	<70
E	Soil profile consisting of a surface alluvium layer with V_s values of type C or D and thickness varying between about 5 m and 20 m, underlain by stiffer material with V_s > 800 m/s.			
S_1 (2)	Deposits consisting, or containing a layer at least 10 m thick, of soft clays/silts with a high plasticity index (PI > 40) and high water content.	<100 (indicative)	–	10–20
S_2 (2)	Deposits of liquefiable soils, of sensitive clays, or any other soil profile not included in types A – E or S_1.			

Notes:

1. The average shear wave velocity $V_{s,30}$ should be computed in accordance with the following expression:

$$V_{s,30} = \frac{30}{\sum_{i=1,N} \frac{h_i}{V_i}}$$

where h_i denotes the thickness (in m) and V_i the shear-wave velocity (at a shear strain level of 10^{-5} or less) of the i-th formation or layer, in a total of N, existing in the top 30 m.

2. For sites with ground conditions matching either one of the two special ground types, S_1 or S_2, special studies for the definition of the seismic action are required. For these types, particularly for S_2, the possibility of soil failure under the seismic action should be taken into account.

Table 3.13 presents the values of the reference peak acceleration and of the seismic magnitude for the two types of seismic action in Lisbon.[1]

The design (horizontal) acceleration at the surface of a type A ground, a_g, can be obtained from the reference peak acceleration by:

$$a_g = a_{gR} \cdot \gamma_I \tag{3.22}$$

where γ_I represents the importance factor, which may be obtained from Table 3.14 as a function of the importance class of the structure and of the type of seismic action. The importance class affects the seismic acceleration value as it depends on the reference period considered. Structures of greater importance are designed for greater reference periods.

---

[1] Seismic action of Type 1 corresponds to earthquakes with large focal distance, generated at the so-called Gorringe Bank, in the contact between the Eurasian and African plates of the earth's crust, in the Atlantic Ocean. Seismic action of Type 2 represents earthquakes with small focal distance, generated in faults on the Iberian Peninsula.

The greater the reference period, the lower the probability of the design acceleration being exceeded in each year.

The maximum surface acceleration for grounds not of type A can be obtained by multiplying a_g by a soil factor, S, to account for acceleration amplification. That factor can be computed according to the following conditions (NP EN 1998-1: 2010, National Annex):

- for $a_g \leq 1$ m/s^2 : $S = S_{max}$
- for 1 m/s$^2 < a_g < 4$ m/s^2 : $S = S_{max} - \left[\left(S_{max} - 1 \right) \left(a_g - 1 \right) / 3 \right]$ (3.23)
- for $a_g \geq 4$ m/s^2 : $S = 1.0$

where S_{max} is given by Table 3.15.

Table 3.16 presents the relation between vertical and horizontal seismic accelerations for the two types of seismic action.

Table 3.13 Values of the reference peak ground acceleration on type A ground, a_{gR}, and of the magnitude for seismic actions in Lisbon (NP EN 1998-1: 2010, National Annex).

Seismic action		Seismic action	
Type I		Type 2	
Magnitude	a_{gR} (m/s^2)	Magnitude	a_{gR} (m/s^2)
7.5	1.5	5.2	1.7

Table 3.14 Importance classes for buildings (EN 1998-1:2004) and values of importance factors for Lisbon, γ_I (NP EN 1998-1: 2010, National Annex).

Importance classes for buildings	Seismic action Type I	Seismic action Type 2
I. Buildings of minor importance for public safety, e.g., agricultural buildings	0.65	0.75
II. Ordinary buildings, not belonging in the other categories	1.00	1.00
III. Buildings whose seismic resistance is of importance in view of the consequences associated with a collapse, e.g., schools, assembly halls, cultural installations, etc.	1.45	1.25
IV. Buildings whose integrity during earthquakes is of vital importance for civil protection, e.g. hospitals, fire stations, power plants, etc.	1.95	1.50

Table 3.15 Values of the factor S_{max} of the foundation soil for Lisbon (NP EN 1998-1: 2010, National Annex).

Ground type	A	B	C	D	E
S_{max}	1.0	1.35	1.6	2.0	1.8

Table 3.16 Relation between the design vertical and horizontal ground accelerations for Lisbon (NP EN 1998-1: 2010, National Annex).

Seismic action	a_{vg}/a_g
Type I	0.75
Type 2	0.95

ANNEX A3.1 SUMMARY OF SOME PROBABILISTIC CONCEPTS

The present annex contains a brief summary of basic probability theory concepts strictly necessary for understanding certain parts of this chapter.

When a given parameter, expressing a certain quantity, is not explicitly fixed, but may instead assume any value within a given series of values, this parameter constitutes a *random variable*.

One of the most common ways for representing random variables is through the so-called *probability density function*, $f(X)$, a sample of which is shown in Figure A3.1.1a. This function expresses the probability that a given value may be assumed by the variable in question in comparison with the other values. The area covered by the probability density function has a unit value.

Another way of presenting the same information is by means of the *cumulative distribution function*, $F(X)$, represented in Figure A3.1.1b. This function is the integral of the former. So, the ordinate corresponding to a given value X_1 in function $F(X)$ is equal to the shaded area in function $f(X)$, representing the probability of X being smaller than X_1.

A probability distribution represented by a probability density function with the bell shape of Figure A3.1.1a is designated as a *normal* or *Gaussian distribution*. Another distribution type common in geotechnical problems is the so-called *lognormal distribution*. A variable X has this type of distribution when $\ln X$ obeys a normal distribution, as shown in Figure A3.1.2.

In a probability density function, there are several parameters to be considered:

  i)   the *mean value* or *expected value*, X_{mean}, which corresponds to the abscissa of the center of gravity of the probability density function; therefore, in a normal distribution, the mean value is the abscissa of the peak of function $f(X)$; in any type of distribution, there is a 50% probability that X assumes values lower than or equal to the mean;

 ii)   the *variance*, which is the weighted average of the square of the deviation of each value of X relative to X_{mean}, being the weight for each value the respective probability density;

iii)   the *standard deviation*, σ_X, which is the square root of the variance and is expressed in the same units as the random variable itself;

iv)   the *coefficient of variation* (COV_X or V_X), a non-dimensional quantity, that represents the ratio between the standard deviation and the mean value, σ_X/X_{mean}.

Table A3.1.1 summarizes the expressions of the mean value and the variance for the cases of discrete and continuous variables.

The probability density function for the normal distribution may be expressed by:

$$f(x) = \frac{1}{\sqrt{2\pi\sigma_X^2}} e^{-\frac{1}{2}\left(\frac{x-X_{mean}}{\sigma_X}\right)^2}$$

(A3.1.1)

The inflexion points of the Gaussian curve are at a distance $\pm\sigma_X$ from the mean point. Table A3.1.2 shows, for a normal distribution, the probability of X to take values within certain intervals centered on the mean value and measured as a function of the standard deviation.

For the strength properties, it is current to consider the characteristic value, X_k, as the value of the property whose probability of occurrence with a lower value does not exceed

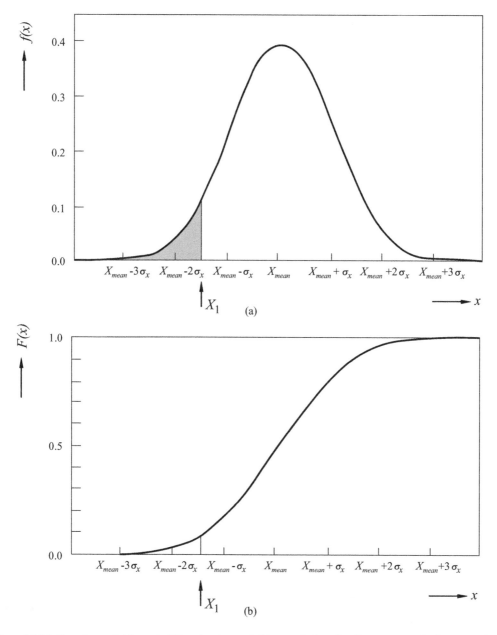

Figure A3.1.1 Functions of (a) probability density and (b) cumulative distribution typical of a normal or Gaussian distribution.

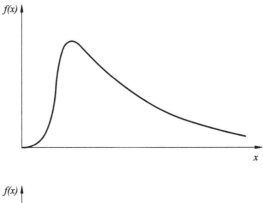

Figure A3.1.2 Lognormal distribution.

Table A3.1.1 Definitions of mean value and variance of random variables.

Parameter	Discrete variable	Continuous variable
Mean value	$X_{mean} = \dfrac{\sum X_i}{n}$	$X_{mean} = \displaystyle\int_{-\infty}^{\infty} x f(x)\,dx$
Variance	$\sigma_X^2 = \dfrac{\sum (X_i - X_{mean})^2}{n-1}$	$\sigma_X^2 = \displaystyle\int_{-\infty}^{\infty} (x - X_{mean})^2 f(x)\,dx$

Table A3.1.2 Probability of X to take values within intervals centered in X_{mean} or below the lower limit of the intervals, for a normal distribution.

Interval centered on X_{mean}	Probability that X takes values in the interval (%)	Probability that X takes values below the lower limit of the interval (%)
$X_{mean} \pm \sigma_X$	68.26	15.87
$X_{mean} \pm 1.645\sigma_X$	90	5
$X_{mean} \pm 2\sigma_X$	95.44	2.28
$X_{mean} \pm 3\sigma_X$	99.73	0.135
$X_{mean} \pm 3.5\sigma_X$	99.978	0.011
$X_{mean} \pm 4\sigma_X$	99.9936	0.0032
$X_{mean} \pm 4.5\sigma_X$	99.99934	0.00033

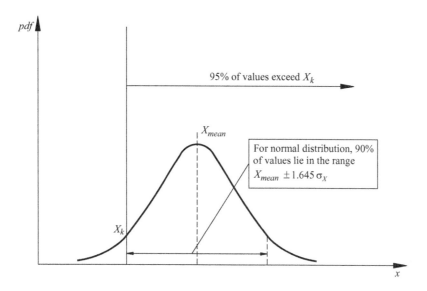

Figure A3.1.3 Probability density function of a normal or Gaussian distribution with indication of the 5% percentile of lower values, normally associated with the definition of the characteristic value of a strength parameter.

5%. As shown in Figure A3.1.3 and Table A3.1.2, for a normal distribution, the characteristic and the mean values are related by:

$$X_k = X_{\mathrm{mean}} - 1.645\sigma_X \qquad\qquad (A3.1.2)$$

ANNEX A3.2 PARTIAL SAFETY FACTORS FOR THE EQU, UPL AND HYD LIMIT STATES, ACCORDING TO EUROCODE 7

See Tables A3.2.1 to A3.2.5.

Table A3.2.1 Equilibrium limit state (EQU): partial factors on actions (γ_F) (EN 1997-1: 2004).

Action	Symbol	Value
Permanent		
Unfavorable[a]	$\gamma_{G;dst}$	1.1
Favorable[b]	$\gamma_{G;stb}$	0.9
Variable		
Unfavorable[a]	$\gamma_{Q;dst}$	1.5
Favorable[b]	$\gamma_{Q;stb}$	0

[a] Destabilizing
[b] Stabilizing

Table A3.2.2 Equilibrium limit state (EQU): partial factors for soil parameters (γ_M) (EN 1997-1: 2004).

Soil parameter	Symbol	Value
Angle of shearing resistance[a]	$\gamma_{\phi'}$	1.25
Effective cohesion	$\gamma_{c'}$	1.25
Undrained shear strength	γ_{cu}	1.4
Unconfined strength	γ_{qu}	1.4
Weight density	$\gamma_{\gamma'}$	1.0

[a] This factor is applied to $\tan \phi'$

Table A3.2.3 Uplift limit state (UPL): partial factors on actions (γ_F) (EN 1997-1: 2004).

Action	Symbol	Value
Permanent		
Unfavorable[a]	$\gamma_{G;dst}$	1.0
Favorable[b]	$\gamma_{G;stb}$	0.9
Variable		
Unfavorable[a]	$\gamma_{Q;dst}$	1.5

[a] Destabilizing
[b] Stabilizing

Table A3.2.4 Uplift limit state (UPL): partial factors for soil parameters and resistances (EN 1997-1: 2004).

Soil parameter	Symbol	Value
Angle of shearing resistance[a]	$\gamma_{\phi'}$	1.25
Effective cohesion	$\gamma_{c'}$	1.25
Undrained shear strength	γ_{cu}	1.40
Tensile pile resistance	$\gamma_{s;t}$	1.40
Anchor resistance	γ_a	1.40

[a] This factor is applied to $\tan \phi'$

Table A3.2.5 Hydraulic heave limit state (HYD): partial factors on actions (γ_F) (EN 1997-1: 2004).

Action	Symbol	Value
Permanent		
Unfavorable[a]	$\gamma_{G;dst}$	1.35
Favorable[b]	$\gamma_{G;stb}$	0.90
Variable		
Unfavorable[a]	$\gamma_{Q;dst}$	1.50

[a] Destabilizing
[b] Stabilizing

ANNEX A3.3 DESIGN APPROACHES 1, 2 AND 3 OF EUROCODE 7 FOR LIMIT STATE TYPES STR AND GEO

Note: the text of this annex is based on Annex B (informative) of Eurocode 7.

A3.3.1 Partial safety factors on actions and the effect of actions

1. In geotechnical design, both partial safety factors γ_F on actions (F_{rep}):

$$E_d = E\{\gamma_F F_{\text{rep}}; X_k/\gamma_M; a_d\} \tag{A3.3.1a}$$

and partial safety factors γ_E on the effect of actions (E) (see Table 3.5) are used:

$$E_d = \gamma_E E\{F_{\text{rep}}; X_k/\gamma_M; a_d\} \tag{A3.3.1b}$$

2. Equations A3.3.1 include X_k/γ_M because ground material properties may affect the values of the effects of actions.
3. In *Design Approach 1*, checks are required for two combinations of sets of factors, applied in two separate calculations.
   In *Combination 1*, factors unequal to 1 are generally applied to actions, with factors equal to 1 on the effects of actions. So, $\gamma_F \neq 1$ and $\gamma_E = 1$ are applied in Equations A3.3.1.
   In *Combination 2*, $\gamma_E = 1$ is always used, with $\gamma_F \neq 1$ used only for variable actions.
   So, in general, in the case of Design Approach 1, only Equation A3.3.1a is applied:

$$E_d = E\{\gamma_F F_{\text{rep}}; X_k / \gamma_M; a_d\} \tag{A3.3.2}$$

4. In *Design Approach 2*, a single calculation is required for each limit state, and the way in which the factors are applied, either to actions or the effects of actions, varies according to the calculation considered, in conformity with national annex.
   Either $\gamma_E \neq 1$ and $\gamma_F = 1$, or $\gamma_F \neq 1$ and $\gamma_E = 1$ are applied. Since $\gamma_M = 1$ is used, Equation A3.3.1a reduces to:

$$E_d = E\{\gamma_F F_{\text{rep}}; X_k; a_d\} \tag{A3.3.3}$$

and Equation A3.3.1b reduces to:

$$E_d = \gamma_E E\{F_{\text{rep}}; X_k; a_d\} \tag{A3.3.4}$$

5. In *Design Approach 3*, a single calculation is required. However, in this design approach, a difference is established between actions F_{rep} from the structure and actions from or through the ground, calculated from X_k. Either $\gamma_E \neq 1$ and $\gamma_F = 1$ or $\gamma_F \neq 1$ and $\gamma_E = 1$ are applied. Thus, Equation A3.3.1a remains:

$$E_d = E\{\gamma_F F_{\text{rep}}; X_k / \gamma_M; a_d\} \tag{A3.3.5}$$

as well as equation A3.3.1b:

$$E_d = \gamma_E E\{F_{\text{rep}}; X_k / \gamma_M; a_d\} \tag{A3.3.6}$$

Note: In calculation procedures, where partial factors are applied to the effects of actions, the partial safety factor for actions, γ_F, is 1.0.

A3.3.2 Partial safety factors on material strengths and resistances

1. In geotechnical design (see Tables 3.6 and 3.8), partial safety factors γ_M are applied to ground properties (X),

$$R_d = R\left\{\gamma_F F_{\text{rep}}; X_k / \gamma_M; a_d\right\} \tag{A3.3.7a}$$

or partial safety factors γ_R are applied to resistances (R),

$$R_d = \frac{1}{\gamma_R} R\left\{\gamma_F F_{\text{rep}}; X_k; a_d\right\} \tag{A3.3.7b}$$

or γ_M and γ_R to both ground properties and resistances,

$$R_d = \frac{1}{\gamma_R} R\left\{\gamma_F F_{\text{rep}}; X_k / \gamma_M; a_d\right\} \tag{A3.3.7c}$$

2. Note that Equations A3.3.7 include $\gamma_F F_{rep}$ in the determination of the design values of resistances, because the actions may affect the values of the geotechnical resistances.
3. In *Design Approach 1*, checks are required for two combinations of sets of factors, applied in two separate calculations.
   In *Combination 1*, factors equal to 1 are applied to ground material properties and resistances. Thus, $\gamma_M = \gamma_R = 1$ in Equation A3.3.7.
   In *Combination 2*, except for piles and anchors, $\gamma_M > 1$ and $\gamma_R = 1$.
   Thus, in most cases, Design Approach 1 adopts Equation A3.3.7a:

$$R_d = R\left\{\gamma_F F_{\text{rep}}; X_k / \gamma_M; a_d\right\} \tag{A3.3.8}$$

   But, in *Combination 2* for piles and anchors, $\gamma_M = 1$ and $\gamma_R > 1$ are used in Equation A3.3.7b, thus:

$$R_d = \frac{1}{\gamma_R} R\left\{\gamma_F F_{\text{rep}}; X_k; a_d\right\} \tag{A3.3.9}$$

4. In *Design Approach 2*, factors equal to 1 are generally applied to material strengths, with factors greater than 1 applied to resistances. Thus $\gamma_M = 1$ and $\gamma_R > 1$ are used in Equation A3.3.7b:

$$R_d = \frac{1}{\gamma_R} R\left\{\gamma_F F_{\text{rep}}; X_k; a_d\right\} \tag{A3.3.10}$$

   When $\gamma_F = 1$ is also used, Equation A.3.3.7b assumes the form:

$$R_d = \frac{1}{\gamma_R} R\left\{F_{\text{rep}}; X_k; a_d\right\} \tag{A3.3.11}$$

5. In *Design Approach 3*, $\gamma_M > 1$ and $\gamma_R = 1$ are generally applied. Thus, Equation A3.3.7a is used:

$$R_d = R\{\gamma_F F_{rep}; X_k / \gamma_M; a_d\} \tag{A3.3.12}$$

In this approach, sometimes there is also the need to have $\gamma_R > 1$ (for piles in tension, for example), so Equation A3.3.7c is then used:

$$R_d = \frac{1}{\gamma_R} R\{\gamma_F F_{rep}; X_k / \gamma_M; a_d\} \tag{A3.3.13}$$

ANNEX A3.4 RESISTANCE FACTORS FROM LRFD BRIDGE DESIGN SPECIFICATIONS (AASHTO, 2012)

Tables A3.4.1 and A3.4.2 include values of the resistance factor for bored and driven pile foundations, respectively, established in the AASHTO code (2012). Table A3.4.3 contains values of the resistance factor for permanent retaining walls established in the same code.

Table A3.4.1 Resistance factors for geotechnical resistance of drilled shafts (AASHTO, 2012).

	Method/Soil/Condition		Resistance factor
Nominal axial compressive resistance of single-drilled shafts, Φ_{stat}	Side resistance in clay	α-method (O'Neill and Reese, 1999)	0.45
	Tip resistance in clay	Total stress (O'Neill and Reese, 1999)	0.40
	Side resistance in sand	β-method (O'Neill and Reese, 1999)	0.55
	Tip resistance in sand	O'Neill and Reese, 1999	0.50
	Side resistance in IGMs	O'Neill and Reese, 1999	0.60
	Tip resistance in IGMs	O'Neill and Reese, 1999	0.55
	Side resistance in rock	Horvath and Kenney, 1979; O'Neill and Reese, 1999	0.55
	Side resistance in rock	Carter and Kulhawy, 1988	0.50
	Tip resistance in rock	Canadian Geotechnical Society (1985) Pressuremeter method (Canadian Geotechnical Society, 1985; O'Neill and Reese, 1999)	0.50
Block failure, Φ_{bl}	Clay		0.55
Uplift resistance of single-drilled shafts, Φ_{up}	Clay	α-method (O'Neill and Reese, 1999)	0.35
	Sand	β-method (O'Neill and Reese, 1999)	0.45
	Rock	Horvath and Kenney (1979) Carter and Kulhawy, 1988	0.40
Group uplift resistance, Φ_{ug}	Sand and clay		0.45
Horizontal geotechnical resistance of single shaft or shaft group	All materials		1.0
Static load test (compression), Φ_{load}	All materials		0.70
Static load test (uplift), Φ_{upload}	All materials		0.60

Table A3.4.2 Resistance factors for geotechnical resistance of driven piles (AASHTO, 2012).

Condition/Resistance Determination Method		Resistance factor
Nominal bearing resistance of single piles – dynamic analysis and static load test methods, Φ_{dyn}	Driving criteria established by successful static load test of at least one pile per site condition and dynamic testing[a] of at least two piles per site condition, but no less than 2% of the production piles	0.80
	Driving criteria established by successful static load test of at least one pile per site condition without dynamic testing	0.75
	Driving criteria established by dynamic testing[a], conducted on 100% production piles	0.75
	Driving criteria established by dynamic testing[a], quality control by dynamic testing[a] of at least two piles per site condition, but no less than 2% of the production piles	0.65
	Wave equation analysis, without pile dynamic measurements or load test but with field confirmation of hammer performance	0.50
	FHWA-modified Gates dynamic pile formula (End of Drive condition only)	0.40
	Engineering news (as defined in Article 10.7.3.8.5) dynamic pile formula (End of Drive condition only)	0.10
Nominal bearing resistance of single pile – static analysis methods, Φ_{stat}	Side resistance and end bearing: clay and mixed soils	
	α-method (Tomlinson, 1987; Skempton, 1951)	0.35
	β-method (Esrig and Kirby, 1979; Skempton, 1951)	0.25
	λ-method (Vijayvergiya and Focht, 1972; Skempton, 1951)	0.40
	Side resistance and end bearing: sand	
	Nordlund/Thurman method (Hannigan et al., 2006)	0.45
	SPT-method (Meyerhof)	0.30
	CPT-method (Schmertmann)	0.50
	End bearing in rock (Canadian Geotechnical Society, 1985)	0.45
Block failure, Φ_{bl}	Clay	0.60
Uplift resistance of single piles, Φ_{up}	Nordlund method	0.35
	α-method	0.25
	β-method	0.20
	λ-method	0.30
	SPT-method	0.25
	CPT-method	0.40
	Static load test	0.60
	Dynamic test with signal matching	0.50
Group uplift resistance, Φ_{ug}	All soils	0.50
Lateral geotechnical resistance of single pile or pile group	All soils and rock	1.0

[a] Dynamic testing requires signal matching, and best estimates of nominal resistance are made from a restrike. Dynamic tests are calibrated to the static load test, when available.

Table A3.4.3 Resistance factors for permanent retaining walls (AASHTO, 2012).

	Wall type and conditions	Resistance factor
Gravity and semi-gravity walls	Bearing resistance	0.55
	Sliding	1.0
Non-gravity cantilevered and anchored walls	Passive resistance of vertical elements	0.75
	Flexural capacity of vertical elements	0.90
Anchors	Pull-out resistance (presumptive ultimate unit bond stress for preliminary design)	
	Granular soils	0.65
	Cohesive soils	0.70
	Rock	0.50
	Pull-out resistance (proof tests conducted on every production anchor to a load equal to or greater than the factored load on the anchor)	1.0
	Tensile resistance of tendon (high-strength steel)	0.80

Note: this is just part of the complete table of the cited document.

Chapter 4

Consolidation theories and delayed settlements in clay

4.1 INTRODUCTION

Geologically recent fine-grained or clayey soils occur close to the ground surface, under the water table, and typically have very high water content (close to their liquid limit) and void ratio. The tendency to experience very high volumetric deformations, and the long time taken for these deformations to develop, cause large difficulties for construction.

It should be noted that the sites where these soils appear – very recent geological areas, namely alluvial valleys on the coast of continents – are the areas where many urban and industrial areas tend to concentrate, and, therefore, where many civil engineering structures are required.

This chapter presents some theories and methodologies that provide: i) the estimation of the magnitude of surface settlement associated with the volumetric deformation of fine-grained soils; ii) the prediction of settlement change over time, from loading to final stabilization stage; and iii) the design of systems to accelerate, where necessary, that change.

In most cases these topics are very relevant to the design of embankments on soft clayey soil, the stability of which was already discussed in Chapter 2 (see Section 2.4). Therefore, some issues treated in Chapter 2 will be reworked here, where necessary.

4.2 STRESS–STRAIN RELATIONS IN SOILS LOADED UNDER CONSTRAINED CONDITIONS

4.2.1 Constrained loading: oedometer tests

The treatment of the questions indicated above will start with a particularly simple case, such as the one illustrated in Figure 4.1a, where a soil mass is loaded at the surface by a uniformly distributed load on an area whose dimensions in plan are much larger than the thickness of the set of compressible layers. These loading conditions may be designated as *constrained loading* or *confined loading* because the horizontal strains are null.

Consider a generic point P in the soil mass and an elementary load Δq_{s1} at the surface, situated to the left of the vertical line that passes through P. As Figure 4.1b suggests, in qualitative terms, the displacement vector δ_1 in P, associated with the load Δq_{s1}, has a downward vertical component and a horizontal component directed to the right. Taking an elementary load Δq_{s2} symmetrical to Δq_{s1} in relation to P, the displacement of this point induced by Δq_{s2}, δ_2, will have a horizontal component symmetrical to that of δ_1, and a vertical component equal to that of δ_1. Therefore, the displacement, δ_{12}, induced by the two elementary loads is vertical.

As the loaded area at the surface is practically infinite, any elementary load at one side of P will have another symmetrical one. Therefore, in the soil mass, only vertical displacements

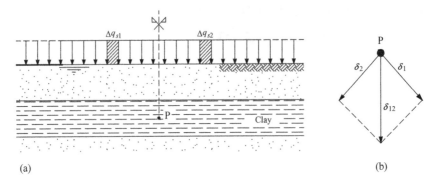

Figure 4.1 Constrained or one-dimensional loading: a) loading scheme; b) displacement vector of a generic point induced by two elementary symmetrical loads.

will occur. The minimum displacement will occur in contact with the bedrock and the maximum at the surface.

For that reason, under the conditions indicated above, the loading will cause only vertical strains, corresponding to a one-dimensional mechanical problem. The volumetric strains are numerically equal to the vertical strains, which result in a reduction in the thickness of the distinct layers, particularly the soft clay layers which have greater *compressibility*. The word "compressibility" is frequently used as synonym of "deformability" when only volumetric deformations are present, i.e., when shear strains are null.

Another consequence of applying the load over an almost infinite area is that, at any point of the soil mass, the total vertical incremental stress coincides with the surface surcharge load, i.e., $\Delta\sigma_v = \Delta q_s$.

From the explanation presented, it is clear that an active sedimentary deposit in Nature, with a regular deposition of new layers over a very large area, corresponds to a constrained loading.

The tests used in soil mechanics to simulate the reported field conditions in the laboratory are the *oedometer tests*, a simplified scheme of which is shown in Figure 4.2.

LEGEND:

1 – Undisturbed and saturated sample

2 – Porous stones

3 – Load (applied by stages)

4 – Rigid ring (it prevents horizontal strains)

5 – Dial gauge (vertical displacements)

6 - Water

7 - Container

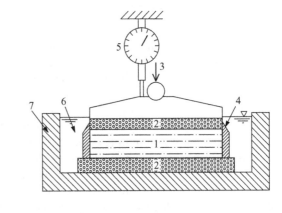

Figure 4.2 Simplified scheme of an oedometer test.

4.2.2 Time effect: hydromechanical analogy

To support our understanding of time-delayed deformations in clayey soils subjected to one-dimensional loading, the hydromechanical analogy represented in Figure 4.3 can be used. It consists of a cylindrical reservoir of rigid walls with a frictionless piston, where there is a very narrow hole. The piston is linked to the reservoir base by means of a spring.

In an initial situation (Figure 4.3a), the force at the spring balances the piston weight, and water is introduced into the reservoir so that its surface coincides with the piston base.

Considering an instantaneous load applied to the piston, ΔW, and bearing in mind that water is almost incompressible, at that precise moment the force at the spring does not change, since the corresponding length has not yet changed. This means that the applied load is therefore supported by the water, whose pressure increases by $u_e(0)$. The product of this overpressure by the piston area is equal to the applied load.

This overpressure will cause the immediate escape of water through the hole, which will enable the piston to come down, with a reduction in spring length, and, consequently, an increase in the force exerted on the spring. In this way, in the following instant t, the equilibrium of the applied load will be shared by both the water and the spring (Figure 4.3c). The proportion supported by the spring, U, increases progressively as the piston comes down, reaching 100% when the water pressure under the piston again becomes equal to the atmospheric pressure. In that instant, the water stops flowing through the hole, the piston stops and an equilibrium situation similar to the initial one is restored (Figure 4.3d).

Table 4.1 presents the analogy between the model and the field situations.

Figure 4.4 represents a soil mass loaded at the surface by a uniformly distributed surcharge over an infinite area (Figure 4.4a). In Figure 4.4b the values of the total and effective vertical stress and pore water pressure in a generic point of the clay layer are illustrated.

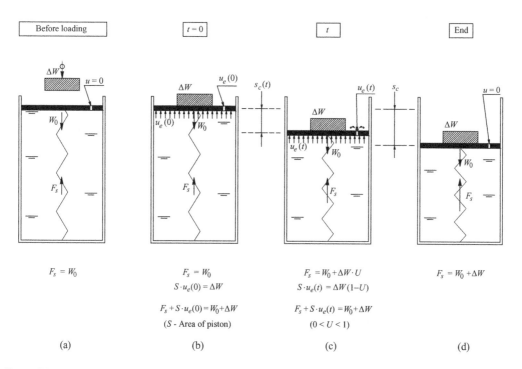

Figure 4.3 Hydromechanical analogy for the one-dimensional consolidation and loading of clay (see meaning of variables in Table 4.1).

Table 4.1 Correspondence between the hydrodynamic analogy and the one-dimensional consolidation.

		Hydrodynamic analogy	*One-dimensional consolidation*
Components		Spring	Soil solid skeleton
		Water	Soil pore water
		Piston	Overlying layers above the clay layer
		Applied load	Weight of the structure built at the surface
		Hole dimensions	Clay permeability
Stresses or forces		Spring force, F_s	Effective stress
		Water pressure, u	Pore water pressure
		Water overpressure, u_e	Excess pore pressure
		Piston weight, W_0	Effective stress at-rest
		Load on the piston, ΔW	Surcharge load applied at the surface
Displacements		Piston displacement at the instant t, $s_c(t)$	Consolidation settlement at the instant t
		Piston displacement at the end of process, s_c	Consolidation settlement (final)

At the time of loading, settlement will occur, known as *immediate settlement* (see Figure 4.4c), associated with the volumetric deformation of the very permeable layers.

If the volumetric deformation of the clay is still null at that instant, its effective stress has not changed, so that the vertical stress increment has instantaneously caused an excess pore pressure, u_e, of the same value. Bearing in mind that in the permeable layers this excess pore pressure is not developed,[1] an hydraulic gradient is established in the clay layer boundaries, which will immediately lead to a flow, with vertical flow lines, of the water from the clay to the draining boundaries.

As far as the water is expelled from the clay, a volumetric deformation occurs, with increasing settlement at the ground surface, an increase in the effective stress, and a decrease of pore pressure. This process, delayed in time, as a result of dissipation of the excess pore pressure, and progressive transfer of the total stress increment from the liquid phase to the solid phase of the soil, is called *consolidation*. The *consolidation settlement*, delayed in time, evolves as illustrated in Figure 4.4c.

The rate at which the excess pore pressure dissipation occurs depends essentially on four issues: i) clay permeability; ii) clay stiffness; iii) clay layer thickness; and iv) drainage conditions at the layer boundaries. These phenomena will be treated in more detail in Section 4.3.

4.2.3 Load diagrams obtained in the oedometer test

Each load increment of the oedometer test will trigger, in the soil specimen, a delayed-in-time volumetric deformation process, described by the hydromechanical analogy presented above. Typically, the increments are applied at 24-h intervals, enough time (given the reduced thickness of the specimen) for the applied load to be transferred to the effective stress, i.e., for complete *excess pore pressure dissipation*.

Figure 4.5 shows the relationships obtained in a clay sample from the Holocene era. These relationships express the soil compressibility, connecting the void ratio with the vertical

---

[1] The permeability of coarse-grained layers is several orders of magnitude higher than that of clay layers. For that reason, in those layers, the volumetric deformation occurs at the same time as surface loading, since the water can easily be expelled from the soil pores.

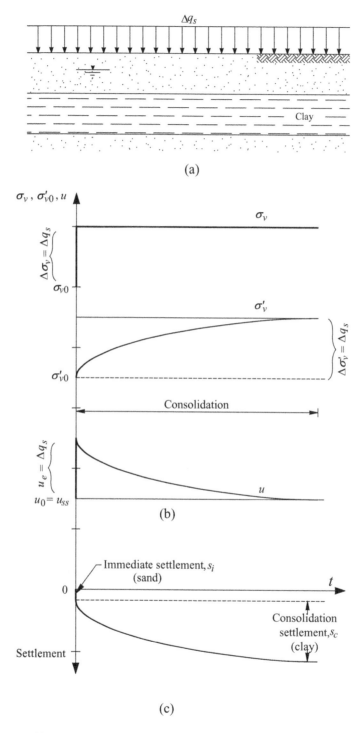

Figure 4.4 Constrained loading: a) ground and loading conditions; b) change with time of the stress state at a generic point of the clay layer; c) change in time (in qualitative terms) of the ground surface settlements.

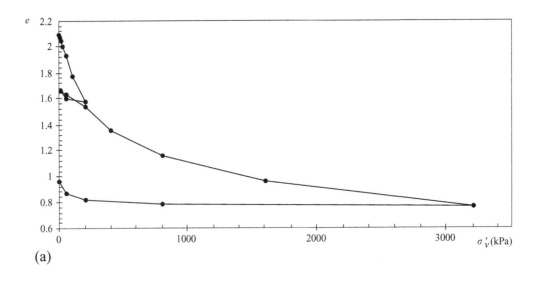

(a)

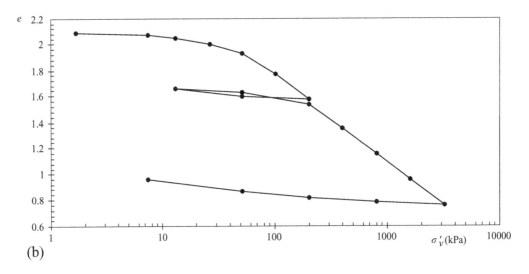

(b)

Figure 4.5 Vertical effective stress *versus* void ratio plot of a Lower Mondego Holocene clay sample submitted to an oedometer test (Coelho, 2000): a) effective stress on an arithmetic scale; b) effective stress on a logarithmic scale.

effective stress, expressed on an arithmetic scale (Figure 4.5a) or on a logarithmic scale (Figure 4.5b). Each point on the graphs relates the vertical effective stress applied to the specimen with the void ratio over 24 h, i.e., immediately before the new increment.

The void ratio is obtained from the change in the specimen thickness measured in the test by the following equation:

$$e_0 - e = \Delta e = \frac{1 + e_0}{h_0} \cdot \Delta h \qquad (4.1)$$

being h_0 and e_0, respectively, the thickness and the void ratio at the start of the test. The deduction of this equation is presented in Annex A4.1, where it is also demonstrated that the ratio $h/(1 + e)$ is constant during loading, with the same occurring, obviously, to its inverse.

The analysis of Figure 4.5a indicates that, when the soil is unloaded, a very important (plastic) fraction of the deformation remains irrecoverable. This happens because the void ratio reduction results from structural rearrangements in the solid skeleton, leading to new inter-particle balances of greater stability. This is also the explanation for the increase in stiffness as loading increases.

In Figure 4.5b, the straight line of the $\log \sigma'_v - e$ relationship, corresponds to the stress range not yet experienced by the sample, hence, by the soil mass from where it was taken, and it is named the *virgin compression line*. On the other hand, the maximum vertical effective stress already experienced by the soil *in situ* (thus, the abscissa of the initial point of the virgin compression line) is called the *pre-consolidation pressure, σ'_p*.

This stress is equal to the vertical effective stress at-rest, σ'_{v0}, in normally consolidated soils. These soils represent, in the majority of the cases, geologically recent soil masses, the stress history of which comprises, since its formation by sedimentation, just a monotonic loading associated with the weight of overlying layers.

In many cases, especially in older soil masses, the vertical effective stress at-rest, σ'_{v0}, is lower or much lower than the maximum vertical effective stress previously installed in the soil, σ'_p. The soil masses in those conditions are said to be *overconsolidated*.

The *overconsolidation* is expressed quantitatively by the *overconsolidation ratio, OCR*, the ratio between the effective pre-consolidation stress and the vertical effective stress at-rest.[1]

$$OCR = \frac{\sigma'_p}{\sigma'_{v0}} \tag{4.2}$$

Table 4.2 includes a classification of soils according to the overconsolidation ratio

As illustrated in Figure 4.5b, in general, the transition from the initial part of the curve, corresponding to the stress range already experienced by the soil *in situ*, to the virgin compression line, is not clear from the laboratory curves. Therefore, the evaluation of the value of the pre-consolidation stress is not obvious in most cases. The literature includes several methodologies for this purpose. The best-known method is the empirical construction proposed by Casagrande (1936), described in Annex A4.2.

4.2.4 Treatment of the compressibility curve

The estimation of the (final) consolidation settlement is made from parameters taken from the curves obtained in oedometer tests, that relate the void ratio with the vertical effective stress. However, to achieve a reliable estimate, the test curves need to be previously treated or adjusted. This treatment generally involves the use of a semi-empirical construction, proposed by Schmertmann (1955), the justification of which is explained in Figure 4.6.

Table 4.2 Classification of clayey soils according to over-consolidation ratio, OCR.

Classification	OCR
Normally consolidated	$\cong 1$
Lightly overconsolidated	$1 - 2$
Moderately overconsolidated	$2 - 5$
Heavily overconsolidated	>5

---

[1] The symbol R_{OC} is also used for overconsolidation ratio.

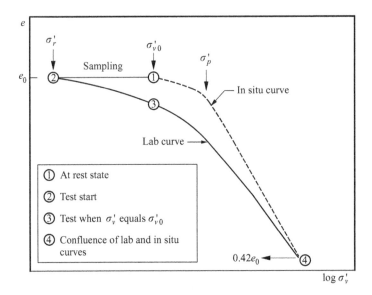

Figure 4.6 Comparison between the curve obtained from an oedometer test with the assumed curve of the soil *in situ*, according to Schmertmann (1955).

In this figure, point 1 represents the situation of the sample in the soil mass, before collection. The undisturbed sampling process, maintaining the sample water content, and thus the void ratio, is represented in the graph by a path from 1 to 2; when the sample is placed in the oedometer and the *in situ* effective stress is applied, the path changes from 2 to 3. Between 2 and 3 the sample is overconsolidated, in that it has low compressibility but is still experiencing deformations. Therefore, at point 3, the sample exhibits a lower void ratio and water content than the *in situ* soil. The difference between points 1 and 3 shows that the laboratory curve can never be coincident with the real behavior, even in very high-quality samples. However, as suggested in the figure, the laboratory and *in situ* curves tend to become closer during loading, progressively reducing the disturbance effects of sampling.

As Figure 4.6 indicates, the great number of experimental curves analyzed by Schmertmann (1955) has shown that curve confluence is generally verified at a void ratio close to $0.42e_0$, where e_0 is the void ratio of the soil in its natural state. Based on that observation, the author proposed the reconstitution of the compressibility curve by means of the constructions illustrated in Annex A4.3, for normally and overconsolidated cases. For these, the construction has to start by the determination of the pre-consolidation pressure, using the Casagrande method or another suitable method.

4.2.5 Parameters for the definition of the stress–strain relationships

It is now important to define some parameters given by the oedometer curves, such as the ones in Figure 4.5, which will allow evaluation of the settlements.

As shown in Figure 4.7a, *the soil compression index C_c,* is defined as the slope (absolute value) of the virgin compression line.

Similarly, the *recompression index, C_r,* represents the slope (absolute value) of the recompression curve of the same graph. These indexes are normally determined on the experimental curves, after the correction explained above.

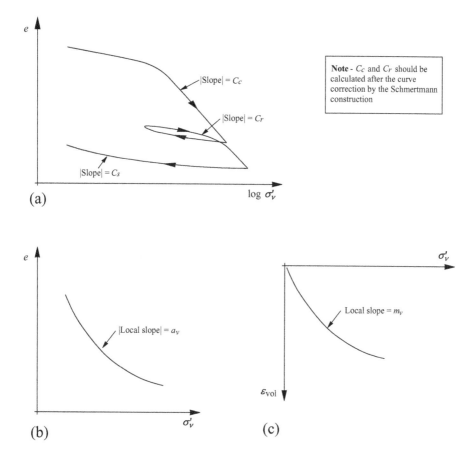

Figure 4.7 Parameters for the definition of the soil compressibility: a) $\log \sigma'_v - e$ plot; b) $\sigma'_v - e$ plot; c) $\sigma'_v - \varepsilon_{vol}$ plot.

As Figure 4.7a suggests, it is not convenient to estimate C_r from the initial part of the experimental curve, being better to estimate it from an unload–reload cycle (generally involving not more than two load increments) performed when the test is clearly in the virgin compression range of the soil.

In certain cases, when the behavior under decreasing vertical effective stress needs to be studied, as happens in a soil mass underneath an excavation, it is necessary to estimate the *swelling index*, C_s. This can be performed, as Figure 4.7a suggests, from the final unloading of the sample, after the maximum load has been achieved.

Usually, the values of C_r and C_s are very close and can range, in the majority of the soils, within 1/5 and 1/10 of the C_c value (Lancellotta, 2008). All these indexes are dimensionless.

According to Figure 4.7b, the parameter that measures the change in void ratio by unit effective stress increment, at each point of the $\sigma'_v - e$ curve, is called the *coefficient of compressibility*, a_v:

$$a_v = \left| \frac{\Delta e}{\Delta \sigma'_v} \right|$$

(4.3)

A similar parameter can also be measured in terms of volumetric deformation, as Figure 4.7c illustrates. The *coefficient of volumetric compressibility* is then defined, m_v:

$$m_v = \frac{\Delta \varepsilon_{vol}}{\Delta \sigma'_v} \tag{4.4}$$

Considering that, in the present case, $\varepsilon_{vol} = \varepsilon_v = \Delta h / h_0$, and, bearing in mind Equation 4.1, the latter two coefficients can be related:

$$m_v = \frac{1}{1 + e_0} a_v \tag{4.5}$$

where e_0 is the initial void ratio of the specimen.

Conversely to what happens with the parameters C_c, C_r and C_s, taken from the experimental curves with effective stresses on a logarithmic scale, a_v and m_v change with the magnitude of the effective stress.

4.2.6 Expressions for calculation of the consolidation settlement

Consider the general case of an overconsolidated soil that, as illustrated in Figure 4.8, will be loaded by a given vertical effective stress increment, $\Delta \sigma'_v$, which, when added to the vertical effective stress at-rest, σ'_{v0}, exceeds the pre-consolidation stress, σ'_p. This means that the soil will experience a recompression between σ'_{v0} and σ'_p, and a virgin compression between σ'_p and $\sigma'_{v0} + \Delta \sigma'_v$.

The variation of the void ratio associated with this recompression is:

$$\Delta e_1 = -C_r \log \frac{\sigma'_p}{\sigma'_{v0}} \tag{4.6}$$

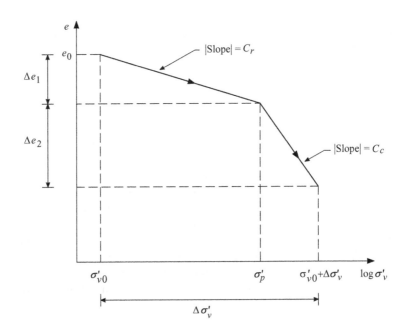

Figure 4.8 Void ratio variation on the loading of an overconsolidated clay in the case where the increment exceeds σ'_p.

while the variation of the void ratio associated with the virgin compression is:

$$\Delta e_2 = -C_c \log \frac{\sigma'_{v0} + \Delta\sigma'_v}{\sigma'_p} \tag{4.7}$$

Bearing in mind Equation 4.1, the variations in layer thickness associated with the changes in void ratio previously mentioned, are.[1]

$$\Delta h_1 = -\frac{h_0}{1+e_0} C_r \log \frac{\sigma'_p}{\sigma'_{vo}} \tag{4.8}$$

$$\Delta h_2 = -\frac{h_0}{1+e_0} C_c \log \frac{\sigma'_{v0} + \Delta\sigma'_v}{\sigma'_p} \tag{4.9}$$

The *consolidation settlement* is obtained by adding the variations in layer thickness and taking the absolute value

$$s_c = \frac{h_0}{1+e_0} \left(C_r \log \frac{\sigma'_p}{\sigma'_{v0}} + C_c \log \frac{\sigma'_{v0} + \Delta\sigma'_v}{\sigma'_p} \right) \tag{4.10}$$

In the case of a normally consolidated soil loading, then $\sigma'_{v0} = \sigma'_p$, and hence, the previous equation becomes:

$$s_c = \frac{h_0}{1+e_0} C_c \log \frac{\sigma'_{v0} + \Delta\sigma'_v}{\sigma'_{v0}} \tag{4.11}$$

On the other hand, in the case of an overconsolidated soil loading, where the pre-consolidation effective stress is not exceeded (i.e., where $\sigma'_{v0} + \Delta\sigma'_v \le \sigma'_p$), the expression to use for the calculation of the consolidation settlement becomes:

$$s_c = \frac{h_0}{1+e_0} C_r \log \frac{\sigma'_{v0} + \Delta\sigma'_v}{\sigma'_{v0}} \tag{4.12}$$

Unless the thickness of the layer in study is small, in order to obtain a more rigorous mathematic calculation, it is convenient to consider the layer divided into several sublayers and, for each one of those, to apply the appropriate settlement expression (one of the three mentioned above, with σ'_{v0} calculated for the center of each sublayer), proceeding then to the corresponding sum.

The consolidation settlement can also be calculated from the parameters a_v and m_v obtained from the curve $\sigma'_v - e$ (Figure 4.7b,c).

In fact, combining Equations 4.3 and 4.1, and considering that s_c is equal to the absolute value of Δh, i.e., to the variation of the layer thickness, the following is obtained:

$$s_c = \frac{a_v}{1+e_0} h_0 \Delta\sigma'_v \tag{4.13}$$

and considering Equation 4.5:

$$s_c = m_v h_0 \Delta\sigma'_v \tag{4.14}$$

---

[1] The reader is reminded that Annex A4.1 demonstrates that the ratio $h/(1+e)$ is constant.

Equations 4.13 and 4.14 are valid for normally consolidated or overconsolidated soils but it should be realized that a_v and m_v are not constant. In a specific case, the values of these parameters should be adopted taking into account the effective stress range that the soil will experience between the at-rest stress state and the end of consolidation, i.e., between σ'_{v0} and $\sigma'_{v0} + \Delta\sigma'_v$.

Equation 4.14 can be rewritten as follows:

$$\Delta\sigma'_v = \frac{1}{m_v} \frac{s_c}{h_0} \tag{4.15}$$

With s_c being a measure of the variation in layer thickness, s_c/h_0 physically represents a vertical strain and, since the horizontal strain is null, it also represents a volumetric strain, associated with the application of a vertical effective stress increment, $\Delta\sigma'_v$.

The interpretation of the previous equation from the elasticity point of view indicates that the inverse of m_v has the physical meaning of an elasticity modulus. It can thus be written that

$$E_{\varepsilon_h=0} = \frac{1}{m_v} \tag{4.16}$$

i.e., that the inverse of m_v represents the soil deformation modulus for loading conditions where the horizontal strains are null, or, in other words, the soil *volumetric deformation modulus*. This parameter is also usually designated as the *oedometer modulus* or the *constrained modulus*.

The oedometer modulus, for which the current symbol is E_{oed}, can be easily related with the deformation modulus for general loading conditions, E, and with the Poisson ratio, ν.

In a constrained loading, $\varepsilon_x = \varepsilon_y = 0$. Then, by Hooke's law (Timoshenko and Goodier, 1951) it can be written that, for those conditions:

$$\sigma_z = \frac{E(1-\nu)}{(1+\nu)(1-2\nu)} \varepsilon_z \tag{4.17}$$

So

$$E_{oed} = \frac{E(1-\nu)}{(1+\nu)(1-2\nu)} \tag{4.18}$$

4.2.7 Some practical issues

When dealing with the construction of embankments over highly compressible soils, with the water level very close to the soil surface, it is not rare that part of the embankment becomes submerged, due to the large magnitude of the settlement. If part of the embankment stays under the effect of buoyancy, then the incremental effective stress will be reduced in relation to the stress initially considered in the calculation, and therefore the consolidation settlement will be overestimated. It is then necessary to do an adjustment to the settlement calculation to take this effect into account, which will require an iterative process.

In embankment works, the elevation of the embankment surface is generally an imposition of the project. As a consequence, the embankment thickness cannot be just the difference between that imposed elevation and the elevation of the ground surface before the intervention. In fact, the consolidation settlement should be added to the thickness corresponding to the difference of elevations already mentioned. Then, in the equations for the

Table 4.3 Correlations between C_c and basic soil parameters.

Equation (w and w_L in %)	Further information (n: number of results; r: correlation coefficient)	References
$C_c = 0.009\ (w_L - 10)$	Normally consolidated clays of low to moderate sensitivity	Terzaghi and Peck (1967)
$C_c = 0.01\ (w - 5)$	Alluvial, marine and aeolian normally consolidated soils from Greece ($n = 717, r = 0.79$)	Azzouz et al. (1976)
$C_c = 0.0093w$	Clays from Alberta, Canada ($n = 109$, confidence level $= 95\%$)	Koppula (1981)
$C_c = 0.013\ (w - 8)$	Soils from Bangladesh ($n = 130$)	Serajuddin (1987)
$C_c = 0.013w$	Holocene Rio de Janeiro marine clays, Brazil (seven sites, $n = 220, r = 0.84$)	Futai (1999) Almeida et al. (2008)
$C_c = 0.013\ (w - 4)$	Holocene Korean clays (South coast, $n = 278$, $r = 0.77$)	Yoon et al. (2004)
$C_c = 0.010\ (w + 3)$	Holocene Korean clays (East coast, $n = 603$, $r = 0.54$)	
$C_c = 0.011\ (w - 11)$	Holocene Korean clays (West coast, $n = 356$, $r = 0.65$)	
$C_c = 0.013\ (w - 10)$	Holocene Portuguese clays ($n = 109, r = 0.93$)	Esteves (2013)
$C_c = 0.014\ (w - 23)$	Soft clays from Ireland ($n = 61, r = 0.93$)	McCabe et al. (2014)

calculation of the consolidation settlement, the incremental effective stress that appears in the second member will depend on the settlement itself; i.e., this will appear in the two members of the equation. The resolution of this equation will allow calculation of the height of embankment to be built.

Table 4.3 presents some empirical equations, relating the compression index with physical and identification properties of the soil. These correlations do not replace oedometer tests on high-quality samples. However, they can be useful in preliminary studies and, also, as a reference with which to compare the test results, in order to identify possible gross mistakes, resulting from low-quality samples or inappropriate interpretation of the experimental data.

4.3 THE TERZAGHI CONSOLIDATION THEORY

4.3.1 Base hypotheses: consolidation equation

The previous section focused on the evaluation of the magnitude of the settlement. In general, it is also very important to evaluate the time that this settlement takes to process. This is the aim of the one-dimensional consolidation theory formulated by Terzaghi (1923a,b). The hydrodynamic analogy presented in Section 4.2.2 explains what happens in a generic point of the soil, according to the theory that will be explained in this section.

Figure 4.9a illustrates a confined loading of a clay layer. Figure 4.9b shows a clay volume element of infinitesimal dimensions. In what follows, the origin of the axis system is situated on the upper boundary of the clay layer.

The Terzaghi consolidation theory is based on the following hypotheses:

  i) the soil is homogeneous and it is saturated;
 ii) the water and particle compressibility are negligible;

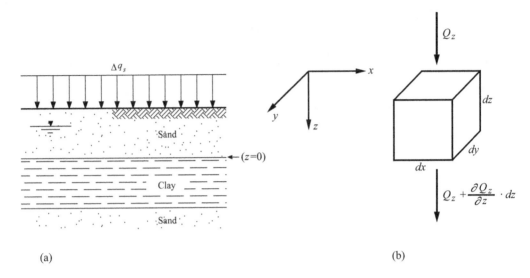

Figure 4.9 One-dimensional consolidation of a clay layer: a) general scheme; b) generic soil element.

iii) the stress and strain states in any horizontal section and at any instant are uniform;
iv) the deformations occur only in the vertical direction;
 v) the flow of water is exclusively vertical and obeys Darcy's law;
vi) the effects, the phenomenon and its development in elements of infinitesimal dimensions are extrapolated to the representative dimensions of a real soil mass;
vii) the permeability coefficient, k, the coefficient of compressibility, a_v, and the coefficient of volumetric compressibility, m_v, remain constant for a small stress and strain interval;
viii) there is a one-to-one relationship between the void ratio and the vertical effective stress at a given point and in a given instant (there is no creep of the solid skeleton);
ix) the small displacement hypothesis is valid (geometric linearity).

Based on these hypotheses, it is possible to obtain the following equation that relates the excess pore pressure, u_e, with space and time (Terzaghi, 1943):

$$\frac{\partial u_e}{\partial t} = c_v \cdot \frac{\partial^2 u_e}{\partial z^2} \tag{4.19}$$

where c_v is the *coefficient of consolidation*, given by:

$$c_v = \frac{k}{m_v \gamma_w} \tag{4.20}$$

expressed in m²/s or m²/year.

For general application of consolidation equation solutions, it is advantageous to work, not with z and t (parameters that represent real time and real space), but with others directly proportional to z and t but of a dimensionless nature, whose expressions are:

$$Z = \frac{z}{H} \tag{4.21}$$

and

$$T = \frac{c_v t}{H^2} \tag{4.22}$$

with Z being the *depth factor*, T the *time factor* and H the longest distance that a water particle needs to travel until reaching a draining boundary; in the case of Figure 4.9a, H is half the layer thickness, bearing in mind that two draining boundaries exist.

Using dimensionless variables, Equation 4.19 becomes:

$$\frac{\partial u_e}{\partial T} = \frac{\partial^2 u_e}{\partial Z^2} \tag{4.23}$$

4.3.2 Consolidation equation solutions

4.3.2.1 Layer with two draining boundaries

Equation 4.23 can be integrated for each case, considering the boundary conditions. For Figure 4.9 conditions, the space and time boundary conditions are the following:

i)   for $T = 0$, $u_e(t) = u_e(0) = \Delta\sigma_v$ for $0 \le Z \le 2$; \hfill (4.24)

ii)   for any $T \neq 0$, $u_e(t) = 0$ for $Z = 0$ and $Z = 2$. \hfill (4.25)

The equation solution is given by the following infinite series:

$$u_e = \sum_{m=0}^{m=\infty} \frac{2u_e(0)}{M} \cdot (\sin MZ) \cdot e^{-M^2 T} \tag{4.26}$$

where

$$M = \frac{\pi}{2} \cdot (2m + 1) \tag{4.27}$$

and m is an integer variable, taking the values 0, 1, 2, 3...

This solution can be presented in a graph, as shown in Figure 4.10, which provides an image of the space and time distribution of the excess pore pressure, represented with dimensionless values in the upper horizontal axis, or the *consolidation ratio*, defined by the equation:

$$U_z(t) = 1 - \frac{u_e(t)}{u_e(0)} \tag{4.28}$$

and represented in the lower horizontal axis.

The physical meaning of the consolidation ratio should be clarified. Bearing in mind that the initial excess pore pressure, $u_e(0)$, is equal to the effective stress increment at the end of consolidation, $\Delta\sigma'_v$ *(final)*, and that the excess pore pressure in a generic instant t, $u_e(t)$, is the

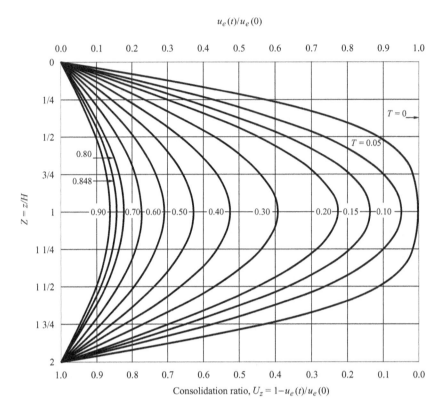

Figure 4.10 Solution of the consolidation equation for the case of two draining boundaries and initial excess pore pressure with constant distribution along the layer thickness.

difference, $\Delta\sigma'_v (final) - \Delta\sigma'_v (t)$, then, carrying out the appropriate replacements in Equation 4.28, it becomes:

$$U_z(t) = \frac{\Delta\sigma'_v(t)}{\Delta\sigma'_v(final)} \qquad (4.29)$$

Therefore, the consolidation ratio, $U_z(t)$, represents in each point and instant, the ratio of the vertical effective stress already installed to the corresponding increment at the end of consolidation.

In this way, considering that the curve represented in Figure 4.11 corresponds to the general solution at a given real instant t for the conditions of a given problem (determined by the soil itself, by means of the coefficient of consolidation, c_v, and by the geometry and prevailing drainage conditions, represented by H), then two related readings can be taken from this figure.

Taking the upper horizontal axis as reference, the lighter part of the figure represents the excess pore pressure that *still* needs to be dissipated, at the instant t. It is clearly seen that, close to the draining boundaries, the dissipation is more advanced while in the layer center it is more delayed.

Taking the lower horizontal axis as reference, the darker part of the figure represents the effective stress increment *already* installed in the same instant. In contrast to the excess pore pressure, the effective stress increment decreases from the draining boundaries to the layer center. At a subsequent instant t, the decrease of the lighter part is associated with the advance of the darker part. The sum $u_e(t) + \Delta\sigma'_v(t)$ is constant at any instant or point, with a progressive decrease of the first term and a consequent increase of the second.

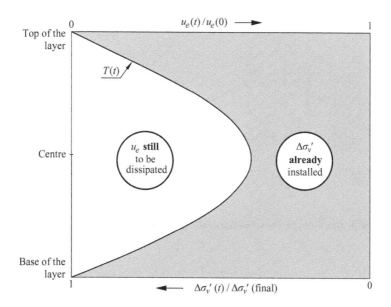

Figure 4.11 Interpretation of Terzaghi's consolidation theory solution.

4.3.2.2 Layer with only one draining boundary

In a case where one of the clay layer boundaries is impervious, the solution of Equation 4.23 is still represented by Equation 4.26 and by Figure 4.10.

In fact, as can be concluded from that figure, when there are two draining boundaries and the initial distribution of the excess pore pressure is symmetrical in relation to the central plan of the layer, this distribution remains symmetrical during the entire consolidation process, hence the hydraulic gradient is null in that plan, with the flow that passes through it also being null. This way, the excess pore pressure distribution in half of the layer will be the same as the distribution that exists in another layer with half the thickness and with drainage by one of the boundaries.

However, it should be pointed out, that, when drainage exists in only one boundary, the parameter H, which is used in the definition of the time factor (Equation 4.22), represents the whole layer thickness. Bearing in mind that H appears squared in the denominator of that expression, in a layer of a certain thickness, a given consolidation ratio will be achieved in a period four times longer if drainage occurs in only one direction.

4.3.3 Consolidation settlement change over time

The consolidation equation solutions presented are not generally used *directly* in engineering. However, based on them, a result of paramount practical interest is achieved: the consolidation settlement changes over time!

From the space and time distribution of the excess pore pressure, $u_e(Z, T)$, and consequently of the distribution of the consolidation ratio, $U(Z, T)$, the average value of this last parameter can be calculated along the layer in consolidation, $\bar{U}_z(T)$. This relationship between $\bar{U}_z$ and T is shown in Figure 4.12.

From this figure, it can be seen that the rate of increase of the average consolidation ratio decreases over time. According to the mathematical solution of the consolidation equation,

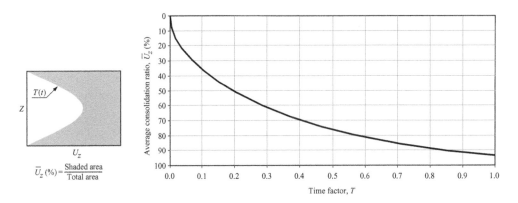

Figure 4.12 Change in the average consolidation ratio with the time factor for the case in the previous figures.

to achieve $\bar{U}_z = 100\%$, an infinite time period would be necessary: the horizontal axis of the figure is an asymptote of the $\bar{U}_z(T)$ curve.

The physical explanation for this result is simple: when the consolidation ratio tends toward 100%, the excess pore pressure tends to zero; then, the same happens to the hydraulic gradient and to the flow rate, which, by Darcy's law, is proportional to the former. In practical terms, it is common to take the end of consolidation for $T = 1$, which corresponds to a $\bar{U}_z$ of approximately 90%.

For values of $\bar{U}_z < 60\%$ the curve presented almost coincides with a parabola, with $\bar{U}_z$ being approximately proportional to $\sqrt{T}$.

It is possible to demonstrate that the relation between $\bar{U}_z$ and T expressed by Figure 4.12 is valid for any linear distribution (constant or not) of the excess pore pressure, for $T = 0$. Table 4.4 illustrates that relationship (case 1) in numerical terms; in the upper part of the table, the average consolidation ratio is expressed as a function of the time factor, whereas, in the lower part, it is the time factor that is expressed as a function of the consolidation ratio.

In the same table, consolidation equation solutions are included for other initial excess pore pressure distributions (cases 2 to 4), which will be commented on further below (see Section 4.4.3.2).

It should be highlighted that the average consolidation ratio represents the average percentage of the total stress increase applied that has already been converted to effective stress up to a given instant (for that purpose, see Equation 4.29). By the fact of assuming parameters a_v and m_v as constants (see Section 4.3.1), $\bar{U}_z(t)$ will also represent the percentages of the total variation in the void ratio and the volumetric strain, processed up to that same instant.

Therefore, the consolidation settlement in that instant can be obtained by the equation

$$s_c(t) = \bar{U}_z(t) \cdot s_c \tag{4.30}$$

where s_c represents the total consolidation settlement, the calculation of which was described in Section 4.2.6.

This equation involves a certain error resulting from the hypothesis that a_v and m_v are constant. However, as seen before, these parameters decrease with the increase in effective stress, since the soil becomes denser. For that reason, as Figure 4.13 illustrates, in response to the average increase of effective stress with t, $\bar{U}_z(t) \cdot \Delta\sigma'(\text{final}) = \Delta\sigma'_v(t)$, a volumetric strain greater than $m_v \Delta\sigma'_v(t)$ will be associated. In other words, the settlement at the end of time t will tend to be higher than the one given by Equation 4.30.

Table 4.4 Numerical solutions of Terzaghi's consolidation equation for different distributions of the initial excess pore pressure (Perloff, 1975).

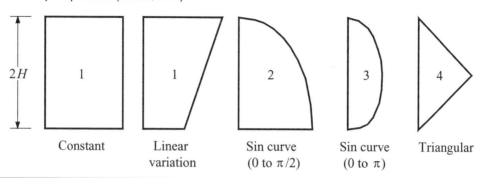

| | Constant | Linear variation | Sin curve (0 to $\pi/2$) | Sin curve (0 to π) | Triangular |

	Average consolidation ratio, $\bar{U}_z$ (%)			
T	*Case 1*	*Case 2*	*Case 3*	*Case 4*
0.004	7.14	6.49	0.98	0.80
0.008	10.09	8.62	1.95	1.60
0.012	12.36	10.49	2.92	2.40
0.020	15.96	13.67	4.81	4.00
0.028	18.88	16.38	6.67	5.60
0.036	21.40	18.76	8.50	7.20
0.048	24.72	21.96	11.17	9.60
0.060	27.64	24.81	13.76	11.99
0.072	30.28	27.43	16.28	14.36
0.083	32.51	29.67	18.52	16.51
0.100	35.68	32.88	21.87	19.77
0.125	39.89	36.54	26.54	24.42
0.150	43.70	41.12	30.93	28.86
0.175	47.18	44.73	35.07	33.06
0.200	50.41	48.09	38.95	37.04
0.250	56.22	54.17	46.03	44.32
0.300	61.32	59.50	52.30	50.78
0.350	65.82	64.21	57.83	56.49
0.400	69.79	68.36	62.73	61.54
0.500	76.40	76.28	70.88	69.95
0.600	81.56	80.69	77.25	76.52
0.700	85.59	84.91	82.22	81.65
0.800	88.74	88.21	86.11	85.66
0.900	91.20	90.79	89.15	88.80
1.00	93.13	92.80	91.52	91.25
1.500	98.00	97.90	97.53	97.45
2.000	99.42	99.39	99.28	99.26

(*Continued*)

Table 4.4 (Continued) Numerical solutions of Terzaghi's consolidation equation for different distributions of the initial excess pore pressure (Perloff, 1975).

$\bar{U}_z$ (%)	Time factor, T			
	Case 1	Case 2	Case 3	Case 4
0	0	0	0	0
5	0.0020	0.0030	0.0208	0.0250
10	0.0078	0.0111	0.0427	0.0500
15	0.0177	0.0238	0.0659	0.0753
20	0.0314	0.0405	0.0904	0.101
25	0.0491	0.0608	0.117	0.128
30	0.0707	0.0847	0.145	0.157
35	0.0962	0.112	0.175	0.187
40	0.126	0.143	0.207	0.220
45	0.159	0.177	0.242	0.255
50	0.197	0.215	0.281	0.294
55	0.239	0.257	0.324	0.336
60	0.286	0.305	0.371	0.384
65	0.342	0.359	0.425	0.438
70	0.403	0.422	0.488	0.501
75	0.477	0.495	0.562	0.575
80	0.567	0.586	0.652	0.665
85	0.684	0.702	0.769	0.782
90	0.848	0.867	0.933	0.946
95	1.129	1.148	1.214	1.227
100	∞	∞	∞	∞

4.3.4 Estimation of c_v from oedometer tests

As mentioned previously, in oedometer tests, the specimen is loaded in stages and lateral strains are prevented. Each load increment triggers a consolidation process similar to the one targeted in Terzaghi's theory. If the change in soil volumetric strain over time were reliably represented by the theoretical curve of Figure 4.12, there would be no difficulty in estimating the consolidation coefficient.

However, the experimental curves have significant deviations from the theoretical curve. The two main reasons for those differences are the following: i) at the start of the deformation process in the oedometer, (almost) immediate settlement occurs due to adjustments of the specimen in relation to the ring and to the compression of some air present in the system; and ii) at the end of the process, where the increase in the theoretical volumetric strain with time tends toward zero, other type of strains occur, due to creep, which are not related to the process described in Terzaghi's theory.

In this way, the experimental curves that express, for each load increment, a volumetric strain change over time, have to be corrected in order to be comparable with the theoretical curves, allowing the estimation of the consolidation coefficient. In Annex A4.4, two methodologies for this purpose are presented, proposed by Taylor (1948) and by Casagrande and Fadum (1940). For soils with relatively fast consolidation, the Casagrande method is more problematic, so Taylor's method is recommended.

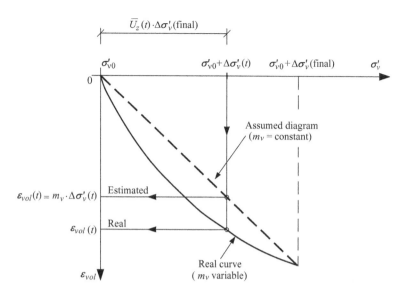

Figure 4.13 Calculation of the volumetric strains in a given instant from the average consolidation ratio obtained from Terzaghi's consolidation theory.

It should be mentioned that c_v changes with vertical effective stress, i.e., from one stage to another, during the test. Its value should then be adopted, taking into account the expected stress range in the specific problem under study.

The variation of c_v with the effective stress has a rather uniform pattern. Beneath the pre-consolidation effective stress, it is relatively high, being substantially reduced when the loading enters the virgin compression line, along which the variations are relatively modest.

This dependence of c_v on the effective stress reflects the variation in permeability and in compressibility. In the transition from the overconsolidated state to the virgin compression, the soil exhibits an abrupt increase in compressibility without significant variation in permeability (since, in that transition, the variation in void ratio is still small). Under those circumstances, according to Equation 4.20, the consolidation coefficient tends to be significantly lower. In virgin compression, the void ratio decrease affects both compressibility and permeability similarly, which makes the variation in the consolidation coefficient smaller.

4.3.5 Settlement change taking into account construction time

In all the previous considerations, the hypothesis that the clay layer loading is instantaneous was implicit, in agreement with the hydrodynamic analogy represented in Figure 4.3 and with the oedometer test.

However, in many cases, particularly in embankment construction, the time needed to achieve the maximum load may be significant when compared with the consolidation time. In such cases the change in settlement over time will differ from the theoretical curve of Figure 4.12.

A solution for this can be found, bearing in mind that the consolidation for each load increment proceeds independently of the processes due to any preceding and succeeding increments (Lambe and Whitman, 1979). That is, the superposition of effects can be applied.

For that purpose, there is a solution based on the following hypothesis (Taylor, 1948): i) during the construction period, the loading rate at the surface is constant, i.e., the load increases linearly with time; ii) the previous point is equivalent, in terms of settlement, to applying the total load at the middle of the construction period; and iii) the settlement is proportional to the applied load.

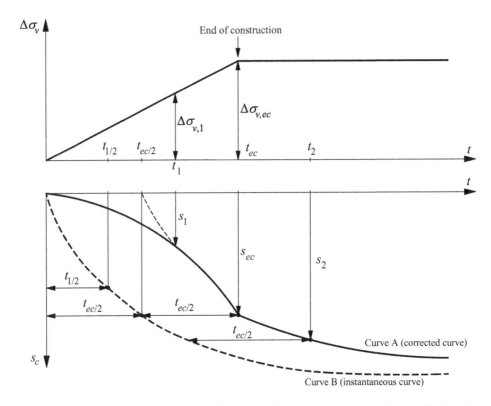

Figure 4.14 Construction of the curve consolidation settlement *versus* time graph, considering the construction period.

Figure 4.14 illustrates the application of the proposed methodology. The upper part of the figure illustrates the change of the pressure applied at the surface until a maximum value, $\Delta\sigma_{v,ec}$, is achieved for the instant corresponding to the end of construction, t_{ec}. The lower part of the figure shows the change in time of the consolidation settlement in the case of an instantaneous application of the loading (curve B, dashed line) and the corrected one, according to the methodology herein presented (curve A, solid line).

The point of curve B for the instant $t_{ec}/2$, corresponds, according to the aforementioned hypotheses, to the corrected settlement for the end of construction, s_{ec}, defining the point of curve A for t_{ec}.

For any instant after t_{ec}, for example t_2, the corrected settlement of curve A, s_2, is equal to the settlement of curve B at instant $t_2 - (t_{ec}/2)$.

For any instant before t_{ec}, for example t_1, the corrected settlement of curve A, s_1, is obtained from curve B settlement at the instant $t_1/2$ multiplied by the factor $\Delta\sigma_{v,1}/\Delta\sigma_{v,ec}=t_1/t_{ec}$. With this last criterion, selecting several instants lower than t_{ec}, other points of the corrected curve during the construction period are obtained.

4.4 LOADING IN GENERAL (NON-CONSTRAINED) CONDITIONS

4.4.1 Introduction: generalization of the hydromechanical analogy

In most practical problems, the loaded area at the ground surface is of the same order of magnitude as the depth and thickness of the loaded layers, as illustrated in Figure 4.15, which shows an embankment of considerable length (Figure 4.15a) and a cylindrical tank with flexible foundation (Figure 4.15b). Bearing in mind the explanations made in Section 4.2.1, it is easy to conclude that, in these cases, for any point outside the axis of symmetry of the loaded area, the displacement vector will have a non-null horizontal component.

In the case analyzed previously in this chapter, the hydromechanical problem is one-dimensional, that is, what happens in the soil points along any vertical line is representative of the whole phenomenon. In the problems represented in Figure 4.15, the hydromechanical problem becomes two-dimensional (Figure 4.15a) or three-dimensional (Figure 4.15b).

The loading in these *non-constrained conditions* consists of several phenomena related to the stress state, strain state, and time, which are substantially more complex than those previously analyzed. In fact, as a consequence of having the load now being applied to a ground surface area of finite dimensions:

i) the total stress increments become variable, with two or three coordinates of the soil mass point;
ii) the excess pore pressure generated therefore becomes variable with respect to two or three coordinates of the clay layer point;
iii) the subsequent flow, associated with the dissipation of that excess, becomes two-dimensional or three-dimensional;
iv) the surface settlement becomes variable with respect to one or two coordinates of the surface point, with reference to Figure 4.15a,b, respectively.

To better understand the new conditions, a generalization of the hydromechanical analogy presented in Section 4.2.2 (see Figure 4.3) was developed. This generalization is represented in Figure 4.16, and the difference, in relation to the primitive model, is that part of the reservoir wall is made by an elastic membrane.

Before loading (Figure 4.16a), the situation coincides with the one that exists in the primitive model, with the water in hydrostatic equilibrium, and the piston weight installed on the spring.

In the instant the additional load is placed on the piston (Figure 4.16b) and excess pore pressure is generated in the water, this leads to the deformation of the lateral membrane, allowing the downward movement of the piston in the same instant, still without any volume of water expelled. This allows the equilibrium of the additional load to be shared by the water – by means of the overpressure $u_e(0)$ – and by the spring, where the reduction of its length (by s_i) leads to an increment of the installed force.

If α (see Figure 4.16b) is the applied load fraction balanced by the water overpressure, then $1-\alpha$ will be the fraction balanced by the spring. What happens next (described in Figures 4.16c and 4.16d) is similar to what occurs in the primitive model. The difference of pressure in the water–atmosphere interface will induce the water to be expelled, with downward

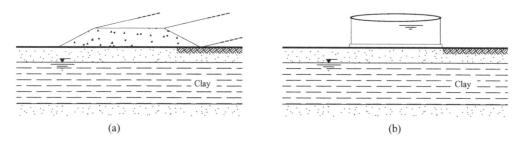

(a)   (b)

Figure 4.15 Examples of non constrained loading of clay layers: a) road embankment with large longitudinal development; b) cylindrical tank with flexible foundation.

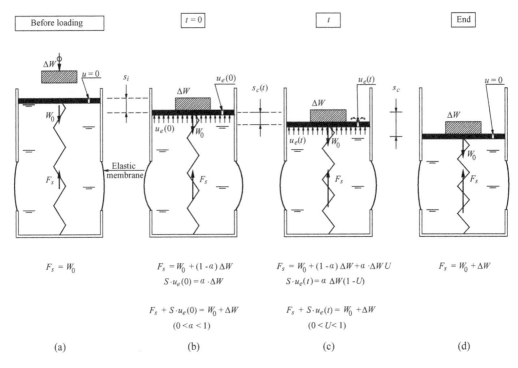

Figure 4.16 Generalisation of the hydromechanical analogy for non-constrained loading and consolidation of clay layers (Matos Fernandes, 2006).

movement of the piston and progressive transfer of the $\alpha\Delta W$ force to the spring. In each instant, t, of the process, the magnitude of U – which varies from 0 to 1 – defines the fraction of $\alpha\Delta W$ already transferred from the water to the spring.

This analogy allows us to better understand what happens in a generic point of a clay layer under non-constrained or general loading. Then:

i) the horizontal strains are not null anymore, allowing an *immediate* settlement concurrent with the load application, associated with a *distortion* and not to a volumetric deformation, which can only occur when the water is expelled, that is, by consolidation;

ii) as a consequence, at each point, the excess pore pressure generated is no longer equal to the total vertical stress increment, since this becomes divided, even in the loading instant, between the two soil phases, giving rise to an excess pore pressure and to a vertical effective stress increment.

4.4.2 Consolidation settlement calculation: classical method of Skempton and Bjerrum

Figure 4.17 illustrates, for the case of the cylindrical tank with flexible foundation described in Figure 4.15b, some issues related to the displacement and incremental stress distribution in the soil mass.

Figure 4.17a shows the typical shape of the vertical displacements of the ground surface and of the horizontal displacements along a vertical line under the limit of the loaded area. Since, at the time of loading, the clay volumetric variation is null (as instantaneous strains are purely distortional), the *immediate settlement*, s_i, under the loaded area is associated

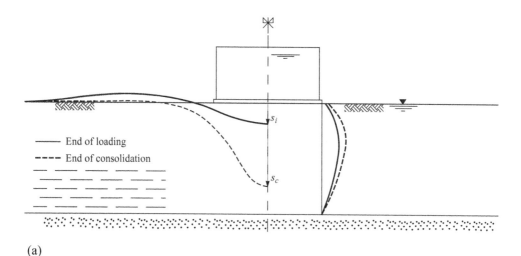

(a)

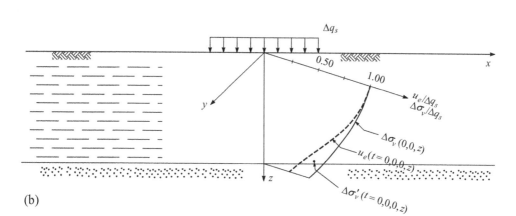

(b)

Figure 4.17 Three-dimensional loading of a clay layer: a) typical shape of surface settlements and horizontal displacements along a vertical line under the limit of the loaded area at the end of loading and at the end of consolidation; b) incremental total and effective vertical stress and excess pore pressure distributions under the center of the loaded area at the time of loading.

with a surface *heave* of the neighboring areas. Subsequently, the settlements tend to experience more or less significant increments with the development of consolidation. The distribution of the horizontal displacements typically has the shape represented, being, in general, more reduced at shallower depths due to the influence of the desiccated surface crust.

As an example, Figure 4.17b shows the incremental total vertical stress distribution under the center of the tank. It also shows, as an example, the distribution of excess pore pressure generated at the loading instant. The difference between the two graphs corresponds to the incremental vertical effective stress distribution at the same instant. From the two parts in which the incremental total stress graph is divided, the part related to the excess pore pressure will lead to *consolidation settlements*, whereas the part related to the effective stresses is associated with *immediate settlements*.

The immediate settlement calculation, s_i, will be dealt with in Chapter 6 (see Section 6.3.2). In what follows, the estimation of the consolidation settlement is considered.

The calculation of the consolidation settlement in the present case is more complex than what was seen for the one-dimensional loading. At present, calculation models based on the finite element method can be used, taking into consideration the stress–strain–time relations of the soil by means of constitutive models more or less sophisticated (Venda Oliveira et al., 2010). These models require user expertise for a successful application.

A classical and simplified solution for the problem was proposed by Skempton and Bjerrum (1957). This solution is based on the hypothesis that – after the undrained loading, which causes distortions to the soil mass, and consequently induces strains on the horizontal direction – the consolidation process is developed without significant horizontal strains, i.e., in way similar to a confined loading. This is suggested in Figure 4.17a, where the horizontal displacements during consolidation along a vertical line under the limit of the loaded area are not very different from the ones caused by undrained loading.

The hypothesis presented is based on rational considerations, discussed by Skempton and Bjerrum (1957), and in agreement with observed results of numerous cases (Tavenas et al., 1979; Magnan et al., 1983; Leroueil et al., 1990; Venda Oliveira et al., 2010). Tavenas et al. (1979) analyzed in detail the lateral strains induced by embankments over soft soils, under short- and long-term conditions. It was concluded that the magnitude and distribution of the lateral displacements are affected by several factors, such as the thickness of the overconsolidated surface crust, secondary consolidation (creep), among others. For the layers where the clay soil is normally consolidated, a distribution, according to the elasticity theory, seems to prevail. Within the behavioral diversity observed in the 21 analyzed case studies, Tavenas et al. (1979) reported that the incremental horizontal displacement of the consolidation process in the first five years represented around 16% of the incremental vertical displacement; this percentage tends to be lower for longer periods.

Considering the hypothesis that, during consolidation, the horizontal displacements are small, the compressibility parameters, defined in Section 4.2.5 and obtained from oedometer tests, will still be appropriate for an estimate of the consolidation settlement in the present case. As explained above, the settlement is calculated for a given point of the ground surface (of coordinates x and y), generally at the center of the loaded area.

In this way, the application of the Skempton and Bjerrum method can be carried out according to the following steps:

1. calculation of the total stress increments for each depth, usually based on the elasticity theory solutions;
2. estimation of the excess pore pressure generated immediately after the loading, Δu (z, $t = 0$); this excess pore pressure also represents the vertical effective stress increment during consolidation, $\Delta\sigma'_v$ (z, t = end of consolidation); and
3. calculation of the consolidation settlement by means of Equations 4.10 to 4.14.

The excess pore pressure generated by the loading can be estimated by means of the Skempton equation:

$$\Delta u(z, t = 0) = \Delta\sigma_3(z) + A\left[\Delta\sigma_1(z) - \Delta\sigma_3(z)\right] \tag{4.31}$$

or from *in situ* observation results, at the end of loading, which would represent a more reliable alternative.

The consolidation settlement change over time will be treated in the following sections.

4.4.3 Two- and three-dimensional consolidation

4.4.3.1 Biot theory results

In the study of the change in time of the consolidation settlement when the loading is not constrained, the hypotheses of Terzaghi's theory are not valid anymore, basically because the flow is no longer exclusively vertical (which is the reason why Terzaghi's theory is also known as the one-dimensional consolidation theory) to become two or three dimensional (in the cases of Figures 4.15a, and b, respectively). In fact, once the excess pore pressure at a certain depth varies from point to point, hydraulic gradients are established also in the horizontal directions. These phenomena are treated by means of Biot theory (1935, 1941a,b) the application of which requires the numerical integration of a differential equation system (for which the finite element or finite difference methods are used).

Although Biot theory is outside the scope of this book, it is interesting to present and discuss some of its solutions.

Figure 4.18 presents the solution for a uniform vertical loading, Δq_s, applied to an infinite strip, considering a half-space consolidation, and limited by a draining upper boundary (Schiffman et al., 1967). For a point under the axis of the strip, the figure illustrates the response with the time factor of: i) the vertical and horizontal total and effective stress increments (Figure 4.18a); and ii) the mean total and effective stresses, and the excess pore pressure (Figure 4.18b). The mentioned stresses are normalized by Δq_s.

In this two-dimensional consolidation problem solved by Biot theory, it is important to notice that, at the time of loading $(T = 0)$:

i) the total vertical stress increment at the mentioned point is divided into a given verti-cal effective stress increment and into a given pore pressure increment (excess pore pressure);

ii) the horizontal total stress increment is lower than the excess pore pressure generated, and, consequently, the horizontal effective stress decreases.

It is curious to note that, conversely to what happens in Terzaghi's theory, during the consolidation process, the vertical total stress does not remain constant at the analyzed point, and that the excess pore pressure increases slightly before tending toward zero. The phenomenon is described in the literature as the *Mandel-Cryer effect* and it has a simple physical explanation (Cryer, 1963). In fact, the initial excess pore pressure decreases with the distance to the symmetry plane, since the incremental mean stress decreases over that distance. On the other hand, for the points at a greater distance from the mentioned axis, the excess pore pressure dissipation is faster and consequently the volumetric strains are faster too. This differential variation of the volumetric strains will generate a redistribu-tion of the total stresses, with an increase in the area that has deformed less in a given instant, that is, the central zone. Naturally, this total stress increment will generate new excess pore pressures beyond those generated initially. With time, the excess pore pressure will tend to zero, and hence, the effective stress increments will tend to be equal to the total stress increments.

In the two- and three-dimensional problems, the key issues or parameters for the change in the consolidation ratio with the time factor, in the comparison with the solution of Terzaghi's consolidation theory, are:

i) the relative dimensions of the loaded area and the thickness of the clay layer;

ii) the permeability anisotropy of the clay layer.

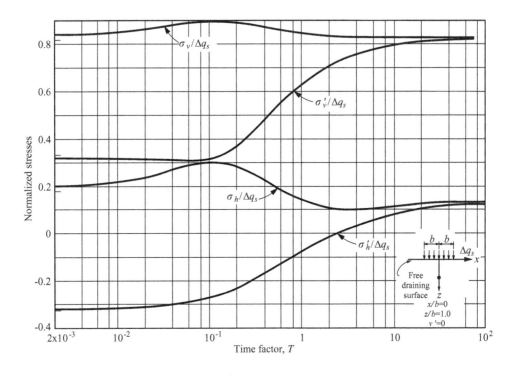

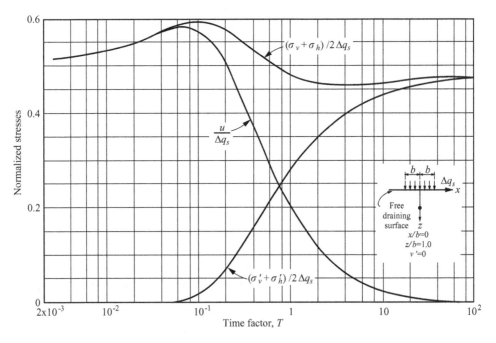

Figure 4.18 Two-dimensional consolidation under a uniformly loaded infinite strip: solution presented by Schiffman et al. (1967) based on Biot theory (reproduced from Lambe and Whitman, 1979).

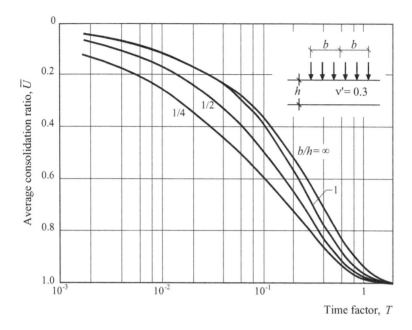

Figure 4.19 Influence of the uniformly loaded infinite strip width on the change of the average consolidation ratio underneath the center of the strip when the base of the clay layer is impervious (adapted from Correia, 1982).

With respect to the first issue, Figure 4.19 illustrates the change in response to time of the average consolidation ratio underneath the central point of an infinite strip under uniform loading, as a function of the ratio of the strip semi-width by the thickness of the layer, with an impervious base (Correia, 1982). The solution for $b/h = \infty$ almost coincides with the one from Terzaghi's theory. As expected, the lower the ratio b/h, the faster the consolidation, because the importance of pore pressure dissipation by drainage in the horizontal direction progressively increases. The figure also shows that, for $b/h \geq 1$, the solution is quite close to the one-dimensional theory solution, especially for low values of the time factor. It should be noted, however, that for other shapes of the loaded area, such as circular areas, the deviation from the one-dimensional solution is very substantial for ratios $b/h \geq 1$ (where b is the radius of the loaded area). Similar results were also obtained by Davis and Poulos (1972).

To evaluate the effect of the permeability anisotropy, Figure 4.20 illustrates, for one of the conditions of the previous figure, the change, relative to time, of the average consolidation ratio underneath the central point of the strip for several ratios between horizontal and vertical permeability. As expected, the permeability anisotropy accelerates the consolidation in a very significant way, and is of greater importance when the base layer is impervious.

It should be mentioned that very high values of the ratio k_h/k_v, such as the ones considered in the figure, are likely to happen in real cases. This aspect is fundamentally due to what can be designated as *macrofabric* typical of many clay deposits. As a consequence of the environmental variations, in which the corresponding sedimentation has been processed, such deposits frequently contain very thin layers of coarser material, which are thus much more permeable, and the presence of which is susceptible to accelerating the consolidation, by drainage through horizontal planes, and by reducing the path of the vertical flow.

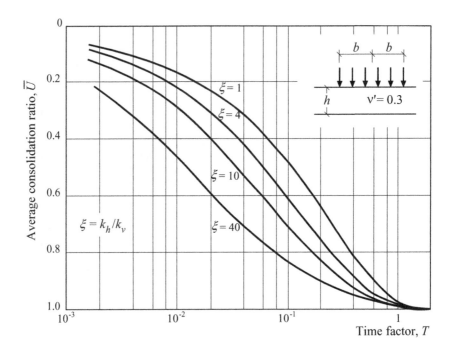

Figure 4.20 Influence of permeability anisotropy on the change of the average consolidation ratio underneath the center of a uniformly loaded infinite strip for $b/h = 1/2$ when the base of the clay layer is impervious (adapted from Correia, 1982).

4.4.3.2 Solutions of Terzaghi's theory for any distributions of the initial excess pore pressure

From what has been mentioned above, it can be concluded that the use of solutions from the classical Terzaghi's consolidation theory for cases where the consolidation phenomenon is two or three dimensional will lead to an overestimation of the consolidation time. However, there are cases where the use of those solutions can be useful, because the geometric conditions are not very different from the ones inherent to that theory, or because the aim is to obtain, as a first approximation, an estimation of *an upper limit* of the consolidation time.

With that perspective, it is interesting to have solutions of Equation 4.23 for non-uniform depth distributions of the initial excess pore pressure. For the case of the double draining boundary, several solutions are included in Table 4.4. Please remember that the solutions of the symmetrical distributions, in relation to the layer center, can still be applied in the case of a single draining boundary, in a way similar to that discussed in Section 4.3.2.2.

Since the differential consolidation equation is linear, the corresponding solutions can be added. Hence, if the distribution of the initial excess pore pressure, for a given case, can be represented by the algebraic sum of simple distributions, the solutions of which are known, then its average consolidation ratio, at a certain instant, can be calculated from the consolidation ratios corresponding to those solutions. This calculation is made by a weighted average of the consolidation ratios for the elementary cases, being the weight that affects each consolidation ratio equal to the area of the corresponding graph, representative of the distribution of the initial excess pore pressure.

The aforementioned procedure is useful, for example, to obtain the average consolidation ratio when a linear initial distribution occurs in a given layer with a single draining boundary, as illustrated in Figure 4.21.

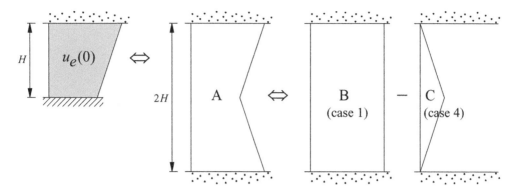

Figure 4.21 Composition of solutions for Terzaghi's consolidation equation (Perloff, 1975).

This distribution can be considered to be the algebraic sum of a uniform (rectangular) distribution (B), with a triangular distribution (C). Therefore, the average consolidation ratio, in a given instant t, for case A, i.e., $\bar{U}_{zA}(t)$, can be calculated from the average consolidation ratios, in the same instant, for cases B and C, namely $\bar{U}_{zB}(t)$ and $\bar{U}_{zC}(t)$, respectively, by the expression:

$$\bar{U}_{zA}(t) = \frac{\bar{U}_{zB}(t) \cdot \text{Area B} - \bar{U}_{zC}(t) \cdot \text{Area C}}{\text{Area A}} \tag{4.32}$$

This procedure can be applied to any number of elementary areas.

4.5 SECONDARY CONSOLIDATION

4.5.1 Introduction

As mentioned in Section 4.3.4, after the load transfer process from the pore water to the solid skeleton, the clay volumetric strains are not completely stabilized. This can be verified, for example, in the experimental graphs of Annex A4.4 (see Figures A4.4.1b and A4.4.2b).

This last deformation process, under constant effective stress, results from the *creep* of the solid skeleton and of the adsorbed water layers, where readjustments continue over a very long period of time. This process is called *secondary consolidation*. To emphasize the difference, the consolidation process treated in previous sections is called *primary* or *hydrodynamic consolidation*.

Secondary consolidation can only occur as a result of the water outlet from the soil. As a consequence, there will necessarily be excess pore pressures and very low hydraulic gradients, corresponding to a flow of very low speed, that is not ruled by Darcy's law.

The settlements resulting from this process can be particularly significant in geologically recent clay deposits with high plasticity and/or high organic matter content. On the contrary, settlements are generally negligible in overconsolidated deposits.

4.5.2 Overconsolidation by secondary consolidation

As Figure 4.22 shows (Bjerrum, 1972), secondary consolidation comprises a progressive reduction of the soil void ratio under constant effective stress. This reduction in void ratio, and thus in water content, leads to greater strength and reduced compressibility. Hence, over time, the clay develops a reserve resistance in relation to a future loading. For that reason,

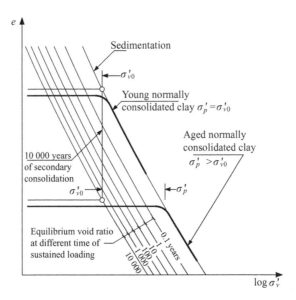

Figure 4.22 Secondary consolidation effect on a clay compressibility curve (Bjerrum, 1972, with permission from ASCE).

as the figure illustrates, the behavior of a given normally consolidated clay deposit will be different, depending on whether the loading has occurred right after its formation or after a significant period of secondary consolidation. In the latter case, the soil will exhibit, in a $\log\sigma'_v - e$ graph, an abrupt increase in compressibility for a σ'_p stress greater than σ'_{v0}. In practice, the clay behaves as if it has been submitted to effective stresses between σ'_{v0} and σ'_p, that is, it behaves as an overconsolidated soil (Bjerrum, 1972).

This phenomenon explains the behavior of many Holocene clay deposits, the stress history of which have never included effective stresses higher than the at-rest ones. Typically, these soils have OCR values lower than 2.0.

Commenting on the figure, the various lines parallel to the one that represents the behavior of the very recent clay (without secondary consolidation), and increasingly distant from it, are equivalent to increasing OCR values, in the context mentioned above.

4.5.3 Secondary consolidation settlement: conventional and Brazilian approaches

Considering that the settlement varies linearly with the logarithm of time during secondary consolidation, the void ratio variation can be expressed by equation:

$$\Delta e = -C_\alpha \log\frac{t_2}{t_1} \tag{4.33}$$

where C_α is the *coefficient of secondary consolidation* or the *coefficient of secondary compression* (dimensionless parameter), t_2 is the instant for which the settlement is required, and t_1 is the instant corresponding to the end of primary consolidation (for t_1, the variable "t_{90}" is sometimes used, i.e., the time corresponding to an average consolidation ratio of 90%).

The *secondary consolidation settlement*, s_d, can then be calculated by the expression:

$$s_d = \frac{h_1}{1+e_1} \cdot C_\alpha \log\frac{t_2}{t_1} \tag{4.34}$$

Table 4.5 Correlation between the secondary consolidation coefficient and the compression index (Mesri, 2001).

Type of soil	C_α/C_c
Granular soils, including rockfills	0.02 ± 0.01
Laminated claystones and siltstones	0.03 ± 0.01
Inorganic clays and silts	0.04 ± 0.01
Organic clays and silts	0.05 ± 0.01
Peat	0.06 ± 0.01

Table 4.5 includes a correlation between the coefficient of secondary consolidation and the compression index (Mesri, 2001).

The approach presented for the secondary consolidation settlement is a simplification of the phenomenon. Firstly, it assumes that the secondary settlement is processed only after the primary one, which is not a reasonable occurrence, although its consideration does not probably lead to gross error in most cases. On the other hand, consolidation tests of very long duration and theoretical studies show that the secondary compression has a really complex (non-linear) evolution with the logarithm of time and, most importantly, it tends toward zero (Martins et al., 1997).

Figure 4.23 supports this statement, showing results of three long-duration tests (around 600 days) on specimens of a Holocene era clay loaded in an oedometer. The applied loads

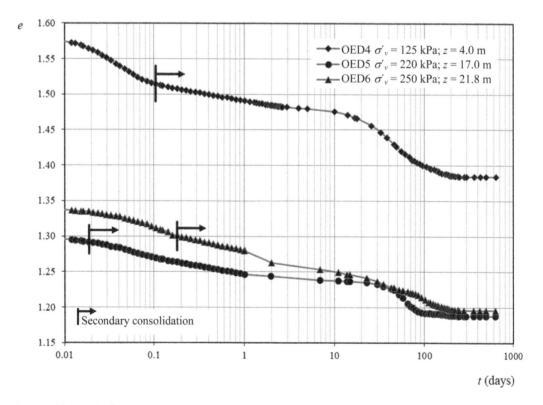

Figure 4.23 Results from three long-duration tests on undisturbed specimens of a Holocene clay loaded in an oedometer (Esteves, 2013).

correspond to a virgin compression for each specimen. It can be verified that, during approximately the second half of the loading period, the volumetric strains were almost null. It can also be observed that the slopes of the curves have a complex evolution until they become null.

From experimental results, such as the ones of this figure, the selection of C_α, to estimate the secondary consolidation settlement along a given time period, is very difficult.

As an alternative, it seems more reasonable to determine an estimate of a *maximum limit* of the secondary consolidation settlement. A proposal for this was presented by Martins (2014) based on theoretical and experimental studies on Holocene Brazilian clays. These studies suggest that the secondary consolidation is complete when the line corresponding to a OCR value of 1.6 is reached. Independently of a specific OCR value (which probably differs from soil to soil), the base idea of this proposal is reasonable, and it has empirical validation. Alonso et al. (2000) state that *"both laboratory and field observations show that there is a sharp reduction in the value of the secondary compression coefficient with only small degrees of overconsolidation".*

Figure 4.24 details the way to identify the line corresponding to OCR equal to 1.6 in the conventional experimental compressibility curve (corresponding to the end of a 24-h of loading in each increment), and, from it, the variation in void ratio corresponding to secondary consolidation:

$$\Delta e_d = -\left(C_c - C_r\right)\log\frac{1.6\sigma_v'\left(final\right)}{\sigma_v'\left(final\right)} = -0.204\left(C_c - C_r\right) \tag{4.35}$$

This leads to the upper limit of the settlement, s_d:

$$s_{d,max} = 0.204\frac{h}{1+e}\left(C_c - C_r\right) \tag{4.36}$$

This methodology can be easily adapted to other OCR values.

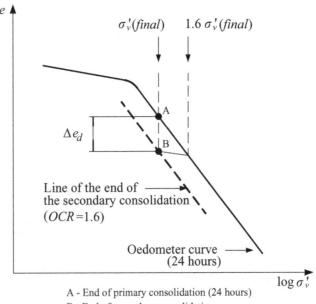

Figure 4.24 Proposal to identify the end of secondary consolidation and the corresponding settlement (adapted from Almeida and Marques, 2018).

In the present state of the art, this perspective of secondary consolidation, although less ambitious, as it does not establish a link between a given settlement and a given time period, seems more reasonable, and, from a practical point of view, satisfactory in most cases. It has an additional important advantage, in that it allows to plan eventual pre-loading operations to eliminate secondary consolidation. This subject will be discussed further below (see 4.6.2.3).

4.6 ACCELERATION OF CONSOLIDATION

4.6.1 Introduction

In most practical cases, delayed consolidation settlements are highly inconvenient. For that reason, processes to *accelerate consolidation*, and thus the settlements, are frequently used.

Sometimes consolidation acceleration can have another objective, complementary or not to the previous one. In fact, as consolidation leads to a reduction in void ratio and water content of the clayey soil, it will stabilize its structure in a new arrangement, which increases the soil shear strength. The sense of the word "consolidation" is due to that fact. For that reason, in certain cases, the use of procedures that allow the acceleration of consolidation is made to enable the application of loads, the magnitude of which is not compatible with the soil strength in its natural state. This issue was already discussed in Chapter 2 (see Section 2.4.3).

4.6.2 Preloading

4.6.2.1 Base scheme

The oldest procedure to accelerate consolidation is, probably, *preloading*. This method consists in building an embankment able to transmit to the ground mass a load higher than the one that will be transmitted in the permanent stage. The temporary excess load will only be removed after settlement is verified of magnitude similar to the one that would be predicted for the intended work to be built. This process requires, obviously, that the ground mass has enough strength for the excess load that is applied to it, a condition that sometimes prevents its use, at least individually.

Figure 4.25 allows a more detailed analysis. Considering a constrained loading of a normally consolidated soil mass, s_c (*perm*) and s_c (*perm + temp*) are the consolidation settlements, respectively induced by the surcharge associated to the structure to be built, and by the same structure and an additional (temporary) surcharge. These loads are associated to incremental vertical stresses on the surface, respectively, Δq_s (*perm*) and Δq_s (*perm + temp*). As Figure 4.25a indicates, assuming by simplification that the consolidation coefficient, c_v, is constant, the evolution in time of the settlements is similar, i.e., a given percentage of the final values, s_c (*perm*) and s_c (*perm + temp*), will happen at the same instant.

A settlement of magnitude s_c (*perm*) will occur for the loading Δq_s (*perm + temp*) in a certain instant t, for which the corresponding consolidation ratio is lower than 100%. Assuming a one-to-one relationship between the volumetric strain (and thus the consolidation settlement) and the effective stress installed in the soil, it can be stated that, at instant t, the incremental vertical stress associated with the permanent structure has already been converted to effective stress. Therefore, as Figure 4.25b suggests, if Δq_s (*temp*) is removed at the instant t, it is expected that, after that, no more settlements are processed.

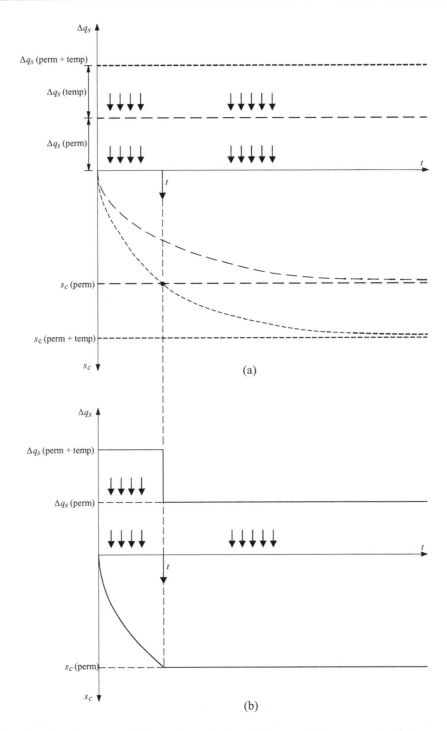

Figure 4.25 Acceleration of consolidation by preloading: a) loading with permanent and temporary surcharges and evolution in time of the corresponding settlements, for the case of loading during enough time for consolidation to complete (in both cases); b) loading really processed and required evolution of the consolidation settlement.

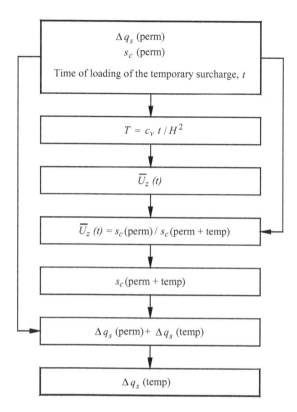

Figure 4.26 Pre-loading flowchart for the calculation of the surcharge load that, in a given period of time, *t*, leads to the required settlement.

4.6.2.2 Calculation of the temporary surcharge

In practical problems, the period *t* for the application of the temporary surcharge is, in general, a problem constraint, being necessary to determine the magnitude of the surcharge (in general, the height of the fill) capable of inducing the desired settlement in that period. Figure 4.26 illustrates the steps necessary to calculate that surcharge magnitude. It is easily recognized that, by an inverse sequence to the one illustrated, it is possible to obtain the necessary application period for a given value of the temporary surcharge.

4.6.2.3 Comment: control of secondary consolidation settlements

The basic idea of the preloading method, described previously, consists of the following: the imposition of a given settlement results in the installation of a certain effective stress increment in the soil, related by Equation 4.14 or similar. With the installation in the soil solid skeleton of the increment equivalent to the weight of the designed work, future consolidation settlements are prevented. Although this idea is not rigorous, the proposed method has been applied with relative success in a great number of works!

The idea described is not rigorous since the *one-to-one relation* between the volumetric strain and the effective incremental stress is only true at the level *of a given point of the soil mass assumed to be normally consolidated* (and ignoring any secondary consolidation) and not to the whole consolidating layer. In fact, as already discussed (for an example, see Figure 4.10), the (primary) consolidation is processed at different rates in different levels of the clay layer. It can be then stated that, at the instant t mentioned, there will be, in fact, zones of the clay mass where the incremental effective vertical stress is higher than Δq_s (*perm*), whereas, in others, it will be lower.

For this reason, a recommended conservative measure is the verification of a consolidation settlement higher than s_c (*perm*) as a criterion for the removal of the temporary surcharge (Jiménez Salas, 1980).

In Section 4.5.3, what was designated as the Brazilian approach was discussed, which indicates the end of secondary consolidation at a point of the log σ'_v –e graph corresponding to a certain *OCR* value (specifically 1.6), and it was highlighted that this basic idea was supported by experience, which showed that secondary consolidation becomes negligible for soils with relatively low *OCR* values (Alonso et al., 2000). Thus, overconsolidation necessary to account for future settlements by secondary consolidation can be induced by preloading.

Figure 4.27 supports an understanding of this issue. Consider a geologically recent normally consolidated clay, which will be loaded by the surcharge Δq_s (*perm*), following in the graph a path from point 0 to point 1 along the virgin compression line. Without preloading, the soil would experience a progressive decrease in the void ratio due to secondary consolidation, becoming progressively more distant from point 1 by means of a vertical path, as represented in the figure (see descending arrows). With the application of an additional surcharge in relation to Δq_s (*perm*), the soil will follow the virgin compression line from 0 to 2. Removing the additional surcharge, the soil changes from 2 to 3, becoming overconsolidated.

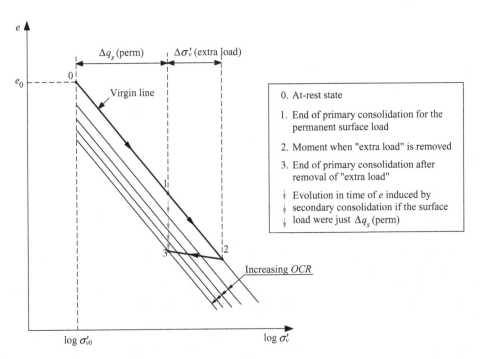

Figure 4.27 Scheme illustrating the pre-loading effect on the anticipation of the secondary consolidation settlement.

If the overconsolidation ratio, $OCR = \sigma'_v(2)/\sigma'_v(3) = \sigma'_v(2)/\sigma'_v(1)$, achieves a certain value, the soil will not experience further settlements by secondary consolidation, and therefore the soil situation becomes, let's say, *frozen* in point 3.[1]

In conclusion: preloading is a method that allows combination of the acceleration of the primary consolidation with the anticipation of secondary consolidation settlements.

After estimating the secondary consolidation settlement for a given period of the structure's service life (or its upper limit, by means of the Brazilian approach), a practical criterion for the removal of the extra load is to take the instant for which the observed settlement is the sum of the expected primary consolidation settlement for the permanent load, $s_c(perm)$, with the secondary consolidation settlement, $s_{d,max}$.

4.6.3 Vertical drains

4.6.3.1 General scheme: types of drains – installation

A process often used in combination with the previous method, consists of the introduction of *vertical drains* of high-permeability material into the soil mass to accelerate consolidation, as illustrated in the scheme of Figure 4.28. The advantages of using vertical drains are mainly the following:

i) in addition to the vertical flow towards the upper and lower draining boundaries of the layer in consolidation, there will be *horizontal flow* to the drains;
ii) the path of the water to leave the clay layer is shortened;
iii) the horizontal flow to the drains runs at a higher velocity than does the vertical flow, since typically k_h is higher than k_v.

As shown in Figure 4.28, before the introduction of the drains, a granular drainage blanket is installed at the ground surface. This layer has a double function: i) to be a working platform for the equipment that will install the drains, with a minimum thickness of around 0.5 m generally being required; and ii) to ensure the free flow of water collected by the drains, so that this layer needs to be made of a highly permeable coarse soil.

In the first applications, the vertical drains were made by *sand columns* but, in most cases, these were quickly replaced by synthetic drains. Figure 4.29 describes schemes and useful information about construction issues. The most up-to-date synthetic drains (Figure 4.29a) are designated as wick drains (or band drains) (with a typical cross section of 95 mm by 4

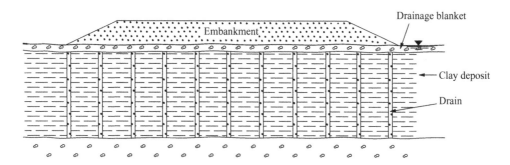

Figure 4.28 General scheme of a vertical drain system to accelerate consolidation.

---

[1] According to the so-called Brazilian approach, the minimum OCR value should be 1.6.

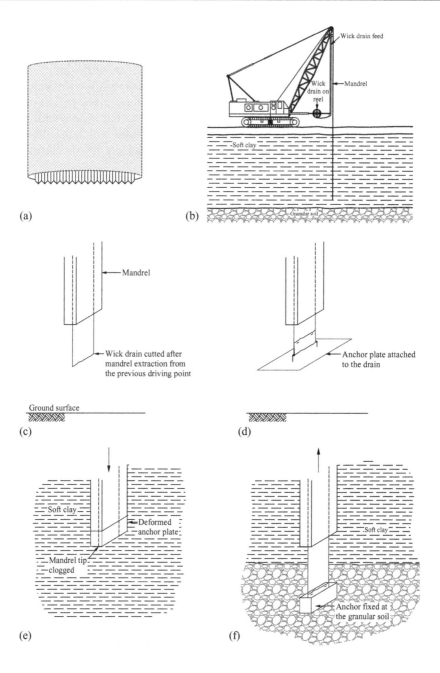

Figure 4.29 Execution of synthetic vertical drains: a) wick drain; b) driving equipment; c) mandrel tip with the drain; d) drain and anchor before driving; e) drain and anchor during driving; f) drain and anchor at the start of mandrel withdrawal.

mm), made of a plastic core wrapped in a geotextile which works as a filter; the core is waved creating in the two faces a certain number of grooves. Figure 4.29b shows, in a general way, the equipment that installs the drains. The vertical steel piece that is (generally) statically driven into the ground is called a mandrel and it consists of a hollow, slender bar, rectangular in cross section, with the drain inside (Figure 4.29c). Before introducing the drain into the ground, it is necessary to apply an anchor plate on its end, as shown in Figure 4.29d. The anchor, generally made of a fine steel rectangular plate slightly larger than the mandrel cross section, has two main roles: i) at the start of the driving, the anchor is deformed, closing the mandrel mouth, so that the soil is prevented from intruding into the mandrel (Figure 4.29e); ii) after reaching the layer that underlies the soft clay layer and before removing the mandrel, the anchor is detached from the mandrel, keeping the drain in position (Figure 4.29f). At the end of mandrel withdrawal, the drain is cut close to the ground surface and the equipment is moved to the new driving point (Figure 4.29c).

4.6.3.2 Radial consolidation

The mathematical formulation of the consolidation problem, involving vertical drains, was developed by Barron (1944, 1947).

Figure 4.30a illustrates two current plan setups of the drains: a square mesh and a triangular mesh. To analyze the problem, it is assumed, in a simplified way, that each drain has a cylindrical zone of influence, the volume of which is equal to the real zone of influence, as suggested by Figure 4.30a,b. In each cylinder of soil around a drain, a consolidation process independent of the neighboring cylinders takes place. This process consists, in fact, of two coupled processes: i) vertical consolidation, as studied before, which develops as if no drains exist, directed to the upper and lower draining boundaries that limit the clay layer; and ii) radial consolidation, in each horizontal section of the cylinder, directed to the drain, which coincides with the cylinder axis.

The equation ruling the vertical consolidation:

$$c_v \cdot \frac{\partial^2 u_e}{\partial z^2} = \frac{\partial u_e}{\partial t} \tag{4.37}$$

was already treated and discussed in Section 4.3.1.

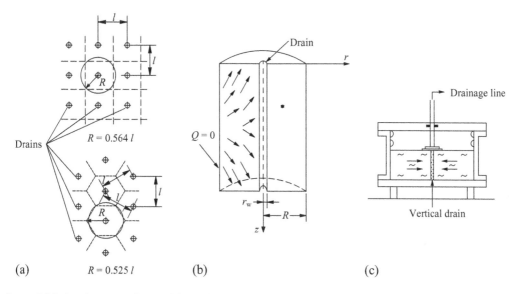

Figure 4.30 Acceleration of consolidation by vertical drains: a) drain arrangement in plan layout; b) flow inside a drain influence zone; c) Rowe cell for evaluation of c_h.

To develop the equation determining the radial consolidation, Barron (1944, 1947) introduces three additional hypothesis to Terzaghi's one-dimensional consolidation theory (see Figure 4.30b):

i) at any $t \neq 0$, $u_e(t) = 0$ for $r = r_w$ $\qquad\qquad\qquad\qquad\qquad\qquad\quad$ (4.38)

        thus, the excess pore pressure in the drain is null;

ii) at any $t \neq 0$, $\partial u_e / \partial r = 0$ for $r = R$ $\qquad\qquad\qquad\qquad\qquad\qquad$ (4.39)

        thus, there is no flow crossing the external surface of the cylinder;

iii) the hydraulic resistance of the drains is null, i.e., its capacity to remove the water coming from the soil is unlimited.

Bearing in mind these hypotheses, the ruling equation of radial consolidation is the following:

$$c_h \cdot \left[\frac{\partial^2 u_e}{\partial r^2} + \frac{1}{r} \cdot \frac{\partial u_e}{\partial r} \right] = \frac{\partial u_e}{\partial t} \qquad\qquad\qquad\qquad\qquad (4.40)$$

where

$$c_h = \frac{k_h}{\gamma_w \, m_v} \qquad\qquad\qquad\qquad\qquad\qquad\qquad\qquad\qquad\quad (4.41)$$

is the *coefficient of horizontal or radial consolidation*. This coefficient, which is typically higher than c_v due to the anisotropy of the soil permeability, can be estimated by the so-called Rowe cell (Rowe and Barden, 1966), represented in Figure 4.30c. This is a constrained compression test with flow to a central (or peripheral) drain.

The phenomenon, illustrated in Figure 4.30b, is, as a whole, ruled by a combination of Equations 4.37 and 4.40:

$$c_v \cdot \frac{\partial^2 u_e}{\partial z^2} + c_h \left[\frac{\partial^2 u_e}{\partial r^2} + \frac{1}{r} \cdot \frac{\partial u_e}{\partial r} \right] = \frac{\partial u_e}{\partial t} \qquad\qquad\qquad (4.42)$$

In order to solve a specific problem, Equations 4.37 and 4.40 are used separately, and then their solutions are combined, using the following equation (Carrillo, 1942; Scott, 1963):

$$1 - \bar{U}(t) = \left[1 - \bar{U}_z(t) \right] \cdot \left[1 - \bar{U}_r(t) \right] \qquad\qquad\qquad\qquad (4.43)$$

where $\bar{U}(t)$ is the average consolidation ratio at a given instant t, and $\bar{U}_z(t)$ and $\bar{U}_r(t)$ are, respectively, the average consolidation ratios for vertical and radial consolidation at the same instant.

Table 4.6 includes the solution of Equation 4.40, providing the average radial consolidation ratio, $\bar{U}_r$, as a function of the ratio between the radius of the cylindrical influence zone and the drain radius:

$$n = \frac{R}{r_w} \qquad\qquad\qquad\qquad\qquad\qquad\qquad\qquad\qquad\quad (4.44)$$

Table 4.6 Solution of the radial consolidation equation.

Average consolidation ratio $\overline{U}_r$ (%)	Time factor, T_r										
	$\dfrac{R}{r_w} = 5$	10	15	20	25	30	40	50	60	90	100
0	0	0	0	0	0	0	0	0	0	0	0
5	0.024	0.040	0.051	0.058	0.063	0.068	0.075	0.081	0.086	0.090	0.093
10	0.049	0.083	0.104	0.119	0.130	0.140	0.155	0.167	0.176	0.184	0.191
15	0.076	0.128	0.160	0.183	0.201	0.216	0.239	0.257	0.272	0.284	0.295
20	0.104	0.176	0.220	0.251	0.276	0.296	0.328	0.353	0.373	0.390	0.405
25	0.135	0.227	0.284	0.324	0.356	0.382	0.423	0.455	0.481	0.503	0.523
30	0.167	0.281	0.352	0.402	0.441	0.474	0.525	0.564	0.597	0.624	0.648
35	0.202	0.340	0.425	0.485	0.533	0.572	0.634	0.681	0.721	0.754	0.782
40	0.239	0.403	0.503	0.576	0.632	0.678	0.751	0.808	0.854	0.894	0.928
45	0.280	0.472	0.589	0.674	0.740	0.794	0.879	0.946	1.000	1.046	1.086
50	0.325	0.547	0.683	0.781	0.858	0.920	1.019	1.096	1.159	1.213	1.259
55	0.374	0.630	0.787	0.900	0.988	1.060	1.174	1.263	1.336	1.397	1.450
60	0.429	0.723	0.903	1.033	1.134	1.216	1.348	1.449	1.533	1.603	1.664
65	0.492	0.828	1.035	1.183	1.299	1.394	1.544	1.661	1.756	1.837	1.907
70	0.564	0.950	1.187	1.357	1.490	1.598	1.771	1.904	2.014	2.107	2.187
75	0.649	1.094	1.366	1.562	1.715	1.840	2.039	2.193	2.319	2.426	2.518
80	0.754	1.270	1.586	1.814	1.991	2.137	2.367	2.546	2.692	2.816	2.923
85	0.888	1.497	1.870	2.138	2.347	2.519	2.790	3.001	3.173	3.319	3.446
90	1.078	1.817	2.269	2.595	2.849	3.057	3.386	3.642	3.852	4.029	4.182
95	1.403	2.364	2.953	3.376	3.706	3.977	4.406	4.739	5.011	5.242	5.441
99	2.156	3.634	4.539	5.190	5.698	6.114	6.773	7.285	7.703	8.058	8.365
100	∞	∞	∞	∞	∞	∞	∞	∞	∞	∞	∞

and the time factor for radial consolidation:

$$T_r = \frac{c_h t}{R^2} \tag{4.45}$$

The following approximate solutions can also be used (Hansbo, 1981):

$$\bar{U}_r = 1 - e^{\frac{-2T_r}{F(n)}} \tag{4.46}$$

and

$$T_r = -\frac{1}{2} F(n) \ln\left(1 - \bar{U}_r\right) \tag{4.47}$$

where

$$F(n) = \frac{n^2}{n^2 - 1} \ln(n) - \frac{3n^2 - 1}{4n^2} \approx \ln(n) - 0.75 \tag{4.48}$$

In the case of the band or wick drains, in Equation 4.44, a value of r_w, corresponding to a cylinder whose lateral surface is coincident with the one of the drain, can be taken (Hansbo, 1981):

$$r_w = \frac{b + t}{\pi} \tag{4.49}$$

where b and t are the width and the thickness of the wick drain, respectively.

On the hydraulic resistance of the drains, Barron (1947) has also developed a radial consolidation equation where that resistance is considered. This equation is not presented here because the synthetic drains available on the market are designed to work as if, in practice, their draining capacity was unlimited (Hansbo, 2004).

4.6.3.3 Smear effect

The radial consolidation equation presented above does not take into account the so-called smear effect caused by mandrel driving. In fact, during driving, there is a soil zone immediately around the mandrel that is remolded and, for that reason, it can become less permeable, thus delaying consolidation.

Barron (1947) and Hansbo (1981, 2004) have dealt with this issue, generalizing the radial consolidation equation. Figure 4.31 compares both situations, with and without the smear effect, with the cylinder radius corresponding to the remolded zone designated as r_s and the permeability coefficient of the same zone k_s.

An approximate solution to the problem involves the employment of two equations as alternatives to Equations 4.46 to 4.48 (Hansbo, 1981; Almeida and Marques, 2018):

$$\bar{U}_r = 1 - e^{\frac{-2T_r}{F(n,s)}} \tag{4.50}$$

and

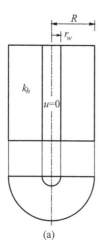
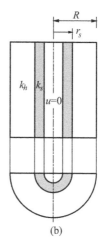

(a)   (b)

Figure 4.31 Scheme showing a generic vertical drain with its influence zone: a) ignoring the smear zone; or b) considering the smear zone.

$$T_r = -\frac{1}{2}F(n,s)\ln(1-\bar{U}_r)$$ (4.51)

where

$$F(n,s) \approx \ln(n) + \left(\frac{k_h}{k_s} - 1\right)\ln(s) - 0.75$$ (4.52)

and

$$s = \frac{r_s}{r_w}$$ (4.53)

The difficulty of dealing with the smear effect lies in which values to consider for the two characterizing parameters, r_s and k_s.

The value of r_s depends on the mandrel cross-section dimensions. The selection of this and the anchor type should be carefully considered in order to reduce the dimensions of the remolded zone. It is usual to consider r_s at between 1.5 and 3 times r_m, with r_m being the radius of a circle whose area coincides with the cross-sectional area of the mandrel (Indraratna et al., 2005).

The issue of the permeability coefficient has, obviously, a relationship with the smear zone dimensions, since, in this zone, the soil disturbance is not uniform, but decreases with increasing distance to the drain. Therefore, the change in the permeability coefficient in relation to the intact soil is greater closer to the drain. In this more disturbed zone caused by drilling, it can be imagined that the soil was bent at a 90° angle, and thus the horizontal permeability coefficient will tend to be close to the vertical permeability coefficient of the intact soil, $k_s \approx k_v$ (Bergado et al., 1992). However, taking this relationship into consideration in the entire smear zone can be too conservative (Hird and Moseley, 2000; Hansbo, 2004).

4.6.3.4 Calculation of the vertical drain network

In practice, it is generally required that the consolidation process is almost concluded at the end of a certain period, t, which therefore becomes one of the project requirements.

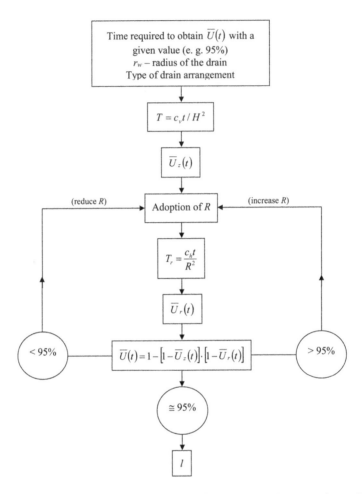

Figure 4.32 Vertical drain flowchart for the calculation of the distance between drains for a given mesh arrangement in order to achieve a certain average consolidation ratio after a certain period, *t*.

Figures 4.32 illustrates the steps necessary for the calculation of the spacing, *l*, of a drain network.

4.6.4 Vacuum preloading

The vacuum preloading system was initially conceived by Kjellman (1952), and it is illustrated in Figure 4.33. In practice, this system combines pre-loading with the use of vertical drains.

The method consists of applying a vacuum to the clay soil mass, the consolidation of which needs to be accelerated. This application will induce a negative excess pore pressure. If the total stress remains constant, the effective stresses will suffer an isotropic increment, equal (in absolute terms) to the excess pore pressure, forcing the water to withdraw through the vertical drains.

To sustain the vacuum application inside the soil mass, the system comprises the following components (assembled in the order listed):

i) a sand blanket placed at the soil surface;
ii) synthetic vertical drains, without reaching the bottom of the clay layer, to avoid contact with any permeable layer beneath;

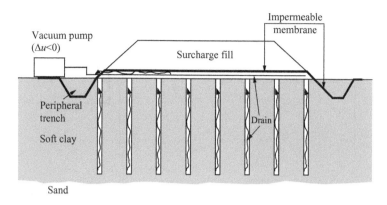

Figure 4.33 Vacuum preloading scheme.

iii) horizontal drains (tubes), placed in longitudinal and transverse directions, inside the sand blanket, at the level of the vertical drain heads;
iv) an impervious membrane, totally covering the sand blanket, and extending to a peripheral trench where it is sealed with water and impermeable material;
v) vacuum pumps capable of extracting air and water, linked to the horizontal drains;
vi) membrane protection.

The impermeable membrane protection is essential, to avoid punctures that may lead to vacuum loss. In the situations where the additional embankment is not applied, it is currently standard practice to cover the membrane with water, to protect against eventual damage by people or by wildlife. In the case where the vacuum application is combined with an embankment (Figure 4.33), the first layers have to be carefully controlled to avoid potentially damaging coarse angular particles.

A particular advantage of this solution is that inducing an immediate and isotropic effective stress increment in the soil does not cause stability problems. For that reason, it is particularly appropriate in very soft soils. In addition, it generally allows much shorter execution and consolidation times (Indraratna et al., 2010). As a limitation, it should be noted that the existence of sandy sublayers in the middle of the clay layer can compromise its application.

In a system without losses, the vacuum application would be equivalent to a surface pressure equal to the atmospheric pressure, i.e., around 100 kPa. It is common to obtain efficiencies of 70% to 80%, which correspond approximately to 4 m of fill. In cases where a greater surcharge is needed, the vacuum application may be combined with the fill, as mentioned above. In this latter case, the vacuum reduces the fill height by about 4 m, in comparison to what would be necessary in a conventional preloading. Figure 4.34 shows a comparison between the two situations (Indraratna et al., 2010).

When the necessary settlement has been achieved, the vacuum system is deactivated without major precautions.

4.6.5 Note on the use of stone columns

In Chapter 2 (see Section 2.4.3.3), stone columns were presented as one of the methods by which to reinforce a soft clay soil mass when loaded at the surface by an embankment or other structure. At that time, it was mentioned that, in addition to its contribution to

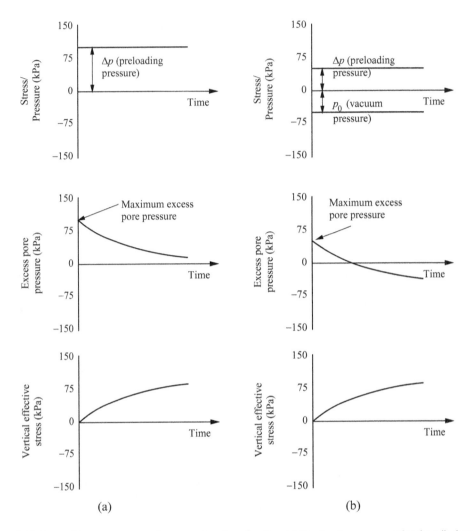

Figure 4.34 Consolidation process: a) conventional preloading; b) idealized vacuum preloading (Indraratna et al., 2010).

stability, stone columns have a very relevant role in the soil mass behavior in terms of deformation: i) being more rigid than the surrounding soil, the stone columns concentrate the vertical stresses, relieving the soft soil, which results in a reduction in the consolidation settlement in comparison with a non-reinforced soil mass; ii) since the stone columns are made of a high-permeability material, they also work as vertical drains.

With regard to the influence of the stone columns on the magnitude of the consolidation settlement, from Equation A2.2.3, the incremental total stress on the soil, $\Delta\sigma_{soil}$ is given by:

$$\Delta\sigma_{soil} = \frac{\Delta\sigma}{1+(m-1)a_s} \tag{4.54}$$

where $\Delta\sigma$ is the total stress increment applied at the ground surface, a_s is the area replacement ratio, expressed by Equation 2.37, and m is the stone column stress concentration factor, expressed by Equation A2.2.1. This factor can be estimated through Figure A2.2.1.

Therefore, the consolidation settlement of the reinforced soil with stone columns may be estimated by introducing $\Delta\sigma_{soil}$, given by Equation 4.54, in the equations presented in Section 4.2.6.

With regard to its behavior as a vertical drain, the calculation of the consolidation time can be made in a similar way as described in Section 4.6.3. In the stone columns, two phenomena need to be taken into account: i) the migration of clayey soil particles to the interior of the columns, creating a peripheral clogged zone of low permeability; and ii) the presence of a smear zone around the column, similar to the one described for the synthetic drains, although with different dimensions, and dependent on the technology used for the construction of the columns.

A simplified way of taking into account the effects of the mentioned phenomena, particularly the first, is by considering a decrease in the effective drain radius (Barksdale and Bachus, 1983). Even introducing a very severe decrease, which may correspond to 50% of the original radius, bearing in mind the diameters and spacing typical of stone columns, the relation, n, given by Equation 4.44, is for those works substantially lower than the one occurring in synthetic drains. As a consequence, in the works involving stone columns, the consolidation time is not generally a critical issue in design. For that reason, this subject will not be developed here. Readers may satisfy their interest in this subject by reading the works of Han and Ye (2000, 2002).

4.7 OBSERVATION OF EMBANKMENTS ON SOFT SOIL

4.7.1 Installation points, equipment, and parameters to monitor

The complexity of natural ground masses and the theoretical limitations that support the predicting methods (for settlements and their change over time) frequently lead to important biases between estimates and reality. This is particularly applicable to settlement change in time, which is generally significantly faster in reality than predicted by the model theories.

This fact results, in many situations, from the soft soil *macrofabric*. In fact, the existence of sublayers of higher permeability can considerably reduce the consolidation time, either by the increment of the average horizontal consolidation coefficient or by the reduction of the flow path associated with the vertical consolidation process.

As a consequence, constructions on soft soils need to be monitored in order to adjust the predicted calculations, and to take decisions (if necessary) that lead to the achievement of the design requirements in the most economical and safe way. The measured parameters, the type of equipment used and the point where it is installed are summarized in Table 4.7. Figure 4.35 illustrates a typical instrumented section of an embankment.

The ground surface settlements are generally measured by means of *settlement plates*. These consist of a base (usually a small square reinforced concrete slab or a steel plate) to which a vertical metallic tube is attached, which can be successively complemented with new portions coupled to the portion immediately underneath. The plates are placed at the

Table 4.7 Equipment, installation points, and parameters to measure for monitoring of soft clay soils.

Measured parameters	Type of equipment	Installation point
Surface settlements	Settlement plates/topography	Ground surface in the loaded area and its neighborhood
Ground horizontal displacements	Inclinometers/inclinometer tubes	Installed in vertical boreholes close to the limit of the loaded area
Pore water pressure	Electric piezometers	Clay layer at several depths

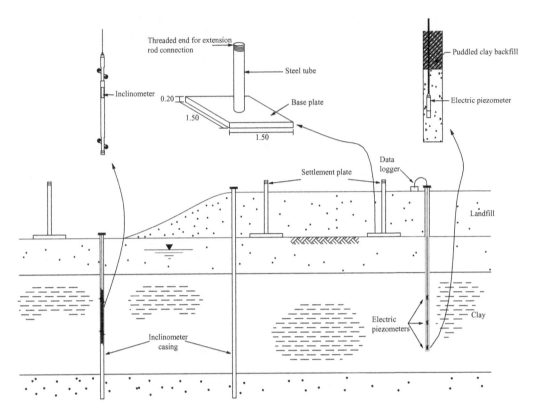

Figure 4.35 Typical instrumented section of an embankment on soft soil.

initial ground surface and the elevation of the tube top is determined by topographic leveling. After the construction of the embankment and the start of the consolidation process, periodic topographic leveling provides the changes in the tube top elevation, and therefore settlement of the ground surface at each point.

From this description, it is easy to understand that this observation process is quite simple and economic. However, its isolated use is not recommended since it is difficult to do an accurate interpretation of the soil mass behavior based only on the settlement observations, in particular when these are still very far from being stabilized. For instance, the observation of much higher settlements than predicted during a given period after loading can result from very different situations: i) underestimation of the compressibility parameters of the soil; ii) underestimation of the coefficient of consolidation; iii) errors in the evaluation of the boundary conditions of the problem, namely in terms of drainage; iv) existence of significant secondary consolidation settlements; v) the linearity between effective stress and volumetric strain adopted in Terzaghi's theory (see Figure 4.13); or vi) combinations of two or more of the aforementioned issues.

It is, therefore, highly convenient to combine these measurements with those of pore water pressures in the clay layer at different depths, using, for that purpose, *electric piezometers* due to the very low permeability of the soil. This matter will be treated in more detail in Chapter 9 (see Annex A.9.3).

In the case of loading in areas of large dimensions, horizontal displacements in the clay mass are only significant under the zones close to the embankment limit. Their measurement is relevant in the context of safety relative to rotational slides of the embankment and foundation soil (as an example, see Figure 2.1). In general, these failures are preceded by quite large horizontal displacements in the foundation soil. The horizontal displacements can be measured along a vertical hole, in which *inclinometer tubes* are installed. The introduction into these tubes of a device – the *inclinometer* - allows the measurements of horizontal displacements at each point of the tube. This type of measurement and the inclinometer operation mode will be discussed in Chapter 9 (see Annex A.9.2).

4.7.2 The Asaoka method

Asaoka (1978) developed a recognized method of easy application, which provides an estimate of the final consolidation settlement based on the settlement change observed over a certain time period. The theoretical basis of the method, which will not be presented here, is in agreement with Terzaghi's one-dimensional consolidation theory.

The method comprises the following steps: i) the change over time of the observed settlement is plotted (Figure 4.36a); ii) a certain instant, t_1, and a certain time interval, Δt, are selected, and, from the previous graph, the settlement values are calculated for t_1 and for all the other subsequent points, spaced for Δt, for which observation results are available; iii) from the previously calculated settlements, a graph is constructed, marking on the x-axis the settlement s_{i-1} and on the y-axis the settlement s_i (Figure 4.36b); iv) in this graph, the line that best fits the plotted points is drawn and its slope, β_1, is calculated; v) in the same graph, a second line at 45° from the origin is drawn which represents the points where $s_{i-1} = s_i$; and vi) the abscissa of the crossing point of these two lines represents the final consolidation settlement estimate.

The coefficient of consolidation can be obtained from the following equation:

$$c_v = -\frac{5H^2}{12\Delta t}\ln\beta_1 \tag{4.55}$$

where H has the same meaning as in Terzaghi's theory of consolidation and β_1 is expressed in radians.

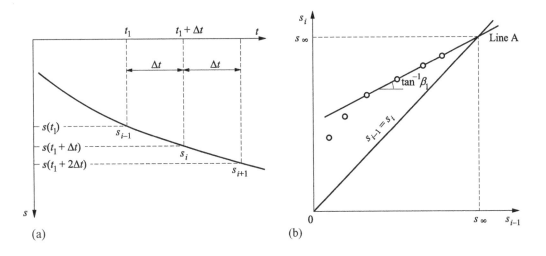

Figure 4.36 Consolidation settlement estimate from observation results (Asaoka, 1978): a) settlement *versus* time curve; b) construction to obtain the final consolidation settlement.

ANNEX A4.1 DEDUCTION OF EQUATIONS

A4.1.1 Equation 4.1

See Figure A4.1.1

$$\left.\begin{array}{l} e_0 = \dfrac{V_{v0}}{V_s} = \dfrac{S \cdot h_{v0}}{S \cdot h_s} = \dfrac{h_{v0}}{h_s} = \dfrac{h_0 - h_s}{h_s} = \dfrac{h_0}{h_s} - 1; \quad h_s = \dfrac{h_0}{1 + e_0} \\[4mm] e = \dfrac{V_v}{V_s} = \dfrac{S \cdot h_v}{S \cdot h_s} = \dfrac{h_v}{h_s}; \quad \Delta e = \dfrac{\Delta h_v}{h_s} = \dfrac{\Delta h}{h_s}; \quad h_s = \dfrac{\Delta h}{\Delta e} \end{array}\right\} \Delta h = \dfrac{h_0}{1 + e_0} \cdot \Delta e \qquad (A4.1.1)$$

A4.1.2 Constancy of the ratio $h/(1 + e)$

Consider e_1 and h_1 to be the void ratio and the thickness after the Δe and Δh variations, respectively. Hence:

$$h_1 = h_0 - \Delta h \qquad (A4.1.2)$$

$$e_1 = e_0 - \Delta e \qquad (A4.1.3)$$

In what follows, it is demonstrated that the ratio $h/(1 + e)$ is constant:

$$\frac{h_1}{1 + e_1} = \frac{h_0 - \Delta h}{1 + e_0 - \Delta e} = \frac{h_0 - \dfrac{h_0}{1 + e_0} \cdot \Delta e}{1 + e_0 - \Delta e}$$

$$= \frac{h_0 \left(1 - \dfrac{\Delta e}{1 + e_0}\right)}{1 + e_0 - \Delta e} = \frac{h_0 \dfrac{1 + e_0 - \Delta e}{1 + e_0}}{1 + e_0 - \Delta e} = \frac{h_0}{1 + e_0} \qquad (A4.1.4)$$

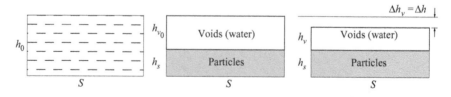

Figure A4.1.1 Saturated clay sample with the volumes of the voids and solid particles before and after consolidation.

ANNEX A4.2 EVALUATION OF σ'_p BY THE CASAGRANDE CONSTRUCTION

The construction consists of the following steps (see Figure A4.2.1):

1. locate the point of the curve log $\sigma'_v - e$ with the least curvature radius;
2. from this point, draw a horizontal line and another line tangential to the curve;
3. draw a bisector line between the two previous lines;
4. extrapolate the virgin compression line upward to cut the bisector;
5. the abscissa of the intersection point corresponds to the pre-consolidation pressure.

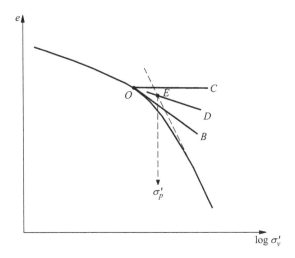

Figure A4.2.1 Casagrande construction for the evaluation of the pre-consolidation stress.

ANNEX A4.3 TREATMENT OF LOG σ'_v–e CURVES BY THE SCHMERTMANN CONSTRUCTION

This construction has to be made after the determination of σ'_p. If $\sigma'_p = \sigma'_{v0}$ (for a normally consolidated soil), as shown in Figure A4.3.1, the steps are as follows:

1. plot point B of coordinates $\left(\log \sigma'_{v0}, e_0\right)$;
2. plot point C, belonging to the test curve and at 0.42 e_0;
3. plot the corrected compressibility curve (virgin line) coinciding with the line B-C.

If $\sigma'_p > \sigma'_{vo}$ (for an overconsolidated soil), as shown in Figure A4.3.2, the steps are as follows:

1. plot point D of coordinates $\left(\log \sigma'_{v0}, e_0\right)$;

2. plot point E, with abscissa log σ'_p, which is on the line that starts at D and is parallel to the recompression curve (characterized by the unload–reload cycle);
3. plot point C, belonging to the test curve at 0.42 e_0;
4. the treated compressibility curve coincides with the line D-E (recompression line) and with the line E-C (virgin line).

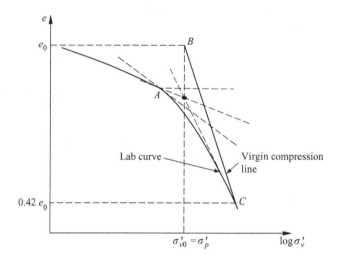

Figure A4.3.1 Reconstitution of compressibility curves according to Schmertmann for normally consolidated soils.

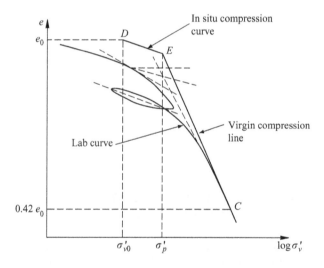

Figure A4.3.2 Reconstitution of compressibility curves according to Schmertmann for overconsolidated soils.

ANNEX A4.4 EVALUATION OF THE VERTICAL CONSOLIDATION COEFFICIENT, c_v

A4.4.1 The Taylor method

With the Taylor method to determine c_v the results are represented by plotting the square root of time on the x-axis, and the vertical displacements registered in the transducer linked to the top of the specimen (and thus, the settlements) on the y-axis.

Figure A4.4.1a shows the graph corresponding to Terzaghi's theory (note that T is proportional to the real time and $\overline{U}_z$ is proportional to the settlement in each instant), whereas Figure A4.4.1b shows the results of a real test.

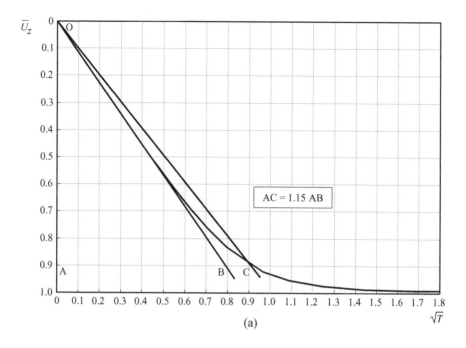

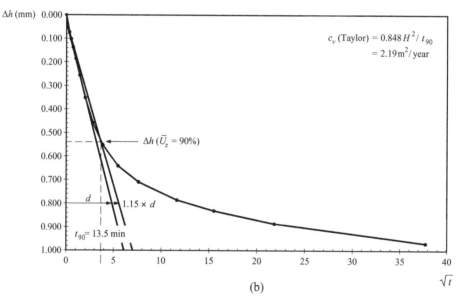

Figure A4.4.1 Evaluation of c_v by Taylor method: a) theoretical curve; b) experimental curve of a soft clay from Lower Mondego region (Coelho, 2000).

It should be noted that the initial part of the theoretical curve is almost linear (given that, as Figure 4.12 shows, the curve that relates T and $\overline{U}_z$ is, at the start, very close to a parabola whose axis is parallel to the x-axis).

Figures A4.4.1a shows two lines, one (OB) tangential to the initial part of the curve and the other (OC) linking the origin with the point of the curve at $\overline{U}_z = 90\%$.

The abscissas of points B and C (both with $\overline{U}_z = 90\%$) are related, as the figure indicates (i.e., the abscissa of C is 1.15 times the one of B).

Bearing this fact in mind, the procedure for the determination of c_v from the experimental curve, as shown in Figure A4.4.1b, consists of the following steps:

1. draw a line (OB) tangential to the linear part of the experimental curve and another line (OC) whose abscissas are 1.15 times larger than the first;
2. the intersection of the second line with the experimental curve gives the point corresponding to $\overline{U}_z = 90\%$, and thus the value of t_{90};
3. for $\overline{U}_z = 90\%$, the theoretical value of T is 0.848 (see Table 4.4); as the initial height of the sample is $2H$, it becomes that:

$$c_v = \frac{T_{90} \cdot H^2}{t_{90}} = \frac{0.848\,H^2}{t_{90}} \tag{A4.4.1}$$

It should be noted that the backwards extrapolation of the tangent to the experimental curve cuts the y-axis at a point of non-null ordinate (generally, positive). This is due to immediate settlements, as mentioned before.

A4.4.2 The Casagrande method

In the Casagrande method to determine c_v the results are represented by plotting the logarithm of time on the x-axis and the vertical displacements registered in the transducer linked to the top of the sample on the y-axis.

Figure A4.4.2a shows the graph according to Terzaghi's theory, whereas Figure A4.4.2b shows the results of a real test.

On the theoretical curve, it can be observed that the intersection of the asymptote with the tangent to the central part of the curve coincides with the point corresponding to 100% of consolidation.

Bearing this fact in mind, the procedure for the evaluation of c_v from an experimental curve, such as the one in Figure A4.4.2b, consists of the following steps:

1. draw two lines tangential to the experimental curve, one in the intermediate part of the curve (at the inflexion point) and another in the final linear part; by analogy to the theoretical curve, the intersection point of the two tangent lines corresponds to the end of consolidation $\left(\overline{U}_z = 100\% \right)$;
2. the initial part of the curve is corrected in a way to remove the immediate settlements, thus evaluating the "real" s_0 record corresponding to $\overline{U}_z = 0$; considering that for $\overline{U}_z < 60\%$, the consolidation ratio is very approximately proportional to the square root of t, taking two instants t_A and $t_B = 4t_A$ in the initial part of the consolidation

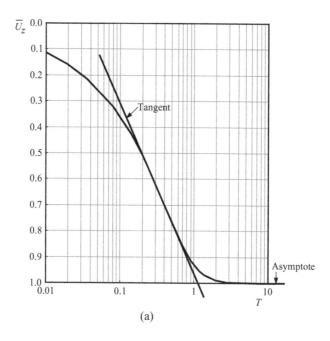

(a)

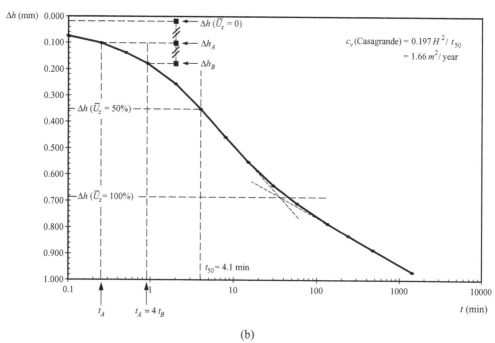

(b)

Figure A4.4.2 Evaluation of c_v by Casagrande method: a) theoretical curve; b) experimental curve of a soft clay from Lower Mondego region (Coelho, 2000)

process, $\Delta h_B = 2\Delta h_A$ should be obtained; in this way, the corrected value of $\Delta h\left(\overline{U}_z = 0\right)$ is obtained by plotting upward from Δh_A upward the distance between Δh_B and Δh_A;

3. then:

$$\Delta h\left(\overline{U}_z = 50\%\right) = \left[\Delta h\left(\overline{U}_z = 0\right) + \Delta h\left(\overline{U}_z = 100\%\right)\right]/2 \qquad (A4.4.2)$$

is calculated as well as the corresponding abscissa, t_{50};

4. for $\overline{U}_z = 50\%$, the theoretical value of T is 0.197 (see Table 4.4); as the initial height of the sample is $2H$, it becomes that:

$$c_v = \frac{T_{50} \cdot H^2}{t_{50}} = \frac{0.197H^2}{t_{50}} \qquad (A4.4.3)$$

ANNEX E4 EXERCISES (RESOLUTIONS ARE INCLUDED IN THE FINAL ANNEX)

E4.1 – Consider the soil profile of Figure E4.1, in which are indicated the initial conditions for a given site. Take $\gamma_w = 9.8$ kN/m³.

a) Calculate the settlement of the normally consolidated clay layer induced by permanent lowering of the water table to elevation –10.0. Assume that the soil above the water table remains saturated by capillarity.

b) A general excavation of the site to elevation –4.0 is performed for the construction of the basement and foundations of a large industrial building, as indicated on the right-hand side of the figure. Assume that the consolidation associated with the lowering of the water table has ended. Estimate the swelling (i.e., the upward displacement) of the ground surface due to the excavation.

c) Assume that, after the swelling has ended, the building is constructed, the foundation of which applies a uniform vertical pressure of 150 kPa to the ground. Evaluate the settlement of the foundation due to the consolidation of the clay layer.

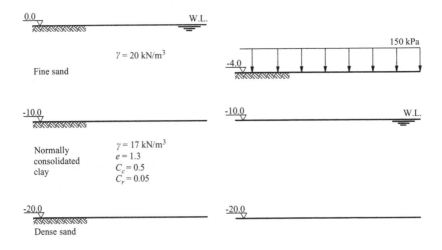

Figure E4.1

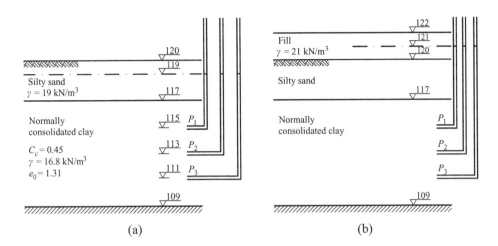

Figure E4.2

d) Using a log σ'_v *versus* e graph, represent the at-rest condition and the changes induced by the above-mentioned operations in a point at elevation –15.0.

E4.2 – Figure E4.2a represents the at-rest state of an alluvial ground mass. A dam will be constructed downstream at some distance from this site, which will cause the water table to rise 2 m, from elevation 119 to 121.

In order to prevent submersion of the site, a fill layer 2 m thick will be placed upon a very large area, so that the ground surface will rise from elevation 120 to 122, as shown in Figure E4.2b. This fill will be constructed in a very short time.

In both figures, three piezometers are represented, the head pressure readings of which are registered in Table E4.2. Taking $\gamma_w = 10.0$ kN/m³, consider the following situations:

i) t_0, initial state, corresponding to Figure E4.2a;
ii) t_1, immediately after the placement of the fill;
iii) t_2, one year after t_1 and immediately before the filling of the reservoir;
iv) t_3, immediately after the filling of the reservoir, which may be considered practically instantaneous;
v) t_4, at the end of consolidation (Figure E4.2b).

a) Sketch the graphs of excess pore pressure in the clay layer for times t_0, t_1, t_2, t_3 and t_4.
b) Does the layer underneath the clay provide a free-draining boundary?
c) Evaluate the coefficient of consolidation, c_v, of the clay.
d) Compute the decrease in thickness of the clay layer from t_0 to t_2.

Table E4.2

Piezometer	Elevation of the base (m)	Pressure head (m)				
		t_0	t_1	t_2	t_3	t_4
P1	115	4.00	8.20	6.68	6.68	6.00
P2	113	6.00	10.20	9.61	9.61	8.00
P3	111	8.00	12.20	10.68	10.68	10.00

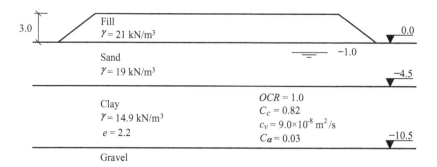

Figure E4.3

E4.3 – Consider the soil profile presented in Figure E4.3, where a large embankment for the construction of a highway will be built. The relationship between the average width of the embankment and the thickness of the clay layer allows for the consideration of a confined loading. Groundwater level is at 1 m depth. Assume that the soil above the water table is saturated by capillarity. Consider $\gamma_w = 9.8$ kN/m³.

The pavement will be constructed when 90% of the expected consolidation settlement has been observed.

a) Calculate the settlement of the normally consolidated clay layer induced by the construction of the embankment, and the expected time necessary to start the paving operations.

b) Calculate the additional height of the fill that will allow paving to start 12 months after the construction of the embankment.

c) Sketch the responses of pore pressure in the clay layer for the following phases: i) at-rest; ii) immediately after the construction of the embankment and application of the temporary load; iii) 12 months later, immediately before the removal of the temporary load; iv) immediately after the removal of the temporary load; v) at the end of consolidation.

d) Estimate the expected settlement 25 years after the construction of the pavement.

E4.4 – On the soil profile presented in Figure E4.4, a large landfill is going to be built. Take $\gamma_w = 9.8$ kN/m³.

a) Calculate the total height of the landfill that is necessary to ensure a +5.0 surface elevation after primary consolidation.

b) Calculate the time necessary to reach 50% of the settlement by primary consolidation.

E4.5 – A 2.5-m thick fill with large dimensions in plan is going to be placed over the ground, the geotechnical profile of which is represented in Figure E4.5. As can be observed, the ground profile includes a soft clay layer 10 m thick.

The results of site investigation suggest the possible existence of a continuous thin sand layer at 6.0 m depth. Nevertheless, to be on the safe side, it has been decided not to consider its effect in the project estimate for the consolidation time.

In order to allow for the start at an earlier date of pavement construction on top of the fill, the consolidation process has been accelerated by the introduction of a network of synthetic vertical drains ($r_w = 0.10$ m), in a square layout with spacing $l = 2.65$ m. Take $\gamma_w = 9.8$ kN/m³.

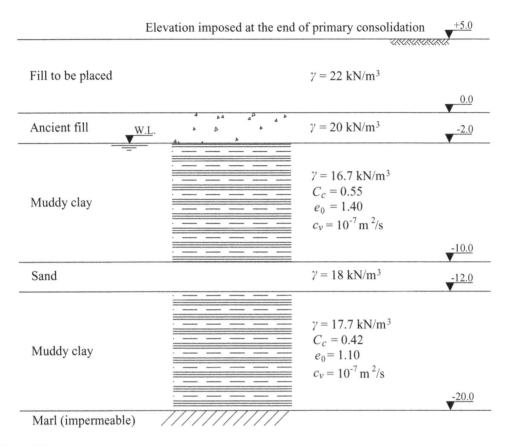

Figure E4.4

a) Calculate the settlement by primary consolidation, due to the construction of the fill.
b) Calculate the average radial consolidation ratio ($\bar{U}_r$) expected after six months (180 d).
c) At the end of the six-month period, a surface settlement of 66 cm has been registered. Assuming that all parameters of the clay layer used in the computations are correct, check whether that settlement value confirms that the thin sand layer works as a draining boundary.

E4.6 – Figure E4.6 presents the geological–geotechnical profile of a sedimentary soil where a 20-m diameter water tank will be built. The design demands that the base of the tank be above the soil surface, through the construction of a 2.0-m thick landfill occupying a considerably large area. The self-weight of the tank is negligible. The tank will be built only after the end of primary consolidation of the clay layer, due to the placement of the landfill.

a) Calculate the consolidation settlement due to the construction of the landfill.
b) Applying the method of Skempton and Bjerrum, calculate the consolidation settlement of the center of the tank filled with 5.0 m of water. Estimate the excess pore water pressure with the pore pressure parameter A.
c) Applying Terzaghi's consolidation theory, calculate the settlement that will occur one year after filling the tank. Considering the conditions of Figure E4.6, is the result obtained in the previous question an overestimate or an underestimate?

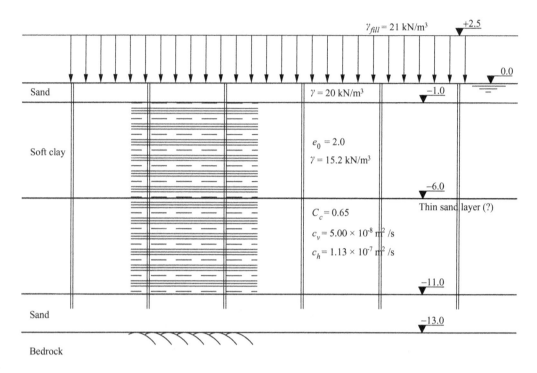

Figure E4.5

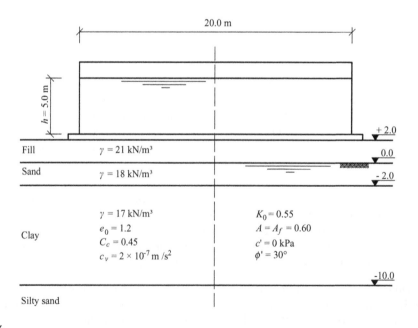

Figure E4.6

Chapter 5

Earth pressure theories

5.1 INTRODUCTION

There are numerous cases where civil engineering structures interact with soil masses by means of a vertical (or almost vertical) structural interface. As a result of that interaction, forces designated as soil *thrusts* are applied by the soil to the structure. These interaction problems can be classified into two main groups.

The first group concerns problems where *the structure supports the soil mass*. The soil mass is considered to be supported when the slope angle of its lateral surface is greater than that which would be adopted in the absence of the given structure, which is designated as a *retaining structure*. Figure 5.1 illustrates examples of this type of structure. In these cases, the soil "pushes" the structural face, by applying a thrust that may be classified as an *action*.

In the second group, *the structure is supported by the soil mass*. In this case, the structural face is "pushed" against the soil, the response of which is represented by a thrust that may be classified as a *reaction*. An example of this situation is depicted in Figure 5.2: a foundation block, loaded by a force with a high horizontal component, which is applied to the soil through the lateral block face.

There is a great number of structures where the soil–structure interaction comprises both groups of interactions simultaneously, some of which are represented in Figure 5.3. These include laminar-shaped structures, made of steel or reinforced concrete, used to support excavations. A relevant issue to the support function is the wall extension below the base of the excavation. It is on the two faces of this embedded part that reactive earth pressures are mobilized, being essential to the support of the soil behind the cut face.

In the majority of these soil–structure interaction cases, the actual interactive force cannot be calculated with static equilibrium equations only, since it depends on the soil stress–strain–strength relationships and on the structure itself. In structural terms, they form an indeterminate or hyperstatic system (Peck, 1972). At present, these problems can be treated by numerical analysis methods, using the finite element method.

However, there are situations in which the design can be properly addressed by evaluating the *minimum* and/or *the maximum soil–structure interaction forces*, herein designated as *active thrust* and *passive thrust*, respectively. The evaluation of these forces can be done by applying scientifically based solutions, as in the Coulomb theory (1773) and the Rankine theory (1857), developed prior to the formulation of the effective stress principle (Terzaghi, 1925). For that reason, these theories were formulated for soils above the water table, omitting pore pressures. However, their generalization to submerged soils, based on the effective stress principle, does not present any difficulty, as will be discussed below.

The aim of this chapter is to present these theories and others developed in the sequence. These theories will later be applied in Chapter 8, dedicated to the design of earth-retaining structures.

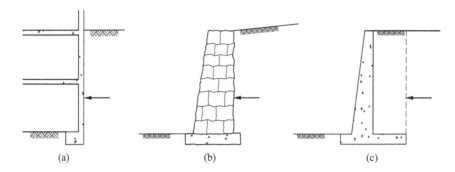

Figure 5.1 Examples of earth-retaining structures: a) peripheral walls in a building basement; b) masonry gravity wall; c) reinforced concrete cantilever wall.

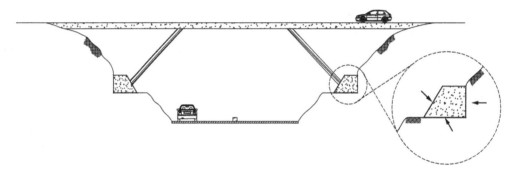

Figure 5.2 Example of a structure (foundation block) where the earth pressures on the vertical face are of the reactive type.

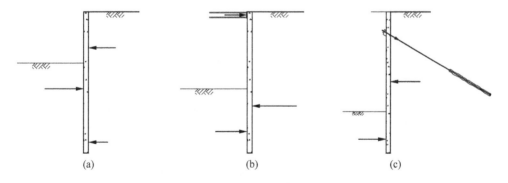

Figure 5.3 Structures where both groups of pressures presented in the previous figures act on both wall faces: a) flexible cantilever embedded wall; b) single-propped wall; c) single-anchored wall.

5.2 RANKINE'S ACTIVE AND PASSIVE LIMIT EQUILIBRIUM STATES

5.2.1 Introduction of the concept

As mentioned before, Coulomb's theory was developed before Rankine's theory. However, from a pedagogical point of view, it is more convenient to start from the latter. In fact, Rankine introduced the concepts of *active limit equilibrium state* and *passive limit equilibrium state*, which will be the *Alpha* and the *Omega* of the present chapter.

Figure 5.4 Starting conditions for the mobilization of the Rankine active and passive states: a) granular soil at-rest with horizontal surface; b) replacement of the left part of the soil mass by a vertical, rigid, and smooth wall face, without introducing any deformations in the remaining soil mass.

Figure 5.4a presents a soil mass at-rest with a horizontal surface. It is assumed that the water level is at a great depth and it does not interfere with the following considerations.

As suggested by Figure 5.4b, the left half of the soil mass was replaced by means of a vertical, rigid, and smooth wall face. If it is assumed that this operation was done without introducing any deformation on the remaining soil mass, the stress state in this mass is still the at-rest stress state. Therefore, the stress imposed on the wall by the soil in a point at a certain depth z is:

$$\sigma'_{h0} = K_0 \sigma'_{v0} \tag{5.1}$$

where σ'_{v0} and σ'_{h0} are the vertical and horizontal effective stresses at-rest, respectively, and K_0 is the coefficient of earth pressure at-rest. Graphically, this stress state can be represented by the Mohr circle of diameter $\overline{AB}$ represented in Figure 5.5.

Considering that the wall moves progressively to the left in the horizontal direction, the minimum and maximum principal stresses will continue to be the vertical and horizontal stresses, respectively, keeping the first constant while the second decreases continuously. (It should be noted that, being a smooth wall, there are no tangential stresses in the contact with the soil). The Mohr circles corresponding to the successive stress states have in common point A, whereas the point representative of the minimum principal stress moves to the left, resulting in the progressive increase of the maximum shear stresses mobilized in the soil. This process has a limit corresponding to the situation in which the soil resistance is fully mobilized (Mohr circle tangential to the failure envelope, with σ'_h represented by point B_1). This situation is designated the *active limit equilibrium state*, and the horizontal stress, corresponding to σ'_{ha}, is designated the *active horizontal stress*.

Starting from the initial situation, if the wall displacement is in the opposite sense, i.e., if the wall is pushed against the soil mass, compressing it, then the horizontal stress will increase, keeping the vertical stress constant. The Mohr circles, corresponding to the successive stress states, have in common point A, whereas the point representative of the horizontal stress moves to the right. At a certain stage, this stress equals the vertical stress, $\sigma'_h = \sigma'_{v0}$, which corresponds to a hydrostatic or isotropic stress state. In the following stages, the horizontal stress becomes the maximum principal stress, shear stresses become progressively larger, until the soil resistance is fully mobilized; this corresponds to the situation where Mohr circle is tangential to the failure envelope, with σ'_h represented by point B_2. This situation is designated the *passive limit equilibrium state*, and the horizontal stress, corresponding to σ'_{hp}, is designated the *passive horizontal stress*.

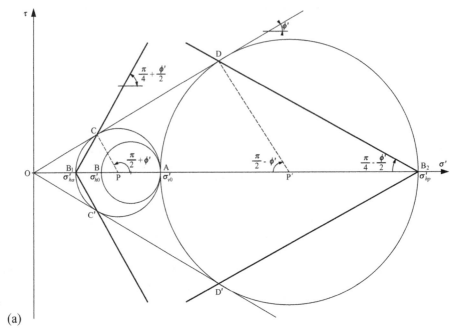

(a)

(b)

Deduction of K_a expression:

$$K_a = \frac{\sigma'_{ha}}{\sigma'_{v0}}$$

$$r \text{ (Mohr circle radius)} = \frac{\sigma'_{v0} - \sigma'_{ha}}{2} = \frac{\sigma'_{v0} - K_a \sigma'_{v0}}{2} = \frac{1 - K_a}{2} \sigma'_{v0}$$

$$\overline{OP} = \sigma'_{v0} - r = \sigma'_{v0} - \frac{1 - K_a}{2} \sigma'_{v0} = \frac{1 + K_a}{2} \sigma'_{v0}$$

$$r = \overline{OP} \sin\phi' ; \frac{1 - K_a}{2} \sigma'_{v0} = \sigma'_{v0} \frac{1 + K_a}{2} \sin\phi' ; (1 - K_a) = (1 + K_a) \sin\phi'$$

$$1 - \sin\phi' = K_a (1 + \sin\phi') \; ; \; K_a = \frac{1 - \sin\phi'}{1 + \sin\phi'}$$

Deduction of K_p expression:

$$K_p = \frac{\sigma'_{hp}}{\sigma'_{v0}}$$

$$r \text{ (Mohr circle radius)} = \frac{\sigma'_{hp} - \sigma'_{v0}}{2} = \frac{K_p \sigma'_{v0} - \sigma'_{v0}}{2} = \frac{K_p - 1}{2} \sigma'_{v0}$$

$$\overline{OP'} = \sigma'_{v0} + r = \sigma'_{v0} + \frac{K_p - 1}{2} \sigma'_{v0} = \frac{1 + K_p}{2} \sigma'_{v0}$$

$$r = \overline{OP'} \sin\phi' ; \frac{K_p - 1}{2} \sigma'_{v0} = \sigma'_{v0} \frac{1 + K_p}{2} \sin\phi' ; (K_p - 1) = (1 + K_p) \sin\phi'$$

$$K_p (1 - \sin\phi') = 1 + \sin\phi' \; ; \; K_p = \frac{1 + \sin\phi'}{1 - \sin\phi'}$$

Figure 5.5 Rankine active and passive limit equilibrium states: a) Mohr circles representative of at-rest, active, and passive states; b) deduction of K_a and K_p expressions.

The active and passive horizontal stresses represent, therefore, the *lower bound* and the *upper bound*, respectively, of the interaction stresses of the soil with the structure, at the given depth. This means that, after being installed, the active and passive states, even if the wall were still moving (away from or against the soil, depending on the case), the horizontal stress would be *frozen* at σ'_{ha} and σ'_{hp} values, respectively. The opposite would mean Mohr circles secant to the envelope, which is physically impossible.

The analysis of Figure 5.5 allows the definition of the directions of the surfaces where the soil shear resistance is fully mobilized, in the active and passive limit equilibrium states. (It should be noted that the active and passive Mohr circle poles are, respectively, points B_1 and B_2). In both cases, those surfaces form an angle of $\pi/4 - \phi'/2$ with the direction of the maximum principal stress (which is the vertical stress in the active case and the horizontal stress in the passive case).

The limit equilibrium states are frequently designated *Rankine's active and passive limit equilibrium states*.

5.2.2 Earth pressure coefficients: coefficients of active and passive earth pressure, according to Rankine

In soil mechanics, the term *coefficient of earth pressure* is currently used, generally represented by the letter K, for the relation between the horizontal and vertical effective stresses acting on a given point of a soil mass. Then,

$$K = \frac{\sigma'_h}{\sigma'_v} \tag{5.2}$$

When the effective stresses of Equation 5.2 correspond to the at-rest state, the coefficient is named the *coefficient of earth pressure at-rest*, as indicated above (Equation 5.1). For normally consolidated soils, the coefficient of earth pressure at-rest can be obtained by the following relationship, which is an approximation to a more complex theoretical solution (Jaky, 1944):

$$K_0 = 1 - \sin\phi' \tag{5.3}$$

For overconsolidated soils, the following relationship between the value of the coefficient of earth pressure at-rest for a normally consolidated soil, $K_0(NC)$, and the value of the same coefficient induced by overconsolidation, $K_0(OC)$, is currently used:

$$\frac{K_0(OC)}{K_0(NC)} = (OCR)^m \tag{5.4}$$

where OCR is the overconsolidation ratio and m is a dimensionless coefficient, generally in the range 0.4 to 0.5 (Mayne and Kulhawy, 1982).[1]

It is now time to present two new earth pressure coefficients.

The *active earth pressure coefficient*, K_a, is the ratio between the horizontal effective stress in the active state and the corresponding vertical effective stress:

$$K_a = \frac{\sigma'_{ha}}{\sigma'_{v0}} \tag{5.5}$$

---

[1] For highly overconsolidated soils, this equation may lead to estimates quite far from reality. In the cases where a more accurate estimate is important, as in underground structures, an experimental *in situ* evaluation is highly recommended. As presented in Chapter 1, the self-boring pressuremeter and the flat dilatometer are the most convenient tests for this purpose.

On the other hand, the *passive earth pressure coefficient*, K_p, is the ratio between the horizontal effective stress in the passive state and the corresponding vertical effective stress:

$$K_p = \frac{\sigma'_{hp}}{\sigma'_{v0}} \tag{5.6}$$

The vertical effective stress, indicated in Equations 5.5 and 5.6, is not necessarily the vertical effective stress at-rest. As will become clear later on, in soil–structure interaction problems, where the active and passive state concepts are used, there may be cases where the vertical stress state at-rest is modified due to surface loading. On the other hand, the stresses in the numerator of K_a and K_p, in general, are not horizontal, since the wall face is often rough and/or inclined to the vertical.

In Figure 5.5b the deduction of K_a and K_p expressions is presented. In agreement with that deduction:

$$K_a = \frac{1-\sin\phi'}{1+\sin\phi'} = \tan^2\left(\frac{\pi}{4} - \frac{\phi'}{2}\right) \tag{5.7}$$

and

$$K_p = \frac{1+\sin\phi'}{1-\sin\phi'} = \tan^2\left(\frac{\pi}{4} + \frac{\phi'}{2}\right) \tag{5.8}$$

5.2.3 Active and passive thrusts according to Rankine

Rankine's method allows, under the postulated conditions, the calculation of the pressure (horizontal stress) on a wall that interacts with the soil mass in limit equilibrium state at a specific depth. As initially formulated, Rankine's method is based on the following hypotheses: i) the soil is non cohesive (the shear strength is null for null effective stress) and the water table does not interfere; ii) the ground surface is horizontal; iii) the wall is vertical and rigid; and iv) the friction between the soil and the wall is null.

The pressures on the wall at a given depth, z, are, as indicated in Figure 5.6:

$$\sigma'_{ha}(z) = K_a\sigma'_v(z) = K_a\gamma z \tag{5.9}$$

$$\sigma'_{hp}(z) = K_p\sigma'_v(z) = K_p\gamma z \tag{5.10}$$

for the active and passive cases, respectively, being the corresponding earth pressure coefficients given by Equations 5.7 and 5.8, respectively. It can therefore be concluded that, if the soil mass is homogeneous (Figure 5.6), the earth pressure graphs are triangular in shape.

The active and passive earth pressure resultants at a given depth, h, are the *active thrust* and the *passive thrust*, respectively, per linear meter of wall longitudinal development, being calculated by integration of the earth pressures between the surface and the given depth:

$$P_a = \int_0^h K_a\,\sigma'_v\,dz = \int_0^h K_a\gamma\,z\,dz = \frac{1}{2}K_a\gamma\,h^2 \tag{5.11}$$

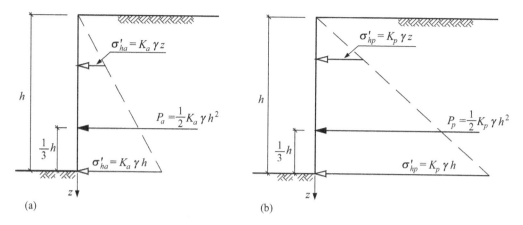

Figure 5.6 a) Active and b) passive earth pressure diagrams and thrusts (Rankine conditions).

$$P_p = \int_0^h K_p \, \sigma'_v \, dz = \int_0^h K_p \, \gamma \, z \, dz = \frac{1}{2} K_p \gamma \, h^2 \qquad (5.12)$$

If the soil mass is homogeneous, the thrust application point is at a depth of $(2/3)h$.

Therefore, the active and passive thrusts represent, respectively, the lower bound and the upper bound of the interaction forces between the soil and the wall. With regard to the K_a and K_p expressions, it is easy to conclude that the value of the first decreases whereas the second increases with increasing angle of shearing resistance.

5.3 DISPLACEMENTS ASSOCIATED WITH ACTIVE AND PASSIVE STATES

5.3.1 Terzaghi experiments

Terzaghi (1920, 1934) conducted model tests in the laboratory with the aim of quantifying the magnitude of structural wall displacements, to which the active and passive limit equilibrium states are mobilized in the adjacent soil masses. As shown in Figure 5.7a, these experiments were performed using a tank or container full of sand, carefully deposited in layers, where one of the lateral walls was hinged at the base. There were also sensors at different heights capable of measuring the earth pressures, and, therefore, the thrust.

During sand filling, the wall was kept in the vertical position; then, at the end of filling the thrust was registered, being considered to be the at-rest thrust, P_0. In fact, it seems reasonable to imagine that the pressures installed against the wall would be similar to the ones that would be installed in that vertical plane if the limit of the tank were located at a great distance behind.

Starting from the situation just described, in certain tests, the wall (hinged at the base) was moved against the soil; the registered thrust increased until a certain maximum value, considered the passive thrust, P_p, was reached. In other tests, the wall was moved in the opposite direction, and therefore the thrust decreased to values lower than P_0, until a certain minimum value, considered the active thrust, P_a, was reached.

Figure 5.7b shows the typical shape of the curves that relate the thrust change with the top wall displacement, d. The tests clearly indicate that substantial displacements are needed

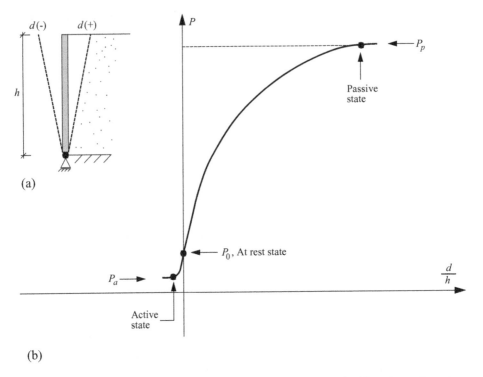

Figure 5.7 Terzaghi's experiment concerning the displacements associated with active and passive states: a) test apparatus; b) typical shape of the graphs relating thrust with the top wall displacement.

for mobilization of the passive state. On the contrary, very small displacements lead to the mobilization of the active state. In terms of orders of magnitude obtained from this and other similar tests, the displacements for the passive state can exceed 5% or even 10% of the wall height, while for the active state 0.1% to 0.5% of the wall height will usually be enough.

5.3.2 Stress paths: typical results of triaxial tests

The explanation for the results presented can be found in the stress–strain–strength relationships of soils, such as sandy soils in the present case. Figure 5.8a shows the effective stress paths corresponding to the mobilization of active and passive limit states. These active and passive states can be obtained by performing, with specimens consolidated to the effective stress state at-rest, triaxial compression tests (with decrease in the horizontal stress, keeping the vertical stress constant), and triaxial extension tests (with increase in horizontal stress, keeping the vertical stress constant), respectively.

Figure 5.8b shows the typical curves from tests performed on those stress paths over loose or dense sand specimens. The results show the changes in the specimens' horizontal strain against the deviator stress. It can be observed that: i) as expected, for the same type of test, the strength is greater and the strain at failure is lower in the dense sand; ii) for both sands, the strain at failure is much lower for the compression test than for the extension test; and iii) the contrast between these deformations is greater in the loose sand.

The explanation for the deformation contrast arises from two different issues. The first is due to the fact that the incremental stresses, corresponding to the transition from the at-rest

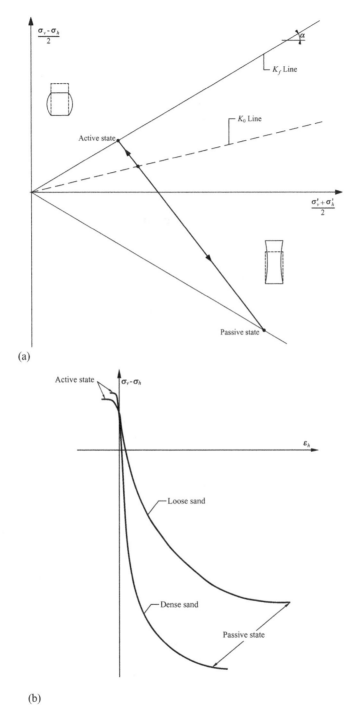

(a)

(b)

Figure 5.8 Active and passive Rankine states: a) effective stress paths; b) typical shape of graphs relating maximum shear stress to horizontal strain on specimens of loose and dense sand.

stress state to the active state, are much less than those necessary to pass from the at-rest stress state to the passive state; compare, for this purpose, the magnitude of $\left|\sigma'_{h0} - \sigma'_{ha}\right|$ and of $\left|\sigma'_{h0} - \sigma'_{hp}\right|$ in Figure 5.5, or the length of the corresponding stress path in Figure 5.8a.

Moreover, the soil stiffness depends on the stress paths. Typically, the soils seem to show greater stiffness in the stress paths where the mean effective stress is reduced – as in the case of the active state – than in those where the mean effective stress is increased – as in the case of the change to the passive state.

The reasons for the behavior observed in the experiments is therefore understood: the mobilization of the active state needs much lower displacements since it involves much smaller incremental stresses and, for the type of loading involved, the soils have greater stiffness.

5.3.3 Consequences for the design of civil engineering structures

From the previous section, it becomes clear that the magnitude of the earth pressures depends on the displacements of the wall. This is, in fact, a critical issue in the design of structures that interact with soil masses by means of vertical (or close to vertical) walls, which is the aim of this chapter.

When the structure is supported by the soil mass, the wall moves against the soil. In general terms, the equilibrium pressures resulting from such an interaction will be greater than the at-rest stresses and equal or less than the passive stresses, being the corresponding integral, i.e., the mobilized thrust, a *support reaction*. The values that this thrust can assume, therefore, cover a very wide range (see Figure 5.7). However, it should be taken into consideration that the displacements needed for the mobilization of the maximum thrust value, i.e., the passive thrust, are very high, and generally incompatible with the ability of the structure to deform. In the design, it is, therefore, usual to limit the passive mobilized resistance.[2] The level of that limitation depends on the type of structure (of higher or lower ductility) and the type of soil (of higher or lower density).[3]

Moving on to another type of interaction, in which the structure supports the soil, the resultant of the interaction stresses, the mobilized thrust, is between the at-rest thrust and the active thrust.

If the structure displacement is null, the thrust will be equal to the one at-rest. This is generally the case with respect to the peripheral walls of building basements, as represented in Figure 5.1a, due to the restraint provided by the beams and slabs of the internal structure.

In terms of gravity walls, like the one represented in Figure 5.1b, the wall movement is essentially controlled by its foundation. As represented in Figure 5.9, the wall foundation receives an inclined force with a certain eccentricity, which is the resultant of its weight and the thrust of the retained soil mass. The equilibrium of this force will require the mobilization of tangential and normal stresses on the wall base, as the figure suggests. The normal stresses, with the distribution indicated, entail growing settlements from the foundation interior border to the exterior border. On the other hand, the tangential stresses require a (very small) tangential displacement between the wall and the foundation.

---

[2] These considerations are done in a purely deterministic context. The uncertainties related to the actions and to the soil strength parameters, that would normally advise against the adoption in design of the maximum value of the thrust, even if the displacements needed for its mobilization were acceptable for the structure itself, were not taken into account.

[3] In traditional geotechnical practice, this is taken into account by adopting high global safety factors. As discussed in Chapter 3, these factors are also used as a semi-empirical method of deformation control.

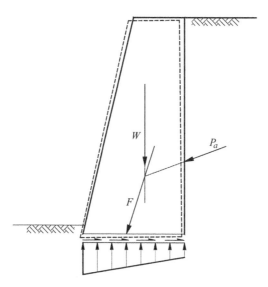

Figure 5.9 Gravity retaining wall, force transmitted to the foundation, stresses mobilized in the wall base, and associated typical movement.

In summary: i) the movement that the wall experiences is essentially due to the deformation of the foundation soil; and ii) that movement can be simply characterized as the combination of a rotation with a horizontal displacement, both directed away from the supported soil mass.

The reader has probably noticed that the thrust represented in Figure 5.9 is the active thrust. This is, in fact, the thrust used in the external design of gravity walls. With the active thrust being the *minimum thrust*, as reported before, this is a very special situation in the context of the design of civil engineering structures.

Contrary to what would be expected, this option is not unsafe, since the active thrust is mobilized for very small displacements of the wall. For that reason, no failure situation (by insufficient bearing capacity of the foundation soil, by base sliding, etc...) can develop without enough wall displacement to the outer side, which ensures that the thrust is reduced to its minimum value, the active thrust. In other words, the soil resistance behind the wall will be mobilized for much smaller deformations than the ones corresponding to the mobilization of the resistance of the foundation soil. If the backfill thrust exceeds the active thrust, this is a good symptom: it means that the wall foundation soil is more resistant than necessary.

In Chapter 8, these issues will be analyzed in detail. In any case, the previous considerations will help introduce the reader in the following context, where the evaluation of the active and passive earth pressure will be treated in a more general way.

5.4 GENERALIZATION OF RANKINE'S METHOD

5.4.1 Introduction

In Section 5.2, Rankine's method was presented for highly idealized, very simple conditions: a granular, homogeneous soil, with a horizontal surface, above the water table, interacting with a vertical, rigid, and smooth wall. In the following sections, this theory will be applied to more complex situations, usually associated with practical engineering cases.

5.4.2 Vertical, uniformly distributed surcharges applied to the ground surface

If a vertical, uniformly distributed surcharge, q, is applied to the ground surface, as Figure 5.10 shows, a vertical effective stress at any point of the soil mass increases consequently by the same amount. So,

$$\sigma'_v(z) = \gamma z + q \tag{5.13}$$

If the soil is in limit equilibrium state, the pressure (active or passive) on the wall at a certain depth, z, becomes,

$$\sigma'_h(z) = K\sigma'_v(z) = K\gamma z + Kq \tag{5.14}$$

where K is K_a or K_p, depending on the case.

It can be therefore concluded that a vertical, uniformly distributed surcharge on the ground surface leads, in a limit equilibrium Rankine state, to a uniform earth pressure acting on the wall, defined by the product of the surcharge by the corresponding earth pressure coefficient. This should be added to the earth pressure graph.

5.4.3 Stratified ground

A similar reasoning can be used to calculate the pressures in a stratified ground, if each layer has a certain value of unit weight and angle of shearing resistance. As Figure 5.11 indicates, the pressure in the point immediately above the layer separation surface is calculated as explained in Section 5.2.3, being $K_1\gamma_1h_1$ (where K_1 is the active or passive coefficient of earth pressure).

In the calculation of the pressures for depths corresponding to layer 2, layer 1 can be considered a uniformly distributed surcharge, the value of which is γ_1h_1, leading to a uniform

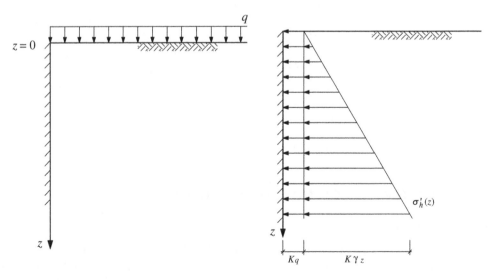

Figure 5.10 Application of Rankine's method when there is a vertical uniformly distributed surcharge applied to the ground surface.

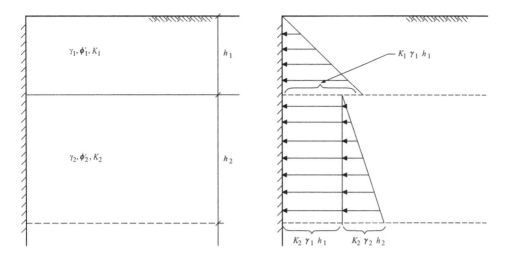

Figure 5.11 Application of Rankine's method to stratified ground.

graph of $K_2\gamma_1h_1$. This graph is added to the earth pressures associated with layer 2, which, at a depth of h_2 below the layer separation surface, is $K_2\gamma_2h_2$.

It should be noted that, since ϕ_1' and ϕ_2' are different, K_1 and K_2 will also be, and therefore the resulting graph (represented in the right-hand part of Figure 5.11) has a discontinuity at the separation depth.

5.4.4 Ground with water table

If the ground includes a stationary water table, the problem can be handled as if two layers existed, one above the water table, of unit weight γ, and another below the water table, of unit weight γ' (submerged unit weight). This means that, if the soil mass, or just part of it, is submerged, the earth pressures decrease. However, these pressures should be added to the hydrostatic pressures, thereby making the total thrust (soil plus water) substantially higher (in the active case) than in the case where the water table does not exist. The opposite occurs in the passive case.

Figure 5.12 represents the calculation process. Graph 1 corresponds to the soil above the water level, increasing from the surface to that level, and being constant from then on, since that soil can be considered to represent a uniform surcharge of value $\gamma(h - h_w)$. Graph 2

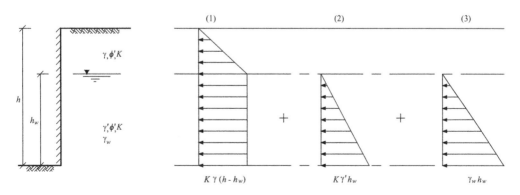

Figure 5.12 Application of Rankine's method to partially submerged soil masses.

corresponds to the soil below the water table, whereas graph 3 represents the hydrostatic pressures.

Being the same soil (with the same ϕ', and consequently with the same K, above and below the water level), the resulting graph presents, at the water level depth, a breakpoint rather than a discontinuity.

5.4.5 Extension of Rankine's theory to soils with cohesion

Figure 5.13 illustrates the Mohr circles representative of the stresses at a depth z in a horizontal surface ground with cohesion and friction angle, in the limit equilibrium Rankine states. It can therefore be verified that the slopes of the failure surfaces correspond to the ones of Figure 5.5.

The deduction of the equations of the coefficients of earth pressures, for the present case, does not offer any difficulty. These equations, called the Rankine-Résal equations, are as follows :

$$K_a = \frac{\sigma'_{ha}}{\sigma'_v} = \frac{1-\sin\phi'}{1+\sin\phi'} - \frac{2c'}{\sigma'_v}\frac{\cos\phi'}{1+\sin\phi'}$$

$$= \tan^2\left(\frac{\pi}{4} - \frac{\phi'}{2}\right) - \frac{2c'}{\sigma'_v}\tan\left(\frac{\pi}{4} - \frac{\phi'}{2}\right)$$

(5.15)

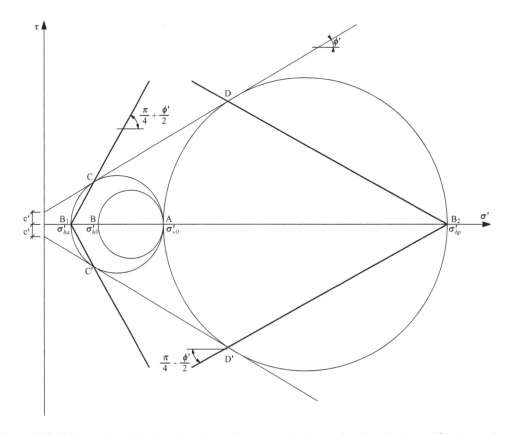

Figure 5.13 Mohr circles in the at-rest, active and passive states in a soil with cohesion and friction angle.

$$K_p = \frac{\sigma'_{hp}}{\sigma'_v} = \frac{1+\sin\phi'}{1-\sin\phi'} + \frac{2c'}{\sigma'_v}\frac{\cos\phi'}{1-\sin\phi'}$$

$$= \tan^2\left(\frac{\pi}{4}+\frac{\phi'}{2}\right) + \frac{2c'}{\sigma'_v}\tan\left(\frac{\pi}{4}+\frac{\phi'}{2}\right)$$

(5.16)

These coefficients are not constant, becoming depth dependent. They both tend to the values of the non-cohesive soils (Equations 5.7 and 5.8) when σ'_v tends toward infinity.

Figure 5.14 shows the Mohr circles representative of the limit equilibrium states at a depth z in a purely cohesive soil mass, $\phi = 0$. As is well known, the concept of purely cohesive ground is meaningful only when the analysis is made in total stresses. So, unlike the situation described in previous figures, in Figure 5.14 the stresses represented are total stresses.

As can be observed, in this case the slopes of the failure surfaces are the same for the two limit situations, the value of which is 45°. At any particular depth, the passive and active pressures differ by $4c_u$.

The deduction of K_a and K_p equations is for this case almost immediate:

$$\sigma_{ha} = \sigma_v - 2c_u; \quad K_a = \frac{\sigma_{ha}}{\sigma_v} = 1 - \frac{2c_u}{\sigma_v}$$

(5.17)

$$\sigma_{hp} = \sigma_v + 2c_u; \quad K_p = \frac{\sigma_{hp}}{\sigma_v} = 1 + \frac{2c_u}{\sigma_v}$$

(5.18)

The earth pressure coefficients in this case are defined by ratios between total stresses. As in the previous case, the coefficients are not constants, but they tend toward the same value, the unit, as depth increases.

It should be pointed out that, in non-cohesive soils, it is always possible to create a limit equilibrium situation (active) due to the decrease in the lateral stress. In cohesive soils, however, this is not always possible. As discussed in Chapter 2 (see Section 2.2.2) for a purely cohesive soil, if the vertical stress is lower than $2c_u$, even if σ_h is null, which happens, for example, at the face of an unsupported excavation, the Mohr circle is not tangential to the envelope. It can therefore be concluded that, for depth z values lower than z_0 given by (see Figure 2.2):

$$z_0 = \frac{2c_u}{\gamma}$$

(5.19)

it is only possible to install the active state if tensile stresses are applied to the soil mass.

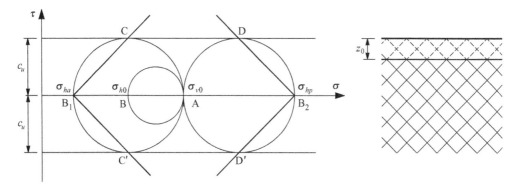

Figure 5.14 Limit equilibrium states in a purely cohesive soil mass: the Mohr circles in the at-rest, active and passive states.

5.4.6 Extension of Rankine's theory to soil masses with inclined surface interacting with vertical wall

Figure 5.15 represents a non-cohesive soil mass with the surface inclined of an angle β to the horizontal. Consider the equilibrium of an infinitesimal prism of vertical faces, the upper side is the ground surface and the lower side is at a depth dz, with ds being the area of the base. Please note that, as ds is infinitesimal, the forces that act on the prism lateral faces have to be equal and symmetrical.

Figure 5.15 illustrates, in general, the forces applied to the element. Considering the axes represented in the figure, two equilibrium equations can be defined, corresponding to the projection of the forces in the two axes:

$$\sum F_x = 0; \quad dP - dT - dP = 0; \quad dT = 0 \tag{5.20}$$

$$\sum F_y = 0; \quad dV + dW - dR - dV = 0; \quad dR = dW \tag{5.21}$$

and an equilibrium equation in terms of the moments of the forces in relation to a given point P:

$$\sum M_P = 0; \quad dV = 0 \tag{5.22}$$

In conclusion: i) the forces that act on the faces parallel to the surface are strictly vertical; and ii) the forces acting on the vertical faces are parallel to the surface. On the other hand

$$dW = \gamma \ dz \ ds \ \cos\beta \tag{5.23}$$

Projecting $dW(= dR)$, in the perpendicular and parallel directions to the base of the prism, and dividing these projections by the area (ds) of that base, it is possible to obtain the stress

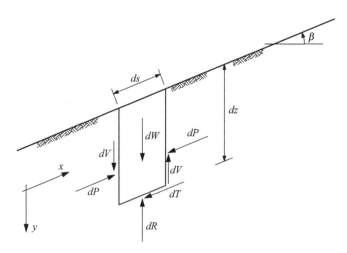

Figure 5.15 Non-cohesive soil mass of inclined surface and element of infinitesimal dimensions, vertical faces and base parallel to the surface.

components in that same base, as well as the stress inclination in relation to the perpendicular direction:

$$\sigma' = \gamma \, dz \cos^2 \beta \tag{5.24}$$

$$\tau = \gamma \, dz \sin \beta \cos \beta \tag{5.25}$$

$$\frac{\tau}{\sigma'} = \tan \beta \tag{5.26}$$

If

$$dp = \frac{dP}{dz} \tag{5.27}$$

then the stress in the lateral faces, its components, and its inclination in relation to the perpendicular direction are, respectively:

$$\sigma' = dp \cos \beta \tag{5.28}$$

$$\tau = dp \sin \beta \tag{5.29}$$

$$\frac{\tau}{\sigma'} = \tan \beta \tag{5.30}$$

In conclusion: the Mohr circle points, representative of the stresses in the two faces (the one parallel to the surface and the one vertical) belong to lines that pass through the origin and have an inclination β in relation to the horizontal.

It is now assumed that part of the soil mass is removed and replaced by a rigid, smooth, and vertical wall face. In this case, limit equilibrium states (active and passive) can be installed in the soil mass, by moving the wall progressively away from the remaining soil or by pushing it against the soil.

In Figure 5.16, the Mohr circles corresponding to the active and passive limit equilibrium states at a point P at a certain depth z are represented. Please note that point A, which is common to both circles, is not on the abscissa axis, as in the previous cases, but in a line of inclination angle β. The position of the circle poles and of the points which are representative of the stresses in the vertical planes adjacent to the wall are, therefore, clear from this figure. In those planes, the stresses are parallel to the surface of the soil mass.

The magnitude of the active stress in the wall at a generic depth z is the segment OB_1, represented in the figure by p_a, while the magnitude of the passive stress is the segment OB_2, represented by p_p. It is possible to relate the magnitude of the segments OB_1 and OB_2 with segment OA, as this is known (see Equations 5.24 and 5.25):

$$p_a = \cos \beta \, \frac{\cos \beta - \sqrt{\cos^2 \beta - \cos^2 \phi'}}{\cos \beta + \sqrt{\cos^2 \beta - \cos^2 \phi'}} \gamma \, z \tag{5.31}$$

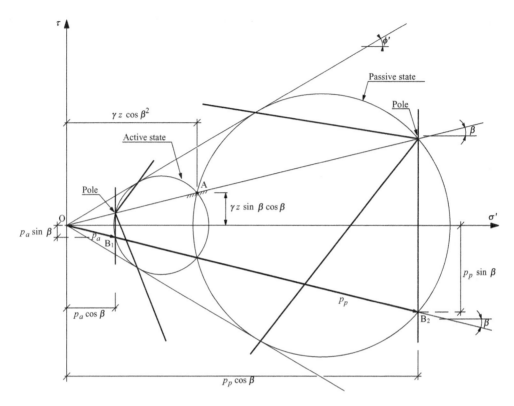

Figure 5.16 Mohr circles in the active and passive states for the case of a non-cohesive soil mass of inclined surface in contact with a vertical wall.

and

$$p_p = \cos \beta \frac{\cos \beta + \sqrt{\cos^2 \beta - \cos^2 \phi'}}{\cos \beta - \sqrt{\cos^2 \beta - \cos^2 \phi'}} \gamma z \qquad (5.32)$$

Extending the concept of the active earth pressure coefficient, K_a, and the passive earth pressure coefficient, K_p, to the ratio of the active stress, p_a, and the passive stress, p_p, respectively, by γz, the previous equations become:

$$p_a = K_a \gamma z \qquad (5.33)$$

and

$$p_p = K_p \gamma z \qquad (5.34)$$

where K_a and K_p are:

$$K_a = \cos \beta \frac{1 - \sqrt{1 - \dfrac{\cos^2 \phi'}{\cos^2 \beta}}}{1 + \sqrt{1 - \dfrac{\cos^2 \phi'}{\cos^2 \beta}}} \qquad (5.35)$$

and

$$K_p = \cos\beta \, \frac{1 + \sqrt{1 - \dfrac{\cos^2\phi'}{\cos^2\beta}}}{1 - \sqrt{1 - \dfrac{\cos^2\phi'}{\cos^2\beta}}} \tag{5.36}$$

The integral of the active and passive stresses in a wall of height h, that is, the active and passive thrusts, are similar to the ones presented previously:

$$P_a = \frac{1}{2} K_a \gamma h^2 \tag{5.37}$$

and

$$P_p = \frac{1}{2} K_p \gamma h^2 \tag{5.38}$$

being, as the corresponding stresses, parallel to the ground surface, and consequently acting on the wall with an inclination angle β in relation to the normal to the wall face.

Figure 5.16 also shows the directions of the faces where the soil fails by shear, whose angles to the horizontal are $\pi/4 + \phi'/2 + (\varepsilon - \beta)/2$ and $\pi/4 + \phi'/2 - (\varepsilon - \beta)/2$, where $\varepsilon = \sin^{-1}(\sin\beta / \sin\phi')$.

When $\beta = \phi'$, the soil mass is in limit equilibrium in the at-rest state, that is, before experiencing strains due to the interaction with a wall. As Figure 5.17 illustrates, in this situation,

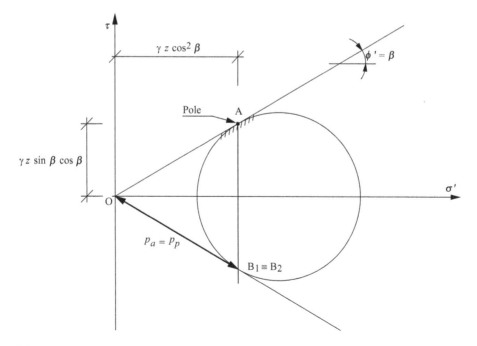

Figure 5.17 Mohr circle for a generic point of a non-cohesive soil mass, the inclination of which is equal to the angle of shearing resistance.

the faces at failure are the vertical and the ones parallel to the ground surface. The active and passive stresses coincide and are (see figure):

$$p_a = p_p = \sqrt{\left(\gamma z \cos^2 \phi'\right)^2 + \left(\gamma z \sin \phi' \cos \phi'\right)^2} = \cos \phi' \gamma z \qquad (5.39)$$

an expression that can also be obtained if $\beta = \phi'$ in Equations 5.31 and 5.32.

5.4.7 Extension of Rankine's theory to soil masses with inclined surface interacting with non-vertical wall

It is now assumed that the soil mass of inclined surface treated in Section 5.4.6 is interacting with a wall inclined at a certain angle, λ, in relation to the vertical and that the soil mass is in active or passive limit equilibrium state.

Figure 5.18a represents the active and passive Mohr circles, similar to the ones represented in Figure 5.16. Similarly to what was seen above, the points representative of the active and passive stresses in a point adjacent to the inclined wall are the points B_1 and B_2, respectively, obtained by drawing lines that pass through the poles of the corresponding Mohr circles and are parallel to the wall face. The magnitude of the active and passive stresses, p_a and p_p, are the segments OB_1 and OB_2, which act on the wall with an inclination that depends on ϕ', λ, and β.

Relating the magnitude of these two segments with the magnitude (already known) of the segment OA, the following equations are obtained (Chu, 1991):

$$p_a = \cos \beta \frac{\sqrt{1 + \sin^2 \phi' - 2 \sin \phi' \cos \theta_a}}{\cos \beta + \sqrt{\sin^2 \phi' - \sin^2 \beta}} \gamma z \qquad (5.40)$$

and

$$p_p = \cos \beta \frac{\sqrt{1 + \sin^2 \phi' + 2 \sin \phi' \cos \theta_p}}{\cos \beta - \sqrt{\sin^2 \phi' - \sin^2 \beta}} \gamma z \qquad (5.41)$$

where:

$$\theta_a = \sin^{-1}\left(\frac{\sin \beta}{\sin \phi'}\right) - \beta + 2\lambda \qquad (5.42)$$

and

$$\theta_p = \sin^{-1}\left(\frac{\sin \beta}{\sin \phi'}\right) + \beta - 2\lambda \qquad (5.43)$$

From Figure 5.18b, it is clear that the base of the wall is at a depth below the ground surface, z_{max}, which is:

$$z_{max} = \frac{\cos(\beta - \lambda)}{\cos \beta \cos \lambda} h \qquad (5.44)$$

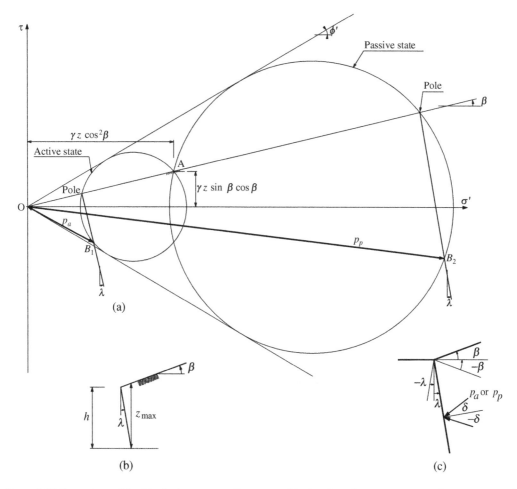

Figure 5.18 Extension of Rankine's theory to soil masses of inclined surface in contact with a non-vertical wall (adapted from Chu, 1991): a) Mohr circles for active and passive states; b) depth of the foot of the wall; c) sign convention.

where h is the height of the wall.

So, $p_{a,max}$ and $p_{p,max}$, the active and passive stresses at the base of the wall, respectively, can be obtained by replacing z with z_{max} in Equations 5.40 and 5.41, respectively:

$$p_{a,max} = \frac{\cos(\beta - \lambda)}{\cos\lambda} \frac{\sqrt{1 + \sin^2\phi' - 2\sin\phi'\cos\theta_a}}{\cos\beta + \sqrt{\sin^2\phi' - \sin^2\beta}} \gamma h \tag{5.45}$$

$$p_{p,max} = \frac{\cos(\beta - \lambda)}{\cos\lambda} \frac{\sqrt{1 + \sin^2\phi' + 2\sin\phi'\cos\theta_p}}{\cos\beta - \sqrt{\sin^2\phi' - \sin^2\beta}} \gamma h \tag{5.46}$$

Bearing in mind that the stress distribution through the wall varies linearly from zero at the surface to the values given by the last two equations, and that the length of the wall is $h/\cos\lambda$, the active and passive thrusts are:

$$P_a = \frac{1}{2}\frac{\cos(\beta-\lambda)}{\cos^2\lambda}\frac{\sqrt{1+\sin^2\phi'-2\sin\phi'\cos\theta_a}}{\cos\beta+\sqrt{\sin^2\phi'-\sin^2\beta}}\gamma h^2 \tag{5.47}$$

and

$$P_p = \frac{1}{2}\frac{\cos(\beta-\lambda)}{\cos^2\lambda}\frac{\sqrt{1+\sin^2\phi'+2\sin\phi'\cos\theta_p}}{\cos\beta-\sqrt{\sin^2\phi'-\sin^2\beta}}\gamma h^2 \tag{5.48}$$

or

$$P_a = \frac{1}{2}K_a\gamma h^2 \tag{5.49}$$

and

$$P_p = \frac{1}{2}K_p\gamma h^2 \tag{5.50}$$

where

$$K_a = \frac{\cos(\beta-\lambda)}{\cos^2\lambda}\frac{\sqrt{1+\sin^2\phi'-2\sin\phi'\cos\theta_a}}{\cos\beta+\sqrt{\sin^2\phi'-\sin^2\beta}} \tag{5.51}$$

and

$$K_p = \frac{\cos(\beta-\lambda)}{\cos^2\lambda}\frac{\sqrt{1+\sin^2\phi'+2\sin\phi'\cos\theta_p}}{\cos\beta-\sqrt{\sin^2\phi'-\sin^2\beta}} \tag{5.52}$$

These expressions become Equations 5.35 and 5.36 for $\lambda=0$ and Equations 5.7 and 5.8 for $\lambda=\beta=0$.

Conversely to the situation analyzed in Section 5.4.6, the active and passive stresses are no longer parallel to the ground surface, and their angle with the wall is no longer the same for the two stresses. This can be confirmed by Figure 5.18a, where the ratio between the y value and the x value is no longer the same for points B_1 and B_2. It can be demonstrated that the inclination of the stresses and thrusts in relation to the normal to the wall face is given by the expressions:

$$\delta_a = \tan^{-1}\left(\frac{\sin\phi'\sin\theta_a}{1-\sin\phi'\cos\theta_a}\right) \tag{5.53}$$

and

$$\delta_p = \tan^{-1}\left(\frac{\sin\phi'\sin\theta_p}{1+\sin\phi'\cos\theta_p}\right) \tag{5.54}$$

All previous expressions (of Section 5.4.7) should be applied, bearing in mind the sign convention of Figure 5.18c.

5.5 BOUSSINESQ, RÉSAL, AND CAQUOT THEORIES FOR THE CONSIDERATION OF SOIL WALL FRICTION

5.5.1 Introduction

As was just concluded, the active and passive stresses, when evaluated by Rankine's generalized theory, have, in the most complex cases (that is, when the ground surface is inclined and/or when the wall is not vertical), a direction that *is not perpendicular to the wall and that results from the application of the method itself.* This is a limitation of generalized Rankine's theory: it does not allow the angle defining the inclination of the thrust *to be given* (that is, to be assumed as one of the design parameters, such as, for example, the soil angle of shearing resistance), being instead *a result* of the thrust evaluation method.

The inclination angle of the stress in relation to the normal to the wall requires the mobilization of a certain tangential strength in the soil–structure interface. This strength depends on the shear resistance of the soil and on the wall roughness, depending therefore on the specific conditions of the problem to be analyzed. In general, the interface strength is defined through an *angle of shearing resistance*, often designated as the soil–wall friction angle, represented by the symbol δ. This is, in general, assumed to be a fraction of ϕ', the angle of shearing resistance of the soil. For very rough walls, δ will tend to be equal to ϕ'.

To better understand the consequences of this question, Figure 5.19 compares the interaction without soil–wall friction (the Rankine hypothesis) and with soil–wall friction, for the active and passive states, in the case originally studied by Rankine, that is, a soil mass of horizontal surface interacting with a vertical wall.

When the wall moves away from the soil, the soil tends to descend in relation to the wall. In a rough wall, the tangential stresses that the soil applies to the wall are directed

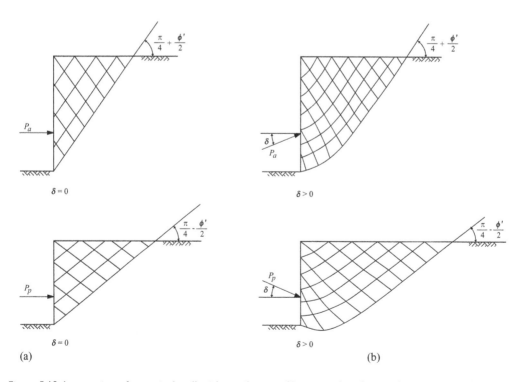

Figure 5.19 Interaction of a vertical wall with a soil mass of horizontal surface in the active state (above) or passive state (below): a) smooth wall – Rankine's conditions; b) rough wall.

downwards, and so the active thrust lies above the normal to the wall at an angle δ. On the contrary, when the wall is pushed against the soil, the soil tends to rise in relation to the wall, applying tangential stresses directed upwards, with the passive thrust lying below the normal to the wall at an angle δ. (Figure 5.19b represents the forces that the wall applies to the soil, that is, forces equal but opposite in direction to the active and passive thrusts).

The tangential stresses mobilized on the wall face cause not only the inclination of the thrust but also, as suggested by Figure 5.19b, a curvature of the failure surfaces. In fact, those tangential stresses cause *a rotation of the principal stresses* in relation to the vertical and horizontal positions (which are the prevailing positions for the conditions of Figure 5.19a). This rotation is maximum next to the wall – since its cause is located there – and tends toward zero at the approximation to the ground surface. Bearing in mind that the failure surfaces have an angle of $\pi/4 - \phi'/2$ with the maximum principal stress (as illustrated in Figure 5.5), those surfaces become curved when the direction of this principal stress varies from point to point.

As will be detailed further on, the existence of friction between the soil and the wall favors the stability of the structures that interact with the soil. In fact: i) the magnitude of the passive thrust increases substantially with the value of δ (see Table A5.1.8); ii) the inclination of the active thrust in relation to the normal to the wall face, reduces the moments and horizontal forces that must be equilibrated by the retaining structure.

5.5.2 The Boussinesq theory: the Caquot-Kérisel tables

The problem of calculating the stresses corresponding to active and passive states, when there is soil–wall friction, was formulated initially by Boussinesq (1885). An exposition of the Boussinesq theory exceeds the scope of this book. However, a brief idea of the starting hypotheses will be given, to introduce the reader to the context that led to the Caquot-Kérisel tables, at present used by engineers.

Figure 5.20 represents a wall with a certain inclination to the vertical, in contact with a homogeneous mass of granular soil, of which the angle of shearing resistance is ϕ' and the unit weight is γ, limited by a surface with a certain inclination to the horizontal. Defining generically the stress tensor in a point M function of r and θ (polar coordinates), Boussinesq assumed the following hypotheses:

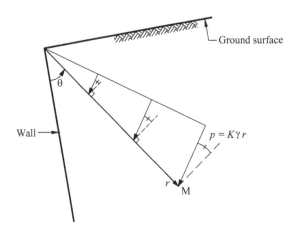

Figure 5.20 Boussinesq theory hypotheses.

i) the ratio (τ/σ') of the tangential and normal stresses on a surface element, contained by a polar radius, is constant along the radius, and therefore, independent of r;

ii) the value of that stress is proportional to r when θ is constant, i.e., along the same radius;

iii) the value of the referred stress is proportional to the soil unit weight, γ.

From these hypotheses, Boussinesq formulated a system of differential equations, imposing the static equilibrium of the soil mass (equilibrium equations for each point), imposing the limit equilibrium condition (through the Mohr-Coulomb law, $\tau/\sigma' = \tan\phi'$), and, finally, imposing adequate boundary conditions (null stresses at the ground surface and stresses at the wall with an inclination δ in relation to the normal).

Due to the complexity of these equations, analytical integration was not accomplished by Boussinesq. After complementary studies made by Résal (1903, 1910), the numerical integration of the system of equations was achieved by Caquot and Kérisel (1949). From that resolution, it is possible to assess the stress state in soil mass under active or passive limit equilibrium, namely the stresses acting on the wall, as well as the field of failure planes (illustrated schematically in Figure 5.19).

Caquot and Kérisel have condensed the result of their work into tables – known by the names of their authors – which give the active and passive earth pressure coefficients as a function of four angles, as shown in Figure 5.21: the angles that define the soil and interface strength, ϕ' and δ, respectively, and the angles that define the geometry, β and λ, respectively (indicated in the figure with the positive sign). In each point P of the wall at a distance l from the vertex, O, the active and passive stresses are given, respectively, by:

$$p_a = K_a \gamma l \tag{5.55}$$

and

$$p_p = K_p \gamma l \tag{5.56}$$

and the active and passive thrusts, corresponding to the stress integral between O and P, by:

$$P_a = \frac{1}{2} K_a \gamma l^2 \tag{5.57}$$

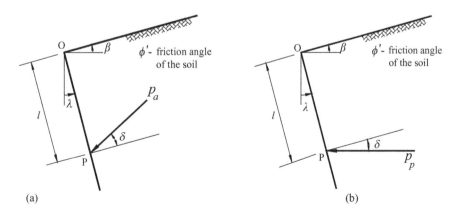

(a)

(b)

Figure 5.21 Conventions used in Caquot-Kérisel tables (all the marked angles are positive): a) active state; b) passive state.

and

$$P_p = \frac{1}{2} K_p \gamma l^2 \tag{5.58}$$

If the soil mass is homogeneous, the application point of the thrust is located at a distance equal to $2/3l$ from O.

In Annex A5.1 an extract of the Caquot and Kérisel tables is included.

5.5.3 Cohesive soil masses: theorem of corresponding states

Up to this point, the generalization of Rankine's theory has been presented to take into account the soil–wall friction, but only with respect to non-cohesive soils. It is time now to discuss how the active and passive stresses can be determined for soils with cohesion and friction angle. This is the purpose of the theorem of corresponding states, also formulated by Caquot and Kérisel. The presentation will be done for the active case and for a soil mass where the surface is inclined by β in relation to the horizontal, interacting with a wall inclined by λ in relation to the vertical.

Figure 5.22 represents, with a solid line, the determination of the active stress on the wall face (represented by the coordinates of point Q), for a soil with cohesion and friction angle in the active limit equilibrium state. It can be seen that the method is analogous to Figure 5.18. A Mohr-Coulomb envelope is then drawn corresponding to a hypothetical, non-cohesive soil, with the angle of shearing resistance equal to that of the real soil (dashed line). Consider now a Mohr circle (also denoted by a dashed line) with the same diameter of the circle previously drawn for the real soil, but moved to the right by a stress

$$H = c' \cotan \phi' \tag{5.59}$$

It can be seen that this circle is tangential to the Mohr-Coulomb envelope of the non-cohesive soil. Considering, for this circle, the point (P*) corresponding to the one that, for the

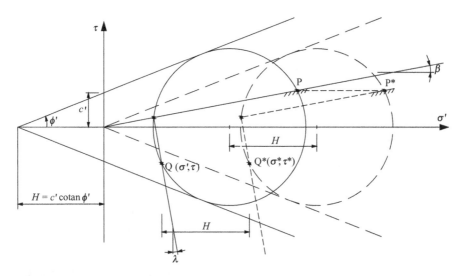

Figure 5.22 Theorem of corresponding states.

real soil, represents the stress on a plane parallel to the free surface (P), the circle pole can be found. This allows the active stress on the wall (Q*) to be determined for the hypothetical, non-cohesive soil.

Then, it is possible to verify that, between the components of this stress and those of the stress in the wall for the real soil (Q), the following relationships apply:

$$\tau = \tau^* \tag{5.60}$$

$$\sigma' = \sigma^* - H \tag{5.61}$$

This allows the conclusion that the stresses inside the cohesive soil mass in limit equilibrium are equal to the stresses in the non-cohesive soil, with the same angle of shearing resistance, with the same geometry, under the same external forces, and also in limit equilibrium, if a normal pressure H, given by Equation 5.59, is considered to act on the surface of the virtual soil; the application of this pressure or surcharge corresponds to the translation to the right operated in the Mohr circle along the xx axis). To obtain the real stress acting on a given plane of a given point, the normal stress that acts on the same plane of the same point of the hypothetical soil mass should be subtracted from a stress H given by Equation 5.59.

Figure 5.23 illustrates the calculation procedure for the active stresses on the back of a gravity wall, applying the theorem of corresponding states. It is possible to pass to the

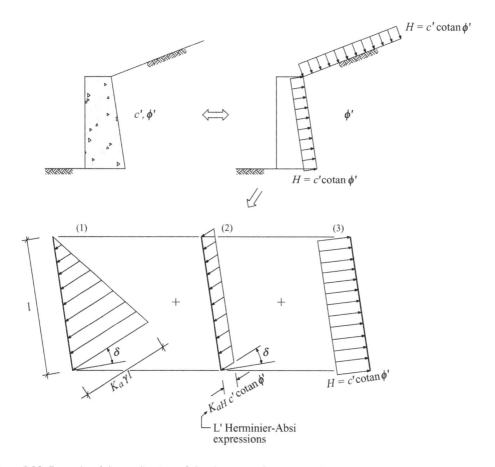

Figure 5.23 Example of the application of the theorem of corresponding states.

analysis of a non-cohesive soil problem, adding, at the surface and at the wall face, the pressures represented in the top right sketch. If surcharges exist at the surface, they would, obviously, be considered.

The stresses on the wall result, then, of the sum of three graphs, as shown in the lower part of the figure:

1) graph 1 for the non-cohesive soil;
2) graph 2 corresponds to the surcharge $H = c'\cotan \phi'$, perpendicular to the surface;
3) finally, graph 3 refers to the pressure H that the application of the theorem of corresponding states requires to be subtracted from each point to obtain the real values of the normal stress components (Equation 5.61); for that reason, the pressure is directed to the interior of the soil mass and perpendicular to the wall.

The evaluation of graph 2 associated with the uniform surcharge applied perpendicular to the surface (and, therefore, only vertical when the ground surface is horizontal) can be made using the expressions of L'Herminier-Absi, included in Figure 5.24 (Absi, 1962; L'Herminier, 1967). These expressions are derived from Prandtl theory that studies the plastic equilibrium of a purely frictional soil mass without weight, limited by a rough wall and by a surface, inclined or not, where a uniform surcharge is applied with any inclination in relation to the direction perpendicular to the surface. The exposition of this theory exceeds the scope of this book. However, it may be noticed that its development has some similarities with the Boussinesq theory referred to above.

5.6 COULOMB'S THEORY

5.6.1 General presentation and hypotheses

This section concerns the evaluation of the active and passive thrusts according to Coulomb (1773), who formulated the first scientific theory to evaluate earth pressures in limit

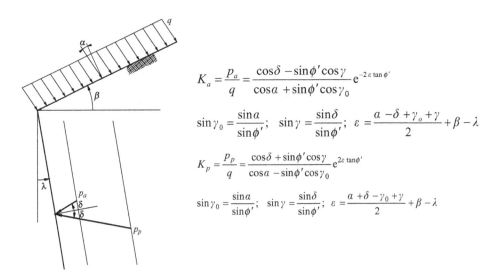

$$K_a = \frac{p_a}{q} = \frac{\cos\delta - \sin\phi'\cos\gamma}{\cos a + \sin\phi'\cos\gamma_0} e^{-2\varepsilon\tan\phi'}$$

$$\sin\gamma_0 = \frac{\sin a}{\sin\phi'}; \quad \sin\gamma = \frac{\sin\delta}{\sin\phi'}; \quad \varepsilon = \frac{a - \delta + \gamma_o + \gamma}{2} + \beta - \lambda$$

$$K_p = \frac{p_p}{q} = \frac{\cos\delta + \sin\phi'\cos\gamma}{\cos a - \sin\phi'\cos\gamma_0} e^{2\varepsilon\tan\phi'}$$

$$\sin\gamma_0 = \frac{\sin a}{\sin\phi'}; \quad \sin\gamma = \frac{\sin\delta}{\sin\phi'}; \quad \varepsilon = \frac{a + \delta - \gamma_0 + \gamma}{2} + \beta - \lambda$$

Figure 5.24 Expressions of L'Herminier-Absi.

equilibrium. Since the designations of active and passive thrust, as well as the corresponding earth pressure coefficients, which are now generally adopted, were in fact introduced by Rankine almost one century later, the original Coulomb method consisted in the evaluation of *minimum and maximum limit soil–wall interaction forces*, and therefore in the actual determination of active and passive thrusts[4].

Essentially, Coulomb assumes that the soil wedge – a non-cohesive and homogeneous soil above the water level – responsible for the limit interaction force with the structural wall, is limited by a plane surface intersecting the toe of the wall face, as shown in Figure 5.25. To determine that force, the wedge condition is assumed to be imminent sliding along the surface mentioned and along the wall itself. The sliding will be of descending or ascending, depending on whether the structure is retaining the soil or the soil is supporting the struc-ture. Knowing the angles of shearing resistance of the soil and of the soil–wall interface, the

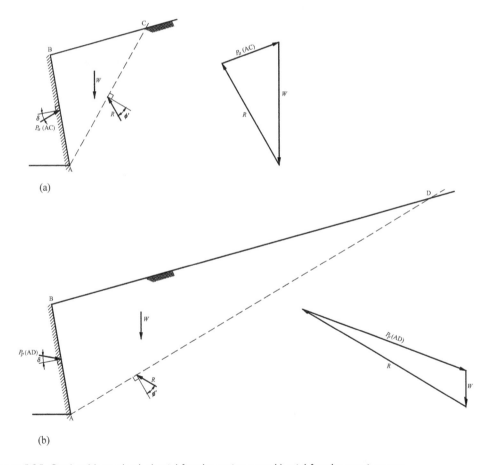

Figure 5.25 Coulomb's method: a) trial for the active case; b) trial for the passive case.

[4] Charles Augustin Coulomb (1736–1806) developed his earth pressure evaluation method in the scope of for-tress design as a military engineer in the Kingdom of France. He was also notable for his scientific work on electricity and electromagnetism.

hypothesis of imminent sliding means that the problem is statically determined, allowing the calculation of the limit interaction force.

The consideration of distinct soil wedges, limited by sliding surfaces, allows the identification of a series of interaction forces, P_1, P_2, ..., P_n.

In the problems where the structure retains the soil, the force to select – called *active thrust* since Rankine – will be the *highest* of that series. On the contrary, in those problems where the soil supports the structure, the force to select – called *passive thrust* since Rankine – will be the *lowest* of that series.

Coulomb's method was then designed as a graphical trial method. Figure 5.25 represents one such trial for each of the active and passive states. Please note the orientation of the forces P and R in each of the cases, which is determined by the sense of the relative movement of the wedge with the remaining soil and the structure.

5.6.2 Culmann construction

A method, such as Coulomb's method, requires several attempts to determine the thrust, which is very laborious. It is therefore convenient to reduce the number of lines to be drawn to the minimum. This is the advantage of the construction, represented in Figure 5.26 for the case of the active thrust, due to Culmann (1875). The construction sequence is as follows:

1. it starts by drawing the horizontal line bb';
2. the line bf, that makes an angle ϕ' with the previous line, is drawn;
3. the line bg, that makes an angle ψ with the previous line is drawn (see in the figure the meaning of ψ);
4. the first sliding surface b_{c1} is selected, the corresponding soil weight (and surcharges) is calculated and point d_1 is marked on the line bf, so that, in a force scale randomly selected, the segment b_{d1} represents the weight of the abc_1 wedge (and the corresponding surcharges);
5. the segment d_1e_1, parallel to bg, represents, in the force scale selected, the reaction that the wall has to apply to avoid the sliding of the abc_1 wedge.

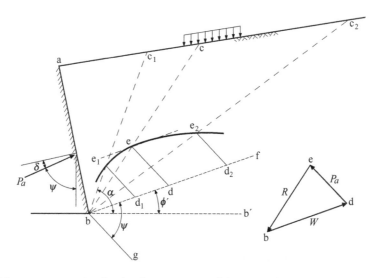

Figure 5.26 Culmann construction for the determination of the active thrust.

Repeating the operation for other wedges, it can readily be determined the point e, corresponding to the highest interaction force, which is the desired active thrust. The identification of the surface that limits the *critical wedge*, abc, is then obvious.

This construction is based on an artifice that allows a much faster calculation: the force polygons ($\mathrm{bd_1e_1}$, $\mathrm{bd_2e_2}$, etc.) are rotated counter-clockwise by an angle $\pi/2+\phi'$ in relation to the normal position (see, for instance, Figure 5.25a).

For the passive thrust, an analogous construction can be applied.

5.6.3 Determination of the thrust application point

Coulomb's method leads directly to the value of the thrust, i.e., to the resultant of the earth pressures from the surface to a depth, h. As will be shown below, for the stability analysis of structures that interact with the soil, it is necessary to know, not only the magnitude of the thrust, but also the corresponding application point. For that purpose, it is necessary to know the distribution of pressures at different depths.

Figure 5.27 illustrates the process that, in general, is followed to solve the problem (Terzaghi, 1943). First, the corresponding thrusts are determined at a series of depths lower than h, which allows the depiction of the graph representative of the change of the normal component of the thrust with respect to depth (Figure 5.27b). In an elementary area dz, situated at a generic depth z, a normal pressure, p_n, can be defined by the following procedure:

$$p_n = \frac{dP_n}{dz} \tag{5.62}$$

The distribution of p_n with respect to depth is then deduced (Figure 5.27c). The application point of the thrust is then situated at the same depth as the centre of gravity of the p_n graph.

5.6.4 Analytical solution

As the problem of the determination of the thrust is, according to Coulomb, a problem of determining maximum and minimum values, it can be treated analytically for the simple cases (linear wall face, linear ground surface, and without non-uniform surcharges acting on the surface).

Figure 5.28 concerns the determination of the active thrust (see the direction of R and P). Coulomb's methodology is no more than the search for the angle α, which defines the sliding

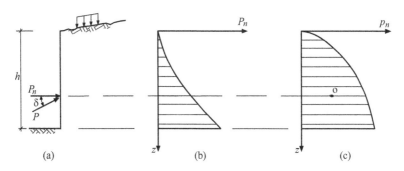

Figure 5.27 Methodology for the determination of the pressure graph and of the thrust application point when this is calculated by Coulomb's method.

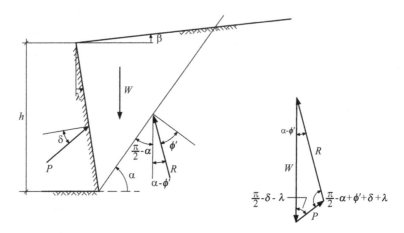

Figure 5.28 Coulomb's method – case susceptible to analytical solution.

surface orientation, which, in turn, leads to the maximum value of the force P. Based on the force polygon represented, it can be written that:

$$\frac{P}{\sin(\alpha - \phi')} = \frac{W}{\sin\left(\dfrac{\pi}{2} - \alpha + \phi' + \delta + \lambda\right)} \tag{5.63}$$

resulting in:

$$P = \frac{W \sin(\alpha - \phi')}{\cos(\alpha - \phi' - \delta - \lambda)} \tag{5.64}$$

Please note that W is a function of the soil unit weight, γ, and of the angles that define the wedge geometry, namely α, λ, and β, and also of the total height of the wall, h:

$$W = W(\gamma, \alpha, \lambda, \beta, h) \tag{5.65}$$

For a given soil and a given problem, γ, h, and all the angles are constant, except α, then:

$$P = P(\alpha) \tag{5.66}$$

Calculating the derivative of P in relation to α and making it equal to zero:

$$\frac{\partial P}{\partial \alpha} = 0 \tag{5.67}$$

the value of α is determined and replaced in Equation 5.64, allowing the calculation of the maximum value of the function, now designated as P_a, the expression of which is:

$$P_a = \frac{1}{2} K_a \gamma h^2 \tag{5.68}$$

where K_a is, naturally, a function of ϕ', δ, β, and λ:

$$K_a = \frac{\cos^2\left(\phi' - \lambda\right)}{\cos^2 \lambda \cos(\delta + \lambda)\left[1 + \left(\frac{\sin(\phi' + \delta)\sin(\phi' - \beta)}{\cos(\beta - \lambda)\cos(\delta + \lambda)}\right)^{1/2}\right]^2} \qquad (5.69)$$

Using a similar procedure for the passive thrust, the following expression is obtained:

$$P_p = \frac{1}{2}K_p \gamma b^2 \qquad (5.70)$$

where:

$$K_p = \frac{\cos^2\left(\phi' + \lambda\right)}{\cos^2 \lambda \ \cos(\delta - \lambda)\left[1 - \left(\frac{\sin(\phi' + \delta)\sin(\phi' + \beta)}{\cos(\beta - \lambda)\cos(\delta - \lambda)}\right)^{1/2}\right]^2} \qquad (5.71)$$

Figure 5.29 illustrates the sign convention for the angles of Equations 5.69 and 5.71.

The expressions of the angles α_a and α_p that the sliding surface makes with the horizontal for the active and passive cases, respectively, are as follows:

$$\cot an\left(\alpha_a - \beta\right) = -\tan\left(\phi' + \delta + \lambda - \beta\right)$$
$$+ \sec\left(\phi' + \delta + \lambda - \beta\right)\sqrt{\frac{\cos\left(\lambda + \delta\right)\sin\left(\phi' + \delta\right)}{\cos\left(\beta - \lambda\right)\sin\left(\phi' - \beta\right)}} \qquad (5.72)$$

$$\cot an\left(\alpha_p - \beta\right) = \tan\left(\phi' + \delta - \lambda + \beta\right)$$
$$+ \sec\left(\phi' + \delta - \lambda + \beta\right)\sqrt{\frac{\cos\left(\lambda - \delta\right)\sin\left(\phi' + \delta\right)}{\cos\left(\beta - \lambda\right)\sin\left(\phi' + \beta\right)}} \qquad (5.73)$$

5.6.5 Thrust calculation due to uniform surcharge at the surface to be used in Coulomb's method

In the original formulation of Coulomb's method, presented in Section 5.6.1, the consideration of any type of surcharge at the surface (concentrated or distributed, uniform or

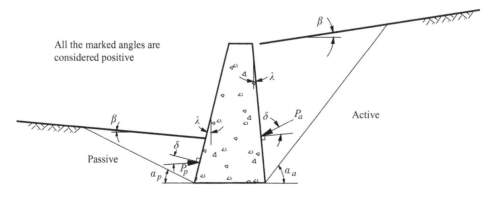

Figure 5.29 Sign convention for the active and passive cases (Equations 5.69 and 5.71, respectively).

non-uniform) does not present further issues: it is added to the weight of the soil wedge, being an additional known force in the force polygon. In this context, Coulomb's method has the advantage of allowing any type of surcharges, while the previous methods (those of Rankine and Caquot-Kérisel) only allow the consideration of uniform surcharges distributed across the whole surface.

In the same way, the analytical version of Coulomb's method, which has just been presented, only allows the consideration of this last type of surcharge. The way Coulomb's analytical solution can deal with the consideration of these surcharges can be understood with the help of Figure 5.30.

As Figure 5.30b suggests, the surcharge can be replaced by an extra soil layer of thickness h_{eq}, with the same unit weight, which would require the vertical extension of the upper part of the wall by a height, h^*.

From Figure 5.30c, the relationship between the surcharge q and the equivalent height is obtained:

$$h_{eq} = \frac{q}{\gamma \cos \beta} \qquad (5.74)$$

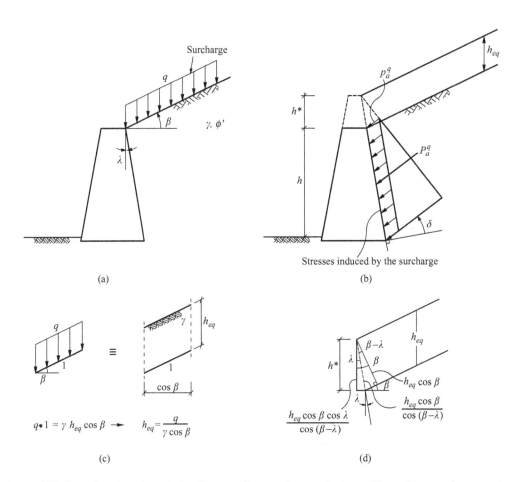

Figure 5.30 Consideration of vertical uniform surcharges: a) general scheme; b) surcharge replacement by a virtual layer of equivalent weight and corresponding wall extension in the upper part; c) definition of equivalent height; d) auxiliary scheme for relationship deduction between h^* and h_{eq}.

By Figure 5.30d, the relationship between h^* and h_{eq} is obtained:

$$h^* = \frac{h_{eq}\cos\beta\cos\lambda}{\cos(\beta-\lambda)} = \frac{q\cos\lambda}{\gamma\cos(\beta-\lambda)} \tag{5.75}$$

From the application of Coulomb's theory, the active stresses graph on the wall of height $h + h^*$ is represented in Figure 5.30b. The part of the graph above h is not considered. The remaining part is divided in two. The part corresponding to the surcharge, is defined by a uniform stress given by:

$$p_a^q = K_a\gamma h^* = \frac{K_a q\cos\lambda}{\cos(\beta-\lambda)} \tag{5.76}$$

Then, the active thrust associated with the surcharge q is:

$$P_a^q = \frac{K_a q h\cos\lambda}{\cos(\beta-\lambda)} \tag{5.77}$$

5.6.6 Comparison of the Rankine, Caquot-Kérisel, and Coulomb methods

Coulomb's theory, on one side, and Rankine's theory and its following developments that led to the Caquot-Kérisel tables, on the other, have substantially different basic philosophies, which became clear during the exposition and corresponding formulations.

Rankine's theory and its developments are included in the philosophy of the lower bound theorem, discussed in Chapter 2. Starting from a local *statically admissible* stress state, its integration is performed to obtain the thrust. On the other hand, Coulomb's theory postulates failure mechanisms which are *kinematically admissible*, directly estimating the thrust without discussing stresses – this is in agreement with the philosophy of the upper bound theorem, also discussed in Chapter 2.

It is interesting to note, however, that the analytical version of Coulomb's theory also leads to the earth pressure coefficient concept. It is therefore convenient to compare the values of this coefficient (active or passive) obtained by the different methodologies.

In Table 5.1, the active and passive earth pressure coefficients obtained by the methodologies mentioned are compared for the values of the four indicated angles.[5] The Rankine solution is only considered for the situation of $\beta = \lambda = \delta = 0$, since, for the other cases, the Caquot-Kérisel solution, within the same philosophy, is the one applicable. It should be reminded that the values included in the table for Coulomb's and Rankine's methods result from analytical expressions, while the ones from Caquot-Kérisel result from numerical integration.

From the table, it can be observed that:

i) the Caquot-Kérisel tables lead to higher earth pressure coefficients, in the active case, and lower, in the passive case, than the values obtained by Coulomb's method; this result is due to the theorems referred to above: the method based on the lower bound

---

[5] Please note that, to carry out such comparisons, it is necessary to change the Coulomb equations, as these give the pressures as a function of the depth h below the top of the wall, while the Caquot-Kérisel tables give the pressures as a function of the distance l measured from the same point. Bearing in mind that $h = l\cos\lambda$ (see both Figure 5.21 and 5.29 notations), the comparison with the earth pressure coefficients retrieved from the Caquot-Kérisel tables has to be made, adjusting the values of K_a and K_p from Equations 5.69 and 5.71 by multiplication by $\cos^2\lambda$.

Table 5.1 Comparison between the active and passive earth pressure coefficients obtained by the Rankine, Caquot-Kérisel and Coulomb methods.

$\phi' = 35°$	\multicolumn{5}{c}{Active earth pressure coefficient, K_a}
	$\beta = 0$	\multicolumn{2}{c}{$\lambda = 0$}	$\beta = 14°$	$\lambda = 10°$
δ	*(1)*	*(2)*	*(3)*	*(2)*	*(3)*
0	0.271	0.271	0.271	0.403	0.393
$\phi'/3$	–	0.252	0.251	0.393	0.377
$2\phi'/3$	–	0.247	0.244	0.383	0.382
ϕ'	–	0.260	0.250	0.409	0.406
$\phi' = 35°$	\multicolumn{5}{c}{Passive earth pressure coefficient, K_p}
	$\beta = 0$	\multicolumn{2}{c}{$\lambda = 0$}	$\beta = 14°$	$\lambda = 10°$
δ	*(1)*	*(2)*	*(3)*	*(2)*	*(3)*
0	3.7	3.7	3.7	4.4	4.5
$\phi'/3$	–	5.4	5.7	6.7	7.5
$2\phi'/3$	–	8.0	10.0	9.8	14.8
ϕ'	–	10.5	23.0	13.2	42.2

Notes:

1. Rankine's method;

2. Caquot-Kérisel tables;

3. Coulomb's method (see footnote 5).

theorem leads to solutions on the safe side, in comparison with those based on the upper bound theorem;

ii) consequently, as Figure 5.31 shows, the exact solution of an active thrust problem is higher than the value obtained by Coulomb's method and lower than the value calculated by the Caquot-Kérisel method; in the passive case, the opposite situation occurs;

iii) the studied methods lead to the exact solution in the case where the surface is horizontal, the wall face is vertical, and the soil–wall friction is null ($\beta = \lambda = \delta = 0$): in fact, the Rankine and Coulomb expressions are identical for these conditions; in the table for the mentioned conditions, the Caquot-Kérisel solution provides a value close to the other two methods, since it is a numerical solution.

Analyzing the differences between the Coulomb and Caquot-Kérisel solutions in quantitative terms, it can be observed that:

iv) the active earth pressure coefficients obtained by the two methods are very similar;

v) for the passive case, the earth pressure coefficients for low values of δ are relatively similar whereas for higher values of δ, the coefficients become very different.

The differences between the passive earth pressure coefficient estimates that have been observed are essentially due to errors in the Coulomb solution. As seen in Section 5.6.1, one of the hypotheses of Coulomb's theory is that the sliding surface that limits the soil wedge is plane. However, the discussion made in Section 5.5.1 shows that this hypothesis

ACTIVE THRUST		
COULOMB	EXACT	CAQUOT-KÉRISEL
$\leq$	$\leq$	
SOLUTION	SOLUTION	SOLUTION
PASSIVE THRUST		
CAQUOT-KÉRISEL	EXACT	COULOMB
$\leq$	$\leq$	
SOLUTION	SOLUTION	SOLUTION

Figure 5.31 Relative positions of Caquot-Kérisel and Coulomb solutions in relation to the exact solution, for the active and passive cases.

is only correct if the tangential stresses on the soil–wall interface are null, whereas the surface is curved when those stresses develop.[6] Then, it becomes clear that the differences increase as δ increases and that they are greater in the passive case. In fact, the higher the tangential stresses, the more pronounced will be the rotation of principal stresses (in relation to the vertical and horizontal directions), and consequently the greater will be the contrast between the soil wedge geometry in limit equilibrium and that assumed in Coulomb's theory.[7]

The following conclusions can then be derived: i) the estimation of the active earth pressure coefficient by Coulomb's method is acceptable and very convenient, with the calculation facilities available at present; and ii) for the evaluation of the passive earth pressure coefficient, it is advisable to use the Caquot-Kérisel tables, since its values err on the safe side.

For the evaluation of the passive thrust, as an alternative to the Caquot-Kérisel tables, Lancellota (2002) developed the following analytical solution, based on the lower bound theorem, also on the safe side (valid for a vertical wall and horizontal surface):

$$K_p = \left[\frac{\cos\delta}{1 - \sin\phi'} \left(\cos\delta + \sqrt{\sin^2\phi' - \sin^2\delta} \right) \right] e^{2\eta\tan\phi'} \tag{5.78}$$

where:

$$2\eta = \sin^{-1}\left(\frac{\sin\delta}{\sin\phi'} \right) + \delta \tag{5.79}$$

---

[6] Note that the sliding surface in the scope of Coulomb's theory corresponds to the planes where the shear strength is fully mobilized in the theory of Rankine and its developments.

[7] Regarding the comparison between Coulomb's and Rankine's methods, it is interesting to observe that the expressions of the earth pressure coefficient presented in Sections 5.4.6 and 5.4.7, in the scope of Rankine's theory generalization to soil masses with inclined surface in relation to the horizontal and/or interacting with inclined walls in relation to the vertical, and the expressions of the same coefficients in Coulomb's theory, are mathematically identical if, in the latter, the angle δ is taken as the one corresponding to the thrust inclination (i.e., β, in the case of Section 5.4.6, and Equations 5.53 and 5.54, in the case of Section 5.4.7). The sign convention in each method should be carefully taken into consideration.

5.7 ACTIVE AND PASSIVE THRUSTS UNDER SEISMIC CONDITIONS: MONONOBE-OKABE THEORY

5.7.1 Introduction

The first theory for the evaluation of active and passive thrusts under seismic conditions resulted from the works of Okabe (1926) and from Mononobe and Matsuo (1929), and can be considered an extension of Coulomb's theory.

The Mononobe-Okabe theory is based, in fact, on the same hypotheses as Coulomb's theory (see Section 5.6.1), adding only one other hypothesis specific to the situation being dealt with: during the earthquake, the soil wedge which interacts with the structure behaves as a rigid body, being therefore uniform the acceleration field.

Considering that hypothesis, and applying the D'Alembert's principle, as Figure 5.32 suggests for the retaining wall case, the seismic action is considered by adding the so-called *inertial forces* to the force system representative of the static situation. The inertial forces are obtained by multiplying the weight of the body, W, by dimensionless factors known as *horizontal seismic coefficient, k_h,* and *vertical seismic coefficient, k_v,* which represent the ratio between the corresponding component of the seismic acceleration and the acceleration of gravity.[8] It should be noted that the wall itself also suffers the action of inertial forces, as defined by seismic coefficients coincident with those applied to the soil.

In terms of what concerns the horizontal, it is important to consider the inertial force $k_h W$ pointing towards the wall, corresponding to a seismic acceleration equal to $k_h g$ in the opposite direction. In the vertical, it should be considered that inertial forces may point up or down, that is, $-k_v W$ or $k_v W$, respectively. In fact, although the inertial force facing down $(k_v W)$ leads to the greater thrust, for that case, the wall resistance is also increased, since the seismic acceleration and the corresponding inertial force (facing down) also apply to the wall. So, the greater thrust may not lead to the most critical situation in terms of stability.

As shown in Figure 5.32, the force resulting from the weight of the soil wedge and the horizontal and vertical components of the inertial forces, W_e, becomes inclined at an angle θ in relation to the vertical, given by:

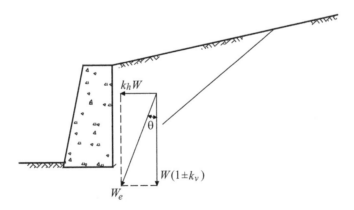

Figure 5.32 Soil wedge under seismic action.

---

[8] A horizontal seismic coefficient of 0.2 means that the seismic action results in a horizontal inertial force of 0.2 W applied to the center of gravity of the body in study, resulting from a horizontal seismic acceleration of 0.2 g in the opposite direction.

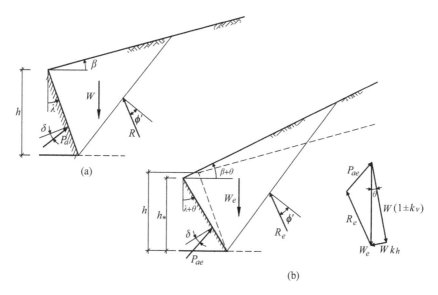

Figure 5.33 Soil wedge in active state: a) static conditions; b) seismic conditions.

$$\theta = \tan^{-1} \frac{k_h}{1 \pm k_v} \tag{5.80}$$

The selection of the seismic coefficients is treated in Annex A8.1.

5.7.2 Expression deduction

The Mononobe-Okabe expression for the calculation of the seismic active thrust considers that the effect of the seismic acceleration in the direction of the resultant of the mass forces, W_e, is taken into consideration by rotating the horizontal and vertical reference planes of an angle θ, as illustrated in Figure 5.33.

Then, W_e will stay vertical and the angles β and λ will become $\beta + \theta$ and $\lambda + \theta$, respectively. In this way, the seismic active thrust may be calculated using Coulomb's theory:

$$P_{ae} = \frac{1}{2} K_{a*} \gamma_* h_*^2 \tag{5.81}$$

where K_{a*} is obtained from Equation 5.69, replacing β with $\beta + \theta$ and λ with $\lambda + \theta$ and γ_* and h_* are obtained from γ and h, respectively, in a very simple way.

In fact, if l is the length of the face of the retaining structure, it can be written that:

$$l = \frac{h}{\cos \lambda} = \frac{h_*}{\cos(\lambda + \theta)} \tag{5.82}$$

which allows for h_* determination:

$$h_* = h \frac{\cos(\lambda + \theta)}{\cos \lambda} \tag{5.83}$$

On the other hand:

$$\frac{\gamma_*}{\gamma} = \frac{W_e}{W} = \frac{W(1 \pm k_v)}{W\cos\theta} = \frac{(1 \pm k_v)}{\cos\theta} \tag{5.84}$$

becoming then,

$$\gamma_* = \gamma \frac{(1 \pm k_v)}{\cos\theta} \tag{5.85}$$

Replacing the Equations 5.83 and 5.85, and also the expression of K_{a*}, as indicated above, in Equation 5.81, the following expression is obtained:

$$P_{ae} = \frac{1}{2} K_{ae} \gamma (1 \pm k_v) h^2 \tag{5.86}$$

with K_{ae} being the *seismic active earth pressure coefficient*, with the expression:

$$K_{ae} = \frac{\cos^2(\phi' - \lambda - \theta)}{\cos\theta\cos^2\lambda\cos(\delta + \lambda + \theta)\left[1 + \left(\frac{\sin(\phi' + \delta)\sin(\phi' - \beta - \theta)}{\cos(\beta - \lambda)\cos(\delta + \lambda + \theta)}\right)^{1/2}\right]^2} \tag{5.87}$$

Similar expressions may be deduced for the seismic passive thrust case:

$$P_{pe} = \frac{1}{2} K_{pe} \gamma (1 \pm k_v) h^2 \tag{5.88}$$

with K_{pe} being the *seismic passive earth pressure coefficient*, with the expression:

$$K_{pe} = \frac{\cos^2(\phi' + \lambda - \theta)}{\cos\theta\cos^2\lambda\cos(\delta - \lambda + \theta)\left[1 - \left(\frac{\sin(\phi' + \delta)\sin(\phi' + \beta - \theta)}{\cos(\beta - \lambda)\cos(\delta - \lambda + \theta)}\right)^{1/2}\right]^2} \tag{5.89}$$

The Mononobe-Okabe solution for the seismic passive thrust has similar limitations to the ones described before for the static passive thrust of Coulomb's solution. As presented before for this latter case, Lancellota (2007) developed the following analytical solution, based on the lower bound theorem, also on the safe side (valid for a vertical wall):

$$K_{pe} = \left[\frac{\cos\delta}{\cos(\beta - \theta) - \sqrt{\sin^2\phi' - \sin^2(\beta - \theta)}} \times \left(\cos\delta + \sqrt{\sin^2\phi' - \sin^2\delta}\right)\right] e^{2\eta\tan\phi'} \tag{5.90}$$

where

$$2\eta = \sin^{-1}\left(\frac{\sin\delta}{\sin\phi'}\right) + \sin^{-1}\left[\frac{\sin(\beta - \theta)}{\sin\phi'}\right] + \delta + (\beta - \theta) + 2\theta \tag{5.91}$$

In agreement with the methodology presented in Section 5.6.5, the expression of the seismic active thrust associated with a uniformly distributed surcharge at the ground surface, q, is:

$$P_{ae,q} = \frac{K_{ae}q(1 \pm k_v)h\cos\lambda}{\cos(\beta - \lambda)} \tag{5.92}$$

5.7.3 Critical horizontal acceleration

The angle $\phi' - \beta - \theta$, whose sine appears inside a square root in Equation 5.87, has to be greater than or equal to zero so that K_{ae} becomes a real number, i.e., to make equilibrium physically possible.[9] (Similarly, for static conditions, the angle $\phi' - \beta$, which appears inside a square root in Equation 5.69, has to be greater than or equal to zero, which corresponds to the well-known result that a non-cohesive soil mass cannot have a surface with an inclination greater than the angle of shearing resistance).

For seismic conditions, it should then be verified that:

$$\phi' - \beta - \theta \geq 0 \tag{5.93}$$

Introducing Equation 5.80 in this last expression, the following is obtained:

$$\theta = \tan^{-1}\frac{k_h}{1 \pm k_v} \leq \phi' - \beta \tag{5.94}$$

and

$$k_h \leq (1 \pm k_v)\tan(\phi' - \beta) \tag{5.95}$$

Therefore, there is a *critical horizontal acceleration* corresponding to a horizontal seismic coefficient equal to

$$k_{h,cr} = (1 \pm k_v)\tan(\phi' - \beta) \tag{5.96}$$

which cannot be exceeded. In other words, a soil mass, with friction angle ϕ' and surface angle β, cannot transmit shear forces produced by acceleration levels greater than the critical value, this being expressed by Equation 5.96.

As the seismic acceleration increases, the active thrust increases and the passive thrust decreases, tending both to the same value when the horizontal acceleration achieves the value previously named as critical. For soil masses with horizontal surface in contact with vertical walls, when the acceleration achieves the critical value, in this case expressed by:

$$k_{h,cr} = (1 \pm k_v)\tan\phi' \tag{5.97}$$

the seismic active and passive earth pressure coefficients achieve their maximum and minimum values, respectively, given by the equation:

$$K_{ae,cr} = K_{pe,cr} = \frac{1}{\cos\phi'\cos(\phi' + \delta)} \tag{5.98}$$

---

[9] The indicated situation is not unusual when a significant inclination of the ground surface appears together with high seismic coefficients. In those situations, the earth pressure coefficient can be calculated, assuming null to be the result of the expression inside the square root. This means that, if the conditions corresponding to the seismic coefficients assumed in the calculations are verified, the soil situated between the planes inclined at $\phi'-\theta$ and β will slide over the top of the retaining wall.

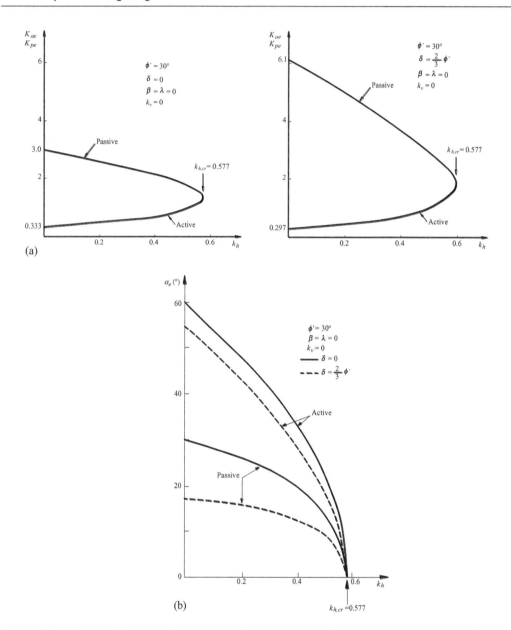

Figure 5.34 Influence of horizontal acceleration: a) on the active and passive earth pressure coefficients; b) on the inclination with the horizontal of the surfaces limiting thrust wedges (Davies et al., 1986).

Figure 5.34a illustrates the change of the active and passive earth pressure coefficients, with k_h between zero and the critical value for a soil mass of horizontal surface and angle of shearing resistance, $\phi' = 30°$.

5.7.4 Inclination of the surfaces that limit the thrust wedges

The inclination of the surfaces that limit the active and passive wedges is reduced by the seismic actions, that is to say that these wedges are not the same for the static conditions (see

Equations 5.72 and 5.73) and for the seismic conditions, these latter being of high volume, with the difference growing with the seismic intensity.

The expressions of the angles, α_{ae} and α_{pe}, with a horizontal sliding surface for the seismic active and passive cases, respectively, are the following:

$$
\begin{aligned}
\cot an\left(\alpha_{ae} - \beta\right) = &-\tan\left(\phi' + \delta + \lambda - \beta\right) \\
&+ \sec\left(\phi' + \delta + \lambda - \beta\right)\sqrt{\frac{\cos\left(\lambda + \delta + \theta\right)\sin\left(\phi' + \delta\right)}{\cos\left(\beta - \lambda\right)\sin\left(\phi' - \beta - \theta\right)}}
\end{aligned}
\tag{5.99}
$$

$$
\begin{aligned}
\cot an\left(\alpha_{pe} - \beta\right) = &\tan\left(\phi' + \delta - \lambda + \beta\right) \\
&+ \sec\left(\phi' + \delta - \lambda + \beta\right)\sqrt{\frac{\cos\left(\lambda - \delta - \theta\right)\sin\left(\phi' + \delta\right)}{\cos\left(\beta - \lambda\right)\sin\left(\phi' + \beta - \theta\right)}}
\end{aligned}
\tag{5.100}
$$

Figure 5.34b shows the evolution of α_{ae} and α_{pe} as a function of the horizontal seismic coefficient for the indicated set of parameters. For moderate levels of acceleration, the inclination of the passive wedge does not differ much from the static case but the same is not observed for the active wedge. It is curious to note that the inclination of the surfaces reduces as the acceleration increases, becoming equal to zero when the critical acceleration is reached, as described above.

5.7.5 Graphical solution of the problem

The problem analyzed has a graphical solution similar to the Culmann one for Coulomb's method, but now working with the force, W_e, the resultant of the wedge weight and the horizontal and vertical inertial forces. Similar to what happens under static conditions, this methodology solves the problem when the geometric conditions, the surface load, or the cohesion of the soil mass do not allow the use of the analytical solution.

It is simple to adapt the Culmann construction to the present case. In fact, taking as reference Figure 5.26, the line bf now has an angle of $\phi' - \theta$ to the horizontal and the line bg an angle $\psi-\theta$ with the first line.

The graphical construction that follows is identical to that described in Section 5.6.2. The maximum vector obtained (corresponding to ed in Figure 5.26) must be multiplied by $(1 \pm kv)/\cos\theta$ to obtain the seismic active thrust.

5.7.6 Thrust application point

In cases where the problem has an analytical solution, the formulation of Mononobe-Okabe, like the one of Coulomb, leads to a thrust that is the resultant of a triangular pressure diagram. The application point coincides with the thrust in static conditions. However, results from different studies – using truly dynamic and not pseudo-dynamic analytical methods, such as Mononobe-Okabe, as well as tests on small-scale physical models – show that the resultant of the incremental stresses of seismic origin has an application point situated frequently above the one for static thrust (Seed and Whitman, 1970; Sherif et al., 1982; Steedman and Zeng, 1990). For that reason, the seismic action effect becomes even more severe, and therefore the thrust, P_{ae}, has to be separated, as explained as follows.

The thrust, P_{ae}, can be considered the result of two components: the thrust that was already acting before the earthquake, P_a, and the thrust increment associated with the seismic action, ΔP_{ae}:

$$P_{ae} = P_a + \Delta P_{ae} \tag{5.101}$$

Bearing in mind that:

$$P_a = \frac{1}{2} K_a \gamma h^2 \tag{5.102}$$

and considering Equation 5.86, it can be written that

$$\Delta P_{ae} = \frac{1}{2} \left[\left(1 \pm k_v \right) K_{ae} - K_a \right] \gamma h^2 \tag{5.103}$$

Taking

$$\Delta K_{ae} = \left(1 \pm k_v \right) K_{ae} - K_a \tag{5.104}$$

it will become finally:

$$\Delta P_{ae} = \frac{1}{2} \Delta K_{ae} \gamma h^2 \tag{5.105}$$

Figure 5.35 includes the solution that, at present, is the most consensual, where the thrust increment associated with the seismic action is applied at the wall mid-height. This position constitutes a conservative option, taking into account the studies cited above, for the typical deformation conditions of gravity walls (Matsuzawa et al., 1985).[10]

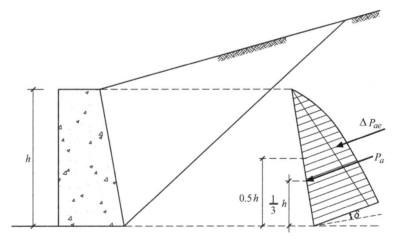

Figure 5.35 Splitting of the active thrust in seismic conditions, where the thrust increment associated to the seismic action is applied at the wall mid-height.

---

[10] The rigorous determination of the position of the application point would be only possible for a *certain retaining wall under the action of a given seismic record*, and it would involve studies of great complexity in order to take into consideration the seismic action characteristics, including the acceleration on the vertical direction, the stiffness of the soil–structure system, the foundation soil characteristics, etc. These studies are only justified for structures of exceptional importance.

As the expressions of K_{ae} and K_a are rather complex, it is not easy to obtain a prompt evaluation of how the seismic coefficients, and, in particular, the horizontal seismic coefficient, influence the value of ΔK_{ae}. Therefore, it is useful to refer to the following approximate proposal from Seed and Whitman (1970) for the case of vertical wall and horizontal ground surface ($\lambda = \beta = 0$):

$$\Delta K_{ae} \approx 0.75 k_b \tag{5.106}$$

This expression gives reasonably approximate values, in general on the safe side, for k_b values up to 0.3, values of the angle of shearing resistance equal to or greater than 30°, and values of δ between 0 and (2/3) ϕ'.

5.7.7 Generalization to submerged soil masses

5.7.7.1 Introduction

One of the hypotheses of the Mononobe-Okabe theory is that the soil mass is above the water table. This hypothesis applies to a great number of gravity-retaining structures, as it is well known that one of the basic conditions for their good performance is the drainage of the supported soil mass, avoiding the establishment of a water level in the backfill. There are, however, situations where the soil mass and the structure itself are partially (or even totally) submerged, as happens, for instance, in quay walls. In these situations, it is necessary to estimate the seismic thrusts of the submerged soil mass and corresponding water table, as well as the thrust of the free water mass in front of the structure.

In very permeable soil masses, it is reasonable to assume that the water and the solid skeleton of the soil behave independently during the earthquake. For these cases, it is not particularly complex to extend the previously described theory, as will be shown below. For non-cohesive soils of medium to low permeability, the seismic actions may result in a loading under undrained conditions. The excess pore pressure generated reduces the effective stresses, and consequently the soil resistance, leading, in extreme cases, to liquefaction. The evaluation of the thrust in these particularly complex conditions exceeds the scope of this book.

5.7.7.2 Seismic thrust of a free water mass on a vertical wall

For the thrusts related to the free water mass in front of the structure, the theoretical solution of Westergaard (1933) can be used. This was developed to estimate the effect of hydrodynamic thrusts in concrete dam walls during earthquakes. According to this solution (see Figure 5.36), the incremental water pressure on the wall at a depth z below the water surface having a total height h_w, is:

$$\Delta p_{we}(z) = \frac{7}{8} \gamma_w k_b \sqrt{h_w z} \tag{5.107}$$

The total thrust is obtained by integrating the previous equation up to the depth h_w:

$$\Delta P_{we} = \int_0^{h_w} \Delta p_{we}(z) dz = \frac{7}{12} \gamma_w k_b h^2{}_w \tag{5.108}$$

where the corresponding application point is located at $0.6 h_w$ from the water surface.

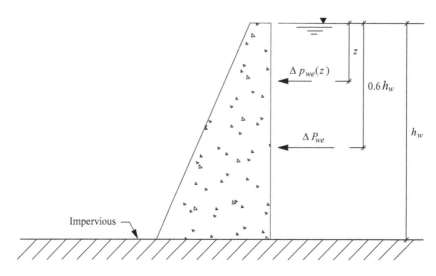

Figure 5.36 Westergaard solution for hydrodynamic thrusts.

5.7.7.3 Thrust of a very permeable submerged soil mass

In very permeable soil masses, like coarse sands, gravels or rockfills with few or no fines (with permeability coefficients around 10^{-3} m/s or higher) it is generally assumed that the water and the soil solid skeleton behave independently during the earthquake, so that the corresponding thrusts are calculated separately. The seismic active thrust of the soil below the water table will be then calculated, like the static one, based on the submerged unit weight by the following equation (Matsuzawa et al., 1985):

$$P_{ae} = \frac{1}{2} K'_{ae} \gamma' \left(1 \pm k_v\right) h_w^2 \tag{5.109}$$

with K'_{ae} being now used for the seismic active coefficient of earth pressure. This different symbol highlights the fact that the calculation is now based on a seismic angle θ' (angle between the resultant of the mass forces and the vertical direction) which is different from the one expressed in Equation 5.80. In fact, as illustrated in Figure 5.37, the horizontal inertial forces per unit volume in the soil below the water table are applied only to the solid

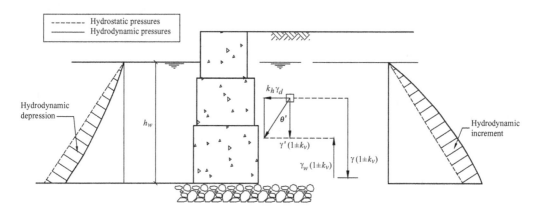

Figure 5.37 Seismic thrusts in very permeable submerged soil masses.

skeleton, i.e., to the dry unit weight, whereas the vertical force results from the difference between the total gravitational force, $\gamma(1 \pm k_v)$, and the buoyancy force, given by $\gamma_w(1 \pm k_v)$, as the water is also subjected to the seismic action. So, K'_{ae} is calculated by Equation 5.87, using θ' instead of θ:

$$\theta' = \tan^{-1} \frac{\gamma_d k_h}{\gamma'(1 \pm k_v)} = \tan^{-1} \frac{G_s k_h}{(G_s - 1)(1 \pm k_v)} \tag{5.110}$$

where G_s is the solid particle density. Taking $G_s = 2.65$, it results in:

$$\tan\theta' \approx 1.6 \tan\theta \tag{5.111}$$

which is equivalent, if $k_v = 0$, to a horizontal seismic coefficient, k'_h, equal to:

$$k'_h \approx 1.6 k_h \tag{5.112}$$

The increment of the seismic active thrust, resulting from the difference between Equations 5.109 and 5.68, should be combined with the incremental hydrodynamic thrust given by Equation 5.108. Since the soil and water act independently, it is unlikely that the corresponding thrusts are in phase, i.e., achieve the maximum values simultaneously. As a consequence, its combination in design needs to be balanced in the most reasonable way. For that purpose, some authors suggest calculating the square root of the sum of the squared incremental thrusts.

It should be noted that, in quay walls or similar structures, as illustrated in Figure 5.37, the hydrodynamic depression in front of the wall needs to be taken into account, the maximum value of which is given (in absolute terms) by Equation 5.108.

5.7.8 Note about the seismic thrusts on walls with null displacements

For structures with restricted displacement (peripheral walls in building basements, bridge abutments, etc.), the Mononobe-Okabe theory is no longer valid for the same reason that, for static conditions, Coulomb's theory loses validity. In static conditions, these structures are designed to support the earth pressure at-rest, defined by the coefficient K_0.

The reference solution to estimate the seismic incremental pressures was developed by Wood (1973) and it is a solution in the elasticity domain.

As shown in Figure 5.38a, Wood studied an elastic and homogeneous ground mass, limited by two lateral boundaries and one rigid bottom boundary, under the action of horizontal inertial forces distributed by the domain, constant in time and space, and associated with an acceleration of equal direction and opposite sense. The Poisson ratio, ν, was considered to be in the range 0.2–0.5.

As shown in Figure 5.38b, for ground masses of large development behind the rigid vertical wall, the thrust (exclusively associated with the acceleration) is approximately:

$$\Delta P_{0e} \approx k_h \gamma h^2 \tag{5.113}$$

and the corresponding application point is located at a distance of $0.5h$ to $0.6h$ from the wall base (Figure 5.38c).

This result corresponds to a coefficient of seismic incremental thrust for null displacement of:

$$\Delta K_{0e} \approx 2 k_h \tag{5.114}$$

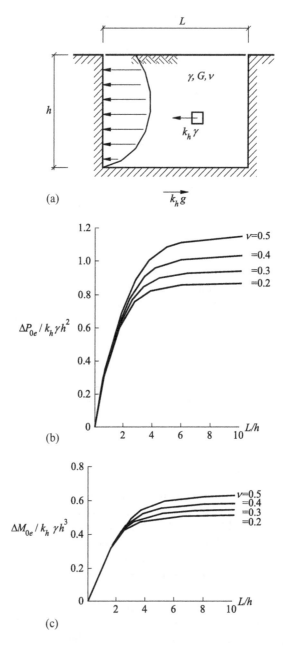

Figure 5.38 Elastic solution for seismic pressures on a fixed wall (Wood, 1973): a) scheme and basic hypotheses; b) seismic incremental thrust; c) distance between the wall base and the thrust application point.

Comparing this expression with the simplified Equation 5.106 for ΔK_{ae}, it can be concluded
that, according to the elastic solution of Wood, the seismic incremental thrust on a wall with
null displacement can achieve values two or three times greater than in a wall where it is
valid to assume the applicability of the Mononobe-Okabe theory.

ANNEX A5.1 - CAQUOT–KÉRISEL TABLES (EXTRACT)

See Tables A5.1.1 to A5.1.8.

Table A5.1.1 K_a values for $\delta=0$ (see sign convention in Figure 5.21a).

$\phi'(°)$	$\lambda(°)$	β/ϕ' values						
		−0.2	0	+0.2	+0.4	+0.6	+0.8	+1.0
	+10	0.52	0.55	0.58	0.62	0.68	0.77	–
20	0	0.47	0.47	0.52	0.55	0.60	0.67	0.98
	−10	0.41	0.41	0.44	0.47	0.51	0.56	0.81
	+10	0.44	0.47	0.50	0.54	0.60	0.70	–
25	0	0.39	0.41	0.43	0.46	0.51	0.59	0.92
	−10	0.32	0.34	0.36	0.38	0.41	0.47	0.73
	+10	0.37	0.40	0.43	0.47	0.53	0.62	–
30	0	0.31	0.33	0.36	0.39	0.43	0.50	0.85
	−10	0.25	0.26	0.28	0.30	0.33	0.38	0.63
	+10	0.31	0.34	0.37	0.40	0.46	0.55	–
35	0	0.25	0.27	0.29	0.32	0.35	0.42	0.77
	−10	0.20	0.21	0.22	0.23	0.26	0.30	0.54
	+10	0.26	0.28	0.31	0.34	0.39	0.47	–
40	0	0.20	0.22	0.23	0.25	0.29	0.34	0.68
	−10	0.15	0.16	0.17	0.18	0.20	0.23	0.45
	+10	0.22	0.24	0.26	0.28	0.33	0.40	–
45	0	0.16	0.17	0.18	0.20	0.23	0.27	0.58
	−10	0.11	0.12	0.12	0.13	0.14	0.17	0.35

Table A5.1.2 K_a values for $\delta=(2/3)\phi'$ (see sign convention in Figure 5.21a).

$\phi'(°)$	$\lambda(°)$	β/ϕ' values						
		−0.2	0	+0.2	+0.4	+0.6	+0.8	+1.0
	+10	0.47	0.50	0.53	0.58	0.64	0.73	–
20	0	0.42	0.44	0.47	0.50	0.55	0.63	0.93
	−10	0.36	0.37	0.39	0.42	0.46	0.52	0.76
	+10	0.40	0.43	0.46	0.50	0.57	0.67	–
25	0	0.34	0.36	0.39	0.42	0.47	0.55	0.87
	−10	0.28	0.29	0.31	0.34	0.37	0.43	0.68
	+10	0.34	0.37	0.39	0.44	0.50	0.60	–
30	0	0.28	0.30	0.32	0.35	0.40	0.47	0.82
	−10	0.22	0.23	0.25	0.27	0.30	0.35	0.60
	+10	0.29	0.31	0.34	0.38	0.44	0.55	–
35	0	0.23	0.25	0.27	0.29	0.33	0.40	0.76
	−10	0.17	0.18	0.19	0.21	0.23	0.28	0.52
	+10	0.25	0.27	0.30	0.33	0.38	0.48	–
40	0	0.19	0.20	0.22	0.24	0.27	0.33	0.68
	−10	0.13	0.14	0.15	0.16	0.18	0.21	0.43
	+10	0.21	0.23	0.25	0.28	0.33	0.41	–
45	0	0.15	0.16	0.18	0.19	0.22	0.27	0.60
	−10	0.10	0.11	0.11	0.12	0.14	0.16	0.35

Table A5.1.3 K_a values for $\delta = \phi'$ (see sign convention in Figure 5.21a).

$\phi'(°)$	$\lambda(°)$	β/ϕ' values						
		−0.2	0	+0.2	+0.4	+0.6	+0.8	+1.0
20	+10	0.47	0.50	0.54	0.58	0.64	0.74	–
	0	0.42	0.44	0.47	0.50	0.55	0.63	0.94
	−10	0.35	0.37	0.39	0.42	0.46	0.52	0.76
25	+10	0.40	0.43	0.47	0.52	0.58	0.69	–
	0	0.35	0.37	0.39	0.43	0.48	0.56	0.91
	−10	0.28	0.30	0.31	0.34	0.37	0.43	0.70
30	+10	0.35	0.38	0.41	0.46	0.52	0.63	–
	0	0.29	0.31	0.33	0.36	0.41	0.49	0.87
	−10	0.22	0.24	0.25	0.27	0.30	0.36	0.63
35	+10	0.31	0.33	0.37	0.41	0.47	0.58	–
	0	0.24	0.26	0.28	0.31	0.35	0.42	0.82
	−10	0.18	0.19	0.20	0.22	0.25	0.29	0.55
40	+10	0.27	0.29	0.33	0.37	0.43	0.53	–
	0	0.20	0.22	0.24	0.26	0.30	0.36	0.77
	−10	0.14	0.15	0.16	0.17	0.19	0.23	0.48
45	+10	0.24	0.26	0.29	0.33	0.38	0.49	–
	0	0.17	0.18	0.20	0.22	0.25	0.31	0.71
	−10	0.11	0.12	0.13	0.14	0.15	0.18	0.41

Table A5.1.4 K_p values for $\delta = 0$ (see sign convention in Figure 5.21b).

$\phi'(°)$	$\lambda(°)$	β/ϕ' values						
		−1.0	−0.8	−0.6	−0.4	−0.2	0	+0.2
20	+10		–	–	1.28	1.54	1.76	1.96
	0	–		1.30	1.60	1.83	2.05	2.25
	−10		1.24	1.56	1.83	2.05	2.30	2.50
25	+10	–		–	1.34	1.73	2.05	2.40
	0		–	1.34	1.76	2.10	2.45	2.80
	−10		1.24	1.70	2.10	2.50	2.90	3.30
30	+10		–	–	1.40	1.90	2.40	2.90
	0	–		–	1.37	1.94	2.45	3.00	3.50
	−10		1.20	1.85	2.45	3.00	3.60	4.30
35	+10		–	–	1.50	2.15	2.85	3.60
	0	–		–	1.40	2.10	2.85	3.70	4.60
	−10		1.20	2.00	2.75	3.70	4,70	5.90
40	+10		–	–	1.55	2.35	3.40	4.60
	0	–		–	1.40	2.30	3.40	4.60	6.20
	−10		1.15	2.05	3.20	4.50	6.20	8.40
45	+10		–	–	1.55	2.60	4.10	5.90
	0	–		–	1.40	2.40	3.90	5.80	8.40
	−10		1.05	2.10	3.60	5.50	8.30	12.40

Table A5.1.5 K_p values for $\delta = (1/3)\phi'$ (see sign convention in Figure 5.21b).

$\phi'(°)$	$\lambda(°)$	β/ϕ' values						
		−1.0	−0.8	−0.6	−0.4	−0.2	0	+0.2
	+10		−	1.20	1.54	1.82	2.05	2.30
20	0	−	1.16	1.56	1.86	2.10	2.40	2.65
	−10		1.48	1.85	2.10	2.15	2.70	3.00
	+10		−	1.23	1.73	2.15	2.55	2.95
25	0	−	1.15	1.73	2.20	2.60	3.10	3.50
	−10		1.56	2.10	2.60	3.10	3.60	4.20
	+10		−	1.30	1.94	2.55	3.20	3.90
30	0	−	1.23	1.87	2.55	3.20	4.00	4.90
	−10		1.70	2.45	3.20	4.0	5.00	6.10
	+10		−	1.34	2.20	3.10	4.10	5.30
35	0	−	1.12	2.10	3.00	4.10	5.40	7.00
	−10		1.73	2.80	4.00	5.40	7.10	9.20
	+10		−	1.38	2.45	3.80	5.40	7.60
40	0	−	1.10	2.20	3.60	5.30	7.60	10.60
	−10		1.80	3.20	5.00	7.30	10.50	15.00
	+10		−	1.50	2.80	4.70	7.40	11.20
45	0	−	1.20	2.40	4.20	7.0	11.00	17.00
	−10		2.00	3.90	6.60	10.50	17.00	26.00

Table A5.1.6 K_p values for $\delta = (2/3)\phi'$ (see sign convention in Figure 5.21b).

$\phi'(°)$	$\lambda(°)$	β/ϕ' values						
		−1.0	−0.8	−0.6	−0.4	−0.2	0	+0.2
	+10		−	1.47	1.76	2.05	2.30	2.60
20	0	−	1.40	1.76	2.10	2.40	2.75	3.10
	−10		1.65	2.05	2.45	2.75	3.10	3.50
	+10		−	1.62	2.10	2.55	3.00	3.60
25	0	−	1.53	2.05	2.55	3.10	3.70	4.30
	−10		1.90	2.50	3.10	3.80	4.50	5.20
	+10		−	1.78	2.50	3.20	4.10	5.00
30	0	−	1.58	2.40	3.20	4.20	5.30	6.50
	−10		2.15	3.10	4.10	5.30	6.60	8.20
	+10		−	1.98	3.00	4.20	5.70	7.50
35	0	−	1.65	2.80	4.10	5.70	8.00	10.00
	−10		2.40	3.80	5.60	7.80	10.50	13.80
	+10		−	2.20	3.60	5.70	8.50	12.10
40	0	−	1.80	3.40	5.40	8.82	12.00	17.50
	−10		2.80	4.80	7.90	11.90	17.60	25.00
	+10		−	2.30	4.80	8.10	13.10	21.00
45	0	−	1.70	4.00	7.10	12.50	20.00	33.00
	−10		3.00	6.00	11.50	19.90	33.00	53.00

Table A5.1.7 K_p values for $\delta = \phi'$ (see sign convention in Figure 5.21b).

$\phi'(°)$	$\lambda(°)$	-1.0	-0.8	-0.6	-0.4	-0.2	0	$+0.2$
					β/ϕ' values			
	$+10$	–	1.28	1.60	1.92	2.25	2.55	2.90
20	0	–	1.54	1.93	2.30	2.70	3.10	3.40
	-10	1.16	1.80	2.25	2.70	3.10	3.50	3.90
	$+10$	–	1.35	1.85	2.35	2.95	3.50	4.20
25	0	–	1.72	2.30	2.95	3.60	4.40	5.10
	-10	1.22	2.20	2.90	3.70	4.40	5.30	6.20
	$+10$	–	1.42	2.15	2.95	3.90	5.00	6.20
30	0	–	1.92	2.85	3.90	5.10	6.50	8.10
	-10	1.30	2.60	3.80	5.10	6.60	8.40	10.40
	$+10$	–	1.45	2.55	3.80	5.50	7.60	10.20
35	0	–	2.10	3.60	5.40	7.60	10.50	14.00
	-10	1.40	3.00	5.00	7.50	10.50	14.20	19.00
	$+10$	–	1.60	3.00	5.10	8.0	12.20	17.80
40	0	–	2.50	4.60	7.70	12.00	18.00	26.00
	-10	1.50	3.70	7.00	11.30	18.00	26.50	39.00
	$+10$	–	1.80	4.00	7.50	12.60	21.50	35.00
45	0	–	2.90	6.20	12.00	21.00	35.00	56.00
	-10	1.55	5.00	10.00	19.00	34.00	56.00	90.00

Table A5.1.8 K_p values for the case of $\beta = \lambda = 0$ (see sign convention in Figure 5.21b).

$\phi'(°)$	$\delta = 0$	$\delta = (1/3)\phi'$	$\delta = (2/3)\phi'$	$\delta = \phi'$
20	2.04	2.40	2.75	3.10
22	2.20	2.65	3.10	3.55
24	2.37	2.95	3.50	4.10
26	2.56	3.30	4.00	4.75
28	2.77	3.65	4.60	5.60
30	3.0	4.0	5.3	6.5
32	3.25	4.6	6.2	8.0
34	3.5	5.2	7.4	9.75
36	3.8	5.8	8.6	11.0
38	4.2	6.6	10	14
40	4.6	7.6	12	18
42	5.0	8.8	15	25
44	5.5	10.3	18	32
46	6.1	12	22	40
48	6.7	14	28	53
50	7.2	15	35	70

ANNEX E5 – EXERCISES (RESOLUTIONS ARE INCLUDED IN THE FINAL ANNEX)

E5.1 - Consider a rigid, smooth plate of 3.0 m height, represented in Figure E5.1, which interacts on the right side with a loose sand mass, the characteristics of which are provided in the figure. Force F equilibrates the resultant of the soil pressure acting on the plate and d is the distance from its application point to the ground surface. Calculate the range of values within which force F and the corresponding d value can vary, for the two following situations:

a) no surcharge load applied at the ground surface;
b) uniform surcharge of 15 kN/m² applied at the surface.

E5.2 - Figure E5.2 shows two granular soil masses in contact with two vertical, rigid, and smooth plates (of nearly infinite length) connected by evenly spaced hinged struts placed 1.0 m apart. Assume that there is no friction between the plates and the base platform. The force on each strut can be adjusted by a mechanical system that increases or decreases the strut length. Take $\gamma_w = 9.8$ kN/m³.
  Determine the maximum and minimum values of the force on each strut, assuming that:

a) both soils are dry;
b) the soil on the left is submerged and the one on the right is dry.

E5.3 - Figure E5.3 displays two granular soil masses in direct contact with two vertical, rigid, and smooth plates (of nearly infinite length) connected by evenly spaced hinged struts

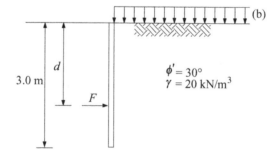

Figure E5.1

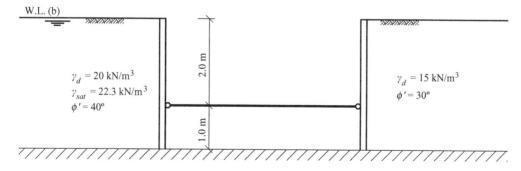

Figure E5.2

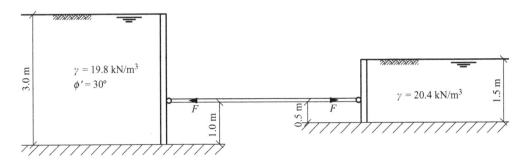

Figure E5.3

placed 1.0 m apart. Both soils are submerged. Assume that there is no friction between the plates and the horizontal base platforms. Take $\gamma_w = 9.8$ kN/m³.

It is possible to introduce small length variations in the struts by means of a mechanical system and also to measure the strut force, F. It has been verified that force F is equal to 59 kN and this value remains constant for (positive or negative) variations of strut length.

a) Determine the friction angle of the soil interacting with the plate on the right side of the figure. How are the stress states installed on the two soils designated? Justify.
b) Draw the Mohr circles which represent the total and effective stress states at a point in the soil adjacent to each plate at 1.0 m depth.

E5.4 - Figure E5.4 represents two vertical, rigid, and smooth plates, connected by steel tendons located 1.0 m apart in the longitudinal direction. It is assumed that no friction exists between the plates and the horizontal base. Dry sand will be placed layer by layer in the space between the plates. As Figure E5.4a shows, during the sand-filling operation, the plates are kept fixed by means of auxiliary struts, while the steel tendons remain inactive. The tendon force can be regulated by a mechanical system at the right end of each tendon, while the left end remains fixed to the plate. Assume that there is no friction between the plates and the base platform.

For the sand, take the following characteristics: $\gamma_d = 16$ kN/m³, $\gamma_{sat} = 19.8$ kN/m³ and $\phi' = 30°$. Consider $K_0 = 1 - \sin\phi'$. Take $\gamma_w = 9.8$ kN/m³.

a) After the sand-filling operation of the space between the plates (see Figure E5.4b), calculate the force F_i that should be installed on each tendon in order to enable the removal of the struts without any displacement of the plates.
b) Determine the values between which the tendon force can vary.
c) Imagine that the minimum tendon force, F_m, is installed, as illustrated in Figure E5.4c, and that the space between the plates is filled with water until its level reaches the top surface of the sand, as indicated in Figure E5.4d. Calculate the new value of force F.

E5.5 - Consider Figure E5.5, which represents a sand mass with a horizontal surface and a water level coinciding with the ground surface. A vertical, smooth, and rigid plate (of nearly infinite length) has been introduced in this soil, with horizontal tendons connected longitudinally spaced 1.0 m apart. Assume that the introduction of the plate and tendons did not modify the soil at-rest stress state. Then, identical forces, F, were applied to

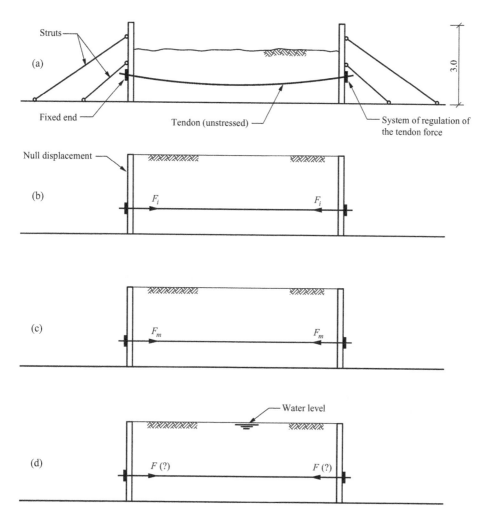

Figure E5.4

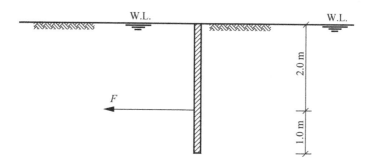

Figure E5.5

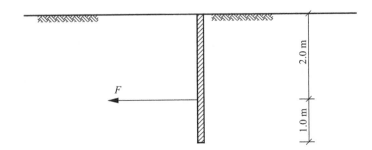

Figure E5.6

all tendons, progressively increasing. Ignore the friction between the plate base and the soil. Take $\gamma = 18.6$ kN/m³ and $\gamma_w = 9.8$ kN/m³.

a) Applying Rankine's theory, determine the angle of shearing resistance of the soil, taking into account that, when F equals 26.5 kN, the soil on the right side of the plate reaches a limit equilibrium state, whereas the stress state in the soil on the left becomes isotropic (the principal stresses are equal). How do you designate the limit equilibrium state installed on the right side?

b) Determine the maximum value that F can reach. How is the corresponding state of limit equilibrium on the left side of the plate designated?

E5.6 - Figure E5.6 represents a dry sand mass with a horizontal surface in which has been introduced a vertical, smooth, and rigid plate (of nearly infinite length), to which are connected horizontal tendons longitudinally spaced 1.0 m apart. Assume that the introduction of the plate and tendons did not modify the soil at-rest stress state and ignore the friction between the plate base and the soil. Take $\gamma_d = 16$ kN/m³, $\gamma_{sat} = 19.8$ kN/m³, $\gamma_w = 9.8$ kN/m³, and $\phi' = 35°$.

Assume that identical forces, F, were applied to all tendons, progressively increasing until mobilizing 65% of the theoretical passive pressure on the left side.

a) When F reaches the value mentioned above, in which situation is the soil on the right side of the plate? Determine the corresponding value of F.

b) Assuming that F remains constant, suppose that the water level rises to the ground surface on both sides of the plate. Calculate the new forces, P_{right} and P_{left}, applied by the ground on either side of the plate. To which side does the plate move?

E5.7 - With the help of the Caquot-Kérisel tables, calculate the horizontal and vertical components of the active thrust on the retaining wall represented in Figure E5.7, under the following conditions:

a) ignoring the cohesion of the supported soil;

b) taking into account the cohesion of the supported soil and applying the corresponding state theorem.

E5.8 - Consider the gabion retaining wall represented in Figure E5.8. Applying the method of Coulomb, estimate the active thrust on the wall backface.

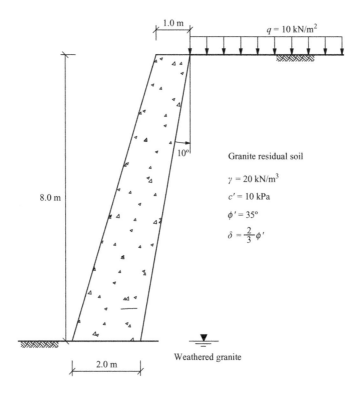

Granite residual soil

$\gamma = 20$ kN/m^3

$c' = 10$ kPa

$\phi' = 35°$

$\delta = \dfrac{2}{3}\phi'$

Figure E5.7

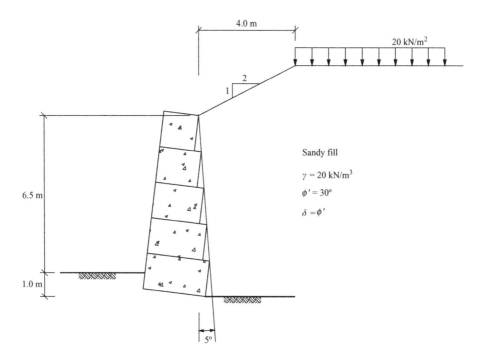

Sandy fill

$\gamma = 20$ kN/m^3

$\phi' = 30°$

$\delta = \phi'$

Figure E5.8

The leaning bell-tower of Pisa

1 THE TOWER

The bell-tower of Pisa Cathedral is admired for its architectural beauty, but it is especially famous by its disconcerting tilt, which had developed over more than eight centuries until being stabilized by a delicate and ingineous intervention concluded in 2001.

As shown in Figures P.1 and P.2, the tower consists of a cylindrical masonry structure, externally decorated by eight orders of sandstone *loggias*. The masonry comprises two concentric rings of square-shaped stones (with 0.4 m to 0.5 m sides), with a filling of gravel and mortar in between. The total thickness of the masonry is 4.1 m in the first *loggia* and 2.6 m

Figure P.1 Engraving of the 17th century by unknown author (AGI, 1991a).

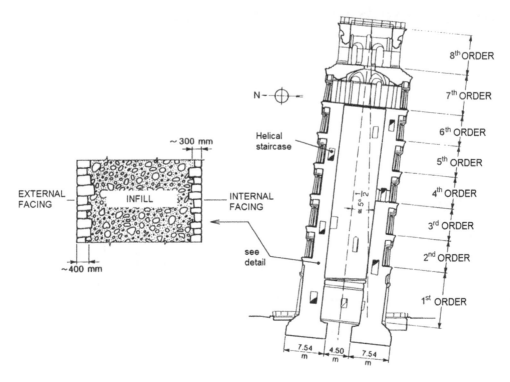

Figure P.2 Structure of the Tower of Pisa (AGI, 1991a).

from the second to the sixth *loggias*. The two upper *loggias* are constructed in brick and pumice stone masonry.

The tower was built in three stages. From 1173 to 1178, the foundation and the first four floors were constructed, reaching 29 m height. The works were then halted for nearly a century. From 1272 to 1278, the tower was built to the seventh *loggia*, rising to 51 m in height. Another stoppage followed, lasting about eighty years. Finally, from 1360 to 1370 the tower was completed, with the addition of the bell-chamber, reaching a height of 58 m.

The foundation, in stone masonry, has a ring shape, with internal and external diameters of 4.5 m and 19.6 m, respectively. The total weight of the tower is 142 MN, applied to the foundation with an eccentricity of 2.25 m measured immediately before the stabilization works were begun. The average pressure transmitted to the ground is approximately 500 kPa.

The average settlement of the tower since the beginning of construction has been estimated to be about 2.5 m to 3.0 m.

2 THE GROUND

The foundation ground of the Tower of Pisa is alluvial in origin and is geologically very recent (Pleistocene and Holocene). Figure P.3a shows the profile of the subsoil under the plane of maximum tower tilt, running practically North-South, and Figure P.3b includes results of a CPT performed at the site. Horizon A, the most recent horizon, comprises a succession of thin strata of silty sand, clayey silt and sand; Horizon B is predominantly clayey, while Horizon C is sandy.

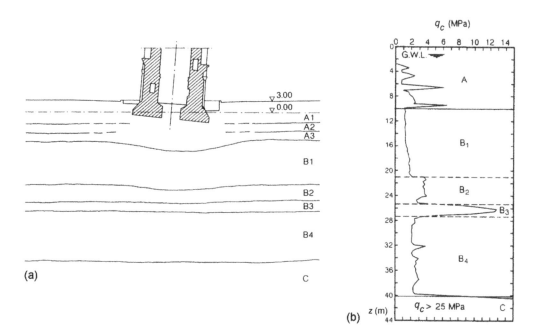

Figure P.3 Foundation ground of the Tower of Pisa: a) stratigraphic profile; b) typical CPT results (AGI, 1991a).

It is curious to note that the boundary between Horizon A and the upper layer (B_1) of the clayey Horizon B is horizontal across the entire *Piazza dei Miracoli*, except under the Tower, where it presents a depression exceeding 2 m. This provides an estimate of the order of magnitude of the consolidation settlement induced by the weight of the monument.

3 THE TILT

Information about the movements of the Tower is essentially derived from measurements of its southward tilt, the evolution of which is represented in Figure P.4.

The tower was already inclined when the second construction phase started. The builders attempted to correct the shift from the vertical along the three subsequent stories by placing more masonry blocks on the south side. The same happened in the third phase, leaving the Tower with a "banana" profile of a North-facing concavity, if observed carefully.

Careful analysis of the corrections introduced during construction permitted the reconstitution of the history of the tower's tilting during those two centuries. It is therefore estimated that, in 1370, at the end of construction, the southward tilt was already close to 2°.

After completion of the Tower, no data were available for more than four centuries. The first reliable observations, by geodetic surveying, date from 1817, when a tilt of 4.8° was registered.

The tilt of the Tower has been continuously monitored by several methods from 1911 onwards. In the early 1990s, the total southward tilt exceeded 5.5° (about 10%) and the rate of inclination was twice that measured in the 1930s, representing a horizontal displacement of the Tower top by 1.5 mm per year.

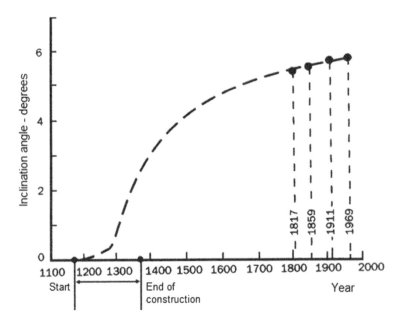

Figure P.4 Evolution of the tilt of the Tower of Pisa (AGI, 1991b).

4 THE PHENOMENON

Despite numerous geotechnical investigations performed on the site during the 20th century, no proof was found that the progressive southward tilt of the Tower was associated with heterogeneity of the foundation ground, i.e., due to weaker mechanical properties of the soil on that side. Overall, the sedimentary soil is actually quite uniform in the *Piazza dei Miracoli* and the two other building located there, the Cathedral and the Baptistery, older than the Tower, do not show signs of differential settlement.

The progressive inclination of the Tower may be explained by the phenomenon of *leaning instability*, which can affect structures in which a relatively deformable foundation soil is combined with a high value for the ratio between the center of gravity elevation above the foundation and the dimensions of the foundation in plan. In order to understand the unfavorable effect of foundation deformability on structures of this type, one may resort to an experience from childhood: it is easier for a *Lego* tower to remain stable on a bare tabletop than on a carpet-covered floor (Puzrin et al., 2010).

For a structure under those conditions, an incipient inclination – motivated by a possible, though small and brief, loading asymmetry or by a minute soil heterogeneity – will produce an overturning moment that the foundation ground has difficulty in equilibrating without experiencing a certain deformation. This allows the inclination to increase, which, in turn, induces the moment to increase, generating a continuous tilt progression at an increasing rate. Given the prolonged time period over which this phenomenon developed in the case of the Tower of Pisa, the deformation produced in the past centuries may be associated with creep (secondary consolidation).

The increased tilt rate during the 20th century represented a strong sign that the collapse of the Tower might eventually occur, unless stabilization works were undertaken.

5 THE STABILIZATION WORKS

In 1990, the Italian government appointed an International Committee for the Safeguard of the Tower of Pisa, led by Professor John Burland (Imperial College of London) and Professor Michele Jamiolkowski (Politecnico di Torino).

The fact that motivated the creation of the Committee had been the collapse of the Civic Tower next to Pavia Cathedral in 1989. This tower, built in the 11th century on a dense formation of sand and gravel and without any apparent inclination, suddenly collapsed to the ground, killing four people. This was similar to the abrupt collapse of Saint Mark's Bell Tower in Venice, occurring at the start of the 20th century. Both towers had some similarities to the structure of the Tower of Pisa, namely infilled stone masonry. These events led to the closure of the Tower of Pisa to the public in 1990.

The Committee soon concluded that the Tower was at serious risk, due not only to its progressive inclination, but also to the risk of practically instantaneous structural collapse.

These two collapse mechanisms, toppling over due to increasing tilt or masonry failure by compression, are closely related. In fact, the continuing inclination increased the compressive stress on the South side, which reached about 8 MPa on the external masonry ring at the second *loggia* level.

Taking this into account, the Committee opted for the proposal of some urgent measures, even though of a provisional and fully reversible nature.

The first was implemented in the summer of 1992. As shown in Figure P.5, lightly tensioned circumferential steel strands were installed at the level of the second *loggia*, in order to prevent compressive failure of the external masonry ring.

The second temporary measure, put into practice between May 1993 and January 1994, consisted of the application of a 600-ton counterweight to the North edge of the foundation, as illustrated in Figure P.6. The operation started with the construction of a reinforced concrete ring around the foundation of the Tower, over which were gradually placed lead ingots on the North edge. The total weight of the concrete ring and the lead ingots corresponded to

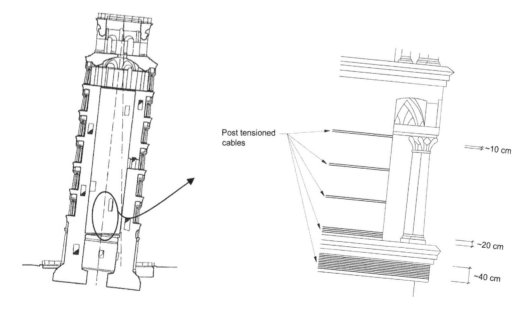

Figure P.5 Circumferential post-tensioned steel cables around the masonry structure of the Tower of Pisa (Burland et al., 1993).

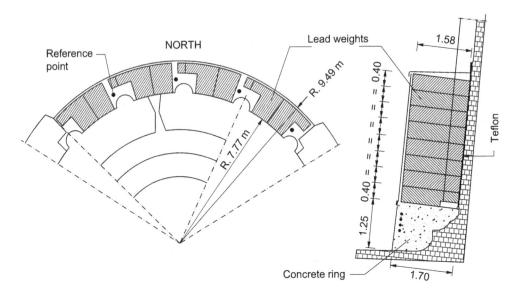

Figure P.6 Counterweights placed on the North edge of the foundation of the Tower of Pisa (Burland et al., 1993).

nearly 4.8% of the Tower's weight, and its placement resulted in a stabilizing moment of about 10% of the moment associated with the eccentricity of the load transmitted to the foundation.

The motion of the Tower during the works was carefully monitored. The results were encouraging: the Tower exhibited a clear tendency toward a reduction of the tilt, rotating northward by about 37 arc seconds (1 second of arc is 1/3600 of a degree). The average settlement associated with the operation was about 2.5 mm (Burland et al., 1994).

The final phase of the stabilization works aimed to achieve a northward rotation of the Tower of about 0.5° (1800 arc seconds), considered appropriate for several reasons: being imperceptible to the naked eye, it would not affect the collective memory of the Leaning Tower, which was imperative to preserve; it ensured a reduction of the compressive stress in the masonry, safeguarding against structural collapse; and it might lead to a stabilization of the progressive southward inclination process.

After detailed studies over a decade, the solution selected was the *under-excavation* technique, already employed with success in the rehabilitation of the Mexico City Metropolitan Cathedral. As shown in Figure P.7a, the method consisted of the localized and progressive removal of soil from inside inclined holes of small diameter; after soil removal, the cavities eventually closed, inducing a gradual and controlled surface settlement.

On February 9, 1999, in an atmosphere of great expectation, the first extraction of soil was initiated on the North side, using 12 holes inclined at 30° on a 6 m wide strip, as shown in the upper part of Figure P.7b. Since that step, the Tower started to rotate slowly northward. When the rotation reached 80 arc seconds, at the beginning of July 1999, this preliminary treatment was halted. The rotation continued, at a decreasing rate, until the following October.

The success of this preliminary phase convinced the Committee that it would be safe to extend the soil extraction to the full width of the foundation. So, between December 1999 and January 2000, as indicated in the lower part of Figure P.7b, 41 holes, spaced 0.5 m apart, were drilled on the North side of the Tower, inclined at 20° and each with its lining tube and internal auger. Soil extraction was started on February 21, 2000, with a positive effect on the Tower behavior, which inclined further northward in a very steady manner.

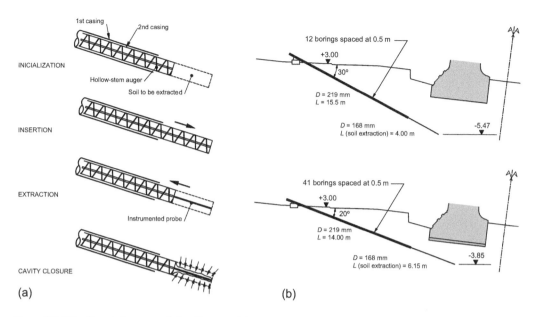

Figure P.7 "Final" stabilization of the Tower of Pisa: a) under-excavation technique (Jamiolkowski, 1999); b) procedure for soil extraction under the foundation on the North side of the Tower (adapted from Burland et al., 2009).

The extraction of soil continued until June 2001, when the Tower rotation had reached the target value (1800 arc s), returning the inclination to that determined in 1844!

One month before, the process of removal of the lead counterweight ingots had been initiated, followed by the demolition of the reinforced concrete base ring.

Complementary stabilizing work involved the construction of a drainage system to attenuate the seasonal fluctuations of the water level in Horizon A during heavy rains. That system consisted of three wells, installed on the North side of the tower and connected to sub-horizontal drains extending toward the limits of the Tower foundation. The water level in the wells was controlled by discharge pipes connected to the sewer system of the Piazza.

After a careful rehabilitation of the surrounding area, in December 2001, more than eleven years after its closure, the Tower was finally reopened to the public, with the city of Pisa in a festive mood.

In 2009, the members of the Committee presented an update on the condition of the Tower (Burland et al., 2009). The results were clearly satisfactory: between 2001 and 2008, an additional northwards inclination of 148 arc seconds had occurred, but at a decreasing rate; at the end of this period, the Tower was practically motionless.

In that publication, the experts draw two scenarios concerning the evolution of the Tower in the future. The *optimistic scenario* corresponded to the permanent stopping of the Tower motion, except for minute seasonal oscillations due to variations in temperature and groundwater conditions. In the *pessimistic scenario*, after a few decades of immobility, the Tower would restart its southward rotation, initially at a very slow rate, but later at an increasing rate. In this latter scenario, about 200 years would be required for the Tower to return the inclination to the point where it was in 1993, immediately before the start of the stabilization works. Nevertheless, even if future generations do not develop novel intervention methods, resorting to the procedure of soil extraction on the north side will permit the inclination of the Tower to be reduced to the required degree at any time.

Chapter 6

Shallow foundations

6.1 INTRODUCTION

It is termed a *shallow foundation*, but also a *spread foundation* or, even, a *footing*, one that transmits the action from the structure to a *load-bearing* layer located close to the ground surface.

In functional terms, what distinguishes a shallow foundation from a deep foundation is that, in its interaction with the soil, to balance out the vertical load, the portion of a shallow foundation load not processed through the base is generally negligible and, therefore, ignored in the design process. As for deep foundations, their interaction with the ground along the lateral surface (also called the shaft) is normally taken into account in the design, given its relevance to the equilibration of the vertical load.

In the great majority of building and bridge foundations, the vertical actions are clearly predominant, with the horizontal forces, normally associated with variable or even accidental actions, such as wind or earthquakes, being of modest or moderate value. It is therefore understandable that, normally, the base of shallow foundations is horizontal. Footings with inclined base are sometimes adopted for gravity retaining walls, as will be seen in the next chapter, in view of the large and permanent nature of the horizontal force transmitted to the foundation.

With respect to the ground, the design of a shallow foundation must satisfy the safety requirements relative to both the ultimate and the serviceability limit states.

The most important ultimate limit state consists of the failure of the ground under the footing, due to insufficient bearing capacity relative to vertical loading. This causes a very large vertical displacement that may also induce an ultimate limit state in the structure.

Another very relevant limit state involves excessive foundation settlement. Normally, this is associated with a serviceability limit state, but, under certain circumstances, it may induce an ultimate limit state in structural elements adjacent to that foundation.

The major objective of the present chapter on shallow foundations is to present methodologies to: i) evaluate the bearing capacity for vertical loading; and ii) estimate the settlement. Based on these procedures, it is then possible to define, for a given practical case, the elevation of the footing base and its dimensions in plan. The further design of the foundation as a structural element (namely establishing its height and the area of steel, in the case of reinforced concrete) is outside the scope of this book.

The sizing of shallow foundations has been made for centuries by means of empirical criteria, with the validity restricted to the soils of a given zone or region, adjusted over time on the basis of the successes and failures achieved. These rules were essentially developed for stone masonry structures.

The generalization of frame structures, associated with the use of steel and concrete as the most current structural materials from the middle of the 19th century onwards, has created additional requirements as to what concerns foundation design. In fact, the more frequent construction of taller structures with larger spans has resulted in higher and more concentrated foundation loadings.

The theoretical tools available for shallow foundation design were developed in the 20th century, in parallel with the development of soil mechanics from the pioneering work of Terzaghi in the 1920s. Nevertheless, due the complexity of some of the questions involved in the design process, many of the methods currently applied are still a mix of theoretical, rationally proved solutions, and semi-empirical, or even empirical, experience-based approaches.

In the present chapter, the empirical component has been limited to the indispensable minimum, with emphasis being placed on the discussion of the mechanical phenomena involved in foundation behavior and on design methods fundamentally based on rational arguments. This will allow the reader, should the will or the need arise, to resort, with critical sense, to the perusal of specialized publications, namely foundation manuals, where other methods are presented (Coduto, 2001; Frank, 2003).

6.2 BEARING CAPACITY UNDER VERTICAL LOADING

6.2.1 Introduction

Consider the footing represented in Figure 6.1a, located at the surface of a soil mass and submitted to an increasing, centered, vertical loading.

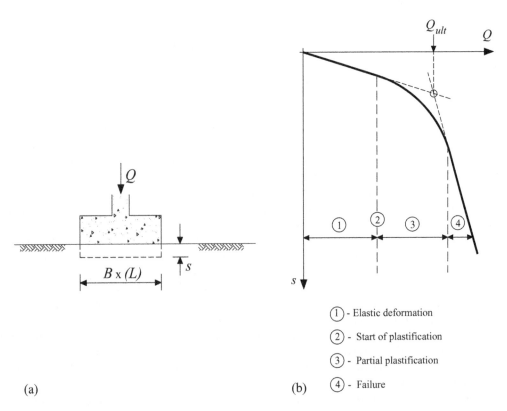

(a)   (b)

① - Elastic deformation

② - Start of plastification

③ - Partial plastification

④ - Failure

Figure 6.1 Bearing capacity of a footing under vertical loading: a) basic layout; b) generic load-settlement diagram.

Figure 6.1b shows the simplified aspect of the vertical load-settlement graph. The initial part, approximately linear and with a small slope, represents the deformation of the foundation layer in an essentially elastic regime. At a certain point, the graph starts to exhibit some curvature, indicating the start, still localized, of a shearing failure in the loaded soil. The slope of the graph becomes more pronounced as the failure zone expands in the ground. The quasi-linear final branch deserves a more detailed comment, as both its slope and extension depend on several factors, particularly the soil type.

In dense sandy soils and in clays loaded under undrained conditions, as shown in Figure 6.2a, that branch is practically vertical and relatively short. Sudden failure occurs, with a clear ascent of the soil surrounding the footing. This means that an unconfined plastic flow develops, i.e., a shear failure situation involving a continuous soil mass under and around the footing, limited by a sliding surface that extends up to the soil surface. Its geometry is sketched on the right side of the figure and will be analyzed in detail in Section 6.2.2. If there is no constraint provided by the connection of the footing to the rest of the structure (isolated footing), a pronounced rotation may occur in the final stages of collapse.[1]

In loose sandy soils and in clays loaded under drained conditions, as shown in Figure 6.2b, the final branch will be significantly inclined but not vertical, and large settlement values may be reached without the occurrence of a generalized failure. The explanation lies in the fact that, as the settlement develops with the increase in loading, so does the soil resistance, which contributes to postponing generalized failure almost indefinitely. In the case of sand, the deformation associated with loading leads to some densification, thus increasing the shearing resistance angle. Additionally, the (large) settlement increases the weight of the soil above the footing base. In the following, this weight will be seen to play a relevant role in the bearing capacity value. The preceding considerations essentially apply also to clays.

Returning to Figure 6.1b, the intersection of the tangents to the approximately linear branches may be considered the theoretical *failure load*, Q_{ult}, which is also known as the foundation *limit load* or *ultimate load*.

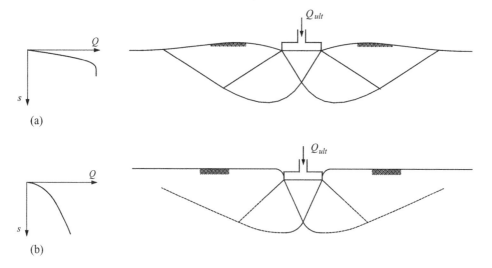

Figure 6.2 Typical footing failure modes (adapted from Vesić, 1963): a) general failure; b) local or partial failure.

---

[1] All soil formations are heterogeneous to some degree. During loading, settlements increase more rapidly on the weaker side and so do the vertical stresses as second order effects become significant, which accelerates the failure process. In conjunction with the settlement, the footing then experiences a rotation towards the weaker soil zone. This has some degree of analogy with the buckling phenomenon studied in Strength of Materials.

The *bearing capacity* of the foundation is the ratio between the ultimate load and the area of the foundation base:

$$q_{ult} = \frac{Q_{ult}}{B\,L} \tag{6.1}$$

with B being the width and L the length.

6.2.2 General expression of the load-bearing capacity under vertical loading

The problem of the evaluation of the load-bearing capacity of a shallow foundation is treated in the context of the theory of plasticity, using methods of limit analysis based on the upper and lower bound theorems (see Chapter 2). As is known, the methods based on the former provide overestimates, which are unsafe, whereas those based in the second theorem lead to underestimates, which are on the safe side. When the solutions produced by both theorems coincide, they are considered to be exact. In these methods, it is assumed that the soil has a rigid-plastic behavior, i.e., the pre-failure deformation is not considered.

Given the analytical and numerical complexity of the problem, a very succinct presentation will be now made of the solution that is generally employed for design purposes, referring to the basic hypotheses and the fundamental contributions. An alternative approach, of a semi-empirical nature, directly based on PMT test results, is presented in Annex A6.1.

Consider then a shallow foundation of width B, like the one represented in Figure 6.3, the base of which is at depth D, on a homogeneous soil with a horizontal surface and a unit weight γ, loaded by a vertical, centered force. Assume that:

i) the footing has infinite length;
ii) the soil obeys the Mohr-Coulomb failure criterion;
iii) the shearing resistance of the soil above the footing base is null, i.e., the soil acts as a uniformly distributed surcharge on the horizontal footing base surface;
iv) the friction and the adhesion between the footing and the foundation soil are null.

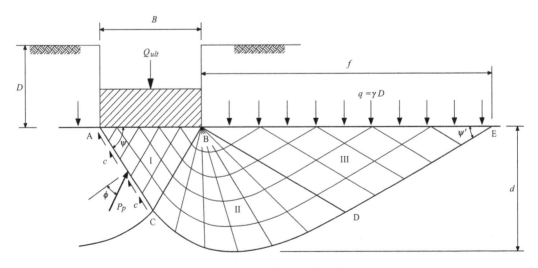

Figure 6.3 Bearing capacity of a shallow foundation: shear wedges and forces opposing failure.

As shown in Figure 6.3, the soil shear failure implies the formation of five blocks or wedges (the problem is symmetric in relation to the center of the footing): wedge I, which descends in Rankine active limit state jointly with the footing, obliging wedges II to move laterally in radial shear, which induces, in wedges III, a lateral ascending displacement in Rankine passive state. As usually happens with mathematical solutions of engineering problems, the failure mechanism considered has been inspired by the observation of real failures as well as results from small-scale laboratory tests.

Line ACDE comprises two straight segments, AC and DE, and one curved part. The rectilinear segments are at angles ψ and ψ' with the horizontal of $\pi/4+\phi/2$ and $\pi/4-\phi/2$, respectively; the curvilinear part is a circular arc for soils with a zero friction angle, or a log spiral in other cases.

Given the mechanical and geometrical connection of wedge ABC with the footing, the equilibrium analysis of the wedge provides a way to analyze the equilibrium of the footing. Besides its self-weight, W, the wedge is subject to the forces included in Figure 6.3. Considering force equilibrium in the vertical direction, we may write:

$$Q_{ult} = 2P_p \cos(\psi - \phi) + 2\overline{AC}\,c\sin\psi - W \tag{6.2}$$

Given that

$$\overline{AC} = \frac{B}{2\cos\psi} \tag{6.3}$$

and

$$W = \frac{\gamma B^2}{4}\tan\psi \tag{6.4}$$

equation 6.2 becomes:

$$Q_{ult} = 2P_p \cos(\psi - \phi) + Bc\tan\psi - \frac{\gamma B^2}{4}\tan\psi \tag{6.5}$$

Dividing this equation by the foundation width, B, we get:

$$q_{ult} = \frac{2P_p}{B}\cos(\psi - \phi) + c\tan\psi - \frac{\gamma B}{4}\tan\psi \tag{6.6}$$

In this equation, the only unknown is the force P_p, in the determination of which resides the difficulty of the problem.

For *weightless* materials ($\gamma = 0$), the problem has a mathematically exact solution. This solution, developed by Prandtl (1921) and Reissner (1924) in the context of metal plasticity, has been adapted to soil mechanics by Caquot (1934), corresponding to the following equation:

$$q_{ult} = cN_c + qN_q \tag{6.7}$$

where

$$q = \gamma D \tag{6.8}$$

is the pressure applied at the footing base surface, on either side, and N_c and N_q are dimensionless coefficients given by:

$$N_c = (N_q - 1)\cotan\phi \qquad (6.9)$$

and

$$N_q = e^{\pi\tan\phi}\tan^2\left(\frac{\pi}{4} + \frac{\phi}{2}\right) \qquad (6.10)$$

For $\phi = 0$, this exact solution provides $N_c = \pi + 2 = 5.14$ and $N_q = 1$.

An exact solution has not yet been found for the bearing capacity of real soils, i.e., *with weight*. The problem was analyzed first by Buisman (1940) and then by Terzaghi (1943), assuming null cohesion and the footing at the soil surface, with zero lateral surcharge ($q = 0$). The proposed expression:

$$q_{ult} = \frac{1}{2}\gamma B N_\gamma \qquad (6.11)$$

involves the dimensionless coefficient, N_γ, the value of which was graphically determined by these authors.

For the general problem of a soil with cohesion, lateral surcharge, and self-weight, Terzaghi (1943) proposed to combine Equations 6.7 and 6.11 into a single one. The general expression for the load-bearing capacity of a shallow footing, known as the Buisman-Terzaghi equation and one of the most used equations in soil mechanics, is then:

$$q_{ult} = c\, N_c + q\, N_q + \frac{1}{2}\gamma\, B\, N_\gamma \qquad (6.12)$$

The second member of this equation can be interpreted in the following way: i) the first term represents the bearing capacity of a material with frictional resistance combined *with cohesion*, but *without self-weight or lateral surcharge*; ii) the second term represents the bearing capacity of a material with frictional resistance combined *with lateral surcharge*, but *without cohesion or self-weight*; and iii) the third term represents the bearing capacity of a material with frictional resistance combined *with self-weight*, but *without cohesion or lateral surcharge*.

Several studies have shown that this *superposition of effects*, though not mathematically correct, involves tolerable errors on the safe side when applied to cases in which all those parameters are involved (Lundgren and Mortensen, 1953; Brinch Hansen and Christensen, 1969).

As mentioned above, a mathematically exact solution is not available for the coefficient N_γ, which is highly dependent on ψ, the angle defining the wedge geometry under the footing base (see Figure 6.3). The expression

$$N_\gamma = 2(N_q - 1)\tan\phi \qquad (6.13)$$

that has had a much-generalized application (having been adopted in Eurocode 7), results from an approximation to the numerical values obtained by Caquot and Kérisel (1953), assuming $\psi = \pi/4 + \phi/2$.

The search for approximate solutions for the N_γ factor has been, over the past decades, a recurrent research topic in soil mechanics, on a par with the development of modern computational methods. A very accurate solution was reported by Hjiaj et al. (2005), combining results from the two limit analysis theorems. The proposed solution is the average of those obtained by each of the two approaches, confining the exact solution to a maximum error of the order of 3%. The authors developed the expression:

$$N_\gamma = e^{\frac{1}{6}\left(\pi + 3\pi^2 \tan\phi\right)} \left(\tan\phi\right)^{\frac{2\pi}{5}} \tag{6.14}$$

which approximates the numerical values obtained.

In Table 6.1 are listed the values of the three load-bearing capacity factors for a practical range of ϕ values. The values of N_c and N_q result from Equations 6.9 and 6.10, respectively. For N_γ are included values computed from Equations 6.13 and 6.14; as can be observed, the differences are very significant, with the values of the former expression exceeding those of the more recent proposal by 25 to 40% for friction angles below 40°. The use of the values supplied by this last equation seems therefore to be recommendable.

Based on Figure 6.4, the depth and width reached by the plastic zone responsible for footing equilibrium can be evaluated (see, in Figure 6.3, the definitions of d and f). It may be seen, for example, that for sand with a friction angle of 30°, the failure surface extends sideways by about three times the footing width, reaching a maximum depth of the order of that width.

6.2.3 Conditions of application of the bearing capacity equation: total and effective stress analyses

Equation 6.12 is a general expression, which can be applied in both effective and total stress analyses.

For effective stress analyses, that equation becomes:

$$q_{ult} = c'N_c + q'N_q + \frac{1}{2}\gamma\, B\, N_\gamma \tag{6.15}$$

where:

i) c' is the effective cohesion;
ii) N_c, N_q, and N_γ are obtained from ϕ', the angle of shearing resistance in terms of effective stress;
iii) q' represents the effective vertical stress at the footing base level;
iv) γ, the unit weight in the third term, which represents the contribution of the weight of the displaced soil wedges, should be taken to equal γ' in cases where the water table is at or above the base of the footing; if the water table is below the footing base but part of the plastic zone is submerged, an intermediate value between γ and γ' should be adopted; for definition of that value, the use of Figure 6.4 is suggested.

For total stress analysis ($\phi = 0$), Equation 6.12 reduces simply to:

$$q_{ult} = c_u N_c + q = c_u\left(\pi + 2\right) + q = 5.14 c_u + q \tag{6.16}$$

with c_u being the soil undrained shear strength and q the total vertical stress at the footing base level.

As is known, total stress analyses are essentially applied in undrained loading of saturated clayey soils. It has already been pointed out (see Section 2.4.1) that the ground bearing capacity is minimum under those conditions, given that, with time, the (positive) excess

Table 6.1 Bearing capacity factors.

ϕ	N_c (1)	N_q (2)	N_γ (3)	N_γ (4)
0	5.14	1.00	0.00	0.00
20	14.83	6.40	3.93	2.86
21	15.81	7.07	4.66	3.37
22	16.88	7.82	5.51	3.97
23	18.05	8.66	6.50	4.67
24	19.32	9.60	7.66	5.50
25	20.72	10.66	9.01	6.46
26	22.25	11.85	10.59	7.60
27	23.94	13.20	12.43	8.94
28	25.80	14.72	14.59	10.52
29	27.86	16.44	17.12	12.40
30	30.14	18.40	20.09	14.62
31	32.67	20.63	23.59	17.26
32	35.49	23.18	27.72	20.42
33	38.64	26.09	32.59	24.19
34	42.16	29.44	38.37	28.71
35	46.12	33.30	45.23	34.16
36	50.59	37.75	53.40	40.75
37	55.63	42.92	63.18	48.75
38	61.35	48.93	74.90	58.49
39	67.87	55.96	89.01	70.43
40	75.31	64.20	106.05	85.11
41	83.86	73.90	126.74	103.27
42	93.71	85.37	151.94	125.85
43	105.11	99.01	182.80	154.10
44	118.37	115.31	220.77	189.66
45	133.88	134.87	267.75	234.72
46	152.10	158.50	326.20	292.25
47	173.64	187.21	399.36	366.25
48	199.26	222.30	491.56	462.24
49	229.92	265.50	608.54	587.85
50	266.88	319.06	758.09	753.79

Notes:

1. Equation 6.9;

2. Equation 6.10;

3. Equation 6.13;

4. Equation 6.14.

pore pressure dissipates, with an increase in effective stress and also in the soil shearing resistance.

6.2.4 Influence of the parameters integrating the bearing capacity expression

In order to comment on the influence of the diverse factors integrating the general bearing capacity expression, it should be stressed, to start with, that the friction angle is the fundamental parameter, since N_c, N_q, and N_γ all increase rapidly with ϕ'.

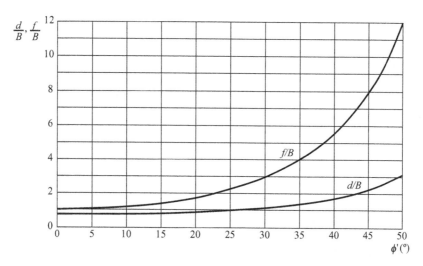

Figure 6.4 Dimensions of the shear wedges under the footing (Meyerhof, 1948; G. A. Leonards. *Les Fondations*, © Dunod, 1968, Paris).

The physical explanation of this fact can be understood with the help of Figures 6.3 and 6.5. As seen in Figure 6.3, the higher the angle of shearing resistance, the greater the angle of inclination of surface AC (and its symmetrical surface), with respect to the horizontal, and the lower the angle of inclination of surface DE (and its symmetrical surface). As shown now in Figure 6.5, this leads to a substantial increase in the dimensions of the soil wedges that ensure bearing capacity. Therefore: i) the cohesive strength can be mobilized along surfaces with larger development; ii) the surcharge lateral to the footing is involved in providing equilibrium along a greater length (segment BE and its symmetrical in Figure 6.3); and iii) naturally, the weight of the wedges also suffers a considerable increment.

With a finer analysis of Table 6.1 and of the equations that have originated it, it is possible to verify that, with the angle of friction, the rate of growth increases substantially from N_c to N_q and from N_q to N_γ. It is interesting to reflect on the reason for this fact. N_q grows more markedly than N_c because the increase of the resultant of the lateral surcharge

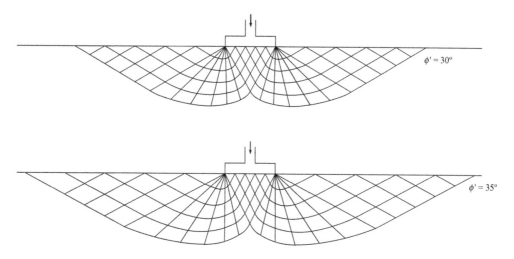

Figure 6.5 Influence of the angle of shearing resistance on the geometry of the wedges that ensure bearing capacity.

leads to a greater frictional resistance of the foundation soil. The growth of N_γ is also involved in an increase in frictional resistance, due to the increase in the wedge weight. But, in this case, the increase is even more pronounced because it is proportional to the wedge *volume*, while that of N_q is proportional to the *area* of the top horizontal surface of those wedges.

Continuing the analysis of the factors that affect the bearing capacity, two of the terms of that same expression are proportional to the soil unit weight. Therefore, in an effective stress analysis, the position of the water table assumes great significance, given that its rise from a great depth to the foundation base, and from here to the soil surface reduces the third and second terms, respectively, by around 50%.

The third term varies proportionally to the foundation width. Therefore, a wide footing resting on a soil with a large friction angle has a high bearing capacity, while a narrow footing on the same soil has a much lower bearing capacity. However, it should be noted that N_γ is zero for $\phi = 0$, which means that, for clayey soils under undrained conditions, the bearing capacity is independent of the foundation width.

The soil cohesion influences only the first term. Given that for $\phi = 0$, we have $N_q = 1$ and $N_\gamma = 0$, the first term becomes dominant with respect to bearing capacity in total stress analyses.

6.2.5 Extension of the bearing capacity expression to cases of practical interest

6.2.5.1 Introduction

The hypotheses assumed for deriving the bearing capacity under vertical loading listed in Section 6.2.2, correspond to a highly idealized and simplified problem. In fact, the following conditions, among others, though current in real problems, are not contemplated in the theoretical solution: i) the footing has finite length, i.e., the longitudinal dimension is of the order of the transverse dimension; ii) the vertical foundation load is applied with a given eccentricity; iii) the load has a certain inclination; iv) the base of the footing is not horizontal; v) the ground surface is not horizontal; and vi) the ground is stratified, or, at a certain depth beneath the foundation base, there occurs a large resistance (firm) layer.

In order to attend to most of the practical conditions listed, it is current to apply *corrective factors* to each of the terms of Equation 6.12. The notation used for these comprises a lowercase letter, generally the initial of the English word for that condition, complemented with an index indicative of the corresponding term. For example, the corrective factor s_q refers to the correction of the second bearing capacity term (relative to the contribution of the lateral surcharge, q) in order to account for the footing *shape*.

Given the complexity of the problem, most corrective factors are of a semi-empirical nature or result from physical tests conducted on small-scale models.

6.2.5.2 Eccentric loads

Within the conditions listed above, there is one requiring a distinct treatment that must be performed beforehand. When applied moments, M_x and M_y, relative to the axes of the footing base plane exist in conjunction with the vertical load, V, , as shown in Figure 6.6a, the generalized force system (V, M_x, M_y) acting on the center of gravity of the foundation base is statically equivalent to force V applied at point P (Figure 6.6b), with coordinates e_x and e_y, such that:

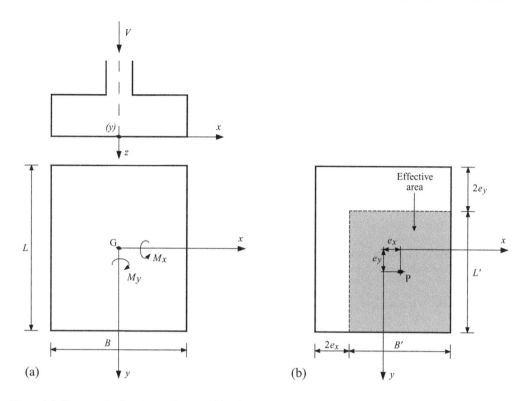

Figure 6.6 Rectangular footing with vertical load and moments: a) force system on the foundation center of gravity; b) effective foundation area.

$$e_x = \frac{M_y}{V} \tag{6.17}$$

$$e_y = \frac{M_x}{V} \tag{6.18}$$

For practical purposes, as indicated in Figure 6.6b, the bearing capacity should be calculated, considering a virtual footing centered on point P, which amounts to taking as the footing *effective dimensions* B' and L', given by

$$B' = B - 2e_x \tag{6.19}$$

and

$$L' = L - 2e_y \tag{6.20}$$

and A_{ef} as the *effective area* of the foundation base,

$$A_{ef} = (B - 2e_x)(L - 2e_y) = B'L' \tag{6.21}$$

If q_{ult} is the bearing capacity calculated taking, for all purposes, B' and L' as the footing dimensions, the foundation ultimate load, Q_{ult}, is:

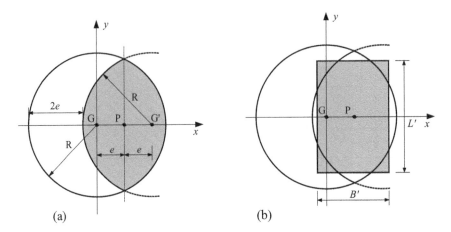

Figure 6.7 Circular footing with vertical load and moments: a) determination of the effective area; b) rectangular footing with equivalent effective area.

$$Q_{ult} = q_{ult} A_{ef} = q_{ult} B'L' \tag{6.22}$$

Figure 6.7 suggests a simplified procedure for applying this methodology to circular footings (Brinch Hansen, 1961a). The so-called footing effective area is the area common to the real footing and to another with an identical diameter but centered at point P, with coordinates e_x and e_y. Then, a rectangular footing is sought, the dimensions B' and L' of which produce an area equal to the effective area A_{ef} and with a proportion that reasonably approximates its shape.

Fang (1991) proposed the following approximate expressions for calculating the effective area A_{ef} and the dimensions B' and L' in accordance with that criterion:

$$A_{ef} = \pi R^2 - 2 \left[e\sqrt{R^2 - e^2} + R^2 \sin^{-1}\left(\frac{e}{R}\right) \right] \tag{6.23}$$

$$B' = L' \left(\frac{R-e}{R+e}\right)^{0.5} \tag{6.24}$$

$$L' = \left[A_{ef} \left(\frac{R+e}{R-e}\right)^{0.5} \right]^{0.5} \tag{6.25}$$

6.2.5.3 Corrective factors for the bearing capacity

Table 6.2 summarizes the practical situations considered, their defining variables, the symbols of the respective corrective factors, as well as the neutrality conditions, for which the corrective factors assume unit value.

Therefore, in the general case, in which those situations occur simultaneously, the expression of the bearing capacity will be:

$$q_{ult} = cN_c s_c i_c b_c g_c f_c + qN_q s_q i_q b_q g_q f_q + \frac{1}{2}\gamma BN_\gamma s_\gamma i_\gamma b_\gamma g_\gamma f_\gamma \tag{6.26}$$

Table 6.2 Practical situations for calculation of the bearing capacity of a shallow foundation and respective corrective factors for the simplified theoretical solution.

Effect	Variables	Corrective factors	Neutrality condition	Scheme
Foundation shape	B, L	s_c, s_q, s_γ	$L = \infty$	
Load inclination	V, H	i_c, i_q, i_γ	$H = 0$	
Inclination of the foundation base	ζ	b_c, b_q, b_γ	$\zeta = 0$	
Inclination of the ground surface	β	g_c, g_q, g_γ	$\beta = 0$	
Proximity of a firm stratum	B, D_f	f_c, f_q, f_γ	$D_f = \infty$	

Taking into account what has been said about how the corrective factors have been obtained, it is understandable that their expressions may vary among the authors who have investigated this subject. In recent years, a relative consensus has been generally reached around the expressions listed in Table 6.3, which have been included in Eurocode 7. By and large, those expressions result from proposals of Vesić (1975), based on his research, and on works of other authors, namely Meyerhof (1953), De Beer (1970) and Brinch Hansen (1961a, 1970). For the corrective factors relative to the first term of the bearing capacity, the expressions for both effective stress and total stress analyses are included.

It should be noted that, both in Equation 6.26 and in the expressions of Table 6.3, the foundation dimensions to be considered are the effective dimensions, B' and L', obtained as discussed in Section 6.2.5.2.

When the foundation base is not horizontal, V and H must be taken as the normal and tangential components of the load, respectively, and Q_{ult} is the ultimate resisting force normal to the foundation base.

When in the presence of a load component tangential to the foundation base, the safety relative to sliding has to be checked. Such verification is processed in a similar manner to what will be seen for retaining wall foundations in Chapter 8. In general, the contribution of the passive thrust for the safety against sliding is disregarded, given that its mobilization, in contrast with that of the interface resistance between the footing base and the foundation soil, demands a displacement magnitude that, in most cases, is not compatible with the deformation capacity of the structure.

6.2.5.4 Shallow foundations on stratified ground

In the context of the evaluation of the bearing capacity of shallow foundations presented so far, it has been assumed that the underlying ground is homogeneous, i.e., the thickness of the soil layer, whose strength data is considered in the preceding expressions, is sufficiently large for the bearing capacity to depend exclusively on that same layer. Taking Figure 6.3 as

Table 6.3 Expressions of the corrective factors for the simplified theoretical solution of the bearing capacity of a shallow foundation.

Effect	1st term (cohesion)	2nd term (surcharge)	3rd term (weight)
Foundation shape (s factors) (see note 1)	Effective stress analysis $s_c = \dfrac{s_q N_q - 1}{N_q - 1}$ Total stress analysis $s_c = 1 + 0.2\dfrac{B}{L}$	$s_q = 1 + \dfrac{B}{L}\sin\phi'$ EC7 (Vesić has $\tan\phi'$ instead of $\sin\phi'$)	$s_\gamma = 1 - 0.3\dfrac{B}{L}$ EC7 (Vesić has 0.4 instead of 0.3)
Load inclination (i factors) (see note 2)	Effective stress analysis $i_c = i_q - \dfrac{1-i_q}{N_c \tan\phi'}$ Total stress analysis $i_c = \dfrac{1}{2}\left[1 + \left(1 - \dfrac{H}{Ac_u}\right)^{0.5}\right]$ with $H \le A\, c_u$	$i_q = \left(1 - \dfrac{H}{V + BLc'\cotan\phi'}\right)^{m}$ (see note 5)	$i_\gamma = \left(1 - \dfrac{H}{V + BLc'\cotan\phi'}\right)^{m+1}$ (see note 5)
Foundation base inclination (b factors) (see note 3)	Effective stress analysis $b_c = b_q - \dfrac{1-b_q}{N_c \tan\phi'}$ Total stress analysis $b_c = 1 - \dfrac{2\zeta}{\pi+2}$ (ζ in radians)	$b_q = \left(1 - \zeta\tan\phi'\right)^2$	$b_\gamma = \left(1 - \zeta\tan\phi'\right)^2$
Ground surface inclination (g factors) (see note 4)	Effective stress analysis $g_c = g_q - \dfrac{1-g_q}{N_c \tan\phi'}$ Total stress analysis $g_c = 1 - \dfrac{2\beta}{\pi+2}$ (β in radians)	$g_q = \left(1 - \tan\beta\right)^2$	$g_\gamma = \left(1 - \tan\beta\right)^2$
Proximity of a firm stratum (f factors)	f_c – see Table 6.4	f_q – see Table 6.4	f_γ – see Table 6.4

Notes:

1. s factors: Vesić adapting and generalizing a proposal of De Beer (1970), later adjusted in Eurocode 7.

2. i factors: Vesić adapting and generalizing a proposal of Brinch Hansen (1961a).

3. b factors: Vesić based on Meyerhof (1953) and Brinch Hansen (1970).

4. g factors: Vesić based on Brinch Hansen (1970).

5. Definition of the exponent m:

$$m = m_B = \frac{2 + (B/L)}{1 + (B/L)} \text{ for } H \text{ parallel to } B;$$

$$m = m_L = \frac{2 + (L/B)}{1 + (L/B)} \text{ for } H \text{ parallel to } L;$$

$$m = m_\theta = m_L \cos^2\theta + m_B \sin^2\theta \text{ for } H \text{ at an angle } \theta \text{ with } L.$$

reference, this means that the sliding surface developed at failure does not extend to a layer other than that immediately underlying the foundation. As already discussed, the maximum depth, d, reached by that surface depends both on the soil friction angle and on the foundation dimensions, and may be estimated with the help of Figure 6.4.

When the bearing capacity depends on more than a single layer, its evaluation requires sound judgement and imagination for adapting the basic solution previously described.

If the contrast in strength between the two (or more) layers involved is not considerable, the computation of the bearing capacity exclusively on the basis of the data from each of them will allow calculation of the *limits of an interval*, within which lies the actual value of the bearing capacity. If this interval is relatively narrow, then a satisfactory solution has been found. Otherwise, the consideration of the relative importance of the layers in question may allow, in many cases, for the extraction of acceptable practical conclusions, concerning the proximity of the real bearing capacity to the upper or to the lower limit of that interval.

A simpler situation, and a particularly interesting one in practical terms, is when a firm stratum, of significantly higher strength, underlies the loaded soil layer. This problem has been tackled by Mandel and Salençon (1969, 1972), who, by applying the theory of plasticity, obtained a numerical solution for the (larger than 1.0) corrective factors, f_c, f_q, and f_γ, for the three terms of the bearing capacity to account for the effect of the lower, rigid boundary of the soil layer. Those factors, included in Table 6.4, are a function of the friction angle of the soil layer and of the ratio between the foundation width, B, and the layer thickness, D_f.

6.2.6 Bearing capacity for vertical loading under seismic conditions

Seismic conditions will give rise to the application of inertial forces to the soil wedges responsible for the bearing capacity, altering in an unfavorable manner their equilibrium conditions, and thereby reducing the ground bearing resistance.

On the other hand, the seismic action applied to the structure will have, as an effect, horizontal forces and moments applied to the foundation, the consideration of which in the evaluation of the bearing capacity to vertical loading is naturally indispensable and has already been treated (see Section 6.2.5).

Similarly, prior to the design of any foundation on a sand mass, the safety against liquefaction phenomena has to be evaluated, which will be discussed at the end of the present chapter (see Section 6.5).

As to the seismic effects referred to in the three preceding paragraphs, only the last two are normally contemplated in the design process of shallow foundations. For the first, only recently have satisfactory mathematical solutions been developed.

The very practical solution proposed by Fishman et al. (2003) to incorporate the effect of inertial forces in the bearing ground allows the continued use of the classical Equation 6.12, but now with the bearing capacity factors for seismic conditions, N_{ce}, N_{qe}, and $N_{\gamma e}$, replacing the homologous factors for static conditions, N_c, N_q, and N_γ. The results of this proposal show that the bearing capacity factors for seismic conditions experience reductions relative to their values for static conditions, which vary in a practically linear way with the horizontal seismic coefficient, k_h, for values of k_h up to 0.3–0.4, an interval that covers the zones of substantial seismic risk. Figure 6.8 shows this result for the factors N_q and N_γ, considering two values for the angle of shearing resistance (the results for the factor N_c are very similar to those for N_q)[2].

An expeditious solution for the problem, suggested by the author of this book, consists of the adoption of a percentage reduction for the bearing capacity factors equal to the value of

---

[2] It should be noted that the curves in the figure tend to the vertical, i.e., to a reduction of 100%, when k_h reaches the value corresponding to the critical horizontal acceleration (Equation 5.97).

Table 6.4 Corrective coefficients f_c, f_q and f_γ of the bearing capacity to account for the presence of a firm stratum at depth D_f from the foundation base.

Values of f_c

ϕ \ B/D_f	0 to 1	1	2	3	4	5	6	8	10
0°	—	1 ($B/D_f \leq 1.41$)	1.02	1.11	1.21	1.30	1.40	1.59	1.78
20°	1 ($B/D_f \leq 0.86$)	1.01	1.39	2.12	3.29	5.17	8.29	22.00	61.50
30°	1 ($B/D_f \leq 0.63$)	1.13	2.50	6.36	17.4	50.20	*	*	*
36°	1 ($B/D_f \leq 0.50$)	1.37	5.25	23.40	*	*	*	*	*
40°	1 ($B/D_f \leq 0.42$)	1.73	11.10	82.20	*	*	*	*	*

Values of f_q

ϕ \ B/D_f	0 to 1	1	2	3	4	5	6	8	10
0°	—	—	—	—	—	—	—	—	—
20°	1 ($B/D_f \leq 0.86$)	1.01	1.33	1.95	2.93	4.52	7.14	18.70	51.90
30°	1 ($B/D_f \leq 0.63$)	1.12	2.42	6.07	16.50	47.50	*	*	*
36°	1 ($B/D_f \leq 0.50$)	1.36	5.14	22.80	*	*	*	*	*
40°	1 ($B/D_f \leq 0.42$)	1.72	10.90	80.90	*	*	*	*	*

Values of f_γ

ϕ \ B/D_f	0 to 1	1	2	3	4	5	6	8	10
0°	—	—	—	—	—	—	—	—	—
20°	—	—	1 ($B/D_f \leq 2.14$)	1.07	1.28	1.63	2.20	4.41	9.82
30°	—	1 ($B/D_f \leq 1.30$)	1.20	2.07	4.23	9.90	24.80	*	*
36°	1 ($B/D_f \leq 0.98$)	1.00	1.87	5.60	21.00	90.00	*	*	*
40°	1 ($B/D_f \leq 0.81$)	1.05	3.27	16.60	*	*	*	*	*

* Values greater than 100.

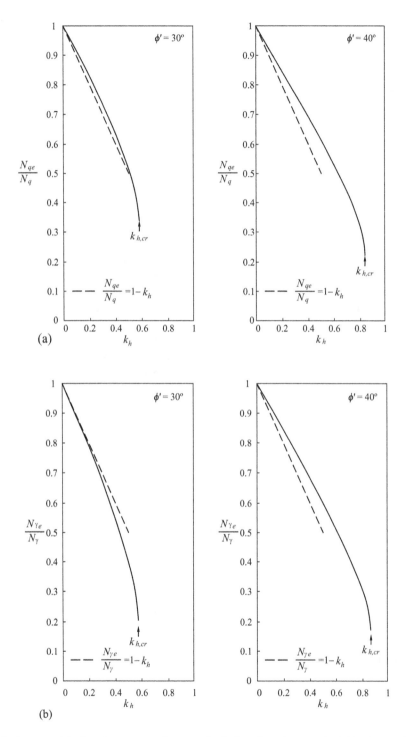

Figure 6.8 Reduction of the bearing capacity factors of a shallow foundation induced by inertial forces in the foundation soil: a) N_q factor; b) N_γ factor (adapted from Fishman et al., 2003).

the horizontal seismic coefficient. This proposal, indicated by broken lines in Figure 6.8, is on the safe side and with a reasonable approximation.

6.2.7 Safety against foundation soil failure due to insufficient bearing capacity for vertical loading

6.2.7.1 Global safety factor

The global safety factor against foundation soil failure, due to insufficient bearing capacity for vertical loading, is usually defined by the ratio:

$$F = \frac{Q_{ult}}{V_S} \tag{6.27}$$

where V_S is the force normal to the foundation base applied by the superstructure. Traditionally, this force is obtained from the diverse combinations of the actions required by the structural design, taking into account the action characteristic values, i.e., it is not affected by partial safety factors.

In general, the minimum value recommended for the global safety factor of shallow foundations is around 3.0.

Lower values are sometimes adopted when evaluating the bearing capacity, assuming undrained conditions, by means of a total stress analysis. This may be understood because: i) the undrained loading hypothesis is, in general, pessimistic for the foundations of buildings, bridges and of many other structures, since, during construction, a partial dissipation of the positive excess pore pressure is likely to occur, the loading in fact occurring under *partially drained conditions*; and ii) the ground resistance substantially increases with the consolidation.

On the other hand, safety factor values higher than those mentioned above are also common in many situations. Such an option is no longer justified by the prevention of failure due to insufficient bearing capacity, but aims fundamentally at limiting the settlement to small, but unspecified, values. In practical terms, this is a semi-empirical approach to ensure safety against both the ultimate and the serviceability limit states. This procedure is acceptable if based on comparable experience, i.e., for the same structural type and foundation soil.

6.2.7.2 Partial safety factors: structural Eurocodes and LRFD codes

In Chapter 3 the partial safety factor method has already been discussed in its essential aspects, as well as the more relevant related features of Eurocode 7 (EN 1997-1:2004) and of the codes based on LRFD.

According to Eurocode 7, with regard to shallow foundations, the following inequality has to be satisfied for all combinations of actions:

$$V_{Sd} \leq V_{Rd} \tag{6.28}$$

where V_{Sd} is the design value, for the verification of ultimate limit states, of the load normal to the foundation base, including the foundation self-weight and that of the fill material above it, and V_{Rd} is the design value of the load-bearing capacity of the foundation (i.e., the design value of Q_{ult}).

V_{Sd} should be obtained from the combination of actions for the diverse design situations, according to Eurocode 0 (EN 1990:2002).

For persistent or transient design situations, the so-called fundamental combinations are expressed by:

$$V_{Sd} = \Sigma_j \gamma_{Gj} G_{kj} + \gamma_{Q1} Q_{k1} + \Sigma_i \gamma_{Qi} \psi_{0i} Q_{ki} \quad (j \geq 1; i > 1) \tag{6.29}$$

For accidental design situations, the combination of actions are expressed by:

$$V_{Sd} = \Sigma_j G_{kj} + A_d + (\psi_{11} \text{ or } \psi_{21}) Q_{k1} + \Sigma_i \psi_{2i} Q_{ki} \quad (j \geq 1; i > 1) \tag{6.30}$$

For earthquake design situations, the combination of actions are expressed by:

$$V_{Sd} = \Sigma_j G_{kj} + A_{Ed} + \Sigma_i \psi_{2i} Q_{ki} \quad (j \geq 1; i > 1) \tag{6.31}$$

where:

- "+" means "to be combined with";
- Σ means "the combined effect of";
- G_{kj} is the characteristic value of permanent action j;
- Q_{k1} is the characteristic value of the leading variable action;
- Q_{ki} is the characteristic values of the accompanying variable action i;
- A_d is the design value of an accidental action;
- A_{Ed} is the design value of the earthquake action;
- γ_{Gj} is the partial safety factor for permanent action j (see Table 3.5);
- γ_{Qi} is the partial safety factor for variable action i (see Table 3.5);
- ψ_0 is the factor for the combination value of a variable action (see, in each case, Eurocode 0);
- ψ_1 is the factor for the frequent value of a variable action (see, in each case, Eurocode 0);
- ψ_2 is the factor for the quasi-permanent value of a variable action (see, in each case, Eurocode 0).

The same is essentially valid for a safety check following the codes based on the LFRD concept.

6.3 ESTIMATION OF FOUNDATION SETTLEMENT

6.3.1 Introduction

The approximate prediction of the settlement of a foundation is, in general, very difficult. It is therefore more appropriate to speak in terms of the *assessment* or *estimate*, and not so much in terms of *calculation* of the settlement. It is certainly useful to discuss with the reader some of the difficulties involved in the process, leaving others to be analyzed later.

In the first place, it should be pointed out that the pressure applied to the soil by a footing is not likely to be evaluated with great accuracy. In fact, differential settlement in adjacent shallow foundations of a statically indeterminate structure will induce a redistribution of internal forces that may be relevant. The load will decrease in those foundations with greater settlement, whereas it increases in those experiencing lesser settlement. On the other hand, by applying any of the expressions to be presented later for settlement estimation, it is implicitly assumed that the applied pressure will not vary, regardless of the actual settlement value.

Still on the subject of ground applied pressure, one should recall that, for settlement estimates, the design value of the actions are employed, along with the respective combinations, corresponding to the so-called serviceability limit states. For the variable actions, these values result from the application of codes of practice, based on probabilistic methods, which aim to cope with situations of various types likely to occur during construction and the service life of the structure. The observed settlement (at a given moment) may be the result of a *real* load (at that moment), significantly different from the one used for design purposes.

On the other hand, the evolution of the settlement in time must be taken into account. The generic time-settlement curve of a shallow foundation is schematically represented in Figure 6.9. The total settlement, s, is the sum of three components:

$$s = s_i + s_c + s_d \tag{6.32}$$

the immediate settlement, s_i, the primary consolidation settlement, s_c, and the secondary consolidation or creep settlement, s_d.

The last two terms have already been addressed in Chapter 4, with a particular emphasis on the case of the loading of clayey soils. The immediate settlement is the component of the total settlement that occurs concomitantly with load application. The curve represented in Figure 6.9 is applicable to all soil formations, if one takes into account that the time scale and the relative value of the three settlement components can vary by several orders of magnitude. Table 6.5 indicates, in general terms, the relative importance of these components for various conditions.

6.3.2 Immediate settlement: elastic solutions

6.3.2.1 General expression

In soil mechanics, it is common to apply solutions from the theory of elasticity to the evaluation of the stresses induced in a soil formation by loads applied at the surface. Such application is valid particularly under the two following conditions:

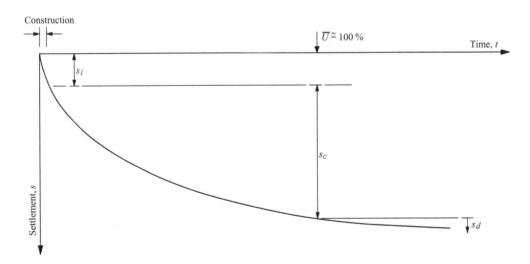

Figure 6.9 Foundation time-settlement generic curve (Perloff, 1975).

Table 6.5 Typical relative importance of the three settlement components as a function of foundation soil type.

Characteristics of the foundation ground and type of loading		s_i	s_c	s_d
Predominantly fine-grained soils, constrained loading	Normally consolidated or lightly overconsolidated	Practically null	Large to very large	Relevant in highly organic soils
	Overconsolidated	Practically null	Small to moderate	Negligible
Predominantly fine-grained soils, non-constrained loading	Normally consolidated or lightly overconsolidated	Variable over a relatively wide interval	Large to very large	Relevant in highly organic soils
	Overconsolidated	Small to moderate	Small	Negligible
Predominantly coarse-grained soils	Load with modest variation	Variable in a relatively wide interval	Null	Generally small, sometimes significant
	Load with significant variation	Variable in a relatively wide interval	Null	Relevant

i) the loading is essentially monotonic, i.e., it increases up to a certain value and then has relatively small variations from that value;
ii) the stress transmitted to the soil is relatively small in relation to the bearing capacity for vertical loading.

Most shallow foundations of civil engineering structures, namely buildings and bridges, satisfy these two conditions. In particular, the second is verified due to the great convenience, discussed later in this chapter, of limiting the ground-induced deformation, by keeping settlements within tight bounds. This circumstance leads to the adoption of relatively low pressure values in the foundation–soil contact surface, which ensure a reasonable proportionality in relation to the deformations produced in the soil mass.

The Annex A6.2 to this chapter contains some of the more useful elastic solutions for the evaluation of shallow foundation settlements (Boussinesq, 1885; Flamant, 1892; Giroud, 1970; Poulos and Davis, 1974). The induced stress distribution is not particularly sensitive to the variation with depth of the soil elastic properties. This means that, for stratified formations, one can employ, with reasonable accuracy, solutions for the stress distribution in homogeneous elastic media.

Consider the formation represented in Figure 6.10, comprising n layers, all with elastic behavior, loaded at the surface by a surcharge Δq_s, uniformly distributed in a given area. If the stress increments induced by that load at the center of layer j, with thickness h_j, and elastic parameters E_j and v_j, are $\Delta\sigma_{zj}$, $\Delta\sigma_{xj}$, and $\Delta\sigma_{yj}$, respectively, the immediate settlement, s_i, at the surface, can be calculated by applying Hooke's law as:

$$s_i = \sum_{j=1}^{n} \Delta\varepsilon_{zj} h_j = \sum_{j=1}^{n} \frac{1}{E_j}\left[\Delta\sigma_{zj} - v_j\left(\Delta\sigma_{xj} + \Delta\sigma_{yj}\right)\right] h_j \qquad (6.33)$$

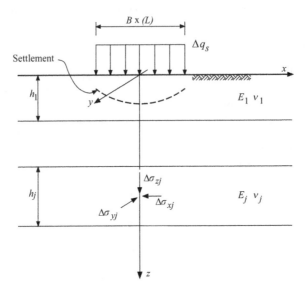

Figure 6.10 Surface loading of a stratified elastic formation.

6.3.2.2 Homogeneous semi-indefinite soil formation

In cases where the elastic parameters are constant with depth, the summation in the previous equation becomes an integral, with the expression:

$$s_i = \int_0^\infty \frac{1}{E}\left[\Delta\sigma_z - v\left(\Delta\sigma_x + \Delta\sigma_y\right)\right]dz \tag{6.34}$$

In the simpler and more current cases, $\Delta\sigma_z$, $\Delta\sigma_x$, and $\Delta\sigma_y$ are analytic functions that depend on the surface applied pressure, Δq_s, on the coordinates of the point, on the dimensions B and L of the loaded area, and also, in the case of $\Delta\sigma_x$ and $\Delta\sigma_y$, on the Poisson ratio. Given that the stress increments are directly proportional to Δq_s, one may write:

$$\Delta\sigma_z = \Delta q_s \, f_{\Delta\sigma_z}\left(B, L, x, y, z\right)$$

$$\Delta\sigma_x = \Delta q_s \, f_{\Delta\sigma_x}\left(v, B, L, x, y, z\right) \tag{6.35}$$

$$\Delta\sigma_y = \Delta q_s \, f_{\Delta\sigma_y}\left(v, B, L, x, y, z\right)$$

Therefore, Equation 6.34 becomes:

$$s_i = \left(\frac{\Delta q_s}{E}\right)\int_0^\infty\left[f_{\Delta\sigma_z} - v\left(f_{\Delta\sigma_x} + f_{\Delta\sigma_y}\right)\right]dz \tag{6.36}$$

The function inside the square bracket on the right-hand side of Equation 6.36, which will be designated as f_s:

$$f_s = \left[f_{\Delta\sigma_z} - v\left(f_{\Delta\sigma_x} + f_{\Delta\sigma_y}\right)\right] = f_s\left(v, B, L, x, y, z\right) \tag{6.37}$$

is a dimensionless function, while the integral has the dimension length. In fact, observing Equation 6.36 and comparing it with the unidimensional Hooke's law, the integral represents what could be designated as the *equivalent thickness* or *depth*, h_s, responsible for the settlement. The result of this integral, for a given surface point with coordinates x and y, is:

$$h_s = \int_0^\infty f_s(v, B, L, z) dz = B(1 - v^2) I_s \tag{6.38}$$

with I_s being a dimensionless factor that depends on the geometry of the loaded area and, naturally, on the location of the surface point.

Using Equation 6.38 in Equation 6.36, the expression of the settlement for an elastic, homogeneous half-space (with E and v as constants) is obtained as:

$$s_i = \Delta q_s B \frac{1 - v^2}{E} I_s \tag{6.39}$$

It is interesting to verify that the above-designated *equivalent thickness* (Equation 6.38), and therefore the settlement (Equation 6.39), are directly proportional to B. This results from the fact that, for a loaded surface area with a given geometric shape, the depth reached by a certain percentage of the surface stress increment increases with the transverse dimension, B (see, for example, Figure A6.2.3).

In Table 6.6 are included the values of I_s as a function of the geometry of the loaded area and for several points in each geometry; as should be expected, it can be seen that I_s is maximal at the center of the loaded area and minimal at the edges.

Strictly speaking, it is important to note that the analytical solution presented does not correspond to the loading of a soil by a *real footing*, because the assumed surface applied pressure, Δq_s, is a *surcharge*, i.e., admitting that each elementary force is applied to the elastic medium without any physical connection between the adjacent forces. Thus, I_s, and, so s_i, are functions of the point under which the integration of the vertical strain was performed. Such solution will correspond to what may be designated as an *infinitely flexible footing*.

However, real foundations are typically *rigid*, with the settlement being practically identical in all its points, if the applied loading is, of course, centered. The solutions for the

Table 6.6 Values of I_s for homogeneous elastic half-space (Perloff, 1975; Milović, 1992).

Foundation shape		I_s, infinitely flexible (surcharge) case					I_s
		Center	Vertex	Middle of the smaller side	Middle of the larger side	Average	Rigid case
Circular		1.00	–	0.64	0.64	0.85	0.79
Square		1.12	0.56	0.77	0.77	0.95	0.92
Rectangular L/B	= 1.5	1.36	0.68	0.89	0.97	1.15	1.13
	= 2.0	1.53	0.77	0.98	1.12	1.30	1.27
	= 2.5	1.67	0.83	1.05	1.25	1.44	1.40
	= 3.0	1.78	0.89	1.11	1.36	1.52	1.51
	= 4.0	1.97	0.98	1.20	1.53	1.71	1.67
	= 5.0	2.10	1.05	1.27	1.67	1.83	1.81
	= 7.0	2.31	1.16	1.38	1.89	2.03	2.01
	= 10.0	2.54	1.27	1.49	2.10	2.25	2.25

settlement of rigid footings on a homogeneous and elastic half-space are no longer exact, analytical solutions, and so, for the more current geometries, numerical solutions are available.

In such cases, the settlement may still be estimated by using Equation 6.39, adopting the I_s values included in the last column of Table 6.6. As can be verified, the I_s values proposed for rigid footings are quite close to the average values for the infinitely flexible (surcharge) case of equal geometry.

It is also interesting to note that the value of I_s in Table 6.6 increases with the longitudinal foundation dimension, L. The explanation for this fact is analogous to what has been stated above, relative to the transverse dimension, B (see, for example, Figure A6.2.3).

6.3.2.3 Homogeneous layer with rigid lower boundary

In case a rigid boundary exists at a depth D_f below the surface of the elastic medium, one way for calculating the settlement would involve integrating the vertical strain as in Equation 6.34, but taking D_f as the integration upper limit. However, this approach has one limitation because the presence of the rigid boundary introduces a significant modification in the induced stress in relation to the elastic solution for the homogeneous half-space. This modification involves essentially the following aspects: i) the vertical stress increment is larger than for the half-space case and now depends on the Poisson ratio (though with a relatively low sensitivity to this parameter); ii) the horizontal stress increment differs even more significantly, with this difference depending on the Poisson ratio value; and iii) the horizontal stress increment, after a similar decrease with depth, increases again toward the lower rigid boundary.

Figure 6.11 illustrates this trend by comparing the finite element result for the vertical and horizontal stress distribution below the center of a rigid circular foundation on an elastic homogeneous medium with a rigid boundary at a depth equal to the footing diameter, against the theory of elasticity solution for the half-space case.

A preferable alternative to the previous questionable procedure consists of applying Equation 6.39 and taking I_s values naturally adapted to the new situation. Table 6.7 presents I_s values applicable to rigid footings on granular soils, as function of the depth of the rigid boundary and the foundation width. The values on that table have been obtained by the finite element method.

6.3.2.4 Foundation rotation associated with moment action

In cases where the foundation is subjected to moments, as illustrated in Figure 6.12, in addition to the settlement previously calculated, the foundation also suffers rotation. The composition of the settlement, due to vertical loading with this rotation, will cause the vertical displacements to be variable from point to point in the foundation base, even if this is rigid.

The foundation rotations, associated with the two moments, M_x and M_y, for the case of homogeneous elastic half-space, may be estimated by the equations:

$$\tan \omega_x = \frac{M_x}{BL^2} \frac{1-v^2}{E} I_{\omega x} \tag{6.40}$$

$$\tan \omega_y = \frac{M_y}{B^2L} \frac{1-v^2}{E} I_{\omega y} \tag{6.41}$$

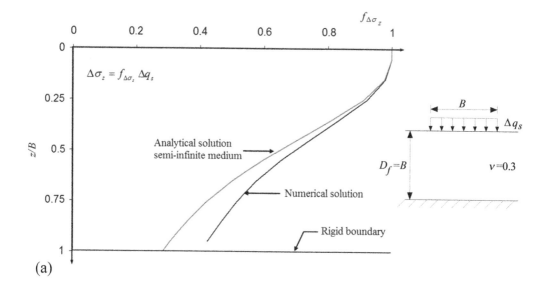

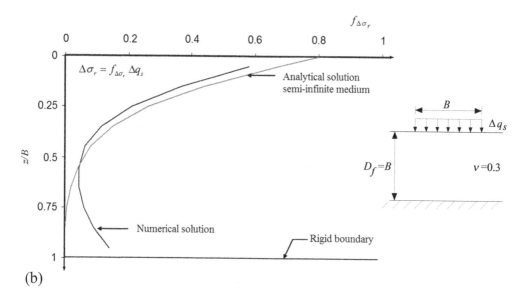

Figure 6.11 Vertical (a) and horizontal (b) incremental stresses under the center of a rigid circular foundation on an elastic medium limited by a rigid boundary: comparison of results from the finite element method and from the theory of elasticity equations for an elastic half-space.

where I_{ω_x} and I_{ω_y} are dimensionless parameters given by the following approximate expressions for a rigid foundation (Taylor, 1967; Bowles, 1997):

$$I_{\omega_x} = \frac{16}{\pi\left(1+\dfrac{0.22L}{B}\right)}$$ (6.42)

and

Table 6.7 Values of I_s for a rigid foundation on an elastic medium with a rigid boundary at depth D_f and $\nu = 0.3$ (Marques and Magalhães, 2010).

D_f/B	Circle diameter = B	Rectangle					
		L/B = 1	L/B = 1.5	L/B = 2	L/B = 3	L/B = 5	L/B = ∞
0.0	0.00	0.00	0.00	0.00	0.00	0.00	0.00
0.5	0.31	0.32	0.32	0.33	0.34	0.35	0.36
1.0	0.47	0.48	0.52	0.54	0.57	0.58	0.63
1.5	0.55	0.57	0.64	0.68	0.72	0.75	0.83
2.0	0.60	0.63	0.72	0.77	0.83	0.87	0.99
2.5	0.63	0.66	0.77	0.83	0.91	0.97	1.12
3.0	0.65	0.69	0.80	0.88	0.97	1.04	1.23
3.5	0.66	0.71	0.83	0.91	1.02	1.10	1.32
5.0	0.69	0.74	0.88	0.97	1.10	1.22	1.54
7.5	0.71	0.77	0.92	1.01	1.15	1.34	1.79
10	0.72	0.77	0.93	1.05	1.17	1.45	1.97

$$I_{\omega_y} = \frac{16}{\pi \left(1 + \dfrac{0.22B}{L} \right)} \tag{6.43}$$

These expressions are valid only when the foundation base is integrally under compression, i.e., when $e_x \leq B/6$ and $e_y \leq L/6$.

6.3.2.5 Shortcomings of the elastic solutions

The settlement estimation by means of elastic solutions involves significant limitations that it is important to discuss.

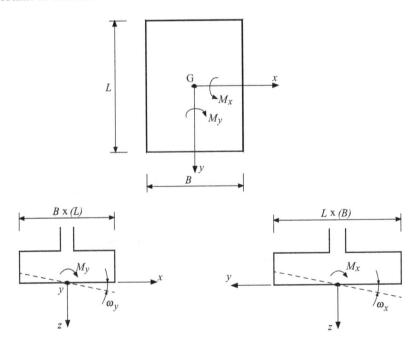

Figure 6.12 Rotation of a shallow foundation on an elastic half-space under moment action.

The form of the function, f_s, defined by Equation 6.37, provides an idea of the distribution with depth of the shear stress increments, and so of the horizons that contribute most to the settlement. Figure 6.13 presents the function f_s for the vertical line through the center of the loaded area, for three geometries of this area (with the same transverse dimension, B) and two Poisson ratio values. Due to the great variation with depth of the shear stress increment, the percentage of the shearing resistance mobilized at each depth is very distinct. Therefore, even in a *physically homogeneous* soil, the deformation modulus exhibited under foundation loading will be different from point to point.

What has just been said also has another consequence: the loading may cause shearing failure in certain regions of the foundation soil, even when the global resistance available is still very considerable. However, the elastic estimates of the stress increments progressively lose validity from the moment when the soil begins to yield, and this extends to ever-growing regions. Testing in physical models and numerical studies based on the finite element method suggest that the plastic yielding progress tends to increase the depth at which the largest deformations occur, i.e., the peak of function, f_s.

Another aspect that may contribute to the gap between the stress increments and the elastic solutions is that, in these, the loading is a pressure or surcharge distributed on a certain surface area, whereas, in real terms, the soil is loaded by practically rigid footings.

It must also be considered that the soil deformation modulus depends not only on the fraction of the shearing resistance that has been mobilized, but also on the average effective stress increasing with it. This explains the well-known favorable effect of foundation embedment on the ground: two identical foundations, with identical applied load, founded on the same homogeneous soil layer at different depths will not experience identical settlement: the one founded closer to the soil surface will settle more.

These preceding considerations help understand the tendency for deviation between predicted and observed settlements, commented on below.

Where the elastic methods seem to differ more from reality is with regard to the depth at which deformation occurs. In fact, the observation of numerous real cases suggests that the soil thickness that affects the settlement is somewhat smaller than what should be expected from the theory of elasticity. Therefore, the settlement is, in fact, neither proportional to the foundation width, contrary to what Equation 6.39 expresses, nor does it increase with the L/B ratio as markedly as illustrated by Table 6.6 (according to which, for instance, the settlement of a rigid footing increases by about 2.5 times when L/B passes from 1 to 10).

It should be recalled that the settlement proportionality, relative to the foundation width, B, inherent in the elastic solution, stems from the fact that this width is, in its turn, proportional to the dimensions of the stress bulb induced in the soil formation, i.e., to the depth until the stress increment is significant. It is precisely for the same reason that the settlement given by the theory of elasticity increases with the foundation length, L (for this, see Figures 6.13 and A6.2.3).

It is worth quoting, by the way, the remarkable work of Burland and Burbidge (1985) who analysed statistically more than 200 well-documented cases of building foundations, tanks and embankments over granular formations (sand and gravel), concluding, among other things, the following:

i) on average, the observed settlement was proportional to $B^{0.7}$;
ii) on average, the immediate settlement increase with the L/B ratio may be expressed by the relationship

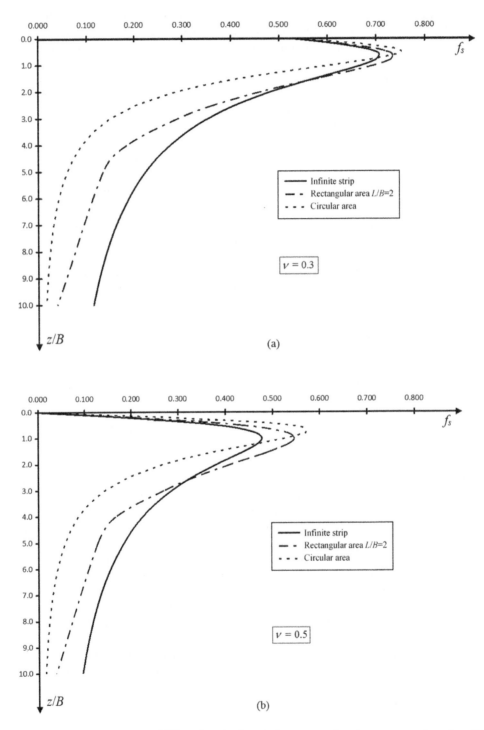

Figure 6.13 Function f_s (Equation 6.37) under the foundation center, for three geometries: a) $\nu = 0.3$; b) $\nu = 0.5$.

$$\frac{s_i\left(L/B>1\right)}{s_i\left(L/B=1\right)}=\left[\frac{1.25L/B}{\left(L/B\right)+0.25}\right]^2 \tag{6.44}$$

whose second term tends to 1.56 when L/B tends to infinity.

These results show that elastic solutions tend to overestimate the immediate settlement. The explanation for this discrepancy between elastic solutions and observed behavior seems to reside in the double dependence of soil stiffness on both the mobilized percentage of soil resistance and the average effective stress, as discussed before. Therefore, for greater depth under the shallow foundation, where the stress increments are modest in relation to the at-rest effective stress, the associated deformation will tend to be negligible.

An expeditious process to account for this deviation consists of adding up the vertical strains in accordance with Equation 6.33, but only to the depth at which the vertical stress increment represents a significant percentage (for example, 20%) of the at-rest effective vertical stress.

An alternative solution, semi-empirical in nature, directly based on PMT results is presented in Annex A6.3.

6.3.3 Immediate settlement in sand

6.3.3.1 Schmertmann's method

Given the limitations of the (elastic) theoretical methods discussed before, it is of no surprise that the specialized literature contains a significant number of empirical and semi-empirical methods for estimating shallow foundation settlements. For example, the work of Burland and Burbidge, quoted above, originated one of these methods, of empirical nature.

Within the referred methods, the well-known semi-empirical Schmertmann's method (1970, 1978) is probably the one most applied to rigid shallow foundations. The expression for immediate settlement is the following:

$$s_i=C_s\Delta q_s\int\left(\frac{I_\varepsilon}{E}\right)dz \tag{6.45}$$

where:

- C_s is a dimensionless corrective factor that accounts for the already-mentioned favorable effect of the foundation embedment, with the expression:

$$C_s=1-0.5\left(\frac{\sigma'_{vb}}{\Delta q_s}\right) \tag{6.46}$$

with σ'_{vb} representing the effective vertical stress at foundation base level;

- Δq_s is the footing-applied pressure after deducting σ'_{vb};
- E is the soil deformation modulus, estimated at each depth from CPT test results, in accordance with Table 6.8;
- I_ε is the so-called *vertical strain influence factor*, the distribution of which is indicated in Figure 6.14.

In practical applications, the integral of Equation 6.45 is replaced by a summation over the n sub-layers into which the soil is divided, with I_{ε_j} and E_j representing the average values

Table 6.8 Correlation between E and q_c (CPT) for application of Schmertmann's method (adapted from Schmertmann, 1970; Robertson and Campanella, 1988).

Soil type	E/q_c
Recent normally consolidated sand	2.5 to 3.5
Aged normally consolidated sand	3.5 to 6.0
Overconsolidated sand	≥ 6.0

of the vertical strain influence factor and of the deformation modulus, respectively, of the generic sublayer, of thickness h_j:

$$s_i = C_s \, \Delta q_s \sum_{j=1}^{n} \frac{I_{\varepsilon j}}{E_j} h_j \qquad (6.47)$$

6.3.3.2 Elastic theoretical method versus Schmertmann's method

The comparison of Equations 6.36 and 6.45 allows understanding that the function f_s of the theoretical solution and the vertical strain influence factor, I_ε, of Schmertmann's method, as

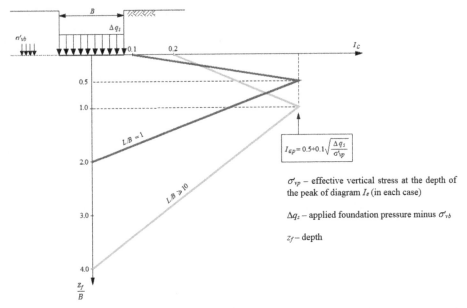

Foundation shape	Depth	Expression for I_ε
Square or circular	$0 \leq z_f \leq B/2$	$I_\varepsilon = 0.1 + (z_f/B)(2\,I_{\varepsilon p} - 0.2)$
	$B/2 \leq z_f \leq 2B$	$I_\varepsilon = 0.667\,I_{\varepsilon p}(2 - z_f/B)$
Continuous strip footing	$0 \leq z_f \leq B$	$I_\varepsilon = 0.2 + (z_f/B)(I_{\varepsilon p} - 0.2)$
$L/B \geq 10$	$B \leq z_f \leq 4B$	$I_\varepsilon = 0.333\,I_{\varepsilon p}(4 - z_f/B)$
$1 < L/B < 10$	$I_\varepsilon = I_{\varepsilon s} + 0.111\,(I_{\varepsilon c} - I_{\varepsilon s})\,(L/B - 1)$	
	$I_{\varepsilon s}$: I_ε value for a square footing	
	$I_{\varepsilon c}$: I_ε value for a continuous strip footing	

Figure 6.14 Vertical strain influence factor of Schmertmann's method.

well as their respective integrals, have the same physical meaning[3]. Comparing Figures 6.13 and 6.14, it can be observed that the latter corresponds to a simplification and adaptation of the former, in order to contemplate some of the aspects, discussed above, responsible for the deviations in elastic solutions. To this end, for example, is the adaptation related to the negligible contribution of the deeper horizons, as well as the greater depth of the graphs' peaks. Given the typical Poisson ratio value for sands, the graphs of Figure 6.14 should be compared, in this context, with those of Figure 6.13a.

In view of the above, the elastic theoretical method and that of Schmertmann frequently lead to not very different results, namely when the latter is used taking a unit value for the C_s factor.

These results will be even closer if, in the theoretical method, other adjustments, already commented on above, are introduced, such as: i) truncating the vertical strain integral below the depth where the incremental vertical stress is lower than a given value, for example $0.20\sigma'_{v0}$; or ii) taking into account Equation 6.44, to assess the influence of the foundation longitudinal dimension.

It seems to the author useful to clarify this aspect because Schmertmann's method is sometimes presented in the literature as a solution with an exclusively practical basis, and therefore empirical. This is not at all correct! As we have seen, Schmertmann's method essentially provides an *adjustment* of the elastic theoretical solution in order to account for certain deviations, with regard to results observed in real cases. For such deviations, a rational explanation has been found, although its quantification by recourse to a general formulation of analytical nature has not been encountered. Therefore, this method may be classified as *semi-empirical*.

The quality of the Schmertmann's method estimates has been the object of several studies, which have come to the conclusion that, in general, the method is conservative, i.e., as a rule, it overestimates the settlement but with a quite acceptable degree of approximation (Frank, 1991). Nevertheless, it should be noted that the elastic theoretical method, if employed with due care, may also provide a reasonable alternative, the application of both therefore being recommended.

The reader will find other empirical and semi-empirical methods in the specialized literature (Coduto, 2001; Frank, 2003).

6.3.3.3 Deformation modulus of sandy soils

Chapter 1 presented the main field tests for characterization of sand formations. It was then stressed that, for some of these tests, – namely SPT and CPT – there is no methodology available for theoretical interpretation of the results by relating them to the soil stiffness parameters. For other tests – such as the plate load test, the cross-hole seismic test and the self-boring pressuremeter test – the theoretical interpretation is known, allowing, in particular, determination of the soil deformation modulus.

Consequently, it might seem reasonable to think that, in general, the latter would be the tests selected for estimating the settlement of shallow foundations. On the contrary, the tests of the first group, the penetration tests, are those most employed for that purpose! The explanation for this apparently paradoxical situation is based on two types of reason.

First, penetration tests are much less costly, and SPT, in particular, is routinely performed along with the geotechnical surveys that constitute the basic site investigation. To this is added the fact that the other tests mentioned, by allowing theoretical interpretation, supply values for E associated with given soil stress and strain conditions that do not necessarily coincide with those prevailing under the shallow foundations being considered (see, *apropos* Section 1.5).

---

[3] However, it should be noted that f_s is obtained for surcharges applied at the surface of the elastic medium and may be defined for any point (x,y) of that surface, while I_ε is defined for rigid footings and for the foundation center.

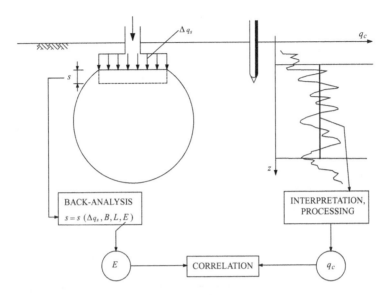

Figure 6.15 Formulation of empirical correlations between *in situ* test results and the soil deformation modulus for the estimation of foundation settlement.

This explains that the value (or values) of that parameter to be introduced into the immediate settlement expressions derive essentially from experience, i.e., from the observation of works, and are obtained from correlations with penetration test data.

The *praxis* is depicted in Figure 6.15 and may be summarized as follows:

1. the observation of foundation behavior allows measurement of the displayed settlement;
2. the application of any of the previously presented settlement expressions, or similar ones, by introducing, in the left-hand side, the observed settlement value, allows the estimation, by *back-analysis*, of the deformation modulus, or (depending on the approach) its change with respect to depth, exhibited by the foundation soil;
3. the comparison of these back-analysis results with those from *in situ* tests permits the establishment of quantitative relationships between the two variables;
4. the accumulated experience, in conjunction with a careful analysis of diverse case studies to identify their common and distinct features, permits a progressive increase in the reliability of these correlations.

Certainly, this methodology can be – and actually has been – used in conjunction with the other *in situ* tests, apart from penetration tests.

For foundations in sandy soils, the correlation of the deformation modulus with CPT results is very popular, usually being expressed by a proportionality relationship of the type:

$$E = \alpha \, q_c \tag{6.48}$$

where q_c is the cone resistance and α is a dimensionless parameter that varies within very wide limits.

Table 6.8, as already mentioned, presents a correlation of this type, recommended for use with Schmertmann's method, but also with the elastic theoretical method.

When only SPT results are available, it is possible to use one of the N_{SPT} *versus* q_c correlations (see Figure 1.27), and subsequently employ a correlation between the CPT results and the deformation modulus.

For completeness, it should be noted that the Poisson ratio of the soil is also involved in the settlement evaluation. Fortunately, the influence of this parameter on the settlement value is relatively small, given its limited range: for sand and residual soil from granite, ν values of 0.2 to 0.3 seem reasonable.

6.3.4 Immediate settlement in clay

6.3.4.1 Settlement expression

The evaluation of immediate settlement in clayey soil is normally based on the elastic theoretical method presented in Section 6.3.2. It is important to recall that the immediate settlement of a foundation in clay corresponds to an undrained loading, the soil elastic parameters being E_u and ν_u. As saturated soils under undrained loading conditions do not experience volume variation, then $\nu_u = 0.5$.

Therefore, the settlement can be obtained by one of the two following equations (adapted from Equations 6.33 and 6.39), in case E_u is either variable or constant with depth, respectively:

$$s_i = \sum_{j=1}^{n} \frac{1}{E_{uj}} \left[\Delta\sigma_{zj} - \nu_u \left(\Delta\sigma_{xj} + \Delta\sigma_{yj} \right) \right] h_j \tag{6.49}$$

$$s_i = \Delta q_s B \frac{1-\nu_u^2}{E_u} I_s \tag{6.50}$$

the I_s factor being given by Table 6.6.

This methodology is also applicable for estimating the immediate settlement in embankment construction on soft soil.

6.3.4.2 Deformation modulus of clayey soils

Contrary to what happens with sandy soils, the collection of undisturbed clay samples for triaxial or other laboratory tests may be considered a widespread practice. However, as already discussed in Section 1.5, this does not mean that the evaluation of immediate foundation settlements in clayey soil is generally based on deformation moduli obtained by recourse to laboratory tests.

The estimation of foundation settlement based on laboratory tests is only justified for foundations of exceptional importance and, in order to have a good probability of success, it requires: i) samples of excellent quality, in particular, those obtained from blocks; ii) carefully conducted tests, with recourse to instrumentation able to accurately determine the specimen stress–strain curves for strain levels below 10^{-4}; iii) correction (scaling) of these curves with the help of field test results, particularly those that provide estimates of G_0.

A simplified procedure for common situations consists of evaluating E_u from a correlation with the undrained shear strength, c_u, which may be determined either in the laboratory, or from *in situ* tests. The correlation between the undrained deformation modulus and the undrained shear strength is expressed by the equation:

$$E_u = M c_u \tag{6.51}$$

where the dimensionless parameter M, may be obtained from Figure 6.16, as a function of the soil plasticity index and the overconsolidation ratio.

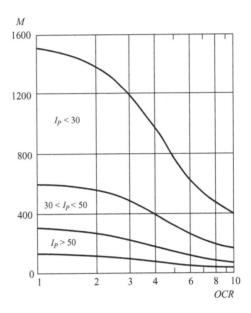

Figure 6.16 Proportionality factor between E_u and c_u (Duncan and Buchignani, 1976).

6.3.5 Note about settlement by secondary consolidation or by creep

With respect to settlement by secondary consolidation or by creep, it is interesting to note that, conventionally, it was considered that these only had to be taken into account in foundations on clayey soil, an aspect to which the considerations presented in Chapter 4 should apply.

Burland and Burbidge (1985) have concluded that foundations on granular soil can also exhibit a non-negligible settlement, delayed in time, particularly in the case of foundations subjected to important variable loads, such as tall chimneys, railway bridges, silos, etc.

The authors suggest the following empirical expression to estimate the settlement, s_t, after a time period t greater than three years:

$$s_t = s_i \left(1 + R_3 + R_t \, \log\frac{t}{3} \right) \tag{6.52}$$

where R_3 and R_t are the delayed settlements, expressed as fractions of s_i, corresponding to the first three years after construction and to each time logarithmic cycle after those three initial years, respectively.

For essentially constant loads, the authors suggest values of 0.3 and 0.2 for R_3 and R_t, respectively; this is equivalent to considering $s_t = 1.5s_i$ for $t = 30$ years. For significantly variable loads, the suggested values are, respectively, 0.7 and 0.8, which corresponds to $s_t = 2.5s_i$ for $t = 30$ years.

6.3.6 Verification of structural serviceability limit states due to foundation movement: actions to consider

The settlement estimate that is being discussed is indispensable for performing the verification of the structural serviceability limit states caused by foundation displacements. In this case, the partial safety factors for the actions are all taken with unit values, i.e., the actions are considered with their characteristic values.

The combinations of actions to be considered are established in Eurocode 0 (EN 1990-1:2002.

For serviceability limit states designated as being irreversible, the combination of actions, called *characteristic combination*, is expressed by:

$$\sum_j G_{kj} + Q_{k1} + \sum_i \psi_{0i} Q_{ki} \quad (j \geq 1; i > 1) \tag{6.53}$$

For serviceability limit states designated as reversible, the combination of actions, called *frequent combination*, is given by:

$$\sum_j G_{kj} + \psi_{11} Q_{k1} + \sum_i \psi_{2i} Q_{ki} \quad (j \geq 1; i > 1) \tag{6.54}$$

For serviceability limit states related with long-term effects and the appearance of the structure, the combination of actions, called *quasi-permanent combination*, is given by:

$$\sum_j G_{kj} + \sum_i \psi_{2i} Q_{ki} \quad (j \geq 1; i \geq 1) \tag{6.55}$$

where:

- "+" implies "to be combined with";
- Σ means "the combined effect of";
- G_{kj} is the characteristic value of a permanent action;
- Q_{k1} is the characteristic value of the dominant variable action;
- Q_{ki} is the characteristic value of the other variable action;
- ψ_0 is the factor for the combination value of the variable action (to obtain, in each case, from Eurocode 0);
- ψ_1 is the factor for the frequent value of the variable action (to obtain, in each case, from Eurocode 0);
- ψ_2 is the factor for the quasi-permanent value of the variable action (to obtain, in each case, from Eurocode 0).

In general, it seems reasonable to adopt the above-designated frequent combination to obtain the vertical load value to be considered for settlement estimation.

6.4 DESIGN OF THE FOUNDATIONS OF A STRUCTURE

6.4.1 Movement of the foundation of a structure: definitions and limits

In the preceding sections, it has been seen that the sizing of a shallow foundation involves both the evaluation of the bearing capacity and the estimation of the settlement. This section will address the question of the design not of *one foundation,* but of the *set of foundations* of a given structure. Before starting with design considerations, it is convenient to clarify some concepts related to foundation movement.

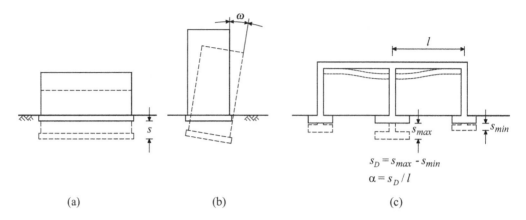

Figure 6.17 Simplified types of foundation movement: a) uniform structural settlement as a rigid body; b) uniform settlement combined with tilt of the structure as a rigid body; c) non-uniform settlement with structural distortion.

When the structure experiences a rigid body motion, a situation depicted in Figures 6.17a and 6.17b, the settlement may be uniform or associated with a rotation (tilt). This is the kind of movement that typically occurs in chimneys, towers or silos. The Tower of Pisa provides an example of settlement associated with a tilt.

The most common situation in buildings with a frame structure or in bridges is sketched in Figure 6.17c, where the various supports experience differential settlements, causing distortional deformation in the structure and its finishing materials.

It is pertinent to recall the following main definitions relative to foundation movements: i) the *total settlement*, s, is the absolute value of the downward vertical displacement of a given foundation; ii) the *rotation* (or *tilt*), ω, is the angle of the rigid body motion around a given point of the structure; iii) the *differential settlement*, s_D, is the difference between the total settlements of two, usually adjacent, foundations; iv) the *angular distortion*, α, is the ratio of the differential settlement between two foundations (or between two points of the same foundation) and their distance. In cases where tilt occurs in conjunction with non-uniform settlement, the angular distortion calculation should be based on the differential settlement obtained after rotating the structure by an angle ω, equal and opposite to the tilt[4].

Considering a structure with uniform settlement, i.e., with coincident total settlement for all foundations, the tolerable limit for it could be substantial. This limit has to deal essentially with the need to safeguard the plumbing (water, sanitation and other) connected to the building, as well as the access.

When the settlement is non-uniform, the parameter usually employed to establish bounds related to serviceability limit states is the angular distortion, the relationship of which with the stress induced in the structural and non-structural elements is more direct than, for example, that of differential settlement.

The subject of the relationship between foundation movement and induced damage to buildings has not undergone significant development recently, with the limits presented in the specialized literature being taken from quite old references (Skempton and McDonald, 1956; Polshin and Tokar, 1957; Bjerrum, 1963; Sowers, 1968; Burland and Wroth, 1974; Burland et al., 1977). The reason resides in the great complexity of the theme, because the value of the differential settlement or of the angular distortion that causes damage in a given building depends on numerous factors, many of them non-quantifiable, related to the

---

[4] Note that by applying this rule to the case of Figure 6.17b, the angular distortion is null.

Table 6.9 Allowable angular distortion for various situations.

Type of structure	Allowable angular distortion, α_a
Simple steel structures. Warehouses with steel structure and external steel walls	5.0/1000
Buildings with steel frame structure:	
• without diagonal braces	2.0/1000
• with diagonal braces	1.5/1000
Buildings with reinforced concrete frame structure	2.5/1000
Buildings with sensitive finishes	1.0/1000
Load bearing walls of non-reinforced masonry – U-shaped deformation (sagging)	
• length/height = 1	0.4/1000
• length/height = 5	0.8/1000
Load bearing walls of non-reinforced masonry – ∩-shaped deformation (hogging)	
• length/height = 1	0.2/1000
• length/height = 5	0.4/1000

(structural and non-structural) building elements, the type and rate of loading, the nature of the foundation soil, etc. For instance, it turns out that the settlement rate significantly affects the structural behavior, causing less relevant damage if very slow.

Additionally, the limits for differential settlement and the respective angular distortion are generally conditioned in most buildings by the cracking of non-structural elements (more sensitive than the structure itself), and even, in some industrial plants, by the operational requirements of particularly sensitive equipment. However, these non-structural elements and equipments are set at a later phase, when part of the foundation load has been applied, and so a significant portion of the settlement has already occurred.

Table 6.9 summarizes the limit values recommended for angular distortion, based on the above-quoted references. Industrial buildings with sensitive equipment will have to obey the limits imposed by the manufacturers. When their dimensions and weight are considerable, it is sometimes convenient to provide a purpose-built foundation, independent from those of the building.

6.4.2 Sizing based on allowable stress or allowable pressure

There is a long tradition in civil engineering of the sizing of the shallow foundations of a structure by empirically setting the maximum pressure to be applied to the ground, designated as *allowable stress or allowable pressure*, based on the experience with a given type of soil. This corresponds, therefore, to satisfying implicitly and simultaneously the ultimate and the serviceability limit states.

When the stress distribution on its base is not uniform, due to the existence of moments transmitted to the foundation, it is also current to adopt as design criterion, for example, that the stress corresponding to three-quarters of the maximum normal stress is lower than the allowable stress (Frank, 2003).

The calculation of footing dimensions is, in this way, based on the allowable pressure and on the applied vertical load. The latter is, in general, defined by combining the permanent and the variable actions without the application of partial safety factors larger than 1.0.

This procedure is not adequate. Not by the fact of being an empirical process – there are numerous empirical methodologies of great utility in geotechnics – but, as shall be seen, for it may lead to unsatisfactory design options, lacking rationality.

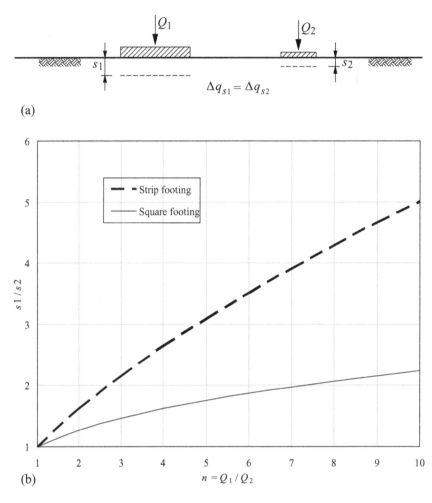

(a)

(b)

$n = Q_1 / Q_2$

Figure 6.18 Sizing based on allowable stress: a) adjacent footings with the same geometric shape under different applied loads, but transmitting identical pressure to the ground; b) ratio of the two foundation settlements (assumed to be proportional to $B^{0.7}$), as a function of the ratio of the applied loads, for two geometric shapes.

Consider, as shown in Figure 6.18a, two adjacent foundations, 1 and 2, of the same structure, which receive significantly different loads. Assume that the geometric shape of the two footings is the same and that the foundation soil does not differ in the two places. If the above-referred sizing methodology is applied, the footings will transmit the same (allowable) pressure, although their dimensions are quite different.

Footing 1, the larger, will have a higher global safety factor relative to failure by insufficient soil-bearing capacity, because this increases with the footing width (see Equation 6.12). Nevertheless, the settlement will also be greater in footing 1. In fact, according to the expression of the elastic method, in footings with the same geometric shape, the settlement is directly proportional to B (see Equation 6.39). As seen, this method tends to overestimate that dependence, and Burland and Burbidge (1985), from the analysis of real cases, have found an average dependence of the type $B^{0.7}$. For this dependence, Figure 6.18b shows the evolution of the two footings settlement ratio as a function of their applied load ratio, for two different geometric shapes.

It should be mentioned that, in the previous context, it is assumed that the soil, homogeneous in plan, will respond with the same stiffness under both footings. Such an hypothesis is not rigorous because this stiffness depends on the mobilized fraction of the soil resistance. With the global safety factor being greater in footing 1, the respective foundation soil will therefore tend to exhibit a somewhat greater stiffness than that of the foundation soil of footing 2. Thus, the settlement ratio expressed by the curves of Figure 6.19 will be, to some extent, overestimated.

What has just been said does not invalidate the conclusion that sizing based on allowable stress will necessarily lead to differential settlement whenever the foundation loads are to some degree dissimilar. If loading variation is now compounded with soil heterogeneity, and softer zones occur under the larger footings, differential settlement will naturally tend to get worse.

The allowable stress method can still be objected to for other reasons.

Normally, the allowable stress is established by soil type, i.e., from its geological description, such as *Boston blue clay*, *residual soil from Porto granite*, etc. For some formations described in this manner, it does not seem reasonable to establish an allowable stress. For example, residual soil from granite occurs with extremely variable resistance and stiffness characteristics.

Another aspect that should be considered is that the successful use of a certain allowable stress value in the past does not guarantee identical success in the future. In fact, it is not reasonable to assign one allowable stress value to a certain soil type regardless of the structure to be founded there. A low-rise building, for instance, will imply relatively low loads, and, as a consequence, needs footings of a relatively small size. A high-rise building on the same soil, with identical allowable stress, will require much larger foundations. Therefore, it will experience greater settlement.

In the preceding explanation, it has been left implicit that the magnitude of the loads on the two foundations of Figure 6.18 are fixed, regardless of the foundation movement. This hypothesis applies only to isostatic structures. On the contrary, in hyperstatic (statically indeterminate) structures, the load distribution on the foundations depends on the displacement that they experience. If, by chance, a given foundation settlement tends to be higher than those of the adjacent foundations, the structural deformation will induce a force redistribution, reducing the load on the former and increasing it on the others.

In other words, the soil–structure interaction in redundant structures provides, in fact, a settlement distribution less discrepant than what would result from their evaluation for each individual footing. The degree of settlement homogeneity will be greater, the greater the structural stiffness, resulting in greater discrepancies between the actual foundation loads and the values considered for their sizing.

So, the criticism of the allowable stress method may be concluded with some irony: this method is not only unable to provide a more or less uniform settlement distribution but is also incapable of ensuring uniformity of the stresses transmitted to the ground!

6.4.3 Design based on identical foundation settlement

With respect to the foundations of a given structure – and assuming that a bearing capacity to vertical loading within the safety margins recommended by the design criterion or code adopted will be ensured – it can be concluded from the previous discussion that the most desirable situation is clearly when:

  i) the foundation settlement assumes a *very close* value for them all;
  ii) this value is small.

In other words, it is preferable to use a criterion based on a *design settlement value, as much as possible common to all foundations.*

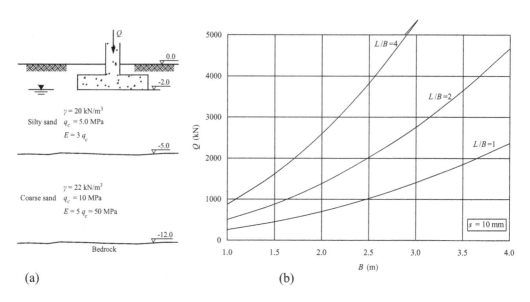

Figure 6.19 Example of a foundation design chart: a) geological–geotechnical model; b) chart providing the transverse dimension B for a predefined settlement (10 mm), given the applied vertical load Q and the L/B ratio, applying the method of Schmertmann.

The method for establishing the value of this settlement will be treated in Section 6.4.4. It is now important to stress that, once the design settlement is selected for a given structure to be founded on a given soil, it is possible to develop design charts like that of Figure 6.19, which relate the applied load with the *B* dimension required to ensure the allowable settlement for various foundation geometric shapes. In the diverse solutions discussed in Section 6.3, in a context of settlement evaluation, this approach corresponds to the *result* of a calculation in which *data* are entered in the form of the deformation modulus of the soil (or of the various sub-layers in which it is divided), the footing dimensions, the respective geometric shape, and the applied load. In the design context now being considered, the perspective is different: by *imposing* a settlement, the objective is to define the foundation dimensions from the same solutions, given the applied load and, naturally, the stiffness characteristics of the foundation soil.

When the settlement equation corresponds to an analytical expression, the design charts are readily obtained. When the settlement evaluation involves an algebraic expression, for example like Equations 6.33 and 6.47, obtaining the design charts requires calculations that are more laborious and the use of spreadsheets is recommended. This is the case of the design chart shown in Figure 6.19b, that arose from the application of the method of Schmertmann to the geological–geotechnical model represented in Figure 6.19a.

It can be readily confirmed from the application of this chart that, for a given geometric shape, the pressure decreases with *B*. For example: for square footings, for $B=1.5$ m the footing would be able to receive a load of approximately 450 kN, which corresponds to a pressure $\Delta q_s = 201$ kPa, whereas, for $B=3.0$ m, the corresponding load amounts to about 1400 kN, which corresponds to a pressure $\Delta q_s = 156$ kPa.

A recurring criticism to this foundation design procedure consists of invoking the difficulties in evaluating the stiffness of the foundation soil, which increase the difficulty in accurately estimating the settlement.

This fact does not limit the merit of this methodology. In fact, it seems reasonable to expect that (possible or probable) errors in the ground stiffness estimation will affect, to a similar extent, the various foundation elements of the structure. Therefore, the objective of minimizing differential settlement has been fulfilled (without requiring redistribution of internal structural forces). However, that difficulty makes it advisable to choose a small value for the target settlement. This relevant issue is discussed in the following section.

6.4.4 Allowable design settlement for structural foundations

If no other settlement-related uncertainties were involved in the design procedure, the value adopted for the settlement common to all foundations could be quite large. However, foundation sizing is frequently affected by uncertainties related both to the spatial heterogeneity of the foundation ground and to the actions applied to the foundation, which make it advisable to limit the allowable design settlement to values which are typically quite small.

The action-related uncertainty has already been addressed in Section 6.3.1, but it will now be further developed. The most relevant feature in the present context has to deal with the variable actions, in particular, the surcharge loads. As suggested in Figure 6.20, the distribution, and its variability in space and time, may cause the load effectively applied to a given foundation to differ significantly (in general, to decrease) from that which has been considered in the design. It is likely that such discrepancy may vary widely between foundations, and this will inevitably result in differential settlement. Naturally, this aspect is more important in structures where variable actions are significant when compared to the permanent ones, as in buildings housing archives and libraries, or in silos.

Even more relevant will be the uncertainties relative to the foundation ground itself, and its effect on all the supported structures. The treatment of this question demands some preliminary considerations to introduce the less experienced reader to the typical *praxis* of a foundation project.

Based on the interpretation of the geotechnical site investigation, it is normally established a *geological–geotechnical model* for foundation design purposes. This model basically comprises: i) the sequence of soil layers, their thickness and soil type; ii) the position of the water table; and iii) the unit weight and the strength and stiffness parameters of the layers that will receive the foundation loading. In the less simple cases and/or when the construction dimensions in the plan are large, there is a *global model* that frequently unfolds into various local models. Each *local model* includes adjustments to one or more of the above components,

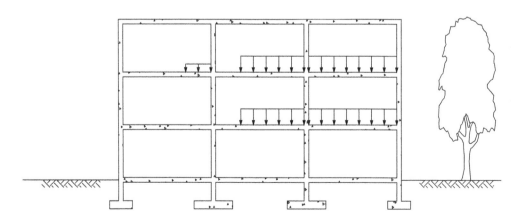

Figure 6.20 Spatial variation in the distribution of variable actions in a building, contributing to differential settlement.

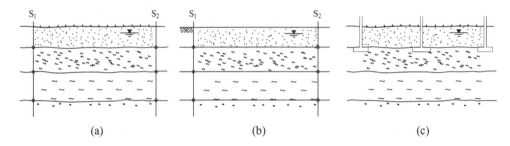

Figure 6.21 Geological–geotechnical scenario, comprising sedimentary soils: a) real formations and boring locations; b) geological–geotechnical model; c) foundation layout.

for distinct zones (in plan) of the foundation ground. The so-called design charts shown in Section 6.4.2 (see Figure 6.19) are developed for each local model.

The referred model concept is based on geotechnical borings and *in situ* test results performed at some locations, and laboratory tests on samples collected there. However, except for structures of exceptional importance, as is the case of bridges and large viaducts, the borings will not cover all the foundation points. Therefore, in general, many local geological–geotechnical models result from the interpolation-extrapolation from models coinciding with boring locations. Hence, the possibility for model differences from real situations.

In this context, there are two very distinct geological scenarios. As shown in Figure 6.21, a generally favorable scenario corresponds to sedimentary soils, where the variations in the horizontal plane are typically smooth, providing less uncertainty in the establishment of local models. The opposite case, represented in Figure 6.22, concerns residual soils, which typically exhibit great horizontal heterogeneity, often with extremely pronounced variation of strength and stiffness over distances of a few meters. As can be understood, the

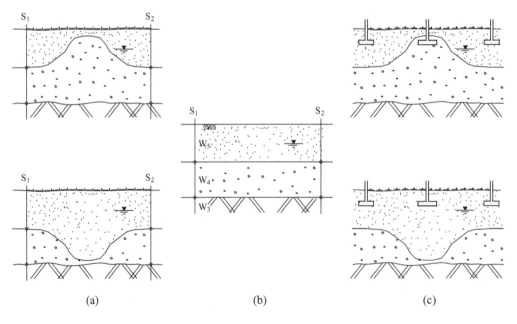

Figure 6.22 Geological–geotechnical scenario comprising residual soils: a) real formations and boring locations; b) geological–geotechnical model; c) foundation layouts.

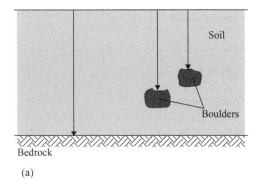
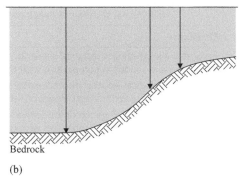

Figure 6.23 Geological–geotechnical scenario formed by residual soils with rock blocks (boulders): a) real profile; b) geological–geotechnical model resulting from the erroneous interpretation of boreholes that intersected the boulders (Mililitski et al., 2005).

settlement of a given foundation in these cases can experience a substantial positive or negative difference in relation to the design prediction (Figure 6.22c).

Geological–geotechnical models quite distant from reality may also result from an erroneous interpretation of geotechnical investigation results. Formations with residual soils from granite or gneiss pose more delicate problems, namely for the correct identification of the location of firm ground, as exemplified in Figure 6.23 (Mililitski et al., 2005).

Therefore, the occurrence of differential settlements is inevitable due to the uncertainties involved in design. Nevertheless, this fact makes even more pertinent the design strategy explained in Section 6.4.3, based on a design settlement common to all foundations. The aim with this strategy is basically to cancel or minimize the differential settlement component that is fully amenable to designer control. The component of these settlements associated with the uncertainties discussed can also be naturally minimized, although not fully suppressed, by means of appropriate design options[5].

It can be understood from these considerations that the strategy for this minimization consists of imposing restrictions to the magnitude of the design settlement, which will be the more severe, the higher the uncertainties in the variables present in the design process. In fact, given these uncertainties, if the total settlements were allowed to reach relatively high values, the differential ones would very likely be outside reasonable bounds.

Table 6.10 contains the author's proposal for the relation between allowable design settlement and allowable differential settlement, s_{Da}.

6.4.5 Design sequence for shallow building foundations

Taking as reference a current building with frame structure, the establishment of the design settlement value and, subsequently, the foundation sizing, assuming identical settlement for all foundations, involves the following essential steps, outlined in Figure 6.24:

1. adoption of the allowable angular distortion, α_a, for example from Table 6.9;
2. calculation of the allowable differential settlement of contiguous foundations for each zone of the structure with similar average column spacing, l, by means of:

$$s_{Da} = \alpha_a l \tag{6.56}$$

---

[5] One way of reducing uncertainty related to soil heterogeneity consists of closer spaced boreholes in the more unfavorable geological scenarios.

Table 6.10 Values of the ratio between allowable design settlement and allowable differential settlement (inspired by the results of Bjerrum, 1963).

Conditions	$s_{a,d}/s_{Da}$
Favorable geological–geotechnical scenario. Minor variable actions in comparison with the permanent ones	$s_{a,d}/s_{Da} \leq 2.0$
Favorable geological–geotechnical scenario. Important variable actions in comparison with the permanent ones	$s_{a,d}/s_{Da} \leq 1.5$
Unfavorable geological–geotechnical scenario. Minor variable actions in comparison with the permanent ones	$s_{a,d}/s_{Da} = 1.0$
Unfavorable geological–geotechnical scenario. Important variable actions in comparison with the permanent ones	$s_{a,d}/s_{Da} = 1.0$ (see note)

Note: It is recommended to perform soil–structure interaction analyses, considering various scenarios for variable action distribution and foundation soil stiffness.

3. adoption of the allowable design settlement, $s_{a,d}$, based on Table 6.10;
4. development of the foundation design charts, relating the applied load with the foundation dimensions for their various geometrical shapes, based on the design settlement and the local geological-geotechnical models;
5. estimation of the dimensions (B, L) for each footing, based on the foundation applied loads, taken from the structural analyses assuming rigid supports and the design charts referred to in 4);
6. verification as to whether the soil-bearing capacity for vertical loading is sufficient, taking for each footing the dimensions obtained in 5); if not, adjust the footing dimensions or re-evaluate the bearing layer; note that the actions to be considered in this verification are distinct from those taken in 5);
7. execution of the structural design of the foundations.

In certain cases, after having fixed the dimensions in plan of the foundations, it may be advisable to perform new structural analyses, considering now deformable (elastic) supports. For example, this will clearly be the case when there is a highly heterogeneous geological–geotechnical scenario in combination with relatively important variable loads. Performing *sensitivity analyses*, by introducing variations in the stiffness distribution of foundation soil, as well as in the spatial layout of the variable loads, will permit assessment of the influence of these uncertainties in the distribution and magnitude of settlements and structural internal forces.

In these analyses, it may be convenient to take jointly into account the changes in the structural geometry and that of the structural and foundation loading. This means that, instead of applying in a single increment the total loading to the final structural geometry, one should analyze the various construction phases (such as the gradual increase in building floors), applying to each geometry only the loads corresponding to the respective construction phase.

In certain situations, the differences between those two analysis procedures may be significant. Note, for example, that, if differential settlements occur in the frame structure of an *n*-story building, the internal forces in the structural members of the top floor will naturally be independent of the settlements induced by the permanent loads of the first *n*-1 stories.

On the other hand, consideration of the changes in the structural geometry allows the evolution of its stiffness to be better taken into account, as is obvious. For example, if, in a high-rise building, the total loading is applied to its final geometry, this will lead to an overestimation of its capacity for redistribution of internal forces induced by foundation settlement.

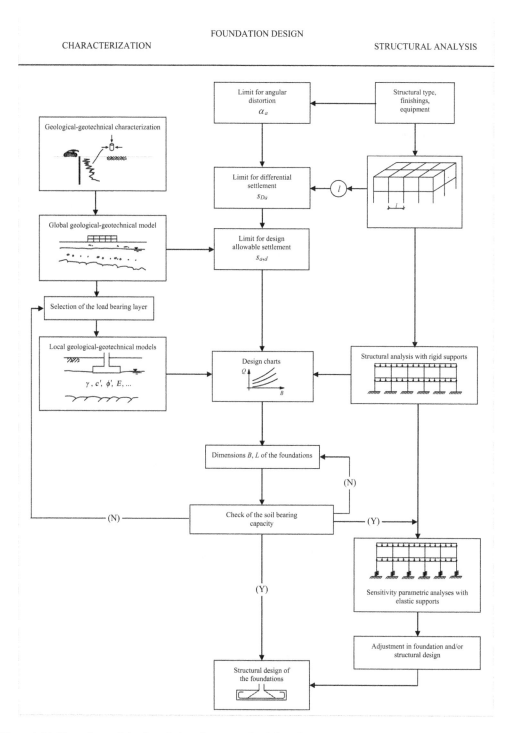

Figure 6.24 Flow chart of the foundation design methodology based on the design settlement value.

The results of these soil–structure interaction analyses may lead in some cases to significant alterations in the envelopes of internal structural forces, and, thereby, to adjustments in the sizing of structural elements and of the foundations. In structures particularly sensitive to settlement, this may even lead to a different solution for the foundation system, or even to the option for pile foundations.

6.5 EVALUATION OF SITE SEISMIC BEHAVIOR: LIQUEFACTION SUSCEPTIBILITY OF SANDY SOILS

6.5.1 Introduction

From soil mechanics, it is known that loose sandy soils are likely to exhibit poor behavior under seismic actions of medium or high intensity. If such soils are submerged, this behavior may result in the liquefaction phenomenon. The cyclic shear stresses associated with seismic action tend to induce volume reduction, which leads to the development of positive excess pore pressures. Their magnitude depends on the amplitude and duration of the cyclic loading, which, in turn, depends on the so-called *earthquake magnitude*.

If the excess pore pressure generated is enough to cancel the soil effective stress, liquefaction is triggered, leading to a drastic reduction in the soil shear strength, which causes serious damage to the structures founded on that ground. Once the seismic event ends, the excess pore pressure dissipates relatively quickly and the soil is rearranged with a higher density, causing a surface settlement. For unsaturated sandy soils above the water table, the seismic loading will tend to induce settlements without the occurrence of liquefaction (Ishihara, 1985, 1993).

As can be readily understood, the evaluation of soil behavior under seismic loading must be done before dealing with the specific aspects of foundation design, previously treated in this chapter. If this seismic assessment leads to the conclusion that a deficient behavior is to be expected, three options can be considered: i) carry out a ground improvement procedure (see Section 6.5.5); ii) change the foundation type to pile foundation, accounting in its design for the deficient seismic behavior of the top soil horizons; or iii) reject the site as being unsuitable for construction.

The base of the liquefaction expertise available today has been developed from the work of Seed and co-workers at the University of Berkeley between the 1960s and the 1980s. To this team is due the best-known methodology for treating the liquefaction phenomenon, known as the cyclic stress procedure, first formulated by Seed and Idriss (1971) and presented below.

The cyclic stress procedure is semi-empirical, because it is based on the observation of cases with and without the triggering of liquefaction in reasonably well-documented situations. Since its initial formulation, almost five decades ago, the database on which the method has been developed has substantially increased, as has the understanding of sand behavior under seismic loading. Therefore, the cyclic stress approach has been the recipient of regular updates. These recent developments tend to make the methodology based on this approach quite complex, and even somewhat less transparent.

Therefore, it was a challenge to reconcile a text meant to provide students with initial guidance on this subject with an up-to-date manual for professional engineers.

In order to overcome this obstacle, an opening section describes the cyclic stress approach in its essence – one might say, in its *original simplicity* – detailing and justifying its essential steps. In the second part, the current methodology based on that approach is summarized. The reader should be aware that certain steps of this methodology are likely to be changed in the near future, as has been observed during the recent past.

The present text will consider only *cyclic liquefaction*, namely, earthquake-induced liquefaction. There is another type of liquefaction, driven by static stresses, called *flow liquefaction*, usually associated with inclined ground surface, which will not be addressed here. For
further reading on that topic, the use of specialized literature is suggested (Kramer, 1996;
Finn, 2001; Jefferies and Been, 2016).

6.5.2 The cyclic stress procedure: general formulation

6.5.2.1 Representation of seismic loading

Consider Figure 6.25a that represents a soil mass with the horizontal surface subjected to a
horizontal acceleration history. Previously, i.e., in the at-rest state, the shear stress was zero
in the horizontal and vertical planes. At the instant being represented, in which the surface
acceleration reaches its peak, the maximum shear stress is acting on any generic horizontal
plane.

Considering a column with vertical lateral faces, height z and unit base area, if it behaved
as a rigid body, it would receive a horizontal inertial force, F_{ih}, equal to:

$$F_{ih} = \frac{a_{\max}}{g} \gamma z \qquad (6.57)$$

inducing, therefore, in its base a shear stress given by:

$$\left(\tau_{\max}\right)_r = \frac{a_{\max}}{g} \gamma z = \frac{a_{\max}}{g} \sigma_{v0} \qquad (6.58)$$

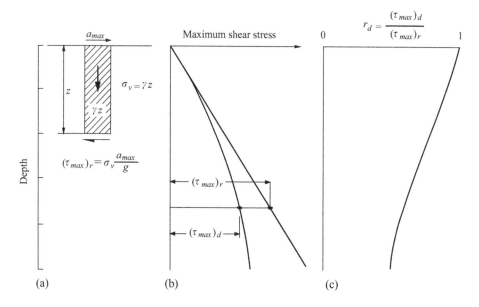

Figure 6.25 Simplified method for evaluating earthquake induced shear stresses (adapted from Seed and
Idriss, 1971): a) soil column scheme; b) change with depth of the shear stress in horizontal
planes for the rigid and deformable body hypotheses; c) typical change with depth of the r_d
coefficient, relating the previous stresses.

As suggested in Figure 6.25b, since the soil column is not a rigid body, the mobilized shear stress is lower than that given by the previous equation, and can be obtained by introducing a factor r_d lower than one and decreasing with depth:

$$\left(\tau_{max}\right)_d = \frac{a_{max}}{g}\sigma_{v0}\,r_d \tag{6.59}$$

The change of r_d with depth has the typical trend expressed in Figure 6.25c, as shown by numerical simulation results of the vertical propagation of earthquake-induced shear waves in sand deposits. Values for r_d will be presented later.

As can be understood, being the seismic record a cyclic loading of random amplitude, the shear stress in any horizontal plane will vary in magnitude and direction from instant to instant. Therefore, the maximum acceleration will not be the most appropriate for characterizing the seismic loading. Furthermore, its duration, i.e., the total number of cycles, is also very relevant.

To study liquefaction, Seed and Idriss (1971) defined the seismic action by means of an *equivalent uniform* cyclic loading, i.e., of constant frequency and amplitude, the latter being equal to 65% of the maximum earthquake amplitude, with the cyclic shear stress then being given by:

$$\tau_{cyc} = 0.65\frac{a_{max}}{g}\sigma_{v0}\,r_d \tag{6.60}$$

Dividing both members by σ'_{v0}, the so-called *cyclic stress ratio*, CSR_M, *for a given seismic magnitude, M*, is given by:

$$CSR_M = \frac{\tau_{cyc}}{\sigma'_{v0}} = 0.65\frac{a_{max}}{g}\cdot\frac{\sigma_{v0}}{\sigma'_{v0}}\,r_d \tag{6.61}$$

The number of uniform loading cycles is given in Table 6.11 as a function of earthquake magnitude. These results have been obtained from numerical simulation studies comparing the excess pore pressure generated in certain sand deposits, by real seismic records (with distinct magnitudes and maximum shear stress values) and by uniform cyclic loadings of amplitude equal to 65% of the maximum amplitude of the corresponding seismic record.

As will be seen next, the liquefaction potential charts were prepared for the reference magnitude $M = 7.5$. Therefore, the cyclic stress ratio given by Equation 6.61 has to be corrected in order to obtain the equivalent cyclic stress ratio for that reference magnitude, $CSR_{M=7.5}$, by means of the expression:

$$CSR_{M=7.5} = \frac{CSR_M}{MSF} \tag{6.62}$$

Table 6.11 Number of cycles of a uniform loading of amplitude $0.65\tau_{max}$, representative of various seismic magnitudes (Seed et al., 1975).

Magnitude, M	Number of cycles with $0.65\,\tau_{max}$
8.5	26
7.5	15
6.75	10
6	5–6
5.25	2–3

where *MSF* is the (dimensionless) *magnitude scaling factor*. Values of *MSF* are presented in Table 6.12.

6.5.2.2 Characterization of liquefaction potential

The selected way to characterize the liquefaction potential has been the study of real, well-documented events with: i) seismic data, from which the above-mentioned *CSR* can be obtained; ii) the proven observation of the *occurrence* or *non-occurrence* of liquefaction (in general, by sand ejecta events and surface cracks); and iii) a parameter characterizing the soil strength *before the earthquake*.

In the great majority of the cases analyzed in the first studies (earthquakes in Japan, China, Alaska, Guatemala and Argentina in the 1960s and 1970s), the available parameter for characterizing the soil strength was the SPT result. Note, by the way, that the increase in parameters that lead to a N_{SPT} increase in a granular medium, namely relative density and average effective stress, is also favorable in terms of liquefaction resistance.

Figure 6.26 shows an example from one of the first studies (Seed et al., 1983), with the representative parameter of ground strength on the horizontal axis and the cyclic stress ratio on the vertical axis. A boundary line has been drawn separating, in a reasonably conservative way, the zone with concentration of symbols representing cases where liquefaction occurred, from the zone where it did not.

The development of studies following this strategy, as well as the accumulation of well-documented data from recent seismic events, led to the development of analogous relations for granular soils with distinct percentages of fines (Seed et al., 1985), for CPT results

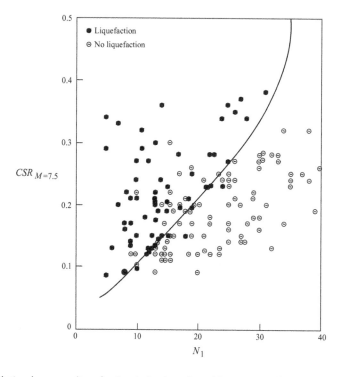

Figure 6.26 Correlation between liquefaction behavior of sand formations ($D_{50} > 0.25$ mm) with horizontal surface under seismic action ($M = 7.5$) and the SPT result N_1 (normalized for an effective vertical stress of 100 kPa, but not yet for the standard energy ratio of 60%) (Seed et al., 1983).

(Robertson and Wride, 1998), and for values of shear wave velocity, V_s (Andrus and Stokoe, 2000).

There is an important detail to be noted: the boundary line of Figure 6.26, and other similar lines for distinct conditions, provide, for a given soil type with a certain resistance (expressed by field test results), the cyclic stress ratio that it can experience without liquefaction. Therefore, the ordinate of Figure 6.26 can also be designated as *cyclic resistance ratio*, $CRR_{M=7.5}$.

6.5.2.3 Evaluation of safety against liquefaction

The evaluation of safety against liquefaction for a given formation involves the following steps:

1. calculate, from Equation 6.61, the change with depth of the cyclic stress ratio, $CSR_M(z)$, taking into account the site seismic safety code, which provides the maximum surface acceleration and its magnitude;
2. correct, with Equation 6.62, the cyclic stress ratio for a reference magnitude of 7.5, $CSR_{M=7.5}$;
3. from a correlation analogous to that of Figure 6.26, determine, at each depth, the value of the cyclic resistance ratio for a magnitude of 7.5, $CRR_{M=7.5}(z)$;
4. calculate, at each depth, the safety factor against liquefaction by means of the ratio:

$$F_L(z) = \frac{CRR_{M=7.5}(z)}{CSR_{M=7.5}(z)} \tag{6.63}$$

Eurocode 8 (EN 1998-5, 2004) recommends, for example, a minimum value of 1.25 for this safety factor.

It is important to stress that this methodology for safety evaluation is applicable to Holocene sand deposits, of horizontal or almost horizontal surface, and for depths up to 20 m[6].

6.5.3 The cyclic stress approach: more recent application proposals

6.5.3.1 Results from the NCEER Workshops, USA (1996, 1998)

Within the recent versions of the cyclic stress procedure, it is referred in the first place, due to its importance, to the one resulting from meetings under the patronage of the United States National Center for Earthquake Engineering Research (NCEER), to review the expertise about liquefaction. Youd and Idriss (2001) published the summary reports of the 1996 and 1998 meetings.

The following equations (Liao and Whitman, 1986) were considered appropriate for the evaluation of the factor r_d:

$$r_d = 1.0 - 0.00768 \cdot z \quad \text{for } z \leq 9.15 \text{ m}$$

$$r_d = 1.174 - 0.0267 \cdot z \quad \text{for } z > 9.15 \text{ m} \tag{6.64}$$

These equations provide a change of r_d with depth very close to that resulting from numerical simulations of the vertical propagation of shear seismic waves in agreement with the initial formulation of Seed and Idriss (1971).

---

[6] The database, updated in the Boulanger and Idriss (2014) report quoted below, includes a mere ten case histories for $\sigma'_v > 120$ kPa, four of which coincided with the occurrence of liquefaction, for the SPT-based chart.

Table 6.12 Values of MSF.

| Magnitude, M | $MSF = CRR_M/CRR_{M=7.5}$ | |
	Youd and Idriss (2001)*	EC8 – Part 5 (EN 1998-5:2004)
5.5	2.20	2.86
6.0	1.76	2.20
6.5	1.44	1.69
7.0	1.19	1.30
7.5	1.00	1.00
8.0	0.84	0.67
8.5	0.72	–

* The values of this column are approximated by the expression $MSF = 174/M^{2.56}$.

With respect to the *MSF* parameter, Table 6.12 presents the values considered appropriate in the NCEER Workshop, together with the values adopted by Eurocode 8 – Part 5 (EN 1998-5, 2004), which derive from a proposal developed by Ambraseys (1988).

Figure 6.27 shows the so-called *liquefaction charts*, based on SPT and on CPT results, recommended in the report. Complementary to each are presented equations adjusted to

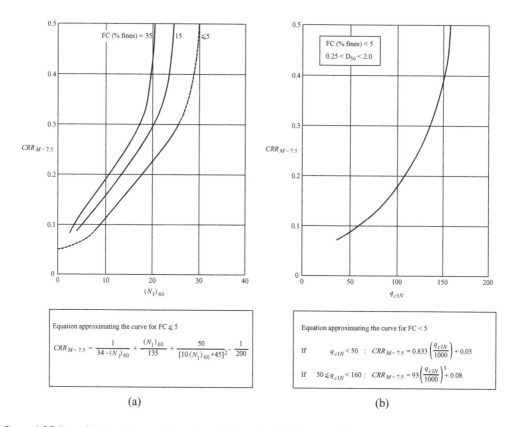

(a)   (b)

Figure 6.27 Liquefaction charts: a) based on SPT results (Seed et al., 1985, corrected, see Youd and Idriss, 2001); b) based on CPT results (Robertson and Wride, 1998).

the corresponding curves for *clean sands*. The chart of Figure 6.27a has been included in Eurocode 8 – Part 5.

It is indispensable to mention that, in the development of Figure 6.27a: i) the C_R rod length correction factor (see Equation 1.14 and Table 1.4) has not been considered in the calculation of $(N_1)_{60}$ for depths between 3 m and 10 m; and ii) the $(N_1)_{60}$ value has been obtained from N_{60} based on Equations 1.12 and 1.13.

As can be seen in Figure 6.27a, the curves for sand with fines correspond to greater liquefaction resistance, for a given value of $(N_1)_{60}$. It may be questioned whether this fact arises from an actually higher liquefaction resistance conferred by the fines or from a decrease in penetration resistance induced by them. As an alternative to the use of curves for sands with fines, as in Figure 6.27a, the meeting of experts mentioned above recommended the adoption of the following methodology (Youd and Idriss, 2001):

1. calculate from the actually measured resistance, $(N_1)_{60}$, a penetration resistance equivalent to that of clean sand, $(N_1)_{60cs}$, using the equation:

$$(N_1)_{60cs} = \alpha + \beta (N_1)_{60} \qquad (6.65)$$

   where α and β are given in Table 6.13;
2. with the values obtained from equation 6.65, use the curve for clean sand of Figure 6.27a.

This procedure, though providing an evaluation of $CRR_{M=7.5}$ for any fines content value between 5% and 35%, does not correspond rigorously to the curves of Figure 6.27a. In fact, the result of its application to 15% fines leads to a curve slightly to the right of the one represented for that percentage, being therefore based on a more conservative evaluation criterion for the cyclic resistance ratio. Such criteria resulted from a reassessment of the data then available.

The CPT-normalized resistance, q_{c1N}, which appears on the abscissa of the graph of Figure 6.27b, has been calculated from Equation 1.21. Those resistance values represent *average values* in the sand strata that suffered liquefaction. However, it is common to directly import CPT data (typically at 1-cm or 2-cm depth intervals) into a spreadsheet and calculate the safety factor over depth. In deposits with marked variability, this may lead to the identification of thin horizons with safety factors lower than 1.0, even though its average value in the layer may be markedly higher.

In such circumstances, this does not necessarily mean (in a purely deterministic perspective) that liquefaction will occur. Sometimes lower values of q_c are influenced by closely interspersed thin soft clay layers and are not representative of the sand layer. It is, therefore, recommended to complement the automatic processing of test data with a careful and detailed analysis of the change, with depth, of CPTu results. Robertson (2009) notes that the analysis of I_c change may be very useful. A value of $I_c = 2.6$ corresponds to an approximate boundary between

Table 6.13 Value of the nondimensional parameters α and β of Equation 6.65.

Fines content, FC (%)	α	β
$FC \leq 5$	0	1
$5 < FC < 35$	$\exp[1.76 - (190 / FC^2)]$	$0.99 + (FC^{1.5}/1000)$
$35 \leq FC$	5.0	1.2

sand-like and clay-like behavior (in sandy soils I_c is below and in clayey soils is above that value). Therefore, when the rate of change of I_c is fast and it crosses the mentioned boundary, it is a symptom that the cone is in transition from sand to clay or *vice versa*.

Contrary to Figure 6.27a, Figure 6.27b does not contain graphs for sand with fines. Robertson and Wride (1998) proposed a methodology, whereby it is possible to distinguish a graph for sand with fines from that for clean sand through the correlation between the soil behavior type index I_c and the fines content. However, the correlation between I_c and the fines content is not particularly strong.

6.5.3.2 New developments from Idriss and Boulanger

Idriss and Boulanger have been developing relevant research within the cyclic stress approach, with the aim of improving the safety predictions relative to liquefaction, taking, as a fundamental tool, the case history database, which has benefited from a sizeable expansion in the 21st century. The main results of this research are available in an open access University of California report (Boulanger and Idriss, 2014), as well as in several published papers (Idriss and Boulanger, 2006; Boulanger and Idriss, 2012, 2015, 2016).

In the scope of this research, the above cited authors aimed at refining the cyclic stress procedure by means, among other aspects, of a better characterization of the r_d and *MSF* parameters, which in the classical formulation are dependent only on the depth (for r_d) and on the magnitude (for *MSF*).

For r_d the authors have adopted a new formulation where, besides depth, the earthquake magnitude is also included:

$$r_d = \exp\left[\alpha(z) + \beta(z).M\right] \tag{6.66}$$

where α and β are given by:

$$\alpha(z) = -1.012 - 1.126\sin\left(\frac{z}{11.73} + 5.133\right) \tag{6.67}$$

$$\beta(z) = 0.106 + 0.118\sin\left(\frac{z}{11.28} + 5.142\right) \tag{6.68}$$

with z being the depth below ground surface in meters and the sine arguments expressed in radians. The resulting variations of r_d with depth and earthquake magnitude are illustrated in Figure 6.28.

With respect to the *MSF* parameter, Boulanger and Idriss (2014, 2015) present equations that, in addition to the earthquake magnitude, also take into account the strength of the sand deposit, expressed by the $(N_1)_{60}$ and q_{c1N} results of SPT and CPT, respectively:

$$MSF = 1 + (MSF_{max} - 1)\left(8.64\exp\left(\frac{-M}{4}\right) - 1.325\right) \tag{6.69}$$

with MSF_{max} given by:

$$MSF_{max} = 1.09 + \left(\frac{q_{c1N_{cs}}}{180}\right)^3 \le 2.2 \tag{6.70}$$

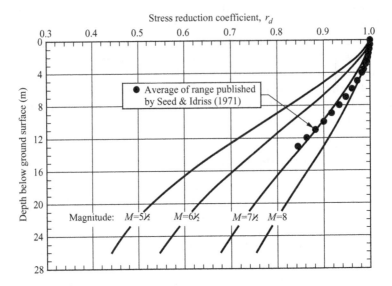

Figure 6.28 Shear reduction factor, r_d, relationship (Idriss and Boulanger, 2006).

or

$$MSF_{\max} = 1.09 + \left(\frac{(N_1)_{60cs}}{31.5}\right)^2 \leq 2.2 \tag{6.71}$$

This proposal corresponds to a stronger dependence of *MSF* on the earthquake magnitude for more competent soils.

In the above-mentioned report, Boulanger and Idriss (2014) present new liquefaction charts, based on SPT and CPT results. The curves for clean sands (fines content $\leq 5\%$) are expressed by the following equations, developed by the cited authors:

$CRR_{M=7.5,\sigma'_v=1\,\mathrm{atm}}$

$$= \exp\left(\frac{(N_1)_{60cs}}{14.1} + \left(\frac{(N_1)_{60cs}}{126}\right)^2 - \left(\frac{(N_1)_{60cs}}{23.6}\right)^3 + \left(\frac{(N_1)_{60cs}}{25.4}\right)^4 - 2.8\right) \tag{6.72}$$

$CRR_{M=7.5,\sigma'_v=1\,\mathrm{atm}}$

$$= \exp\left(\frac{q_{c1Ncs}}{113} + \left(\frac{q_{c1Ncs}}{1000}\right)^2 - \left(\frac{q_{c1Ncs}}{140}\right)^3 + \left(\frac{q_{c1Ncs}}{137}\right)^4 - 2.8\right) \tag{6.73}$$

For sand with fines, the authors propose the following procedure:

1. calculate from the measured penetration resistance, $(N_1)_{60}$, and q_{c1N}, an equivalent resistance for clean sand, $(N_1)_{60cs}$ or q_{c1Ncs}, with the equations:

$$(N_1)_{60cs} = (N_1)_{60} + \Delta(N_1)_{60} \tag{6.74}$$

and

$$q_{c1Ncs} = q_{c1N} + \Delta q_{c1N} \tag{6.75}$$

where $\Delta(N_1)_{60}$ and Δq_{c1N} are given by the equations:

$$\Delta q_{c1N} = \left(11.9 + \frac{q_{c1N}}{14.6}\right) \exp\left(1.63 - \frac{9.7}{FC+2} - \left(\frac{15.7}{FC+2}\right)^2\right) \tag{6.76}$$

and

$$\Delta(N_1)_{60} = \exp\left(1.63 + \frac{9.7}{FC+0.01} - \left(\frac{15.7}{FC+0.01}\right)^2\right) \tag{6.77}$$

with FC being the percentage of fines;
2. introduce the values calculated from Equations 6.74 and 6.75 in Equations 6.72 and 6.73.

Equations 6.72 and 6.73, expressing the cyclic resistance ratio, should be used together with the new equations for r_d and MSF.

The difference between the liquefaction chart of Figure 6.27a and the new proposal, based on SPT, is relatively small. Conversely, the difference between the chart of Figure 6.27b and the new proposal based on CPT is significant, the latter being more conservative[7].

All the previous proposals represent a deterministic approach. The same authors, using results from SPT (Boulanger and Idriss, 2012) and CPT (Boulanger and Idriss, 2016), have proposed analogous charts associated with a given probability of liquefaction triggering.

6.5.4 Settlement evaluation in sand formations above the water table

As mentioned before, in loose, unsaturated, sandy soils above the water table, seismic loading is likely to cause surface settlements of substantial magnitude. The evaluation of these settlements is even more difficult than that of those associated with static loading, already discussed. The most employed method for this purpose is, probably, a simplified procedure developed at the University of Berkeley by Tokimatsu and Seed (1987).

Taking into account Equation 6.60, which estimates cyclic shear stress, the cyclic shear strain may be obtained from the equation:

$$\gamma_{cyc} = \frac{\tau_{cyc}}{G} = 0.65 \frac{a_{max}}{g} \sigma_{v0} r_d \frac{1}{G} \tag{6.78}$$

where G is the soil shear modulus.

As is known, G depends on the strain level, being usually obtained from charts like that of Figure 1.51, as a function of G_0 and the cyclic shear strain. Therefore, the application of Equation 6.78 requires a small iterative procedure: i) adoption of a starting value for G, representing a given fraction of G_0; ii) calculation of γ_{cyc} from Equation 6.78; iii) calculation of the G/G_0 ratio as a function of γ_{cyc}, using Figure 1.51, or a similar chart; and iv) in situations where this result is not compatible with the value adopted in i), adjust it until agreement is reached.

Once the shear strain is calculated for each sublayer of the sand stratum, it is necessary to relate it to the volumetric strain. This relationship is highly dependent on sand dilatancy, and, therefore, on its density index. Figure 6.29a shows experimental results expressing

---

[7] It is possible that such difference may partly result from the way of calculating r_d and MSF. The reader may observe the comparison of the proposed curves with the database in Figures 4.1 and 6.2 of the Boulanger and Idriss (2014) report.

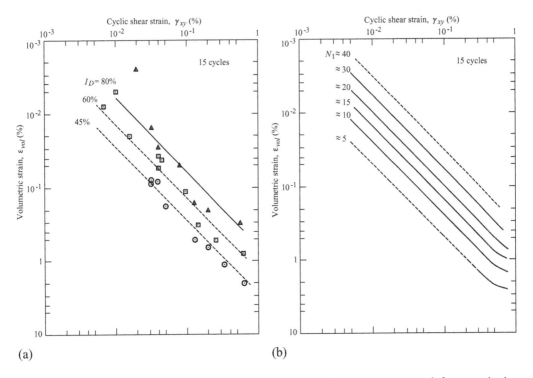

Figure 6.29 Relationship between the cyclic shear strain and the volumetric strain in sands for several values of: a) relative density (Silver and Seed, 1971); and b) $(N_1)_{60}$ (Tokimatsu and Seed, 1987).

such dependence, obtained in cyclic triaxial tests (with 15 complete cycles, corresponding to an earthquake of magnitude 7.5). Figure 6.29b has been obtained from it by using the relationship between $(N_1)_{60}$ and I_D expressed in Figure 1.14.

Once the volumetric strain is obtained from Figure 6.29 for the various sublayers, in case the earthquake magnitude is different from 7.5, those results must be corrected in accordance with Table 6.14. After that, the calculation of sublayer thickness variation is immediate, and its summation provides the total settlement.

As the cyclic horizontal stresses are applied in two directions, according to the results of Pyke et al. (1975), it is convenient to multiply the obtained results by a factor of 2.0. Also in agreement with those authors, the additional effect of cyclic stress associated with the vertical accelerations may increase the settlements by 50%, relative to those caused by the horizontal loading components, the only ones considered in the approach presented.

Table 6.14 Corrective factor to be applied to the volumetric strain taken from Figure 6.29 for magnitudes other than 7.5 (Tokimatsu and Seed, 1987).

Magnitude, M	$\varepsilon_{vol,M}/\varepsilon_{vol,M=7.5}$
8.5	1.25
7.5	1.00
6.75	0.85
6	0.60
5.25	0.40

6.5.5 Ground improvement for liquefaction mitigation

6.5.5.1 Most common densification methods for sand strata

When susceptibility to liquefaction or to significant settlement under seismic conditions is identified, the feasibility of using shallow foundations requires previous ground treatment. The treatment types currently available are quite numerous and involve a wide range of technologies (JGS, 1998). However, for the scenario being considered – near horizontal surface with the presence of one or more loose sand layers – the most common technique is *densification*, in order to reduce their void ratio. In this way, the tendency for soil contraction is reduced or cancelled. Therefore, when the soil is saturated, seismic loading will not induce positive excess pore pressure, and the liquefaction phenomenon is mitigated. The post-earthquake settlements are also minimized, which may occur, as already seen, regardless of full soil saturation.

The most common densification methods for sand strata in the liquefaction context are: i) vibro-compaction (sometimes also known as vibro-flotation); b) dynamic compaction; and iii) vibro-replacement. Table 6.15 summarizes some essential aspects of those methods.

The ideal soils for the application of *vibro-compaction* are clean sands, since the technique is not recommended when the fines content exceeds 15 to 20%. Sometimes, even though the soil to be treated is essentially clean sand, the presence of interspersed thin clay or silty clay strata severely affects the efficiency of the treatment. In that case, it is better to adopt the *vibro-replacement* technique, which consists of the installation of stone columns.

It is important to clarify the designation of the vibro-replacement technique and, in the context of the present chapter, its inclusion in the densification methods. When the method is applied to improve the bearing capacity of soft clayey soils (as discussed in Chapter 2), the stone columns actually *replace* the soft soil but do not *densify* it. When the same method is applied to sandy soils (though with fines), the densification effect predominates over that of substitution. However, since the technology is essentially the same, the name "vibro-replacement" prevails.

Dynamic compaction has quite a wide applicability range in terms of the soils which can be treated, being therefore a satisfactory alternative to vibro-compaction when the fine fraction is relatively significant (Mayne et al., 1984). However, it can only be used in places

Table 6.15 Most common methods of soil treatment to mitigate liquefaction.

Method	Parameters	Effects	Application conditions
Vibro-compaction or vibro-flotation	Spacing of the treatment points Vibration frequency Vibration amplitude Power of the vibrating probe	Reduction of void ratio Increase of the mean effective stress (especially σ'_{h0})	Applicable to about 30 m depth. Applicable to soils with % fines < 15%.
Vibro-replacement (stone columns)	Spacing of the columns Vibration frequency Vibration amplitude Power of the vibrating probe	Soil reinforcement (column of more resistant material). Improved drainage Reduction of void ratio between columns Increase of the mean effective stress (especially σ'_{h0})	Applicable to about 30 m depth. Applicable to all kinds of soil (with or without fines).
Dynamic compaction	Spacing of the impact points No. of impacts/point Height of fall Falling weight	Reduction of void ratio Increase of the mean effective stress (especially σ'_{h0})	Applicable to about 10 m to 12 m depth. Applicable in large areas away from constructions and inhabited zones.

located away from constructions and inhabited areas, due to the associated noise and intense vibrations. It is also a procedure that only becomes economically attractive when the area to be treated is relatively large.

When the treatment is applied in a large area, it is highly convenient to carry out a previous adjustment of the treatment parameters (see Table 6.15) in a given restricted experimental area, in order to ensure the accomplishment of the objectives with maximum economy.

6.5.5.2 Degree of improvement to be achieved and area to be treated

The effectiveness of the vibro-compaction and dynamic compaction techniques is assessed by field tests, namely SPT and CPT. In general, the treatment project specifies the minimum level to be reached by those field test results, i.e., the increment achieved relative to the tests results performed in the same location before treatment. This minimum level is directly obtained from the liquefaction analysis methods previously presented.

It is recommended to locate the verification tests inside the polygons of the grid of vibrator penetration points (in the case of vibro-compaction) or of impact points (in the case of dynamic compaction), so that the evaluation of the new conditions is made on the safe side. Figure 6.30 summarizes part of the Japanese experience with improvement procedures similar to vibro-compaction, illustrating the relationship between SPT results before and after treatment in points inside that grid, for several values of the ratio between the volume of the densified "columns" around the penetration point and the total volume[8]. For practical purposes, it seems reasonable to assume that the diameter of the densified "columns" is three times the diameter of the vibrator (this being equal to 0.3 m to 0.5 m).

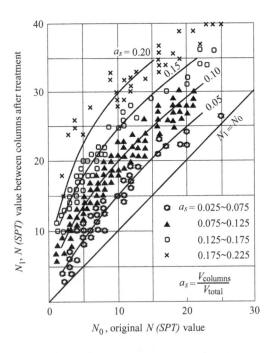

Figure 6.30 Relation between SPT results before and after vibro-compaction treatment in internal points of the penetration grid, for several values of the ratio between the volume of the densified "columns" and the total volume (JGS, 1998).

---

[8] This ratio is equal to the so-called replacement ratio used for stone columns, expressed by Equation 2.37, which will be referred to below.

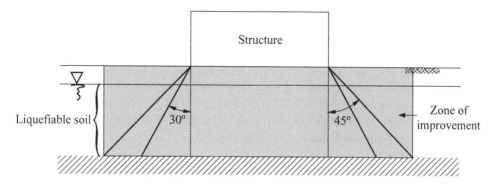

Figure 6.31 Definition of the volume of ground to be treated in the vicinity of the shallow foundations of a structure (Kramer, 1996).

For the vibro-replacement strategy, no methods have yet been developed that assess, in a satisfactorily fundamental way, the contribution of its three effects to resistance to liquefaction: i) the reinforcement of the original ground mass; ii) the densification of this mass between columns; and iii) the drainage effect, which attenuates the earthquake-induced excess pore pressure.

The execution of SPT or CPT tests between stone columns may also be instrumental in evaluating the improvement introduced by vibro-replacement in the soil *in situ*. The assessment of the reinforcement effect is frequently performed by means of the following empirical procedure: i) let N_{soil} be the SPT value at the centre of the areas between columns after treatment; ii) let N_{col} be the SPT value at the column axis; and iii) the resulting average SPT value for the reinforced soil, N_{res}, will then be:

$$N_{res} = \left(1 - a_s\right)N_{soil} + a_s N_{col} \tag{6.79}$$

where a_s is the replacement ratio given by Equation 2.37. For the evaluation of N_{col}, the relationship between $(N_1)_{60}$ and the density index given by Figure 1.14 can be applied, considering, in complement, that the column density index exceeds 80%.

The literature does not provide consistent criteria for defining the extension of the area surrounding the structure, within which the treatment should be performed.

In the case of a foundation with a relatively low safety factor against failure under vertical loading and, taking as reference Figure 6.3, the area to be treated should, at least, extend laterally to a distance f on either side of the foundation. This criterion may lead to unreasonable solutions in some cases, as, for instance, in large mat foundations.

However, as stressed before, the foundation reserve of resistance relative to global failure – with liquefaction being naturally safeguarded against – is typically very considerable. Figure 6.31 illustrates the most common criteria referred to in the literature. Before studies on this issue are available that provide properly fundamented solutions, the author recommends the adoption of the most conservative criterion indicated in the figure.

ANNEX A6.1 EVALUATION OF THE BEARING CAPACITY FOR VERTICAL LOADING BY DIRECT APPLICATION OF PMT RESULTS (MELT, 1993)

The present annex describes a method for the evaluation of the bearing capacity of shallow foundations, subject to vertical loading, directly based on the limit pressure, pl, determined by the Ménard pressuremeter test, PMT (see Section 1.3.10). This method is essentially empirical. For additional information, the book by Frank (2003) is recommended.

The so-called *net limit pressure*, p_ℓ^*, is defined by:

$$p_\ell^* = p_\ell - \sigma_{h0} \tag{A6.1.1}$$

where p_ℓ is the limit pressure measured by the PMT, and σ_{h0} is the at-rest total horizontal stress at the same level.

The so-called *equivalent net limit pressure*, $p_{\ell e}^*$, is calculated in two distinct forms: i) if the soil is reasonably homogeneous at a depth $1.5B$ under the shallow foundation (i.e., it belongs to the same layer), a straight line is drawn representing the change with depth of p_ℓ^*, and $p_{\ell e}^*$ is then taken to be equal to the value at a depth $0.67B$ under the footing, as shown in Figure A6.1.1; or ii) otherwise, and still at a depth of $1.5B$ under the footing, after eliminating the more discrepant p_ℓ^* values, $p_{\ell e}^*$ is taken as the geometric mean of p_ℓ^*, i.e.:

$$p_{\ell e}^* = \sqrt[n]{p_{\ell 1}^* \times p_{\ell 2}^* \times \ldots \ldots p_{\ell n}^*} \tag{A6.1.2}$$

Taking into consideration Figure A6.1.1, the *equivalent embedment depth of the foundation*, D_e, is given by:

$$D_e = \frac{1}{p_{\ell e}^*} \int_0^D p_\ell^*(z)\,dz \tag{A6.1.3}$$

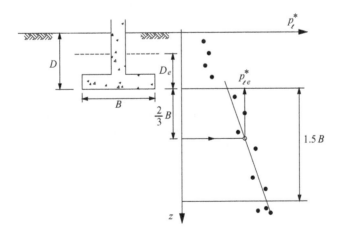

Figure A6.1.1 Criteria for determining the equivalent net limit pressure under the foundation and for calculating the equivalent depth of the foundation.

Table A6.1.1 Values of factor k_p from PMT results for application of Equation A6.1.4

Soil type	p_ℓ category	p_ℓ (MPa)	k_p	$k_{p,max}$ Square footing	$k_{p,max}$ Strip footing
Clay and silt	A	< 0.7	$0.8\left[1+0.25\left(0.6+0.4\dfrac{B}{L}\right)\dfrac{D_e}{B}\right]$	1.30	1.10
	B	1.2 – 2.0	$0.8\left[1+0.35\left(0.6+0.4\dfrac{B}{L}\right)\dfrac{D_e}{B}\right]$	1.50	1.22
	C	> 2.5	$0.8\left[1+0.50\left(0.6+0.4\dfrac{B}{L}\right)\dfrac{D_e}{B}\right]$	1.80	1.40
Sand and gravel	A	< 0.5	$\left[1+0.35\left(0.6+0.4\dfrac{B}{L}\right)\dfrac{D_e}{B}\right]$	1.88	1.53
	B	1.0–2.0	$\left[1+0.50\left(0.6+0.4\dfrac{B}{L}\right)\dfrac{D_e}{B}\right]$	2.25	1.75
	C	> 2.5	$\left[1+0.80\left(0.6+0.4\dfrac{B}{L}\right)\dfrac{D_e}{B}\right]$	3.00	2.20
Chalk	B and C	1.0–2.5 > 3.0	$1.3\left[1+0.27\left(0.6+0.4\dfrac{B}{L}\right)\dfrac{D_e}{B}\right]$	2.18	1.83
Marl and limestone Weathered rock	–	> 1.5 2.5–4.0	$\left[1+0.27\left(0.6+0.4\dfrac{B}{L}\right)\dfrac{D_e}{B}\right]$	1.68	1.41

The foundation bearing capacity under vertical loading can be calculated by the expression:

$$q_{ult} = \sigma_{v0} + k_p\, p_{\ell e}^* \tag{A6.1.4}$$

with:

σ_{v0} being the at-rest total vertical stress at the foundation base level;

$p_{\ell e}^*$ being the equivalent net limit pressure, defined above;

k_p being the bearing resistance factor given in Table A6.1.1.

ANNEX A6.2 INCREMENTAL STRESS IN AN ELASTIC, HOMOGENEOUS HALF-SPACE LOADED AT THE SURFACE

A6.2.1 Solutions of Boussinesq and Flamant and "stress bulbs"

See Figures A6.2.1 to A6.2.3.

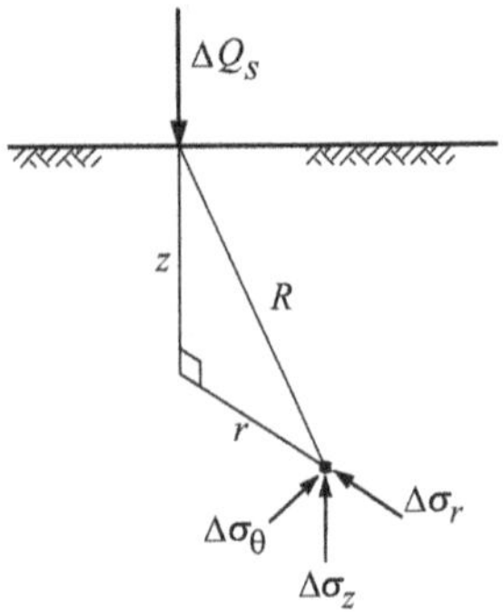

$$\Delta\sigma_z = \frac{3\Delta Q_s z^3}{2\pi R^5} \tag{A6.2.1}$$

$$\Delta\sigma_r = -\frac{\Delta Q_s}{2\pi R^2}\left[\frac{-3r^2 z}{R^3} + \frac{(1-2v)R}{R+z}\right] \tag{A6.2.2}$$

$$\Delta\sigma_\theta = -\frac{(1-2v)\Delta Q_s}{2\pi R^2}\left[\frac{z}{R} - \frac{R}{R+z}\right] \tag{A6.2.3}$$

$$\Delta\tau_{rz} = \frac{3\Delta Q_s r z^2}{2\pi R^5} \tag{A6.2.4}$$

Figure A6.2.1 Stresses induced in an elastic half-space by vertical point load at the surface (Boussinesq, 1885).

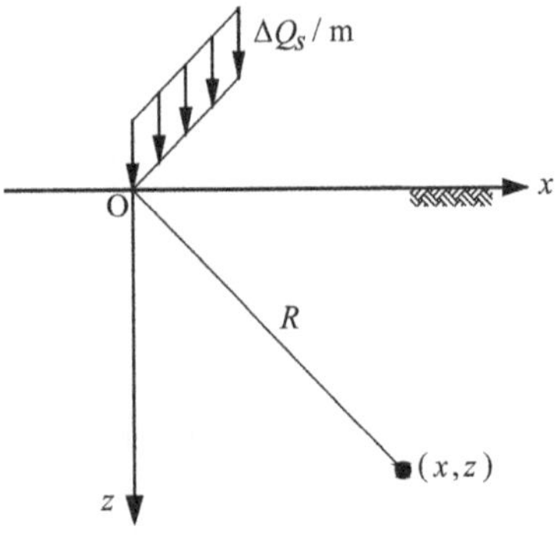

$$\Delta\sigma_x = \frac{2\Delta Q_s}{\pi}\frac{x^2 z}{R^4} \tag{A6.2.5}$$

$$\Delta\sigma_z = \frac{2\Delta Q_s}{\pi}\frac{z^3}{R^4} \tag{A6.2.6}$$

$$\Delta\sigma_y = \frac{2\Delta Q_s}{\pi}v\frac{z}{R^2} \tag{A6.2.7}$$

$$\Delta\tau_{xz} = \frac{2\Delta Q_s}{\pi}\frac{xz^2}{R^4} \tag{A6.2.8}$$

Figure A6.2.2 Stresses induced in an elastic half-space by a linear vertical load at the surface (Flamant, 1892).

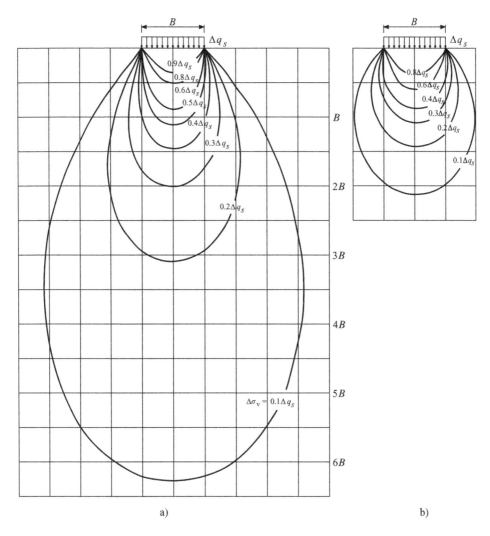

Figure A6.2.3 Contours of vertical incremental stress ("stress bulbs") on elastic half-space beneath uniformly loaded (a) infinite strip and (b) square area.

A6.2.2 Uniform pressure on infinite strip

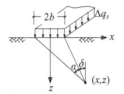

Note: δ is positive if the point is located outside the vertical lines that delimit the loaded area and negative otherwise.

$$\Delta\sigma_z = \frac{\Delta q_s}{\pi}\left[\alpha + \sin\alpha\cos\left(\alpha + 2\delta\right)\right] \qquad (A6.2.9)$$

$$\Delta\sigma_x = \frac{\Delta q_s}{\pi}\left[\alpha - \sin\alpha\cos\left(\alpha + 2\delta\right)\right] \qquad (A6.2.10)$$

$$\Delta\sigma_y = \frac{2\Delta q_s}{\pi}v\alpha \qquad (A6.2.11)$$

$$\Delta\tau_{xz} = \frac{\Delta q_s}{\pi}\sin\alpha\sin\left(\alpha + 2\delta\right) \qquad (A6.2.12)$$

(Poulos and Davis, 1974)

A6.2.3 Triangular pressure on infinite strip

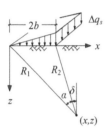

Note: δ is positive for $x > 2b$.

$$\Delta\sigma_z = \frac{\Delta q_s}{2\pi}\left(\frac{x}{b}\alpha - \sin 2\delta\right) \qquad (A6.2.13)$$

$$\Delta\sigma_x = \frac{\Delta q_s}{2\pi}\left(\frac{x}{b}\alpha - \frac{z}{b}\ln\frac{R_1^2}{R_2^2} + \sin 2\delta\right) \qquad (A6.2.14)$$

$$\Delta\tau_{xz} = \frac{\Delta q_s}{2\pi}\left(1 + \cos 2\delta - \frac{z\alpha}{b}\right) \qquad (A6.2.15)$$

(Poulos and Davis, 1974)

A6.2.4 Uniform pressure on circular area

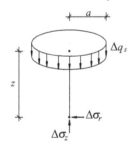

$$\Delta\sigma_z = \Delta q_s\left\{1 - \left[\frac{1}{1 + \left(b/z\right)^2}\right]^{3/2}\right\} \qquad (A6.2.16)$$

$$\Delta\sigma_r = \Delta\sigma_\theta = \frac{\Delta q_s}{2}\left[\left(1 + 2v\right) - \frac{2\left(1 + v\right)z}{\left(b^2 + z^2\right)^{1/2}} + \frac{z^3}{\left(b^2 + z^2\right)^{3/2}}\right] \qquad (A6.2.17)$$

(Poulos and Davis, 1974)

A6.2.5 Uniform pressure on rectangular area

Equations A6.2.18 to A6.2.20 and Tables A6.2.1 to A6.2.5 provide the normal stress increments under the vertex of a rectangle with sides b and l (Giroud, 1970; Poulos and Davis, 1974). The stress increments under the center of the rectangle with sides $2b$ and $2l$ are four times greater.

$$\Delta\sigma_z = \Delta q_s K_1 \qquad (A6.2.18)$$

$$\Delta\sigma_x = \Delta q_s\left[K_2 - \left(1 - 2v\right)K_2'\right] \qquad (A6.2.19)$$

$$\Delta\sigma_y = \Delta q_s\left[L_2 - \left(1 - 2v\right)L_2'\right] \qquad (A6.2.20)$$

Table A6.2.1 Values of K_1

z/b \ l/b	0	0.1	0.2	1/3	0.4	0.5	2/3	1	1.5	2	2.5	3	5	10	∞
0	0.000	0.250	0.250	0.250	0.250	0.250	0.250	0.250	0.250	0.250	0.250	0.250	0.250	0.250	0.250
0.2	0.000	0.137	0.204	0.234	0.240	0.244	0.247	0.249	0.249	0.249	0.249	0.249	0.249	0.249	0.249
0.4	0.000	0.076	0.136	0.187	0.202	0.218	0.231	0.240	0.243	0.244	0.244	0.244	0.244	0.244	0.244
0.5	0.000	0.061	0.113	0.164	0.181	0.200	0.218	0.232	0.238	0.239	0.240	0.240	0.240	0.240	0.240
0.6	0.000	0.051	0.096	0.143	0.161	0.182	0.204	0.223	0.231	0.233	0.234	0.234	0.234	0.234	0.234
0.8	0.000	0.037	0.071	0.111	0.127	0.148	0.173	0.200	0.214	0.218	0.219	0.220	0.220	0.220	0.220
1	0.000	0.028	0.055	0.087	0.101	0.120	0.145	0.175	0.194	0.200	0.202	0.203	0.204	0.205	0.205
1.2	0.000	0.022	0.043	0.069	0.081	0.098	0.121	0.152	0.173	0.182	0.185	0.187	0.189	0.189	0.189
1.4	0.000	0.018	0.035	0.056	0.066	0.080	0.101	0.131	0.154	0.164	0.169	0.171	0.174	0.174	0.174
1.5	0.000	0.016	0.031	0.051	0.060	0.073	0.092	0.121	0.145	0.156	0.161	0.164	0.166	0.167	0.167
1.6	0.000	0.014	0.028	0.046	0.055	0.067	0.085	0.112	0.136	0.148	0.154	0.157	0.160	0.160	0.160
1.8	0.000	0.012	0.024	0.039	0.046	0.056	0.072	0.097	0.121	0.133	0.140	0.143	0.147	0.148	0.148
2	0.000	0.010	0.020	0.033	0.039	0.048	0.061	0.084	0.107	0.120	0.127	0.131	0.136	0.137	0.137
2.5	0.000	0.007	0.013	0.022	0.027	0.033	0.043	0.060	0.080	0.093	0.101	0.106	0.113	0.115	0.115
3	0.000	0.005	0.010	0.016	0.019	0.024	0.031	0.045	0.061	0.073	0.081	0.087	0.096	0.099	0.099
4	0.000	0.003	0.006	0.009	0.011	0.014	0.019	0.027	0.038	0.048	0.055	0.060	0.071	0.076	0.076
5	0.000	0.002	0.004	0.006	0.007	0.009	0.012	0.018	0.026	0.033	0.039	0.043	0.055	0.061	0.062
10	0.000	0.000	0.001	0.002	0.002	0.002	0.003	0.005	0.007	0.009	0.011	0.013	0.020	0.028	0.032
15	0.000	0.000	0.000	0.001	0.001	0.001	0.001	0.002	0.003	0.004	0.005	0.006	0.010	0.016	0.021
20	0.000	0.000	0.000	0.000	0.000	0.001	0.001	0.001	0.002	0.002	0.003	0.004	0.006	0.010	0.016
50	0.000	0.000	0.000	0.000	0.000	0.000	0.000	0.000	0.000	0.000	0.000	0.001	0.001	0.002	0.006

Table A6.2.2 Values of K_2

z/b \ l/b	0	0.1	0.2	1/3	0.4	0.5	2/3	1	1.5	2	2.5	3	5	10	∞
0	0.000	0.250	0.250	0.250	0.250	0.250	0.250	0.250	0.250	0.250	0.250	0.250	0.250	0.250	0.250
0.2	0.000	0.069	0.116	0.149	0.159	0.169	0.177	0.184	0.187	0.188	0.188	0.188	0.188	0.188	0.188
0.4	0.000	0.031	0.058	0.085	0.095	0.106	0.118	0.128	0.133	0.134	0.134	0.134	0.135	0.135	0.135
0.5	0.000	0.022	0.043	0.064	0.073	0.083	0.094	0.105	0.110	0.112	0.112	0.112	0.113	0.113	0.113
0.6	0.000	0.017	0.032	0.049	0.056	0.065	0.075	0.086	0.091	0.093	0.093	0.094	0.094	0.094	0.094
0.8	0.000	0.009	0.018	0.029	0.034	0.040	0.047	0.057	0.062	0.064	0.064	0.065	0.065	0.065	0.065
1	0.000	0.006	0.011	0.018	0.021	0.025	0.030	0.037	0.042	0.044	0.045	0.045	0.045	0.045	0.045
1.2	0.000	0.003	0.007	0.011	0.013	0.016	0.020	0.025	0.029	0.031	0.032	0.032	0.032	0.032	0.032
1.4	0.000	0.002	0.004	0.007	0.008	0.010	0.013	0.017	0.020	0.022	0.023	0.023	0.023	0.023	0.023
1.5	0.000	0.002	0.004	0.006	0.007	0.008	0.011	0.014	0.017	0.019	0.019	0.020	0.020	0.020	0.020
1.6	0.000	0.001	0.003	0.005	0.006	0.007	0.009	0.012	0.015	0.016	0.017	0.017	0.017	0.017	0.017
1.8	0.000	0.001	0.002	0.003	0.004	0.005	0.006	0.008	0.011	0.012	0.012	0.013	0.013	0.013	0.013
2	0.000	0.001	0.001	0.002	0.003	0.003	0.004	0.006	0.008	0.009	0.009	0.010	0.010	0.010	0.010
2.5	0.000	0.000	0.001	0.001	0.001	0.002	0.002	0.003	0.004	0.005	0.005	0.005	0.006	0.006	0.006
3	0.000	0.000	0.000	0.001	0.001	0.001	0.001	0.002	0.002	0.003	0.003	0.003	0.003	0.003	0.003
4	0.000	0.000	0.000	0.000	0.000	0.000	0.000	0.001	0.001	0.001	0.001	0.001	0.001	0.002	0.002
5	0.000	0.000	0.000	0.000	0.000	0.000	0.000	0.000	0.000	0.000	0.001	0.001	0.001	0.001	0.001
10	0.000	0.000	0.000	0.000	0.000	0.000	0.000	0.000	0.000	0.000	0.000	0.000	0.000	0.000	0.000
15	0.000	0.000	0.000	0.000	0.000	0.000	0.000	0.000	0.000	0.000	0.000	0.000	0.000	0.000	0.000
20	0.000	0.000	0.000	0.000	0.000	0.000	0.000	0.000	0.000	0.000	0.000	0.000	0.000	0.000	0.000
50	0.000	0.000	0.000	0.000	0.000	0.000	0.000	0.000	0.000	0.000	0.000	0.000	0.000	0.000	0.000

Table A6.2.3 Values of K_2^I

z/b \ l/b	0	0.1	0.2	1/3	0.4	0.5	2/3	1	1.5	2	2.5	3	5	10	∞
0	0.000	0.234	0.219	0.199	0.189	0.176	0.156	0.125	0.094	0.074	0.061	0.051	0.031	0.016	0.000
0.2	0.000	0.059	0.097	0.118	0.121	0.122	0.118	0.103	0.082	0.067	0.056	0.048	0.030	0.016	0.000
0.4	0.000	0.026	0.048	0.069	0.075	0.082	0.086	0.083	0.071	0.060	0.051	0.045	0.029	0.015	0.000
0.5	0.000	0.019	0.036	0.054	0.060	0.067	0.073	0.074	0.066	0.056	0.049	0.043	0.028	0.015	0.000
0.6	0.000	0.015	0.028	0.043	0.049	0.056	0.062	0.066	0.061	0.053	0.047	0.041	0.028	0.015	0.000
0.8	0.000	0.009	0.018	0.029	0.033	0.039	0.046	0.052	0.052	0.047	0.043	0.038	0.026	0.015	0.000
1	0.000	0.007	0.013	0.021	0.024	0.029	0.035	0.042	0.044	0.042	0.039	0.035	0.025	0.014	0.000
1.2	0.000	0.005	0.009	0.015	0.018	0.022	0.027	0.034	0.037	0.037	0.035	0.032	0.024	0.014	0.000
1.4	0.000	0.004	0.007	0.012	0.014	0.017	0.021	0.027	0.032	0.033	0.032	0.030	0.023	0.014	0.000
1.5	0.000	0.003	0.006	0.010	0.012	0.015	0.019	0.025	0.029	0.031	0.030	0.029	0.022	0.014	0.000
1.6	0.000	0.003	0.006	0.009	0.011	0.013	0.017	0.023	0.027	0.029	0.028	0.027	0.022	0.013	0.000
1.8	0.000	0.002	0.005	0.007	0.009	0.011	0.014	0.019	0.024	0.025	0.026	0.025	0.021	0.013	0.000
2	0.000	0.002	0.004	0.006	0.007	0.009	0.012	0.016	0.020	0.023	0.023	0.023	0.020	0.013	0.000
2.5	0.000	0.001	0.002	0.004	0.005	0.006	0.008	0.011	0.015	0.017	0.018	0.019	0.017	0.012	0.000
3	0.000	0.001	0.002	0.003	0.003	0.004	0.006	0.008	0.011	0.013	0.015	0.015	0.015	0.011	0.000
4	0.000	0.000	0.001	0.002	0.002	0.002	0.003	0.005	0.007	0.008	0.010	0.011	0.012	0.010	0.000
5	0.000	0.000	0.001	0.001	0.001	0.002	0.002	0.003	0.004	0.006	0.007	0.008	0.009	0.009	0.000
10	0.000	0.000	0.000	0.000	0.000	0.000	0.001	0.001	0.001	0.002	0.002	0.002	0.003	0.005	0.000
15	0.000	0.000	0.000	0.000	0.000	0.000	0.000	0.000	0.001	0.001	0.001	0.001	0.002	0.003	0.000
20	0.000	0.000	0.000	0.000	0.000	0.000	0.000	0.000	0.000	0.000	0.000	0.001	0.001	0.002	0.000
50	0.000	0.000	0.000	0.000	0.000	0.000	0.000	0.000	0.000	0.000	0.000	0.000	0.000	0.000	0.000

Table A6.2.4 Values of L_2

z/b \ l/b	0	0.1	0.2	1/3	0.4	0.5	2/3	1	1.5	2	2.5	3	5	10	∞
0	0.000	0.250	0.250	0.250	0.250	0.250	0.250	0.250	0.250	0.250	0.250	0.250	0.250	0.250	0.250
0.2	0.000	0.010	0.045	0.094	0.112	0.134	0.158	0.184	0.201	0.208	0.211	0.214	0.217	0.218	0.219
0.4	0.000	0.002	0.010	0.032	0.045	0.064	0.091	0.128	0.156	0.169	0.176	0.179	0.186	0.188	0.189
0.5	0.000	0.001	0.006	0.020	0.029	0.044	0.068	0.105	0.136	0.151	0.159	0.164	0.172	0.175	0.176
0.6	0.000	0.000	0.003	0.013	0.019	0.031	0.051	0.086	0.118	0.134	0.144	0.149	0.158	0.163	0.164
0.8	0.000	0.000	0.001	0.006	0.009	0.016	0.029	0.057	0.087	0.106	0.117	0.124	0.135	0.141	0.143
1	0.000	0.000	0.001	0.003	0.005	0.009	0.017	0.037	0.064	0.083	0.095	0.103	0.116	0.123	0.125
1.2	0.000	0.000	0.000	0.002	0.003	0.005	0.011	0.025	0.047	0.065	0.077	0.085	0.100	0.108	0.111
1.4	0.000	0.000	0.000	0.001	0.002	0.003	0.007	0.017	0.035	0.051	0.062	0.071	0.087	0.095	0.099
1.5	0.000	0.000	0.000	0.001	0.001	0.003	0.005	0.014	0.030	0.045	0.056	0.064	0.081	0.090	0.094
1.6	0.000	0.000	0.000	0.001	0.001	0.002	0.004	0.012	0.026	0.040	0.051	0.059	0.076	0.085	0.089
1.8	0.000	0.000	0.000	0.000	0.001	0.001	0.003	0.008	0.020	0.031	0.041	0.049	0.066	0.077	0.081
2	0.000	0.000	0.000	0.000	0.001	0.001	0.002	0.006	0.015	0.025	0.034	0.041	0.058	0.069	0.074
2.5	0.000	0.000	0.000	0.000	0.000	0.000	0.001	0.003	0.008	0.014	0.021	0.027	0.043	0.055	0.061
3	0.000	0.000	0.000	0.000	0.000	0.000	0.000	0.002	0.004	0.008	0.013	0.018	0.032	0.045	0.051
4	0.000	0.000	0.000	0.000	0.000	0.000	0.000	0.001	0.002	0.003	0.006	0.008	0.018	0.031	0.039
5	0.000	0.000	0.000	0.000	0.000	0.000	0.000	0.000	0.001	0.002	0.003	0.004	0.011	0.022	0.031
10	0.000	0.000	0.000	0.000	0.000	0.000	0.000	0.000	0.000	0.000	0.000	0.000	0.001	0.006	0.016
15	0.000	0.000	0.000	0.000	0.000	0.000	0.000	0.000	0.000	0.000	0.000	0.000	0.000	0.002	0.011
20	0.000	0.000	0.000	0.000	0.000	0.000	0.000	0.000	0.000	0.000	0.000	0.000	0.000	0.001	0.008
50	0.000	0.000	0.000	0.000	0.000	0.000	0.000	0.000	0.000	0.000	0.000	0.000	0.000	0.000	0.003

Table A6.2.5 Values of L_2^1

z/b \ l/b	0	0.1	0.2	1/3	0.4	0.5	2/3	1	1.5	2	2.5	3	5	10	∞
0	0.000	0.016	0.031	0.051	0.061	0.074	0.094	0.125	0.156	0.176	0.189	0.199	0.219	0.234	0.250
0.2	0.000	0.013	0.025	0.041	0.049	0.060	0.076	0.103	0.130	0.148	0.160	0.169	0.188	0.203	0.219
0.4	0.000	0.010	0.020	0.032	0.039	0.047	0.061	0.083	0.106	0.122	0.133	0.141	0.159	0.174	0.189
0.5	0.000	0.009	0.017	0.029	0.034	0.042	0.054	0.074	0.096	0.111	0.121	0.129	0.146	0.161	0.176
0.6	0.000	0.008	0.015	0.025	0.030	0.037	0.048	0.066	0.086	0.100	0.110	0.118	0.134	0.149	0.164
0.8	0.000	0.006	0.012	0.020	0.023	0.029	0.037	0.052	0.069	0.082	0.091	0.098	0.114	0.127	0.143
1	0.000	0.005	0.009	0.015	0.018	0.023	0.029	0.042	0.056	0.067	0.075	0.082	0.097	0.110	0.125
1.2	0.000	0.004	0.007	0.012	0.015	0.018	0.024	0.034	0.046	0.056	0.063	0.069	0.083	0.096	0.111
1.4	0.000	0.003	0.006	0.010	0.012	0.015	0.019	0.027	0.038	0.046	0.053	0.058	0.072	0.084	0.099
1.5	0.000	0.003	0.005	0.009	0.011	0.013	0.017	0.025	0.035	0.043	0.049	0.054	0.067	0.079	0.094
1.6	0.000	0.002	0.005	0.008	0.010	0.012	0.016	0.023	0.032	0.039	0.045	0.050	0.062	0.074	0.089
1.8	0.000	0.002	0.004	0.007	0.008	0.010	0.013	0.019	0.027	0.033	0.039	0.043	0.055	0.066	0.081
2	0.000	0.002	0.003	0.006	0.007	0.008	0.011	0.016	0.023	0.029	0.033	0.038	0.048	0.059	0.074
2.5	0.000	0.001	0.002	0.004	0.005	0.006	0.007	0.011	0.016	0.020	0.024	0.027	0.036	0.047	0.061
3	0.000	0.001	0.002	0.003	0.003	0.004	0.005	0.008	0.012	0.015	0.018	0.021	0.028	0.038	0.051
4	0.000	0.000	0.001	0.002	0.002	0.002	0.003	0.005	0.007	0.009	0.011	0.013	0.018	0.026	0.039
5	0.000	0.000	0.001	0.001	0.001	0.002	0.002	0.003	0.005	0.006	0.007	0.009	0.013	0.019	0.031
10	0.000	0.000	0.000	0.000	0.000	0.000	0.001	0.001	0.001	0.002	0.002	0.002	0.004	0.007	0.016
15	0.000	0.000	0.000	0.000	0.000	0.000	0.000	0.000	0.001	0.001	0.001	0.001	0.002	0.003	0.011
20	0.000	0.000	0.000	0.000	0.000	0.000	0.000	0.000	0.000	0.000	0.000	0.001	0.001	0.002	0.008
50	0.000	0.000	0.000	0.000	0.000	0.000	0.000	0.000	0.000	0.000	0.000	0.000	0.000	0.000	0.003

ANNEX A6.3 SETTLEMENT EVALUATION BY DIRECT APPLICATION OF PMT RESULTS (MELT, 1993)

The present annex describes a method for the evaluation of the settlement of shallow foundations directly based on the pressuremeter modulus, E_{PMT}, determined by the Ménard pressuremeter test, PMT (see Section 1.3.10). This method is essentially empirical. For additional information, the book by Frank (2003) is recommended.

According to this method, the foundation settlement is considered to have two components: i) a settlement, $s_{c,PMT}$, predominantly attributed to the volumetric deformation of the immediately underlying layer of thickness $B/2$, therefore loaded in conditions close to lateral confinement; and ii) a settlement, $s_{d,PMT}$, mostly associated with distortional deformation down to depth $8B$. So:

$$s_{PMT} = s_{c,PMT} + s_{d,PMT} \tag{A6.3.1}$$

with:

$$s_{c,PMT} = \left(q - \sigma_{v0}\right)\frac{\lambda_c\,B\,\alpha}{q\,E_c} \tag{A6.3.2}$$

and

$$s_{d,PMT} = \left(q - \sigma_{v0}\right)\frac{2\,B_0}{9\,E_d}\left(\frac{\lambda_d\,B}{B_0}\right)^{\alpha} \tag{A6.3.3}$$

where:
- σ_{v0} is the at-rest total vertical stress at the foundation base level;
- q is the foundation applied pressure;
- B_0 is the reference width equal to 0.6 m;
- B is the foundation width;
- λ_c and λ_d are shape coefficients given by Table A6.3.1;
- α is the rheological coefficient given by Table A6.3.2;
- E_c is the value of E_{PMT} between the foundation base and a depth $B/2$ below (modulus E_1 in Figure A6.3.1);
- E_d is calculated from the E_{PMT} values down to a depth $8B$ below the foundation base, by means of the following equation (see also Figure A6.3.1):

$$\frac{4}{E_d} = \frac{1}{E_1} + \frac{1}{0.85E_2} + \frac{1}{E_{3,5}} + \frac{1}{2.5E_{6,8}} + \frac{1}{2.5E_{9,16}} \tag{A6.3.4}$$

where $E_{i,j}$ is the harmonic mean of the E_{PMT} values measured in the layers located between depths $iB/2$ and $jB/2$. So, for example:

$$\frac{3.0}{E_{3,5}} = \frac{1}{E_3} + \frac{1}{E_4} + \frac{1}{E_5} \tag{A6.3.5}$$

Table A6.3.1 Values of the shape coefficients, λ_c and λ_d, for use in Equations A6.3.2 and A6.3.3.

L/B	Circular	Square	2	3	5	20
λ_c	1	1.1	1.2	1.3	1.4	1.5
λ_d	1	1.12	1.53	1.78	2.14	2.65

Table A6.3.2 Values of the rheological coefficient α for use in Equation A6.3.2.

Soil type	Description	E_{PMT}/p_ℓ	α
Peat			1
Clay	Overconsolidated	>16	1
	Normally consolidated	9–16	0.67
	Underconsolidated or remolded	7–9	0.5
Silt	Overconsolidated	>14	0.67
	Normally consolidated	8–14	0.5
	Underconsolidated or remolded	5–8	0.5
Sand	Very dense	>12	0.5
	Medium	7–12	0.33
	Loose	5–7	0.33
Sand and gravel	Very dense	>10	0.33
	Medium	6–10	0.25
Rock	Very little fractured	–	0.67
	Fairly fractured	–	0.5
	Extensively fractured	–	0.33
	Weathered	–	0.67

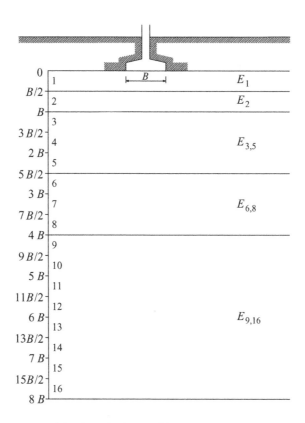

Figure A6.3.1 Pressuremeter modulus for calculation of foundation settlement.

ANNEX E6 EXERCISES (RESOLUTIONS ARE INCLUDED IN THE FINAL ANNEX)

E6.1 – Figure E6.1 shows three identical isostatic structures with the same applied loads but with different foundation conditions, as indicated in the figure. Consider that the footings have a square shape. The Poisson ratio has the same value in all foundation soils.

a) Sort, in ascending order, the ratio of the left and right footing settlements for each structure. Complement it with as much numerical data as possible.
b) Sort, in ascending order, the settlement of the six footings.

E6.2 – Figure E6.2 presents four pairs of footings over the same ground. The longitudinal footing dimension is practically infinite and transversally they are sufficiently away from

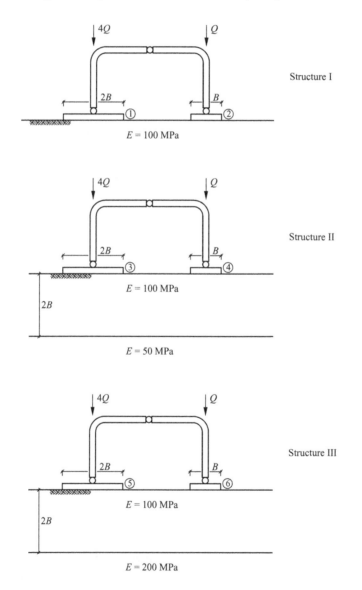

Figure E6.1

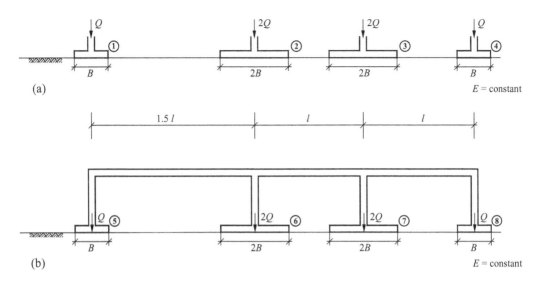

Figure E6.2

each other that the intersection is negligible between their stress bulbs. Assume that the deformation modulus of the foundation ground does not vary with depth.

The footings of Figure E6.2a are structurally disconnected, whereas those of Figure E6.2b belong to a structure of great stiffness. The loads indicated in Figure E6.2b were obtained from a structural analysis assuming rigid supports.

a) Establish quantitative relations between the foundation settlements, s.
b) Sort, in ascending order, the eight foundation settlements.
c) Sort, in ascending order, the ratios s_5/s_1, s_6/s_2, s_7/s_3, and s_8/s_4. Include in that ordering numerical values, where appropriate.

E6.3 – Figure E6.3 displays three shallow foundations (with a square shape in plan) under four different conditions relative to the foundation ground and to the superstructure. So: i) Structure I is isostatic while the other three are hyperstatic (Structures III and IV are identical); and ii) the elastic modulus of the foundation soil is constant for the first three cases, while increasing linearly with depth in the fourth case. The foundation loads indicated resulted from analyses admitting rigid supports. Consider an identical Poisson ratio in the four cases.

a) Sort, in ascending order, the footing settlement ratios s_2/s_1, s_4/s_3, s_6/s_5, and s_8/s_7. Add numerical values for their upper and lower bounds.
b) In case the foundation ground is a sand stratum with a given grain size distribution and density index, which will be the more plausible scenario: that of the three first cases or that of the last one?

E6.4 – Figure E6.4 displays four pairs of square footings over the same sandy soil, the deformation modulus, E, of which increases linearly with depth. The footings of Figure E6.4a are structurally disconnected, whereas those of Figure E6.4b belong to a structure with great stiffness. Assume that the foundation dimensions in plan have been defined using the loads from an analysis of the structure with rigid supports, and taking as a criterion that the pressure (Δq_s) transmitted to the ground is identical for all footings.

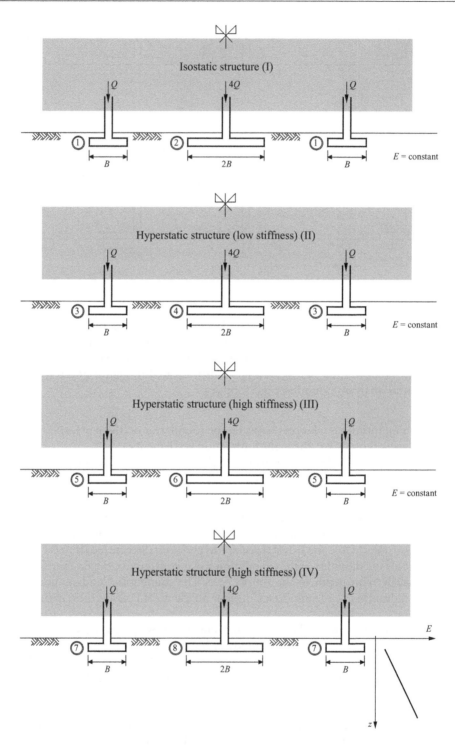

Figure E6.3

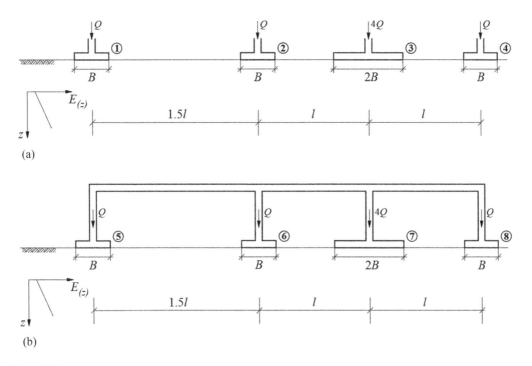

Figure E6.4

a) Order the four foundation settlements, s, of Figure E6.4a by increasing value.
b) Order the four foundation settlements, s, of Figure E6.4b by increasing value.
c) Order the eight foundation settlements, s, of Figure E6.4 by increasing value.
d) Order the ratios s_5/s_1, s_6/s_2, s_7/s_3, and s_8/s_4 by increasing value.
e) Now, assume that underlying the sandy soil there is a very firm layer at a depth of about 3B. How will the differential settlements change in relation to the conditions of Figure E6.4b?

E6.5 – Figure E6.5 shows a cylindrical storage tank with 24.0-m diameter, founded on a 12.0-m thick lightly overconsolidated clay layer with the water table at the surface. The self-weight of the tank and of the stored material correspond to uniform pressures on the foundation soil of 10 kPa and 90 kPa, respectively. The figure includes the geotechnical parameters of the clay. Take $\gamma_w = 9.8$ kN/m³.

a) Calculate the global safety factor relative to a bearing resistance failure of the foundation immediately after the first tank-filling operation, which will occur immediately after the tank construction.
b) Calculate the global safety factor relative to a bearing resistance failure of the foundation under long-term conditions (end of the consolidation).
c) Estimate the immediate settlement of the foundation center for the first filling, considering a flexible foundation.

E6.6 – Figure E6.6 represents a wind tower with a square, rigid footing on a thick and homogeneous sand layer. The water level coincides with the base of the foundation. Take $\gamma_w = 9.8$ kN/m³.

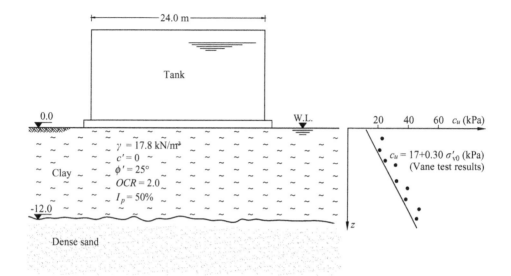

Figure E6.5

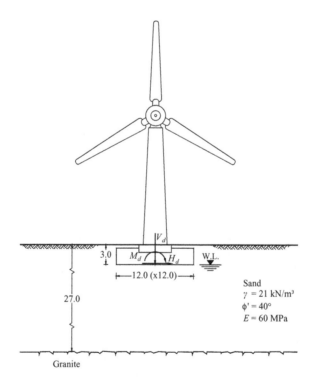

Figure E6.6

a) Check the safety of the foundation in relation to a bearing resistance failure due to vertical loading, according to Eurocode 7 (Design Approach 1 – Combination 2).
b) Calculate the foundation settlement and verify if its rotation exceeds the 0.1° limit imposed by the wind tower manufacturer.

The applied force systems, referred to the center of the footing base, for Ultimate Limit State (ULS) and Serviceability Limit State (SLS), with wind as the leading variable action, are the following:

i) ULS: $V_d = 15000$ kN; $M_d = 60000$ kN.m; $H_d = 1500$ kN
ii) SLS: $V_d = 15000$ kN; $M_d = 30000$ kN.m; $H_d = 750$ kN

E6.7 – Figure E6.7 represents a reinforced concrete silo founded at 1.0 m depth on a rigid mat. In the figure, the succession of soil strata overlying the schist bedrock is displayed, as well as the mechanical parameters required for estimating the bearing capacity and the settlement of the foundation. The water table coincides with the base of the foundation. Assume that the soil above the water table is saturated. Take $\gamma_w = 9.8$ kN/m³.

a) Determine an upper and a lower limit, as close as possible, for the global safety factor with regard to a bearing resistance failure of the foundation.

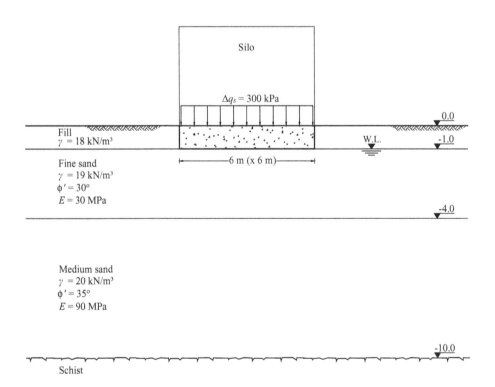

Figure E6.7

b) Using the analytical expression for the immediate settlement of a footing on an elastic foundation with constant E value in depth, provide an estimate of the immediate settlement of the foundation. Assume the Poisson ratio to be $\nu = 0.30$.

c) Applying the Schmertmann's method, estimate the immediate settlement of the mat. Take the same dimensions, E and ν values, and loading as in question b).

E6.8 – Figure E6.8 represents part of the reinforced concrete hyperstatic structure of an industrial building, founded on rigid footings. For design reasons, the width of the central footing is 4.00 m. The footing forces, relative to the center of gravity of each footing, are given in Table E6.8.

An SPT campaign provided results that, after application of correction factors, led to the following average normalized values: fine sand – $(N_1)_{60} = 15$; medium sand – $(N_1)_{60} = 30$. Consider the unit weight of all layers to be $\gamma = 20.0$ kN/m³ and take $\gamma_w = 9.8$ kN/m³.

a) Estimate the angle of shearing resistance of the two sands, using a correlation with the SPT results. Indicate in your answer which correlation has been applied.

b) Calculate the length, L, of the central footing in order to obtain a global safety factor of 4.0, relative to bearing capacity.

c) Using the analytical expression for the immediate settlement of a footing on an elastic half-space with constant E value in depth, establish an interval, as narrow as possible,

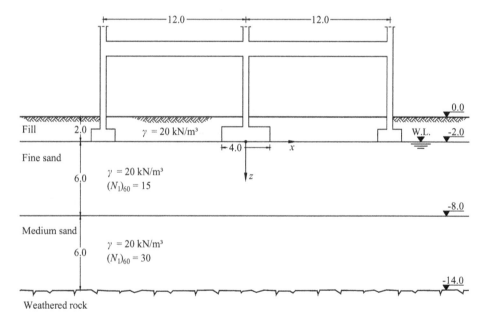

Figure E6.8

Table E6.8

	Ultimate Limit States			Serviceability LS
	V (kN)	*Hx* (kN)	*My* (kN.m)	*V* (kN)
Central footing	6300	630	2520	4000
Lateral footings	3000	300	1200	1900

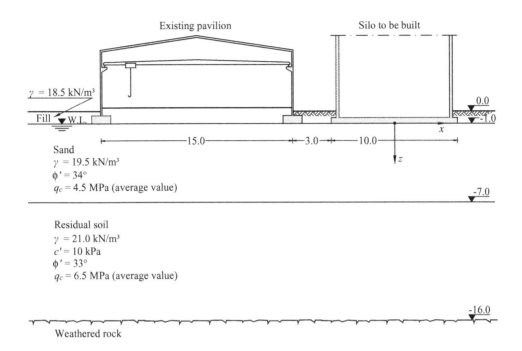

Figure E6.9

for the settlement of the central footing. Adopt the following values for the deformation modulus of the two sands: fine sand, $E = 30$ MPa; medium sand, $E = 60$ MPa. In
both cases, take the value $\nu = 0.30$ for the Poisson ratio. Consider only the vertical load
V corresponding to the Serviceability Limit State.

d) Using the Schmertmann's method, estimate the immediate settlement of the central
footing. Take the same dimensions, E values, and loading as in question c). Comment
on the result obtained by comparing it with the interval determined in c).

e) Assume once again that the dimensions of the central footing are $B = L = 4.00$ m. With
respect to the lateral footings, which criterion, relative to the central footing, would
you adopt for their design: i) the same safety factor; ii) the same vertical pressure
applied to the ground; or iii) the same settlement?

E6.9 –Figure E6.9 represents the ground conditions at the site of an industrial unit where a
silo is going to be constructed, close to an existing pavilion. This pavilion has a steel frame
structure on shallow foundations. The silo has a 10 m by 60 m reinforced concrete rigid
foundation.

Estimate the value of the differential settlement of the steel frame footings due to the silo
construction in the vicinity. Consider only the vertical load $V = 146$ MN (corresponding to
the serviceability limit state). Adopt a value of 0.35 for the Poisson ratio.

Chapter 7

Pile foundations

Paulo Pinto and Pedro Alves Costa
University of Coimbra, Portugal
University of Porto, Portugal

7.1 INTRODUCTION

Piles are slender structural elements used for transferring loads, coming from the structure, to deep ground layers. In comparison with shallow foundations, the equilibrium of a vertical load by a pile foundation typically involves interaction with the soil through both the shaft and the base, as shown in Figure 7.1a. The maximum load equilibrated through the shaft is called *shaft resistance*, whereas the maximum load carried by the base is called *base resistance*.[1] The top of the pile, where it receives the load from the structure, is called the pile head.

The most common situation that requires a pile foundation is when the resistance and/or the stiffness of the layers close to the ground surface are not adequate to support the compressive load applied by the structure. Moreover, pile foundations are the normal solution when the load transmitted to the foundation is a tensile load. In this case, as illustrated in Figure 7.1b, the load equilibrium requires the mobilization of the shaft resistance. As for the pile–soil interaction induced by horizontal loads and/or moments applied at the pile head, it produces horizontal pile displacements, which lead to the mobilization of normal reactive stresses on the adjacent ground, as suggested by Figure 7.1c. Such interaction induces bending moments and shear stresses in the pile along a certain length. This length, as well as the magnitude of the bending moments and shear stresses, is dependent on the soil and pile stiffnesses.

Beside the scenarios described above, due to their versatility, piles are a convenient solution under very different conditions and for different purposes, when:

i) ground layers with sufficient bearing capacity are at great depth;
ii) surface layers are very deformable, although they may have enough bearing capacity, or when their stiffness varies greatly from point to point;
iii) layers immediately below the structure may undergo internal erosion or instability due to seepage forces or liquefaction;
iv) superficial ground layers are likely to experience seasonal movements (expansive soils);
v) the structure to be supported is particularly sensitive to differential settlements;
vi) the structure transmits unusually high vertical and/or horizontal loads;
vii) the structure transmits tensile loads to the foundation or moments, the magnitude of which give rise to great eccentricity of the vertical load.

---

[1] Shaft resistance is also commonly designated as skin friction. The base resistance is often called end bearing or tip resistance.

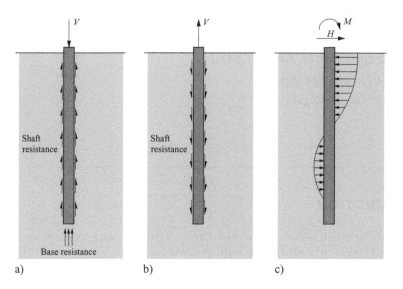

Figure 7.1 Pile–soil interaction for (a) compressive vertical loading, (b) tensile vertical loading, and (c) for horizontal forces or bending moments at the pile head.

A reason that further contributes to the increasing use of piles is the fact that their construction involves more industrialized (mechanized) processes than the execution of shallow foundations. This can reduce, in some cases, the time required for the execution and/or the cost of the work.

Despite the technological advances verified in the last decades in piling engineering, it should be pointed out that the basic concept inherent in this type of foundations is very ancient. In river valleys and in locations with poor geotechnical characteristics, piles have been used since prehistoric times. However, until the beginning of the last century, the piles were typically of timber, with diameters of around 30 cm and lengths that did not exceed 10 to 15 m. At present, piles can reach diameters ranging from 3 to 4 m, or even 6 m in the case of mono-pile wind turbines, and lengths that can be greater than 60 m, consisting of reinforced concrete, pre-stressed concrete or steel.

7.2 TYPES OF PILE FOUNDATIONS AND CONSTRUCTION METHODS

Piles can be classified in several ways. Based on the installation method, piles are usually classified into two categories, namely displacement and non-displacement piles. Displacement piles are installed without removal of soil during construction, requiring soil movement to permit pile penetration into the ground. On the other hand, when the soil is removed to create the hole in which to accommodate the pile, this is classified as a non-displacement pile. Commonly, these piles are made of concrete, which is the material used to fill the cavity drilled into the ground.

This classification is quite broad, since different techniques can be adopted to construct both displacement and non-displacement piles. Table 7.1 presents the most usual types of piles within each category.

Installation of displacement piles implies movement of the soil as the pile is pushed into the ground, causing a very significant increase of horizontal stress in the pile vicinity, as well as changes in the strength properties of the surrounding soil.

Table 7.1 Pile classification and construction technique description.

Pile type		Material	Construction technique description/Observations	
Displacement piles	Pre-fabricated driven piles	Timber	The pre-formed element is pushed into the ground using different techniques such as vibration or an impact hammer.	Traditional piles until the 19th century. Usually not longer than 10 to 15 m.
		Steel sections or tubes		Solution of choice for moderate loads imposed by the structure, and when it is necessary to minimize soil movements and ground-borne vibrations induced during pile driving
		Prestressed concrete		Usually with square cross section. Piles adopted when dealing with relevant loads imposed by the structure. The concrete elements are prestressed in order to avoid cracking due to the tensile stress wave generated during pile driving.
	Cast-in-place displacement-piles	Steel and reinforced concrete	A tube with sealed tip is driven in the ground using vibration or an impact hammer. After reaching the required depth, the reinforcement steel cage is placed inside the tube and then fully filled with concrete.	
Non displacement piles	Bored piles with permanent steel casing	Steel and reinforced concrete	A steel open-ended tube is driven to the desired depth. The soil inside the tube is removed. The reinforcement steel cage of the pile is then inserted and, finally, concrete is poured into the tube using a tremie (see Figure 7.2).	
	Bored piles concreted against the ground	Reinforced concrete	A hole is drilled to the desired depth. To guarantee its stability, temporary stabilization techniques, if necessary, are adopted, such as a temporary steel casing or a stabilization fluid. The construction technique, when adopting temporary steel casing, is similar to the case described in the paragraph above, with the only difference that concrete is delivered concurrently with tube withdrawal. If stabilization fluids are adopted, during ground drilling, the hole is filled with stabilization slurry. After reaching the required depth, the steel reinforcement cage is inserted into the hole. Concrete is placed from the bottom of the hole, replacing the slurry (see Figure 7.3)	
	Continuous flight auger piles	Reinforced concrete	A hollow-stem auger is used to drill the ground to the required depth. Then, concrete is injected from the hollow auger, which is withdrawn smoothly while rotating to form the pile and remove the soil (the pile is formed at the same time that concrete is delivered). After completion of concreting, the reinforcement steel cage is inserted into the fresh concrete (see Figure 7.4)	

Pile driving is an attractive solution when dealing with deep foundations in ground with soft or loose shallow layers, since drivability can be easily achieved. If the soil is stiff, there is a need to increase the energy required for pile driving. Despite the advances in pile driving equipment, which provide drivability in such soils, some other issues may arise. In fact, when increasing pile driving energy, the amplitude of ground-borne vibrations may be enough to

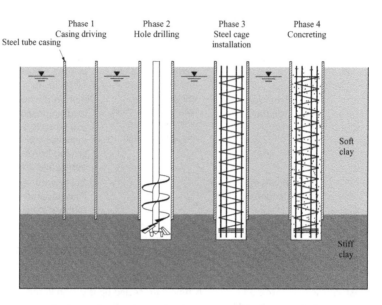

Figure 7.2 Construction of piles with permanent steel casing: 1) driving of casing; 2) soil extraction; 3) steel reinforcement cage installation; 4) concreting.

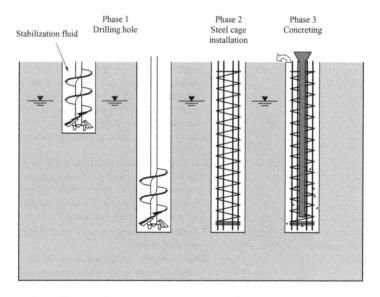

Figure 7.3 Construction of bored piles with fluid stabilization of the borehole.

induce damage in nearby buildings and facilities or, at least, to cause annoyance among the inhabitants, and to cause malfunction in sensitive equipment.

Non-displacement piles are constructed by drilling the ground and placing the pile in the hole. In general, piles are cast-in-place and are usually called bored piles or drilled shafts. Different technologies and construction methods can be adopted for pile installation, where the most popular are: i) bored piles with permanent steel casing; ii) bored piles concreted against the ground; and iii) continuous flight auger piles. Table 7.1 summarizes the main

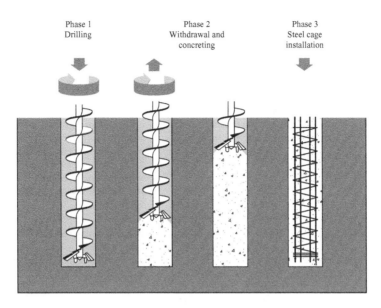

Phase 1
Drilling

Phase 2
Withdrawal and
concreting

Phase 3
Steel cage
installation

Figure 7.4 Continuous flight auger pile construction.

aspects of each technology. A brief description of the main steps of the construction techniques is also given.

It should be highlighted that the change in the ground stress state during the installation of non-displacement piles from each one of these three groups is distinct. In the first, the at-rest stress state remains practically unchanged or the horizontal stress increases slightly due to casing driving. On the other hand, in the second and third cases, there is, probably, a reduction in the at-rest stress state, of variable magnitude, depending on the technique adopted for borehole stabilization.

The selection of the most suitable pile type for a specific foundation scenario is dependent on several factors, such as the soil type, the loading intensity, environmental concerns, namely in terms of noise and vibration, the logistic chain required for the construction, etc.

Table 7.2 summarizes the main advantages and drawbacks associated with the different pile installation techniques.

As will be discussed in later sections, pile performance and behavior are dependent on the installation technique. Thus, the selection of the installation technique must be defined in the early stages of foundation design, since it has significant implications with respect to the pile bearing capacity.

7.3 SINGLE AXIALLY LOADED PILE

7.3.1 Shaft and base resistance mobilization

As mentioned above, when a pile is subjected to a compressive axial force, two distinct interaction mechanisms are activated: i) mobilization of resistance along the shaft; and ii) mobilization of base resistance. Therefore, the ultimate bearing capacity, V_{ult}, of the pile under axial compressive load is given by the sum of both components, i.e., the shaft resistance, R_s,

Table 7.2 Advantages and drawbacks associated to pile type.

Pile type		Advantages	Drawbacks
Displacement piles	Pre-formed driven piles	Increase of the horizontal stresses in the ground with consequent increase of shaft resistance. Soil improvement in the pile vicinity (mainly in loose sandy soils). Borehole stabilization not required. Faster installation with reduced construction staff.	The pile cross section is reduced when compared with the cross section of regular displacement piles. Vibration and noise associated to driving may prevent the usage of this technique in urban environment, close to buildings and other facilities. Soil movement can induce damage if piles are driven close to existing structures. Difficult to apply when it is required to cross stiff soil layers or layers with cobbles or boulders.
Non displacement piles	Bored piles with permanent steel casing	Recommended when it is required to cross very soft soil layers. Reduced disturbance of the soil properties. Large pile cross sections may be adopted. Can cross layers with obstacles (cobbles or boulders). Can be used when rock sockets are required.	The steel casing reduces shaft unit resistance. Higher cost. Base resistance may be compromised when installed in soils with artesian pressure.
	Bored piles concreted against the ground	Large pile cross sections may be adopted. Large depths can be reached. Can cross layers with obstacles (cobbles or boulders). Can be used when rock sockets are required.	Longer construction time. Base resistance may be compromised when installed in soils with artesian pressure.
	Continuous flight auger piles	Fast construction.	Limited dimension of the cross section and, above all, limited pile length because the height of the hollow-stem auger must be at least equal to the pile length, and also because introduction of the steel reinforcement cage into fresh concrete is difficult for longer piles. The quality and integrity of the pile is greatly dependent on the skills of the contractor (operations such as auger withdrawal can seriously affect the integrity of the pile since concrete can become contaminated with soil). Base resistance may be compromised when installed in soils with artesian pressure.

and the base resistance, R_b. The mobilization of both resistances is the result of a complex pile–soil interaction.

To better illustrate the load transfer mechanism, consider a pile installed in a homogeneous half-space subjected to an increasing compressive load, V, as shown in Figure 7.5a. Figure 7.5c shows the typical load–settlement graph for a pile under such load conditions.

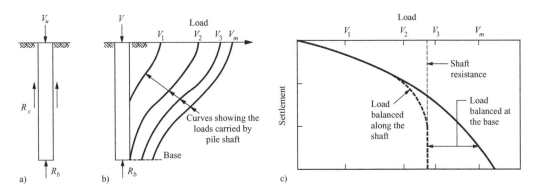

Figure 7.5 Mechanism of load transfer for piles subjected to compressive axial forces: a) and b) load transfer mechanism; c) load-settlement graphs.

In the latter figure, the change in shaft resistance mobilization with the pile settlement is also depicted. Due to the axial pile deformability, settlement is highest at the pile head and diminishes along the pile length. Therefore, the mobilization of tangential stresses along the shaft is also larger close to the pile head, decreasing with depth, as shown in Figure 7.5b. As can be seen in Figure 7.5c, when the applied load is low, the settlement is low and the balance condition is achieved only by shaft resistance mobilization.

For subsequent loading stages, which induce further increases in pile settlement, the shaft resistance mobilization continues to increase, and base resistance starts to be mobilized. After a certain loading stage, the shaft resistance is fully mobilized, and further loading increments must be balanced by the development of base resistance.

This effect is justified by the nature of the pile–soil interaction mechanisms. Shaft resistance is due to friction/adhesion phenomena along the pile shaft, which can be mobilized for very small soil–shaft relative displacements (typically, a few millimetres). Conversely, base resistance mobilization requires much greater settlements, which can be as large as 20% of the pile diameter for non-displacement piles.

The previous considerations apply to geotechnical scenarios, where no pronounced contrast exists between the soil enveloping the shaft and the base. However, it is also frequently the case where the pile tip is resting on a much stiffer material (e.g. unweathered rock) than the overlying layers. Under these conditions, the pile tip practically does not settle and the shaft resistance mobilization will be highly dependent on the pile axial stiffness.

As a conclusion, under service conditions where the load is considerably lower than the ultimate state design load, the contribution of the shaft resistance to the global balance is typically very substantial. On the other hand, for ultimate limit states, the balance usually demands the contribution of the base resistance.

7.3.2 Influence of the construction process on pile axial resistance

Contrary to what happens in shallow foundations, pile construction plays a relevant role in its bearing capacity, as it may seriously change the properties of the surrounding soil. In fact, soil properties evaluated prior to pile construction are distinct from the properties after pile installation. Moreover, the level of soil disturbance is also complex, since it depends on the pile installation technique, soil properties, pile properties, etc. (Coduto, 2001).

Regarding displacement piles, significant changes in surrounding soil properties occur because piles push the soil aside as they are driven into the ground, thus increasing the horizontal stresses in comparison with the at-rest stress state.

When piles are driven into clayey soils, the soil around the pile is remolded within a zone that extends to at least one pile diameter from the pile. If the soil is saturated, the resulting pore pressure in the vicinity of the pile may reach, or even locally exceed, the total overburden pressure (Hansbo, 1994). This high pore pressure dramatically decreases the soil shear strength. With the dissipation of excess pore pressure, pile load capacity increases. For single isolated piles, it is expected that the consolidation process does not take more than one month, whereas, when dealing with pile groups, it can last more than one year (Coduto, 2001). This phenomenon, usually called *pile setup* or *pile freeze*, may have implications for the construction schedule, since full bearing capacity becomes time dependent.

The excess pore pressure generated by pile driving in sandy soils dissipates very quickly, so that the full pile bearing capacity develops almost instantaneously. However, this does not mean that pile installation does not affect granular soil properties! The pile penetration implies particle rearrangement, resulting in a denser structure in loose soils and increasing the bearing capacity.

The soil alteration induced by non-displacement pile installation is less evident, since the pile is inserted in a borehole previously created in the ground. Contrary to the situation with displacement piles, the temporary hole needed for construction allows a horizontal stress relief, which is dependent on the construction technique. For instance, the reduction of horizontal stresses is almost negligible when dealing with bored piles with permanent casing, but it can be significant when fluid stabilization is adopted. In addition, when a pile is driven into the ground, the soil around the tip is highly densified due to the impact and vibration required for pile penetration. This effect, favorable for the mobilization of base resistance, is absent when dealing with non-displacement piles.

7.4 ULTIMATE RESISTANCE OF A SINGLE AXIALLY LOADED PILE

7.4.1 Bearing capacity

The pile bearing resistance for compressive loads is equal to the sum of the base resistance, R_b, and the shaft resistance, R_s. Notice that part of these resistances is spent to equilibrate the pile self-weight, W, and therefore the bearing capacity is equal to:

$$R_c = R_b + R_s - W \tag{7.1}$$

If the pile supports tensile loads, the base resistance cannot be mobilized, therefore the tensile capacity is equal to:

$$R_t = R_s + W \tag{7.2}$$

Alternatively, when performing safety verification, the pile self-weight can be considered to be a load and added to the load from the structure.

The pile base resistance is equal to:

$$R_b = q_b A_{\text{base}} \tag{7.3}$$

where q_b is the unit base resistance and A_{base} is the pile base cross section.

The pile shaft resistance, R_s, results from the mobilization of the shear resistance at the soil–shaft interface and is given by:

$$R_s = \int_0^D q_s(z)C(z)dz \cong \sum_i q_{s;i} A_{s;i} \tag{7.4}$$

where q_s is the unit shaft resistance and C is the pile perimeter, and both may vary with depth, z. The soil profile can be divided into layers and the shaft resistance is computed by evaluating a constant unit shaft resistance in each layer, $q_{s;i}$, and multiplying it by the corresponding shaft area, $A_{s;i}$. The determination of q_b and q_s can be performed by analytical methods, semi-empirical methods and load tests.

7.4.2 Assessment from analytical methods

7.4.2.1 Base resistance

The pile unit base resistance, q_b, may be determined based on the bearing capacity theory for shallow foundations, dealt with in Chapter 6. As the length/diameter (D/B) ratio for piles is much larger than for shallow foundations, the third term in Equation 6.15 becomes negligible when compared with the first two. Therefore, the bearing capacity expression simplifies to:

$$q_b = c'N_c' + q'N_q' \tag{7.5}$$

where N_c' and N_q' are the bearing capacity factors, modified to account for the increase of resistance due to the depth and to the shape of the failure surfaces near the pile base, respectively, c' is the effective cohesion, and q' is the vertical effective stress at the pile tip depth.

Figure 7.6 presents a summary of the methods proposed by different authors to determine N_q' as a function of the angle of shearing resistance (Terzaghi, 1943; De Beer, 1945; Brinch Hansen, 1951, 1961b; Meyerhof, 1953; Skempton et al., 1953, Caquot-Kérisel, 1956; Berezantzev et al., 1961, Vesić, 1963). The influence of the shape of the failure surfaces is notoriously large, and the value of N_q' can vary by one order of magnitude, for example, when the method proposed by Terzaghi (1943) is compared with the one by De Beer (1945).

Vesić (1977) concluded that the resistance of the soil at the base of the pile depends not only on the shear strength but also on the ground stiffness. The general failure mode would occur only for shallow foundations placed near the ground surface on dense soils, and all other cases would correspond to a punching failure, the resistance of which can be evaluated through the cavity expansion theory. Figure 7.7 depicts this mode of failure: a conical wedge, with a base angle $\psi = \pi/4 + \phi'/2$, moves with the pile tip and pushes laterally the radial-shear zone II into the plastic zone III.

Vesić considered that the base resistance depends on the mean normal effective stress:

$$\bar{\sigma}' = \frac{\sigma_v' + 2\sigma_h'}{3} = \frac{q'(1 + 2K_0)}{3} \tag{7.6}$$

and rewrote Equation 7.5 as:

$$q_b = c'N_c' + q'[(1 + 2K_0)/3]N_\sigma \tag{7.7}$$

where

$$N_\sigma = \frac{3}{3 - \sin\phi'} \cdot \exp\left((\pi - \phi') \cdot \tan\phi'\right) \cdot \tan^2\left(\frac{\pi}{4} + \frac{\phi'}{2}\right) \cdot I_{RR}^{\left(\frac{4\sin\phi'}{3(1+\sin\phi')}\right)} \tag{7.8}$$

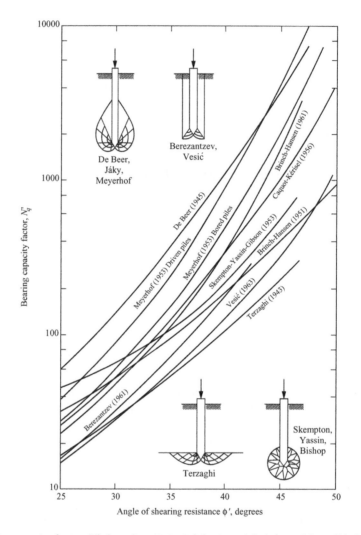

Figure 7.6 Bearing capacity factor, N'_q, based on various failure models (adapted from Kézdi, 1975).

The influence of the soil stiffness is introduced by the reduced rigidity index of the soil, I_{RR}. The rigidity index, I_R, is the ratio between the shear modulus and the shear strength, and is computed at the depth of the pile tip:[1]

$$I_R = \frac{G}{c' + \sigma'_v \tan\phi'} = \frac{E}{2(1+v)(c' + \sigma'_v \tan\phi')} \tag{7.9}$$

As the behavior of the soil in the plastic zone III, near the pile tip, is influenced by the volumetric strain, ε_v, the reduced rigidity index is equal to:

$$I_{RR} = \frac{I_R}{1 + I_R \varepsilon_v} \tag{7.10}$$

---

[1] In the original report, Vesić (1977) uses the mean normal stress, $\bar{\sigma}'$, instead of the vertical effective stress, σ'_v, to compute I_R. Many of the subsequent authors use the latter.

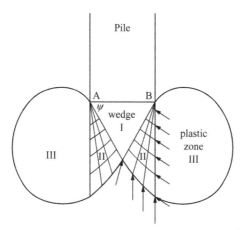

Figure 7.7 Punching failure mode at the pile tip (Vesić, 1977; reproduced with permission from the National Academy of Sciences, courtesy of the National Academies Press, Washington, D.C.).

The values of N'_q computed as:

$$N'_q = [(1 + 2K_0)/3]N_\sigma \tag{7.11}$$

are depicted in Figure 7.8. Kulhawy et al. (1983) reviewed this procedure, maintained the failure mode and rigidity index approach and proposed a slightly different formulation where

$$N'_q = N_q \, \xi_{qr} \, s_q \, d_q \tag{7.12}$$

The bearing capacity factor, N_q, has already been presented in Equation 6.1; ξ_{qr} is the corrective factor introduced to account for the dependence of the failure mode on the rigidity

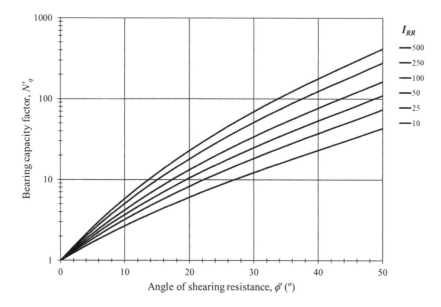

Figure 7.8 Bearing capacity factor N'_q (Vesić, 1977).

index, and s_q and d_q are the shape and depth corrective factors, respectively. The value of the critical rigidity index I_{RC}, that separates the generalized failure mode from the punching failure mode, is:

$$I_{RC} = 0.5 \exp\left[2.85 \cot an\left(\frac{\pi}{4} - \frac{\phi'}{2}\right)\right] \tag{7.13}$$

When the reduced rigidity index I_{RR} is greater than I_{RC}, a generalized failure mode is predicted, and $\xi_{qr} = 1$. Otherwise, the corrective factor is computed as:

$$\xi_{qr} = \exp\left[-3.8 \tan\phi' + \frac{(3.07 \sin\phi')(\log 2I_{RR})}{1 + \sin\phi'}\right] \tag{7.14}$$

The corrective factors s_q and d_q are equal to:

$$s_q = 1 + \tan\phi' \tag{7.15}$$

$$d_q = 1 + 2\,\tan\phi'\left(1 - \sin\phi'\right)^2\,\tan^{-1}\left(\frac{D_b}{B}\right) \tag{7.16}$$

where D_b is the depth of the pile tip. Figure 7.9 presents N'_q according to the Kulhawy et al. (1983) approach and considers $D_b/B = 5$. It can be easily demonstrated that larger values of this ratio have a limited influence on the resulting N'_q.

When the pile tip is embedded in clay, the short-term base resistance is controlled by the soil undrained behavior and is computed through a total stress analysis. Equation 7.5 is written as:

$$q_b = c_u N'_c + q N'_q \tag{7.17}$$

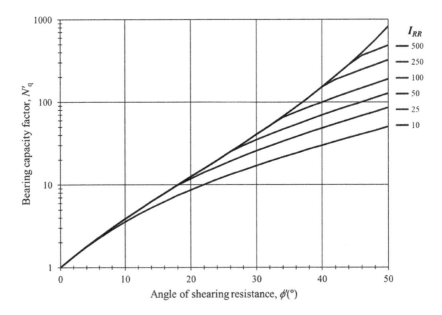

Figure 7.9 Bearing capacity factor N'_q (Kulhawy et al., 1983).

where c_u is the undrained shear strength, q is the total vertical stress at the pile tip depth, and the bearing capacity factor is $N'_q = 1$. The factor N'_c can be computed as:

$$N'_c = N_c \xi_{cr} s_c d_c \tag{7.18}$$

where $N_c = 5.14$. As stated previously, the rigidity correction factor ξ_{cr} is equal to unity if $I_{RR} > I_{RC}$ and, if the opposite occurs, the correction factor for a square or circular pile is equal to:

$$\xi_{cr} = 0.44 + 0.60 \log I_{RR} \tag{7.19}$$

For undrained conditions and a square or circular cross section, the corrective factors s_c and d_c are equal to:

$$s_c = 1.2 \tag{7.20}$$

$$d_c = 1 + 0.33 \tan^{-1}\left(\frac{D_b}{B}\right) \tag{7.21}$$

Considering $D_b/B > 5$ and a general failure mode, the resulting bearing capacity factor is $N'_c = 9$ (Skempton, 1951) and the expression for the unit base resistance becomes:

$$q_b = 9c_u + q \tag{7.22}$$

7.4.2.2 Shaft resistance

The pile shaft resistance, R_s, results from the mobilization of the shear resistance at the soil-shaft interface, q_s. In effective stress analysis:

$$q_s = c_a + K\sigma'_v \tan\delta \tag{7.23}$$

where c_a is the adhesion, K is the horizontal earth pressure coefficient, and δ is the angle of friction at the shaft–soil interface. The range of values for K and δ, summarized by Kulhawy et al. (1983), are presented in Tables 7.3 and 7.4. The influence of the construction process on the horizontal stress may be observed in the range of values proposed for K. The roughness of the pile shaft varies with the pile material and constructive process. Precast concrete piles have smoother shaft surfaces than bored piles concreted against the soil, partially compensating for the stress reduction caused by boring.

Table 7.3 Coefficient of horizontal earth pressure (Kulhawy et al., 1983).

Pile type	K/K_0
Driven piles with jetting	0.5 to 0.7
Bored piles	0.7 to 1.0
Piles driven with small soil displacement	0.8 to 1.3
Piles driven with large soil displacement	1.0 to 2.0

Table 7.4 Ratio δ/ϕ' (Kulhawy et al., 1983).

Interface	δ/ϕ'
Sand/rough concrete (bored pile)	1.0
Sand/smooth concrete (driven pile)	0.8 to 1.0
Sand/rough steel (corrugated steel)	0.7 to 0.9
Sand/smooth steel	0.5 to 0.7
Sand/timber	0.8 to 0.9

For piles installed in clayey soils, the calculation is performed in total stress analysis, and the adhesion is computed as a fraction of the undrained shear strength:

$$q_s = \alpha c_u \tag{7.24}$$

The adhesion factor, α, has been determined by pile load tests and, as shown in Figure 7.10, there is a wide range of values. It can be seen that α decreases with increased undrained shear strength ratio. For driven piles, API (2005) adopted the following expressions, proposed by Randolph and Murphy (1985):

$$\alpha = \frac{1}{2\left(c_u/\sigma'_{v0}\right)^{0.5}} \quad \text{for } \frac{c_u}{\sigma'_{v0}} < 1 \tag{7.25}$$

$$\alpha = \frac{1}{2\left(c_u/\sigma'_{v0}\right)^{0.25}} \quad \text{for } \frac{c_u}{\sigma'_{v0}} > 1 \tag{7.26}$$

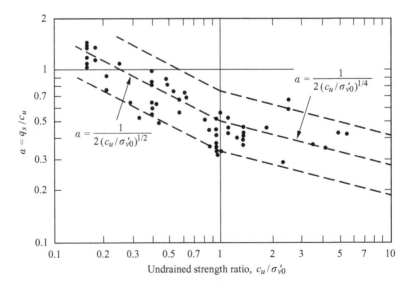

Figure 7.10 Values of the adhesion factor $\alpha = q_s/c_u$ (Randolph and Murphy, 1985; used with permission of Offshore Technology Conference).

7.4.3 Assessment from semi-empirical methods based on field tests

7.4.3.1 General aspects

The equations presented in the previous section have shown a wide scatter of results and often are not able to include the influence of the construction procedures on the bearing capacity. It is not only the values of N'_q which vary widely with the theoretical methods, but bored piles also often exhibit a much smaller (30 to 50%) base resistance than do driven piles. These difficulties are augmented when the soil parameters are determined by correlations of *in situ* test results, which may increase the results scatter. An alternative to these expressions is to correlate the results from field tests directly with the shaft and base unit resistances measured in pile load tests. In the following sections, some of these methods will be presented.

7.4.3.2 Method based on CPT and PMT results (French or LCPC method)

Cone penetrometer (CPT) and pressuremeter (PMT) test results can be used to estimate the bearing capacity of piles. Bustamante and Gianeselli (1981) correlated CPT tip resistance, q_c, and the limit pressure, p_L, from PMT with base and shaft resistances determined in pile load tests, and proposed the method detailed in MELT (1993). This method has been continuously updated and the most recent version can be found in AFNOR (2012). Soil type is classified, based on CPT or PMT results, as shown in Table A7.1.1, and piles are organized into 20 types and eight classes, according to the construction method (Table A7.1.2). The prediction method will be presented in the current section but comprehensive reading of AFNOR (2012) is recommended before its application is used in design.

7.4.3.2.1 Base resistance based on CPT

The base resistance based on CPT is given by:

$$q_b = k_c \cdot q_{ce} \tag{7.27}$$

where k_c is the penetrometer base resistance coefficient (Tables A7.1.3 and A7.1.4) and q_{ce} is the equivalent cone resistance. For a homogeneous layer, the equivalent cone resistance is determined by:

$$q_{ce} = \frac{1}{b + 3a} \int_{D-b}^{D+3a} q_{cc}(z)\,dz \tag{7.28}$$

$$a = \max\left\{\frac{B}{2}; 0.5m\right\} \tag{7.29}$$

$$b = \min\{a; h\} \tag{7.30}$$

where D is the depth of the pile tip, b and $3a$ are the thicknesses of the ground above and below the pile tip considered in the calculation, respectively, $q_{cc}(z)$ is the corrected cone resistance profile, B is the pile width or diameter, and h is the embedded length in the bearing layer.

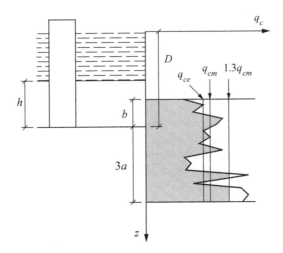

Figure 7.11 Computation of the equivalent cone resistance, q_{ce}.

The corrected cone resistance profile $q_{cc}(z)$ is obtained by computing the average cone resistance, q_{cm}, along the $b + 3a$ length:

$$q_{cm} = \frac{1}{b + 3a} \int_{D-b}^{D+3a} q_c(z)\,dz \tag{7.31}$$

and removing the values greater than 1.3 q_{cm}, as shown in Figure 7.11.

Table A7.1.3 shows how the value of coefficient, k_c, depends on the effective penetration depth, D_{ef}, which is computed as:

$$D_{ef} = \frac{1}{q_{ce}} \int_{D-h_D}^{D} q_c(z)\,dz \tag{7.32}$$

where h_D is equal to 10B. If $D_{ef}/B \leq 5$, the value of k_c is computed as shown in Table A7.1.3. For $D_{ef}/B > 5$, the maximum value of k_c is shown in Table A7.1.4.

7.4.3.2.2 Shaft resistance based on CPT

The unit shaft resistance based on CPT at depth z is determined as a function of the cone resistance, q_c, soil, and pile type:

$$q_s(z) = \alpha_{\text{pile-soil}} f_{\text{soil}} \left[q_c(z) \right] \leq q_{s\max} \tag{7.33}$$

where the values of the dimensionless parameter, $\alpha_{pile\text{-}soil}$, are presented in Table A7.1.5. The value of f_{soil} depends on the soil type and cone resistance as depicted in Figure 7.12. The analytical expression of these curves can also be expressed as:

$$f_{\text{soil}}(q_c) = (a q_c + b)\left(1 - e^{-c q_c}\right) \tag{7.34}$$

where the coefficients a, b and c are presented in Table A7.1.6. The limit shaft resistance, q_{smax}, is presented in Table A7.1.7.

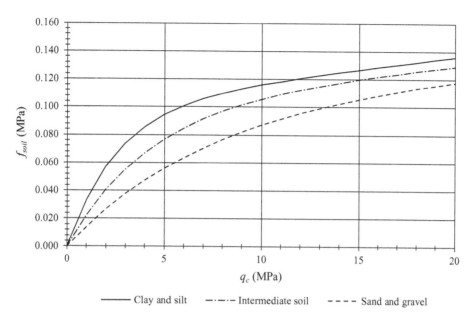

Figure 7.12 Determination of f_{soil}.

7.4.3.2.3 Base resistance based on PMT

Similarly to the procedure followed in the previous section, the base resistance based on PMT can be correlated to the results from pressuremeter tests:

$$q_b = k_p \cdot p_{le}^* \qquad (7.35)$$

where k_p is the pressuremeter base resistance coefficient and p_{le}^* is the equivalent net limit pressure computed as:

$$p_{le}^* = \frac{1}{b+3a} \int_{D-b}^{D+3a} p_l^*(z)\,dz \qquad (7.36)$$

where $p_l^*(z)$ is the profile of net limit pressure considered to be representative, and the remaining dimensions are displayed in Figure 7.11. The value of the coefficient, k_p, depends on the effective embedment depth, D_{ef}, which is computed with a procedure similar to Equation 7.32:

$$D_{ef} = \frac{1}{p_{le}^*} \int_{D-h_D}^{D} p_l^*(z)\,dz \qquad (7.37)$$

The coefficient, k_p, ranges from unity to k_{pmax} (see Table A7.1.8), where the higher value requires that effective embedment depth be greater than $5B$. If $D_{ef}/B \leq 5$, the interpolation is equal to:

$$k_p(D_{ef}/B) = 1.0 + (k_{pmax} - 1.0)(D_{ef}/B)/5 \qquad (7.38)$$

7.4.3.2.4 Shaft resistance based on PMT

The unit shaft resistance based on PMT is computed as:

$$q_s(z) = \alpha_{\text{pile-soil}} \, f_{\text{soil}} \left[p_l^*(z) \right] \le q_{smax} \tag{7.39}$$

where the dimensionless parameter, $\alpha_{pile\text{-}soil}$, is presented in Table A7.1.9. The function, f_{soil}, depends on soil type and limit pressure, as displayed in Figure 7.13, and may be computed by:

$$f_{\text{soil}}(p_l^*) = (a\, p_l^* + b)(1 - e^{-c p_l^*}) \tag{7.40}$$

where the coefficients a, b and c are presented in Table A7.1.10. The values of q_{smax} have been listed in Table A7.1.7.

7.4.3.3 Methods based on SPT results

7.4.3.3.1 General

There are several methods that use SPT blow count directly to estimate pile capacity (Schmertmann et al., 1970; Meyerhof, 1976; Aoki and Velloso, 1975; Décourt and Quaresma, 1978, Décourt, 1998). Most of these methods refer to driven piles and rely on local data bases. Their application to a different region should be performed with care, and local calibration through load tests is recommended.

7.4.3.3.2 Décourt and Quaresma (1978)

The method was originally developed for driven piles by Décourt and Quaresma (1978) and was later updated for other pile types (Décourt, 1998). The base resistance is given by:

$$q_b\,(\text{kPa}) = \alpha K N \tag{7.41}$$

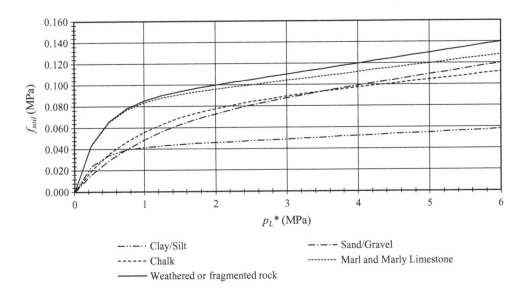

Figure 7.13 Determination of f_{soil}.

Table 7.5 Coefficient α (Décourt, 1998).

Soil type	Pile type Driven piles	Bored piles	Bored piles with bentonite	CFA	Micropiles	Injected under high pressure
Clay	1.0	0.85	0.85	0.3*	0.85*	1.0*
Intermediate soil	1.0	0.60	0.60	0.3*	0.60*	1.0*
Sand	1.0	0.50	0.50	0.3*	0.50*	1.0*

* Values suggested, based on a limited number of tests

Table 7.6 Coefficient K (Décourt and Quaresma, 1978).

Soil type	K (kPa)
Clay	120
Silty clay (residual soil)	200
Silty sand (residual soil)	250
Sand	400

Table 7.7 Coefficient β (Décourt, 1998).

Soil type	Pile type Driven piles	Bored piles	Bored piles with bentonite	CFA	Micropiles	Injected under high pressure
Clay	1.0	0.80	0.9*	1.0*	1.5*	3.0*
Intermediate soil	1.0	0.65	0.75*	1.0*	1.5*	3.0*
Sand	1.0	0.50	0.60*	1.0*	1.5*	3.0*

* Values suggested, based on a limited number of tests

where N is the SPT blow count at the pile toe, α is a coefficient that depends on the pile type (Table 7.5) and soil conditions, and K depends on the soil type (Tables 7.6).

The unit shaft resistance is given by:

$$q_s \left(kPa \right) = 10\beta \left(N/3 + 1 \right) \tag{7.42}$$

where β is a coefficient from Table 7.7.

7.4.3.3.3 O'Neill and Reese (1999)

This method is not based purely on SPT blow count but uses N_{60} to estimate base resistance in cohesionless soil, and its common use in several parts of the world justifies its presentation here. O'Neill and Reese (1999) proposed the following empirical expression to estimate base resistance for bored piles (drilled shafts) in granular soil:

$$q_b \left(kPa \right) = 57.5 N_{60} \leq 2900 \ kPa \tag{7.43}$$

where N_{60} is the average blow count between one diameter above and two diameters below the pile tip.

The base resistance was determined, considering pile failure when the settlement reaches 5% of the pile diameter, and the expression is valid for $N_{60} \leq 50$. Higher values may be used if load tests are performed to support them. Bored piles can be constructed with a large diameter, and the 5% criterion may lead to excessive settlement. To solve this, the authors proposed a reduction of capacity for base diameters, B_b, greater than 1.20 m:

$$q_{br} = \frac{1.20}{B_b} q_b \tag{7.44}$$

Alternatively, a load–settlement analysis may be performed.

The shaft resistance is computed for drained conditions with Equation 7.23 and it can be expressed as the β *method* where the horizontal stress coefficient is combined with the pile–soil angle of friction:

$$q_s = K \tan \delta \, \sigma_v' = \beta \, \sigma_v' \tag{7.45}$$

The factor β can be determined, computing K and δ independently, but it has also been evaluated empirically. For bored piles installed in sand with $N_{60} \geq 15$, O'Neill and Reese (1999) limited the unit shaft resistance to $q_s \leq 190$ kPa and proposed:

$$\beta = 1.5 - 0.245\sqrt{z} \qquad 0.25 \leq \beta \leq 1.20 \tag{7.46}$$

When $N_{60} < 15$, the following expression applies:

$$\beta = \frac{N_{60}}{15}\left(1.5 - 0.245\sqrt{z}\right) \qquad 0.25 \leq \beta \leq 1.20 \tag{7.47}$$

For bored piles installed in cohesive soils, the undrained unit base resistance is given by:

$$q_b = c_u N_c^* + q \leq 3800 \text{ kPa} \tag{7.48}$$

where the bearing capacity factor, N_c^*, depends on the undrained shear strength at the pile base, as shown in Table 7.8. For intermediate values of c_u, the factor may be determined by interpolation. The undrained shear strength, c_u, should be taken as the average value determined from the pile tip to two diameters below. If the pile embedment in the clay layer, D/B, is less than 3, the base resistance must be reduced, according to the equation:

$$q_b = \frac{2}{3}\left[1 + \frac{1}{6}\left(\frac{D}{B}\right)\right]c_u N_c^* + q \tag{7.49}$$

In order to keep settlement within acceptable limits for piles with diameters larger than 1.90 m, O'Neill and Reese (1999) proposed a reduction of q_b, according to the following equations:

$$q_{br} = F_r q_b \tag{7.50}$$

$$F_r = \frac{2.5}{\psi_1 B_b + 2.5\psi_2} \leq 1.0 \tag{7.51}$$

$$\psi_1 = 0.28B_b + 0.083\left(L / B_b\right) \tag{7.52}$$

$$\psi_2 = 0.065\sqrt{c_u} \tag{7.53}$$

The unit shaft resistance is computed by the α method, as shown in equation 7.24, imposing a limit of $q_s \leq 260$ kPa. O'Neill and Reese (1999) adopted $\alpha = 0.55$ if $c_u \leq 150$ kPa, and $\alpha = 0.45$ if $c_u \leq 250$ kPa. Intermediate values can be determined by linear interpolation.

Table 7.8 Bearing capacity factor for bored
piles (O'Neill and Reese, 1999).

c_u (kPa)	N_c^*
25	6.5
50	8.0
≥ 100	9.0

7.4.3.4 Pile resistance in hard soil/soft rock

The transition between soil and rock is continuous, and it is difficult to define a limit for the material behavior between soil and rock. The concept of Intermediate Geomaterial (IGM) has been proposed by O'Neill et al. (1996) to consider materials that behave as either hard soil or soft rock. Notice that this definition is not necessarily in agreement with geological interpretation but depends solely on the shear strength. Tip embedment in IGM is only possible for bored piles, where rock drilling tools are required to open the rock socket. IGM may be defined as:

- cohesive or cemented material, such as clay shales or mudstones, with uniaxial compressive strength, q_u, from 500 kPa to 5000 kPa.
- cohesionlesss material with $N_{60} > 50$.

For cohesive IGM and rock, with Rock Quality Designation, $RQD = 100\%$, with tip embedment of at least $1.5B$:

$$q_b = 2.5 q_u \tag{7.54}$$

If $70\% \leq RQD \leq 100\%$, with closed and nearly horizontal joints:

$$q_b = 4830 (q_u)^{0.51} \tag{7.55}$$

where q_u is expressed in MPa and q_b is determined in kPa.

If the IGM is jointed, with joints primarily horizontal, typical of sedimentary rock masses, spaced at least 300 mm apart, and with apertures not exceeding 6 mm, the unit base resistance may be computed as:

$$q_b = 3 q_u K_{sp} d \tag{7.56}$$

in which:

$$K_{sp} = \frac{3 + \dfrac{s_v}{B}}{10 \sqrt{1 + 300 \dfrac{t_d}{s_v}}} \tag{7.57}$$

$$d = 1 + 0.4 \frac{D_s}{B} \leq 3.4 \tag{7.58}$$

where s_v is the vertical spacing of the discontinuities, B is the rock socket diameter, t_d is the aperture of the discontinuities, and D_s is the depth of the rock socket embedment. The shaft resistance for cohesive IGM may be computed as:

$$q_s = \alpha\, q_u \tag{7.59}$$

Rowe and Armitage (1987) proposed the adhesion factor α for clean sockets with side undulations between 1 mm and 10 mm deep and less than 10 mm wide:

$$\alpha = 0.6\left(q_u\right)^{-0.5} \tag{7.60}$$

and for clean sockets with sidewall undulations greater than 10 mm deep and 10 mm wide:

$$\alpha = 0.75\left(q_u\right)^{-0.5} \tag{7.61}$$

In Equations 7.60 and 7.61, q_u is expressed in MPa.

When using the allowable stress design method, the adopted factors of safety are typically 2.5 to 3.0.

7.4.4 Assessment from axial load tests

7.4.4.1 General

The scatter of bearing capacity values predicted by analytical methods has already been discussed. Semi-empirical methods often predict values with a wide range of variation, and it is recognized that the construction procedure has a strong influence on the pile bearing capacity. These facts explain the need for load testing of piles, in order to determine the safety level, and potentially improve the design. There are three main types of load tests: static load (applied to the top or the pile toe), rapid load testing (Bermingham and Janes, 1989), and dynamic load testing. In the following sections, the static and dynamic load tests will be briefly presented.

7.4.4.2 Conventional static load test (compression)

In a static load test, hydraulic jacks are used to apply load to the pile head, reacting against a frame supported by other piles or anchors, as shown in Figure 7.14. In some cases, kentledge systems may be used, where dead weights on platforms provide the reaction. The load is controlled with load cells, and the pile top settlement is measured with displacement transducers fixed on reference beams.

When the assessment of base and shaft resistance distribution is required, internal instrumentation such as strain gauges or tell tales are installed prior to concreting. By measuring axial strain at several positions along the pile length, the axial force is determined, as shown in Figure 7.15, and its variation, divided by the shaft area between two adjacent sections, yields the average unit shaft resistance. Load tests should be performed on piles with the same diameter and length as the production piles, but, in some cases, where the loads required to test large-diameter piles are very high, this is not possible and the diameter of the test piles can be smaller. It is recommended that the test pile diameter should be at least 50% of the production piles.

Figure 7.15a illustrates the setup of a long pile, instrumented with ten levels of retrievable strain gauges, installed in alluvial soil. The axial force distribution along the pile length

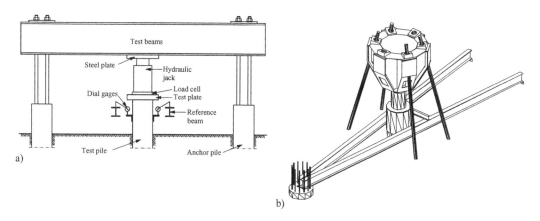

Figure 7.14 Reaction systems for static load test: a) reaction frame supported by piles; b) ground–anchor reaction system.

is shown in Figure 7.15b, and it can be observed how the shaft and base resistances were mobilized for each load increment.

The determination of the failure load is performed with one of several possible criteria (Fellenius, 2019). For small- diameter piles ($B \leq 600$ mm), Davisson (1972) defined failure as the intercept of the load-settlement curve with a line parallel to the elastic deformation line, as shown in Figure 7.16. The settlement at failure s_f is equal to:

$$s_f \left(\text{mm} \right) = \frac{QL}{AE} + 4.0 + 0.008B \tag{7.62}$$

Eurocode 7 recommends the definition of pile failure as when the settlement reaches 10% of the pile diameter. In some cases, when that value is not reached, a careful extrapolation of the load-settlement curve may be performed.

Figure 7.16 shows the application of Davisson's criterion to the results of a static load test on a 800-mm diameter Continuous Flight Auger (CFA) pile.

7.4.4.3 Bi-directional (Osterberg) load test

The bi-directional load test does not require a reaction system, which is a considerable advantage in comparison with the conventional static load test. A hydraulic jack is installed at or near the pile tip, as shown in Figure 7.17. When it is expanded, an upward force pushes the top segment of the pile out of the ground, while the lower segment compresses the soil. Shaft- and base-mobilized resistances are measured separately, having the same value but being linked to different displacements. This limitation can be minimized when two or more bi-directional cells are used at different elevations.

When one of the segments mobilizes its capacity, the test is concluded and the load-settlement curve is obtained by combining the two individual curves, as shown in Figure 7.18. For each selected displacement, the upward and downward forces are added and the curve is determined.

7.4.4.4 Assessment from high-strain dynamic load tests

On a high-strain dynamic load test, the pile is instrumented with two pairs of strain transducers and accelerometers, close to the pile head, as shown in Figure 7.19. A pile driving

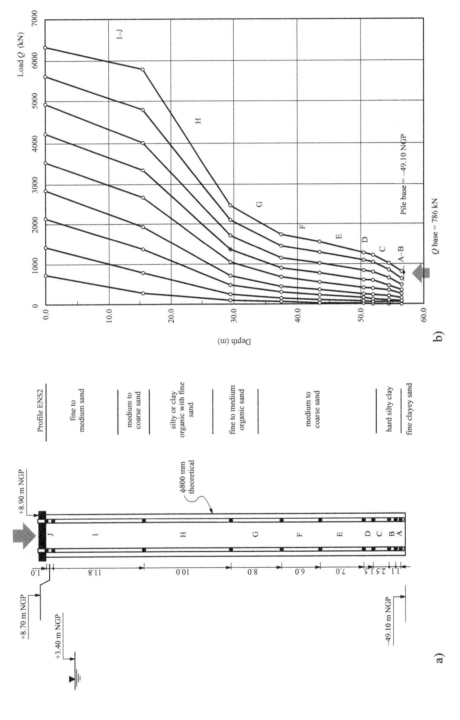

Figure 7.15 Pile static load test (Bridge over River Tagus in Santarém): a) instrumentation with retrievable strain gauges; b) axial force distribution along the pile length, for each load incremente (courtesy of Estradas de Portugal).

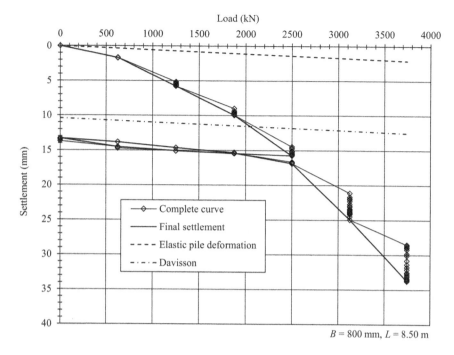

Figure 7.16 Load–settlement curve of a static test on a 800-mm diameter, 8.50-m long, CFA pile.

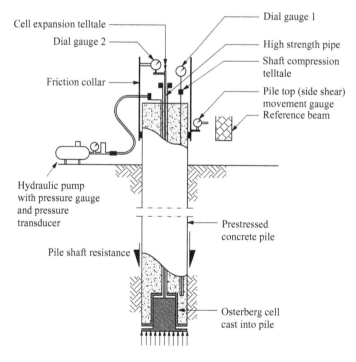

Figure 7.17 Bi-directional test setup.

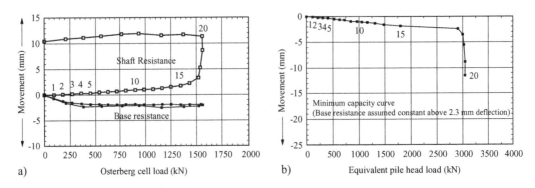

a)   b)

Figure 7.18 Bi-directional test: a) test results; b) total load–settlement curve (Hannigan et al., 2006).

hammer or an equivalent system is used to apply a series of blows and, for each blow, the axial strain and acceleration are measured and the axial force and particle velocity are derived. By interpreting the data through 1-D wave propagation theory, it is possible to evaluate the static pile capacity.

The differential equation governing wave propagation in linear elastic homogeneous bars, without resistance, is:

$$\frac{\partial^2 u}{\partial t^2} - v_c^2 \frac{\partial^2 u}{\partial x^2} = 0 \qquad (7.63)$$

where u is the displacement, x is the coordinate along the bar, t is the time, and v_c is the pressure wave propagation velocity in a linear elastic medium, with Young modulus E and mass density ρ, given by:

$$v_c = \sqrt{\frac{E}{\rho}} \qquad (7.64)$$

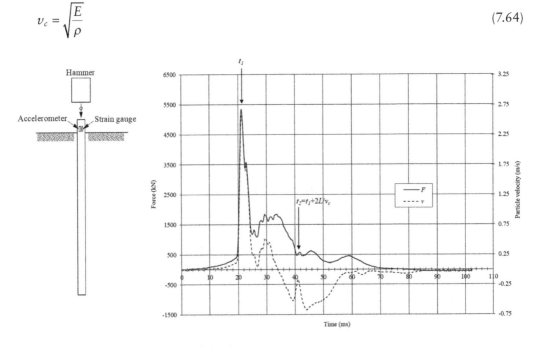

Figure 7.19 Readings from a dynamic pile load test.

In concrete, v_c ranges from 3000 to 4000 m/s and in steel it is equal to 5100 m/s. If there is no resistance to the wave propagation, the force $F(t)$ in a given control section and the particle vibration velocity, $v = \partial u/\partial t$, are related by the pile impedance ($Z = EA/v_c$):

$$F(t) = \frac{EA}{v_c} \cdot v(t) \tag{7.65}$$

where A is the cross-sectional area.

The ground resistance, including static and damping components, can be introduced in Equation 7.63, but the closed-form solution is rather complex to obtain. Alternatively, a numerical solution has been proposed by Smith (1960), which is the fundamental reference of the current analysis procedure. Smith divided the pile into segments, with a given mass and connected with springs and dashpots. The soil static and dynamic resistances are introduced as springs and dashpots, as shown in Figure 7.20.

A simplified solution to determine the pile capacity was developed by Goble and Rausche (1970) and is known as the CASE method. It uses the force and velocity readings for each blow at two instants: t_1, when the pile is hit and maximum force is measured, and t_2, when the wave reflected at the pile tip returns to the instrumented section, traveling twice the pile length below the gauge location, L:

$$t_2 = t_1 + 2L / v_c \tag{7.66}$$

In this method, damping is considered to occur solely at the pile tip. The total dynamic resistance (RTL) is equal to:

$$RTL = \frac{1}{2}\left[F(t_1) + F(t_2)\right] + \frac{1}{2}\left[v(t_1) - v(t_2)\right]\frac{EA}{c} \tag{7.67}$$

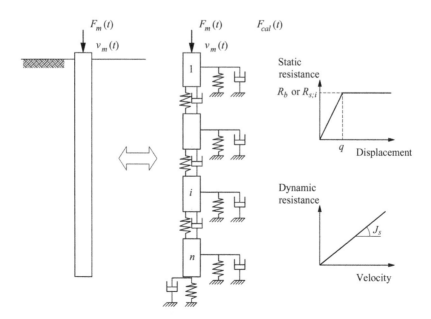

Figure 7.20 Signal matching model to interpret a dynamic load test.

Table 7.9 Indicative values of damping coefficient, J_c (Pile Dynamics, 2004).

Soil type (at tip)	J_c
Clean sand	0.10–0.15
Silty sand	0.15–0.25
Silt	0.25–0.40
Silty clay	0.40–0.70
Clay	0.70 or higher

and the static resistance, R_c, is then computed as:

$$R_c = RTL - J_c \left[v(t_1) \frac{EA}{c} + F(t_1) - RTL \right] \tag{7.68}$$

where J_c is a dimensionless damping coefficient of the soil near the pile tip, with indicative values listed in Table 7.9. These values may be calibrated locally, with signal matching techniques or with static load tests.

A more complete method, used to interpret a high-strain dynamic load test and determine the static resistance, involves performing a signal matching procedure. Software packages such as CAPWAP (Rausche et al., 1985) or TNOWAVE (Reiding et al., 1988) are routinely used for this purpose. In this method, the pile is modeled as a set of masses connected by springs and dashpots, as shown in Figure 7.20. Soil–pile interaction is introduced with a set of bi-linear soil springs and dashpots with constant damping coefficients. Each spring has constant stiffness and is limited by the shaft resistance or base resistance at that elevation. The damping coefficient, J_s, is distinct from the one used in the CASE method, J_c, which is dimensionless. In the signal matching procedure, the Smith (1960) damping factor, J_s, is used and the dynamic resistance is determined by multiplying the damping factor by the particle velocity.

In the analysis, the measured velocity, $v_m(t)$, is inputted at the pile top and the resulting computed force at the same section, $F_c(t)$, is compared with the measured force, $F_m(t)$. Through an iterative procedure, the spring stiffness and maximum resistance values, together with the damping coefficient, J_s, are tuned until a good match is obtained. The static capacity is determined as the sum of all the spring maximum resistances. The final model with the tuned springs can also be used to compute a curve of the static load *versus* displacement relationship.

Being an inversion process, with multiple admissible solutions from a mathematical point of view, the results must be checked to ensure that the properties of the springs are consistent with the ground conditions. It is also recommended that static load tests should be used to confirm the dynamic test results.

Dynamic tests are routinely performed during pile driving and are also becoming more frequently used with bored piles. Eurocode 7 includes specific clauses for this type of load test.

7.5 A NOTE ON NEGATIVE SKIN FRICTION

The mobilization of stresses along the pile shaft requires the development of pile–soil relative displacement. When the pile settles relative to the surrounding ground, upward shear stresses develop along the shaft–soil interface, corresponding to the shaft resistance mobilization

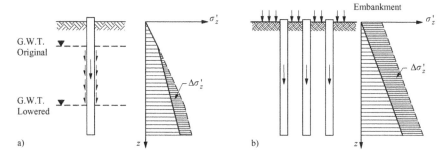

Figure 7.21 Generation of negative skin friction: a) water table dewatering; b) consolidation due to surface loads applied after pile installation.

explained above. However, there are some situations where the opposite occurs, i.e., the surrounding soil settles more than the pile. Thus, the shear stresses that develop along the shaft have a downward direction, i.e., they have the same direction as the compressive load applied on the pile head.

This effect is called negative skin friction and the resulting forces are called *drag forces*. In this case, the force from the structure plus the one induced by negative skin friction must be balanced through the mobilization of base resistance, as well as shaft resistance along the horizons where the pile settles more than the surrounding soil. The location where this force balance occurs is called the *neutral plane* and corresponds to the point where the soil and pile settlements are equal (Fellenius, 1972, 2019)

Negative skin friction loads can be generated fundamentally when the pile crosses soft soils and the tip is resting in a stiffer bearing stratum. In such a scenario, there is potential to develop negative skin friction in the following situations (Figure 7.21):

i) settlement of large areas of soft clays resulting from the increase of vertical effective stresses, due to a regional lowering of the water table (Figure 7.21a);
ii) consolidation of soft layers, due to embankment construction or surcharge application on the ground surface after pile installation (Figure 7.21b) (Fleming et al., 2009).

The assessment of this phenomenon is relevant for several reasons, namely:

i) additional compressive loads, also known as *drag forces*, are generated in the piles, which can generate large pile settlement (*downdrag*);
ii) in situations where the thickness of the strata is not uniform, negative skin friction can cause differential settlements;
iii) for very long and slender piles, the increase of axial load can lead to structural failure due to excessive compressive stress.

The effect of negative skin friction may be observed in Figure 7.22. The soil and pile settlements are shown in Figure 7.22a. From the ground surface to the depth of the neutral plane, the soil settles more than the pile and the negative skin friction, q_n, increases the compressive axial force (Figure 7.22b). From the depth of the neutral plane to the pile base, the inverse occurs. Figure 7.22c depicts the change of the axial force and resistance on the pile. The curve of the ultimate capacity is drawn starting at the pile base, with the base resistance, R_b, and adding the shaft resistance, R_s, along the pile length, to obtain the pile capacity R_c. The load applied to the pile head is Q. When the drag force is added to Q, the axial force

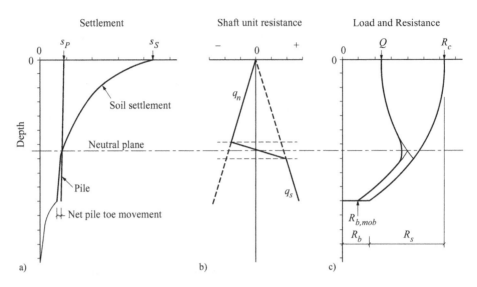

Figure 7.22 Generation of negative skin friction: a) unit shaft resistance; b) load-resistance evolution with depth; c) soil and pile settlement evolution with depth (adapted from Fellenius, 2019).

increases until the neutral plane is reached. At that point, there is force equilibrium and the soil and pile settlements are equal. The figure also shows that, for this load, the base resistance mobilized, $R_{b,mob}$, is smaller than the ultimate base resistance. Its mobilization depends on the net pile toe movement.

It should be noticed that the negative skin friction cannot cause a geotechnical ultimate limit state by exceeding the pile bearing capacity. If the pile settles due to the drag force, the pile length suffering negative skin friction is reduced and the neutral plane rises, warranting additional positive shaft resistance and therefore maintaining the force equilibrium.

The additional load caused by negative skin friction may lead to a service limit state as it increases the pile settlement, and the ability of the structure to withstand this displacement needs to be verified. The structural resistance of the pile needs to be checked with the maximum axial force.

7.6 SINGLE LATERALLY LOADED PILE

7.6.1 Basic phenomenology of the soil–pile interaction for rigid and slender piles

As shown in Figure 7.1c, piles can also be subjected to horizontal loads or to moments applied at their heads. In the past, this issue was often sidestepped by adopting pile groups where some piles were installed at an angle from the vertical in order to absorb the horizontal loads through the mobilization of axial stress. Piles installed in this manner are usually known as batter piles or racker piles. However, due to difficulties found in the installation process, this practice has been abandoned and nowadays vertical piles are more usual. Moreover, the demands for construction of earthquake-resistant or -resilient structures imply the option for piles with larger cross sections, which have a greater capacity to support lateral loads.

A pile subjected to lateral loading constitutes a classical soil–structure interaction problem, where the deflection of the pile and the lateral resistance of the soil are interdependent

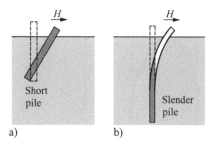

Figure 7.23 Failure modes of laterally loaded piles: a) short piles; b) slender piles.

(Reese et al., 2005). In general, laterally loaded piles can be divided into two major categories:

i) short or rigid piles;
ii) slender or flexible piles.

The failure mechanism of a short pile is primarily governed by soil properties, i.e., the soil fails before depleting the structural capacity of the pile (Figure 7.23a). This is one of the soil–structure interaction problems where the structural unit behaves as a rigid body. Conversely, the failure mechanism of a slender pile is governed by the structural capacity of the pile, i.e., the pile fails prior to the soil, as depicted in Figure 7.23b.

Classification of the pile as short (or stiff) or slender (or flexible) is dependent on several parameters, with the most relevant being: i) pile bending stiffness; ii) pile length; iii) soil stiffness. For engineering purposes, it is usual to assume that, if the ratio D/B is larger than 25, the collapse mechanism is governed by the structural properties, i.e., the pile is classified as flexible.

In regular scenarios, since piles are constructed to reach a deep bearing layer, they are usually long and slender. In the context of the present book, only the case of slender piles is treated. Details about behaviour of short piles can be found in reference books devoted to pile foundation engineering (Poulos and Davies, 1980; Hansbo, 1994; Fleming et al. 2009).

7.6.2 Structural analysis of slender piles

7.6.2.1 Soil–structure interaction

Laterally loaded slender piles constitute a classical soil–structure interaction problem, where the movements and flexural stresses of the pile depend on the lateral stress mobilized in the surrounding soil, and *vice versa*. Thus, geotechnical and structural aspects cannot be separated, demanding an integrated analysis.

Figure 7.24 shows response diagrams of a pile, as a function of depth, when subjected to a generic lateral load (and moment) applied to its head. The lateral load/moment applied to the pile head induces pile deflection and, consequently, lateral pressures are mobilized in the soil in order to counteract the pile movement.

Due to the mobilization of large lateral reaction pressures close to the pile head, pile deflection decreases with depth, and possibly reaches null value at some depth. However, since stresses are not null at that depth, pile deflection changes direction, and soil reaction is oppositely mobilized. Obviously, the magnitude of the diagrams presented above is dependent on several aspects such as: i) pile bending stiffness; ii) soil properties; iii) pile boundary conditions.

As in other soil–structure interaction problems, a comprehensive analysis can only be achieved using advanced numerical modelling techniques, such as tridimensional non-linear

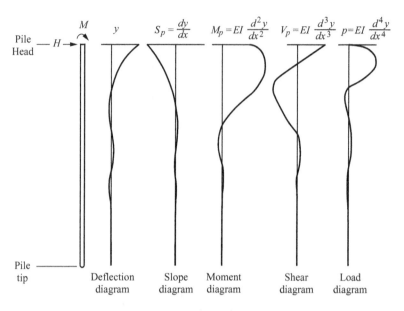

Figure 7.24 Diagrams of displacement, slope, and internal forces for a laterally loaded pile.

finite element method. However, the usage of such advanced resources is not desirable in regular applications, since they are very time consuming and demand knowledge of input parameters that are not easy to quantify. Thus, alternative approaches have been developed, such as the *P-y* method (Reese, 1984). A simplified version of the *P-y* method can be used if the soil is assumed to be homogeneous, elastic, and linear. Under such conditions, the problem is treated using closed-form solutions based on the theory of a beam on an elastic foundation, as proposed by Winkler (1867).

7.6.2.2 Winkler theory: analytical solutions

Consider a beam of infinite length (with width B and bending stiffness EI) subjected to a known load diagram, $Q(z)$, and resting on an elastic medium, simulated through a set of springs, as shown in Figure 7.25. The spring stiffness can be computed based on the sub-grade reaction coefficient,[1] k, expressed in units of force per cubic length (FL^{-3}), as:

$$k = \frac{p}{y} = \frac{P}{By} \tag{7.69}$$

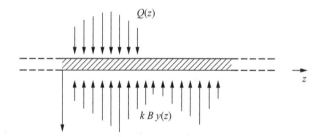

Figure 7.25 Beam on elastic foundation subjected to external loading.

[1] This coefficient is also known as the subgrade modulus.

where p is the soil pressure (FL^{-2}), P is the soil reaction force per unit length (FL^{-1}), and y is the associated displacement. The reaction modulus or Winkler coefficient, k', in units of (FL^{-2}), may be expressed as:

$$k' = \frac{P}{y} = kB \tag{7.70}$$

If discrete springs are used, their stiffness, K, is given by

$$K = k'\Delta L \tag{7.71}$$

where ΔL is the spring spacing along the beam length.

The establishment of a relationship between the beam deflection, $y(z)$, the bending and shear stresses, and the soil reaction, $P(z)$ (force per unit length), is achieved through successive derivatives regarding the spatial variable z.

The differential equation that describes the problem is given by:

$$EI\frac{\partial^4 y(z)}{\partial z^4} = Q(z) - kB \cdot y(z) \tag{7.72}$$

or,

$$EI\frac{\partial^4 y(z)}{\partial z^4} + k' \cdot y(z) = Q(z) \tag{7.73}$$

Among other resolution techniques, Equation 7.73 can be solved by splitting the solution into two parts, the first one referring to the homogeneous solution and the second to the particular solution.

The homogeneous equation is given by:

$$EI\frac{\partial^4 y(z)}{\partial z^4} + k' \cdot y(z) = 0 \tag{7.74}$$

which can be manipulated to give rise to:

$$\frac{\partial^4 y(z)}{\partial z^4} + 4\lambda^4 \cdot y(z) = 0 \tag{7.75}$$

where,

$$\lambda = \sqrt[4]{\frac{k'}{4EI}} = \frac{1}{l_e} \tag{7.76}$$

where l_e is the so-called *characteristic elastic length*.

The characteristic elastic length has a very clear physical meaning, since it gives an idea about the length of the pile along which the displacements, internal stresses, and pressures mobilized in the soil are significant. It should be noted that, if the pile is very flexible in relation to the ground, the elastic length will be small, which means that the internal stresses will affect the pile along a limited extension around the point of application of the horizontal load/moment.

The solution of Equation 7.75 can be written in the following form:

$$y(z) = e^{\lambda z}\left(C_1\cos(\lambda z) + C_2\sin(\lambda z)\right) + e^{-\lambda z}\left(C_3\cos(\lambda z) + C_4\sin(\lambda z)\right) \tag{7.77}$$

where C_1, C_2, C_3 and C_4 are integration constants that are defined as functions of the boundary conditions.

It should be noted that the solution presented in Equation 7.77 is based on the assumption of a beam of infinite length. However, from Figure 7.24, it becomes clear that, beyond a given depth, all the magnitudes represented tend toward negligible values. This length has a direct relationship with the elastic length of the pile. As such, it is reasonable to assume that, if the length of the pile is substantially larger than the elastic length, the assessment of the pile response can be performed by assuming it to be a semi-infinite beam. Therefore, integration constants C_1 and C_2 present in Equation 7.77 must be zero. As for the constants C_3 and C_4, they are defined from the pile head boundary conditions. Table 7.10 gives the closed-form solution in terms of deflection, shear force, and bending moment for the most typical boundary conditions at the pile head.

A key point for an accurate evaluation of the pile response is the correct assessment of the Winkler coefficient, k'. This coefficient is dependent not only on the soil stiffness but also on the pile properties. Actually, due to the soil non-linear behavior, the consideration of a constant value for k', independent of the strain level, corresponds to a rough approximation to the real behavior. Nevertheless, some authors propose that this parameter can be estimated based on elasticity solutions, i.e., assuming for the soil linear elastic behavior. In such conditions, the soil elastic parameters should be chosen by taking into account the expected strain range, adopting a procedure of equivalent linearization as already introduced in Chapter 6 for the estimation of the settlement of shallow foundations. By the way, a simple estimation of k' can be performed by manipulation of Equation 6.39 used for estimation of shallow foundation settlement in elastic homogeneous ground. Thus, after some mathematical manipulation, it is possible to find the following relationship:

$$k' = \frac{E}{\left(1-v^2\right)I_s} \tag{7.78}$$

where E is the linear equivalent soil Young modulus, v is the Poisson ratio, and I_s is a factor that takes into account boundary condition effects, such as the pile diameter.

For regular conditions, the denominator present in Equation 7.78 takes a value close to one (Bowles, 1997), so that, for engineering practice, it is common to assume that:

$$k' \approx E \tag{7.79}$$

A more elaborate approach for the assessment of the reaction modulus was proposed by Vesić (1961), where the pile bending stiffness and diameter are also taken into account:

$$k' = 0.65 \sqrt[12]{\frac{E B^4}{E_p I_P}} \frac{E}{1-v^2} \tag{7.80}$$

From the analysis of Figure 7.24 and also from the equations presented in Table 7.10, it is possible to conclude that the lateral pile–soil interaction is relevant only in the upper region of the pile, along 4 to 5 times the elastic length. Therefore, the Winkler coefficient should be selected by taking into account the soil stiffness distribution in that region.

7.6.2.3 P-y curves approach

In an attempt to overcome the limitations of the Winkler approach, and taking into account the possibilities provided by numerical analysis, more sophisticated formulations for lateral

Table 7.10 Closed-form expression for pile deflection, shear force, and bending moment when lateral loads or moments are applied at the pile head.

Boundary condition at the pile head	Lateral load applied on pile head (H)		Moment applied at the pile head (M)
	Rotation free	Rotation fixed	Rotation free
Deflection	$y(z) = \dfrac{H}{2\lambda^3 EI} e^{-\lambda x}\cos(\lambda z)$	$y(z) = \dfrac{\lambda H}{k'} e^{-\lambda x}\left(\sin(\lambda z) + \cos(\lambda z)\right)$	$y(x) = -\dfrac{M}{2\lambda^2 EI} e^{-\lambda x}\left(\sin(\lambda x) - \cos(\lambda x)\right)$
Shear force	$V(z) = -He^{-\lambda z}\left(\cos(\lambda z) - \sin(\lambda z)\right)$	$V(z) = -He^{-\lambda z}\left(\cos(\lambda z)\right)$	$V(z) = -2M\lambda e^{-\lambda z}\sin(\lambda z)$
Bending moment	$M(z) = -\dfrac{H}{\lambda} e^{-\lambda z}\sin(\lambda z)$	$M(z) = \dfrac{H}{2\lambda} e^{-\lambda z}\left(\cos(\lambda z) - \sin(\lambda z)\right)$	$M(z) = -Me^{-\lambda z}\left(\sin(\lambda z) + \cos(\lambda z)\right)$

pile–soil interaction have been presented. The most popular of these methods is usually designated the *P-y approach*. Actually, this method is no more than a generalization of the Winkler method, but assumes non-linear laws for the discrete springs that simulate the soil response. Usually, these laws are called *P-y* curves and are dependent on the pile displacement.

From a conceptual point of view, the problem can be solved using a numerical approach based on the finite element method. Figure 7.26 shows a schematic representation of this approach. The pile is usually simulated as a Timoshenko or Euler-Bernoulli beam, with the real properties of the pile (finite length). Each node of the beam is linked to a non-linear spring, the behavior of which is represented by the *P-y* curve. A schematic representation of typical *P-y* curves is given in Figure 7.27.

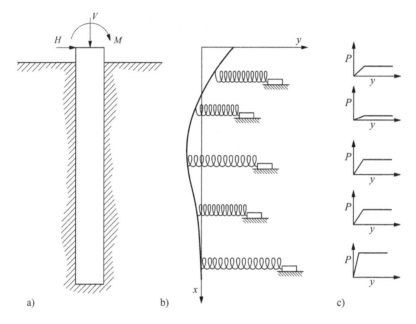

Figure 7.26 P-y analysis: a) pile and loading conditions; b) pile–soil model; c) P-y curves.

Ductile *P-y* curves, such as the curves depicted in Figure 7.27a, are typical of the response of sands under static loading. *P-y* curves of clays are more complex, as shown in Figure 7.27b, where stiff clays can exhibit brittle behavior. Although the shape of the *P-y* curve is

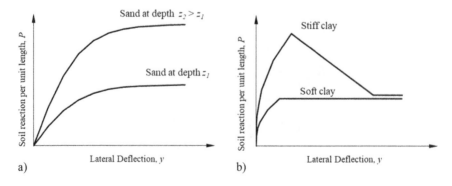

Figure 7.27 Schematic representation of typical P-y curves.

mostly influenced by soil properties, it depends also on the depth, pile width, water table location, and loading conditions (static or cyclic).

Due to the inherent complexity of the soil non-linear behavior, the definition of P-y curves is strongly supported on empirical correlations, derived from full-scale tests. Based on these, API (2005) proposed analytical expressions for definition of P-y curves for the most common types of soil. Alternatively, P-y curves can also be obtained by interpretation of full-scale lateral load tests, only viable in large projects, or by direct correlation from results of *in situ* tests, such as pressuremeter and dilatometer tests (Robertson et al., 1989).

7.7 PILE GROUPS

7.7.1 General aspects

When the loads imposed at the foundation level are not capable of being balanced by a single pile, a set of piles is installed in order to constitute a pile group. The piles are connected at the top by a very rigid element, the pile cap, which has the function of distributing the loads imposed by the structure over the different piles.

Considering the generalized loading system shown in Figure 7.28, composed of vertical and horizontal forces, as well as moments acting in both directions, it is common practice to develop the analysis of the axial and lateral pile response by an uncoupled approach, i.e.,

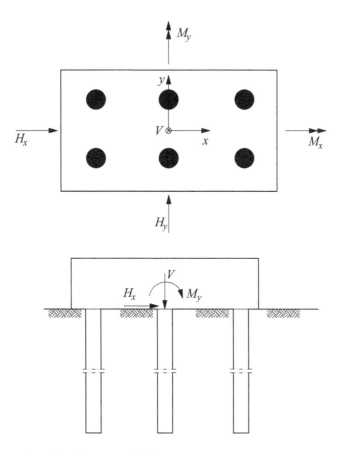

Figure 7.28 Determination of axial force on piles from a group.

assuming that the pile–soil interaction mechanisms are independent of the loading direction (axial or lateral). As can be understood, since the piles are spaced apart, the moments arising from the structure are balanced by the generation of increments (positive and negative) to the axial force applied to each of the piles. Therefore, the axial force on each pile is given by:

$$N_i = \frac{V}{n} + \frac{M_x y_i}{\sum_{j=1}^{n} y_j^2} + \frac{M_y x_i}{\sum_{j=1}^{n} x_j^2} \tag{7.81}$$

where N_i is the axial force in pile i, M_x, M_y are the moments and V the vertical force at the center of gravity at the base of the pile cap, and x_i and y_i are the coordinates of the center of gravity of the cross section of pile i.

When the pile group is subjected to horizontal loads, the distribution of the loads by the different piles is not uniform and it is a consequence of the pile–soil–pile interaction mechanism that will be analysed in Section 7.7.3.

Due to the development of a complex mechanism of load transfer, the response of a pile group can be quite different from that of a single isolated pile, giving rise to the so-called group effect. In following sections, this will be analyzed in detail.

7.7.2 Group effect in axially loaded piles

The interaction between piles belonging to a group and the surrounding soil is complex and the group bearing capacity can be different from the sum of the bearing capacity of the individual piles. Figure 7.29 presents a schematic representation of the increase of the soil loading due to the overlapping of soil stresses induced by each individual pile. As suggested in the same figure, the settlement of a pile group is also larger than the settlement of a single pile subjected to the same axial force as one of the piles from the group, since the volume of soil subjected to large stress increments is also larger.

The pile group bearing capacity, R_{cg}, is given by:

$$R_{cg} = E_g n R_c \tag{7.82}$$

where E_g is the group efficiency factor, n is the number of piles of the group and R_c is the axial bearing capacity of a single isolated pile.

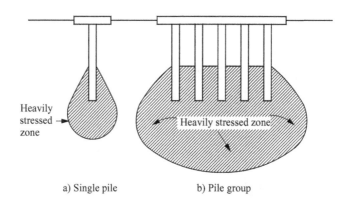

a) Single pile   b) Pile group

Figure 7.29 Heavily stressed zones near single piles and pile groups (Tomlinson and Woodward, 2007).

The group efficiency factor can be lower or greater than 1, depending on several issues, namely (Coduto, 2001):

i)   the number, length, diameter, arrangement, and spacing of piles;
ii) the relative contributions of shaft and base resistances to the ultimate bearing capacity;
iii) the sequence of pile installation, mainly in the case of displacement piles;
iv) the soil type;
v)   the interaction, if any, between the pile cap and the soil;
vi) the direction of the applied load (tension or compression).

The relative lack of experimental data available makes it difficult to validate theoretical models for the estimation of the group efficiency factor.[1] Despite that, there are some main trends that can be readily discerned:

i)   the group efficiency of displacement piles in loose sand is usually greater than 1 (Lo, 1967; Vesić, 1969). This is explained by the higher normal effective stress on the shaft and by the greater compaction of the soil around the piles driven in a group (Fleming et al., 2009);
ii) the group efficiency of displacement piles installed in clayey soils is generally lower than 1 due to soil remolding; another aspect that deserves attention is the time dependence of the group bearing capacity (see Section 7.3.2);
iii) for non-displacement piles, there are even fewer experimental data; accordingly to some authors (Das, 2004; Reese et al., 2005), the group efficiency of bored piles should be slightly lower than 1 for pile spacing less than 3B.

For close-spaced pile groups, an alternative failure mechanism can develop, as shown in Figure 7.30. The bearing capacity of the pile group is given by the sum of the lateral resistances along the block perimeter, with the base resistance being computed by taking into account the area corresponding to the base of the block. It should be noticed that this failure mechanism only governs the group bearing capacity when dealing with close-spaced pile groups installed in clay under undrained conditions. Therefore, the block bearing capacity can be estimated as follows:

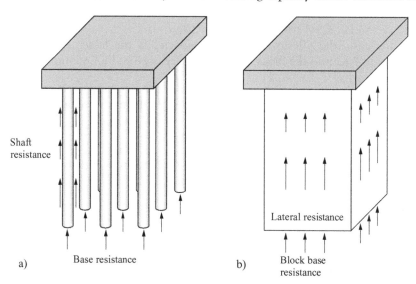

Shaft resistance

Base resistance

a)

Lateral resistance

Block base resistance

b)

Figure 7.30 Failure modes in pile groups: a) isolated single pile failure; b) block-failure (adapted from Reese et al., 2005).

---

[1] Experimental data is scarce because the required loads for pile groups are extremely high, making the testing difficult and expensive. Alternatively, small-scale centrifuge tests have provided useful insights about the topic.

$$R_{ug,block} = 2D(L_B + B_B)c_{u_1} + B_B L_B c_{u_2} N_c' \tag{7.83}$$

where D is the pile length, L_B and B_B are the horizontal cross-sectional dimensions of the block, c_{u1} is the value of the undrained shear strength along the pile shaft, and c_{u2} is the undrained shear strength at the base of the block.

The bearing factor, N_c', is given by:

$$N_c' = 5\left(1 + \frac{D}{5B_B}\right)\left(1 + \frac{B_B}{5L_B}\right) \le 9 \tag{7.84}$$

Obviously, if two distinct failure mechanisms are admissible, the group bearing capacity corresponds to the lowest value from Equations 7.82 and 7.83.

The following guidelines, presented by Hannigan et al. (2016) for driven piles, may be used for design purposes.

For driven piles in cohesionless soils:

i) For piles spaced more than three diameters apart and not underlain by a weak deposit, assume $E_g = 1$.

ii) Jetting or pre-drilling to pile installation should be avoided, since it can result in group efficiencies less than 1.

For driven piles in cohesive soils:

i) When $c_u < 95$ kPa and without pile cap resting on the ground, for pile spacing, s, equal to 2.5B, use $E_g = 0.65$; for $s > 6B$ use $E_g = 1$; linear interpolation should be used for intermediate center-to-center pile spacing.

ii) When $c_u < 95$ kPa, but with the pile cap in firm contact with the ground, $E_g = 1$ may be used if $s > 3B$.

iii) If $c_u > 95$ kPa, $E_g = 1$ may be used if $s > 3B$.

iv) In any case, the pile group bearing capacity is the lesser value obtained from Equation 7.82 and the block failure capacity is given by Equation 7.83.

As already mentioned, experimental results for non-displacement piles are even rarer than for driven pile groups. The following guidelines were presented by Loehr et al. (2011):

For bored piles in cohesionless soils:

The pile group bearing capacity can be estimated from Equation 7.82, and group efficiency factor should be selected as follows:

i) for $s = 2.5B$, take $E_g = 0.65$;

ii) for $s > 4B$, assume $E_g = 1$;

iii) linear interpolation may be applied for intermediate spacings.

The guidelines expressed above are applicable regardless of the contact conditions between the pile cap and the ground.

For bored piles in cohesive soils:

The guidelines presented above for groups of driven piles installed in cohesive soils are also applicable for bored piles.

7.7.3 Group effect in laterally loaded piles

When a pile group, with n piles, joined by a rigid cap, is subjected to a lateral force applied at cap level, it is observed that:

i) the pile head lateral displacement of all piles is equal to the cap displacement;
ii) the lateral displacement of the group, u_g, is greater than the displacement of a single isolated pile, with the same boundary conditions at the head, when subjected to a lateral load corresponding to a uniform load distribution;
iii) the distribution of the load by the piles is not uniform and depends on the location of each pile.

The above findings are the result of the development of pile–soil–pile interaction effects, since the displacement of a given pile contributes to the displacement of the remainder. When a pile group is submitted to a lateral load, each pile pushes the adjacent soil, creating a localized shear zone. With the increase of the load, the shear zone of each individual pile enlarges in order to mobilize the required soil reaction, and overlapping of effects occurs, as depicted in Figure 7.31. The overlapping effect between piles in different rows is called the *shadowing effect* and it is responsible for a lower mobilization of soil reaction in the piles located in the intermediate rows. Since the overlapping effect can also occur due to the proximity of piles of the same row (the *edge effect*), the piles located in the corners usually make a larger contribution to the balance of the global lateral load.

From the representation depicted in Figure 7.31, it is possible to conclude that the reaction offered by the piles located in the peripheral corners is only slightly lower than the reaction given by an isolated pile under the same conditions, whereas the piles located at the group interior provide a significantly lower reaction due to shadowing and edge effects. Since all piles suffer the same head displacement, and the reaction offered by the soil is dependent on pile location, piles in the external rows are subjected to greater shear forces than are piles located in the inner rows.

Experimental evidence of this interaction problem was presented by Brown et al. (1988) Rollins et al. (1998), among others. It should be highlighted that, in pile groups with a large number of piles, this effect can be very significant, and its omission can lead to serious underestimation of mobilized internal forces on the peripheral piles.

As in other geotechnical problems, the complexity of the phenomena under analysis can be understood in detail only through an approach based on a continuous formulation. However, it is also well known that such approaches, taking into account the 3-D character

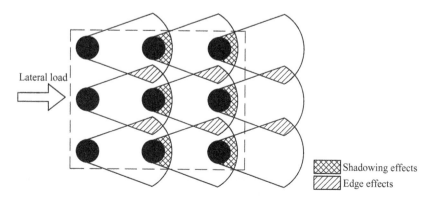

Shadowing effects
Edge effects

Figure 7.31 Laterally loaded pile group: shadowing and edge effects (Walsh, 2005).

of the problem and the soil non-linearity, are very demanding from the computational point of view, discouraging its use as a regular technique for design support. Therefore, simplified methods have been proposed, such as the method of influence factors, proposed by Poulos (1971), and the *P-multipliers* method proposed by Brown and Bollman (1993).

7.7.3.1 Method of the influence factors

The application of the method of influence factors is limited to linear-elastic analysis, since it is based on superposition of effects. The method can be faced as an extension of the Winkler model for the analysis of pile groups, introducing influence coefficients in order to take into account the pile–soil–pile interaction effects. The influence factor is a multiplier of the pile deflection due to the presence of other piles, subjected to similar loading conditions, in its vicinity. Thus, assuming K_t as the horizontal stiffness of an isolated pile, then the displacement of the pile head i belonging to a group with n equal piles is given by:

$$u_i = \frac{1}{K_t} \sum_{j=1}^{n} \alpha_{ij} H_j \tag{7.85}$$

where u_i is the horizontal displacement of the head of pile, α_{ij} is the influence factor between pile i and pile j, and H_j is the lateral load transmitted to pile j.

On the other hand, taking into account the expressions given in Table 7.10, it is easy to find that, for a pile with fixed rotation head, the horizontal stiffness is equal to:

$$K_t = \frac{k'}{\lambda} \tag{7.86}$$

Randolph (1981), through the calibration of the results predicted by the present method with those obtained by means of continuous three-dimensional analysis, proposed a set of analytical expressions for the calculation of the influence factors. Given the generality of practical situations, the most interesting is the expression for influence coefficients related to the horizontal displacement of the pile head with fixed rotation and lateral loading. In this context, and considering the ground to be an isotropic and homogeneous elastic medium, the author proposes the following expression for the evaluation of the influence factors:

$$\alpha_{ij} = 0.6 \zeta_{ij} \left(1 + \cos^2 \psi_{ij} \right) \tag{7.87}$$

where ψ_{ij} is the angle formed by the load direction and the hypothetical line that links pile i to pile j, and ζ_{ij} is a dimensionless coefficient given by:

$$\zeta_{ij} = \left[\frac{E_p}{G_s \left(1 + \frac{3 v_s}{4} \right)} \right]^{\frac{1}{7}} \frac{B}{2 s_{ij}} \tag{7.88}$$

where E_p is the pile Young's modulus, G_s and v are the soil shear modulus and Poisson ratio, respectively, B is the pile diameter, and, finally, s_{ij} is the center-to-center distance between piles i and j.

The distribution of horizontal loads is obtained from the prescription of displacement compatibility and load equilibrium at the pile heads. From a mathematical point of view, the load attributed to each pile is obtained by solving the following system of equations:

$$\begin{cases} u_i = u_j = u_g, \quad \forall i_{ij} \\ \displaystyle\sum_{i=1}^{n} H_i = H_g \end{cases} \tag{7.89}$$

Notice that the pile head displacement is given by Equation 7.85.

After assessing the load distribution by the different piles, the estimation of the pile response (in terms of displacements, internal forces, etc.) is a simple step. Actually, after solving the system of Equations 7.89, the head displacement (u_i, which is the same as u_g due to the displacement compatibility imposed by the pile cap) and the applied load (H_i) are known variables. Thus, the pile response can be obtained directly by application of expressions presented in Table 7.10, but considering a hypothetical Winkler coefficient, which depends on pile location. This hypothetical coefficient, k'_i, is obtained taking into account the following relationship:

$$k'_i u_i = \lambda_i H_i \tag{7.90}$$

where,

$$\lambda_i = \sqrt[4]{\frac{k'_i}{4 E_p I_p}} \tag{7.91}$$

Therefore, after some mathematical manipulation:

$$k'_i = \left(\frac{H_i^4}{4 E_p I_p u_i^4} \right)^{\frac{1}{3}} \tag{7.92}$$

A final comment must be presented about the suitability of the method. This method can be considered to be a balance between simplicity and accuracy. However, there are some drawbacks that the reader must be aware of when applying it:

i)   it is assumed that the soil is elastic, homogeneous, and linear;
ii) the comparison of experimental results with the results provided by the method shows that there is usually an overestimation of the loads in the peripheral piles;
iii) there is no distinction of load for piles located in symmetrical positions relative to the group center of gravity, i.e., the load attributed to front piles is the same as that for rear piles, the adherence of which to experimental evidence is dubious.

7.7.3.2 P-multipliers method

One way to evaluate the group response to applied lateral load is to modify the P-y curves to reflect the pile–soil–pile interaction effects. Brown and Bollman (1993) propose a procedure whereby the P-y curves for an isolated pile are modified as functions of the pile location. An illustration of this concept is depicted in Figure 7.32. Application of the method should follow these steps:

i)   select a P-y curve compatible with the soil and pile characteristics for a single pile;
ii) apply a P-multiplier, P_m, to the selected P-y curve, generating a revised P-y curve as shown in Figure 7.32. For $s = 3B$, the recommended P_m values are 0.8 for the lead row,

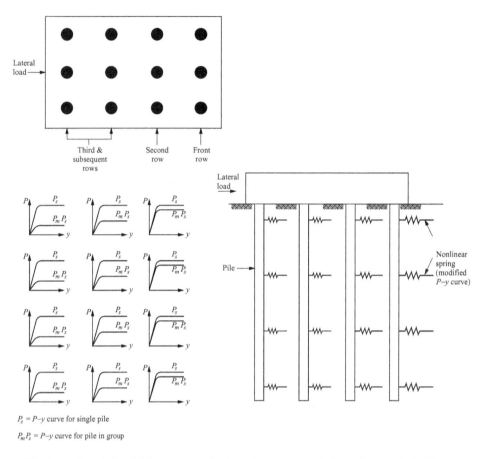

$P_s = P\text{–}y$ curve for single pile

$P_m P_s = P\text{–}y$ curve for pile in group

Figure 7.32 Illustration of *P*-multiplier concept for lateral group analysis (Hannigan et al., 2016).

0.4 for the second row, and 0.3 for the third and subsequent rows. For $s = 5B$, the recommended P_m values are 1.0 for the lead row, 0.85 for the second row, and 0.7 for the third and subsequent rows. For intermediate pile spacing, linear interpolation may be used;

iii) develop a numerical model (using the finite element method, for instance), where the piles are simulated by beam elements linked by a rigid cap and assuming the modified *P-y* curves, as functions of the pile location in relation to the load direction (see Figure 7.32).

Usually, all the piles of the group have the same structural configuration, so the pile design is based on the shear forces and bending moments mobilized in the critical row, i.e., the front row, as represented in Figure 7.32.

7.8 INTEGRITY PILE TESTING

7.8.1 General

Integrity tests are essential to control the quality of the foundations. The constructive process for driven or bored piles can cause pathologies that may hinder their capacity. Pile driving under hard conditions may crush concrete or yield steel if a very stiff layer is reached and the compression stresses exceed the material strength. When a soft layer follows a very stiff one, tension waves may be generated and crack the concrete. Bored piles may suffer from variations of cross section, from "necking", when the section is reduced due to soil caving into the borehole, to "bulging", due to section increase. The concreting process is also important, as soil can fall into the fresh concrete or segregation may occur. To identify these problems, the most common integrity tests are based on the propagation of waves generated by hammer blows, the sonic-echo tests, and the cross-hole tests. These are low-strain dynamic tests, as opposed to the high-strain dynamic load tests presented in Section 7.4.4.4.

7.8.2 Sonic-echo test

The sonic echo test is performed by applying a hammer blow at the pile top and registering the particle acceleration, as shown in Figure 7.33. Knowing the pile length, L, and the pressure wave speed in the pile material (as referred to in Section 7.4.4.4), concrete has a v_c of 3000 to 4000 m/s and steel has a v_c of 5100 m/s), the time required for the wave to travel to the pile tip and back is $t = 2L/v_c$. If reflections are detected earlier than expected, caused by changes in pile impedance ($Z = EA/v_c$), they are symptomatic of integrity problems, such as pile cracking, variations in cross section, or changes in concrete quality.

Figure 7.34 shows the records of an undamaged pile and another two with reflections affected by defects located in the upper part. Figure 7.34a shows a clear wave reflection of the undamaged pile. When the pile has a sharp change of impedance, caused by a change of cross section or a change in wave speed due to material defects, the record shows an early wave reflection. If the section decreases, the early reflection has the same signal as the downward wave (Figure 7.34b), and the opposite occurs when a sudden section increase is found (Figure 7.34c)

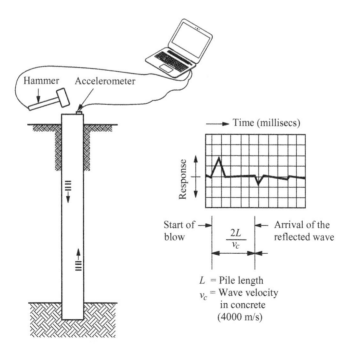

Figure 7.33 Sonic-echo integrity test (adapted from Sliwinski and Fleming, 1983; used with permission of ICE Publishing).

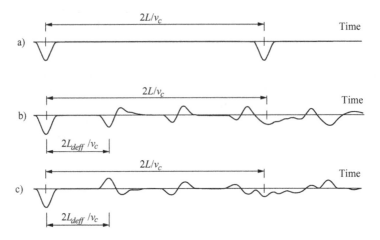

Figure 7.34 Results from pile integrity tests: a) undamaged pile; b) pile with reduction of section; c) pile with increase of section.

This test has limitations when the energy is insufficient to obtain a clear reflection, which may occur for long ($D/B > 30$) or large-diameter piles. It should also be noted that when the pile has a severe section reduction, the energy may not pass, and no interpretation is possible below that section. The analysis of a pile with variable sections is also difficult to perform.

7.8.3 Cross-hole test

The cross-hole test is performed by placing parallel pipes, tied to the longitudinal reinforcement bars, before concreting, with known distances. During the test, a wave generator and a receiver (geophone) are lowered and raised in two pipes and the time required for the wave to travel from one pipe to the other is determined. The resulting speed is compared with the value expected for the material. Defects are detected when the arrival time, and consequently the speed, are different from those expected. Figure 7.35 displays a typical setup and the zones tested. The number of pipes installed depends on the diameter of the pile, with a minimum of three pipes being used for small-diameter piles.

A tomography test may also be performed, where fixed geophones are placed in the pipes and a moving wave generator is used in one pipe.

7.9 INTRODUCTION OF SAFETY

7.9.1 General

When verifying the safety of a foundation, the designer must ensure that no ultimate limit state (ULS) is reached, such as the bearing capacity, and that no serviceability limit state (SLS) is exceeded. One SLS corresponds to excessive settlement that may damage the structure. When working stress design (WSD) is used, the adopted safety factors indirectly warrant that pile settlements are kept within acceptable values and, in most cases, no explicit settlement calculation is performed.

With the transition of codes and regulations from WSD towards Limit State Design (LSD), SLS should be verified. Nevertheless, as stated in Eurocode 7, in typical situations, where the piles are installed in medium-to-dense soil, and for tension piles, the safety requirements

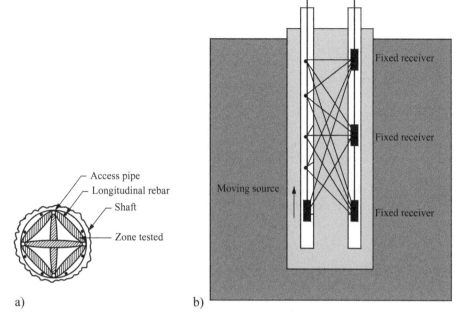

Figure 7.35 a) Cross-hole test setup; b) Pile tomography using cross-hole tests.

for ultimate limit state design are normally sufficient to prevent a serviceability limit state in the supported structure.

7.9.2 Global safety factors

Global factors of safety are applied to the ultimate pile capacity in order to determine the allowable load. They are traditionally used in allowable stress design procedures, where the applied axial force is computed for service load conditions, and characteristic load combinations are used.

The pile service load is then compared with the allowable value. For piles under compression:

$$Q_{all} = \frac{R_{c;ult}}{F} \tag{7.93}$$

$$Q_{all} = \frac{R_{b;ult}}{F_b} + \frac{R_{s;ult}}{F_s} \tag{7.94}$$

The values of the safety factor range from 2 to 6, depending on the ground characteristics, the existence of load tests, the type of loading (compression *vs.* tension), and the type of structure to be supported. A double verification may be imposed by using Equations 7.93 and 7.94 and adopting the lower allowable load value.

For piles loaded in tension:

$$Q_{all} = \frac{R_{s;ult}}{F_s} \tag{7.95}$$

7.9.3 Partial safety factors (Eurocode 7)

The verification of safety is performed for ULS. For a pile under compression load:

$$F_{c;d} \leq R_{c;d} \tag{7.96}$$

and for a pile under tensile force:

$$F_{t;d} \leq R_{t;d} \tag{7.97}$$

where the design axial compression, $F_{c;d}$, or tensile force, $F_{t;d}$, is determined with ULS load combinations, according to EN 1990, which were already presented in Section 6.2.7.2. The pile self-weight is added to the axial force.

The design value of the pile compressive resistance, $R_{c;d}$, is derived from the characteristic compressive resistance, $R_{c;k}$, as follows:

$$R_{c;k} = R_{b;k} + R_{s;k} \tag{7.98}$$

When it is not possible to determine separately the base and shaft resistances, such as on load tests without internal instrumentation, the design resistance, $R_{c;d}$ can be computed as:

$$R_{c;d} = \frac{R_{c;k}}{\gamma_t} \tag{7.99}$$

or, when that separation is possible, as:

$$R_{c;d} = R_{b;d} + R_{s;d} = \frac{R_{b;k}}{\gamma_b} + \frac{R_{s;k}}{\gamma_s} \tag{7.100}$$

The partial factors for total resistance, γ_t, base resistance, γ_b, and shaft resistance, γ_s, are defined in Eurocode 7 and summarized in Table 3.7. They depend on the pile type and the Design Approach. The characteristic resistances, $R_{b;k}$ and $R_{s;k}$, may be determined from static load tests, ground test results, or dynamic impact tests.

7.9.3.1 Ultimate compressive resistance from static load tests

If static load tests are performed on several piles with similar characteristics, the characteristic compressive resistance may be computed as:

$$R_{c;k} = \min\left\{ \frac{\left(R_{c;m}\right)_{\text{mean}}}{\xi_1} ; \frac{\left(R_{c;m}\right)_{\text{min}}}{\xi_2} \right\} \tag{7.101}$$

where ξ_1 and ξ_2 are correlation factors applied to the mean and the lowest of the measured resistances, $R_{c;m}$, respectively. The objective of these correlation factors is to introduce the variability of the ground conditions into the analysis. Their recommended values are presented in Table 7.11.

7.9.3.2 Ultimate compressive resistance from ground test results

Two approaches may be adopted: the "model pile" and the "alternative" or "ground model" approach. In the "model pile" approach, a pile compressive resistance is calculated for each

Table 7.11 Correlation factors ξ to derive characteristic values from static pile load tests (EN 1997-1:2004).

ξ for n^* =	1	2	3	4	≥ 5
ξ_1	1.40	1.30	1.20	1.10	1.00
ξ_2	1.40	1.20	1.05	1.00	1.00

* n = number of tests

test profile, using a semi-empirical method based on *in situ* or laboratory test results. After calculating one resistance, $R_{c;cal}$, per profile, the characteristic value is computed as:

$$R_{c;k} = R_{b;k} + R_{s;k} = \frac{R_{b;cal} + R_{s;cal}}{\xi} = \frac{R_{c;cal}}{\xi} = \min\left\{ \frac{\left(R_{c;cal}\right)_{mean}}{\xi_3} ; \frac{\left(R_{c;cal}\right)_{min}}{\xi_4} \right\} \qquad (7.102)$$

where ξ_3 and ξ_4 are correlation factors applied to the mean and the lowest of the calculated resistances, $R_{c;cal}$, respectively, and depend on the number of test profiles, n, and are applied, respectively:

- to the mean values:

$$\left(R_{c;cal}\right)_{mean} = \left(R_{b;cal} + R_{s;cal}\right)_{mean} = \left(R_{b;cal}\right)_{mean} + \left(R_{s;cal}\right)_{mean} \qquad (7.103)$$

- to the lowest values:

$$\left(R_{c;cal}\right)_{min} = \left(R_{b;cal} + R_{s;cal}\right)_{min} \qquad (7.104)$$

The values of the proposed correlation factors are given in Table 7.12.

In the "alternative" or "ground model" procedure, the characteristic values of the base and shaft resistances are computed, based on ground parameters:

$$R_{b;k} = q_{b;k} A_b \qquad (7.105)$$

$$R_{s;k} = \sum_i q_{s;i;k} A_{s;i} \qquad (7.106)$$

Note that, for calculation of the characteristic values of the unit base and shaft resistances, no partial factor is applied to the ground parameters (except for Design Approach 3) and no correlation factors ξ are applied to the resistances. In this "ground model" alternative procedure, a model factor, γ_{Rd}, greater than 1, defined in the National Annex, is used to

Table 7.12 Correlation factors ξ to derive characteristic values from ground test results (EN 1997-1:2004).

ξ for n^*=	1	2	3	4	5	7	10
ξ_3	1.40	1.35	1.33	1.31	1.29	1.27	1.25
ξ_4	1.40	1.27	1.23	1.20	1.15	1.12	1.08

* n = number of test profiles

increase the partial factors for base and shaft resistances. In this case, Equation 7.100 can be rewritten as:

$$R_{c;d} = R_{b;d} + R_{s;d} = \frac{R_{b;k}}{\gamma_{Rd}\gamma_b} + \frac{R_{s;k}}{\gamma_{Rd}\gamma_s} \tag{7.107}$$

7.9.3.3 Ultimate compressive resistance from dynamic impact tests

When dynamic impact tests are used to determine pile resistance, the characteristic value is given by:

$$R_{c;k} = \min\left\{\frac{\left(R_{c;m}\right)_{\text{mean}}}{\xi_5}; \frac{\left(R_{c;m}\right)_{\text{min}}}{\xi_6}\right\} \tag{7.108}$$

The correlation factors ξ_5 and ξ_6 are listed in Table 7.13.

Table 7.13 Correlation factors ξ to derive characteristic values from dynamic impact tests (EN 1997-1:2004).

ξ for *n	≥ 2	≥ 5	≥ 10	≥ 15	≥ 20
ξ_5	1.60	1.50	1.45	1.42	1.40
ξ_6	1.50	1.35	1.30	1.25	1.25

* n = number of tested piles
a The ξ-values in the table are valid for dynamic impact tests.
b The ξ-values may be multiplied by a model factor of 0.85, when using dynamic impact tests with signal matching.
c The ξ-values should be multiplied by a model factor of 1.10, when using a pile driving formula with measurements of the quasi-elastic pile head displacement during the impact.
d The ξ-values should be multiplied by a model factor of 1.20, when using a pile driving formula without measurements of the quasi-elastic pile head displacement during the impact.
e If different piles exist in the foundation, groups of similar piles should be considered separately when selecting the number n of test piles.

7.9.4 Load and Resistance-Factored Design (LRFD)

The Load and Resistance-Factored Design has been presented in Chapter 3. The verification of safety is performed under the following conditions:

$$P_u \leq \Phi R_n \tag{7.109}$$

where P_u is the factored normal load, computed as the sum of factored overall load effects, for a given load combination, Φ is the resistance factor, and R_n is the nominal normal load capacity.

ANNEX A7.I EVALUATION OF THE PILE AXIAL CAPACITY BY DIRECT APPLICATION OF PMT OR CPT RESULTS

The present annex provides the tables required by the LCPC method to estimate pile capacity based on the direct application of PMT or CPT results (AFNOR, 2012). The definition of soil classes is presented in Table A7.1.1.

Table A7.1.1 Definition of the soil classes.

	Soil class	p_l^* (MPa)	q_c (MPa)
Clay, silt	Very soft to soft	< 0.4	< 1.0
	Firm	0.7 to 1.2	1.0 to 2.5
	Stiff	1.2 to 2.0	2.5 to 4.0
	Hard	≥ 2	≥ 4.0
Sand, gravel	Very loose	< 0.2	< 1.5
	Loose	0.2 to 0.5	1.5 to 4
	Medium dense	0.5 to 1.0	4 to 10
	Dense	1.0 to 2.0	10 to 20
	Very dense	≥ 2	> 20
Chalk	Soft	< 0.7	< 5
	Weathered	0.7 to 3	5 to 15
	Unweathered	≥ 3	≥ 15
Marl and marly limestone	Soft	< 1	< 5
	Compact	1 to 4	5 to 15
	Very compact	≥ 4	≥ 15
Rock	Weathered	2.5 to 4	
	Fragmented	> 4	

Table A7.1.2 Definition of pile type (AFNOR, 2012).

Pile type	Group	Class
1 – Dry drilling	G1	1
2 – Drilling with bentonite	G1	1
3 – Drilling with permanent casing	G1	1
4 – Drilling with temporary casing	G1	1
5 – Dry drilling or with drilling fluid, with grooving	G1	1
6 – CFA simple or with double rotation	G1	2
7 – Screwed, concreted in situ	G1	3
8 – Screw pile with permanent tube	G1	3
9 – Driven, precast concrete or pre-stressed	G1	4
10 – Driven with casing (concrete, mortar, grout)	G2	4
11 – Driven with concrete (Franki)	G1	4
12 – Driven steel pipe, close ended	G1	4
13 – Driven steel pipe, open ended	G1	5
14 – Driven, H	G1	6
15 – Driven, H, grouted IGU or IRS	G2	6
16 – Driven sheet-piles	G1	7
17 – Micropiles type I (gravity pressure)	G2	1
18 – Micropiles type II (low pressure)	G2	1
19 – Piles or micropiles type III (high pressure)	G2	8
20 – Piles or micropiles type IV (high pressure with TAM)	G2	8

Table A7.1.3 Penetrometer capacity coefficient k_c for $D_{ef}/B \leq 5$.

Soil	
Clay and silt	$k_c \left(D_{ef} / B \right) = 0.3 + \left(k_{cmax} - 0.3 \right) \left(D_{ef} / B \right) / 5$
Intermediate soil	$k_c \left(D_{ef} / B \right) = 0.3 + \left(k_{cmax} - 0.3 \right) \left(D_{ef} / B \right) / 5$
Sand and gravel	$k_c \left(D_{ef} / B \right) = 0.1 + \left(k_{cmax} - 0.1 \right) \left(D_{ef} / B \right) / 5$
Chalk, marl and weathered or fragmented rock	$k_c \left(D_{ef} / B \right) = 0.15 + \left(k_{cmax} - 0.15 \right) \left(D_{ef} / B \right) / 5$

Table A7.1.4 Penetrometer capacity coefficient k_{cmax} for $D_{ef}/B > 5$.

Pile class	Clay (%CaCO$_3$ < 30%), silt	Intermediate soil	Sand, gravel	Chalk	Marl and marly limestone	Weathered or fragmented rock
1	0.40	0.30	0.20	0.30	0.30	0.30
2	0.45	0.30	0.25	0.30	0.30	0.30
3	0.50	0.50	0.50	0.40	0.35	0.35
4	0.45	0.40	0.40	0.40	0.40	0.40
5	0.35	0.30	0.25	0.15	0.15	0.15
6	0.40	0.40	0.40	0.35	0.20	0.20
7	0.35	0.25	0.15	0.15	0.15	0.15
8	0.45	0.30	0.20	0.30	0.30	0.25

Table A7.1.5 Determination of $\alpha_{pile\text{-}soil}$.

Pile type	Clay (%CaCO$_3$ < 30%), silt	Intermediate soil	Sand, gravel	Chalk	Marl and marly limestone	Weathered or fragmented rock
1	0.55	0.65	0.70	0.80	1.40	1.50
2	0.65	0.80	1.00	0.80	1.40	1.50
3	0.35	0.40	0.40	0.25	0.85	–
4	0.65	0.80	1.00	0.75	0.13	–
5	0.70	0.85	–	–	–	–
6	0.75	0.90	1.25	0.95	1.50	1.50
7	0.95	1.15	1.45	0.75	1.60	–
8	0.30	0.35	0.40	0.45	0.65	–
9	0.55	0.65	1.00	0.45	0.85	–
10	1.00	1.20	1.45	0.85	1.50	–
11	0.60	0.70	1.00	0.95	0.95	–
12	0.40	0.50	0.85	0.20	0.85	–
13	0.60	0.70	0.50	0.25	0.95	0.95
14	0.55	0.65	0.70	0.20	0.95	0.85
15	1.35	1.60	2.00	1.10	2.25	2.25
16	0.45	0.55	0.55	0.20	1.25	1.15
17	–	–	–	–	–	–
18	–	–	–	–	–	–
19	1.35	1.60	2.00	1.10	2.25	2.25
20	1.70	2.05	2.65	1.40	2.90	2.90

Table A7.1.6 Parameters to determine f_{soil}.

	Clay (%CaCO_3 < 30%), silt	Intermediate soil	Sand, gravel	Chalk	Marl and marly limestone	Weathered or fragmented rock
Curve	Q1	Q2	Q3	Q2	Q2	Q2
a	0.0018	0.0015	0.0012	0.0015	0.0015	0.0015
b	0.1	0.1	0.1	0.1	0.1	0.1
c	0.4	0.25	0.15	0.25	0.25	0.25

Table A7.1.7 Maximum unit shaft resistance, q_{smax}.

Pile type	Clay (%CaCO_3 < 30%), silt	Intermediate soil	Sand, gravel	Chalk	Marl and marly limestone	Weathered or fragmented rock
1	90	90	90	200	170	200
2	90	90	90	200	170	200
3	50	50	50	50	90	–
4	90	90	90	170	170	–
5	90	90	–	–	–	–
6	90	90	170	200	200	200
7	130	130	200	170	170	–
8	50	50	90	90	90	–
9	130	130	130	90	90	–
10	170	170	260	200	200	–
11	90	90	130	260	200	–
12	90	90	90	50	90	–
13	90	90	50	50	90	90
14	90	90	130	50	90	90
15	200	200	380	320	320	320
16	90	90	50	50	90	90
17	–	–	–	–	–	–
18	–	–	–	–	–	–
19	200	200	380	320	320	320
20	200	200	440	440	440	500

Table A7.1.8 Pressiometric coefficient k_{pmax} for a relative embedment $D_{ef}/B > 5$.

Pile class	Clay (%CaCO_3 < 30%), silt, intermediate soil	Intermediate soil, sand, gravel	Chalk	Marl and marly limestone	Weathered and fragmented rock
1	1.15	1.10	1.45	1.45	1.45
2	1.30	1.65	1.60	1.60	2.00
3	1.55	3.20	2.35	2.10	2.10
4	1.35	3.10	2.30	2.30	2.30
5	1.00	1.90	1.40	1.40	1.20
6	1.20	3.10	1.70	2.20	1.50
7	1.00	1.00	1.00	1.00	1.20
8	1.15	1.10	1.45	1.45	1.45

Table A7.1.9 Determination of $\alpha_{pile\text{-}soil}$.

Pile type	Clay (%CaCO$_3$ < 30%), silt, intermediate soil	Intermediate soil, sand, gravel	Chalk	Marl and marly limestone	Weathered and fragmented rock
1	1.10	1.00	1.80	1.50	1.60
2	1.25	1.40	1.80	1.50	1.60
3	0.70	0.60	0.50	0.90	–
4	1.25	1.40	1.70	1.40	–
5	1.30	–	–	–	–
6	1.50	1.80	2.10	1.60	1.60
7	1.90	2.10	1.70	1.70	–
8	0.60	0.60	1.00	0.70	–
9	1.10	1.40	1.00	0.90	–
10	2.00	2.10	1.90	1.60	–
11	1.20	1.40	2.10	1.00	–
12	0.80	1.20	0.40	0.90	–
13	1.20	0.70	0.50	1.00	1.00
14	1.10	1.00	0.40	1.00	0.90
15	2.70	2.90	2.40	2.40	2.40
16	0.90	0.80	0.40	1.20	1.20
17	–	–	1.20	–	–
18	–	–	–	–	–
19	2.70	2.90	2.40	2.40	2.40
20	3.40	3.80	3.10	3.10	3.10

Table A7.1.10 Parameters to determine f_{soil}.

	Clay (%CaCO$_3$ < 30%), silt, intermediate soil	Intermediate soil, sand, gravel	Chalk	Marl and marly limestone	Weathered and fragmented rock
Curve	Q1	Q2	Q3	Q4	Q5
a	0.003	0.01	0.007	0.008	0.01
b	0.04	0.06	0.07	0.08	0.08
c	3.5	1.2	1.3	3	3

ANNEX E7 EXERCISES

E7.1

Consider a precast concrete driven pile, with square cross section of 0.40 m×0.40 m, installed in a sandy soil as depicted in Figure E7.1. Soil properties are also indicated in the same figure.

Compute the pile bearing capacity, applying the analytical approach.

E7.2

A building is to be constructed over sandy ground, consisting of an upper loose layer which overlies a dense layer. Preliminary studies show that the piles should reach 12 m deep. The site geotechnical survey consisted of CPT tests. Representative results of these tests are depicted in Figure E7.2.

a) Considering that the piles will be driven and that they are 0.40 m×0.40 m in cross section, compute the pile bearing capacity through the LCPC method;
b) Assuming that continuous flight auger piles, with a diameter of 0.50 m, are selected, compute the pile bearing capacity through the LCPC method.

E7.3

Consider the pile group, formed by two concrete piles ($E_p = 30$ GPa), 500 mm in diameter, represented in Figure E7.3. The pile group is subjected to the loading cases of Table E7.3.

The ground properties are also depicted in Figure E7.3.

a) For loading case 1, compute the piles shear force and bending moment graphs. Note: ignore group effects.
b) For loading case 2, compute the maximum bending moment in the piles and the pile head displacement.

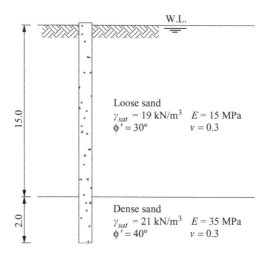

Loose sand
$\gamma_{sat} = 19$ kN/m^3 $E = 15$ MPa
$\phi' = 30°$ $\nu = 0.3$

Dense sand
$\gamma_{sat} = 21$ kN/m^3 $E = 35$ MPa
$\phi' = 40°$ $\nu = 0.3$

W.L.

15.0

2.0

Figure E7.1

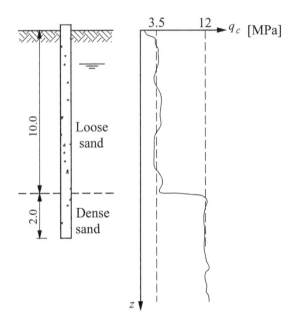

Figure E7.2

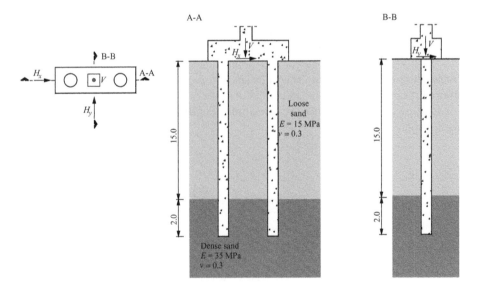

Figure E7.3

Table E7.3 Loading cases.

Loading case	V (kN)	H_x (kN)	H_y (kN)
1	1600	125	0
2	1600	0	100

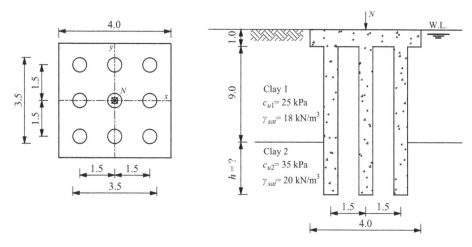

Figure E7.4

E7.4

Consider the pile group of Figure E7.4, with nine bored piles installed with temporary casing, 500 mm in diameter, spaced 1.50 m apart, and connected by a rigid pile cap. A vertical force, N, of 1000 kN is applied by the structure at the center of the pile cap.

Determine the pile embedment in the lower clayey soil required to warrant a global factor of safety, F = 3.0.

E7.5

The foundation of a pier of a special structure is formed by a pile group consisting of eight concrete piles (E_p = 30 GPa), 600 mm in diameter, connected by a rigid cap. Figure E7.5 shows the configuration of the pile group. The ground comprises an upper layer of medium dense sand, 15 m thick, overlying dense sand. The figure also displays the ground properties.

Together with the vertical load, the pile group is subjected to a horizontal load, H_x = 800 kN.

a) Taking into account the pile–soil–pile interaction through the influence coefficient method, compute:
   a1) the pile group lateral displacement and the lateral load distribution between the different piles;
   a2) the maximum bending moment in the piles subjected to the highest lateral load;
b) Compare the maximum bending moment computed in a2) with the values that are obtained if the pile–soil–pile interaction is ignored, i.e., assuming a uniform distribution of the horizontal load on the piles.

E7.6

Figure E7.6 shows the foundation of a wall of an industrial building. The geotechnical survey shows that the ground is composed of three main layers, corresponding to loose sand of

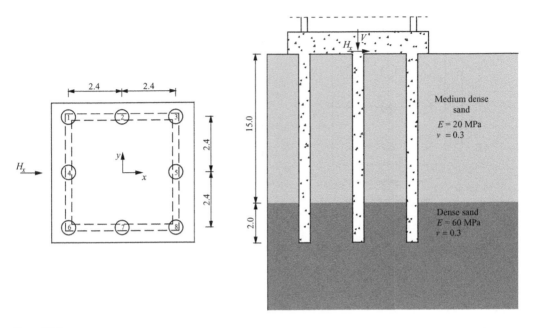

Figure E7.5

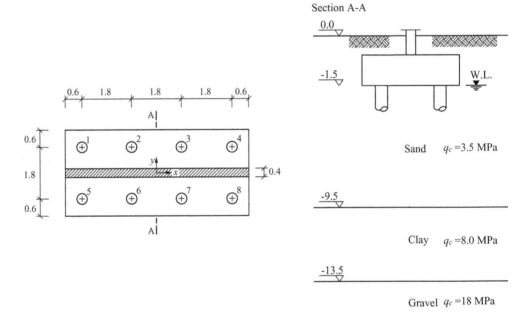

Figure E7.6

Table E7.6 Loading cases

Load type	V (kN)	M_x (kN.m)	M_y (kN.m)	H_x (kN)	H_y (kN)
Permanent	9300	100	1000	100	70
Variable	2500	0	2000	500	30

Materials: concrete C30/37; steel reinforcement S500

8 m thickness, and stiff clay of 4 m thickness, which lies over a dense gravel. A CPT test was conducted and the average value of q_c for each layer is also depicted in Figure E7.6.

The preliminary design proposed that the pile group should consist of eight CFA piles of 600 mm diameter.

The equivalent loads at the barycenter of the pile cap bottom are given in Table E7.6.

a) Compute the axial load in the most stressed pile according to Eurocode 7 – Design Approach 1 and Design Approach 2.
b) Following the LCPC method and EC7, compute the required length for the piles.
c) Compute the longitudinal steel rebars of the piles.

Chapter 8

Earth-retaining structures

8.1 INTRODUCTION

This chapter deals with the analysis and design of earth-retaining structures. A ground mass is said to be retained when the respective lateral surface has an inclination relative to the horizontal greater than it would adopt without the presence of a given structure, designated as a *retaining structure*.

These structures can support either earthfills or excavations, with Figures 8.1 and 8.2, respectively, representing simplified examples for each case.

Assume, for instance, that a platform for a car park is required at a given elevation above the natural ground surface (indicated by the broken line in Figure 8.1a). A retaining wall is first constructed at the boundary of the future platform (Figure 8.1b). After (or simultaneously with) this construction, a fill is placed, layer by layer, until reaching the required elevation (Figures 8.1c,d). In the case of the retaining wall depicted, its self-weight plays a fundamental role in its stability; it is therefore designated as a *gravity wall*.

Consider now that, in the ground represented in Figure 8.2a, an excavation will be performed (as indicated by the broken line) to construct an unlevelled road crossing. In the lateral boundaries of the future excavation, two symmetrical retaining walls are first constructed, the toe of which are located below the base of the intended excavation (Figure 8.2b). Then follows the execution of the excavation between the retaining walls. Depending on the depth of the projected excavation, the ground removal may proceed without the need for any wall support, a solution known as cantilever embedded wall (Figure 8.2c), or it can be accompanied by the introduction of one or more levels of structural supports, such as prestressed anchors (Figure 8.2d), struts (Figure 8.2e), or even a slab (Figure 8.2f, supposing in this case that it is the pavement slab of the upper perpendicular road).

Contrary to the wall described in Figure 8.1, the weight of the retaining walls in Figure 8.2 does not have a significant support effect. The wall thickness is just the one necessary for a satisfactory structural performance under the bending and shearing forces caused by the earth pressures and by the forces applied by the structural supports; hence, their classification as *flexible* structures and their designation as *sheet-pile* wall, in the case of steel, or *diaphragm* wall, in the case of reinforced concrete. Due to the relevant role of their embedded height, such walls also currently come under the generic designation of *embedded* walls.

The constructive and structural solutions for the support of earthfills or excavations are extremely varied and employed in works with the most diverse purposes and scales.

This chapter will focus on gravity walls used in general for backfill support, and cantilever and single-propped embedded walls, both employed for excavation support. As will be seen, all these structures can be designed using analytical methods with theoretical bases. The final part of the chapter will deal with the behavior and design of ground anchors, often

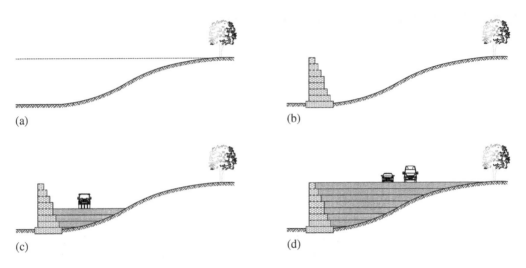

Figure 8.1 Schematic representation of an earthfill supported by a gravity retaining wall.

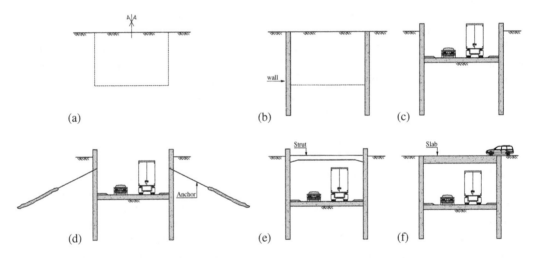

Figure 8.2 Schematic representation of an excavation work supported by embedded cantilever, single-propped or single-anchored walls.

used to complement flexible walls, but also in the stabilization of natural slopes, as will be discussed in Chapter 9.

8.2 GRAVITY WALLS

8.2.1 Wall types: conception

Gravity walls are very diverse in terms of their material, shape, and construction process. Figure 8.3 presents gravity walls of various materials and shapes.

Stone masonry walls (Figure 8.3a) are one of the most ancient man-made constructions. From ancient times, terraces for cultivation on hill slopes have required the construction of masonry retaining walls, with application of empirical rules transmitted and improved through generations.

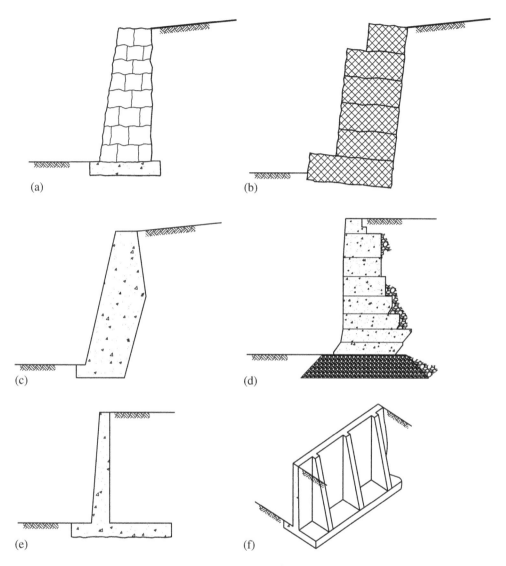

Figure 8.3 Retaining wall types: a) stone masonry; b) gabion; c) cyclopean concrete; d) block quay wall; e) cantilever reinforced concrete; f) reinforced concrete with counterforts.

Gabion walls (Figure 8.3b), made from parallelepipedic boxes of galvanized steel mesh filled with crushed rock, are the modern version of masonry walls. They are currently employed in roadworks, providing good landscape integration.

Concrete walls are also very common, cyclopean concrete for low to moderate height walls (Figure 8.3c) and reinforced concrete for higher walls. Concrete block quay walls are a special case of cyclopean concrete walls (Figure 8.3d). The blocks are pre-fabricated at the construction site, transported to the location and then sunk, their position being fixed with the help of divers. In general, they are laid out on a rockfill base, previously deposited on site.

Reinforced concrete cantilever walls, by extending the heel to the back side, use the supported soil weight to ensure stability (Figure 8.3e). Therefore, they may be included in the gravity wall family.

The thickness of the stem and base slab is determined by the internal design (see Section 8.2.7). From the triangular earth pressure diagram, the bending moment at the stem base increases with the cube of the wall height. Therefore, it is advantageous to include counterforts beyond a certain value of the wall height (Figure 8.3f), in spite of the additional expense in formwork and the increased difficulty in executing the backfill.

Some general guidelines can be issued concerning the shape of the retaining wall:

i) the wall center of gravity should be close to the supported ground and as low as possible, since this reduces the load eccentricity on the foundation, in both static and seismic conditions;

ii) it is desirable to provide the wall base with a certain inclination towards the supported ground to reduce the tangential component and increase the normal component of the foundation load;

iii) the inclination of the two wall faces towards the supported earth improves stability; nevertheless, this inclination should not be excessive, as the inclined wall takes up additional space above its top or in front of its base, and space is wasted with excessive inclination (Jiménez Salas et al., 1976);

iv) in any case, a vertical front face should be avoided, because a subsequent, though small, rotation under soil thrust action may cause an adverse psychological effect, especially in high walls; therefore, it is convenient to provide a slight receding slope to that face, also designated as *batter*.

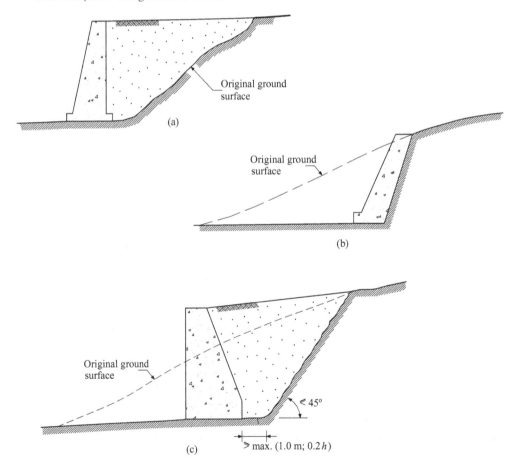

Figure 8.4 Typical cases in the construction of gravity retaining walls: a) fill; b) excavation; c) substitution.

Even though these guidelines seem reasonable, it should be acknowledged that the search for the optimal shape of a gravity wall is quite complex. In a real situation, two experienced designers would be unlikely to arrive at the same solution.

Figure 8.4 shows three typical cases in terms of what is involved in the construction process. Walls of the type of Figure 8.4a correspond to the most common case, which does not involve significant excavation, with backfill being placed behind the wall.

In the case of Figure 8.4b, the work essentially involves excavation to obtain a platform in front of the wall and this is constructed to sustain unexcavated ground, the shear strength of which would be insufficient, in the long-term, to ensure stability on its own. In these circumstances, the wall is typically concreted against the ground, without the need for formwork on the back face. When the slope of the excavation face approaches the angle of shearing resistance of the ground, the wall no longer has a structural function and becomes a facing wall, for protection against ground erosion and weathering. Walls of this type are frequent in the protection of cuts, e.g., in roadworks.

Figure 8.4c illustrates an intermediate situation, in which the natural ground is partially excavated and the backfill replaces part of that soil.

8.2.2 Ultimate limit states of gravity walls by external failure

The expression *external failure* means that the retaining wall is assumed to be rigid, so that ultimate limit states occur by failure of the adjacent or surrounding ground.

Figure 8.5 shows a generic gravity wall that is essentially subjected to the action of two forces: self-weight, W, and the thrust, P_a, applied on the wall back face by the supported soil. As illustrated in the figure, the resultant, F, of these two forces is applied to the wall foundation with a given inclination and a given eccentricity.

As the wall is assumed to be rigid, its displacement will result from the response of the foundation ground to the force F. In a wall with a comfortable margin of safety, the mobilization,

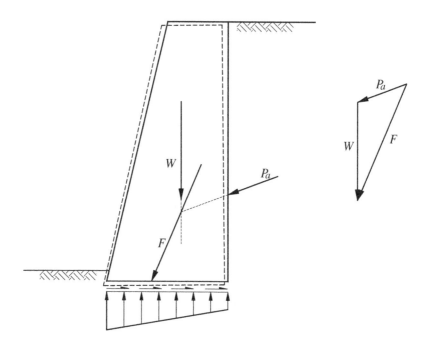

Figure 8.5 Gravity retaining wall, force transmitted to the foundation, stresses mobilized at the wall base and typical associated movement.

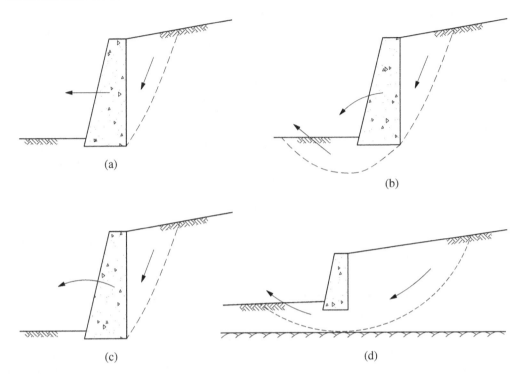

Figure 8.6 Ultimate limit states of gravity walls: a) base sliding; b) foundation soil failure; c) overturning; d) global sliding.

in the wall base–foundation interface, of normal and tangential stresses for equilibrating that force, will require a movement of small magnitude as suggested in the figure (with a dashed line). This movement combines a horizontal translation with a rotation away from the supported soil.

On the contrary, in a wall without suitable stability conditions, the action F applied to the foundation may lead to the following ultimate limit states: i) base sliding (Figure 8.6a); and ii) foundation soil failure, due to insufficient resistance to vertical loading (Figure 8.6b).

In the relevant bibliography, one can also frequently find references to the ultimate limit state by overturning, which consists of wall rotation around the external foundation edge, as illustrated in Figure 8.6c.

This limit state would only be possible on a wall on a rigid base, i.e., with indeformable foundation ground. To better understand this statement, consider Figure 8.7, which shows the normal stress distribution at the wall base for increasing eccentricity, e, of the foundation load, together with the expression of the maximum normal stress value in each case.

It can be seen that (from left to right): i) the normal stress distribution is uniform when the eccentricity is zero and becomes trapezoidal when the eccentricity lies between zero and $B/6$, i.e., when the load is applied within the middle third; ii) the distribution becomes triangular, with the whole base under compression, when the eccentricity reaches $B/6$; iii) the distribution is triangular with the wall base partially under compression when the eccentricity exceeds $B/6$ and is lower than $B/2$; and iv) when the eccentricity is $B/2$, the area where normal stress can be mobilized reduces to zero, so that stress would have to be infinite to ensure equilibrium.

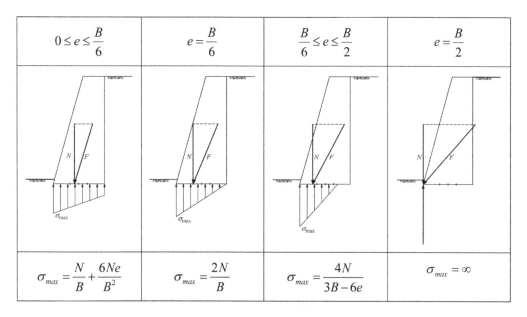

$0 \leq e \leq \dfrac{B}{6}$	$e = \dfrac{B}{6}$	$\dfrac{B}{6} \leq e \leq \dfrac{B}{2}$	$e = \dfrac{B}{2}$
$\sigma_{max} = \dfrac{N}{B} + \dfrac{6Ne}{B^2}$	$\sigma_{max} = \dfrac{2N}{B}$	$\sigma_{max} = \dfrac{4N}{3B - 6e}$	$\sigma_{max} = \infty$

Figure 8.7 Normal stress distribution on the base of a gravity retaining wall for increasing eccentricity, *e*, of the foundation load.

This last situation, when the application point of *R* coincides with the extreme point of the footing base, is precisely when overturning of the wall would be imminent. Thus, it can be understood that this limit state would not be reached, since it would be preceded by failure of the foundation ground, due to the reduction of the effective area of the footing and very high values of the normal stress. Even imagining a retaining wall founded on very high strength bedrock, an internal failure of the wall material would probably occur before failure of the foundation ground did.

The usual reference to the limit state by overturning is certainly associated with the fact that the failure of the foundation soil, due to excessive load eccentricity, is preceded by very large settlement of the external zone of the foundation, causing a pronounced outwards rotation of the retaining wall, in an uncontrollable progression, favored by second-order effects.[1]

Following these considerations, attention should be given to the fact that a competent foundation soil is absolutely critical for the satisfactory behavior of a gravity wall! In fact, permanent and quasi-permanent actions in buildings, bridges, and most civil engineering structures imply foundation loading with small inclination and eccentricity. Only for certain variable actions, of an ephemeral nature, such as wind and seismic actions, are these eccentricities and inclinations significant. On the other hand, permanent actions on the foundation of a gravity retaining wall typically have high eccentricity and inclination. So, foundation failure of gravity retaining walls is much more frequent than foundation failure of other types of structures.

As a conclusion to this topic, the ultimate limit state by global sliding is usually considered, involving both the wall and the surrounding soil formation, as shown in Figure 8.6d.

---

[1] The large rotation of the foundation increases the load eccentricity, which causes an increase in normal stress close to the external edge of the footing, inducing additional settlement at that point, and so on.

It is relevant to draw attention to this limit state because the construction of gravity retaining walls frequently occurs within a wider context of works involving large earth movements, i.e., relevant modifications of the ground geometry and stress state. These modifications are conducive to fostering the destabilization of earth masses, sometimes of large volume and dimensions. In this context, sliding, involving the retaining wall and a circumjacent soil mass, must be considered. The methods for verification of global stability have been the object of Chapter 2 and discussion of this topic will be resumed in Chapter 9.

The failure modes of retaining walls under seismic action are essentially the same as may happen under static conditions. Those limit states may occur due to the unfavorable effect of inertial forces on the ground and on the wall itself, but also due to liquefaction of the backfill and/or the foundation. In situations where this phenomenon occurs in the supported backfill, the soil thrust will reach values scarcely bearable by the structure. A significant number of collapses due to earthquakes has actually occurred in quay walls, due to liquefaction of the submerged sandy backfill. The liquefaction of the wall foundation soil will naturally induce foundation failure or failure by global sliding. Evaluation of the liquefaction potential of a granular deposit has been addressed in Chapter 6.

8.2.3 Soil thrust to be considered for the evaluation of external safety

In Chapter 5, it was demonstrated that the magnitude of the soil thrust acting on a structural face is closely related to the direction and magnitude of the displacement that it experiences. In particular, it has been concluded that a very small outward displacement of the face is sufficient to install an active limit equilibrium state (Figure 5.7).

In gravity retaining walls, as illustrated in Figure 8.5, the movement is essentially controlled by the foundation and implies an outward displacement of the face in contact with the supported soil. This displacement is small in a wall with satisfactory safety conditions. On the other hand, it will be very large if the wall experiences one of the limit states previously discussed.

The two previous paragraphs summarize the reasons that allow the adoption of the active thrust as the action of the supported ground in the external design of gravity retaining walls. In other words, the thrust in Figure 8.5 should be taken as $P = P_a$.

This is a quite curious situation (already mentioned in Chapter 5), and one which is apparently contrary to the philosophy of safety of civil engineering structures: to adopt the active thrust as a design action is to take the *lowest possible thrust*!

Such an option is only apparently imprudent. If the retaining wall and its foundation are able to equilibrate the active thrust, wall stability is guaranteed. In fact, none of the limit states featured above may develop without a (very small!) displacement of the wall back face, which ensures that the soil thrust reaches its minimum value, the active thrust.

It should be clarified that these considerations – when mentioning the *smallest of the possible thrust values* – must be seen in a purely deterministic context! In this case, as well as in the design relative to any other limit state, it is indispensable to use safety factors to account for the uncertainties that affect the magnitude (and also the direction) of the active thrust.

In the general case, there is yet another soil–wall face interaction involved in the stability of the retaining wall. The footing, and sometimes a portion of the lower part of the wall, are buried, so the interaction between the soil and the frontal face of the wall also have to be considered. In this case, the frontal face moves into the soil.

In Chapter 5, it has been seen that, for this type of relative soil–wall movement, the passive thrust may be mobilized. Although the mobilization of this thrust requires substantial displacement, none of the ultimate limit states presented may develop without mobilizing the passive thrust in front of the wall. As the coefficient of passive earth pressure is much larger than that of the active case, even for a small value of the buried height of the frontal wall face, the consideration of the passive thrust may have a significant influence on the design.

8.2.4 Practical questions about design and construction

8.2.4.1 Parameters of the soil and interfaces

The fill material to be placed behind retaining walls must have granular characteristics (non-plastic soil), with a fines content below 5%. In fact, the behavior of a plastic soil is highly dependent on the water content, and so their use in the backfill would cause important seasonal variations, both in the backfill volume and in the magnitude of the soil thrust, with the aggravating circumstances that such variations are very difficult to evaluate and even to control during the wall's lifetime.

Instead, the high permeability of a granular material, with the presence of adequate draining devices (see Section 8.2.4.2), prevents the rise of the groundwater level in the supported ground, which would lead to a very high value for the total thrust (the sum of soil and water thrusts).

Heavy compaction of the backfill must be avoided. Although higher compaction effort increases the soil relative density, and thus its peak friction angle, the associated dynamic effects may increase the pressure on the wall considerably (Aggour and Brown, 1974). Moreover, in order to minimize the compaction-induced pressure, the compaction equipment must always travel in the direction parallel to the wall development and never perpendicularly to it (see Figure 8.1c).

Taking into account these already-described compaction conditions, the calculations should be based on the critical friction angle, ϕ'_{cv}, of the backfill.

As already mentioned, the soil–wall friction angle depends on the angle of shearing resistance of the soil and on the roughness of the wall face. The roughness is larger in stone masonry and gabion walls or in concrete walls poured against the ground, where $\delta = \phi'_{cv}$ can be considered. In concrete walls built with formwork, it is advisable to adopt $\delta = (2/3)\phi'_{cv}$. This value is the maximum admitted in Eurocode 8 – Part 5 for the evaluation of the seismic active thrust (EN 1998-5: 2004).

The preceding considerations apply to the evaluation of the active thrust. For the passive thrust, in the cases when it is explicitly considered in design, it is advisable to adopt lower values for δ, as the magnitude of this thrust increases very much with this angle. An option sometimes adopted in design consists of considering the passive thrust effect only for the verification of safety under seismic conditions (since the probability of seismic action coinciding with a temporary excavation in front of the wall is very low). In these circumstances, a zero value for δ should be taken, according to Eurocode 8 – Part 5 (EN 1998-5: 2004). Note that, as the Mononobe-Okabe theory is an extension of Coulomb's theory, it is also susceptible to overestimating the passive seismic thrust when the soil–wall friction angle is not null.

The shear strength of the contact surface between the wall base and the foundation ground is defined by the parameters adhesion, a_b, and friction angle, δ_b, respectively. For design purposes, it is advisable to ignore the contribution of adhesion. As to the friction angle, δ_b, it is reasonable to assume that it will be close to the angle of shearing resistance of the foundation ground, provided that the footing construction follows the rules of best practice: thorough cleaning of the foundation surface, immediately followed by the placement of blinding concrete, a thin layer of very dry coarse concrete, carefully pounded.

In situations where this procedure is applied, failure by base sliding would imply dragging a certain thickness of the underlying ground, which would lead to mobilization of the soil shearing resistance.

8.2.4.2 Backfill drainage

Backfill drainage constitutes a critical feature for retaining wall safety. In fact, the existence of a water table in the backfill is highly adverse, substantially increasing the total thrust.

Figure 8.8 presents some traditional drainage solutions. In Figures 8.8a and b, a *sub-vertical drain* can be observed, adjacent to the wall back face, made with high-permeability material. This can correspond to larger grain size backfill material or to a man-made product (a draining geocomposite, for example). As shown in Figure 8.8b, where there is no inconvenience to draining water to the wall front face, *transversal drains* or *weep holes* can be used. Figure 8.8c illustrates another particularly convenient draining scheme, which consists of the construction of a *sloping drain* under the backfill, over the excavated surface of the natural ground. The three solutions presented are complemented by a longitudinal perforated drain pipe with fall draining to a soak pit, watercourse, or stormwater system.

The zone adjacent to the drains must be provided with a filter in order to prevent both internal backfill erosion and drain clogging. Geosynthetics have been increasingly used for filtering purposes, due to being quick and easy to place.

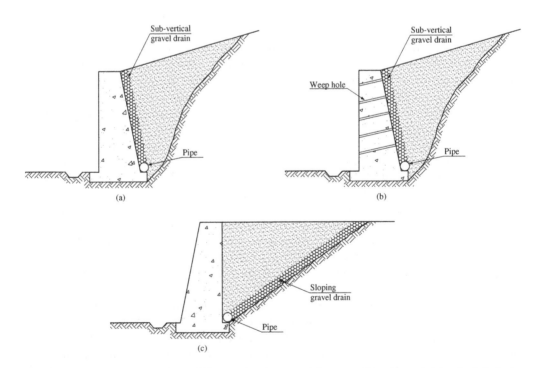

Figure 8.8 Drainage of gravity retaining walls: a) sub-vertical drain connected to a longitudinal drainpipe; b) sub-vertical drain connected to a longitudinal drainpipe and to weep holes; c) sloping underdrain connected to a longitudinal drainpipe.

8.2.5 Evaluation of the soil thrust in some practical cases

8.2.5.1 Consideration of seepage forces

Any of the drainage schemes previously described in Section 8.2.4.2, if in good working order, will prevent accumulation of groundwater and, therefore, the establishment of hydrostatic thrust. Nevertheless, at least in walls whose potential failure involves substantial risk to people and property, it is convenient in design to take into account the effect of seepage forces in the supported ground.

Figure 8.9 represents in a schematic way the flow nets corresponding to steady-state rainwater infiltration at the ground surface for the two drainage solutions presented in Section 8.2.4.2. Note that the drainage schemes of Figures 8.8a and b may be considered to coincide, for the question now being addressed.

The evaluation of the thrust, taking into account the effect of the seepage forces, is discussed in detail by Lambe and Whitman (1979), who demonstrate that the solution with a sloping underdrain is frankly more favorable, leading to a lower active thrust.[2]

8.2.5.2 Retaining walls with a non-rectilinear back face

The back face of stone masonry and gabion retaining walls often has the shape represented in Figure 8.10a. In such cases, for the evaluation of the active thrust, the real back face can be replaced by a rectilinear, notional one, marked with a broken line in the figure, which is equivalent to assuming that the triangular prisms of backfill adjacent to the wall back face behave as an integral part of the wall, in a limit equilibrium situation. Therefore, in the calculation of the thrust, the soil–wall friction angle, δ, should be taken to equal the friction angle of the backfill, ϕ'.

Another situation, quite frequently encountered in cyclopean concrete walls, is depicted in Figure 8.10b, in which the wall back face comprises two plane areas with different inclination. In practice, the problem can be solved by starting with the calculation of ground pressure on part AB of the back face, as if the wall ended in B. Then, considering the extension of line CB to point D at the ground surface, the pressure is computed as if the wall back face was DBC. Naturally, only the portion along BC is taken for design purposes.

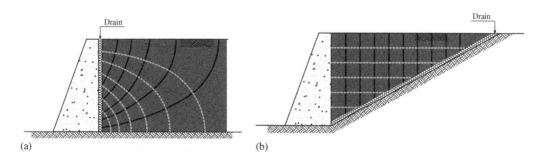

(a) (b)

Figure 8.9 Flow nets for steady-state infiltration at the ground surface: a) drainage at the wall back face; b) sloping underdrain (adapted from Lambe and Whitman, 1979).

---

[2] A curious note: the effect of the seepage forces associated to rainwater flow to the drains, analyzed in this point, was the theme of a paper presented by Karl Terzaghi in the 1st International Conference of Soil Mechanics and Foundation Engineering, held in Cambridge, Massachusetts, USA, in 1936.

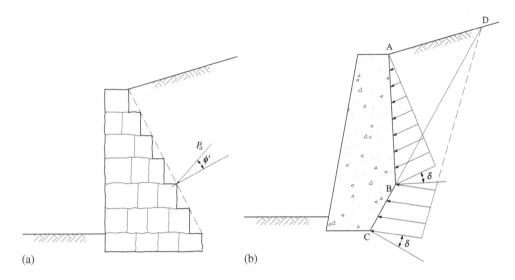

Figure 8.10 Gravity retaining walls with non-rectilinear back face: a) stone masonry wall; b) cyclopean concrete wall.

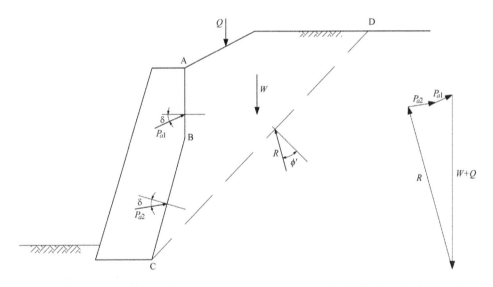

Figure 8.11 Determination of soil thrust in a wall with non-rectilinear back face by the Coulomb method (Jiménez Salas et al., 1976).

In case the problem does not have an analytical solution, due to the geometry of the ground surface and/or to the presence of surcharge point loads applied there, the solution can be obtained by a slightly different method, outlined in Figure 8.11.

Using Coulomb's method, the thrust P_{a1} along AB is first calculated. Then, applying the same method, several trial sliding surfaces, passing through point C at the wall base, are examined, and generically represented in the figure by CD. For each surface, the polygon of the intervening forces is drawn, two of them being known, $W+Q$, the resultant of the self-weight of the soil and the surcharge loads applied between A and D, and P_{a1}, the previously determined thrust applied between A and B. The equilibrium condition (by closure of the force polygon) allows the determination of the magnitude of the remaining forces, R

and P_{a2}, the directions of which are known, being P_{a2}, the thrust applied in BC for the sliding surface under consideration. The active thrust will naturally be the largest of the values obtained for the various trial sliding surfaces.

8.2.5.3 Reinforced concrete cantilever walls

8.2.5.3.1 Wall with wide base slab

The displacement of an L-shaped reinforced concrete cantilever wall jointly involves the ground mass located over its base slab, which ensures its stability. Therefore, it is logical that the problem solution approach consists of separating this ground mass from the rest, considering it as part of the retaining structure, and then applying the earth pressure theories.

As shown in Figure 8.12, the solution to the problem consists of assuming that the wall displacement mobilizes the active limit state in the wedges ABC and ADC, defined, respectively, by the angles with the horizontal $\alpha_1 = \pi / 4 + \phi' / 2 + (\varepsilon - \beta) / 2$ and $\alpha_2 = \pi / 4 + \phi' / 2 - (\varepsilon - \beta) / 2$, with $\varepsilon = \sin^{-1}(\sin \beta / \sin \phi')$, which are both equal to $\pi / 4 + \phi' / 2$ when β is zero (see Section 5.4.6). As a consequence of this hypothesis, the shaded wedge ABGF (Figure 8.12) behaves as an integral part of the wall.

As indicated in Figure 8.13a, the thrust can be calculated along AB, taking $\delta = \phi'$, and using the Caquot-Kérisel tables or Coulomb's analytical solution.

As shown in Figure 8.13b, if the equilibrium of wedge ABC is analyzed, inverting the direction of the previously calculated thrust $P_a(AB)$ and taking its weight W, the thrust $P_a(AC)$ in the vertical plane AC will be obtained. It can be easily verified that this force coincides with the Rankine active thrust in both magnitude and direction.

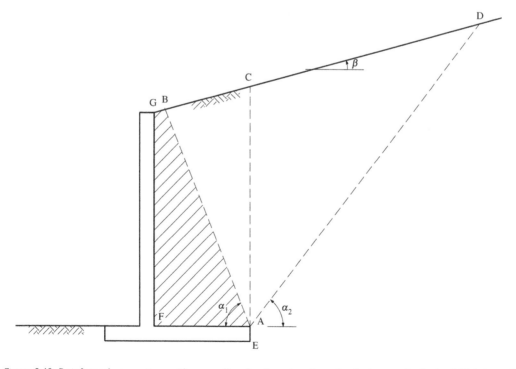

Figure 8.12 Reinforced concrete cantilever wall and soil wedges in active limit state in the backfill (adapted from Jiménez Salas et al., 1976).

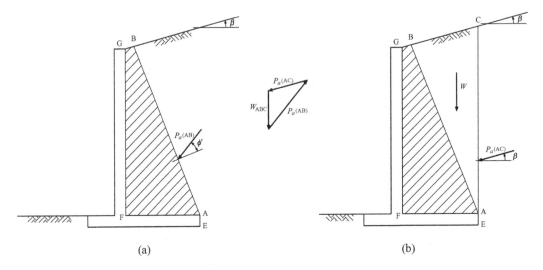

Figure 8.13 Thrust evaluation in reinforced concrete cantilever walls: a) along the delimiting surface of the backfill wedge, moving jointly with the wall; b) along the vertical plane, passing at the heel.

With regard to this approach, the following can also be demonstrated: the orientation of AB previously mentioned maximizes the horizontal component and minimizes the vertical component of the thrust in the plane AC, this situation corresponding to the theoretical Rankine thrust (Greco, 2001).

Therefore, for simplicity, the soil thrust calculation is normally performed by adopting this alternative, taking, in the safety checks, the weight of the soil wedge AFGC jointly with the wall self-weight for the calculation of the resisting forces and moments.[3]

8.2.5.3.2 Wall with short base slab

In walls with short base slabs the thrust calculation becomes more complex because surface AB intersects the wall back face, as shown in Figure 8.14a. In order to apply to this case a calculation procedure compatible with that presented above for walls with long base slab, it seems convenient to keep computing the thrust in plane AC.

This problem has been analyzed by Barghouthi (1990) and Greco (1992, 1999, 2001) who concluded the following: i) the angle α_2, of the right-side boundary of the limit equilibrium soil wedge, keeps, for practical purposes, the value determined for the previous case, i.e., $\alpha_2 = \pi/4 + \phi'/2 - (\varepsilon - \beta)/2$, with $\varepsilon = \sin^{-1}(\sin\beta/\sin\phi')$; ii) the angle α_1, in addition to depending on ϕ' and β, now also depends on the length ratio of AF and GF and on the value of the soil–wall friction angle, δ, between B and G; and iii) the α_1 value that maximizes the horizontal (destabilizing) force component in plane AC coincides with the one that minimizes the vertical (stabilizing) component of that force.

Therefore, the suggested procedure is (see Figure 8.14b):

1. adopt a value for α_1, fixing in this way the location of point B;

---

[3] Some controversy exists concerning the thrust in plane AC. It is sometimes argued that the thrust on that plane should be at an angle $\delta = \phi'$ with the normal. This opinion is not correct, as can be concluded from the preceding explanation. Plane AC is not a failure plane, i.e., one where the soil shearing strength is fully mobilized. In fact, the failure planes are AB and AD. For simplicity, the thrust behind the wall is not usually divided into the thrust on AC plus that on EA, a single thrust value being calculated on EC coinciding with the Rankine thrust, which is on the safe side.

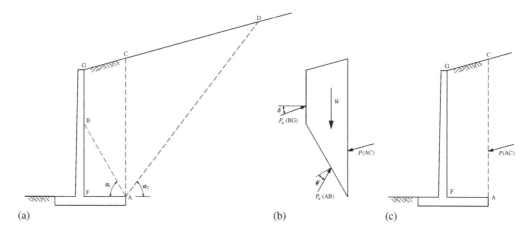

Figure 8.14 Determination of active thrust on a cantilever reinforced concrete wall when the boundary of the limit equilibrium wedge intersects the stem: a) wall and wedge layout; b) free body graph of the wedge part to the left of the vertical plane passing through the heel; c) active thrust on the vertical plane passing through the heel.

2. calculate the active thrusts P_a (BG) and P_a (AB), for example by the Coulomb method, as in the case of a non-rectilinear back face (see Section 8.2.5.2);
3. with these two forces and the weight of the soil block ABGC, calculate the force $P(AC)$;
4. repeat the previous steps with new values for α_1, in order to obtain the maximum value of the horizontal component of $P(AC)$;
5. for that value of α_1, force $P(AC)$ will be the active thrust, $P_a(AC)$;
6. in wall design, the soil block AFGC is considered to integrate the retaining wall, as in the case of walls with long base slab.

In face of this explanation, it is readily anticipated that $P_a(AC)$ will be within the two following values:

i) the Rankine theoretical active thrust, at an angle β with the horizontal, when the intersection of the failure plane with the stem is at point G (for $\alpha_1 = \pi/4 + \phi'/2 + (\varepsilon - \beta)/2$), a situation that coincides with the long base slab case;

ii) the Coulomb theoretical active thrust, at an angle δ with the horizontal, when the base slab length tends to zero.

Therefore, whenever $\delta > \beta$, the adoption of $P_a(AC)$ from the Rankine solution, i.e., the solution corresponding to the wall with long base slab, will be on the safe side.

8.2.5.4 Pressure induced by point loads or uniform surcharges applied on a limited surface area

Chapter 5 addressed the calculation of active stresses and their integral, the active thrust, associated with a uniformly distributed surcharge at the surface. In the current section, the case of surcharges applied on a limited surface area will be described. What follows is applicable to structures supporting backfills or excavations.

Consider the vertical wall face of a generic retaining structure and a uniform line load, Δq_s, parallel to the wall development, applied at the surface, as shown in Figure 8.15a.

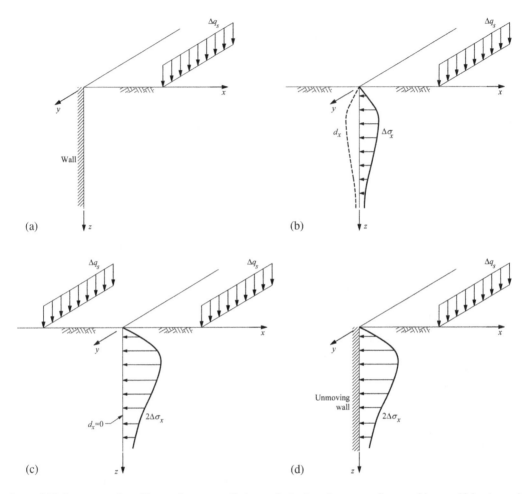

Figure 8.15 Pressure induced by surcharges applied on a limited surface area: a) general layout; b) horizontal stress and displacement induced by the loading on a vertical plane of an elastic half-space; c) effect of two symmetrical loads on the same vertical plane; d) pressure induced on an unmoving wall.

With reference to Figure 8.15b, imagine that the ground where the surcharge is applied is an elastic half-space and consider a vertical plane in a relative position to the surcharge identical to that of the wall face. In that vertical plane, stresses $\Delta\sigma_x$ will be developed under the action of Δq_s. These stresses and the associated displacements, d_x, are shown in Figure 8.15b. Considering now a symmetrical surcharge, Δq_s, as shown in Figure 8.15c, the combined result on the vertical plane of the two surface loads would consist of zero horizontal displacement, and horizontal stresses equal to $2\Delta\sigma_x$.

Given this fact, the following conclusion may be achieved: surcharges applied on a limited area of the supported ground surface, will induce, on the face of an unmoving structure, twice the horizontal stresses that would be applied in a plane with that location in an elastic half-space (Figure 8.15d). This conclusion permits us to obtain an *envelope* of the pressures induced on a retaining wall face under the effect of surface loading. Naturally, if the wall experiences some displacement, the induced stresses will be smaller.

Incidentally, a note of historical character is justified. The conclusion referred to resulted from an interesting and fruitful discussion held during one of the sessions of the 1st International Congress on Soil Mechanics and Foundation Engineering

(Cambridge, Massachusetts, 1936). Spangler (1936) presented experimental results from a physical model: a vertical wall face retaining a backfill, on the horizontal surface of which point or line loads were applied, with stress measurement at the wall face with total stress cells. The form of this pressure distribution was analogous to the elastic stress distribution on a vertical plane under a similar surface-applied loading. Mindlin (1936) showed that, by multiplying the elastic stress distribution by 2, an envelope of the experimental results was obtained, and supplied a rational explanation for this fact, which essentially corresponded to the considerations presented above with the help of Figure 8.15.

Later, Terzaghi (1953), based on a careful interpretation of the results obtained by Spangler and others, found that the experimental pressure values were frequently greater than twice the value of the elastic solution, when the surface loading was applied at a distance from the wall face of less than 40% of its height ($m \leq 0.4$, in Figure 8.16). Therefore, Terzaghi proposed, for such situations, the adoption of the curve corresponding to twice the elastic

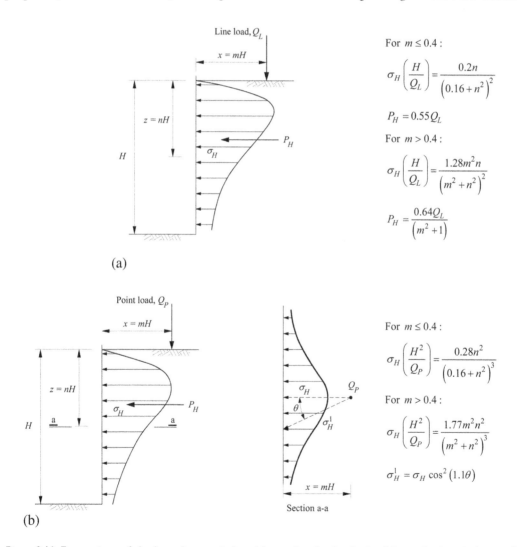

For $m \leq 0.4$:

$$\sigma_H \left(\frac{H}{Q_L} \right) = \frac{0.2n}{\left(0.16 + n^2 \right)^2}$$

$$P_H = 0.55 Q_L$$

For $m > 0.4$:

$$\sigma_H \left(\frac{H}{Q_L} \right) = \frac{1.28 m^2 n}{\left(m^2 + n^2 \right)^2}$$

$$P_H = \frac{0.64 Q_L}{\left(m^2 + 1 \right)}$$

For $m \leq 0.4$:

$$\sigma_H \left(\frac{H^2}{Q_P} \right) = \frac{0.28 n^2}{\left(0.16 + n^2 \right)^3}$$

For $m > 0.4$:

$$\sigma_H \left(\frac{H^2}{Q_P} \right) = \frac{1.77 m^2 n^2}{\left(m^2 + n^2 \right)^3}$$

$$\sigma_H^1 = \sigma_H \cos^2 \left(1.1 \theta \right)$$

Figure 8.16 Expressions of the lateral stress induced by surface loads obtained from elastic solutions and experimental results: a) line load; b) point load (Terzaghi, 1953).

solution calculated for $m = 0.4$, which supplies an adequate envelope to the experimental results. This solution is presented in Figure 8.16.

The procedure presented may be employed with other surcharge types, such as those applied in limited areas, uniformly or non-uniformly (linearly variable with the distance to the wall face). For this purpose, the elastic solutions included in Annex A6.2 may be used.

Before closing this subject, two important caveats are in order.

The procedure presented is reasonable when the resultant of the surcharge-induced pressure is significantly lower than that of the supported ground. Otherwise, a different approach is recommended, namely the analysis of the equilibrium of successive ground wedges, applying Coulomb's theory.

Special attention must be given to repeatedly applied surcharges, which are frequent, for example, in quay walls. The pressure mobilized after a large number of applications may substantially exceed the pressure associated with the first application or the pressure that would reasonably be expected from a (permanent) static surcharge of identical value (Sherif and Mackey, 1977; Budin and Demina, 1979).

As mentioned before, the approaches presented are also useful for structures retaining excavations.

8.2.6 Evaluation of the external safety of gravity retaining walls

8.2.6.1 Global safety factors

The traditional form of safety evaluation of gravity walls is based on global safety factors. Figure 8.17 shows a gravity retaining wall and the applied forces. A general situation has been considered, contemplating not only static but also seismic action, as well as the passive thrust in front of the wall and a non-horizontal base.

In Figure 8.17b, which addresses seismic action, the passive thrust has been assumed to be normal to the wall face, in accordance with the comment made in Section 8.2.4.1. In that part of the figure, an aspect of great relevance for stability can also be observed: at the wall center of gravity are applied vertical and horizontal inertial forces, defined by the same seismic coefficients that were used to calculate the seismic thrust increment, ΔP_{ae}.

The global safety factor against foundation soil failure is:

$$F = \frac{V_R}{V_S} \tag{8.1}$$

where V_R represents the resistance to vertical loading, the evaluation of which has been addressed in Chapter 6, and V_S is the vertical load applied to the wall foundation.

In Chapter 6, it was seen that, when the foundation base is not horizontal, the forces V_S and H_S must be taken as the normal and tangential components of the base load, respectively, and V_R will be the resisting force, normal to the foundation base.

The global safety factor against base sliding is given by:

$$F = \frac{T_R}{T_S} \tag{8.2}$$

where T_R represents the tangential resisting force in the plane of the wall base, and T_S the tangential force applied to the wall base.

The global safety factor against global sliding will not be treated in the present chapter. It has already been discussed in Chapter 2 and will be addressed again in Chapter 9.

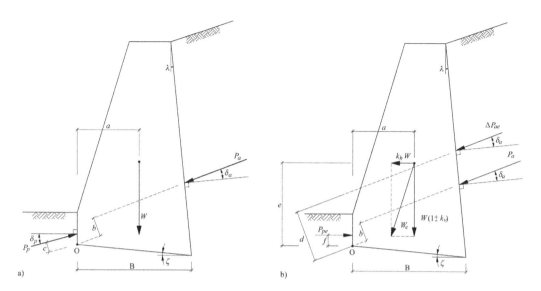

Figure 8.17 Gravity retaining wall with applied forces: a) static conditions; b) seismic conditions.

Tables 8.1 and 8.2 include the expressions of the forces and moments that appear in Equations 8.1 and 8.2, taking as reference Figure 8.17.

Table 8.1 does not include the passive thrust because the resistance of the soil in front (and above) the foundation base is currently ignored when computing the bearing capacity of the foundation soil, as seen in Chapter 6.

In Table 8.2, in order to present expressions with maximum generality, the passive thrust has been considered in the expressions of N_s (except for the case of seismic conditions and horizontal base, in which the corresponding projection onto the plane normal to the base is zero) and, naturally, of T_R. If the passive thrust is discarded, the expressions of N_S and of T_S coincide with those of V_S and of H_S of Table 8.1, respectively.

Table 8.3 includes the minimum values considered acceptable for the safety factor of current retaining walls. Higher values should be adopted for structures whose potential failure involves substantial risk to people and property.

8.2.6.2 *Verification of safety by the partial safety factor method: application of Eurocode 7*

In the context of the so-called *limit state design*, used in the Structural Eurocodes, and in particular in Eurocode 7, partial safety factors are applied, as is known. With these factors, the design value of the resistance and the design value of the effect of the actions are obtained for each limit state. Then, the safety check can be performed, which implies that the first of these values be greater than or equal to the second.

For the limit states of base sliding and foundation soil failure, the expressions of the verifications of safety are, respectively:

$$T_{Sd} \leq T_{Rd} \tag{8.3}$$

$$V_{Sd} \leq V_{Rd} \tag{8.4}$$

Table 8.1 Expressions of the forces involved in the global safety factor against foundation soil failure (Equation 8.1, Figure 8.17).

Conditions	Force	Expression
Seismic, inclined base	V_R	Calculated as seen in Chapter 6
	V_S	$\left(1\pm k_v\right)W\cos\zeta + k_h W\sin\zeta + \left(P_a+\Delta P_{ae}\right)\sin\left(\delta_a+\lambda+\zeta\right)$
	H_S	$\left(P_a+\Delta P_{ae}\right)\cos\left(\delta_a+\lambda+\zeta\right) + k_h W\cos\zeta - \left(1\pm k_v\right)W\sin\zeta$
	$M_S\,(1)$	$P_a\,b + \Delta P_{ae}\,d + k_h We - \left(1\pm k_v\right)Wa + V_S\dfrac{B}{2}$
Seismic, horizontal base	V_R	Calculated as seen in Chapter 6
	V_S	$\left(1\pm k_v\right)W + \left(P_a+\Delta P_{ae}\right)\sin\left(\delta_a+\lambda\right)$
	H_S	$\left(P_a+\Delta P_{ae}\right)\cos\left(\delta_a+\lambda\right) + k_h W$
	$M_S\,(1)$	$P_a\,b + \Delta P_{ae}\,d + k_h We - W\left(1\pm k_v\right)a + V_S\dfrac{B}{2}$
Static, inclined base	V_R	Calculated as seen in Chapter 6
	V_S	$W\cos\zeta + P_a\sin\left(\delta_a+\lambda+\zeta\right)$
	H_S	$P_a\cos\left(\delta_a+\lambda+\zeta\right) - W\sin\zeta$
	$M_S\,(1)$	$P_a\,b - W\,a + V_S\dfrac{B}{2}$
Static, horizontal base	V_R	Calculated as seen in Chapter 6
	V_S	$W + P_a\sin\left(\delta_a+\lambda\right)$
	H_S	$P_a\cos\left(\delta_a+\lambda\right)$
	$M_S\,(1)$	$P_a\,b - W\,a + V_S\dfrac{B}{2}$

The forces and moments that are entered in Equations 8.3 and 8.4 are still those indicated in Figure 8.17 and listed in Tables 8.1 and 8.2, with the index d now added to them all, in order to indicate that the *design value* replaces the characteristic value.

The same is valid for a safety check following the LRFD concept.

8.2.6.3 Gravity retaining walls under seismic action: selection of k_h and k_v

8.2.6.3.1 General considerations

The following features make seismic action particularly unfavorable to gravity retaining walls:

i) in regions of higher seismicity, the seismic thrust increment, ΔP_{ae}, may represent a substantial percentage of the static thrust (Seed and Whitman, 1970; Davies et al., 1986);

Table 8.2 Expressions of the forces involved in the global safety factor against base sliding (Equation 8.2, Figure 8.17).

Conditions	Force	Expression
Seismic, inclined base	N_S	$(1 \pm k_v) W \cos\zeta + k_h W \sin\zeta + (P_a + \Delta P_{ae}) \sin(\delta_a + \lambda + \zeta) - P_{pe} \sin\zeta$
	T_R	$N_S \tan\delta_b + P_{pe} \cos\zeta$
	T_S	$(P_a + \Delta P_{ae}) \cos(\delta_a + \lambda + \zeta) + k_h W \cos\zeta - (1 \pm k_v) W \sin\zeta$
Seismic, horizontal base	N_S	$(1 \pm k_v) W + (P_a + \Delta P_{ae}) \sin(\delta_a + \lambda)$
	T_R	$N_S \tan\delta_b + P_{pe}$
	T_S	$(P_a + \Delta P_{ae}) \cos(\delta_a + \lambda) + k_h W$
Static, inclined base	N_S	$W \cos\zeta + P_a \sin(\delta_a + \lambda + \zeta) - P_p \sin(\delta_p + \zeta)$
	T_R	$N_S \tan\delta_b + P_p \cos(\delta_p + \zeta)$
	T_S	$P_a \cos(\delta_a + \lambda + \zeta) - W \sin\zeta$
Static, horizontal base	N_S	$W + P_a \sin(\delta_a + \lambda) - P_p \sin\delta_p$
	T_R	$N_S \tan\delta_b + P_p \cos\delta_p$
	T_S	$P_a \cos(\delta_a + \lambda)$

Table 8.3 Minimum global safety factor values for gravity retaining walls.

Failure mode	F static	F seismic
Loss of global stability	1.5	1.2
Foundation failure	2.0–3.0	1.5
Base sliding	1.5–2.0	1.1–1.2

ii) the seismic thrust increment, ΔP_{ae}, is applied at a greater distance from the base of the wall face than is the static thrust (see Figure 5.35);

iii) last but not least (!), the inertial forces, defined by the same seismic coefficients that led to the calculation of ΔP_{ae}, are also applied to the wall.

This implies that in regions of high or even moderate seismicity, the design of gravity retaining walls is generally governed by seismic action.

Before analyzing in detail how the seismic coefficients are selected, it is useful to present a theory that relates these coefficients to the displacement that the wall may experience during an earthquake.

8.2.6.3.2 Seismic design involving permanent displacement: theoretical model of Richards and Elms

When base sliding is the critical limit state – which frequently happens, particularly with reinforced concrete cantilever walls – a more economic design may be achieved by accepting

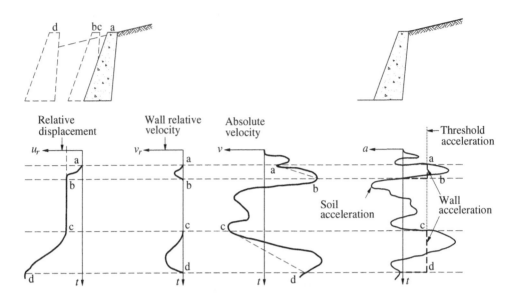

Figure 8.18 Ground and retaining wall motion during an earthquake (Richards and Elms, 1979).

that some relative movement between the foundation soil and the retaining structure may occur during the earthquake, providing that it does not exceed a given value imposed *a priori* to be acceptable. This approach – inspired by methodologies already used for embankment dams, based on the well-known Newmark method (1965) – was developed by Richards and Elms (1979) and permits us to use lower seismic coefficients in design than those adopted in the traditional design method.

Consider Figure 8.18, where a retaining wall is represented with an insufficient margin of safety, relative to base sliding under seismic conditions. In the rightmost graph, (horizontal) ground acceleration is represented by a continuous line and wall acceleration by a broken line. The wall and the soil experience a joint displacement until time *a*, when the seismic acceleration exceeds the *cut-off acceleration*, i.e., that which corresponds to a unit safety factor against sliding. After that moment, the wall acceleration remains constant, leading to different absolute ground and wall velocities. Therefore, the wall experiences increasing relative displacement. This only ceases at time *b* when the absolute velocities become equal again. The joint displacement of wall and ground will be disrupted once more at time *c* and will be re-established at time *d* for similar reasons, naturally with increasing wall displacement. Note that relative motion only occurs when acceleration is directed toward the supported ground, therefore implying destabilizing inertial forces, i.e., toward the wall.

Based on these considerations, Richards and Elms stress that the displacement of a retaining wall with a safety factor against base sliding under seismic action lower than one does not occur *continuously and unlimitedly*, but in increments, the sum of which assumes a *finite* value, which, for a given seismic record, can be *calculated* as a function of the ratio between the cut-off and the peak acceleration of the seismic record. Small-scale tests presented by the same authors essentially validate this theoretical model, including the adequacy of the Mononobe-Okabe theory for predicting the cut-off acceleration (Elms and Richards, 1990).

Nevertheless, the authors underline that both the calculation of this acceleration and that of the final displacement for a given record must be performed by taking into account the strength parameters corresponding to the critical state, although the initiation of wall sliding requires the acceleration to previously exceed a higher value, corresponding to the peak values of the strength parameters. In other words, and taking again as reference Figure 8.18, the wall acceleration plateaux occur at the start of the earthquake for greater acceleration, whereas the ground deformation is still incipient, and, after some cycles, tends to an approximately constant value, corresponding to the acceleration for which the safety factor, calculated with the critical state strength parameters, has a unit value.

This circumstance may explain the phenomenon, revealed by the result of several experimental and numerical studies, whereby a significant part of the incremental pressures remains after the earthquake (Seed and Whitman, 1970). Such a phenomenon will be more likely in dense soils, where the initial structure may be destroyed by the cyclic shear stresses induced by the earthquake, with the soil exhibiting a lower angle of shearing resistance after the seismic event, imparting, therefore, a larger thrust onto the wall.

As referred above, the theory presented permits the selection of seismic coefficients for gravity retaining wall design as a function of the displacement that they may experience.

8.2.6.3.3 Design methodology according to Eurocodes 7 and 8

The selection of the horizontal and vertical seismic coefficients to be applied to the design of earth-retaining walls, according to Eurocode 8, is summarized in Annex A8.1. These coefficients depend on:

i) the seismic zone where the structure is located, which determines the seismic base action;
ii) the characteristics of the wall foundation ground, which will affect the ground surface acceleration;
iii) the importance class of the structure;
iv) the wall-admissible displacement.

This last aspect, i.e., the dependence of the seismic coefficients on the displacement that the structure may experience under seismic conditions, derives precisely from the Richards and Elms theory briefly outlined and from its application to a large number of seismic records, either real or numerically generated.

In summary, the seismic design of gravity retaining walls involves the following steps:

1. establishment of the admissible (permanent) wall displacement, d_r, and, consequently, of the conception of the structure for it to be accommodated without significant damage;
2. selection of the horizontal and vertical seismic coefficients, k_h and k_v, according to Annex A8.1;
3. design of the retaining wall with thrust calculation by the Mononobe-Okabe theory, and taking the partial safety factors in accordance with the design approach adopted.

8.2.7 Internal design

8.2.7.1 General considerations

In Sections 8.2.2 and 8.2.6 the *external design* of retaining walls has been addressed. This implied considering the wall as a rigid body and assessing the ultimate limit states associated with the adjacent ground. The so-called *internal design*, related to the stresses and internal forces mobilized within the structure itself, will now be considered.

The essential question in internal design is the value of the stresses applied by the supported ground to the structure, because – contrary to what might be expected from the initial considerations of the present chapter, particularly in Section 8.2.3 – that ground is generally not in the active limit equilibrium state.

To facilitate the explanation, assume, in what follows, a purely deterministic problem context. Let us start by recalling the justification given for taking account of the active thrust in the external design: *before any of the (external) limit states are able to develop, the wall will experience sufficient displacement for the thrust to decrease to the active thrust value.*

However (and this is the point!), since safety factors are adopted in wall external design, the wall constructed will reach a situation of imminent failure only for a soil thrust value greater than the active value. If, for example, a global safety factor of 2.0 has been adopted in design for each failure mode – in a purely deterministic context, once more – only at *twice the active thrust* will the wall reach, in fact, a limit state.

From the interaction between the soil and the backfill, a given displacement will result – essentially due to foundation ground deformation, as already seen – which is lower than that required for mobilizing the active limit equilibrium state. The supported ground will be in an *intermediate* stress state between the at-rest and the active states.

The more competent the foundation, the smaller that displacement, and the closer the thrust will be to the at-rest value. The more deformable the foundation, the closer the installed thrust will be to the active value.[4]

The relevance of this question is therefore understandable when dealing with the internal design of retaining walls.

8.2.7.2 *Reinforced concrete retaining walls*

Figure 8.19 illustrates the author's proposal for the internal design of reinforced concrete retaining walls.

In the case of walls founded on soil or very weathered rock, the adopted thrust should correspond to the average of the at-rest and active thrusts (Figure 8.19a). For walls with foundation on fresh or slightly weathered rock, the at-rest thrust should be considered, which is equivalent to assuming that the foundation deformation is null (Figure 8.19b).

The option of Figure 8.19a is compatible, for instance, with the recommendation of the American Concrete Institute (Coduto, 2001), which establishes a factor of 1.7 by which to multiply the effect of active stresses.

Let us make a very simple exercise: consider a soil with $\phi' = 30°$ and the Rankine solution, applicable to the case of Figure 8.19. Then:

$$K_a = \frac{1 - \sin\phi'}{1 + \sin\phi'} = 0.33 \tag{8.5}$$

---

[4] The reader, familiar with the design of reinforced concrete structures, will perhaps be able to better understand this issue by considering the design model of the bending reinforcement at a given cross section of a beam. It is assumed that part of the section has tensile cracking, with only the steel reinforcement being active; in the remaining (non-reinforced) part, compression stress is mobilized on the concrete. The combination formed by the reinforcement tensile force and the resultant of concrete compression stresses supplies the so-called resisting moment. The safety check requires that it be greater than or equal to the applied moment. Why are the beams in our structures not cracked, as in the described model? Because, in this model, the resisting moment is calculated from factored (reduced) values of strength parameters of concrete and steel, and the acting moment is obtained from factored (increased) actions. Therefore, the use of safety factors renders the model and the real structure different. In gravity retaining walls, something analogous occurs!

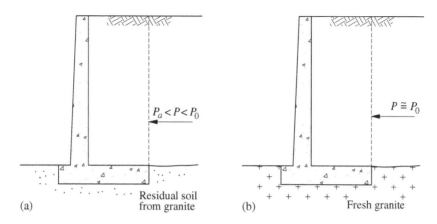

Figure 8.19 Typical situations of reinforced concrete cantilever walls concerning the stresses to be considered for internal design: a) foundation in soil or very weathered rock; b) foundation in fresh or slightly weathered rock.

$$K_0 = 1 - \sin\phi' = 0.50 \tag{8.6}$$

and

$$\frac{K_a + K_0}{2} = 0.42 \tag{8.7}$$

If a safety factor of 1.35 is adopted for the permanent actions, in accordance with Eurocode 7, the design pressure will be factored by $1.35 \times 0.42/0.33 = 1.70$ relative to the active pressure.

Note that, in the preceding considerations, no bending deformation of the wall stem has been taken into account. It might have been invoked that the (internal) failure of the wall itself would have been necessarily preceded by sufficient stem deformation to mobilize the active state in the backfill. However, it is not convenient to take such deformation into account for justifying the adoption of active state, because that may imply significant concrete cracking, with non-compliance with the serviceability limit state of the structure (EN 1992-1-1:2004).

8.2.7.3 Stone masonry, gabion, and cyclopean concrete retaining walls

In stone masonry, gabion, and cyclopean concrete retaining walls, it is necessary to perform certain checks of the stress state in sections parallel to the base.

The action applied in each section can be expressed by the resultant of the normal stresses, N_S, the resultant of the tangential stresses, T_S, and the moment, M_S. These actions, on any section parallel to the base of the wall, are calculated in a similar way to that presented in Section 8.2.6 for the wall base.

In sections corresponding to joints between blocks in masonry or gabion walls, given that the tensile strength is zero or practically zero, the resultant of the applied stresses should pass in the middle third, thus eccentricity must be lower than $B/6$.

In gabion walls, the maximum normal stress (see expressions in Figure 8.7) must be controlled in order to prevent excessive deformation by block compression, as in the example illustrated by Figure 8.20.

Figure 8.20 Front face of a gabion retaining wall revealing internal deformation by excessive compression (photo: Pedro Costa).

The normal admissible stress can be estimated as (Maccaferri, 1987):

$$\sigma_a \left(\text{kPa} \right) = \left(\frac{5\gamma_g}{\gamma_w} - 3 \right) \times 98 \tag{8.8}$$

where γ_g and γ_w are the gabion and the water unit weights, respectively.

In gabion walls, the safety against sliding on the block interfaces must also be verified. The corresponding global safety factor is given by:

$$F = \frac{T_R}{T_S} \tag{8.9}$$

where T_R is the sliding resisting force, given by:

$$T_R = N \tan \phi^* \tag{8.10}$$

with ϕ^* representing the friction angle between gabion baskets, which is a function of the type of filling aggregate and of its density. Its value may be empirically estimated (Maccaferri, 1987) as:

$$\phi^* \left(\degree \right) = 25 \frac{\gamma_g}{\gamma_w} - 10 \tag{8.11}$$

A similar verification is indispensable in concrete block quay walls, the friction angle being determined experimentally.

8.3 EMBEDDED FLEXIBLE WALLS

8.3.1 Introduction

This section is dedicated to certain types of excavation-retaining structures, as in the example of Figure 8.2. In this case, the retaining wall is, in general, installed in the ground prior to excavation. Its weight does not play a significant role in ground support.

In contrast to gravity walls, these structures are designated as *flexible*, experiencing bending deformations in service that may affect the magnitude and distribution of earth pressures and associated internal forces (Rowe, 1952; Peck, 1969, 1972).

In general, these structures have a significant embedded height, i.e., they are extended beneath the excavation base. Earth pressures are mobilized on both faces of the wall embedded height, which is essential for supporting the ground adjacent to the excavation.

When the retaining wall has no structural support above the excavation base, i.e., when its stability is exclusively ensured by the earth pressure exerted on its embedded part, the wall is designated as a *cantilever embedded wall* (Figure 8.2c). This solution is frequently adopted in excavations of relatively low depth.

In other situations, stability is achieved with the additional contribution of one or more levels of structural supports – such as struts, anchors or slabs – complementing the earth pressures on the embedded part and allowing very deep excavations to be performed, cases increasingly frequent in urban environment.

This section will focus on cantilever embedded flexible walls and on walls with a single support close to the top (see Figures 8.2d, 8.2e and 8.2f), the analysis of which may be performed by limit equilibrium methods.

For the analysis of multi-propped or multi-anchored flexible retaining walls, numerical models are required, namely those based on the finite element method. Their comprehensive treatment is beyond the scope of this book. The reader may deepen their knowledge on this topic with works such as those of Puller (1996) and Ou (2006).

8.3.2 Cantilever embedded walls

8.3.2.1 Analysis methods: safety factors

Cantilever embedded flexible walls may provide convenient solutions for supporting temporary or permanent excavations of low to medium depth (typically up to 4–5 m depth), when they are not in the immediate vicinity of structures and infrastructures particularly sensitive to ground movement.

The stability of a cantilever embedded flexible wall exclusively depends on the earth pressures developed along both faces of its embedded height.

The conventional design methods of flexible cantilever walls are limit equilibrium methods, which involve the consideration of active and passive earth pressures in close relation with an assumed movement of the wall as a rigid body. The design mainly consists of calculating the embedded wall height and the bending moment distribution.

Figure 8.21 summarizes the so-called *British method* (Padfield and Mair, 1984; Powrie, 1996) for the simplified case of a homogeneous ground above the water table and without surface surcharges. It is assumed that the wall rotates around a pivot point P close to its tip (Figure 8.21a). Due to this movement, above point P, active stresses are mobilized behind the wall and passive stresses are mobilized in front; below point P, the reverse occurs, with passive stresses acting on the back face and active stresses on the front face.

Figure 8.21b illustrates the base of the British simplified method: the horizontal stresses mobilized below the pivot are replaced by their resultant, R_d, applied at that point. By means

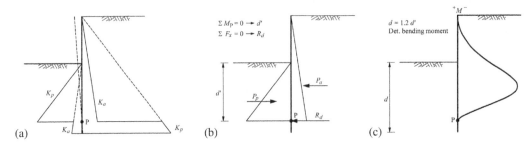

Figure 8.21 British limit equilibrium method for design of embedded cantilever walls: a) wall movement and idealized pressure distribution; b) force system considered and equilibrium equations; c) bending moment graph corresponding to the force system represented in figure b).

of the moment equilibrium equation in relation to the application point of R_d, the distance d' between that point and the bottom of the excavation is determined. Establishing the force equilibrium in the horizontal direction, the value of R_d is calculated. Given the geometry of the structure and the applied loads, the bending moment distribution can be computed, as shown in Figure 8.21c.[5] The wall embedded height, d, is obtained by multiplying d' by 1.2. A final verification should be performed to confirm that the algebraic sum of the active and passive pressures between d' and $1.2d'$ is greater than R_d.

With the help of a spreadsheet, the "exact" solution for the wall embedded height is readily obtained, assuming the pressure distribution represented in Figure 8.21a. Force and moment equilibrium can be achieved by means of the algebraic sum of areas and first-order moments of the various triangles that appear in the pressure diagram. This procedure, which corresponds to the British exact method, may, however, become complex in the case of stratified ground.

The simplified method provides solutions on the safe side, relative to the exact method, for two reasons: i) the vector sum of the passive and active pressures between d' and $1.2d'$ is greater than R_d; and ii) raising the application point of R_d to point P implies a decrease in the stabilizing moment developed in the wall embedded height.

The application of the method presented above must be combined with the use of safety factors. Several ways of introducing these factors are known (Padfield and Mair, 1984; Powrie, 1996):

i) Method 1 – divide the passive pressures in front of the wall by a factor, F_p, ranging from 1.5 to 2.0;
ii) Method 2 – divide the net passive resistance (calculated from K_p-K_a) on the embedded length (that is, below formation level) by a factor, F_r (Burland et al., 1981); values from 1.5 to 2.0 have been used;
iii) Method 3 – multiply the embedded wall length by a factor, F_d; a value of 1.3 is normally adopted;
iv) Method 4 – apply a partial safety factor, F_s, to reduce the soil strength, $\tan \phi'$; $F_s = 1.25$ is recommended.

Safety factors according to Eurocode 7 will also be presented later.

---

[5] The moments compressing the exposed face of the wall (i.e., on the excavation side) are conventionally considered negative.

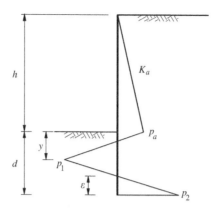

Figure 8.22 American limit equilibrium method for design of embedded cantilever walls.

Figure 8.22 shows the pressure distribution and further features that are the basis of the so-called *American method*, for the simplified case of a homogeneous ground above the water table and without surface surcharges (Bowles, 1988; King, 1995; Day, 1999). This distribution is based on the following assumptions:

i) active pressures are applied behind the wall, except for the zone close to the tip, reaching, at depth h, the value $p_a = K_a \gamma h$;
ii) the passive pressure is fully mobilized immediately below the bottom of the excavation, so that the pressure gradient between p_a and p_1 is equal to $\gamma(K_p - K_a)$;
iii) at the tip of the wall, passive pressure is mobilized behind the wall and active pressure in front, so $p_2 = K_p \gamma(h + d) - K_a \gamma d$.

Taking into consideration the previous hypotheses, a system of two equations can be established, one of moment equilibrium relative to the wall tip and the other of force equilibrium in the horizontal direction, with the distances d and y as unknowns. System solution provides the embedded length, d. With this value and y, the pressure values p_1 and p_2 are obtained. Given the geometry of the structure and the pressure diagram, the bending moment distribution can be computed.

The results provided by the American method are very close to those of the exact British method, whether the embedded height, the bending moment distribution or, in particular, its maximum value.

Safety considerations have been accounted for in the American method by two approaches (Day, 1999): i) multiplying the embedded height by 1.3; or ii) reducing the passive thrust.

8.3.2.2 Brief parametric study

A brief parametric study is now presented in order to compare the application results of the British method, using distinct types of safety factor.

An excavation in a homogeneous soil is considered with the angle of shearing resistance ϕ' varying from 25° to 40°, in order to cover the most common range of shear strengths from plastic clayey soils to dense sands. Values of the soil unit weight have been adjusted following the variation of ϕ' (19 kN/m³ to 22 kN/m³). The water table was assumed to be below the wall tip or coincident with the bottom of the excavation.

For those simple conditions, the ratios d/h and $M_{max}/(\gamma h^3)$ are independent of the value of h (Matos Fernandes et al., 2002). This permits the development of charts like the ones represented in Figure 8.23.

Two conditions were considered for the soil–wall friction: $\delta = 0$ and $\delta = (1/3)\phi'$. The active and passive earth pressure coefficients have been obtained from the theories of Rankine ($\delta = 0$) and of Caquot and Kérisel ($\delta = (1/3)\phi'$).

From the various forms of safety introduction referred to above, the following were considered:

i) a safety factor, F_p, dividing the passive thrust by 2.0;
ii) a safety factor, F_r, dividing the net available passive resistance (calculated from $K_p - K_a$) on the embedded length by 2.0;
iii) the Design Approach 1 of Eurocode 7: Combination 1 implies factoring the active thrust by 1.35; Combination 2 is equivalent to what has been designated above as Method 4, with $F_s = 1.25$; as seen before, the design is governed by the most unfavorable result of these two combinations;
iv) the Design Approach 2 of Eurocode 7; this implies factoring the active thrust by 1.35 and dividing the passive thrust by 1.4.

When dealing with the application of Design Approach 1 of Eurocode 7, the critical combination is indicated in the figure. For an adequate comparison with the results of Eurocode 7, the bending moments corresponding to the application of coefficients F_p and F_r have been multiplied by 1.35.

The examination of the figure permits the identification of some conclusions:

i) differences among the results of the various approaches tend to disappear with increasing angle of shearing resistance;
ii) as expected, the use of F_p is more conservative than the use of F_r, if the same value of the safety factor is adopted;
iii) $F_r = 2.0$ leads to results very similar to Design Approach 2 of Eurocode 7;
iv) when Design Approach 1 of Eurocode 7 is applied, Combination 2 always controls the sizing of the structure, and, in most cases, it also controls the required bending resistance of the wall;
v) Design Approach 1 of Eurocode 7 is less conservative than Design Approach 2 for smaller values of ϕ' and is more conservative for values closer to the upper limit of the range considered.

These trends seem to be essentially independent of the water table conditions and of the resistance of the soil–wall interface.

Comparing the first two pairs of graphs of Figure 8.23, the remarkable influence of the position of ground water surface can be easily evaluated. For $\phi' = 30°$, raising the water level from the wall tip up to the base of the excavation leads to an increase of around 50% on the embedded wall height and 40% on the maximum bending moment.

A comparison of the second and third pairs of graphs of the same figure, which correspond to the same water conditions, permits assessment of the influence of the soil–wall angle of friction. It can be observed that a smooth interface results in a substantial increase in the embedded wall height and in the design bending moments. This issue will be discussed below (see Section 8.3.2.3).

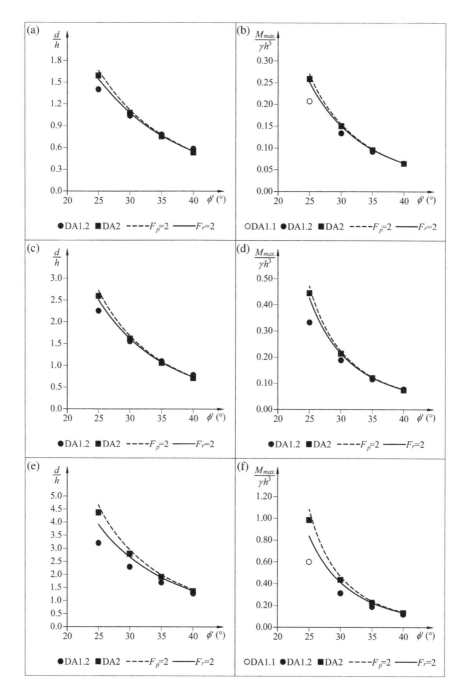

Figure 8.23 Parametric study on the embedded height and maximum bending moment of embedded cantilever wall: (a,b) water level below the wall tip and $\delta = (1/3) \, \phi'$; (c,d) water level at the bottom of the excavation and $\delta = (1/3) \, \phi'$; (e,f) water level at the bottom of the excavation and $\delta = 0$.

8.3.2.3 Vertical equilibrium of cantilever embedded walls

In the particular case of cantilever embedded walls, the soil–wall friction on the front face (on the excavation side) is especially favorable because it substantially increases the magnitude of the passive thrust (see, for example, Table A5.1.8).

However, there is a delicate issue in these structures that requires careful consideration: the tangential friction forces on the soil–structure interface depend not only on the roughness of the structural face and the shear strength of the ground, but also on the equilibrium conditions in the vertical direction, as presented in Figure 8.24.

In fact, considering $\delta \neq 0$ on the passive side implies an upward tangential force applied by the soil to the wall. Since the embedded length is relatively large in this type of structure (see Figure 8.23), so is the tangential force, for relatively high δ values.

Taking as reference Figure 8.24, if that force, P_{pv}, exceeds the sum of the downward tangential force, P_{av}, associated with the active thrust, plus the wall weight, W_w, there are not conditions for the mobilization of the thrust assumed at the front wall face. It will then be necessary to perform an adjustment by reducing the value of δ on that face until the vertical equilibrium is verified.

The following procedure is therefore suggested:

1. adopt initial values for δ as a function of the roughness of the wall face and of the soil with which it interacts; for a homogeneous soil, this will be represented by :

$$\delta_{a,1} = \delta_{p,1} \tag{8.12}$$

  where δ_a and δ_p represent the friction angle of the soil–wall back interface and front interface, respectively;
2. calculate the wall height, using one of the methods presented;
3. check if:

$$P_{pv,1} \leq P_{av,1} + W_{w,1} \tag{8.13}$$

  If this is confirmed, the internal wall design can be performed. Otherwise, the process continues in the following way:

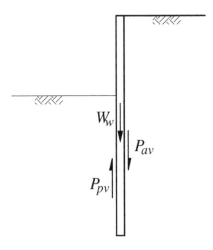

Figure 8.24 Analysis of vertical equilibrium of a cantilever embedded wall.

4. adopt:

$$\delta_{p,2} < \delta_{p,1} \quad \text{and} \quad \delta_{a,2} = \delta_{a,1} \tag{8.14}$$

5. recalculate the wall height;
6. check if:

$$P_{pv,2} \leq P_{av,2} + W_{w,2} \tag{8.15}$$

The steps 4 to 6 should be repeated, if necessary, until verification is obtained.

For reinforced concrete walls (diaphragm walls or pile walls), it is reasonable to assume in the above process:

$$\delta_{a,1} = \delta_{p,1} = \frac{2}{3}\phi' \tag{8.16}$$

where ϕ' is the angle of shearing resistance of the soil in contact with the wall.

8.3.3 Single-propped embedded walls

8.3.3.1 Analysis method: safety factors

Figure 8.25 shows the basis of the so-called Free-earth Support Method (Tschebotarioff, 1951, 1973; Terzaghi, 1953).

Figure 8.25a illustrates the single-propped embedded wall movement; supported at the top and on the embedded part, the wall experiences bending deformation, with maximum displacements above the excavation base. With this movement, active stresses are mobilized behind the wall and passive stresses in front. In the embedded height, displacement is reduced but inversion of the wall curvature is not achieved.

Figure 8.25b displays the force system involved: the resultants of the active and passive pressures and the reaction, R_p, in the structural support at the top of the wall. Moment equilibrium, relative to point P provides a 3rd degree equation with the embedded height, d, as the single unknown. Once the value of d is determined, a horizontal force equilibrium equation supplies the R_p value. Knowing the geometry of the structure and all the applied forces, the bending moment diagram is easily obtained, as represented in Figure 8.25c. Note

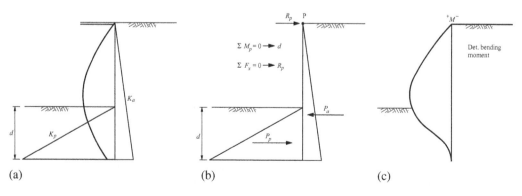

Figure 8.25 Free-earth support method for single-propped embedded walls: a) wall movement and idealized pressure distribution; b) force system considered and equilibrium equations; c) bending moment graph corresponding to the system represented in figure b).

that, in the lower zone, the moment remains positive until reaching a zero value at the tip, as in the extremity of a simple supported beam.

Naturally, the application of this method has to be combined with the introduction of safety factors. As seen for cantilever walls, there are several ways to do it. This subject will be addressed in the next section.

8.3.3.2 Brief parametric study

Similar to what has been done in Section 8.3.2.2, a brief parametric study is now presented but considering a single-propped wall, analyzed by means of the free-earth support method.

The conditions of the ground, water level and soil–wall friction are identical to those considered in Section 8.3.2.2. For these simple conditions, the ratios d/h, $R_p/(\gamma h^2)$ and $M_{max}/(\gamma h^3)$ are independent of the value of h. This permits the development of charts like the ones represented in Figure 8.26 (R_p results are not included). The safety factors have the same symbols and values as discussed in Section 8.3.2.

The comments formulated about the previous parametric study (Section 8.3.2.2) essentially apply also to the results of Figure 8.26.

8.3.3.3 Vertical equilibrium of single-propped and single-anchored walls

Figure 8.27 schematically displays the forces involved in the vertical equilibrium of single-propped (Figure 8.27a) and single-anchored walls (Figure 8.27b).

As can be ascertained by jointly examining Figures 8.26 and 8.23, in single-propped and single-anchored walls, the embedded height is much lower than for cantilever walls, under comparable conditions. Therefore, the verification discussed in Section 8.3.2.3, though still recommended in this case, seldom affects the design.

As the anchors generally present a (descending) inclination relative to the horizontal, the verification concerning friction mobilization on the front wall face now becomes:

$$P_{pv,i} \leq P_{av,i} + W_{w,i} + F_{a,i} \sin \varepsilon \tag{8.17}$$

in iteration i of a process analogous to that described in Section 8.3.2.3.

It is essential to perform another check, concerning failure by insufficient bearing capacity of the wall foundation soil against vertical loading:

$$P_{av,i} + W_{w,i} + F_{a,i} \sin \varepsilon < P_{pv,i} + R_b \tag{8.18}$$

where R_b represents the bearing capacity of the ground adjacent to the wall base.

It is recommended to use the same values for the tangential forces $P_{pv,i}$ and $P_{av,i}$ as in Equation 8.17 in this check. This option is on the safe side. The wall foundation ground should be sufficiently competent for this check to be comfortably satisfied with respect to R_b.

8.3.4 Note on water-related pressure

When the excavation is going to be performed below the pre-existing water table, a fundamental part of design concerns the evaluation of water pressure. In particular, when the water table is close to the surface, a high percentage of the total thrust is due to the water and not to the soil.

As shown in Figure 8.28a, when the wall is impermeable and its toe penetrates a lower impermeable layer, the position of the water table outside the excavation is not altered and

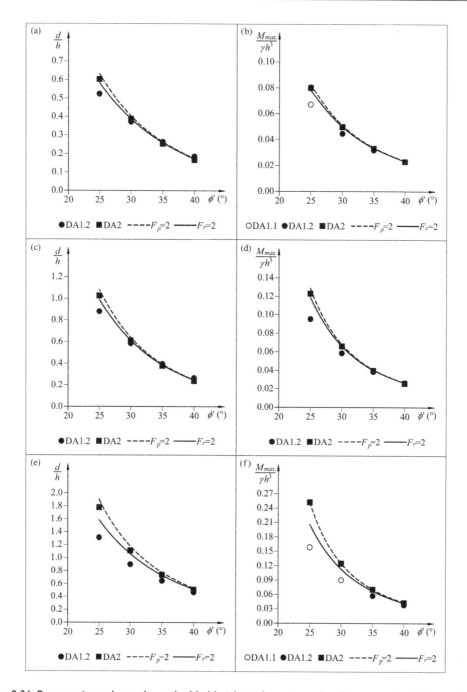

Figure 8.26 Parametric study on the embedded height and maximum bending moment of single-propped embedded wall: (a,b) water level below the wall tip and $\delta = (1/3)\, \phi'$; (c,d) water level at the bottom of the excavation and $\delta = (1/3)\, \phi'$; (e,f) water level at the bottom of the excavation and $\delta = 0$.

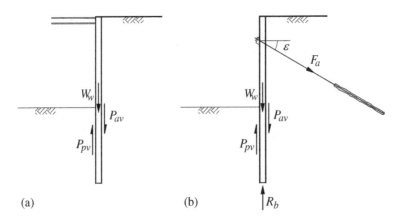

Figure 8.27 Vertical equilibrium analysis: a) single-propped embedded wall; b) single-anchored embedded wall.

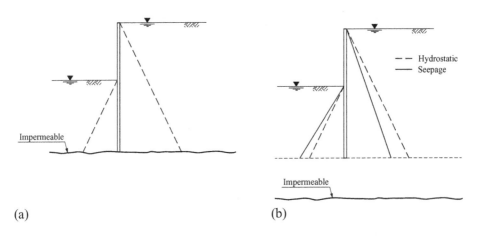

Figure 8.28 Excavation below the water level: a) wall tip reaches lower impermeable boundary (hydrostatic conditions); b) wall tip does not reach lower impermeable boundary (flow towards the excavation).

the water conditions can be considered hydrostatic on both sides of the wall; on the excavation side, the water table will in general be lowered as a result of the excavation process. The earth pressures below the water level must be calculated, taking into account the submerged unit weight (see Section 5.4.4).

When the impermeable boundary is below the wall tip, as shown in Figure 8.28b, there is now flow directed towards the base of the excavation, passing under the wall tip. Due to this flow, the water pressure is less than hydrostatic behind the wall and greater on the opposite side. This variation ensures that, at the wall tip, the water pressure is equal on both sides.

The calculation of water and ground pressures, under the conditions associated with Figure 8.28b, should be based on the flow net. A simplified solution can be obtained without recourse to the flow net by assuming that the total head loss is uniformly distributed along the flow line adjacent to the wall face.

For clarity, consider Figure 8.29. The total head loss, h_w, is processed along the flow line ABC, with length $h_w + 2d$. Assuming the criterion referred to above, the head loss between A and B is:

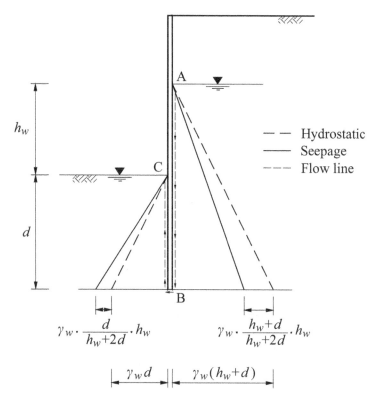

Figure 8.29 Simplified solution for obtaining the water pressure distribution in flow conditions towards the excavation.

$$\Delta h_{AB} = \frac{h_w + d}{h_w + 2d} \cdot h_w \qquad\qquad (8.19)$$

and between B and C is:

$$\Delta h_{BC} = \frac{d}{h_w + 2d} \cdot h_w \qquad\qquad (8.20)$$

These two head losses are proportional to the differences, marked in the figure, between the water pressure, u_B, at the wall tip and the hydrostatic (independent) pressures corresponding to the external water level, $\gamma_w (h_w + d)$, and to the excavation base, $\gamma_w d$.

It is easily confirmed that:

$$u_B = 2\gamma_w d \frac{h_w + d}{h_w + 2d} \qquad\qquad (8.21)$$

which permits calculation of the pressure distribution under flow conditions that correspond to the solid lines in the figure.

Figure 8.30a shows an example of an excavation in a homogeneous sand with permeability isotropy. It is assumed that a thin, superficial layer exists with much higher permeability, which ensures a constant water level on the supported side. The figure displays the flow net,

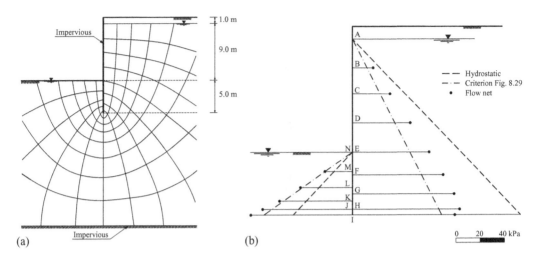

Figure 8.30 Excavation below the water table: a) flow net; b) water pressure distribution on both sides of the wall calculated from the flow net or from the simplified criterion of Figure 8.29.

assuming that the excavation has been concluded and that the water level coincides with the excavation base, due to permanent pumping.

Figure 8.30b shows the water pressure distribution on both sides of the wall, calculated from both the flow net and from the simplified criterion previously described. It can be seen that the differences between the two results are more substantial in the supported side, with the water pressures supplied by the simplified solution being lower than those calculated from the flow net.

The explanation for this deviation may be found by examining the flow net of Figure 8.30a, comparing, in particular, the results for both sides of the wall below the elevation of the excavation base. In points adjacent to the wall at the same elevation, the hydraulic gradient is lower on the upstream side. In other words, the simplified criterion of Figure 8.28 overestimates the head losses on the supported side.

Caution is, therefore, recommended when applying that criterion, particularly when the water level outside the excavation is considerably above its base. In a stratified ground with layers of contrasting permeability, the hypothesis of uniform head loss along the path from the upstream to the downstream boundaries is naturally even less reasonable.

Knowing the water pressure, it is possible to calculate the effective vertical stresses next to the wall and to obtain from them, by applying the active and passive thrust coefficients, the active and passive stress distributions. Taking into account the differences between the water pressures and the (independent) hydrostatic pressures on both sides of the wall, it is readily concluded that, under flow conditions, the active stresses behind the wall increase while the passive pressures in front decrease, in comparison with the hydrostatic conditions. This may be interpreted as a result of seepage force action, which has a downward direction behind the wall (thus increasing the effective vertical stress) and an upward direction in front of the wall (thus decreasing the effective vertical stress).

8.3.5 Construction sequence combining cantilever and single-propped walls in distinct stages

In Sections 8.3.2 and 8.3.3, the sizing of cantilever and single-propped walls has been discussed. As shown in Figure 8.2, the two structural types are often adopted in different locations of the same work, according to the depth to be reached and the clearance to be respected. In roadworks, the minimum clearance, in terms of height, is usually 5.0 m. This means that, in the context of works, such as those of Figure 8.2, it is convenient to use the cantilever solution until depths reach a little more than that value, taking into account that the dimension of the top support cannot be zero, although its impact on the clearance can be reduced, as exemplified in Figure 8.2e. In addition, achieving a depth of 5.0 m in the permanent phase requires that a greater depth be considered in the construction phase, in order to account for the pavement structure construction. Therefore, it may be concluded that the cantilever wall may have to reach a depth close to 6.0 m, about 20% higher than that of the permanent phase.[6]

The situation described is particularly unfavorable because, in a cantilever wall, both the embedded height and the maximum moment (and, *a fortiori*, the displacements) increase very sharply with depth. An interesting perspective of the problem may be obtained by recourse to the design charts of Figures 8.23 and 8.26. Figure 8.31 shows the relation of d/h and of $M_{max}/\gamma h^3$ with the angle of shearing resistance, for both cantilever and single-propped walls, considering the other conditions indicated. A reduction of about 2.5 times exists in the d/h ratio from the first to the second structural type. The reduction is even more pronounced for the $M_{max}/\gamma h^3$ ratio, namely between 3 to 4 times!

This contrast is curious because it stresses, in a simple way, the much lower structural efficiency of the cantilever solution compared to the single-propped solution.

A particularly convenient solution for the problem outlined – that of having cantilever walls for relatively large excavation depths and, in particular, larger in the temporary phase than in the permanent one – may be encountered with a construction phasing, as illustrated in Figure 8.32, which combines the two types of structure analyzed before.

The advantages of this solution are diverse and relevant:

i) allows a reduced excavation depth, for which the wall functions as a cantilever;
ii) achieves a more favorable wall behavior, because the passive thrust in front of the wall may be supplied by a rigid pavement or an improved soil layer (stage 6 of Figure 8.32);

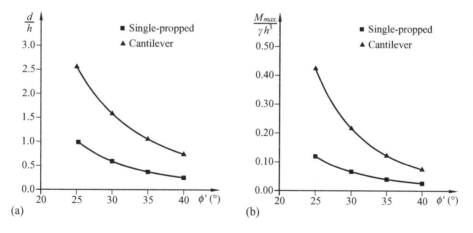

Figure 8.31 Relations (a) ϕ' vs. d/h and (b) ϕ' vs. $M_{max}/\gamma h^3$ for cantilever and single-propped embedded walls (homogeneous soil, water level at the excavation base, $\delta = (1/3)\,\phi'$ and design with safety factor $F_r = 2.0$).

---

[6] As a parenthesis, it should be noted that the solution with anchors (Figure 8.2d) may enable a clearance for an excavation of any depth. However, the use of permanent anchors in urban environment is frequently prevented, either by the client or by the local authority.

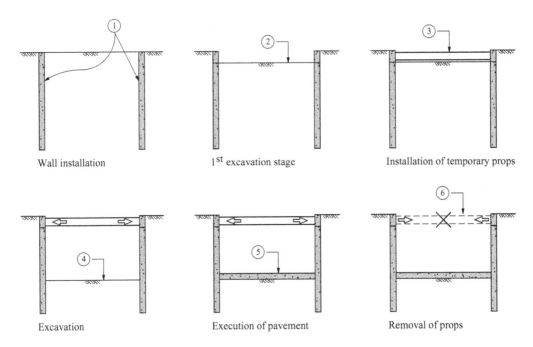

Figure 8.32 Sequence combining a single-propped wall in the construction phase with a cantilever wall in the permanent phase.

iii) last but not least, the construction phasing introduces a stress state in the wall (associated with positive moments) in sections where, in the permanent phase, negative moments will be mobilized, i.e., the construction phasing and the temporary shoring provide the wall with a pre-stress relative to the permanent phase.

8.4 ANCHORS

8.4.1 Introduction: terminology and execution

In geotechnics, an anchor may be defined as any structural element capable of transmitting an applied tensile force to a bearing stratum. Consideration will be given exclusively to anchors of high-strength steel sealed with cement grout under pressure in soil or rock ground (bond anchor).

Within the normative documents on anchors, special reference is given to the standards EN 1537:2013 on execution, EN ISO 22477-5:2018 on testing and also EN 1997-1:2004/A1 (2013), i.e., Eurocode 7, on design. These three documents have recently been updated and form a coherent set. They constitute, therefore, a relevant reference for what follows. The notation adopted here is essentially that used in those standards.

Figure 8.33 includes the scheme of a pre-stressed bond type anchor with the main designations normally employed. The essential anchor element is the steel tendon, which encompasses the three fundamental anchor zones: i) the head, where the steel tendon is connected to the structure to be anchored; ii) the fixed length grout body, which is the resistant zone, through which the tendon tensile force is transmitted to the bearing ground; and iii) the free length between the two previous ones, along which there is no interaction between the tendon and the ground.

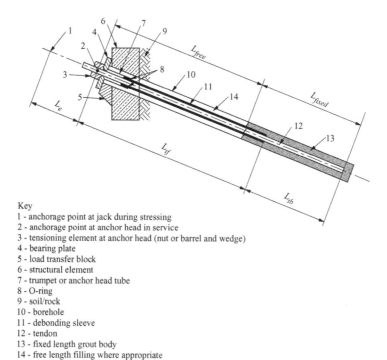

Key
1 - anchorage point at jack during stressing
2 - anchorage point at anchor head in service
3 - tensioning element at anchor head (nut or barrel and wedge)
4 - bearing plate
5 - load transfer block
6 - structural element
7 - trumpet or anchor head tube
8 - O-ring
9 - soil/rock
10 - borehole
11 - debonding sleeve
12 - tendon
13 - fixed length grout body
14 - free length filling where appropriate

Figure 8.33 Sketch of a bond type ground anchor (EN 1537:2013; ISO 22477-5:2018).

The main execution phases of an anchorage are: i) drilling of the borehole, if necessary, accompanied by internal lining; ii) introduction of the tendon and its positioning at the borehole axis with the help of spacers; iii) connection between the tendon and the ground along the bond length, by means of grout injection under pressure; and iv) tendon prestressing and fixing at the anchor head.

Several features, such as the drilling process, the grout composition, the grouting process and pressure, the tendon, and its corrosion protection, are likely to vary, whether with the employed technology, or with the site conditions, and the type of work.

The drilling should disturb the natural ground conditions as little as possible, because the ground conditions will play a decisive role in the anchor performance. An even more important operation for that performance is grouting. The grout is a mixture of water and cement and, possibly, some admixture (a setting accelerator, for example).

Normally, one starts by filling the borehole, from the bottom to the top, by gravity. Filling shall continue until the consistency of the emerging grout is the same as that being introduced (typically, $0.3 <$ water/cement ratio < 0.4). Then the fixed length is grouted (typically, $0.4 <$ water/cement ratio < 0.6) under pressure (up to 3–4 MPa).

The injection of the bond length is performed along limited extents, from bottom to top, in order to increase the effectiveness of the procedure. In less favorable ground, after a primary injection, additional injection phases may be required in order to improve the conditions of the surrounding ground, and the connection between the grout bulb and the ground. This *selective reinjection* technology (i.e., repeated in limited stretches), has a decisive importance in the anchor performance, being able to substantially increase the bulb pull-out force for a given ground and bond length.

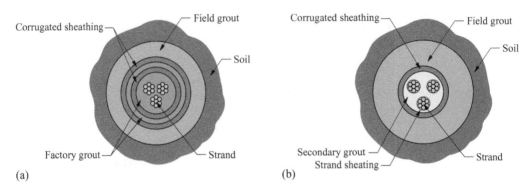

Figure 8.34 Double protection against corrosion of a permanent anchor: a) bond length; b) free length.

The borehole diameter is usually between 10 and 15 cm, corresponding to the diameters of the most current drilling equipment. However, the final average diameter of the grout bulb, when performed in soil, is normally greater than the drilling diameter due to the injection pressures utilized for bonding. Littlejohn (1990) references values for the ratio between bulb and borehole diameters of 1.5 to 2.0 in medium dense sands and of 1.2 to 1.5 for very dense sands. In the case of coarse sands and gravels, where grout permeation is possible, this ratio may reach twice those values.

From this brief description of the construction operations, it is obvious that, in works involving anchors, the quality of workmanship is extremely important for the final result.

As to what concerns tendon protection, the requirements depend on the expected service life of the anchors. They are considered temporary when that period does not exceed two years, and permanent otherwise. In permanent anchors, protection against corrosion assumes vital importance. Figure 8.34 shows an example of protection against corrosion for anchors of this type. Temporary anchors should also be provided with protection against corrosion, given the local site conditions.

Utilizing the most current execution technologies, anchors can be constructed in a wide range of sandy and clayey soils of sedimentary origin and residual soils, with the exception of soft and medium clays. Current values of service tensile forces range from 400 to 600 kN in stiff and very stiff clays, and from 500 to 800 kN in dense to very dense sands and also in residual soils from granite.

8.4.2 Mechanical behavior of anchors

8.4.2.1 Limit states: force-displacement graph

In anchors, both ultimate and serviceability limit states need to be considered. The ultimate limit states are basically two: i) tendon failure; and ii) failure by grout bulb sliding relative to the surrounding ground.

The most important serviceability limit state corresponds to excessive deformation, with the corresponding loss of pre-stress, due to creep of the soil surrounding the grout bulb and of the soil–grout bulb interface. In general, this limit state is critical only in the design of permanent anchors in fine-grained soils.

Consider the generic anchor of Figure 8.35a, in which points 1, 2 and 3 represent, respectively, the head, the boundary between the free and fixed lengths, and the extremity of the

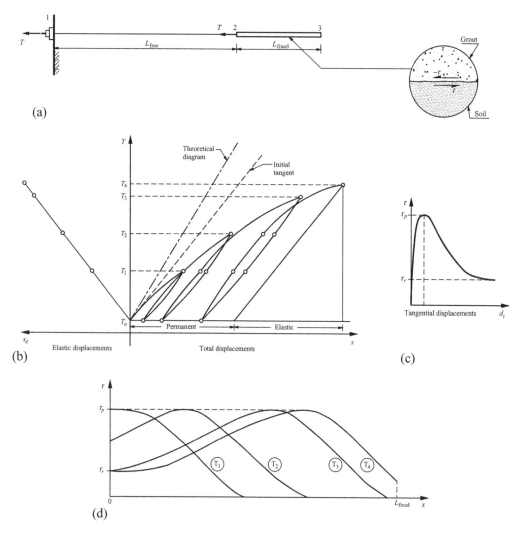

Figure 8.35 Anchor mechanical behavior: a) simplified layout; b) typical tensile force-displacement graph; c) behavior of the grout–soil interface; d) tangential stress distribution on the grout–soil interface for several load stages.

fixed length. Let L_{free} and L_{fixed} be, respectively, the free and fixed anchor lengths and E_t and A_t the elastic modulus and the cross sectional area of the tendon, respectively. Assume, for simplicity, that the tendon free and bond lengths, L_{tf} and L_{tb}, respectively, are equal to L_{free} and L_{fixed} (see Figure 8.33).

An analysis is presented here of the shape of the graph that relates the tensile force, T, applied to the tendon close to the anchor head, with the absolute tendon displacement, s, measured at the same point, as shown in Figure 8.35b. The graph starts at a point with non-zero ordinate, T_a, the datum load, which is the tensile load required for stretching the tendon (in general, assumed to be equal to $0.1T_p$, with T_p being the maximum test proof load).

If the anchor conditions of the grout bulb and of the surrounding ground guarantee zero displacement for point 2, the graph obtained for T values lower than the tendon elastic limit would be a straight line with the equation:

$$T - T_a = \frac{E_t A_t}{L_{free}} \cdot s \qquad (8.22)$$

This straight line, represented by the dash-dot line in Figure 8.35b, is designated as the *theoretical anchor diagram* and its slope, K_{theo}, as the anchor theoretical stiffness:

$$K_{theo} = \frac{E_t A_t}{L_{free}} \qquad (8.23)$$

The theoretical diagram represents, therefore, the contribution for the measured displacement from the elastic deformation of the tendon free length.

If, to this displacement is added the displacement caused exclusively by the elastic deformation beyond point 2, i.e., of the grout bulb and the surrounding ground, the dashed line is obtained, whose slope is lower than K_{theo}.

The real loading diagram is given by the continuous line, which starts tangential to the dashed line. The progressive gap existing between both lines essentially corresponds to the contribution from the relative displacement (slip) at the contact between the grout bulb and the ground, and thus to predominantly plastic deformations. The graph tends toward a plateau corresponding to failure by sliding of the bond length relative to the ground (the so-called *pull-out anchor resistance*), in case tendon failure is not reached for a lower tensile load value.

If, after raising the tensile load to a certain value, an unloading-reloading cycle is performed, as shown in Figure 8.35b, the corresponding diagram is approximately linear (though displaying a small hysteresis). Its slope is practically independent from the tensile load level at the start of the cycle, and is, in general, very close to the slope of the initial tangent (dashed line). Therefore, the elastic displacement diagram is, in general, very close to this line.

8.4.2.2 Mobilization of the grout bulb–ground interface resistance

The equilibrium of force T applied at point 2 (in Figure 8.35a) is achieved with the mobilization of tangential stresses on the grout bulb–ground interface. This problem has an obvious analogy with that of the mobilization of shaft resistance, discussed in Chapter 7. However, there are relevant differences between the two problems, namely: i) in the present case, the axial stiffness of the structural element (the grout bulb) is much lower, so that significant (tensile) axial deformation will occur; ii) this deformation causes tangential interface displacements of considerably different magnitude between point 2 (maximum) and point 3 (minimum); iii) the grout bulb–ground interface has a tangential displacement *versus* tangential stress law with a significant post-peak strength loss (as shown in Figure 8.35c), due to the grout bulb injection process previously described and the soil types in which anchors are constructed; and iv) consequently, the tangential stress distribution on the interface is typically non-uniform and varies substantially with the load level, as illustrated in Figure 8.35d.

In the initial load stages, the peak strength is mobilized only in the zone closer to point 2. Subsequently, this effect is gradually extended along the fixed length, as the relative displacement along the interface reaches the one corresponding to the peak strength. Therefore, when the peak strength is mobilized to more distant zones from point 2, the relative displacement is very high in zones closer to this point, and so the mobilized tangential stress corresponds to residual strength.

As a new load increment is being applied, failure by bulb pull-out becomes apparent when continuing tendon displacement occurs at the anchor head, preventing the tendon from being locked off and the new total load from being reached. This situation will correspond to the mobilization of the residual interface strength practically along the entire fixed length.

The considerations presented explain the two subsequent features typical of anchor mechanical behavior:

i) the tensile force that causes pull-out failure does not increase proportionally to the bond-to-ground length, whose increase becomes ineffective beyond a certain value (see, for example, Figure 8.36);

ii) the portion of the tendon displacement at the anchor head, due to the mobilization of the grout bulb–ground interface resistance, is, in general, substantially lower than that associated with the tendon elastic deformation.

8.4.3 Anchor design

8.4.3.1 Limit State Design

In anchored structures, the anchors have to provide forces that prevent both an ultimate limit state, $F_{ULS;d}$, and a serviceability limit state, $F_{SERV;d}$. In anchored or multi-anchored embedded walls, anchor pre-stress is currently established in order to control the excavation-associated movements and its value is frequently higher than that necessary for preventing the ultimate limit state of the anchored structure.

Therefore, the ULS design force to be resisted by the anchor is given by:

$$E_{ULS;d} = \max\left(F_{ULS;d}; F_{SERV;d}\right) \tag{8.24}$$

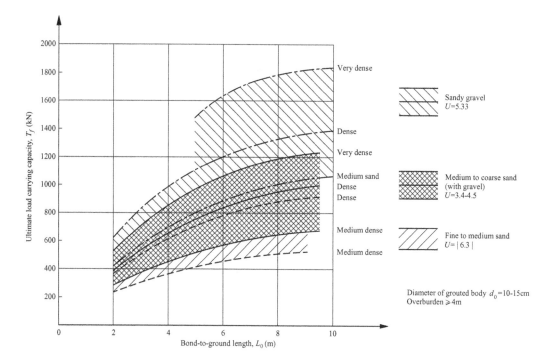

Figure 8.36 Carrying capacity of anchors in non-cohesive soils showing influence of soil type (coefficient of uniformity), density and bond-to-ground length (Ostermayer, 1974; Ostermayer and Scheele, 1977).

with

$$F_{SERV;d} = \gamma_{SERV} \cdot F_{SERV;k} \qquad (8.25)$$

where γ_{SERV} represents the partial safety factor. EN 1997-1:2004/A1 (2013) recommends that $\gamma_{SERV} = 1.35$, for all design approaches.

The safety check with regard to a ULS in the anchor is established through the following pair of equations:

$$E_{ULS;d} \leq R_{ULS;d} \qquad (8.26)$$

and

$$E_{ULS;d} \leq R_{t;d} \qquad (8.27)$$

where:

- $R_{ULS;d}$ is the geotechnical design value of the anchor resistance complying with ULS criteria, usually designated by pull-out resistance;
- $R_{t;d}$ is the structural design value of the anchor resistance complying with ULS criteria, that is, the tensile resistance of the steel tendon.

To prevent a fragile system collapse, it is convenient that these two resistances confirm the following relationship:

$$R_{ULS;d} < R_{t;d} \qquad (8.28)$$

8.4.3.2 Estimation of anchor resistance: anchor load tests and the role of comparable experience

As discussed above, it is necessary to estimate $R_{ULS;d}$, which represents the geotechnical design value of the anchor resistance complying with ULS criteria.

In Chapter 7, the difficulties in estimating the bearing capacity of pile foundations by analytical methods were pointed out. These difficulties are even greater in the case of anchor resistance, taking into account the enormous influence of execution conditions and quality of workmanship. Therefore, there is no reliable analytical method by which to calculate that resistance, based on anchor geometry and on the parameters of the stress state and strength of the surrounding ground.

The anchor resistance estimation may be performed in two ways:

i) by load tests, the most reliable and recommended alternative;
ii) by comparable experience.

As to what concerns load tests, they are detailed in the ISO 22477-5:2018 standard, with the following designations and definitions:

i) investigation test: a load test to establish the geotechnical ultimate resistance of an anchor and to determine the characteristics of the anchor in the working load range;

ii) suitability test: an on-site load test to confirm that a particular anchor design will be adequate under particular ground conditions;

iii) acceptance test: an on-site load test to confirm that an individual anchor conforms with its acceptance criteria.

The anchors submitted to investigation or to suitability tests must have the same geometrical characteristics, be sealed in the same geological formations, and be executed with the same technology of the so-called production or working anchors, which will be the object of acceptance tests in the project under study.

Commenting now on the comparable experience approach, it is curious to observe that estimating anchor bearing capacity is one of the geotechnical areas where that concept has been applied in a most justified and effective way. On one hand, the profession has nowadays more than sixty years of experience in soil anchors. On the other hand, most works incorporating anchors are performed in large urban centers. This explains the improvement of empirical rules that associate: i) a given geotechnical formation; ii) the anchor type; iii) the value of the bond length; and iv) the magnitude of the resistance or, at least, the service tensile load range corresponding to a good performance.

In what concerns bond length, values below 3.0 m and above 10.0 m are not recommended, according to experience. The considerations presented in Section 8.4.2.2 permit an understanding of the reason for the upper limit of that interval. To justify the lower limit, Littlejohn (1990) states that, although it may seem excessive for certain rock anchors, if a very short fixed length is used, any sudden reduction of the ground quality is likely to induce a serious reduction in the anchor resistance. In retaining structures of urban excavations and in stabilization works of natural slopes, when dealing with anchors in competent soils, in most cases the bond length lies between 6.0 m and 8.0 m.

In situations where it is legitimate to invoke comparable experience, one may proceed on this basis for the design of the anchored structure. In the construction phase, this design is verified by means of suitability tests on sacrificial anchors constructed in advance (for example, during the construction of the peripheral retaining wall) and, later, by means of acceptance tests on all the production anchors.[7] If comparable experience is correctly employed and execution is competent, it is highly plausible that these tests confirm the design predictions or, at most, prescribe minor adjustments that do not invalidate the general conception of the anchored structure.

The investigation tests must be performed whenever the anchors are to be constructed in ground conditions that have not been subjected to previous investigation tests, or when the service loads are significantly higher than those adopted under similar ground conditions.

What essentially distinguishes investigation tests from suitability tests is that the former have the declared objective of reaching the anchor pull-out resistance. Therefore, the anchors that are to be submitted to investigation tests normally have higher tendon resistance and may be constructed with smaller bond length. In this case, it will always be necessary to perform supplementary suitability tests with the same bond length as the production anchors.

Another distinctive feature of those two tests is that anchors submitted to investigation tests cannot subsequently be used as production anchors, while the opposite is true for those submitted to suitability tests.

---

[7] Suitability tests may be also performed on production anchors. In many cases, performing acceptance tests on all the working anchors is considered adequate and no suitability tests are carried out.

8.4.3.3 Estimation of anchor resistance from tests

According to Eurocode 7 (EN 1997-1:2004/A1), the measured geotechnical ultimate limit state resistance of an anchor, $R_{ULS,m}$, shall be determined from a number (n) of investigation or suitability tests carried out in accordance with EN ISO 22477-5.[8] This last standard establishes three distinct test methods. In what follows, only the so-called Method 1 will be considered.

As mentioned before, the anchors submitted to investigation and suitability tests generally have higher tendon cross-sectional area, in order to be able to reach the estimated value for the proof load, T_p, with respect to the following conditions:

$$T_p \leq 0.80 \cdot f_{tk} \cdot A_t \tag{8.29}$$

and (depending on the steel quality)

$$T_p \leq 0.95 \cdot f_{t0.1k} \cdot A_t \quad \text{or} \quad T_p \leq 0.95 \cdot f_{t0.2k} \cdot A_t \tag{8.30}$$

where f_{tk} is the tensile steel strength and $f_{t0.1k}$ and $f_{t0.2k}$ are both yield stresses at 0.1% and 0.2% yield strain, respectively.

The referred tests consist of the incremental application of tensile tendon loads with displacement measurement at the anchor head. Unloading–reloading cycles are performed for measuring permanent displacements. When the maximum load value is reached in each cycle, the tensile force is kept constant during a given time period, the displacement evolution being registered for studying creep. The anchor is always unloaded to the datum load, T_a, before being reloaded to the next, higher load cycle.

Table 8.4 presents the periods during which the load should be kept constant, while registering the displacement evolution.

At each load stage, the tendon end displacement *versus* log time is plotted and the slope of the straight line-of-best-fit through these points is determined, which is designated as the creep coefficient, α. The value of α corresponding to each load stage is then represented as a function of the respective load, as shown in Figure 8.37. The load corresponding to a value

Table 8.4 Loading sequence for suitability tests and minimum observation period at maximum load for each cycle (ISO 22477-5:2018).

Cycle	Maximum load	Minimum observation period at maximum load for each cycle (min)			
		Temporary anchor		Permanent anchor	
		Coarse soil and rock	Fine soil	Coarse soil and rock	Fine soil
0	T_a	1	1	1	1
1	$0.40\,T_p$	1	1	15	15
2	$0.55\,T_p$	1	1	15	15
3	$0.70\,T_p$	5	10	30	60
4	$0.85\,T_p$	5	10	30	60
5	$1.00\,T_p$	30	60	60	180

---

[8] Eurocode 7 recommends $n = 3$. Nevertheless, the number of tests should depend on several aspects, such as the number of anchors in the project and the degree of comparable experience.

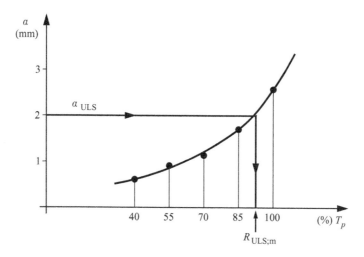

Figure 8.37 Anchor load test (ISO 22477-5:2018): plot of α versus anchor load to calculate $R_{ULS;m}$ (pull-out resistance was reached before target value of T_p).

of $\alpha = 2$ mm (as established in EN 1997-1:2004/A1) is considered to be the pull-out resistance measured in the test, $R_{ULS;m}$.[9]

In cases where α does not exceed that limiting value during the test, the measured pull-out resistance, $R_{ULS;m}$, is defined as the maximum test load, T_p. Then:

$$R_{ULS;m} = \min\{R_m(\alpha_{ULS}); T_p\} \tag{8.31}$$

According to Eurocode 7, if all the anchors are submitted to acceptance tests, the characteristic value of the ultimate limit state geotechnical resistance, $R_{ULS;k}$, may be considered:

$$R_{ULS;k} = \min(R_{ULS;m}) \tag{8.32}$$

The design value of the ultimate limit state geotechnical resistance, $R_{ULS;d}$, is obtained through the equation:

$$R_{ULS;d} = \frac{R_{ULS;k}}{\gamma_{a;ULS}} \tag{8.33}$$

where $\gamma_{a;ULS}$ is the partial safety factor given by Tables 3.7 and A3.2.4.

According to Eurocode 7, the measured serviceability limit state resistance, $R_{SLS;m}$, should be determined from test results, using the following relationship:

$$R_{SLS;m} = \min\{R_m(\alpha_{SLS}); T_p\} \tag{8.34}$$

This allows determination of the characteristic and design values of the serviceability limit state resistance:

$$R_{SLS;k} = \min(R_{SLS;m}) \tag{8.35}$$

$$R_{SLS;d} = R_{SLS;k} \tag{8.36}$$

---

[9] This creep coefficient corresponds to a displacement of about 12 mm between 30 minutes and 50 years.

The limit value of the creep rate (α_{SLS}) recommended in Eurocode 7 is:

$$\alpha_{SLS} = 0.01 \times F_{SERV;k} \cdot \frac{L_{tf}}{E_t A_t} \tag{8.37}$$

which represents 1% of the elastic displacement of the free tendon length under service tensile load.

At each loading stage, the separation of the elastic displacements, s_e, recoverable on unloading, as illustrated on the left-hand side of Figure 8.35b, allows the calculation of the so-called *apparent tendon free length*, L_{app}, as:

$$L_{app} = \frac{E_t A_t s_e}{T - T_a} \tag{8.38}$$

According to ISO 22477-5:2018, the value of the apparent tendon free length should be in the following range:

$$0.8L_{tf} + L_e \leq L_{app} \leq L_{tf} + L_e + 0.5L_{tb} \tag{8.39}$$

where L_e is the external length of tendon measured from the tendon anchorage at the anchor head to the anchorage point in the stressing jack.

8.4.3.4 Confirmation of the anchorage resistance from the results of the acceptance tests

According to the so-called Method 1, an acceptance test consists of applying the anchor load in, at least, five stages until reaching the proof load, T_p, with the objective of checking if the anchor satisfies the design requisites. Figure 8.38 shows the load stages and the period of time for observation of the displacement.

The tendon displacement at the anchor head is registered for each load stage. When the proof load, T_p, is reached, the load is kept constant over a given time period, with measurement of the tendon displacement. After this period, the anchor is unloaded along these stages, being then reloaded to the pre-stress loading, T_0, and locked off.

According to Eurocode 7, the proof load, T_p, to be applied in the acceptance test shall be:

$$T_p \geq \gamma_{a;acc;ULS} \times E_{ULS;d} \tag{8.40}$$

or

$$T_p \geq F_{SERV;k} \tag{8.41}$$

where the safety factor $\gamma_{a;acc;ULS}$ for Method 1 is defined as 1.1.

The option for the verification of the service load (Equation 8.41) is more frequently encountered in retaining structures of deep urban excavations, where the anchor pre-stress loads are established in order to minimize the excavation-induced displacements, being usually much greater than the forces strictly necessary to ensure wall stability.

It should be noted that the pre-stress force (in anchor terminology, the lock-off force) is not the F_{SERV} force. In fact, in the excavation phases performed below each anchor level, the anchor forces typically increase above the lock-off force due to the movement of the wall towards the excavation. If the pre-stressing forces are adequately established, these variations are relatively small. These and other aspects of the behavior of multi-anchored

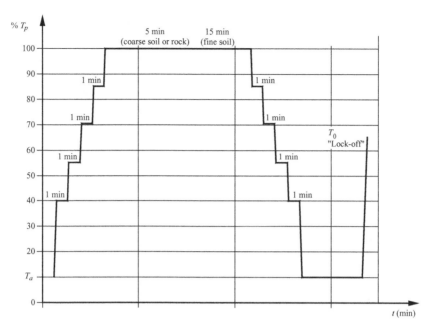

Figure 8.38 Loading sequence for acceptance tests (Method I, ISO 22477-5:2018).

embedded flexible walls can be analysed using numerical models based on the finite element method (Clough and Tsui, 1974; Matos Fernandes and Falcão, 1988; Raposo et al., 2017).

For anchor acceptance, it is necessary to:

i) satisfy the limit for the creep rate under the proof load (2 mm for ULS and Equation 8.37 for SLS); and
ii) satisfy the limits for the apparent tendon free length (see Equation 8.39).

When the apparent tendon free length lies outside the limits, Littlejohn (1990) recommends performing two further load cycles up to proof load, in order to check the repeatability of the load-displacement graph (the results of the 2nd cycle differ by less than $\pm 5\%$ from those of the 1st cycle). If the behavior is consistent and elastic, the anchor should be accepted (see also EN 1537:2013, 9.8.3). Littlejohn (1990) further notices that the elastic modulus, E_t, of the (very long and multi-strand) tendon may be up to 10% lower than the modulus provided by the manufacturer (which is obtained in a laboratory test on a tendon with a length of 610 mm tensioned between two rigid platens). This fact should also be considered in the diagnosis.

8.4.3.5 Aspects concerning geometry

The design of a pre-stressed anchor also involves the establishment of a series of geometry-related issues.

In functional terms, the anchors would be most conveniently horizontal, given that the effective pre-stress component is, in general, only the horizontal. However, execution-related problems disallow inclinations lower than 10° to 15°.[10]

---

[10]In certain works, horizontal anchors are used and even anchors with negative inclination, i.e., with the grout bulb at an elevation higher than that of the anchor head. However, this very rarely occurs in retaining structures of urban excavations and in slope-stabilization works.

Nevertheless, in certain cases, the adopted inclination is substantially higher, sometimes even exceeding 45°. In excavations, this is fundamentally the result of two factors. The first is the presence of obstacles in the close vicinity of the anchored work, such as foundations, buried pipelines, metro tunnels, and others, that lead to inclining the anchors, especially those closer to the ground surface.

The second factor is the depth of the stratum for location of the anchor bond length relative to the elevation of the anchor head. This aspect is indissolubly connected to the value of the total length. As is obvious, in order to reach the load-bearing stratum at a certain depth, the inclination can be increased, decreasing the total length, or, conversely, the option can be to increase the length, decreasing the inclination.

In the establishment of the total length, other exigences have also to be respected, as summarized in Figure 8.39 (Littlejohn, 1972, 1990; Ostermayer, 1976; Hanna, 1983):

i) the grout bodies must be located outside the active wedge of the soil supported by the anchored wall;
ii) the depth below ground surface must be greater than 6.0 m, in order to ensure a given minimum value of the effective stresses at the interface with the ground, and also to prevent damage at the ground surface induced by grout under pressure;
iii) the depth of the grout bodies below the foundations of neighbouring buildings must be greater than or equal to 3.0 m, in order to prevent damage induced by grouting;
iv) the distance between grout bodies must be greater than or equal to 1.5 m, in order to minimize the interference between anchors, and hence possible decrease the respective carrying capacity;
v) the free length should not be less than 6.0 m, so that the force directed to the anchor head applied on the grout body (point 2 of Figure 8.35a) will be degraded inside the ground instead of being "short-circuited" back to the wall.

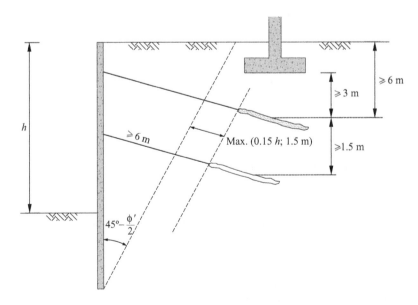

Figure 8.39 Conditions to be respected in the location of anchor fixed length grout bodies.

8.4.3.6 Verification of global stability

Locating the grout bulb outside the active wedge (see Figure 8.39) is a necessary (and obvious) stability condition, but it may not be sufficient. A supplementary stability check is therefore required.

Figure 8.40a presents the proposal of Jelinek and Ostermayer (1967a, 1967b), for the global stability analysis of a single-anchored wall, inspired by the Kranz method (1953), developed for waterfront anchored bulkheads with plate anchors.

It involves the stability analysis of block ABCD, which encloses the anchor bond length (the block boundary passes through the center point of that length, for safety). The idea is to calculate the force, $F_{a,ult}$, applied by the anchor to the block, which leads it to a situation of imminent sliding along plane BC.

Assuming a limit equilibrium situation, the magnitude and direction of the following forces are known: i) the resultant, W, of the block weight and surface surcharges (if their effect is unfavorable); ii) the active thrust, $P_{a,CD}$, on the vertical plane CD, which includes the effect of the surface surcharge to the right of plane CD (the Rankine active thrust should be considered, as discussed in Section 8.2.5.3); and iii) the force, P_a, applied by the wall to the block, which is equal and opposite to the active thrust of the supported soil. The direction is known of the two remaining forces, the reaction, R, in plane BC and the anchor force, $F_{a,ult}$. Closure of the force polygon allows the determination of their magnitude, as shown in Figure 8.40a.

As described by the Kranz classic method, the safety factor:

$$FS_K = \frac{F_{a,ult}}{F_{a,serv}} \qquad (8.42)$$

where $F_{a,serv}$ represents the anchor service load, should be equal to or greater than 1.5.

This approach is susceptible to criticism because it may suggest the (wrong) idea that in these structures it is convenient to have anchors with reduced service load (Littlejohn, 1976).[11]

Figure 8.40b shows an alternative procedure (Broms, 1968), whereby the wall is now incorporated in the block under analysis. Therefore, the active thrust and the anchor force are no longer explicitly involved in the equilibrium, because they are now internal forces to the block, while the passive thrust in front of the wall appears in the equilibrium equations. In this case, the aim is to calculate the safety factor, FS_B, that is, the ratio of the mobilizable passive thrust in front of the wall, P_p, and its mobilized value, $P_{p,mob}$, i.e., the value required for equilibrium of the block:

$$FS_B = \frac{P_p}{P_{p,mob}} \qquad (8.43)$$

Broms (1968) suggests 1.5 for the minimum value of this safety factor.

Whatever the adopted safety factor definition, if its value is considered insufficient, it is necessary to increase the free anchor length, moving the grout bulb away from the wall. This allows reduction of the inclination angle α of plane BC, which is critical for the stability check. In fact, if $\alpha < \phi'$, the force R, the magnitude of which is close to that of the block weight, and thus very large, has a stabilizing horizontal component. Naturally, the opposite occurs when $\alpha > \phi'$.

---

[11] Note, by the way, that the Kranz method has been developed for passive, plate anchorages, i.e., non pre-stressed anchors.

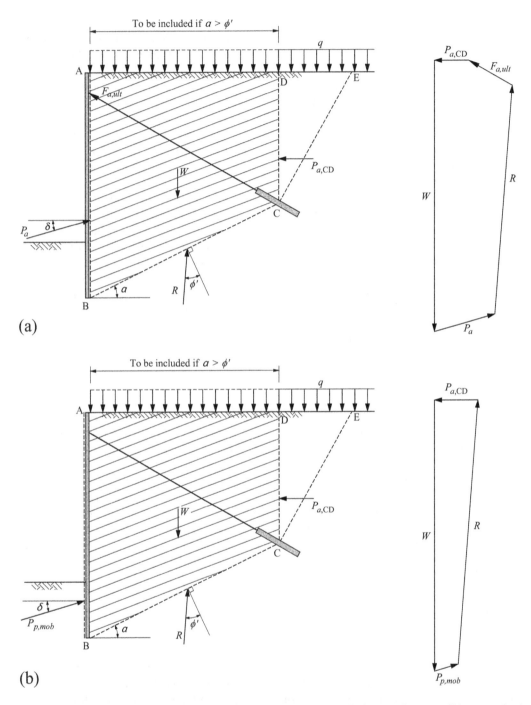

Figure 8.40 Global stability of a single-anchored embedded flexible wall: a) generalization of Kranz method (Jelinek and Ostermayer, 1967a; 1967b); b) method considering the joint soil-wall equilibrium (adapted from Broms, 1968).

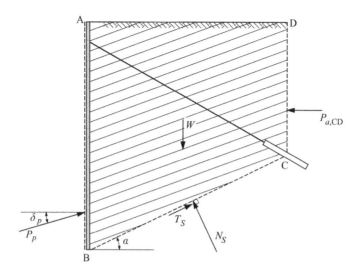

Figure 8.41 Block in limit equilibrium in the global stability analysis of a single-anchored embedded flexible wall.

The essential of the Broms method can be applied but establishing the safety factor in a different way. In fact, since the failure mechanism, as shown in Figure 8.41, involves a block that can slide along its base, under the active thrust action on its back face and a resisting force of the passive type on its front face, the stability analysis can be performed in a way similar to that of the base sliding of a gravity retaining wall, as seen in Section 8.2.6.

Therefore, the force, T_S, that induces sliding along plane BC is:

$$T_S = W \sin\alpha + P_{a,\mathrm{CD}} \cos\alpha \tag{8.44}$$

The normal force, N_S, to that plane is:

$$N_S = W \cos\alpha - P_{a,\mathrm{CD}} \sin\alpha + P_p \sin(\alpha - \delta_p) \tag{8.45}$$

which allows the calculation of the resisting force to sliding, T_R, along plane BC:

$$\begin{aligned}
T_R &= N_S \tan\phi' + P_p \cos(\alpha - \delta_p) \\
&= \left[W \cos\alpha - P_{a,\mathrm{CD}} \sin\alpha + P_p \sin(\alpha - \delta_p) \right] \tan\phi' + P_p \cos(\alpha - \delta_p)
\end{aligned} \tag{8.46}$$

Then, the global safety factor F can be obtained:

$$F = \frac{T_R}{T_S} \tag{8.47}$$

Alternatively, a safety check within the philosophy of Eurocode 7 leads to:

$$T_S \le T_R \tag{8.48}$$

where the several variables in Equations 8.44 to 8.46 should now be taken with their design values, applying the appropriate partial safety factors. The same is valid for a safety check following the LRFD approach.

ANNEX A8.I SEISMIC COEFFICIENTS FOR DESIGN OF RETAINING WALLS, ACCORDING TO EUROCODE 8

This annex summarizes the items of Eurocode 8 (EN 1998-1:2004 and EN 1998-5:2004) that lead to the establishment of the seismic coefficients for use in the design of earth-retaining walls.

The horizontal seismic coefficient is given by:

$$k_h = \left(\frac{a_g}{g} \right) S \frac{1}{r} \tag{A8.1.1}$$

where:

- a_g represents the design value of the (horizontal) seismic acceleration on type A ground, and results from the product of the reference peak ground acceleration, a_{gR}, in the seismic zone where the structure is located, by the importance factor of the structure, γ_I;
- S is a factor which accounts for possible acceleration amplification between the bedrock and the surface;
- r is a factor that depends on the admissible displacement of the retaining wall;
- g is the acceleration due to gravity.

The determination of S has been discussed in Chapter 3 (see Section 3.8). Table A8.1.1 contains the author's proposal of importance classes for retaining walls inspired in Eurocode 0 (Annex B (informative) – Management of Structural Reliability for Construction Works) and in Eurocode 8 - Part 1 (Table 4.3 – Importance classes for buildings) (EN 1990-1:2002 and EN 1998-1:2004), which, jointly with Table 3.14, permits the determination of the structure importance factor, γ_I.

The values of factor r are indicated in Table A8.1.2.

The vertical seismic coefficient is given by (EN 1998-5:2004):

$$k_v = \pm 0.5 k_h \quad \text{if} \quad \frac{a_{vg}}{a_g} > 0.6 \tag{A8.1.2}$$

$$k_v = \pm 0.33 k_h \quad \text{if} \quad \frac{a_{vg}}{a_g} \le 0.6 \tag{A8.1.3}$$

Table A8.1.1 Proposal of importance classes of retaining walls.

Importance class	Retaining wall
I	Walls of minor importance for public safety and whose collapse would have minor economic, social and environmental consequences, e.g., agricultural walls, etc.
II	Ordinary walls, not belonging in the other categories.
III	Walls whose seismic resistance is of importance for people safety in view of the considerable economic, social or environmental costs associated with a collapse.
IV	Walls whose seismic resistance is vital, walls whose collapse would affect hospitals, fire stations, power plants, main railways, highways and bridges.

Table A8.1.2 Values of factor r for the calculation of the horizontal seismic coefficient in function of the admissible displacement of the retaining structure (EN 1998-5:2004).

Type of retaining structure	r
Free gravity walls that can accept a maximum displacement up to $d_r = 300\ a_g\ S/g$ (mm)	2.0
Free gravity walls that can accept a maximum displacement up to $d_r = 200\ a_g\ S/g$ (mm)	1.5
Flexural reinforced concrete walls, anchored or braced walls, reinforced concrete walls founded on vertical piles, restrained basement walls and bridge abutments	1.0

Table A8.1.3 Values of k_h and k_v for use in Lisbon for seismic design of retaining walls (importance factor $\gamma_I = 1$ and factor $r = 1$).

Seismic action	a_g (m/s^2)	$\alpha = a_g/g$	a_{vg}/a_g	S Soil A/B/C	k_h	k_v	Soil
Type 1	1.5	0.153	0.75	1.0/1.29/1.5	0.153	±0.076	A
					0.197	±0.099	B
					0.229	±0.115	C
Type 2	1.7	1.173	0.95	1.0/1.27/1.46	0.173	±0.087	A
					0.220	±0.110	B
					0.253	±0.126	C

As an example, Table A8.1.3 presents the seismic coefficients to be used in Lisbon, given the following conditions:

i) the structure is of importance class II, which corresponds to a unit importance factor;
ii) the admissible displacement, d_r, is null;
iii) the foundation soils are of types A, B and C;
iv) $k_v = \pm 0.5\ k_h$ (according to Table 3.16 and Equation A8.1.2).

For structures with a different importance factor, the values of the table must be multiplied by the values of γ_I given by Table 3.14. For structures that may experience permanent displacement induced by the earthquake, the values in the table must be divided by the values of r given by Table A8.1.2. The soils of Classes D and E are not recommendable for direct foundation of retaining walls in seismic regions unless specialized studies demonstrate its viability.

ANNEX E8 EXERCISES (RESOLUTIONS ARE INCLUDED IN THE FINAL ANNEX)

E8.1 – Figure E8.1 shows a gravity retaining wall, long in length and of constant cross section, supporting a homogeneous fill, and founded on granitic formations.

The wall has been designed for the conditions of case 1 (foundation on a residual soil from granite with the water table at the surface), imposing a minimum global safety factor of 2.0. There are no surface surcharges.

Granitic formations are typically very heterogeneous, and distinct foundation conditions are present in the other three sections, as specified in the figure.

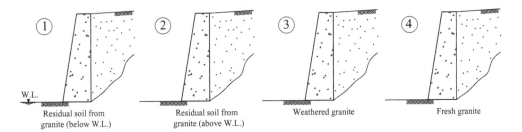

Figure E8.1

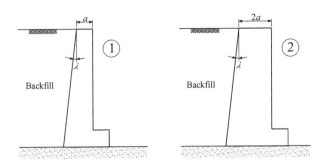

Figure E8.2

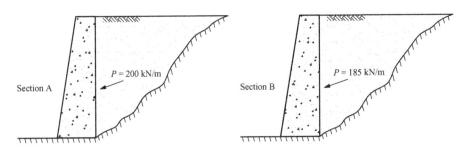

Figure E8.3

Sort in ascending order the thrust P from the supported ground in each of the four cases and relate them, as clearly as possible, to the reference values corresponding to the at-rest state, P_0, and to the active limit state, P_a.

E8.2 – Consider the two gravity retaining walls represented in Figure E8.2. The walls have the same height, foundation soil (sand with medium density), backfill material and back face inclination and roughness). No surcharge loads are applied at the surface.

Sort in ascending order the soil thrust for the two walls and include also the reference values corresponding to the at-rest state, P_0, and to the active limit state, P_a.

E8.3 – Figure E8.3 shows two cross sections of a long gravity retaining wall. The wall and the supported backfill are identical in both cases. No surcharges are applied at the ground surface. The water table is well below the wall base. In these conditions, the theoretical value of the active thrust is $P_a = 156$ kN/m.

In the figure are indicated the values of the total thrust for the two sections, which have been measured by total stress cells installed on the back face of the wall.

Read the following statements and state your opinion with an X, adding a brief explanatory comment.

	Agree	Disagree
1 – It would have been more prudent to design the wall for the passive thrust, which is the larger, rather than the active thrust, which is the smaller.		
2 – The fact that the measured thrust is, in both cases, larger than the active thrust is reasonable, because safety factors have been introduced in the design process.		
3 – The wall is in danger of collapsing because the thrust due to backfill is larger than that considered in design, the active thrust.		
4 – The resistance and stiffness of the foundation soil in section B are lower, providing larger wall displacement, so lower earth pressures.		
5 – The thrust being larger in section A means that, in A, the wall displacement was larger than in section B.		
6 – The soil thrust in both sections is lower than the at-rest thrust.		

E8.4 – Consider the reinforced concrete retaining wall represented in Figure E8.4.

a) Verify the safety relative to base sliding and to the foundation-bearing capacity, applying Eurocode 7 (Design Approach 1). For the calculation, it is recommended the use of the Rankine theory considering the thrust parallel to the ground surface and applied on the vertical plane passing in the right end of the wall base. Ignore the thrust of the soil mass in front of the wall. Take $\gamma_w = 9.8$ kN/m³.

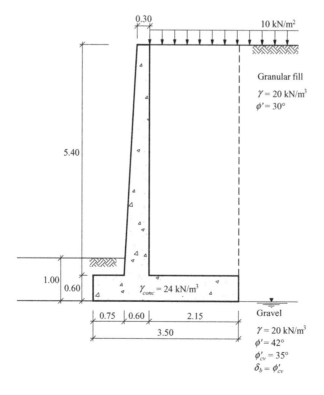

Figure E8.4

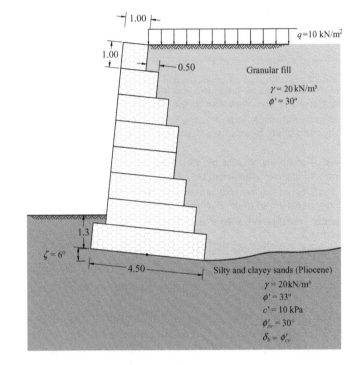

(a)

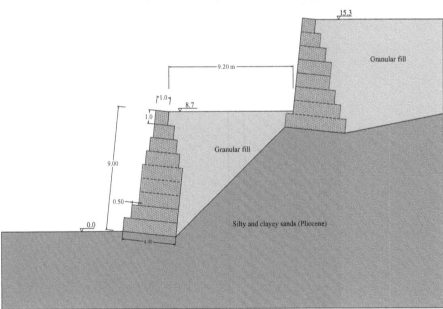

(b)

Figure E8.5

b) Calculate the internal stresses in the wall and the required steel reinforcement, apply-ing Eurocode 2.

E8.5 – Consider the gabion wall represented in Figure E8.5a, supporting a granular back-fill and founded on Pliocene silty and clayey sands. Take 16.25 kN/m³ for the gabion unit weight.

a) Calculate the global safety factors relative to base sliding and to foundation-bearing capacity for static conditions.
b) Calculate the global safety factors relative to base sliding and to foundation-bearing capacity for seismic conditions, defined by $k_h = 0.11$ and $k_v = \pm 0.056$.
c) Check the safety regarding the internal stresses in the wall for static conditions.
d) Suppose that the wall analyzed above is part of a major work involving significant alteration of the natural ground conditions and geometry, as shown in Figure E8.5b. Calculate the safety factors relative to a global sliding for static and seismic conditions. The parameters indicated in Figure E8.5 should be used for the backfill and for the natural ground.

E8.6 – Figure E8.6 represents a cross section of an excavation supported by a reinforced con-crete cantilever embedded wall. Take the wall thickness equal to 0.6 m and $\gamma_{conc} = 25$ kN/m³.

a) Calculate the embedded wall length, by considering:
    a.1 - a safety factor applied to the net passive resistance on the embedded part of the wall, $F_r = 2.0$;
    a.2 - safety factors according to Eurocode 7 – Design Approach 1.

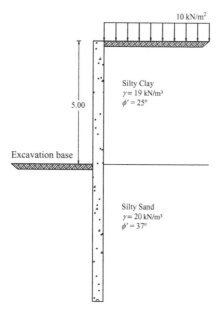

Figure E8.6

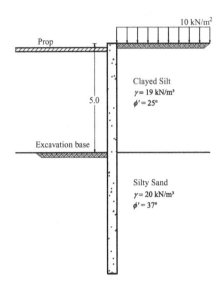

Figure E8.7

b) Calculate the internal stresses in the wall and the required steel reinforcement using Eurocode 7 – Design Approach 1, and Eurocode 2.

E8.7 – Figure E8.7 represents a cross section of an excavation supported by a reinforced concrete single-propped embedded wall. Take the wall thickness equal to 0.4 m and $\gamma_{conc} = 25$ kN/m^3.

a) Calculate the embedded wall length and the prop force, by considering:
   a.1 - safety factors according to Eurocode 7 – Design Approach 1;
   a.2 - safety factors according to Eurocode 7 – Design Approach 2.
b) Calculate the internal stresses in the wall and the required steel reinforcement using Eurocode 7 and Eurocode 2.

The Vajont catastrophe

The Vajont river valley is located in the Italian Alps, less than 100 km to the north of the city of Venice. In a region where the river passes through a narrow gorge of dolomitic limestones from the Jurassic, a 276-m high concrete arch dam was built between 1957 and 1960. Figure V.1 shows a recent downstream view of the dam from the town of Longarone.

In 1963, the dam reservoir was filled in stages, under constant monitoring. This monitoring was required, following a potential instability of the slopes on the left bank, consisting of a slow displacement, which had already been observed during the first partial filling stage, in 1960.

During the night of 9 October 1963, when the water level in the reservoir was at 22.5 m below its maximum level, the Mount Toc hillside, on the left bank of the canyon and

Figure V.1 Downstream view of the Vajont dam taken from the town of Longarone (photo: Alberto Sayão).

immediately upstream of the dam, suffered an abrupt landslide over approximately 2 km, with a failure surface of 2 km². The failure mass, of approximately 280,000,000 m³ volume, poured into the reservoir at a speed of approximately 30 m/s (around 100 km/h), hitting the slope on the opposite bank, and blocking the entire valley.

The water in the reservoir was initially thrown against the right bank, rising 260 m above the initial elevation (700 m) up to the edge of the village of Casso, rushing over the crest of the dam, forming a wave more than 200 m high. This wave, after reaching the riverbed, 500 m below, destroyed Longarone and other downstream villages, taking the lives of 2,040 people.

The sudden change in the displacement rate of the slope, as well as the nature and the scale of the landslide, took the monitoring team by surprise, with the on-site team being unable to save their own lives or to warn the families located in the nearby town of Longarone.

Figure V.2 shows an aerial view of the valley, following the landslide, with the following being observed: in the top right-hand corner, the dam, which remained almost intact, with a small lake in the back; on the left-hand side, the sliding surface (dark gray area); in the center, the soil mass from the landslide (in a lighter colour); at the center bottom, the beginning of the reservoir, now 2 km upstream from the dam.

Figure V.3 shows a geological profile of the landslide area.

Throughout the period of more than 50 years since the disaster, the Vajont landslide catastrophe has been the subject of several studies. It has been widely established that what

Figure V.2 Aerial view of the Vajont river valley following the landslide (reproduced from Engineering Geology, vol. 24, no. 1–4, Muller-Salzburg, L., The Vajont catastrophe – a personal review, pp. 423–444, 1987, courtesy of Elsevier).

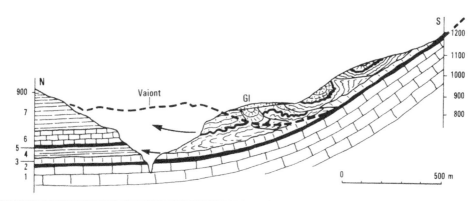

GI – Moraine; 7 – Marl; 6 – Compact limestone; 5 – Marly limestone; 4 – Marl; 3 – Thin layered limestone; 2 – Platy limestone; 1 – Compact limestone.

The sliding surface corresponds to level 2. The present topography is represented by the dashed line.

Figure V.3 Geological profile of the Vajont river valley in the landslide area (Letourneur, J. and Michel, R., La Géologie du Génie Civil, © Armand Colin, 1971, Paris).

happened between 1960 and 1963, combined with the reservoir filling, was the reactivation of an older landslide, which had probably formed tens of thousands of years previously. Along the failure surface, corresponding to formations from the Upper Jurassic, clays and very plastic marls have been identified. In these types of soil, shear deformations, associated with previous landslides, resulted in a strong reduction in strength, designated as residual strength (Skempton, 1985). On the other hand, the safety factor of the soil masses that had previously experienced such landslides is frequently quite small, and in some cases marginally above 1.0.

Figure V.4 Downstream view of the dam, showing the soil mass that suffered the landslide in the background (photo: Anna Laura Nunes).

Figure V.5 Panorama taken from the right bank, with the soil mass that suffered failure in the foreground, with Mount Toc in the background and the dam on the right-hand side (photo: Alberto Sayão).

Figure V.6 Panorama taken from the right bank, showing the village and the Casso cemetery in the foreground and Mount Toc in the background (photo: Alberto Sayão).

The high geological complexity of the ground conditions, the influence of tectonic stress in the total stress state, the difficulty of interpreting the previous landslide, the scale of the soil mass that suffered the landslide, and the limited knowledge, at the time, regarding the residual strength of clayey soils, are all fundamental aspects pointed out by specialists, who have studied the catastrophe and wondered how it could have been prevented (Leonards, 1987; Muller-Salzburg, 1987; Semenza, 2001; Alonso et al., 2010).

Figure V.4 shows a recent image of the dam taken from downstream, with part of the soil mass that suffered the landslide observed in the background, the surface of which reached a level of 160 m above the crest of the dam.

Figures V.5 and V.6 represent two recent panoramic views taken from the right bank. Figure V.5 shows, in the foreground, the soil mass that rushed down the valley, with Mount Toc being shown in the background, on the left bank, and, in the right, the top of the upstream face of the dam. Figure V.6, taken from a higher level than the previous picture, shows the village and the Casso cemetery in the foreground, with Mount Toc in the background, and the failure surface observed on the hillside.

One last remark: in the dialect from the Veneto region, where Vajont is located, Toc means "what moves"!

Chapter 9

Stability and stabilization
of natural slopes

9.1 INTRODUCTION

This chapter is dedicated to the stability and stabilization of natural slopes. Given the discussion on the equilibrium of soil masses, the present chapter is strongly linked to Chapter 2, in which the global stability of ground masses was studied. Some methods previously discussed will be reviewed here, namely the method of slices.

It is well known from geology that the so-called external geodynamic agents – water, under its most varied forms, such as rain, rivers, oceans, glaciers, and others, wind, and temperature variations– tend to flatten the Earth's crust, due to chemical and physical weathering of rock and soil masses. These weathering and weakening actions on ground masses, combined with other forces of Nature, such as gravity, seepage forces, and earthquakes, can sometimes result in great mass movements or landslides.

One of the most common natural causes of mass movements is the occurrence of earthquakes. For instance, the 1971 San Fernando (California) earthquake led to over 6,000 slope failures (Nilsen and Brabb, 1975). Figure 9.1 shows landslides in volcanic formations in Faial Island, Azores, triggered by the earthquake of July 9, 1998.

If it is true that failures of natural slopes have always existed, since before human civilization, it is also unquestionable that human actions have directly contributed to failures, as they are often adversely combined with the action of natural agents. Examples of these unfavorable human actions are numerous.

In this regard, several types of construction, such as modern railways and roads, but also the expansion of densely populated urban areas, lead to important changes in site geometry and, thus, in changes in the stress state of ground masses, in particular as a result of large excavations. Excavated soils are often located in areas that play a key stabilizing role to soil masses located at higher elevations, so that such excavations can lead to landslides typically involving greater volumes than that of the excavated soil.

In certain tropical regions, increasing deforestation has removed part of the protection of slopes against torrential rainfall, common in these regions, promoting both soil erosion and deep infiltration. Therefore, soil degradation is strongly accelerated, and very unfavorable water conditions are created. A number of slope failures have taken place in these regions, which are frequently dominated by constructions in risk areas (that are often the cause of deforestation), leading to great loss of life.

Figure 9.2 shows an image of the Po Shan Road landslide, in a densely populated area of Hong Kong, on June 18, 1972. The landslide took place after a period of heavy rain and involved approximately 40,000 m^3 of material, destroying two residential buildings and causing the death of 67 people. Due to popular commotion in the aftermath of the disaster, the local government (Hong Kong was under British rule at that time) decided to establish

Figure 9.1 Natural slope failures in Faial Island, Azores, July 9 1998, in the aftermath of an earthquake (photo: Carlos Fraga).

Figure 9.2 Po Shan Road landslide, in Hong-Kong, on June 18 1972 (courtesy of GEO, 1992).

the Geotechnical Engineering Office (GEO), which was founded in 1977, aimed at specifically managing and supervising slopes in order to prevent similar disasters.

The Brazilian state of Rio de Janeiro, in the Western hemisphere, is probably the region with the highest concentration of frequent and catastrophic failures of natural slopes. This frequency and intensity of events are due to the particularly adverse natural conditions: very complex orography, intense precipitation over short periods of time, and the predominance of soils and rocks (particularly granite and gneiss, with residual soil cover) with unfavourable behavior (Lacerda, 2004). These adverse natural conditions are also associated with deforestation as well as with disorderly human occupation in areas at high risk of landslide.

Figures 9.3 and 9.4 illustrate two slope failures, with great loss of human life and property, in the State of Rio de Janeiro.

Figure 9.3 depicts the Bananal Bay landslide, in Ilha Grande, in the area of Angra dos Reis, that took place on New Year's Eve of 2009–10, leading to 32 deaths and the destruction of properties at the bottom of the hillside. The failure surface developed in the contact between the soil cover, consisting of residual soil and colluvial deposits, the latter closer to the base, with the underlying rock mass. As pointed out by Nunes et al. (2013), the hillside instability was caused by exceptional rainfall intensity (143 mm in 24 hours, equivalent to the average monthly rainfall).

Figure 9.4 illustrates one of several slope failures that took place in the Serrana Region of the State of Rio de Janeiro, between January 11 and 12 2011, representing the greatest landslide tragedy in the history of Brazil, with approximately one thousand deaths (the real figure was never confirmed). Nunes et al. (2013) state that the landslides were caused by strong storms: precipitation of 297 mm in the night between January 11 and 12 2011, preceded by 388 mm monthly rainfall in December 2010.

Figure 9.3 Natural slope failure in Angra dos Reis, Ilha Grande, Bananal Bay, January 1 2010 (photo: António Lacerda).

Figure 9.4 Natural slope failure in Nova Friburgo, State of Rio de Janeiro, January 2011 (courtesy of the Department of Mineral Resources, State of Rio de Janeiro, Brazil).

Another classical example of particularly serious human action in the context of landslide triggering is the construction of dams, especially tall dams. Such construction involves a rise in upstream water level, leading to unfavorable ground water conditions in the hillsides upstream of the dam. One of the most tragic landslide events of the twentieth century, the Vajont landslide, which took place in 1963 in Italy, and is described in a previous text, is related precisely to this fact (Leonards, 1987; Alonso et al., 2010).

The problems related to the stability of natural slopes are probably among the most complex problems in geotechnical engineering, with several factors contributing to this.

Frequently, these problems involve large ground masses, sometimes reaching millions of cubic metres, and large-scale potential slip surfaces. As will be explained in due course, extrapolation of the soil shear resistance from that measured in small-sized samples to that mobilized in extensive failure surfaces is not at all straightforward.

Slopes that exhibit instability have commonly experienced slides throughout their geological history, subsequently acquiring a geometry favorable to stability, which can be maintained for thousands or millions of years. Despite certain symptoms of instability currently displayed by these slopes, their history and the slip surface are often unknown.

These pre-existent slip surfaces are usually controlled by thin layers of high-plasticity clayey soils, with unfavorable arrangement or inclination. Along these surfaces, in which large shear deformations have occurred, the residual shear strength is mobilized, which is much lower than the peak shear strength.

Therefore, the adequate treatment of these issues requires complex and detailed geological and geotechnical investigations, carried out by experienced and competent multidisciplinary teams.

As will be discussed later, these investigations are essential for slope stabilization projects, i.e., construction projects aimed at increasing the safety margin in the slopes.

Since the problems involving the stability of natural slopes are very diverse, it is difficult to establish methods of general application. The present chapter addresses some of the methods used, in addition to those presented in Chapter 2, familiarizing the reader with the philosophy behind slope stability problems and providing the framework to help decide which of these methods is more appropriate in the face of a concrete problem.

In the final part of the chapter, typical methodologies for the development of the stabilization projects will be discussed, with an introduction to stabilization measures and to the key role of slope observation, as a support to the geological–geotechnical characterization before the conception of the stabilization works, as an essential mean of verifying the efficiency of built solutions and for safety control.

9.2 INFINITE SLOPES

9.2.1 Introduction: geological-geotechnical scenarios

One of the slope classifications usually applied to simplify the choice of analysis methodology is the division between infinite and finite slopes.

Figure 9.5 presents two geological–geotechnical scenarios, whose conditions can be associated with infinite slopes: i) sedimentary geological units with the attitude of bedding parallel to the ground surface; or ii) granite or gneiss formations overlaid by a surface layer of residual soil.

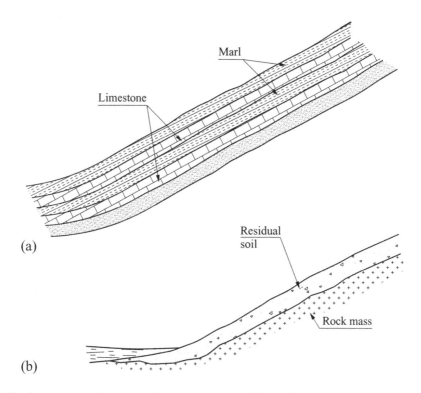

Figure 9.5 Geological–geotechnical scenarios associated with infinite slopes: a) sedimentary geological units with the attitude of bedding parallel to the ground surface; b) granite or gneiss rock formations overlaid by a surface layer of residual soil.

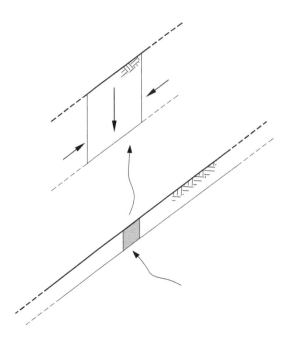

Figure 9.6 Prism representative of the infinite slope for the purpose of a stability analysis.

As inferred by the examples shown, a slope can be characterized as infinite when the properties of the mass localized at a certain depth of the ground surface remain approximately constant and when the thickness of the potentially unstable surface soil layer is small when compared to the extension of the respective slope. In these conditions, a prism with vertical faces and of arbitrary width, as represented in Figure 9.6, can be considered to be representative for the stability analysis of the slope.

9.2.2 Infinite slope in granular material, above the ground water level

The analysis of infinite slopes can begin with a very simple case, which will serve as reference for subsequent developments.

Consider an infinite and homogeneous ground mass, above the ground water level, of granular material, whose surface is tilted by an angle β and with an angle of shearing resistance equal to ϕ'. Also, consider an elemental prism bounded by vertical faces, by the ground surface, and by a plane parallel to the ground surface at a depth z, as illustrated in Figure 9.7. Given that the dimension of the prism parallel to the ground surface is of arbitrary dimensions, this will be considered equal to a unity, with the forces and stresses at the base of the prism being coincident (both the normal and the tangential components).

Equilibrium considerations, similar to those analyzed in Chapter 5 (see Section 5.4.6), show that the reaction at the base of the prism must result in an equal and directly opposed force to the weight of the prism. Thus,

$$W = \gamma \, z \cos \beta \tag{9.1}$$

being the normal and tangential stress components equal to:

$$\sigma' = \gamma \, z \cos^2 \beta \tag{9.2}$$

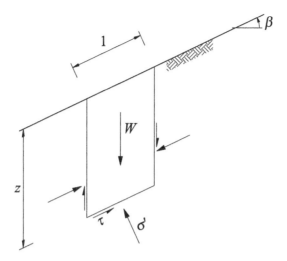

Figure 9.7 Infinite slope in granular material, above the ground water level.

and

$$\tau = \gamma z \, \sin\beta \, \cos\beta \tag{9.3}$$

On the other hand, the available shearing resistance at the prism base is equal to:

$$\tau_f = \sigma' \tan\phi' = \gamma z \cos^2\beta \tan\phi' \tag{9.4}$$

The global factor of safety is defined by the ratio between the available shearing resistance and the mobilized shear stress, i.e., the stress necessary to ensure static equilibrium, thus:

$$F = \frac{\tau_f}{\tau} = \frac{\gamma z \cos^2\beta \tan\phi'}{\gamma z \sin\beta \cos\beta} \tag{9.5}$$

or:

$$F = \frac{\tan\phi'}{\tan\beta} \tag{9.6}$$

Therefore, it can be concluded that, on an infinite slope, above the ground water level in a granular material, the global factor of safety is independent of depth and is equal to the ratio between the tangent of the angle of shearing resistance and the tangent of the slope angle.

With this, it can be said that the soil mass will be in a limit equilibrium state when the slope angle is equal to the angle of shearing resistance of the material.

9.2.3 Infinite slope in a material with cohesion and angle of shearing resistance, above the ground water level

Now consider the ground mass with a shearing resistance characterized by the effective stress parameters c' e ϕ'. Thus, Equations 9.4 to 9.6 should be re-written as follows:

$$\tau_f = c' + \sigma' \tan\phi' = c' + \gamma z \cos^2\beta \tan\phi' \tag{9.7}$$

$$F = \frac{\tau_f}{\tau} = \frac{c' + \gamma z \cos^2 \beta \tan\phi'}{\gamma z \sin\beta \cos\beta} \tag{9.8}$$

$$F = \frac{c' + \gamma \ z \cos^2 \beta \ \tan\phi'}{\gamma \ z \ \sin\beta \ \cos\beta} \tag{9.9}$$

Therefore, it can be concluded that the factor of safety becomes a function of z, decreasing as z increases.

Designating the critical depth, h_{cr}, as the value of z at which the global factor of safety is equal to unity, the following equation can be easily obtained:

$$h_{cr} = \frac{c'}{\gamma} \cdot \frac{1}{\cos^2 \beta \ (\tan\beta - \tan\phi')} \tag{9.10}$$

This expression shows that in a soil with cohesion and angle of shearing resistance the slope angle, β, can be greater than ϕ', when the thickness of the potentially unstable soil mass is lower than the critical depth.

This situation is illustrated in Figure 9.8, in which, despite the slope angle being greater than the angle of shearing resistance of the soil, the slope can be considered stable, as the residual soil layer has a limited depth, lower than the respective critical value.

9.2.4 Infinite slope with seepage parallel to the ground surface

9.2.4.1 Introduction

In the previous considerations, the effect of water was not included. Figure 9.9a shows a possible schematic representation of the conditions created by intense rainfall on a slope considered to be infinite. A seepage regime, which can be assumed to be parallel to the ground surface, is likely to be created, with a phreatic line at a depth z_w. This depth, under extreme

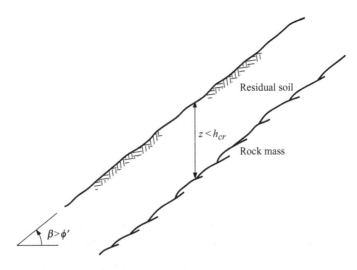

Figure 9.8 Infinite slope in a soil with non-zero cohesion, in which the slope angle is greater than the angle of shearing resistance ($\beta > \phi'$).

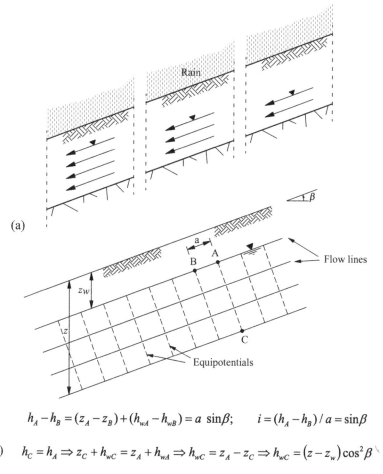

$$h_A - h_B = (z_A - z_B) + (h_{wA} - h_{wB}) = a \, \sin\beta; \qquad i = (h_A - h_B)/a = \sin\beta$$

(b) $h_C = h_A \Rightarrow z_C + h_{wC} = z_A + h_{wA} \Rightarrow h_{wC} = z_A - z_C \Rightarrow h_{wC} = (z - z_w)\cos^2\beta$

Figure 9.9 Seepage in a natural slope: a) schematic diagram showing that, in the region close to the toe of the slope, seepage approaches the ground surface; b) flow net with seepage parallel to the slope face, with the phreatic line at depth z_w.

precipitation conditions and in the region close to the toe of the slope, can be approximately zero.

Figure 9.9b illustrates the flow net for a seepage regime as described above. As shown in the figure, the hydraulic gradient in the ground mass is constant and equal to:

$$i = \sin\beta \tag{9.11}$$

and the pore water pressure at depth z can be calculated according to the following expression:

$$u = \gamma_w (z - z_w)\cos^2\beta \tag{9.12}$$

In order to analyze the equilibrium of the prism representing the slope, two approaches can be considered, as represented in Figure 9.10.

In Figure 9.10a, the total weight of the prism is considered. Thus, water pressures on the boundaries of the prism, the resultant force of which is equal to the vector sum of the buoyant force with the resultant of the seepage forces, should be accounted for. As an alternative, for the part of the prism underneath the phreatic surface, the respective submerged weight combined with the resultant of the seepage forces can be considered, as shown in Figure 9.10b. This last alternative is more convenient, so it will be contemplated in the following analyses.

The weight above the phreatic line, the submerged weight, and the resultant of the seepage forces are respectively equal to:

$$W_1 = \gamma z_w \cos \beta \tag{9.13}$$

$$W_2 = \gamma'(z - z_w) \cos \beta \tag{9.14}$$

and

$$J_2 = \gamma_w(z - z_w) \sin \beta \cos \beta \tag{9.15}$$

Thus, it is possible to calculate the normal effective stress and the tangential stress at the base of the prism, respectively:

$$\sigma' = \gamma z_w \cos^2 \beta + \gamma'(z - z_w)\cos^2 \beta \tag{9.16}$$

and

$$\tau = \gamma z \sin \beta \cos \beta \tag{9.17}$$

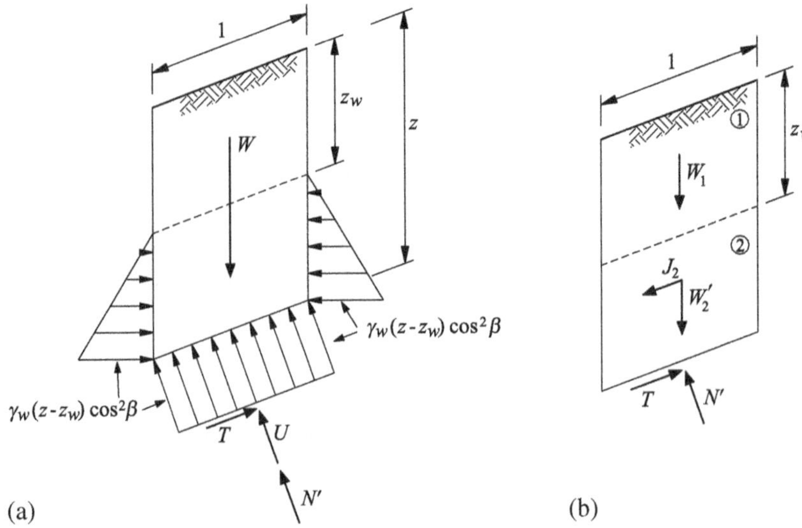

Figure 9.10 Forces applied to an elemental prism: a) when considering water pressures on the boundaries of the prism and the total prism weight; b) when considering submerged weight below the phreatic line combined with the resultant of seepage forces.

When considering the calculation of the factor of safety, for a granular material, after some mathematical manipulation, the following equation can be obtained:

$$F = \frac{(\gamma_w z_w + \gamma' z)}{\gamma z} \frac{\tan \phi'}{\tan \beta} \tag{9.18}$$

In cases where the phreatic line coincides with the surface, z_w is zero, thus obtaining:

$$F = \frac{\gamma'}{\gamma} \frac{\tan \phi'}{\tan \beta} \tag{9.19}$$

Now considering a material with cohesion and angle of shearing resistance, the factor of safety for seepage, with the phreatic line coinciding with the surface, corresponds to the following expression:

$$F = \frac{c' + \gamma' z \cos^2 \beta \tan \phi'}{\gamma z \sin \beta \cos \beta} \tag{9.20}$$

Designating h_{cr} as the value of z at which the factor of safety is equal to unity, the following expression can be obtained under these conditions:

$$h_{cr} = \frac{c'}{\gamma} \cdot \frac{1}{\cos^2 \beta \, (\tan \beta - \dfrac{\gamma'}{\gamma} \tan \phi')} \tag{9.21}$$

The comparison between this equation and Equation 9.10 allows the conclusion that the existence of seepage dramatically reduces the critical depth.

Both of these equations correspond to two extreme conditions, with regard to stability: the first (Equation 9.10) represents an upper limit of the critical depth, in practice intangible, as seepage related to precipitation will still naturally occur on the superficial layer. The second (Equation 9.21) represents the lower limit of this critical depth, given that it corresponds to limit conditions regarding the unfavorable effects of seepage. It seems reasonable to say that, in the case shown, the slope will start to exhibit symptoms of instability associated with periods of exceptional precipitation, when the thickness of the residual superficial layer exceeds the value of h_{cr} given by Equation 9.21, with slippage occurring along the contact with the underlying rock mass.

9.2.5 Infinite slope under seismic action

In the problems of slope stability under seismic action, it is common to adopt D'Alembert's principle, similar to what was discussed in Chapters 5 and 8, in which the effects of horizontal and vertical seismic accelerations are represented by inertial forces.

As shown in Figure 9.11, in the present case it is still possible to adopt the approach of Mononobe-Okabe, imagining a counter-clockwise rotation of the figure by an angle θ, defined by the seismic actions. The ground surface then has a slope $\beta + \theta$ and the vertical force, W_e, applied at the center of gravity of the representative slice, is a resultant of the gravitational force and the vertical and horizontal inertial forces.

Considering the case of a granular material above the ground water level with

$$W_e = \frac{W(1 \pm k_v)}{\cos \theta} = \frac{\gamma(1 \pm k_v) z \cos \beta}{\cos \theta} \tag{9.22}$$

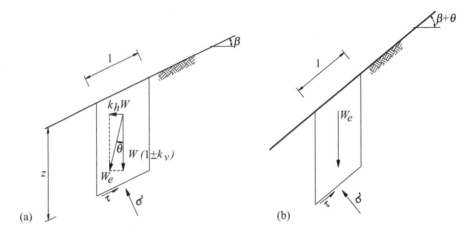

Figure 9.11 Application of the Mononobe-Okabe approach for considering an infinite slope under seismic actions.

the normal and tangential stress components are equal to:

$$\sigma' = W_e \cos(\beta + \theta) = \frac{\gamma(1 \pm k_v)z \cos \beta \cos(\beta + \theta)}{\cos \theta} \tag{9.23}$$

and

$$\tau = W_e \sin(\beta + \theta) = \frac{\gamma(1 \pm k_v)z \cos \beta \sin(\beta + \theta)}{\cos \theta} \tag{9.24}$$

The factor of safety can be expressed as:

$$F = \frac{\sigma' \tan \phi'}{\tau} = \frac{\tan \phi'}{\tan(\beta + \theta)} \tag{9.25}$$

an equation that reveals an obvious similarity to Equation 9.6. In practice, it results from replacing the angle that defines the inclination of the ground surface by the sum of this angle and angle θ, similar to the one used in the method of Mononobe-Okabe.

Now considering a material with cohesion and angle of shearing resistance above the ground water level, the equation of the factor of safety can be written as:

$$F = \frac{c' \cos \theta + \gamma(1 \pm k_v)z \cos \beta \cos(\beta + \theta) \tan \phi'}{\gamma(1 \pm k_v)z \cos \beta \sin(\beta + \theta)} \tag{9.26}$$

As seen previously, by designating the value of z corresponding to a factor of safety equal to unity at the critical depth, h_{cr}, the following equation is obtained:[1]

$$h_{cr} = \frac{c' \cos \theta}{\gamma(1 \pm k_v)} \cdot \frac{1}{\cos \beta \cos(\beta + \theta)\left[\tan(\beta + \theta) - \tan \phi'\right]} \tag{9.27}$$

---

[1] This expression can be understood as derived from Equation 9.10 with the following adjustments, by analogy with the Mononobe-Okabe solution studied in Chapter 5: i) β is replaced by $\beta + \theta$; ii) γ is replaced by the resultant of mass forces during seismic action, so that, for a volume unit, $\gamma(1 \pm k_v)/\cos \theta$; and iii) for a rotation of an angle of θ, h_{cr} should be replaced by $h_{cr} \cos \beta / \cos(\beta + \theta)$.

Analyzing Equation 9.25, applicable when the effective cohesion is null, it can be concluded that the vertical seismic coefficient $-k_v$ (vertical inertial force directed upwards) is always critical because it leads to a greater value of the angle θ. If the effective cohesion is not null, which corresponds to the application of Equation 9.26, both $-k_v$ and $+k_v$ may be critical.

The combination of seismic action and seepage under the conditions described in Section 9.2.4 does not impose any difficulties, although the expressions of the factor safety and critical depth are more complex. The specific conditions to consider in this situation, particularly with respect to the depth of the phreatic line, z_w, should be appropriately pondered, considering the content of Chapter 3.

Annex A9.1 summarizes the procedure for selecting the horizontal and vertical seismic coefficients, according to Eurocode 8 (EN 1998-5: 2004).

9.3 GENERAL METHOD FOR SLOPE STABILITY ANALYSIS: THE MORGENSTERN AND PRICE METHOD

9.3.1 Introduction

In this section and the following one, slope stability analysis methods are discussed for which it is necessary to consider *the entire mass* susceptible to sliding and not only the mass of a representative element, as in the case of infinite slopes.

Chapter 2 presented the method of slices, in the versions designated by the Fellenius method, the simplified Bishop method and the Spencer method. All these methods consider that, in a cross section of the slope, slip surfaces are an arc of a circle. In this section, the Morgenstern and Price method will be discussed, which falls within the philosophy of the method of slices but considers slip surfaces of any shape. This aspect is relevant as the slip surface in deep landslides of natural slopes is frequently very different from a circular arc.

Indeed, in large natural slopes, it is common to find a wide range of geological formations with contrasting mechanical characteristics. As will be seen in the following sections, layers of high-plasticity clays and claystones are often responsible for the observed situations of slope instability. In these conditions, slip surfaces are entirely or mostly developed along the contact surface between these materials and other, more resistant materials, consisting of approximately rectilinear sections, possibly combined with other curved sections.

Figure 9.12 illustrates a geological–geotechnical scenario that is favorable to the development of this type of instability, with the presence of a thin clayey layer that presents a given, although reduced, inclination. The execution of an excavation in the slope area into which that layer dips may trigger the destabilization of a soil mass above this same layer.

With this, it is possible to understand the relevance of using methods, such as the Morgenstern and Price method, for analyzing more complex problems. This method has the additional advantage of being a mathematically correct method, that is, consisting of a system of equilibrium equations with the same number of equations as the number of unknowns, as will be seen in the following sections. This method requires computational application, given the iterative process involved.

9.3.2 Hypotheses

Consider the slope illustrated in Figure 9.13a containing a potentially unstable mass limited by the slip surface described by a general equation, $y(x)$. In turn, Figure 9.13b

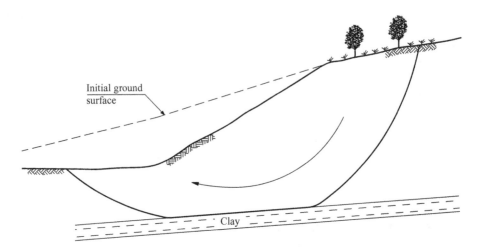

Figure 9.12 Scenario with a deep landslide along a slip surface consisting of curved sections and a rectilinear section.

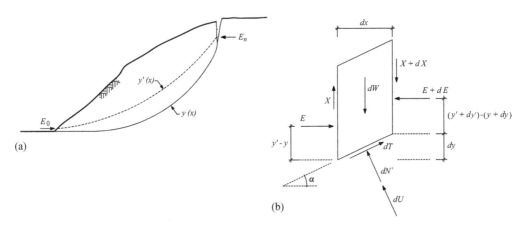

(a)

(b)

Figure 9.13 The Morgenstern and Price method: a) slope in analysis with potentially unstable mass limited by a surface of any shape; b) generic slice of infinitesimal width with forces applied.

represents a generic slice of vertical faces and infinitesimal width, dx, with the following forces applied: i) weight dW; ii) normal effective component of the reaction at the base of the slice, dN'; iii) tangential component of the reaction at the base of the slice, dT; iv) resultant of the pore water pressure at the base of the slice, $dU = udx/\cos\alpha$, supposedly known from the flow net, or any other process; and v) the normal and tangential components of the interaction forces on the left of the slice, E and X, respectively, and on the right, $E + dE$ and $X + dX$, respectively. The line that joins the application points of the interaction forces, called the *line of thrust*, is described by the equation $y'(x)$. E_0 and E_n are normal components applied at the boundary of this mass, considered to be known; as a general rule, the former will be equal to zero while the latter may correspond to the resultant of the water pressure in a tension crack.

At this point it may be useful to remind the reader that Chapter 2 discussed the indeterminate nature of the problem in analysis. It was then observed that, whereas the number of equations available is $3n$ (see Table 2.1), the number of unknown variables is $4n-2$, with n being the number of slices.

The Morgenstern and Price method makes the problem statically determined, by assuming a relationship between the tangential and normal components of the interaction forces of type:

$$X = \lambda f(x) E \qquad (9.28)$$

where $f(x)$ is a shape function – arbitrated in the application of the method – and λ is a dimensionless scale factor, to be calculated.

By arbitrating $f(x)$, the authors are, in fact, establishing n-1 hypotheses regarding the interaction forces. At the same time, the introduction of a scale factor, λ, as an additional unknown variable results in a number of unknowns equal to $4n-2-(n-1)+1 = 3n$, that is, the same number as the number of equilibrium equations.

Therefore, the Morgenstern and Price method leads to a *mathematically correct* solution and a balanced system of forces and moments on all slices. Nonetheless, the solution must be *physically reasonable* or realistic. This matter will be discussed later.

9.3.3 Equilibrium equations for a slice of infinitesimal width

Taking into account that the width at the base of the slice is infinitesimal, the forces dN', dU, and dW will generate null moments in relation to the base midpoint, with the moment equilibrium equation being given as follows:

$$E\left[y'-y-\frac{dy}{2}\right]-(E+dE)\left[(y'+dy')-(y+dy)+\frac{dy}{2}\right]$$
$$+X\frac{dx}{2}+(X+dx)\frac{dx}{2}=0 \qquad (9.29)$$

By simplifying the expression and making dx tend to zero, the following is obtained:

$$(y'-y)\frac{dE}{dx}+E\frac{dy'}{dx}-X=0 \qquad (9.30)$$

Thus, the forces' equilibrium equations are as follows:

$$dN-dW\cos\alpha-dX\cos\alpha+dE\sin\alpha=0 \qquad (9.31)$$

$$dT-dE\cos\alpha-dX\sin\alpha-dW\sin\alpha=0 \qquad (9.32)$$

Force dT can be expressed in terms of the available shearing resistance and from the factor of safety, F, through the equation:

$$dT=\frac{c'\dfrac{dx}{\cos\alpha}+\left(dN-u\dfrac{dx}{\cos\alpha}\right)\tan\phi'}{F} \qquad (9.33)$$

where c' and ϕ' are the effective strength parameters of the soil at the base of the slice.

By substituting this equation in Equation 9.32, the following is obtained:

$$dN = \frac{dE\cos\alpha + dX\sin\alpha + dW\sin\alpha - \dfrac{c'}{F}\dfrac{dx}{\cos\alpha} + u\dfrac{dx}{\cos\alpha}\dfrac{\tan\phi'}{F}}{\dfrac{\tan\phi'}{F}} \qquad (9.34)$$

Now substituting this equation in Equation 9.31, with dx tending to zero, we obtain:

$$\frac{c'}{F}\left[1+\left(\frac{dy}{dx}\right)^2\right] + \frac{\tan\phi'}{F}\left\{\frac{dW}{dx} + \frac{dX}{dx} - \frac{dE}{dx}\frac{dy}{dx} - u\left[1+\left(\frac{dy}{dx}\right)^2\right]\right\}$$
$$= \frac{dE}{dx} + \frac{dX}{dy}\frac{dy}{dx} + \frac{dW}{dx}\frac{dy}{dx} \qquad (9.35)$$

9.3.4 Solution of the system of equations

Equations 9.30 and 9.35 are differential equations that govern the equilibrium of moments and the equilibrium of forces in the slice, respectively. In these equations, the components of the interaction forces between slices, E and X, and the position of the respective application point, y', are unknown variables. By joining Equation 9.28 to these equations, the problem becomes statistically determined, as long as λ and F are known.

The procedure proposed can be described as follows:

1. divide the soil mass in slices of finite width;
2. adopt a function $f(x)$ and values for the forces E_0 and E_n (with the former being zero as a rule);
3. adopt a pair of values for λ and F;
4. integrate the system of Equations 9.28, 9.30 and 9.35, for every slice, starting from the first slice and attributing to it the value of E_0, until obtaining the value of E_n for the last slice;
5. repeat the process, changing both λ and F, until obtaining the correct value of E_n (assumed in point 2);
6. adopt a new function $f(x)$ and repeat steps 3 to 5, until new convergence of the process is reached.

Therefore, the results correspond to a set of values of λ and F that, for every function $f(x)$, lead, in the iterative process, to a balanced system of forces, respecting the boundary conditions, i.e., a *mathematically correct solution*.

Nevertheless, the results obtained still lack a verification of their *physical reasonability*. For such verification, it is necessary to analyze the interaction forces between slices, E and X, as well as the line of thrust, $y'(x)$, according to the following criteria: i) the line of thrust must be inside the soil mass being analyzed; and ii) the interaction forces must respect the failure criterion in each boundary between slices. For the latter, the resultant of the normal effective stresses, E', should be calculated from E and from the resultant of the pore water pressures, that are assumed to be known.

After presenting the formulation of the method, the reader can now have a better physical understanding of the meaning of both $f(x)$ and λ: i) $f(x)$ represents the manner in which the ratio between the tangential and normal components of the interaction forces, X/E, varies

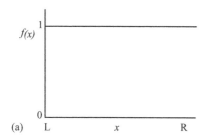 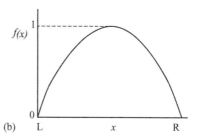

Figure 9.14 Examples of $f(x)$ functions usually adopted in the application of the Morgenstern and Price method: a) $f(x) =$ constant; b) $f(x) = \sin(\eta)$ where $\eta = \pi\left(x - x_L\right) / \left(x_R - x_L\right)$.

along the slope; and ii) λ represents the fraction or percentage of the function that is used to calculate the factor of safety. The values of F obtained are typically not sensitive to the function $f(x)$ adopted, which indicates that the numerical process is quite robust. Figure 9.14 shows examples of $f(x)$ functions that have led to satisfactory results in most cases.

It is important to note that the function represented in Figure 9.14a assumes a constant inclination of the interaction forces, equal to λ, which corresponds to the hypothesis of the Spencer method (1967), presented in Chapter 2. The Spencer method can therefore be considered to be a specific case of the Morgenstern and Price method, by considering the slip surfaces as arcs of a circle with the shape function $f(x) =$ constant. In this case, the tangent of angle ρ defined in the Spencer method to describe the inclination of the resultant of the interaction forces coincides with λ.

The Morgenstern and Price method can be found in most commercial software packages for stability analysis available on the market. In this work, it is, therefore, unnecessary to include the equations used in the iterative process described, adapted from the previous equations for slices of finite width. In these software packages, the method is adapted in order to consider seismic actions (simulated through inertial forces applied at the center of gravity of the slices) and external forces, such as those applied by ground anchors.

9.4 METHOD OF SLIDING WEDGES

In the previous section, the Morgenstern and Price method was presented, which is one of the general approaches of the method of slices and enables consideration of slip surfaces of any shape, requiring computational application.

When the geometry of the soil mass is relatively simple, it is possible to carry out a simplified analysis of the problem through the so-called *method of wedges*. The method consists, for each specific case, in imagining, with some common sense, tempered by the study of other similar cases, several sets of blocks or wedges and analyzing the respective static equilibrium.

For instance, when considering the case of Figure 9.12, the potentially unstable soil mass can be split into three blocks, as represented in Figure 9.15. Consider that the upper soil unit will exhibit an essentially drained behaviour, with a shear strength characterized by c' and ϕ', while the sub-horizontal clayey layer will exhibit, in a possible landslide, undrained behavior with an undrained shear strength c_u.

The global factor of safety can be calculated using a trial-and-error procedure. This factor of safety is defined here as the value by which the available shearing resistance of

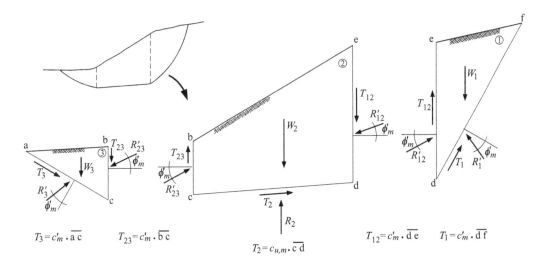

$T_3 = c'_m \cdot \overline{a\,c}$ $\qquad$ $T_{23} = c'_m \cdot \overline{b\,c}$ $\qquad\qquad\qquad$ $T_{12} = c'_m \cdot \overline{d\,e}$ $\qquad$ $T_1 = c'_m \cdot \overline{d\,f}$

$T_2 = c_{u,m} \cdot \overline{c\,d}$

Figure 9.15 Forces applied to the three wedges resulting from the simplification of the destabilization of Figure 9.12.

the soil should be divided in order to obtain the mobilized shear resistance. Therefore, in general:

$$F = \frac{c'}{c'_m} = \frac{\tan\phi'}{\tan\phi'_m} = \frac{c_u}{c_{u,m}} \tag{9.36}$$

an equation extended to the different materials involved in the analysis of slope stability, with m corresponding to the mobilized shear resistance components.

By analysing Figure 9.15, six unknown variables are identified in the problem (R'_1, R'_{12}, R_2, R'_{23}, R'_3 and F) and the same number of equilibrium equations (two for each wedge). One way of calculating the value of F, using a trial-and-error procedure, is:

  i) to assume a value for F;
 ii) calculate c'_m, ϕ'_m and $c_{u,m}$;
iii) determine, by equilibrium analysis of wedges 1 and 3, the interaction forces of these wedges with the central wedge 2, R'_{12} and R'_{23};
 iv) verify if both equilibrium equations of the central wedge are indeed satisfied (note that R_2 is the unknown in both these two equations); if confirmed, the value of F initially adopted is considered to be the global factor of safety; if the equilibrium equations are not satisfied, a new process is initiated, with a new value of F assumed.

It is convenient to carry out other trial-and-error calculations, taking into account distinct directions of the contact surfaces between the wedges, in order to identify the most critical situation (Lambe and Whitman, 1979).

The advantage of this method is the simplicity of calculations involved, when compared with the Morgenstern and Price method, which requires computational application. The drawback of the method is the fact that it is not a general method, as the calculated factor of safety is only applicable to the wedge system considered.

9.5 CONTEXT OF SLOPE STABILITY STUDIES

9.5.1 Introduction

The stabilization of a natural slope consists of a combination of interventions of a constructive nature, aimed at increasing the safety of the slope or, at least, preventing or attenuating any reduction in safety. With this in mind, it can be inferred that, as a general rule, natural slopes exhibit signs of instability without necessarily leading to immediate catastrophic landslide events.

The identification of such signs of instability will lead to studies aimed at building stabilization solutions. In some cases, identification of the signs of instability leads to an immediate interpretation of the phenomenon, that is, to an identification of the unstable soil mass and of the potential failure mechanism. However, in other cases, particularly in deep failures of large natural slopes, it may be very difficult to interpret the potential failure mechanism from the observation, on the ground surface and, at certain points, of the signs of instability.

In the subsequent sections, the following aspects will be discussed: i) how the failure of a natural slope begins and develops; ii) how the failure starts to manifest itself; and iii) how to identify the unstable soil mass, in more complex cases, i.e., how to interpret the developing failure mechanism.

9.5.2 Active zone, passive zone, and neutral line in a natural slope

Consider a large natural slope under complex geological conditions, as illustrated in Figure 9.16, as well as the potential slip surface represented, and the mass limited by this surface divided into slices of vertical faces.

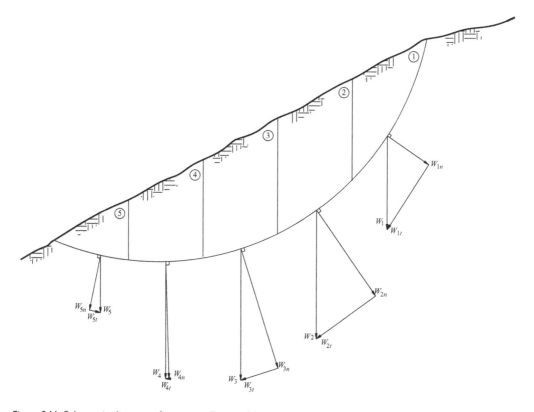

Figure 9.16 Schematic diagram of a potentially unstable mass on a natural slope, showing the weight decomposition of each slice, according to the normal and tangential directions in relation to the respective base.

As previously discussed, for the analysis of slope stability, the weight of each slice is decomposed into a normal and a tangential component in relation to the base of the slice. The tangential component will contribute to the *destabilizing* or *acting moment* (or *force*), while the normal component will contribute to the *resisting moment* (or *force*). From the figure, it is easy to understand that, in relative terms, the upper slice (slice 1) will present the greatest tangential component and the lowest normal component. The further away from the top of the slope the slice is, the greater the normal component. An opposite trend is exhibited by the tangential component, with the lower slice (slice 5) having a favourable direction with respect to stability.

Therefore, by considering the moment or force equilibrium of *each slice isolated,* that is, without considering the interaction forces between slices – that would provide the so-called *local factor of safety* – this factor would be lowest in slice 1, increasing progressively for the lower slices. In a situation like the one represented, the local factor of safety is typically lower than unity for a group of slices in the upper part of the slope and greater than unity for the other slices.

From the previous considerations, it follows that, on a slope similar to the one represented, a *deficit* of resistance can be observed in the upper zone of the slope, while a *surplus* of resistance is seen in the lower zone. As long as the value of the latter is greater than the former, the stability of the whole soil mass is ensured. In fact, the interaction forces between slices are *physically* responsible for enabling that *surplus* to compensate for the *deficit* in resistance. The upper zone of the slope, with a deficit of resistance, is referred to as the *active zone*, while the lower zone, with a surplus of resistance, is referred to as the *passive zone*.

A simple way of identifying the boundary between both active and passive zones, for a given soil mass, consists of applying a moving vertical surcharge to the ground surface, as shown in Figure 9.17a, similar to drawing the line of influence of a structure, and then calculating the global factor of safety for each surcharge position (Hutchinson, 1977). By applying the surcharge to the upper limit of the slope, in the slices from the active zone, the factor of safety obtained is lower than the value calculated without surcharge. As the surcharge is applied progressively at a lower position, this reduction in factor of safety becomes increasingly smaller. From a certain position, the value of the factor of safety obtained will be higher than that without the surcharge; this means that this position is already in the passive zone of the slope, and the boundary between the active and passive zones has been crossed. This boundary is designated the *neutral line*, which can be understood as the vertical plane that contains the point of application of the vertical surcharge, such that the factor of safety remains unchanged, regardless of the presence or absence of the surcharge. By imagining a very thin slice, with the weight applied along the neutral line, the local factor of safety in this slice is equal to 1.0. Figure 9.17b supports this explanation.

9.5.3 Circumstances and symptoms of slope instability

Consider Figure 9.17 again. If the soil mass under consideration consisted of a rigid body, a movement of this mass would generate a tangential displacement of the same magnitude at any point on the potential slip surface. Therefore, soil shear resistance would be mobilized in a reasonably uniform manner along this slip surface.

However, the soil mass under consideration is not a rigid body! Imagine the slices with the larger deficit at the top of the slope pushing the slices immediately below, still in deficit, with these pushing the ones in the middle zone, with a local factor of safety close to 1.0, and so on, until the lower limit of the unstable soil mass is reached; this interaction under the force of gravity will generate compressive deformations. These deformations are greater in the upper region of the active zone and lower in the lower limit of the passive zone. This

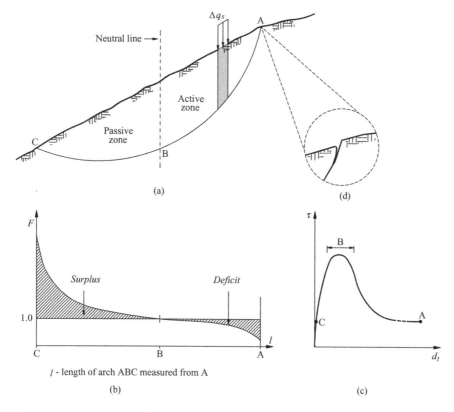

Figure 9.17 Generic potentially unstable soil mass: a) search for the neutral line and the active and passive zones separated by the neutral line; b) evolution of the local factor of safety along the potential slip surface; c) generic diagram of shear stress *versus* tangential displacement on the potential slip surface and levels of mobilized shear resistance at three different points; d) detail representing the upper limit of the unstable mass.

leads to a maximum relative tangential displacement on the potential slip surface at point A (Figure 9.17a) and a minimum displacement at point C (Figure 9.17a). As the surface can be hundreds of meters long, considerable relative displacements can occur at the top (of a few decimeters or even of a few meters), whereas, close to the lower limit, these displacements might still be relatively small or even null.

Thus, the mobilization of shearing resistance begins at the upper limit of the slip surface and progresses towards the lower limit, as the stability conditions deteriorate.

Given that these materials are typically stiff overconsolidated clayey soils, exhibiting pronounced strength loss after large tangential displacements, as illustrated in Figure 9.17c, the residual shear strength is probably mobilized in the upper zone. In turn, peak shear strength will be mobilized in an intermediate zone, still with low shear deformations. As for the lower zone, the mobilized shearing resistance will be small, corresponding to a low percentage of the peak shear strength, as shear deformations in this section of the potential slip surface are (still) very small.

The above considerations explain the appearance of cracks, with considerable openings, on the ground surface, close to the upper limit of the slip surface, as suggested in Figure 9.17d. If this zone is inhabited, more or less severe damage can occur in constructions, roads, or other infrastructures, which will lead to the initiation of studies on the given slope.

In most cases, these signs of instability are linked to exceptionally adverse ground water conditions, caused by (more or less prolonged) periods of intense precipitation. This precipitation will tend to move up the phreatic surface in the slope, reducing effective stresses on the potential slip surface, and thus the available shearing resistance. On the other hand, a greater volume of potentially unstable soil mass will be exposed to the action of seepage forces, in an unfavorable direction.

Accordingly, the *deficit* of the upper zone of the slope deteriorates, whereas the *surplus* of the lower zone is reduced. The neutral line translates towards the base of the slope. Further compressive and shear deformations occur, leading to new displacements and more damage at the ground surface in the upper zone.

As soon as these adverse water conditions are attenuated or dissipated, due to the natural reduction of precipitation in the region, with no global landslide having occurred, the slope reassumes the conditions of stability, or lower deformation rates over a longer or shorter period of time, depending on the return period of the conditions that led to the mentioned damage.

Figure 9.18 illustrates what has been described, showing a very clear correlation between the periods of higher precipitation (registered at a nearby weather station) at a natural slope, with the larger increments of horizontal displacement at a given point of the slope, over a period of eighteen years (Barradas, 1999).

Fortunately, large natural slopes exhibit very clear symptoms of instability before a global landslide is imminent, which allows, in most cases, the assessment and execution of suitable stabilization works.

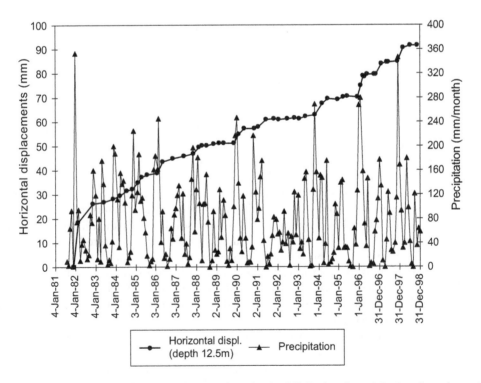

Figure 9.18 Evolution of horizontal displacements (at a depth of 12.5 m) and precipitation throughout eighteen years in a natural slope in the region of Coimbra, Portugal (Barradas, 1999).

There are, however, exceptions, with sudden landslides often being associated with great loss of human life and material damage. Such catastrophes are or have been linked to the following contexts:

i) slopes in inhospitable or underdeveloped zones, which, despite showing symptoms of instability, were not observed/or interpreted as such;

ii) slopes that developed significantly more unfavorable conditions than those previously experienced.

Examples corresponding to this last context include, among others: i) the occurrence of a large earthquake; ii) the removal, by excavation, of part of the passive zone of the slope, for instance, for the construction of a new road; and iii) a rise in the phreatic surface, caused by the construction of a tall dam downstream.

Regarding the last two examples, it is important to point out that the project of such new structures (such as a new road or a new dam) must include the assessment of the stability of the slopes affected by the construction. As this assessment is not easy to be carried out in a reliable way, especially when the geology is complex, careful instrumentation of the slopes affected is a common procedure for preventing catastrophic landslides, such as the ones mentioned.

9.5.4 Interpretation of the potential failure mechanism

In the previous considerations, the potentially unstable soil mass, shown in Figure 9.16, was assumed as a starting point for the discussion. However, it is not too early to make the following (important) remark: in large natural slopes of complex geology that reveal signs of instability, neither the geometry of the unstable mass is entirely known, nor is the respective slip surface!

What can be noticed are certain symptoms on the ground surface at certain points (cracks, certain movements, structural damage in constructions and road systems, etc.), usually concentrated in the upper part of the hillside. However, the recognition of such signs is not enough to understand the *depth* and *extension* reached by the entire potentially unstable soil mass.

As the reader may understand, the interpretation of the failure mechanism is an essential condition for a successful stabilization project. In fact, this interpretation is the core of the first stage of such project. With this in mind, Figure 9.19 summarizes the typical stages of studies of stabilization of natural slopes.

The interpretation studies will involve a careful geological–geotechnical characterization of the slope, which requires several boreholes to be carried out. These boreholes must be of extended depth, in order to reach a certain level, for which there is no reasonable doubt that it is well below the potential slip surface. In similar studies, drilling equipment that allows continuous sampling of the soil is usually preferable, as this will ensure a full and reliable geological interpretation.

Piezometers are subsequently installed into some of the boreholes in order to characterize the ground water conditions, while *inclinometer casing tubes* are placed in other boreholes, along their full height. Such tubes have this designation as a device denominated as *inclinometer* is introduced into them, in order to measure the horizontal ground displacement at each depth. In Annex A9.2, the reader can find the explanation on how an inclinometer operates and how to interpret its measurements.

As an example, Figure 9.20 presents the displacements measured in an inclinometer tube at a certain slope section, at several dates following an initial or reference reading. The

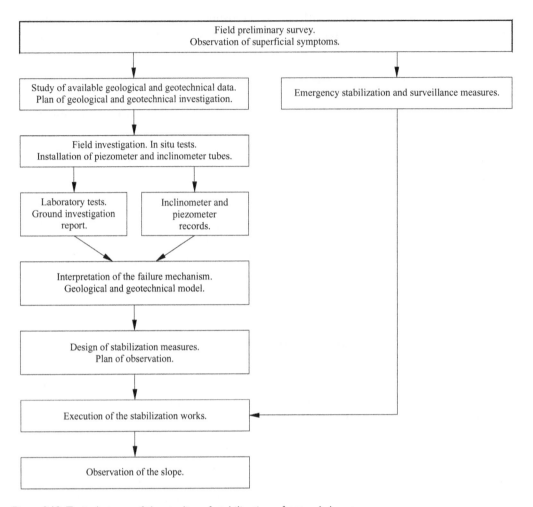

Figure 9.19 Typical stages of the studies of stabilization of natural slopes.

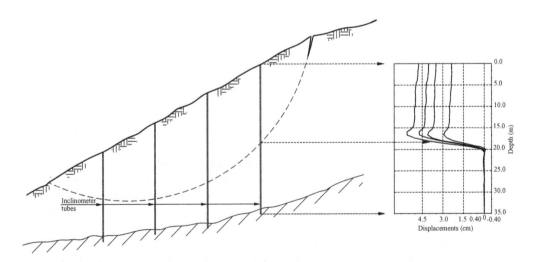

Figure 9.20 Typical inclinometer records in a natural slope that lead to the identification of the slip surface.

inclinometer records typically present almost null displacements in the lower zone, a very limited zone with a high displacement gradient, and an upper zone where the evolution of displacements is variable (practically constant, increasing or decreasing in height), though with a relatively small displacement gradient. The short section where the displacement gradient is high (at approximately 18.0 m depth, in the figure) corresponds to the intersection of the potential slip surface, where relative displacements between the unstable zone and the remaining mass are concentrated. The characterization of the position of this surface on several inclinometer tubes placed at different points on the same slope cross section enables the identification, with reasonable precision, of the unstable soil mass. For the reasons previously discussed, displacements are usually higher when registered in inclinometer tubes that intersect the unstable mass closer to the top of the hillside.

It is particularly useful to compare the location of the potential slip surface with the soil intersected in each borehole, at the same depth where this surface was found. Such comparison often allows the identification of the geological unit (layer or stratum) responsible for the instability. This source of instability is commonly associated with a small thickness layer, which cannot easily be identified as such only by observation of the material collected from the boreholes (see Figure 9.25). The identification of such a layer is crucial as it can allow the "prediction" or "extrapolation" of the progress of the surface in the lower zone, in case the inclinometer records located in this zone are unclear. On the other hand, with new boreholes, high-quality samples of the soil in this layer can be obtained for a more detailed mechanical characterization through laboratory tests.

The geological–geotechnical model of the slope, that will serve as a support to the stabilization project, is then finalized or adjusted, taking into account the interpretation of the failure mechanism and the careful evaluation of the shear strength soil parameters relevant for that mechanism.

9.5.5 Stages of slope movements: reactivation of landslides

Consider once again the example of Figure 9.18. The average return period that leads to the largest incremental displacements registered in the figure is very short (approximately 3 years). For the same slope or an analogous one, more severe ground water conditions would be associated with longer return periods. These conditions may have led to large movements or even to a landslide, with substantial changes to the slope geometry. This event may have taken place before our age, and its evidence may have been, in the meantime, erased or hidden by Nature.

If, at any moment, such unfavorable ground water conditions are repeated or exceeded, the slope will experience new displacements or even a new landslide, with these occurrences being the *reactivation* of similar ones in the past.

Leroueil (2001) proposes dividing natural slope movements into the following four stages, according to Figure 9.21:

i. The pre-failure stage, including all the deformation processes leading to failure. This stage is controlled mostly by deformations due to changes in stresses, creep, and progressive failure.

ii. The onset of failure, characterized by the formation of a continuous shear surface through the entire soil mass.

iii. The post-failure stage, which includes movement of the soil mass involved in the landslide, from just after failure until it essentially stops. It is generally characterized by an increase of the displacement rate, followed by a progressive decrease in velocity.

iv. The reactivation stage, when a soil mass slides along one or several pre-existing shear surfaces.

When observing the results of Figure 9.18, it is not clear to which of these stages the movement corresponds. More specifically, it is unclear whether it is linked to a situation of *pre-failure* or to a situation of *reactivation*. Note that this reactivation can correspond to movements of very different magnitude: from new slides, large enough to change the slope geometry, to limited displacements.

A geological scenario where the occurrence of frequent movements is common, as seen in Figure 9.21, is classed as an *active landslide*. An example consists of hillsides with colluvial deposits, also known as colluvium, accumulated at the bottom. These materials result from the transport by gravity of detritus created by the physical disintegration and chemical weathering of formations that outcrop at higher elevations, as represented in Figure 9.22. This figure is inspired by a real situation, regarding the landslide of colluvial deposits over Jurassic clays, in the region of Torres Vedras, Portugal (Matos Fernandes, 2011).

During its formation, as a result of progressive accumulation of particles by gravity, this detritus mass settles on a position in which the resisting forces are those strictly necessary to ensure static equilibrium. Thus, the factor of safety against sliding is very close to 1.0,

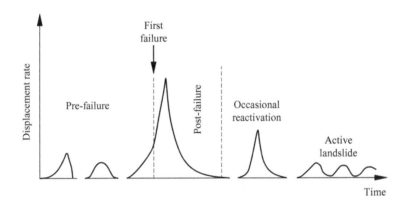

Figure 9.21 Different stages of slope movements (Leroueil et al., 1996).

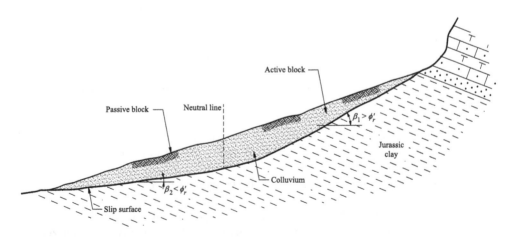

Figure 9.22 Unstable hillside with colluvial deposits overlying an older clayey formation.

with the possibility of, even temporarily, being marginally lower than this value, a situation in which the mass experiences movement, at a more or less variable rate.

In scenarios such as the one in Figure 9.22, typically there is a part of the contact surface of the colluvial deposits with the underlying stratum, the inclination of which is greater than the angle of shearing resistance in the contact surface, while the opposite occurs in the lower part. As suggested in the figure, the soil mass can be divided into two blocks: an *active block* in the upper part, corresponding to the most inclined section of the contact surface of the colluvial deposits with the underlying stratum, and a *passive block*, in which the contact with the underlying stratum is less inclined. The sliding of new materials from the top of the hillside increases the weight of the active block, inducing the movement of the whole soil mass, which tends to stabilize when part of the weight of the active block is transferred to the passive block.

Therefore, throughout geological time, the mass of the colluvial deposits increases in volume (and thus, in weight), sliding along the hillside. It is understandable that the displacement rate increases in periods of more intense precipitation, as this leads to a growth in the unit weight of the material, a reduction of the effective stress on the slip surface, and the generation of seepage forces.

On the surfaces corresponding to such landslides, the relative displacements can be of a few meters or even of tens of meters. In a scenario such as the one of Figure 9.22, these displacements lead to a progressive re-orientation of the particles in the underlying clayey formations, parallel to the direction of the movement, with a consequent reduction of the available shear strength, which is known as *residual shear strength* (Skempton, 1970, 1985).

Figure 9.23 shows a photograph of the shear surface of the slope from the previous figure. The destabilization occurred after digging part of the passive block for building a new road. The polished surface of the clay results from the phenomenon previously discussed. The visible grooves, a result of the landslide, are known in the literature as *slickensides*. In the case presented, the residual angle of shearing resistance, ϕ_r', mobilized in the slip surface ranged between 11° and 14° (with an almost null residual effective cohesion, c_r').

The example presented suggests that the choice of strength parameters to be used in the stability analyses of slopes that have shown periods of instability has to be carefully

Clay

Figure 9.23 Slide of colluvial deposits overlying a Jurassic clay, showing the clay surface, in which the residual shear strength was mobilized (photo: Carlos Sacadura).

pondered, otherwise serious safety issues might occur, namely in the design of certain retaining structures, or it may lead to the misinterpretation of the identified destabilizations. In fact, in many cases, the residual shear strength is mobilized in the slip surface and, in the case of soils with a large clay fraction, the residual angle of shearing resistance can be much lower than the peak and critical state values (Lupini et al., 1981). Thus, the characterization of residual shear strength, through laboratory shear tests on samples collected from the geological formations that control the slip surface, is extremely important.

9.5.6 Note on factor of safety

Before concluding this section, it is worth making some remarks on factors of safety. Within the context discussed in Chapter 3, factors of safety must cover the uncertainties related to the strength parameters of materials, the actions, the geometrical data, and some limitations of the calculation models. In large natural slopes, with the exception of soil strength parameters, the uncertainties associated with the remaining aspects mentioned are mostly low.[2]

It is true that, in problems involving large natural slopes, the variable actions (with the exception of earthquake actions) are negligible when compared to the permanent actions, in this case the weight of soil masses. The uncertainties related to the weight of the slope depend, on one hand, on the unit weight of the different soils involved and, on the other hand, on the geometric characterization of the potentially unstable soil mass. Regarding the first aspect considered, a cautious assumption of the characteristic value of the unit weight is recommended. As for the second aspect, the means currently available, in terms of topography, allow a precise geometric characterization of the slope surface, in a 3-D perspective. Assuming that the potential slip surface is identified – an absolutely critical aspect for assessing safety – the geometry of the unstable soil mass can, therefore, be well defined, reducing the uncertainty regarding the weight of this mass.

In terms of the calculation models, the models most frequently applied and available through well-established commercial programs – such as the Bishop's simplified, the Spencer and the Morgenstern and Price methods – ensure good mathematical approximation and lead to results on the safe side. Additionally, and this is a fundamental aspect to take into account, the use of those methods, through computational application, enables the analysis of a great number of slip surfaces. This fact strongly limits the possibilities of error associated with the selection of the surfaces to analize, a not uncommon occurrence prior to the use of computers.

Regarding the shear strength parameters of the soils involved, uncertainties can be very high. This can be explained with reference to the previous considerations stated in this chapter, which can be summarized as follows: unlike a foundation or a retaining structure, in the present case what is at stake is the available shear resistance along surfaces that can run for hundreds of meters. In addition, this resistance depends on which of the stages of Figure 9.21 the slope is at, although this stage is often unknown.

Considering a possible application of Eurocode 7, ignoring any live loads, both in Design Approach 1 – Combination 2 (this combination is the most unfavorable within DA1 in this context) and in Design Approach 3, an effective stress analysis would require the application of the coefficients $\gamma_M = 1.25$ to the strength parameters of all soils (as $\gamma_G = 1.0$ in both approaches). Therefore, the adoption of these approaches results in the application of a global factor of safety equal to 1.25. This value corresponds to a substantial reduction relative to the value adopted in traditional practice for this type of problems, which tends to consider global factors of safety of 1.5 or higher (Salgado, 2008).

---

[2] The uncertainties related to the geometry are fundamentally reflected in the weight of the slope.

In many situations, the issue of the acceptable factor of safety to be adopted is raised in studies of slope stabilization, for slopes that have shown signs of instability or experienced localised failures. In these cases, when the destabilizations analyzed are interpreted as reactivations along a pre-existing failure surface, the strength parameters can be obtained through back-analyses (assuming factors of safety equal to 1.0 or slightly higher), in combination with shear tests of the soils located along the failure surface.

On this basis, on a large natural slope, it may be extremely costly to undertake stabilization works that allow an increase of the factor of safety from a value close to 1.0 to, for instance, 1.5. However, in these cases, it seems reasonable to consider that the strength parameters were established with a considerable level of confidence, so that it is acceptable to adopt a lower value for the global factor safety in the stabilization project. The case study represented ahead, in Figure 9.25, is an excellent example of this *praxis* (Alonso et al., 1993).

When pondering this issue of the lower acceptable factor of safety, it seems appropriate to remind the reader of a well-known statement: complex matters rarely have simple answers!

9.6 STABILIZATION MEASURES

9.6.1 Introduction

Table 9.1 shows a summary of the most common stabilization measures, which can be divided into three major groups: i) change in geometry; ii) hydraulic measures; and iii) structural measures.

A more thorough discussion of these stabilization measures can be found in works on this specific field (Ortigão and Sayão, 2004). In the following sections, some brief remarks on these stabilization measures will be considered, in the context of its application to natural slopes and not to the stabilization of cut slopes or excavations. Quite often, stabilization projects of large natural slopes involve combinations of the three types of measures listed above.

9.6.2 Alteration of the slope geometry

Figure 9.24 shows the two types of geometry alteration: excavation of the upper part and placement of fill at the slope toe.

The change in slope geometry, in particular the removal of part of the active zone and reduction of the average slope angle, illustrated in Figure 9.24a, is often a convenient solution, being frequently less costly than solutions based on structural measures, even when it implies the relocation of constructions and services.

In an unstable slope, the application of a stabilization measure, involving the removal of part of the slope mass, has to be undertaken from top to bottom. Otherwise, due to the reasons discussed in Section 9.5.2, it could cause a reduction in the factor of safety existing before the intervention. Such reduction, even if temporary, could lead to global failure of the slope.

A particularly convenient solution consists of combining excavation with fill, using the material from the former to be used in the latter. Figure 9.25 illustrates one concrete example of a large natural slope in the region of Valencia, Spain, in Cretaceous limestone and marl formations (Alonso et al., 1993; Alonso, 2019). The landslide was reactivated by quarry excavations, and the geometry of the unstable soil mass was determined by the presence of a thin clayey marl layer, as indicated in the figure. The stabilization measure consisted of

Table 9.1 Summary of the stabilization measures of natural slopes.

Type of measures	Description	Advantages	Observations
Alteration of the slope geometry	Removal of soil mass from the active zone and/or reduction of slope angle.	Reduces the weight in the active zone; reduces the mobilized resistance.	May be unfeasible in built areas; must be undertaken from top to bottom; it is usually followed by superficial drainage and planting of vegetation.
	Placement of fill at the base of the slope.	Increases the weight of the passive zone.	May be unfeasible in built areas; may be part of an emergency measure; the material placed has to be more permeable than the material of the slope.
Hydraulic measures	Superficial drainage. Network of surface drains to reduce rainwater infiltration.	Reduces pore water pressure, as well as seepage forces and the weight of the soil mass; counteracts the reduction in shear resistance of very plastic soils.	Usually complemented with vegetation protection measures.
	Deep drainage. Draining galleries, trenches and wells.	See cell immediately above; maintains the phreatic surface away from part or the entire unstable zone.	Drainage can be by gravity or by pumping.
	Crack filling in the active zone with impermeable material, such as clay or cement grout.	Prevents water infiltration to the inside of the slope and, specially, to the slip surface.	In case of emergency, can be replaced by ground cover with synthetic and impermeable material.
	Protection of the slope face with vegetation, sprayed concrete or gabions.	Protects the ground from superficial erosion.	Sprayed concrete is not very convenient in terms of landscape; gabions have the disadvantage of increasing the weight of the active zone.
Structural measures	Application of external anchor forces. Pre-stressed ground anchors, connected to walls or reinforced concrete beams on the face of the slope.	Normal component increases the resistant force on the slip surface; the tangential component acts as an external stabilizing force applied to the unstable mass.	Ground anchors heads are connected to reinforced concrete elements, which distribute the anchor forces onto the face of the slope; grout bodies have to be placed outside the unstable zone.
	Ground reinforcement of the slip surface with steel-nailing reinforcement, reinforced concrete piles, or jet-grout columns.	Increases shear resistance on the slip surface.	The reinforcements have to penetrate a certain length beyond the slip surface.

simply changing the slope geometry, as indicated, with a consequent movement of 800,000 m³ of material from the active zone to the passive zone.

Modern roadworks in complex orography typically involve large earthworks (excavations and fills). These changes are often sources of destabilization of natural slopes in the vicinity. In cases in which there is an excess in the excavation volume in relation to the fill volume, placing the excess material at the toe of the potential unstable slopes is a very convenient solution from a technical, economic, and environmental point of view.

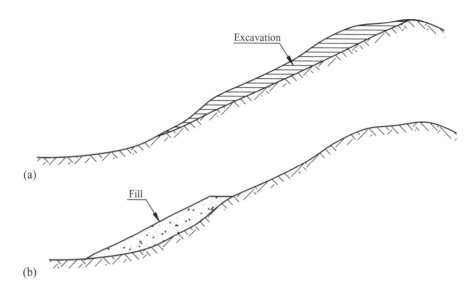

Figure 9.24 Schematic illustration of geometry change in natural slope: a) excavation in the upper part; b) placement of fill in the lower part.

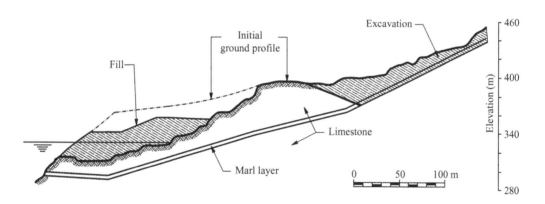

Figure 9.25 Stabilization of the Cortes de Pallas landslide (adapted from Alonso, 2019).

9.6.3 Superficial and deep drainage

With water being the cause of a large proportion of the landslides on natural slopes, it is understandable that drainage, either superficial or deep, consists of one of the most effective stabilization measures.

Superficial drainage is aimed at minimizing rainwater infiltration. Accordingly, a network of superficial drains must be designed in order to avoid water, that has fallen on a certain point of the slope, from running over significant distances on the surface of the slope, before being collected and drained by gravity to the base of the slope. In non-built areas, the protection of this surface with vegetation is an essential complement to the drainage system. Figure 9.26 shows the partial aspect of a superficial drainage system on a natural slope.

Figures 9.27 and 9.28 show two measures involving deep drainage of natural slopes. The solution illustrated in Figure 9.27 consists of the opening from the slope face, in the area close to the toe of the slope, of *draining galleries*. This opening can be executed with the

Figure 9.26 Superficial drainage system of a natural slope.

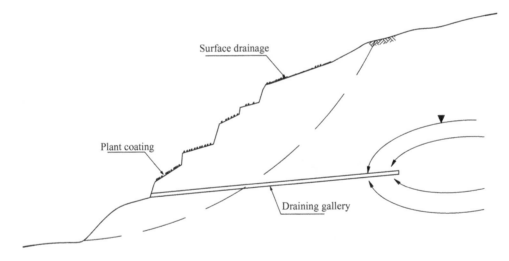

Figure 9.27 Deep drainage in natural slopes - sub-horizontal drainage galleries.

same equipment used in micro-tunnelling (tunnels of small diameter, usually accessible for inspection or maintenance, such as in the present case), with the galleries being lined with a permeable support. The inclination of these galleries allows drainage by gravity to the slope face, where it connects with the superficial drainage system. In general, a system of galleries is designed, in terms of implantation on the slope face, length and inclination, to maintain the phreatic surface as far away as possible from the slope surface, and hence from the potentially unstable zone.

An alternative solution, but with a similar goal, consists of constructing *draining trenches*. As illustrated in Figure 9.28, these consist of a deep trench filled with highly permeable material (gravel, for instance), which causes the lowering of the upstream ground water level, preventing the water from reaching the unstable zone. The water that flows into the

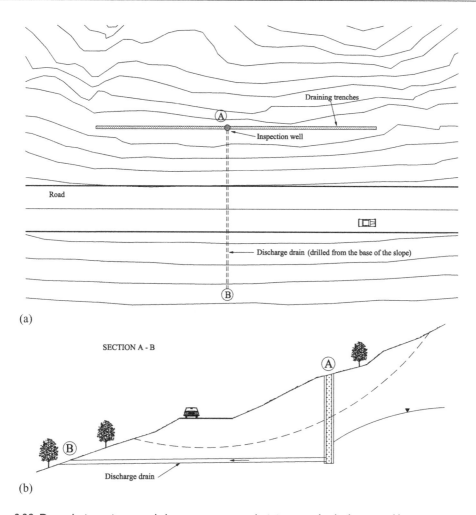

(a)

(b)

Figure 9.28 Deep drainage in natural slopes – upstream draining trench: a) plan view; b) cross section.

trench is drained by gravity through transverse discharge drains, constructed from a lower point on the hillside, using appropriate drilling technology that can ensure high precision.

A solution with an equivalent effect to that of the draining trenches is a set of *draining wells*, with the layout on the plan view and in the cross section being analogous to a draining trench. Drainage can be carried out by pumping, being automatically activated when the water in the well reaches a level considered to be dangerous, which, however, has the inconvenience of requiring maintenance of the system. The drainage by gravity is also feasible by connecting the wells at the base by a collector, constructed in a similar manner to the transverse drains mentioned above. This collector is able to conduct the water to a central well, from where it will be discharged in a similar way to the draining trench (Collotta et al., 1988).

From the description presented, the reader may conclude that the deep drainage system is expensive, and its use is justified only on large natural slopes, which is, in fact, the context of this chapter. In addition, changes to the local hydrological regime, induced by the deep drainage measures, may lead to environmental impact, the evaluation of which would be necessary.

9.6.4 Structural solutions

9.6.4.1 Ground anchors connected to reinforced concrete walls

Among the structural measures, the most frequently used are pre-stressed ground anchors combined with retaining walls on the slope face. The discussion of ground anchors was carried out in Chapter 8.

Figure 9.29, inspired by a real situation, shows a schematic illustration of a stabilization solution combining deep and superficial drainage, ground anchors, and changes in the geometry, i.e., the three types of stabilization measures presented in Table 9.1. In the stabilization works, the ground anchor heads are usually connected to reinforced concrete walls concreted against the slope face, aimed at distributing the loads applied by the ground anchors (usual loads range between 400 kN and 1000 kN). The opposite ends (the grout bodies) are in the ground beyond the failure surface, that is, in the area of the soil mass not involved by the destabilization, as suggested in the figure.

Figure 9.30 shows an image of a retaining wall, such as the one mentioned above, being able to observe the ground anchor heads. At a final stage, the heads are protected against corrosion, normally by using concrete. It is important to note that the reinforced concrete structure is exclusively used with the purpose of distributing the forces of the ground anchors to the ground, with its thickness being strictly necessary to support the internal stresses associated with the concentrated forces received at each point corresponding to the ground anchor head.

The stabilizing effect of the ground anchors can be described as follows: i) the component of the ground anchor forces parallel to the slip surface runs in the opposite direction to the landslide itself; and ii) the component perpendicular to the slip surface increases the resisting forces on this surface.

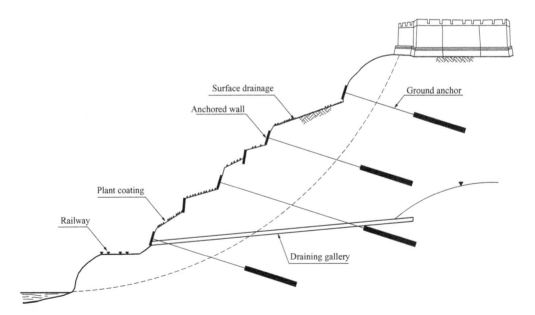

Figure 9.29 Stabilization measure combining deep and superficial drainage, ground anchors and changes in the geometry (schematic illustration inspired by the Portas do Sol Hillside, Santarém, Portugal).

Figure 9.30 Retaining wall for the distribution of the forces of pre-stressed ground anchors for stabilization of the Santarém Castle Hillside (courtesy of Teixeira Duarte).

9.6.4.2 *Ground reinforcement with inclusions that intersect the failure surface*

Another family of structural solutions involves ground reinforcement through structural elements installed in order to intersect the slip surface, as shown in Figure 9.31. The most robust reinforcement elements are reinforced concrete piles or diaphragm wall panels. The shear resistance of these reinforcement elements, which is much greater than the resistance of the ground replaced, increases the total available shear resistance on the slip surface.

However, it is important to note that this type of reinforcement, designed to work under shear, is used less frequently in stabilization works of natural slopes with deep failure surfaces. In fact, in these cases, in order to add significant increments to the total shear resistance, it is necessary to resort to concrete elements with large and highly reinforced sections,

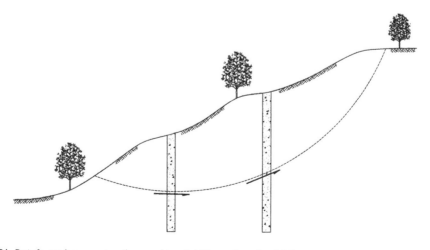

Figure 9.31 Reinforced concrete piles used to stabilize a deep landslide.

which makes it a costly solution when compared to others, such as those involving ground anchors and ground-anchored retaining walls. In addition, the latter requires light and versatile equipment for their construction, while the former needs heavy equipment, which is difficult to place and to maneuver on certain natural slopes.

An interesting example, in which the use of piles combined with ground anchors was revealed to be particularly convenient, is shown in Figure 9.32. It concerns the stabilization of a relatively shallow landslide of colluvial deposits overlying Jurassic clays, triggered by an excavation for a new road on the slope. In the present case, the piles have a dual role: i) to distribute the forces of ground anchors; and ii) to intersect the failure surface.

For the stabilization of shallow landslides, a common reinforcement solution is soil nailing. Figure 9.33 shows the stabilization of an infinite slope, with a shallow failure surface, with steel nails, which are arranged in a square mesh.

Soil nailing consists of a very simple technology: i) execution of a hole, usually perpendicular (or approximately perpendicular) to the slope face or to the cutting face (standard diameter of 7.5–10 cm) ii) hole filling by gravity with cement grout; iii) introduction of a standard steel bar, similar to the ones used in conventional reinforced concrete structures; and iv) attachment of a steel plate, next to the slope face, to the extremity of the steel bar (in order to better distribute the force that will be mobilized in the soil nails). These soil nails are usually complemented by a facing of sprayed concrete and a steel mesh, placed on the slope face.

9.6.5 Methods of analysis of the stabilization solutions

To conclude, it is important to highlight that the methods of stability analysis previously studied in this chapter and in Chapter 2 play a key role in the design of any stabilization system. In fact, these methods provide quantitative indications in terms of the efficiency of

Figure 9.32 Anchored piled wall for the stabilization of a landslide in Lisbon, 1960s (courtesy of Teixeira Duarte).

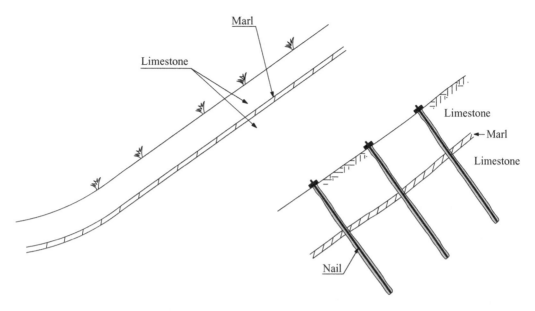

Figure 9.33 Stabilization of an infinite slope in sedimentary soils, using steel nails working under shear.

the stabilization solutions, by comparing the factors of safety of the slope before and after these works were carried out.

It is interesting to observe that the benefit of this comparison is not compromised, even though those factors of safety may be (and probably will be!) affected by errors resulting from the limitations of ground geotechnical characterization, which is especially complex on many natural slopes. In fact, it seems reasonable to expect that these limitations will affect the results of stability analyses, with or without stabilization works, in a similar manner.

9.7 OBSERVATION OF NATURAL SLOPES

9.7.1 General remarks

From the previous remarks, the reader would have understood the relevance of the observation or monitoring of natural slopes in the development of stabilization studies. At this stage, such observation essentially plays two key roles. First of all, it constitutes a basic measure to control the safety of the slope! If the symptoms of instability worsen substantially, foreshadowing impending failure, monitoring and/or observation will allow the development of safety measures to safeguard people and property. On the other hand, as previously discussed, observation is fundamental for interpreting the potential failure mechanism, as well as being important in obtaining specific information required for the stabilization project, namely the position and changes of the phreatic surface(s) in the body of the slope.

After carrying out stabilization works, observation remains of the greatest importance. It will permit assessment of the efficacy of these works and identify whether they need to be adjusted or reinforced. In practice, in the most complex cases, the project is progressive, that is, it comprises several phases driven by the behavior revealed by the slope. This procedure has a long tradition in geotechnical engineering, being known as the *observational method*

(Peck, 1969). Therefore, it can be agreed that the monitoring plan is a fundamental part of the stabilization project.

It is also important to recall that most of the destabilizations observed on natural slopes are associated with exceptionally adverse ground water conditions, which occur with certain – shorter or longer – return periods. This explains why a slope that exhibited instability in one exceptionally rainy year can regain stability during some years, even without any stabilization work having been carried out.

From what has been stated, it can be understood that, after certain stabilization works have been implemented, their efficiency will only be seen in the *long-term*. Thus, long-term observation of slopes is essential. Additionally, this observation will permit assessment of whether any relevant reinforcement element has deteriorated or is no longer active, requiring maintenance or repair works. Hence, the following rule: a slope that underwent intervention following signs of instability must remain under observation.

9.7.2 Observation plan

Table 9.2 summarizes the elements of a typical instrumentation plan of a natural slope, in terms of the quantities to be measured and the respective measuring methods. This table does not exhaustively describe all items. Accordingly, in some special cases, it may be necessary to make other measurements in the ground (such as the measurement of internal deformations at various levels, using extensometers, for instance) or in the constructions, located on the slope, that experienced damage (measuring crack width with a crackmeter, for example).

The usefulness of the measurement of horizontal movements along a vertical line inside of a slope (using an inclinometer) has already been emphasized, with the operation of an inclinometer being addressed in Annex A9.2. Regarding movements at ground surface, their usefulness is also evident. However, in some cases it may be difficult to find a convenient project datum in the surroundings of the slope, to be used in topographic surveys.

The measurement of forces in the ground anchors is very important for several reasons. As the ground anchors are sealed away from the unstable soil mass, the movement of the mass will lead to an increase in the anchor's forces. Therefore, the combined analysis of the variation of these forces and of the displacements in the inclinometer tubes is valuable for interpretation of the behavior of the slope and its evolution. Reduction in the ground

Table 9.2 Basic observation plan of a natural slope: measurements to be carried out, measurement methods and equipment, as well as respective installation locations.

Measurements to be carried out	Equipment/method	Installation sites	Observation
Horizontal displacements in depth	Inclinometer	Inclinometer tubes installed in boreholes	See Annex A9.2
Superficial displacements	Survey pins and targets	Points on the surface of the slope or on structures located on the slope	–
Forces	Load cells	Ground anchor heads	–
Pore pressure/ground water level	Hydraulic or electric piezometers	Installed in boreholes at various depths	See Annex A9.3
Precipitation levels	Pluviometers	Point in the area surrounding the slope and on the slope itself	Data from the national meteorological network can also be used

anchor's forces can result from the relaxation of the steel tendon, due either to deficiencies on the anchor head or by creep in the ground surrounding the bond length, or even due to corrosion of the steel tendon. Under these circumstances, in order to maintain the stabilizing role of the ground anchors, these must be prestressed again (in the case of relaxation) or replaced (in the case of corrosion).

The ground water levels in the slope and the respective pore water pressures can be measured with either hydraulic or electric piezometers. It is worth noting that different aquifer levels, without vertical continuity, can be found in large slopes with complex geology. Annex A9.3 includes schematic diagrams of piezometers.

Finally, a remark regarding precipitation levels may be relevant. The relationship between the displacements registered in a given slope throughout time, in response to the precipitation in the region, is always enlightening, as this comparison can demonstrate the enormous influence that unfavourable ground water conditions have on the instability of natural slopes. This influence can be observed in Figure 9.18, as previously discussed.

However, it is important to explain the meaning of precipitation, that is, which is the most appropriate time interval for computing a given rainfall volume? Accordingly, very intense precipitation over a short period of time is mostly associated with relatively shallow landslides (see Section 9.2, regarding the scenarios with infinite slopes), while deep landslides, involving large slopes, are typically related to relatively long periods of time, with high values of average precipitation (Leroueil, 2001).

Given the complexity of this problem, the relationship, in quantitative terms, between precipitation and slope failures can only be established based on a local or regional experience. This has been explored with success for public safety in cities like Rio de Janeiro and Hong Kong, where annual average precipitation is high, and there are also very high values of precipitation occurring over relatively short periods of time (Brand, 1985; Ortigão and Sayão, 2004).

The precipitation in a given slope can be measure by devices called pluviometers.

ANNEX A9.I SEISMIC COEFFICIENTS TO BE USED IN SLOPE STABILITY ANALYSES, ACCORDING TO EUROCODE 8

This annex presents a summarized description of the aspects considered in Eurocode 8 (EN 1998-1: 2004; EN 1998-5:2004), which establishes the seismic coefficients to be used in the slope stability analyses. As seen in the other chapters, the city of Lisbon was taken as reference.

The horizontal seismic coefficient is given by the following equation:

$$k_h = 0.5 \left(\frac{a_g}{g} \right) S\, S_T \qquad\qquad (A.9.1.1)$$

where:

- a_g represents the design value of (horizontal) seismic acceleration on a type A ground, which results from the product between maximum reference acceleration, a_{gR}, and an importance factor, γ_I;
- g is the acceleration due to gravity;
- S is the coefficient that takes into account the possible amplification of acceleration between the bedrock and the ground surface;
- S_T is the topographic amplification factor, which should be applied whenever there are structures with an importance factor greater than 1.0 over or near cliffs, slopes longer than 30 m, and inclinations greater than 15°.

The calculation of a_g and S has been already discussed in Chapter 3 (see Section 3.8). The values of S_T are shown in Table A9.1.1, with the topographic effects being particularly relevant in shallow landslides. On the other hand, in deep landslides, in which the failure surface passes through the toe of the slope, the topographic effect can be ignored, as the seismic amplification factor decreases rapidly with depth.

In turn, the vertical seismic coefficient is given by the following equations (NP EN 1998-5: 2004):

$$k_v = \pm 0.5 k_h \quad \text{if} \quad \frac{a_{vg}}{a_g} > 0.6 \qquad\qquad (A9.1.2)$$

Table A9.1.1 Values of the topographic amplification factor, S_T (adapted from EN 1998-5:2004).

Conditions of the slope	S_T
Slope angle less than 15°	1.0
Sites near the top edge of isolated cliffs and slopes	≥1.2
Ridges with crest width significantly less than the base width, and average slope angle between 15° and 30°	≥1.2
Ridges with crest width significantly less than the base width, and average slope angle greater than 30°	≥1.4

Notes:
1. In the presence of a loose surface layer, the values of the three lines above should be increased by at least 20%.
2. The value of S_T may be assumed to decrease as a linear function of the height above the cliff or ridge, and to be unity at the base.

$$k_v = \pm 0.33 k_h \quad \text{if} \quad \frac{a_{vg}}{a_g} \le 0.6 \tag{A9.1.3}$$

Considering what has been established in Table 3.16, the following is obtained:

$$k_v = \pm 0.5 k_h \tag{A9.1.4}$$

ANNEX A9.2 INCLINOMETER MEASUREMENTS

Figure A9.2.1 shows an inclinometer and the rest of the apparatus. The inclinometer, with the shape of a torpedo, contains a pendulum which emits an electric signal proportional to the inclination (hence the name "inclinometer") between the device axis and the vertical direction.

When it is intended to measure horizontal displacements in a soil mass along a vertical line between the ground surface and a certain depth, a borehole is executed along this line. A tube (usually made of aluminum), the inclinometer casing, is installed in this borehole with the space between the tube and the walls of the borehole being filled with cement grout or sand.

The inclinometer casing has four grooves forming angles of $90°$ between them. These grooves are used for guiding the inclinometer probe to the appropriate depth, as the probe has two sets of wheels, as shown in Figures A9.2.1 and A9.2.2a. This allows the device to descend along the tube, suspended by a cable, without experiencing any rotation.

In order to carry out the readings, the inclinometer probe is lowered to the base of the hole and, from there, it is successively raised, with the respective inclination, $\delta\omega$, being measured at regular intervals of L (typically, 0.60 m). As shown in Figure A9.2.2b, from the successive inclination readings, it is possible to calculate the horizontal distances at the points at heights L, $2L$, $3L$, etc., above the base of the tube, in relation to the vertical that passes through the base of the hole, in the plane that contains the pair of grooves along which the

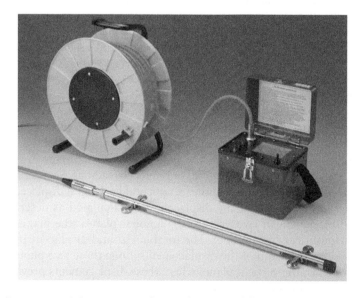

Figure A9.2.1 Inclinometer and the respective devices (courtesy of Geokon Inc.).

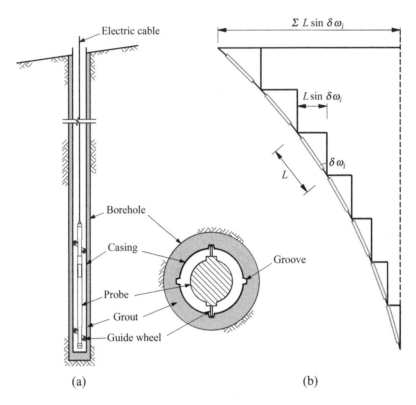

Figure A9.2.2 Working principle of an inclinometer: a) schematic diagram of the inclinometer; b) interpretation of the inclinometer readings.

inclinometer probe has travelled. Having finished this operation, a similar procedure is carried out, by introducing the inclinometer probe in the other set of grooves, in order to obtain the horizontal movements in the vertical orthogonal plane.

Following the installation of the inclinometer casing tubes, initial or reference readings (zero reading) are carried out. With readings being performed on a subsequent date, the difference between the distances measured, for each height and vertical plane, and the initial measurement represents the horizontal displacement of the soil mass between the two dates.

Two important additional remarks can be made.

Firstly, it is important to note that the horizontal displacement obtained is not given in absolute terms, but *in relation to the vertical plane that passes through the base of the casing*. However, as this base is sealed in the ground outside the unstable zone, its displacement can be taken as being zero, which means that the horizontal displacements measured by the inclinometer are, in practical terms, absolute displacements. These displacements (absolute in *space*) are relative over *time*; that is, they are relative to the initial or reference readings. Successive measurements, following the first, enable an evaluation of the incremental horizontal displacements in depth.

Secondly, it should be noted that the displacements, calculated as described above, correspond to the displacement in two vertical orthogonal planes (the planes containing the grooves of the inclinometer casing tube). The total horizontal displacement will, therefore, be obtained from the vector sum of the displacements along these two planes. In the case of slope stability problems, the vertical plane, where these displacements prevail, is reasonably clear. Accordingly, when installing the inclinometer casing in the borehole, it is important to place one pair of grooves along the direction with (supposedly) greater ground mass displacement, presumably along the dip direction.

The installation of inclinometer casings is also common in some structural elements used for slope stabilization. For instance, when using reinforced concrete piled walls (see Figure 9.32), the casings are installed together with the pile steel cages; it is essential that these are carefully sealed, in order to prevent them being filled with concrete, during pile concreting.

A9.3 *IN SITU* MEASUREMENT OF PORE WATER PRESSURE

The basic working principle of a piezometer is very simple: i) a porous body is placed at a point in the ground where the pore water pressure is to be measured; ii) the water in the soil flows through the porous body, entering a compartment within the device; and iii) subsequently, either the water level or the water pressure in this compartment can be measured, with both corresponding to the pore water pressure at the given point, when the process is stabilized.

A given volume of water must flow from the adjacent soil into the piezometer to pressurize the system, with the variation in pressure being proportional to this volume of water. Taking into account that all types of soil exhibit finite permeability, there will be always a *time lag* between the variation in pore water pressure in the soil and this being registered in the piezometer. This time lag, also designated *the piezometer response time*, determines the type of device to be used in each situation.

In soil masses with medium to high permeability, namely those consisting of granular soils, the most common form of measuring water pressures is through a piezometer tube installed in a vertical hole, as shown in Figure A9.3.1. A pipe of synthetic material (polyvinyl chloride, PVC), with the lower section perforated and lined with a geotextile, is inserted in

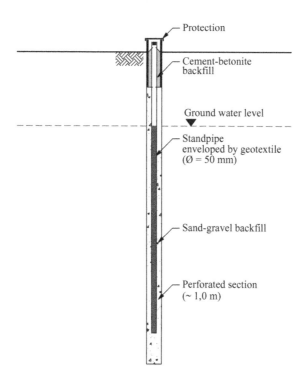

Figure A9.3.1 Open standpipe piezometer.

the hole. In addition, the space between the pipe and the wall of the hole is filled with sand or fine gravel, with the top of the pipe being protected at ground surface.

The water level reached in the tube corresponds to the pressure head at the point where the piezometer is placed, with the product between this head and the water unit weight resulting in the pore water pressure at this given point.

From the description given, it can be stated that this type of *open standpipe piezometer*: i) requires a significant water volume to flow into the device, in order to measure the water pressure; and ii) establishes a hydraulic connection between the various ground layers that are intersected by the piezometer. Consequently, the application of this device is restricted to medium- to high-permeability soils, exhibiting permeability values greater than 5×10^{-6} m/s. In these soil masses, the first point raised above (i) does not lead to an extended response time, with the second point (ii) having no practical relevance, as the phreatic surface in the ground is common to the various soil layers.

In order to reduce the time lag, it is common to use a *Casagrande hydraulic piezometer*, such as the one represented in Figure A9.3.2. This piezometer consists of a porous ceramic tip (with a diameter of approximately 40 mm), installed at the level where water pressure is to be measured, and connected to a pipe of reduced diameter (usually 10 mm), made of impermeable synthetic material. With this system, the volume of water that flows into the device is substantially reduced, which corroborates the fact that it is usually acceptable to use this device in soils with a permeability greater than 10^{-7} m/s.

As shown in Figure A9.3.2, the test section is sealed at the top with clayey material and cement-bentonite grout, isolating the layer where the water pressure is to be measured from

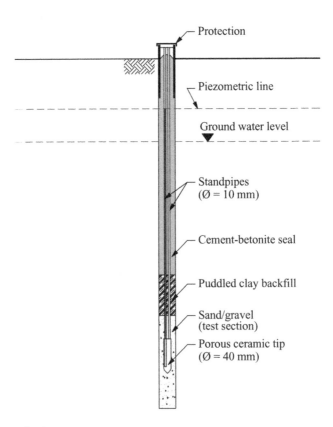

Figure A9.3.2 Casagrande piezometer.

the overlying layers. The water level measured will, thus, represent the pressure only in the soil adjacent to the test section.

The small diameter tube can operate as an open standpipe, with water in contact with the atmosphere, or as a closed tube, connected to a pressure transducer. The latter can be very convenient in situations where the pressure registered corresponds to a water column that exceeds the ground surface.

The reader may have already noticed that Figure A9.3.2 shows two pipes of small diameter, instead of only one. One of the pipes extends to the base of the porous ceramic tip, while the other ends at its top. This set of pipes is used in order to expel the air present in the system, both at the time when the piezometers are installed and during the exploration phase, when necessary. This is seen to be necessary, as the presence of air bubbles in the system can substantially reduce permeability, namely of the porous ceramic tip, hampering the measurements. In order to expel the air, water is injected from the surface into one of these tubes, with the water and the expelled air flowing to the surface through the other. Following this procedure, the shorter pipe is sealed.

In soil with very low permeability, such as clayey soils, the reduction of the time lag requires other types of piezometers, with *electric piezometers* being most commonly used. Figure A9.3.3 shows a schematic diagram of one of these devices. This type of piezometer also includes a porous ceramic tip through which the water will flow until reaching a closed chamber. In this chamber, the water will exert pressure on an electric transducer, connected to the ground surface by a cable, which, in turn, is connected, permanently or not, to the data logger.

Certain transducers are detachable, which allows them to be replaced in case of malfunction, or for the injection of water in order to extract the air in case the porous body is occluded.

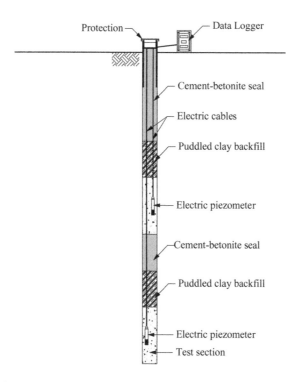

Figure A9.3.3 Electric piezometer.

This type of piezometer can measure pressures lower than atmospheric pressure. As shown in Figure A9.3.3, by sealing the test section with impermeable material, it is possible to measure pore water pressure at different depths, with additional piezometers installed in the same hole.

ANNEX E9 EXERCISES (RESOLUTIONS ARE INCLUDED IN THE FINAL ANNEX)

E9.1 – Consider the infinite slope represented in Figure E9.1. The slope has shown some signs of instability during periods of heavy rainfall. Given that a highway passes at the toe of the slope, stabilization work is being considered. One of the pondered solutions consists of reducing the thickness of the residual soil layer (this reduction, by excavation, will be combined with vegetation planting, both for landscaping purposes and for protecting the soil surface from erosion).

Determine the thickness, d, to be removed from the topsoil layer in order to achieve a minimum safety factor of 1.5, assuming that the water table coincides with the soil surface after excavation.

Take for the residual soil: $\gamma = 20$ kN/m³, $c' = 20$ kPa, $\phi' = 31°$.

E9.2 – Figure E9.2a shows a natural slope of tertiary limestone formations within which occur some fine layers of clayey marl. An excavation will be carried out, as indicated in the figure.

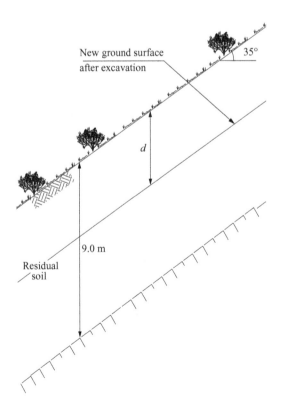

New ground surface after excavation

35°

d

9.0 m

Residual soil

Figure E9.1

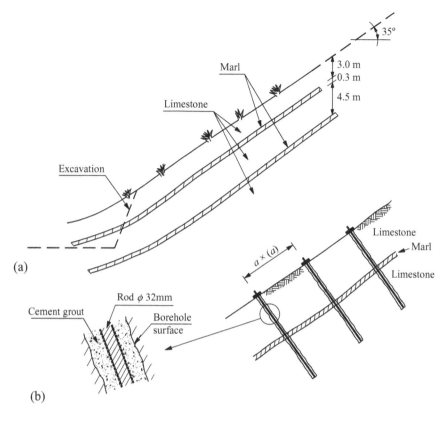

Figure E9.2

Table E9.2

Formation	γ (kN/m³)	c' (kPa)	ϕ' (°)
Limestone	23.0	200	40
Clayey marl	21.5	10	20

a) Using the parameters of Table E9.2 (and considering them as characteristic values), prove that the soil mass will probably experience slipping as a result of the excavation. As the thickness of the potentially unstable ground is very small, compared with the slope dimension, consider this to be an infinite slope.

b) In order to make the excavation viable, a set of nails will be installed perpendicularly to the slope in a square layout, as shown in Figure E9.2b. Each nail comprises a 32 mm diameter steel rod inserted in a 7.5 cm diameter pre-drilled hole filled with cement grout. As the nails intersect the potential slip surface, their shearing resistance is added to that of the soil along that surface. Using partial safety factors in accordance with Eurocode 7 (Design Approach 1 – Combination 2), determine the distance, a, between nails, given that the design value of the shearing resistance of each nail is 75 kN.

E9.3 – Figure E9.3 represents a natural slope where a road will be built. The excavation required for the construction of the road may cause instability to the slope, causing sliding of the limestone layer at the contact with the underlying formation, consisting of clayey sandstone. Table E9.3 includes the geotechnical parameters of the two materials.

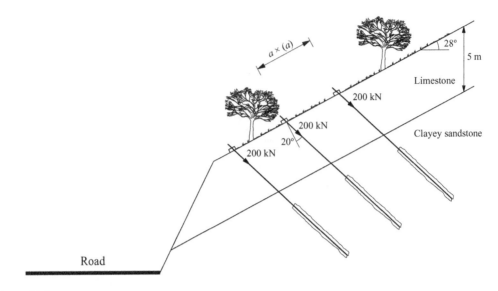

Figure E9.3

Table E9.3

Material	γ (kN/m³)	c' (kPa)	ϕ' (°)
Limestone	23	200	38
Clayey sandstone	22	8	25

In order to stabilize the slope, the construction of a grid of ground anchors in a square layout is being considered, with an inclination of 20° in relation to the perpendicular to the slope, pre-stressed at 200 kN each.

Define the value of the spacing between anchors, a, in order to satisfy a global safety factor of:

a) 1.5 in static conditions;
b) 1.15 in seismic conditions defined by the seismic coefficients $k_h = 0.12$ and $k_v = \pm\, 0.06$.

E9.4 – Figure E9.4a represents a natural slope with a railway running along its base. The slope has a layer of marl with an unfavorable inclination. Moreover, a fault has been identified through the limestone with a filling material (mylonite) with relatively low strength. The characteristic values of their geotechnical parameters are indicated in Table E9.4.

The slope location has relatively high seismicity, which may raise concern about the stability of the limestone mass delimited by the marl layer and the fault.

In the stability analyses, consider the potentially unstable mass divided into two blocks, as indicated in Figure E9.4b, and assume that the ground shearing resistance is fully mobilized on the vertical separation surface.

a) Calculate the safety factor of the slope in static conditions. Consider the safety factor according to Equation 9.36, as the value by which the available shearing resistance of the soil should be divided in order to obtain the mobilized shearing resistance:

$$F = \frac{c'}{c'_m} = \frac{\tan\phi'}{\tan\phi'_m}$$

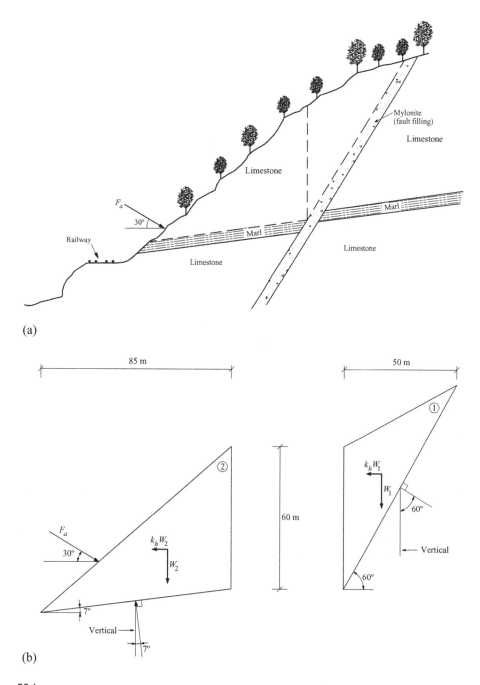

(a)

(b)

Figure E9.4

Table E9.4

Material	γ (kN/m³)	c' (kPa)	ϕ' (°)
Limestone	24	100	40
Mylonite (fine-grained metamorphic rock)	–	0	30
Marl	–	0	22

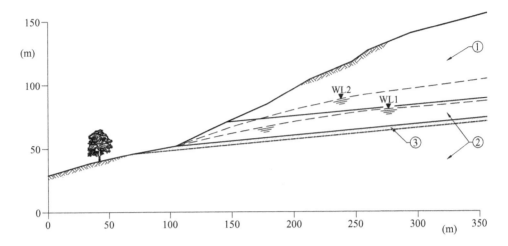

Figure E9.5

Table E9.5

Material	γ (kN/m³)	c' (kPa)	ϕ' (°)
Upper dolomitic limestone (1)	23	50	35
Lower dolomitic limestone (2)	24	100	35
Clay (3)	22	0	16

b) Calculate the safety factor of the slope under a seismic action defined by $k_h = 0.10$.

c) To improve the stability conditions under the seismic action considered in b), anchored walls will be constructed at the surface of the lower block (block 2). The ground anchors will form an angle of 30° with the horizontal, as indicated in Figures E9.4a and b. Calculate the anchor force per meter of longitudinal dimension of the slope, in order to obtain a safety factor of 1.15.

E9.5 – Figure E9.5 shows the cross section of a natural slope of Jurassic sedimentary formations. The main material is a dolomitic limestone, which can be divided into two parts, the upper part being of lower resistance. Interbedded in the lower and stiffer part, there is a thin layer of plastic clay, inclined 5° in relation to the horizontal. The geotechnical parameters of the three main materials are presented in Table E9.5.

Applying a program capable of performing stability analysis by the method of Morgenstern and Price, evaluate the safety factor of the slope for the three subsequent conditions regarding the water level:

i)   the water level is very deep and does not interfere with the stability of the slope;
ii) the water level is in position 1, as indicated in the figure;
iii) the water level is in position 2, as indicated in the figure.

Compare the result obtained for condition ii) with the application of the simplified Bishop method. Comment on the results obtained.

Chapter 10

Introduction to earthworks

Soil compaction

10.1 INTRODUCTION

The use of soil as a construction material is as ancient as human civilization itself. Among the uses of soil as a construction material, *earthfill works* stand out in terms of their variety, technical relevance, and the volume of soil used. These works involve earthfill and rockfill dams and embankments for transport infrastructures (roads, railways, and airports), among others.

Figures 10.1 and 10.2 show cross sections of the Oroville dam in the U.S. state of California, and of the Beliche dam in the Algarve, Portugal, respectively. Figure 10.3 shows the cross section of an embankment for a road in Northern Portugal.

The conventional process for the construction of earthfills usually consists of the following essential phases: i) removal, by excavation, of the soil or rock to be used as fill material from its natural place of deposit, the so-called *borrowing area*; ii) transport of the material to the construction site; iii) spreading of the material in layers of specified thickness; and iv) compaction of the layers by *compaction rollers* for a pre-established number of passes.

The key phase for the construction of earthfills is the compaction phase. Compaction is the process applied to a partially saturated soil mass – consisting of three phases, namely solid (particles), liquid (water), and gas (air) – aimed at increasing its density by reducing the volume of air, through the repeated application of loads. Thus, compaction involves expelling air without causing a significant change in the amount of water present in the soil. Since the amount of air is reduced without variation in water content, the saturation degree increases with compaction. Nevertheless, the air cannot be entirely removed by compaction, so soil saturation is not reached.

The properties of soils that are important to modify through compaction, when they are used as earthfill materials, are shear strength, stiffness, and permeability. In general, compaction increases the shear strength and stiffness, while it causes a reduction in permeability.

In some cases, such as earthfills for road or railway systems, certain areas are built using crushed rock instead of mineral particles from natural soils. The reuse of industrial waste material has become increasingly common, as is the case of metallurgical slag. The focus of this chapter, however, is on earthfill works that make use of natural soils.

The first part of this chapter is essentially dedicated to the topic of soil compaction and to the behavior of compacted soils. The second part outlines the circumstances for the design and construction of earthfill works. This aspect is discussed in more detail in Section 10.3.

Before finishing this introduction, it is worth pointing out that this chapter approaches compaction only in its *conventional* sense, that is, the densification process described above that occurs during the construction, by layers, of the earthworks. Other types of compaction can be applied to natural soils, such as the cases of vibro-compaction and dynamic compaction, which have already been discussed (see Section 6.5.5).

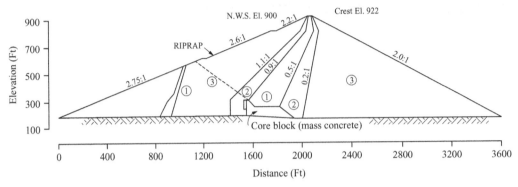

Zone 1 - Impervious (Clayey sandy gravel)
     2 - Transition (Amphibolite gravel, finer distribution than Zone 3)
     3 - Pervious (Amphibolite gravel)

Figure 10.1 Oroville dam maximum height cross section (Kulhawy and Duncan, 1972).

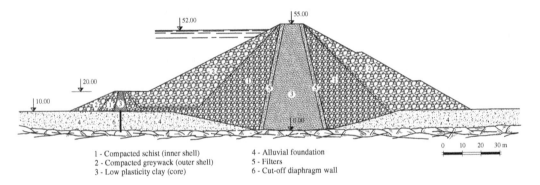

1 - Compacted schist (inner shell)   4 - Alluvial foundation
2 - Compacted greywack (outer shell)   5 - Filters
3 - Low plasticity clay (core)   6 - Cut-off diaphragm wall

Figure 10.2 Beliche dam maximum height cross section (Veiga Pinto, 1983).

10.2 COMPACTION

10.2.1 Compaction curves

10.2.1.1 The w–γ_d relationship in soils with a significant fraction of fines

When comparing soils with a significant fraction of fines or coarse granular soils, such as clean sands and gravels[1], the aspects of compaction are significantly different. Taking into account the Unified Soil Classification system, Hilf (1991) includes the groups CH, CL, MH, SC, SM, GC, and GM in the first type, with the second including the groups SW, SP, GW, and GP. This separation into two main groups has to be understood as an approximation and simplification, as there is, naturally, a number of borderline cases.

The first relevant aspect to be pointed out, regarding compaction involving soils with a significant fraction of fines, is the fact that the result, in terms of density reached when using a particular equipment and a certain compaction procedure, depends strongly on the water content of the soil to be compacted (Proctor, 1933).

---

[1] In the literature, it is common to designate these two types of soils as cohesive soils and cohesionless soils, respectively. In the current text, however, it was considered more appropriate to use the terms *soils with a significant fraction of fines* and *granular clean soils*, respectively.

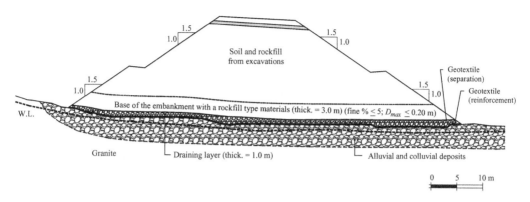

Figure 10.3 Embankment for the IC5 road (courtesy of Estradas de Portugal).

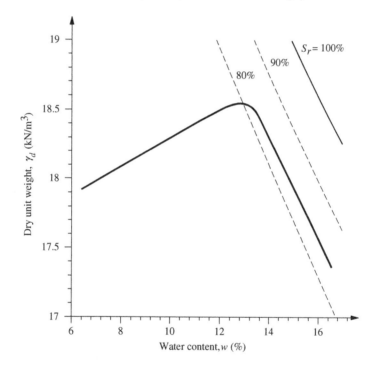

Figure 10.4 Compaction curve of a soil with a significant fraction of fines (granite residual soil).

For instance, Figure 10.4 shows the relationship between water content, w, and the dry unit weight, γ_d, of a soil involving coarse and fine particles, compacted by a specific process and equipment. Graphs such as the one presented here are known as *compaction curves*. It can be seen that there is a certain value of water content that leads to a maximum value of dry unit weight, for the compaction process and equipment used. This value of water content is termed the *optimum water content, w_{opt}*. The branch of the compaction curve to the left of the optimum value is considered to be the *dry side*, whereas the branch to the right is considered to be the *wet side*. The ordinate of the point corresponding to w_{opt} is termed the *maximum dry unit weight*, represented by the symbol, $\gamma_{d,max}$.

This last designation ("maximum dry unit weight"), adopted by most authors and standards, may lead to some confusion. In soil mechanics, when considering sedimentary granular soils, the concepts of maximum void ratio, e_{max}, and minimum void ratio, e_{min}, are introduced. The maximum dry unit weight, $\gamma_{d,max}$, is the dry unit weight corresponding to e_{min}; thus, it exists as a *single value* for each granular soil. This interpretation will be the one used when regarding the compaction of clean granular soils (see Section 10.2.1.3).

In the context of the present section of this chapter, dedicated to soils with a significant fines fraction, $\gamma_{d,max}$ is considered to be the maximum dry unit weight of a given soil, compacted by a certain process, using specific compaction equipment. Therefore, as the reader may understand, it is no longer a single value for each soil.

A simplified explanation, which attracts wide consensus, for the change of γ_d in response to variation in w, can be summarized as follows (Proctor, 1933; Hilf, 1991).

It is reasonable to admit that a mass of moist soil, prepared to be compacted, in the field or laboratory, is composed of "lumps" of particles kept together due to the effect of suction associated with capillarity. Accordingly, the drier the soil, the harder these lumps will be. The compaction process tends to deform these lumps, transforming them into a uniform and homogeneous mass. A certain compaction effort will more easily have that effect when the lumps are softer, that is, when water is added, than when the lumps are harder, due to a lack of water.

From a certain point, the positive effect of this softening, due to the higher water content, becomes less important than another unfavorable effect: removing air becomes increasingly more difficult, as a growing fraction of the air present in the soil is no longer in continuity with the atmosphere. From this point onwards, an increase in the water content inevitably causes a reduction in soil density and, consequently, in its dry unit weight. The value of water content, beyond which the second effect is predominant when compared with the first, corresponds to the peak of the compaction curve, and is known as the *optimum water content*.

It can be noted that, on the wet side, the compaction curve shown in Figure 10.4 approaches the so-called *saturation line*, that is, the curve that represents the relationship between the water content and the dry unit weight for the soil considered, if all the air has been expelled. The equation of this line is expressed as follows:

$$\gamma_d = \frac{G_s\,\gamma_w}{1 + G_s\,w} \tag{10.1}$$

where G_s is the density of the solid particles, and γ_w the water unit weight. This line, for a given G_s, i.e., for a given soil, represents an equilateral hyperbola in the axis system (w, γ_d).

In the same figure, the dashed lines represent the relationship between the water content and the dry unit weight for several values of saturation degree, also being equilateral hyperbolas. The coordinate point $(w_{opt}; \gamma_{d,max})$ is usually close to the curves corresponding to an S_r between 80% and 90%.

Figure 10.5 illustrates the curve representing the relationship between w and γ_d of the soil in Figure 10.4, as well as two other curves corresponding to soils with greater plasticity than the first but compacted by the same process and equipment. By observing the figure, it becomes evident that each soil has its own compaction curve. In addition, it can be noted that soils with greater plasticity exhibit wider compaction curves, with higher values of optimum water content and lower values of maximum dry unit weight.

It is common to express the results of compaction in terms of a so-called *relative compaction (RC)*, given by the ratio:

$$RC = \frac{\gamma_{d,fill}}{\gamma_{d,max}} \times 100(\%) \tag{10.2}$$

where $\gamma_{d,fill}$ represents the dry unit weight of the compacted soil, and $\gamma_{d,max}$ represents the respective maximum value corresponding to the process and compaction equipment used in

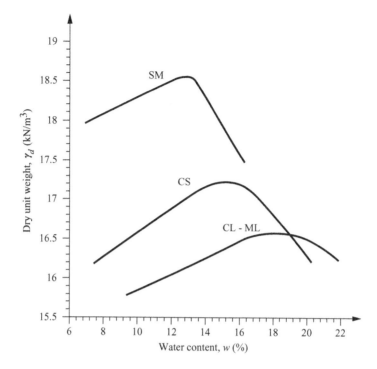

Figure 10.5 Compaction curves of three soils with significant fraction of fines but of different plasticity.

the laboratory. The term *degree of compaction* is also commonly used, as an alternative to *relative compaction*.

10.2.1.2 Effect of the compaction effort

The curve of Figure 10.4, as well as each curve in Figure 10.5, refers to compaction when using *a given equipment and a given procedure*, that is, transmitting to the soil a certain *compaction energy or effort*. If distinct compaction efforts are applied to the soil, the final state will not be the same. In other words, there is a $w-\gamma_d$ curve for each compaction effort applied to the soil.

Figure 10.6 shows schematically the compaction curves of the same soil as described in Figure 10.4 for two normalized compaction efforts, which will be presented in Section 10.2.2.1. It is observed that the increase in compaction effort leads to a decrease in the optimum water content and to an increase in the corresponding value of dry unit weight. Furthermore, the compaction curves become closer to the saturation line, on the wet side, although not reaching this line.

However, the increase in the value of the maximum dry unit weight with the increase in compaction effort is limited; from a certain level of compaction energy, which is different for each soil, the increase in compaction effort will tend to have a null effect on $\gamma_{d,max}$.

10.2.1.3 The $w-\gamma_d$ relationship in clean granular soils

It is relevant to remind the reader that the previous considerations refer to soils with a significant fines fraction. As for clean granular soils (sands and gravels without a significant fraction of fines), given their high permeability, they are less sensitive to the value of water content they have when compacted. Thus, the $w-\gamma_d$ curve, with a more or less expressive peak, typical of soils with a significant fraction of fines, is poorly defined or does not exist at all in clean sands and gravels.

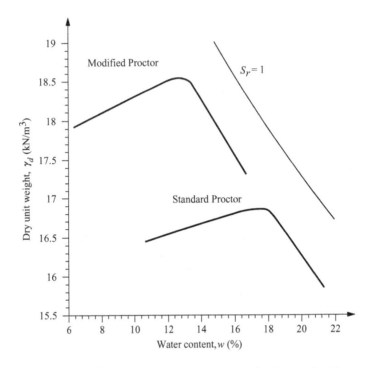

Figure 10.6 Effect of compaction effort on the compaction curve of a given soil with a significant fraction of fines.

As shown in Figure 10.7, it can be observed that, in this type of soil, the dry unit weight, achieved by a given process and compaction equipment, reaches its maximum value when the soil is completely dry or close to saturated, with values slightly lower for intermediate water content (Lambe and Whitman, 1979). This response is a consequence of the effect of suction, which occurs at low water content values, hampering the rearrangement of the grains.

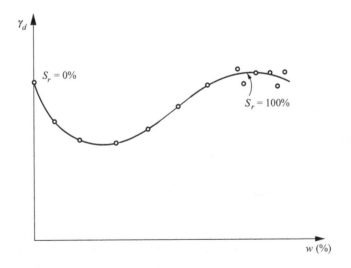

Figure 10.7 Compaction curve for a clean granular soil (Lambe and Whitman, 1979).

In granular fill materials, it is useful to consider the concept of *compactibility,* introduced by Terzaghi (1925), and expressed as:

$$C = \frac{e_{max} - e_{min}}{e_{min}} \tag{10.3}$$

This C factor is higher in well-graded soils in which, as is known, the interval between the maximum and minimum void ratios (e) is relatively high and the minimum void ratio is low. Indeed, well-graded granular soils are easily compactable materials and, when placed with a void ratio close to the minimum, constitute fills with excellent mechanical properties. On the other hand, uniform or poorly graded granular soils are not easily compacted.

In these granular soils, particularly those that are poorly graded, with a small $e_{max} - e_{min}$ interval, the ratio $\gamma_{d,min}/\gamma_{d,max}$ is close to unity. Thus, the ratio expressed by Equation 10.2 becomes close to 100%, hampering the sensitivity of the analysis of the results from compaction. With this in mind, it is preferable to express the compaction to be reached in terms of the density index, I_D:

$$I_D = \frac{e_{max} - e}{e_{max} - e_{min}} \times 100\% \tag{10.4}$$

The density index can also be expressed in terms of the dry unit weight (which is related to the void ratio, for particles with a given density), according to the following equation:

$$I_D = \frac{\gamma_{d,max}}{\gamma_d} \times \frac{\gamma_d - \gamma_{d,min}}{\gamma_{d,max} - \gamma_{d,min}} \times 100(\%) \tag{10.5}$$

10.2.2 Laboratory compaction tests

10.2.2.1 Proctor tests

Compaction curves, such as the ones presented in the previous figures, can be obtained in the laboratory through the so-called Proctor tests (Proctor, 1933), established by the ASTM D 698 (2002) and ASTM D 1557 (2002) standards.

These tests consist of compacting, in a cylindrical mold, a soil sample which had previously been air dried and subsequently mixed with a fixed weight of water in order to obtain a given homogeneous water content. Compaction is then carried out in layers, with each layer being compacted with blows by a hammer dropped from a fixed height.

The tests are performed with one of two types of mold (small or large), depending on the soil particle distribution, and are also performed with two specific values of *compaction effort*, being expressed by the following equation:

$$E_C = \frac{W_h \, h \, n \, c}{V} \tag{10.6}$$

where W_h is the weight of the hammer, h the hammer falling distance, n the number of blows per layer of soil, c the number of layers, and V the volume of the mold.

The test which employs the lower compaction effort is designated the Standard Proctor test (or light compaction test), while the other is designated the Modified Proctor test (or heavy compaction test). Tables 10.1 and 10.2 summarize the equipment data and the types of tests, with Figure 10.8 providing a schematic illustration of the equipment.

Table 10.1 Characteristics of Proctor compaction tests.

Items	Quantity (units)	Standard effort ASTM D 698 (2000)		Modified effort ASTM D 1557 (2002)	
		Small mold	Large mold	Small mold	Large mold
Mold	Inside diameter (mm)	101.6	152.4	101.6	152.4
	Height (mm)	116.4		116.4	
	Volume (cm³)	944	2124	944	2124
Hammer	Weight (kgf)	2.50		4.54	
	Falling distance (cm)	30.5		45.7	
Blows	Blows/layer	25	56	25	56
	Layers	3		5	
Effort	(kN.m/m³)	600		2700	

Table 10.2 Criteria related to gradation in Proctor tests (ASTM D 698 (2002); ASTM D 1557 (2002)).

Method	Mold	Material	Use
A	Small	Passing through N. 4 (4.75-mm) sieve	May be used if 20% or less of the mass of the material is retained on the N. 4 (4.75-mm) sieve. (1) (3)
B	Small	Passing through 3/8-inch (9.5-mm) sieve	Shall be used if more than 20% of the mass of the material is retained on the N. 4 (4.75-mm) sieve and 20% or less by mass of the material is retained on the 3/8-inch (9.5-mm) sieve. (2) (3)
C	Large	Passing through ¾-inch (19.0-mm) sieve	Shall be used if more than 20% of the mass of the material is retained on the 3/8 inch (9.5-mm) sieve and less than 30% by mass of the material is retained on ¾ in. (19.0-mm) sieve. (3)

Notes:

1. Other use: if this method is not specified, materials that meet these gradation requirements may be tested using Methods B and C.
2. Other use: if this method is not specified, materials that meet these gradation requirements may be tested using Method C.
3. If the test specimen contains more than 5% by mass of material retained in the ¾-inch sieve (coarse gravel) and the material will not be included in the test, corrections must be made to the unit mass and water content of the specimen or to the appropriate field in place density specimen using ASTM D 4718.

As illustrated in Figure 10.9, having compacted the soil following any of the test procedures described above for several values (a minimum of five) of water content, the compaction curve can be obtained to assess $\gamma_{d,max}$ and w_{opt}, for the specific value of compaction effort employed. The water content values shall be such that w_{opt} lies near the middle of the range.

It is important to point out that the compaction efforts used in the laboratory tests are not arbitrarily adopted. These values of compaction effort are supposed to reproduce in the laboratory the compaction that will be carried out in the field by construction equipment. The use of the modified Proctor test, developed more recently than the standard Proctor test, in fact, reflects the need for simulating the greater compaction efforts provided by the increasingly more powerful and heavier field equipment used by contractors.

Thus, two remarks are still relevant to be made regarding Proctor tests.

The first concerns that all the compacted specimens, required to establish the compaction curve, should be derived from similarly disturbed samples, with none originating from a specimen previously compacted and subsequently disaggregated to be reused. Massad (2010) argues that such reuse is not to be recommended, given that: i) compaction may lead to particle

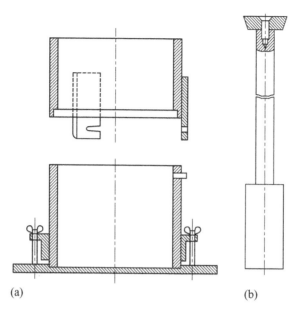

(a)   (b)

Figure 10.8 Schematic illustration of equipment used for Proctor compaction tests: a) vertical cross section of the cylindrical mould and extension; b) hammer.

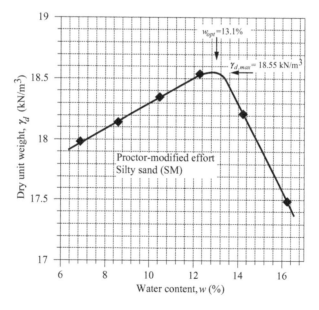

Figure 10.9 Results of a Proctor test.

breakage, thus, altering the particle size distribution in the reused sample; and ii) the drying–wetting cycle, associated with this reuse, may alter the characteristics of certain soils, either by agglomerating particles or by irreversible transformation of clay minerals (such as halloysite).

With this, the second remark may now be stated. In certain types of soil, particularly in residual and tropical soils, previous drying can cause the soil compacted in the laboratory to have characteristics significantly different from those in the earthfill in the field. Pinto

(2002) notes that, in earthfill works, the soil to be used is brought to the construction site, from a borrowing pit, with a water content close to the natural water content or with a water content already close to that specified for compaction. Therefore, the author recommends, reasonably, that the process is carried out in a similar manner in the laboratory, without excessive drying of the material.

10.2.2.2 Vibration compaction tests

There are several laboratory tests in which the compaction process involves vibration, rather than impact, as in the Proctor tests. These vibration tests are used, in particular, with coarse granular materials to be used in earthfills for road or railway works.

As an example, the test considered in the British Standard 1377: Part 4 (1990), using a vibrating hammer, will be presented.

The test is suitable for soils containing no more than 30% by mass of particles retained on the 20-mm test sieve, which may include some particles (up to 10%) retained on the 37.5-mm test sieve. The test uses the same cylindrical mold as the one employed in the California Bearing Ratio (CBR) test.

Table 10.3 presents the details of this test.

The test displays obvious similarities with the Proctor test: compaction is carried out in (three) layers in a cylindrical mold. The essential difference lies in the compaction process itself. A circular tamper is placed on the surface of the soil, which is compacted by the vibrating hammer for 60 s. During this period, a steady downward force of 300 N to 400 N (including the weight of the hammer) is applied.

Following the compaction of the sample in the mold, the water content and the dry unit weight are determined similarly to the methodology described in Proctor tests. The compaction curve is also drawn similarly, from a minimum of five points.

10.2.2.3 Compaction parameters of soils with coarse gravel (material retained in the 19-mm sieve)

The compaction tests previously discussed have limitations in terms of the coarser fraction of the soil to be tested. When this material is beyond the range of the admissible particle size distribution described above, there are essentially two ways of overcoming this difficulty:

  i) to carry out special tests with molds of larger dimensions;
 ii) to carry out conventional tests with material where the particle size distribution has been adjusted to exclude the soil particles which are larger than the maximum

Table 10.3 Apparatus for the vibrating hammer compaction test (BS 1377: Part 4: 1990).

Items	Quantities (units)	Vibrating hammer compaction test
Mold	Inner diameter (mm)	152
	Height (mm)	127
Layers to be compacted	–	3
Hammer	Power (W)	600–750
	Frequency (Hz)	25–45
	Time of vibration per layer (s)	60 ± 2
	Steady downward force (N)	300–400
Tamper	Mass (kg)	≤ 3
	Diameter (mm)	145
	Thickness (mm)	10

admissible particle size, referred to, from now on, as *oversize particles*; this is combined with the application of a correction to the experimental results, in order to achieve the compaction parameters of the total sample.

This correction can be approached as follows.

If the percentage of oversize particles is not too high, it may be reasonable to assume that their presence does not interfere with the compaction of the remaining, finer particles in the field (Ziegler, 1948). Thus, in the laboratory compaction tests on samples with the oversize particles removed, the maximum dry unit weight and optimum water content will be equal to those that would be reached by that same (finer) fraction in a total soil sample, following the same compaction process.

Considering this hypothesis to be valid, the corrections to be made to the compaction parameters obtained in the laboratory on samples containing only the finer fraction, can be understood by observing Figure 10.10.

The upper part of Figure 10.10a shows a volume element of a total sample neighboring an oversize particle, whereas the upper part of Figure 10.10b depicts a volume element equivalent to the previous one, but of a sample of only the finer fraction. If the hypothesis previously stated were valid, the density of the finer fraction would be the same in both samples.

The lower part of the figure illustrates a unit volume of the total sample (Figure 10.10a) and of the sample with the finer fraction (Figure 10.10b). In the first, there is a certain volume of solids corresponding to the oversize particles. In the second, the same volume corresponds to compacted finer soil, therefore, consisting of solid particles and voids.

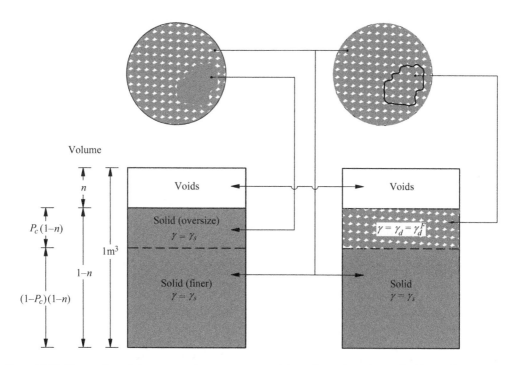

Figure 10.10 Correction of the compaction parameters: a) hypothetical total sample; b) sample with oversized particles removed, according to Ziegler's hypothesis.

Now, consider the symbols γ_d^T and γ_d^F for the designation of the dry unit weights of the total sample and the finer sample, respectively, and with P_C being the proportion by weight of the oversize particles.

Considering, by simplification, a single value for the unit weight of the solid particles, γ_s, P_C then also represents a percentage in volume, as shown in Figure 10.10a. From this, the dry unit weights of the total and the finer samples can be expressed as follows:

$$\gamma_d^T = \frac{W_s}{V} = (1-n)\gamma_s \tag{10.7}$$

and

$$\gamma_d^F = \frac{W_s}{V} = (1-P_C)(1-n)\gamma_s + P_C(1-n)\gamma_d^F = (1-n)\big[(1-P_C)\gamma_s + P_C\gamma_d^F\big] \tag{10.8}$$

Dividing the two previous equations, member by member, the following expression is found:

$$\gamma_d^T = \frac{1}{\dfrac{1-P_C}{\gamma_d^F} + \dfrac{P_C}{\gamma_s}} \tag{10.9}$$

This relation is also, naturally, applicable to the maximum values of the dry unit weight of the two samples $\gamma_{d,\max}^T$ and $\gamma_{d,\max}^F$, respectively.

As expected, the equation provides a value of $\gamma_{d,\max}^T$ higher than $\gamma_{d,\max}^F$, with this difference increasing with the percentage of oversize particles, P_C. Obviously, $\gamma_{d,\max}^T = \gamma_{d,\max}^F$ when P_C is equal to zero.

This equation becomes meaningless, that is, no longer applicable, for values of P_C that lead to results of $\gamma_{d,\max}^T$ higher than the value corresponding to a sample composed only of oversize particles and with a relative density equal to 100% (USBR, 1998).

The water content correction, still in line with the hypothesis stated, is carried out through the equation:

$$w_T = (1-P_C)w_F + P_C w_C \tag{10.10}$$

where:

- w_T is the water content of the total sample;
- w_F is the water content of the finer sample, obtained in the compaction test;
- w_C is the water content (water absorption value) of the coarse oversize particles; the latter is typically very low, and even close to zero for quartz particles.

This methodology is included in the standard ASTM D 4718 (2007), as well as the standard for the Proctor tests mentioned above, that recommend its application for situations in which the percentage retained on the N 4 (4.75-mm) sieve is lower than 40% and the percentage retained on the ¾-inch (19.0-mm) sieve is lower than 30%. This is what is essentially expressed, as well, in the Earth Manual of the USBR (1998) and in the AASHTO procedures (2018a,b).

Experimental studies, comparing the results of compaction tests on total samples (using large-dimension molds) and on samples with the oversized particles removed, under the same compaction effort allow an analysis of the real influence of the presence of coarser oversize particles on the density reached by the finer soil fraction as a result of compaction (Donaghe and Townsend, 1976; Torrey and Donaghe, 1994; Winter et al., 2016). These studies confirm, as expected, that, when the percentage of oversize particles is high, their influence is negative, leading to lower values of $\gamma_{d,\max}^{F}$.

In order to correct for the detrimental effect of the presence of oversize particles on the density reached by the finer soil fraction in the total sample, $\gamma_{d,\max}^{F}$, a reduction or density correction factor, r, can be introduced into Ziegler's equation:

$$\gamma_{d}^{T} = \frac{1}{\dfrac{1-P_{C}}{r\,\gamma_{d}^{F}} + \dfrac{P_{C}}{\gamma_{s}}} \tag{10.11}$$

However, there is no consensus in the literature regarding the most suitable values for this density correction factor. When needed, these values can be experimentally determined, requiring compaction tests in large-dimension molds to be carried out.

10.2.3 Field compaction equipment

A detailed discussion of the characteristics of field compaction equipment goes beyond the scope of this book. Therefore, only the main types of equipment, and the conditions under which their use is most appropriate, will be examined.

Sheepfoot rollers will be the first to be discussed, being represented in Figure 10.11a. Sheepfoot rollers can be found of various dimensions, with regard to weight, drum size (diameter and length) and tamping foot (length, area, and shape). They are available with weight that can exceed 50 tons and may be self-propelled or towed by a tractor. Their weight can be increased by filling the drum with ballast (wet sand or water).

The "sheepfeet" are cylindrical protrusions uniformly distributed over the surface of the drum. The most important features of sheepfoot rollers are the weight and pressure transmitted to each "foot". This high and localized pressure causes the penetration by the foot into the layer to be compacted, improving the characteristics of the lower part of the layer, which becomes progressively more compacted from the bottom to the top, as a result of successive passes of the equipment. Therefore, the superficial zone is only entirely compacted when compaction is performed on the layer placed immediately above.

With this type of equipment, contrary to other rollers referred to later, the compaction efficiency of sheepfoot rollers increases with the travel speed. In fact, the energy transmitted to the soil by each foot increases with the impact velocity, which increases with the travel speed of the cylinder.

These compacting rollers are ideal for clayey soils or for soils with a minimum of 20% fines. In these type of soil, sheepfoot rollers are very effective at preventing *lamination*, that is, the tendency (which can occur when, for instance, pneumatic rollers or smooth wheel rollers are used on these type of soils) for the surface of the layer that has just been compacted to become practically smooth, which hampers the connection to the layer immediately above. This phenomenon causes a decrease in the global resistance of the earthfill – which is divided by horizontal surfaces of low resistance – and a significant increase in the coefficient of horizontal permeability of the fill (Hilf, 1991).

A common variant of the sheepfoot compacter is the tamping or padfoot roller, in which the protrusions are tapered blocks, such as the one represented in Figure 10.11b.

(a)   (b)

Figure 10.11 a) Towed sheepfoot roller; b) self-propelled padfoot roller.

Equipment such as the one represented in Figure 10.12 is known as a *pneumatic roller*, and can reach up to 100 tons in total weight. They are appropriate for a wide range of soils, from clean sands to low-plasticity silty clays.

The contact area of each tire with the soil and the pressure transmitted through this area are determinant for compaction. The given area and pressure depend not only on the load transmitted per tire, but also on the tire inflation pressure. This load per tire can be adjusted by using ballast inside an appropriate container, as shown in Figure 10.12.

One advantage of pneumatic rollers lies in their mobility. The compaction with this type of roller is typically faster than when carried out with a sheepfoot roller. The travel speed of a pneumatic roller is usually less than 10 km/h.

The use of pneumatic rollers combined with sheepfoot rollers at the same construction site is often favorable. For instance, the occurrence of rainfall on a layer that has just been compacted using a sheepfoot roller will lead to infiltration of water up to a certain depth into the fill, delaying the execution of the works. In a scenario of imminent rainfall, this can be avoided by carrying out one or two passes with a pneumatic roller over this layer, which will seal the compacted surface and, thus, retard infiltration.

The previous types of compaction rollers can be combined with vibrators, which may increase compaction efficiency. A third type of compaction roller, which is frequently used in combination with a vibrator, is smooth wheel roller, such as the one represented in

Figure 10.12 Pneumatic tired compaction roller.

Figure 10.13 Towed smooth drum vibratory roller.

Figure 10.13. In this type of roller, the static weight of the drum is usually greater than or equal to 10 tons, and typically the travel speed does not exceed 3 to 4 km/h.

These compaction rollers lead to good results in the compaction of clean granular materials or materials with small percentages of fines, and in rockfills. They are suitable for the compaction of granular materials of any particle size, in which a considerable reduction in void ratio, associated with particle rearrangement, requires soil vibration.

For this vibration to be appropriate, a significant amount of load (self-weight plus vibratory force) and a certain amplitude and frequency are required. For instance, heavy rollers with low frequencies are suitable for gravels and rockfills, whereas light- to medium-weight rollers with high frequencies are more appropriate for sands and (non-plastic) silts. In order to reach a high compaction efficiency with these rollers, the materials should be used with water content above the optimum. This promotes the rearrangement of the particles as a result of vibration and thus increases soil density.

10.2.4 Mechanical behavior of compacted soils

10.2.4.1 Introduction: soils with significant fines content and the effect of soil suction

For soils with significant fines content (passing through the N. 200 (75-μm) sieve), the Atterberg limits, namely the liquid limit, and the plastic limit, are crucial reference points for various aspects of soil behavior. In particular, comparison of the water content of soil samples in terms of those limits enables characterization of the consistency of the soil, providing a simplified and qualitative assessment of the soil strength.

Although this comparison is more informative and powerful for natural sedimentary soils, it seems useful to use it also when considering the optimum water content of compacted soils. As an example, Figure 10.14 shows a comparison between the optimum water content, and the liquid and plastic limits for residual and colluvial soils from Brazil (Cruz, 1996). It can be observed that the optimum water content lies clearly below the plastic limit, a position which, in a sedimentary soil, would correspond to a consistency index greater than 1.0, i.e., to a high consistency.

As previously pointed out, compacted soils are partially saturated soils. In such soils with a significant fines content, suction plays a key role in the soil mechanical behavior.

Figure 10.15 depicts the failure envelopes of a soil with a significant fines content, compacted under a given compaction effort, at a water content slightly below the optimum value

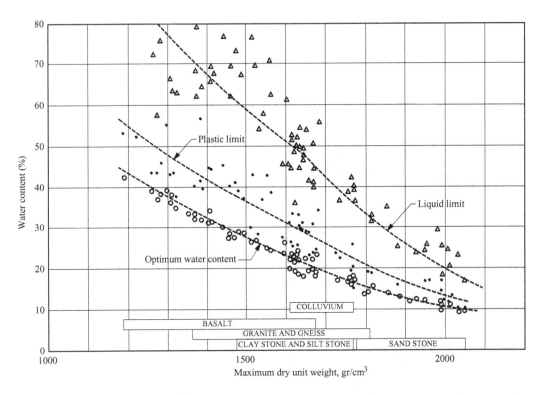

Figure 10.14 Optimum water content (standard Proctor), liquid limit, and plastic limit *versus* dry unit weight for residual and colluvial soils from Brazil (Cruz, 1996).

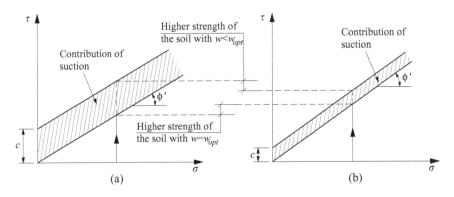

Figure 10.15 Failure envelope for a soil with significant fines content compacted with a given compaction effort: a) with water content slightly below the optimum; b) at the optimum water content.

(Figure 10.15a) and at the optimum water content (Figure 10.15b). These failure envelopes can be obtained through laboratory shear tests at constant suction, equal to that developed during compaction. The figure also includes, as a reference, the failure envelopes for zero suction for each of the cases above.

As shown, the angle of shearing resistance, ϕ', is higher in the soil compacted at the optimum water content as, by definition, the optimum content corresponds to the maximum density. Nevertheless, the increment in shear strength associated with suction, which is significantly higher for the soil compacted on the dry side, leads to a higher shear strength in the soil compacted on the dry side, for the conditions prevailing at the end of compaction. The same can be stated with regard to stiffness.

Nonetheless, benefit from the higher shear strength and stiffness of soil compacted on the dry side is seriously limited by the fact that, in many cases, the fill shifts to a saturated state during its lifespan, with dams being considered a prime example. Therefore, it is paramount to consider each of these conditions with regard to the mechanical behavior of the fill and, in particular, to consider the transition between the partially saturated and fully saturated states. This subject will be discussed below.

10.2.4.2 Dam failures during construction and the practice of compacting on the dry side of the optimum

In the first half of the 20th century, several failure events took place in earthfill dams during the construction phase, particularly in the United Kingdom, as a result of slope failures (Penman, 1986). These failures occurred in fills with significant clay fraction and were caused, or at least facilitated, by the generation of positive excess pore water pressures in the deeper fill layers due to the loading imposed by the upper layers.

Consider the fill compacted with a water content equal to or above the optimum, i.e., with a high saturation degree and, thus, with a small air volume that is not in contact with the atmosphere. With the increase in mean stress, the air can be progressively dissolved in the water of the pores and the degree of saturation reaches approximately 100%. As the height of the earthfill increases, the pore water pressure becomes positive and steadily increases, without any significant pore water pressure dissipation occurring during construction.

In these situations, the shear strength of the fill corresponds to the shear strength under undrained conditions, which is much lower than that under drained conditions. This explains why the most critical period for slope failure in an earthfill dam is during the construction phase or immediately after the completion of the earthfill.

The scenario described and the assessment of the soil response can be studied in the laboratory, using unconsolidated undrained shear tests (UU tests), with the load being applied to the samples in the state they present after compaction and under undrained conditions.

Penmam (1986) notes (and regrets!) that these setbacks led to the practice, which is still advocated by some designers, of compacting the fill on the dry side of the optimum. Compaction on the dry side of the optimum, in fact, leads to a favorable response from the earthfill during the construction phase, though it may cause an unfavorable response during saturation or inundation, as will be discussed later.

Failures during the construction phase have become increasingly rare in recent decades, despite the tendency for both the height of dams and the rate of construction to increase. This reduction in failures was promoted by changes in the criteria adopted for the selection of earthfill materials and the design of the earthworks, as well as by the careful monitoring of these works from the construction phase, namely in terms of the pore water pressures developed (Penman, 1986).

10.2.4.3 Volume change in a soil saturated by inundation. The phenomenon of soil collapse

As previously stated, it is also important to consider the behavior of the soil in other situations, rather than only at the end of compaction. The soil often becomes saturated during the lifetime of the earthfill, and it is essential to take account of the behavior of the soils in the transition to saturation and afterward.

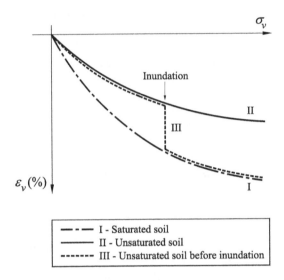

Figure 10.16 Phenomenon of collapse by inundation.

This inundation can cause a certain contraction or reduction in volume, resulting in a settlement. This phenomenon, which occurs under constant total stress, is designated in the literature as *collapse*, with the soils that are susceptible to exhibiting such a phenomenon being known as *collapsible soils* (Alonso et al., 1987; Ng and Menzies, 2007; Alonso, 2014).

Consider Figure 10.16, which shows a graph relating vertical stress with the volumetric strain obtained in one-dimensional compression tests (or oedometer tests) on three specimens of the same soil with similar initial void ratio. Curve I illustrates the behavior of the saturated soil.

Now imagine a partially saturated specimen of the same soil, with a certain value of suction. Carrying out a similar oedometer test would result in curve II, as shown in the same figure. Due to the beneficial effect of suction, the curve of the partially saturated soil exhibits lower compressibility (or greater stiffness).

Consider a third partially saturated specimen of the same soil, with the same suction as the previous sample, which is loaded to a certain level of vertical stress and subsequently inundated. The figure shows what would be the behavior of a collapsible soil: the test curve (curve III) exhibits a reduction in the void ratio associated with inundation, tending towards the curve of the saturated soil (curve I). Following subsequent loading, it would tend to behave in a manner similar to that of the saturated soil, with curve III joining curve I.

As previously pointed out, this contractive volumetric strain is designated by collapse. The increase in the degree of saturation up to a value close to 100% practically annuls the suction, which leads to a rearrangement of the soil structure to a state of higher density, leading to the occurrence of surface settlement.

From the explanation given, it can be understood that this phenomenon is particularly relevant in the case of the *first inundation*. Indeed, as it may seem reasonable to think, the soil structure, more dense after the collapse associated with the first inundation, remains largely unchanged, regardless of future drying–wetting cycles (Alonso et al., 1990; Ng et al., 1998).

Therefore, this phenomenon is considered to be particularly relevant in the compacted soils that constitute earthfill dams, with collapse taking place during the first filling of the reservoir, or in the earthfill of road and railway works, with such collapses tending to occur during the first intense rainfall events.

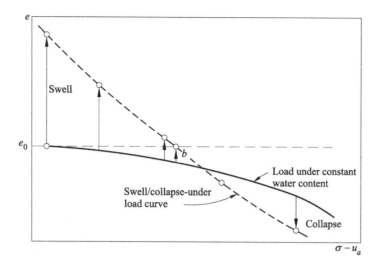

Figure 10.17 Graph of vertical stress (log) *versus* void ratio, showing the loading curve at constant water content up to inundation, at different values of vertical stress, and the swelling–collapse curve under constant net stress (adapted from Alonso et al., 1987).

The previous explanation is a simplification of a complex phenomenon, in which the change in volume is caused by the increased saturation of a partially saturated soil. In fact, the inundation of a given soil can lead to either swelling or collapse, depending on the conditions of the soil prior to the inundation event. The brief remarks that follow aim to detail these matters (Alonso et al., 1987).

For a compacted soil with a given void ratio and water content (and therefore a given suction), Figure 10.17 shows the variation of the void ratio in a series of oedometer tests, in which the samples are loaded to certain values of vertical total stress, and then inundated, thereafter experiencing a variation in void ratio under constant vertical stress.

It can be observed that, when loading is carried out until relatively low values of vertical stress, the soil experiences swelling, following inundation. However, the greater the vertical stress, the lower this swelling will be. There is a value of vertical stress for which the soil regains the initial void ratio (point b of Figure 10.17). Considering samples similar to the previous ones and loading them until the vertical stress exceeds the value of point b, volumetric deformation, due to inundation, becomes contractive, so that collapse will occur.

By joining the points from the several tests representing the conditions when the volumetric deformations associated with inundation have stabilized, the dotted line curve of the figure is obtained. This curve expresses the relationship between the void ratio resulting from swelling or collapse and the vertical stress under which these phenomena have occurred. Therefore, it can be concluded that the manifestation of swelling or collapse will occur depending on the value of vertical stress applied before inundation.

It is worth noting that the curve expressing the relationship between the void ratio, resulting from either swelling or collapse, and the vertical stress under which these responses occurred is not unique for a given soil, but depends on the initial void ratio and on the suction operating before inundation.

An approach which comprises the behavior of partially saturated soil, expressed by an analytical model able to capture, among other aspects, the qualitative considerations expressed in Figure 10.17, was developed by Alonso and collaborators in the Politecnic

University of Catalonia (Alonso et al., 1990). This model will not be presented here, given the introductory scope of this chapter.

From what has been pointed out, it can be understood that, in an earthfill dam, there will be an upper region, with relatively low vertical stresses, which experiences swelling as a result of inundation, and another, lower region, with higher vertical stresses, which undergoes collapse. Taking into account that high vertical stresses are usually encountered in medium to large dams, collapse is typically predominant over swelling, then resulting in a settlement of the body of the dam.

The occurrence of settlements due to collapse in earthfill dams can lead to development of important cracks in the body of the dam, which can favor internal erosion phenomena. This issue is particularly critical in narrow valleys, where differential settlements tend to be larger. In turn, in roadworks, settlements due to collapse may cause great damage to pavements.

With this in mind, there is nowadays considerable interest in adopting compaction conditions to minimize settlements due to collapse. This can be reached by carrying out the compaction at a value of water content very close to or even above the optimum content. Under these circumstances, the degree of saturation of the fill becomes high, typically close to 90%, with the corresponding suction being reduced; as a consequence, deformation by collapse also decreases.

A particularly delicate case concerns the clayey core of earthfill dams, such as the ones represented in Figures 10.1 and 10.2. This area of the dam is aimed at reducing seepage flow through the body of the dam, while stability is guaranteed by the soil masses that surround the core, the so-called shoulders, which typically consist of coarser material.

An aspect that can threaten the safety of an earthfill dam is the appearance of cracks in the core, induced either by collapse, as stated above, or by other events (Casagrande, 1950). These cracks, once formed, can lead to internal erosion of the finer particles under the high hydraulic gradients common in dams (although dams are equipped with filters to prevent these erosion effects). The compaction of the core at a water content well above the optimum is, from this viewpoint, to be highly recommended, given that it favours ductility, inhibiting the formation of these cracks (Cruz, 1996).

Another example concerns rockfills, in which wetting is thoroughly carried out during the construction phase, anticipating deformation due to collapse (Maranha das Neves and Veiga Pinto, 1989). The same can be said of earthfills of clean granular soils, in which the high water content, combined with vibration, facilitates the rearrangement of particles, necessary to achieve the required relative density.

10.2.4.4 Compaction charts

Pinto (2002) presents a very convenient illustration of the influence of the characteristics of compaction on the mechanical and hydraulic properties of soils. Several samples are made following the Proctor test procedures (standard or modified) and subsequently submitted to appropriate laboratory tests (permeability, oedometer, triaxial, etc.). Over a soil compaction curve, the (w, γ_d) co-ordinates of each sample tested are marked, next to which are the values of the respective parameter (strength, stiffness, permeability), the response of which is to be studied. By examining this figure, it is possible to intuitively draw over the compaction curve lines of equal value, as well as to identify the change in the respective parameter with the compaction characteristics. Figure 10.18 illustrates the result of this exercise for Brazilian soils used in earthfill dams.

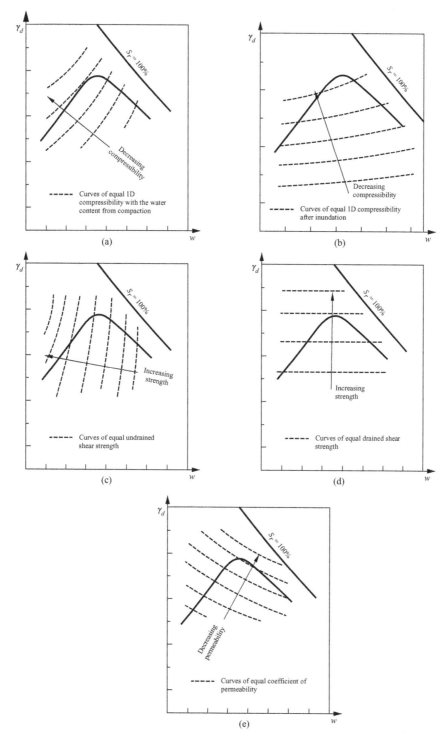

Figure 10.18 Compaction charts of Brazilian soils for the following soil parameters: a) compressibility under confined compression of samples with the compaction water content; b) compressibility under confined compression of samples following inundation; c) undrained shear strength of samples with the compaction water content; d) drained shear strength of samples with the compaction water content; e) permeability (adapted from Pinto, 2002).

10.3 INTRODUCTION TO EARTHWORKS

10.3.1 The peculiar context of earthwork design and construction

This section will introduce the methodology and typical phasing of design and construction of earthworks, mainly considering earthfill dams. This is a broad topic, as will be later understood, which involves specialized aspects, with detailed discussion being beyond the scope of this book.

In order to understand what will be presented, it is worth highlighting that the subject of the present chapter is based on a very peculiar context, different from those underlying *all* the previous chapters.

In fact, in the previous chapters, the implicit idea was the following: there is a natural soil mass on which a civil engineering structure is to be built. Therefore, it is seen to be necessary to characterize this soil mass in order to predict its response to the actions transmitted by the structure. Generally speaking, such a prediction will require carrying out laboratory tests on samples collected on the site and/or *in situ* tests on elements of the soil (samples, in a broad sense). In brief, the soil mass is a *pre-existence*, from which certain points (or samples) are selected (or extracted) for testing.

In earthworks, the opposite takes place: *the "samples", elements of the fill to be tested, precede the existence of the earthfill itself*! Therefore, field compaction aims to *replicate*, that is, to copy, at the scale of the earthfill and with the contractor's equipment, the samples created in the laboratory, through Proctor tests or other similar tests, under pre-established conditions defined in relation to the optimum values of the respective compaction curve.

With this in mind, the essence of earthwork construction is how the samples on which the design is based are *replicated* in the field, as well as how this process is controlled and verified. This point is subsequently developed to introduce the typical phasing in large earthwork projects, shown schematically in Figure 10.19.

10.3.2 Selection of materials

With regard to road and railway works, the *rule-of-thumb* concerning fill materials is to take advantage, as much as possible, of soils resulting from excavations on the construction site, in order to avoid either using borrowing areas outside those of the construction site, or having to dispose of the excess material in a spoil dump. If this is not avoided, the economic and environmental costs can be severe or even unaffordable.

In terms of earth or rockfill dams, it is convenient to borrow from areas that are located within the area of the future reservoir. This prevents additional cost and environmental impact. A common saying in this type of works is that the best fill material is the one that is the closest! The use of borrowing areas in the reservoir may lead, in some cases, to dams with larger volumes of earthfill (when compared with solutions relying on more distant borrowing areas, where materials with better properties can be found), though with a lower final cost (Cruz, 1996).

The selection of borrowing areas involves collecting (either directly from pits or through boreholes executed with light augers) disturbed samples to be submitted to laboratory characterization tests (particle size distribution and Atterberg limits). With these tests, it is possible to establish the soil classification according to the Unified Soil Classification system. Experience has made it possible to associate certain expected behaviors with each one of the 15 groups from this classification when used as fill materials, as Table 10.4 shows (Lambe and Whitman, 1979). The Unified Soil Classification system can lose its meaning in the case of residual soils. In these cases, Cruz (1996) suggests that it is preferable to identify them as saprolites or laterites.

For rock material used in rockfills, it is common, in this phase, to carry out tests to determine the wet and dry unit weights, as well as the Los Angeles abrasion test, and to assess its strength through the point load test and the uniaxial compression test.

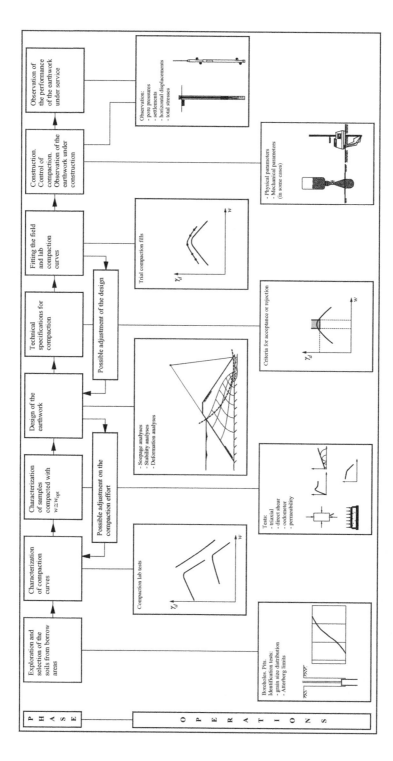

Figure 10.19 Typical phasing in large earthworks projects

Table 10.4 Expected behavior of the Unified Classification groups when used in earthworks (adapted from Lambe and Whitman, 1979).

Group symbols	Permeability when compacted	Shearing strength when compacted and saturated	Compressibility when compacted and saturated	Workability as a construction material
GW	Pervious	Excellent	Negligible	Excellent
GP	Very pervious	Good	Negligible	Good
GM	Semipervious to impervious	Good	Negligible	Good
GC	Impervious	Good to fair	Very low	Good
SW	Pervious	Excellent	Negligible	Excellent
SP	Pervious	Good	Very low	Fair
SM	Semipervious to impervious	Good	Low	Fair
SC	Impervious	Good to fair	Low	Good
ML	Semipervious to impervious	Fair	Medium	Fair
CL	Impervious	Fair	Medium	Good to fair
OL	Semipervious to impervious	Poor	Medium	Fair
MH	Semipervious to impervious	Fair to poor	High	Poor
CH	Impervious	Poor	High	Poor
OH	Impervious	Poor	High	Poor
Pt	–	–	–	–

When considering large dams, this process typically consists of several phases, which involve, following selection of the most convenient soils, a thorough assessment of the volumes available, which may require the application of geophysical methods. However, there will rarely be an available volume of the material marked as being the most appropriate type by the identification tests. Therefore, the project will usually have to compromise, dealing with several types of materials (in addition to the simple distinction between coarse materials for the upstream and downstream shoulders, clayey material for the core and the granular soils to be used as the filters) and combine them in the best way possible.

A careful analysis of the volumes of material available in the borrowing areas is of paramount importance. Interrupting the construction works due to lack of fill material will have severe implications for the cost and deadline of the project, besides leading to a number of technical difficulties (Cruz, 2018).

A borrowing area within the future reservoir must comply with a minimum distance from the toe of the upstream face. This distance must, at least, fulfil the following requirements:

a) the excavations must not jeopardise the bearing capacity of the foundation of the dam;
b) the excavations must not significantly change the seepage conditions in the foundation of the dam during the exploration phase.

As an example to portray this second requirement, Figure 10.20 illustrates a scenario in which a borrowing area excavation causes a high permeability layer to be in direct contact with the water of the reservoir, which can cause the phenomenon of piping. In this case, it

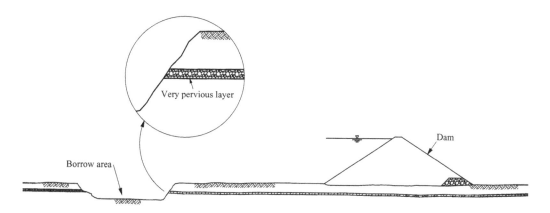

Figure 10.20 Scenario in which the exploration of a borrowing area in the area of the future reservoir can significantly alter the seepage conditions in the foundation of the dam.

would be essential to carefully seal the face of the excavation with a low-permeability fill before filling the reservoir.

10.3.3 Assessment of the compaction characteristics: mechanical and hydraulic characterization

Following selection of the materials, characterization of the respective compaction curves is carried out.

After identifying the sensitivity of each soil to water content when compacted, the same compaction procedure is then employed (in the Proctor molds, or in larger molds, when vibration compaction tests are carried out, in the case of coarser materials) to produce samples for carrying out mechanical and hydraulic characterization tests. These samples are naturally compacted with the effort that is anticipated to be used in the field, with the water content close to the respective optimum.

For these samples, submitted to triaxial, direct shear, constrained compression (oedometer), and permeability tests, among others, the strength, stiffness, and permeability parameters of the various soils are estimated.

Often, depending on the results obtained, a first decision is made regarding the compaction effort to be employed. Actually, the compaction effort to be adopted depends on the mechanical and permeability properties desired, which, in turn, depends on the soils available and on the type of earth structure to be built.

For example, an earthfill for a high dam or for the iron ore storage area of a steelworks plant will experience much higher stresses than a fill for the construction of a canal. Thus, for the same soil, it is necessary that the compaction effort employed is higher in the first two cases than in the latter. This issue is of great importance, given that a higher compaction effort will lead to higher costs and, often, to longer execution periods.

10.3.4 Phase of conception and design

Having selected the various types of soils, ranked from excellent to fair, the design will make the best use of them. Soils with better mechanical behavior are preferably used in areas where loading is more unfavorable: in dams, in the downstream shoulder and deeper layers; in roadworks, in the pavement foundation and in the lower parts, in potential flood areas. Soils, ranked as fair, are preferably used in other, less demanding areas.

It is in this phase that the stability analyses are carried out, using methods such as those studied in Chapters 2 and 9. For dams, several scenarios are analyzed, namely: i) the construction phase (it may be prudent to assume loading under undrained conditions in some cases); ii) the exploration phase at full reservoir level, with the corresponding permanent seepage regime installed, an often critical scenario for the downstream slope; and iii) the rapid drawdown of the reservoir, an often critical scenario for the upstream slope. The combination of one or more of these scenarios, particularly the second, with a seismic action is usually imperative. Most of these stability analyses require considering seepage through the body of the dam. Seepage analyses are equally necessary for estimating the flow rate through the dam.

There are specific software packages available on the market, based on the finite element method, which are able to simulate the relevant construction and exploration phases of the project in great detail, combining the analysis of deformation, stability, and seepage, under static and seismic conditions.

Based on the results of the design analyses, it may be necessary to revise the compaction effort to be employed and even to reconsider the selected materials.

10.3.5 Specifications for field compaction

The technical specifications provided to the contractor will identify the materials to be used, the respective borrow area, the minimum relative compaction required, and the range of water content values of the compacted soil, in relation to the optimum water content.

For instance, the following specifications are common: each layer must be compacted to obtain a value for γ_d of at least $0.98\gamma_{d,max}$ determined with the modified Proctor test, with the water content in the material before and after compaction having to be controlled in order to lie within the range $w_{opt}\pm 1\%$. Figure 10.21 illustrates, in a graph of w versus γ_d, the acceptable zone for the compaction results corresponding to this example. For the reasons previously discussed, this type of specification depends on the type of earthfill structure, on the location of the fill within the structure, and on the fill material itself.

In certain cases, an upper limit is also established for γ_d (for instance, 102%) in order to avoid over-compaction. This phenomenon can cause unwanted structural alterations in the compacted soils, namely lamination, as well as an increase in the swelling potential of soils with significant fines content.

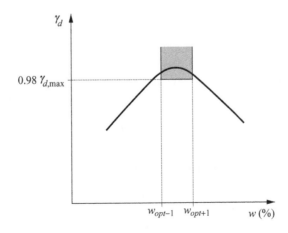

Figure 10.21 Laboratory compaction curve and an example of the acceptable zone for the results of field compaction.

It is worth noting that specifying the values of γ_d and w, such as in the example above, is essentially aimed at ensuring the *similarity* between the fill to be built and the samples created in the laboratory, the properties of which formed the basis of the design. Strictly speaking, the specification of the soils to be used and the specification of these two physical parameters, for each soil, ensure a *physical similarity* between the samples and the fill. With this similarity safeguarded, it is reasonable to expect that the *mechanical* (in terms of strength and stiffness) *and hydraulic* (in terms of permeability) *similarities*, which are actually determinants for the performance of the fill, are maintained.

Hilf (1991) notes:

> The use of a percentage of a standard laboratory maximum dry density, rather than a stated value for the dry density, stems from the inherent variability of soils. Small changes in size and gradation from place to place within a borrow area will result in significant changes in absolute values of maximum dry density and optimum moisture content. (...) It is more desirable, therefore, to require a fill density equal to the maximum density (or 98% of it) of a laboratory standard than to specify a numerical value of dry density. This procedure ensures that a compactative effort, comparable to that of the laboratory standard, is achieved in the fill. Since the designs are usually based on values of compressibility and shear strength obtained from laboratory tests on samples compacted at the standard density, this procedure is valid.

In case the fill material corresponds to clean sands and gravels, as an alternative to stipulating certain values of dry unit weight and water content, it is more appropriate to establish a minimum value of the density index, I_D, such as 75%.

In certain situations, the project can specify that field tests should be carried out to measure the mechanical and hydraulic parameters of the fill and/or undisturbed samples of the compacted fill should be collected for the evaluation of these parameters through laboratory tests (see Section 10.3.8).

This mechanical characterization of the compacted fill is particularly important when potentially *evolutive materials* are used. This designation is applicable to fill materials which, with excavation, transport, and consequent exposure to certain weather conditions, or when compacted on the site, experience more or less significant changes in the particle size distribution, when compared to the samples tested during the design phase.

10.3.6 Field compaction

Considering now the construction phase itself, the crucial question is reproducing in the field the compaction adopted in the laboratory, on which the project is based.

First of all, the material in the borrowing pit will exhibit a given water content that, only by chance, will coincide with the specified value. If the soil is drier than that required, it should be wetted, preferably in the area of the borrowing pit (Hilf, 1991). If necessary, a final correction in terms of the amount of water can be carried out when spreading the lift (or layer) on site; this procedure is aimed at obtaining a homogeneous moisture content throughout the lift. Otherwise, if the soil is wetter than required (which naturally depends on the precipitation events that took place in the recent past or at the time of spreading), partial drying has to be ensured. Following rainfall events, the soil is usually turned over by disc harrows in order to accelerate the evaporation of any excess water.

From this explanation, it is understood that earthworks depend strongly on the meteorological conditions. Thus, wherever possible, the compaction phase should be planned to be carried out during the season when average precipitation is low.

For a given soil, placed in lifts, with a certain water content, compaction will essentially depend on the characteristics of the compaction equipment used (see Section 10.2.3), on the layer thickness and on the number of passes per layer of the equipment, as well as on the travel speed of the compaction equipment. With this, *the adjustment of the field compaction curve to the laboratory compaction curve* will have to be carried out, using a trial and error procedure.

Previous experience with certain equipment and a type of soil greatly facilitates this adjustment. In large earthworks, *field compaction trials* are performed in order to choose the most appropriate compaction equipment to be used, as well as to determine the thickness of the lifts, and the number of passes that will result in the desired level of compaction, with the lowest cost per unit volume of constructed fill.

For a given combination of equipment and soil, there is an optimum combination of the thickness of the lifts and the number of passes of the compaction equipment that will lead to the desired level of compaction, with the greatest savings. The number of passes usually does not exceed 10 to 12 passes.

In general, a minimum thickness of the lift of 1.5 times the maximum particle size of the fill material is imposed. However, there is a maximum lift thickness which allows adequate compaction by a given equipment, regardless of the number of passes. Thus, the adopted thickness usually lies close to this maximum value, *broadly*, ranging between more or less 20 cm, for light equipment compacting fine-grained soils, and between 1.5 m to 2.0 m, for heavy equipment in large rockfill construction works.

Field compaction trials usually consist of the execution of several bands of fill at different values of water content. The preparation of this fill must comply with all the rules specified for the works and be carried out on one or more identical lifts, and never on another surface, even if it is the natural ground. Following the spreading and wetting of the lifts of the field compaction trial, compaction with a certain number of passes is carried out, with this number being presumably lower than the one initially expected for the work, immediately followed by the measurement of the control parameters. The number of passes is then gradually increased, with new trials being performed, and so on.

From this phase, a curve similar to the one presented in Figure 10.22 can usually be drawn for each soil, in which the laboratory and field compaction curves are compared, with the latter being associated with a given procedure essentially defined by the equipment used, the thickness of the lifts and the number of passes per lift.

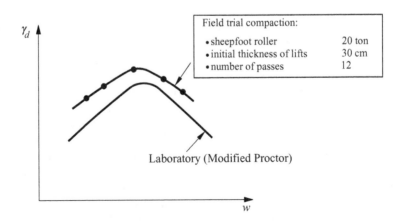

Figure 10.22 Example of laboratory and field compaction curves.

10.3.7 Control of compaction through physical characteristics

10.3.7.1 Conventional process

As previously stated, the soil compaction carried out during construction must be controlled in order to ensure the correct reproduction, at the scale of the construction work, of the samples on which the design was based.

The conventional process of compaction control involves determining the unit weight and water content. Thus, the dry unit weight is given by the following equation:

$$\gamma_d = \frac{\gamma}{1+w} \tag{10.12}$$

The measurement of the unit weight and water content can be carried out by a method based on radioactivity, as represented in Figure 10.23. This equipment essentially consists of a radioactive source and of a receiver. Two radioactive materials are present in the source: one for the emission of gamma radiation, which allows assessment of the unit weight of the soil, and another for the emission of neutrons, from which water content is determined.

In order to determine the unit weight, the source emits gamma radiation, either from the ground surface (indirect transmission) or from its interior (direct transmission); in the latter case, the source is introduced in a hole previously made. The amount of gamma radiation that will be captured per unit time by the Geiger-Muller meter placed inside this nuclear densometer is inversely proportional to the density of the material through which the radiation passes. In the measurement of the water content, the source emits neutrons from the ground surface (in this case, only indirect transmission is used). The intensity of the response detected in the receptor is proportional to the water content in the soil.

The readings are shown in a digital display, with the equipment being capable of recording a great amount of data, to be subsequently analyzed. This data handling involves in particular the analysis of the statistical distribution of the results. The technical specifications usually

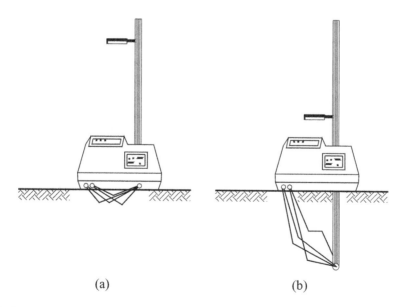

(a)   (b)

Figure 10.23 Control of the compaction using a nuclear densometer for the determination of the water content and the unit weight: a) indirect transmission mode; b) direct transmission mode.

stipulate the admissible percentage of results that can lie outside the prescribed zone in terms of the graph w versus γ_d. It is worth pointing out that these instruments require frequent calibration, with the operator requiring special protection from the radioactive emissions.

It must be noted that radioactive instruments still raise some doubts among specialists, regarding the reliability of the results. However, the alternative experimental methods available to analyse the unit weight and water content – the sand replacement method and oven drying, respectively – have the disadvantage of being more time-consuming (the oven-drying method requiring 24 h of testing). This matter can be overcome with Hilf's method, which is described below.

For more rapid determination of the water content, drying in a microwave oven can be used. The drying time depends on the type of soil and on the mass of the sample. Nonetheless, this method requires calibration by comparing the results with those from the conventional method of oven drying. Detailed information on the method can be obtained from the ASTM D4643 (2017) standard.

The determination of a control unit weight through the use of a radioactive instrument is impossible when the fill contains large particles, such as in the case of rockfills. In these construction works, the measurement of the unit weight can be done using the water replacement method, carried out in a test cavity. Figure 10.24 shows an example of this method.

The cavity varies in diameter, depending on the particle size distribution, and may reach up to 5.0 m in diameter. By dividing the weight of the material removed by the volume of water required to fill the cavity excavated, the unit weight of the fill is obtained. The details of the test procedures can be consulted in the ASTM D 5030 (2013) standard. When large particles are present, the determination of the water content requires some special corrective procedures to take into account that oven drying is applicable only to part of the material removed from the cavity.

10.3.7.2 Control by Hilf's Rapid Method

10.3.7.2.1 Introduction: simplified procedure

Hilf's Rapid Method (1959) was developed to provide a swift compaction control, understood as the deviation of the water content and dry unit weight of the compacted fill relative

Figure 10.24 Water replacement method for the determination of the unit weight of a rockfill (photo: Eduardo Fortunato).

to the optimum water content and the maximum dry unit weight, respectively, with the absolute value of the former being unknown.

The method essentially consists of the following steps: i) a certain mass of fill that has just been compacted is collected and appropriately protected from evaporation in order to preserve its water content; ii) this mass is divided into n specimens, with the total weight of each specimen being determined; iii) to each specimen is added a known mass of water, followed by mixing to form a homogeneous mass, similar to the procedure carried out in a compaction test; and iv) each specimen is then compacted in a Proctor mold, applying a compaction effort similar to that in the field, and the unit weight of these cylindrical samples is determined.

10.3.7.2.2 Evaluation of the relative compaction

Consider Z, the ratio (expressed as a percentage) of the weight of the added water, ΔW_w, and the initial weight, W, of each soil specimen. From the results of the procedure stated above, a curve relating Z (along the x-axis) and the unit weight, γ, of the several compacted specimens can be obtained, as shown in Figure 10.25. Designating W_s as the weight of the solid phase (particles) of each specimen, w_{fill} as the water content of the mass removed from the fill, and w as the water content of each specimen, the following can be obtained:

$$Z = \frac{\Delta W_w}{W} = \frac{w\,W_s - w_{fill}\,W_s}{W_s\left(1 + w_{fill}\right)} = \frac{w - w_{fill}}{1 + w_{fill}} \tag{10.13}$$

For any value of Z, the general equation is valid for correlating the unit weight with the dry unit weight:

$$\gamma_d\left(1 + w\right) = \gamma \tag{10.14}$$

By dividing both members of this equation by $1+Z$ and subsequently replacing Z by the second member of Equation 10.13, the following expression is obtained after some developments:

$$\gamma_d\left(1 + w_{fill}\right) = \frac{\gamma}{1 + Z} \tag{10.15}$$

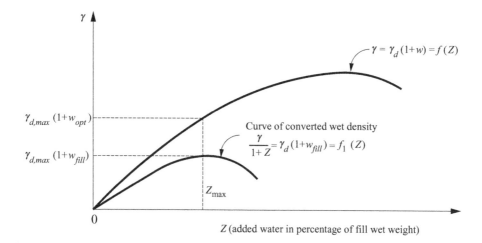

Figure 10.25 Hilf's method: added water *versus* the unit weight and the converted unit weight.

When analyzing this equation, it is possible to observe that the $(1 + w_{fill})$ factor of the first member is constant, although unknown. Thus, if the curve $\gamma = f(Z)$ in Figure 10.25 is divided by $(1 + Z)$, another expression can be obtained:

$$f_1(Z) = \frac{\gamma}{1 + Z} \tag{10.16}$$

which is the so-called *curve of converted unit weights,* which, apart from a constant and unknown scale factor, consistently represents the relationship between γ_d and Z.

This relationship is particularly valid at the maximum point of the curve:

$$\gamma_{d,\max}\left(1 + w_{fill}\right) = f_{1,\max}(Z) \tag{10.17}$$

Therefore, the relative compaction (RC) of the fill, that is, the ratio between the dry unit weight of the fill, $\gamma_{d,fill}$, and the maximum dry unit weight obtained in the Proctor test (see Equation 10.2) can be expressed as:

$$RC = \frac{\gamma_{d,fill}}{\gamma_{d,\max}} = \frac{\gamma_{d,fill}\left(1 + w_{fill}\right)}{\gamma_{d,\max}\left(1 + w_{fill}\right)} \tag{10.18}$$

or, considering Equations 10.14 and 10.17:

$$RC = \frac{\gamma_{fill}}{f_{1,\max}(Z)} \tag{10.19}$$

In conclusion, the relative compaction can be obtained by dividing the unit weight of the fill, γ_{fill}, by the ordinate of the maximum point of the curve of converted unit weights.

It is worth noting that the Z value on the x-axis can be either positive or negative. Negative values are associated with a certain weight of water being removed from the fill, which makes the method less immediate. However, this can be crucial to obtain an adequate definition of the curve of converted unit weights, particularly in fills compacted relatively close to the optimum. As a rule, four points are recommended to establish the curve, with three being the minimum, in combination with the hypothesis that the curve has the shape of a parabola close to the peak.

10.3.7.2.3 Assessment of the deviation of the water content from the optimum

By designating Z_{max} as the abscissa of the peak of the curve of converted unit weights, it corresponds to the optimum water content. Thus, considering Equation 10.13:

$$w_{opt} - w_{fill} = Z_{max}\left(1 + w_{fill}\right) \tag{10.20}$$

When taking Equation 10.15 into account, the following can be obtained:

$$1 + w_{fill} = \frac{\gamma}{\gamma_d} \frac{1}{1 + Z} \tag{10.21}$$

or:

$$1 + w_{fill} = \frac{1 + w}{1 + Z} \tag{10.22}$$

which, for $Z = Z_{max}$, can be expressed as:

$$1 + w_{fill} = \frac{1 + w_{opt}}{1 + Z_{max}}$$ (10.23)

Combining this equation with Equation 10.20, the following can be obtained:

$$w_{opt} - w_{fill} = Z_{max} \frac{1 + w_{opt}}{1 + Z_{max}}$$ (10.24)

In this equation, in addition to w_{fill}, w_{opt} is also, strictly speaking, unknown, as it expresses the optimum water content of the field compaction curve instead of the laboratory curve. Nevertheless, the difference between the abscissas of the peaks of both curves is relatively small. An error of a few percentage points in the value of w_{opt} introduced in the second member of the equation has little effect on the value of the first member, i.e., in the deviation of the water content of the fill from the respective optimum value.

The accuracy of this estimation can be further improved by using the chart represented in Figure 10.26 which, based on the results on 1,300 soils, relates the optimum water content with the unit weight at the same water content.[1]

The author of the method stresses one additional advantage of its use: the fact that this information obtained and treated for a swift control for compaction – though not including,

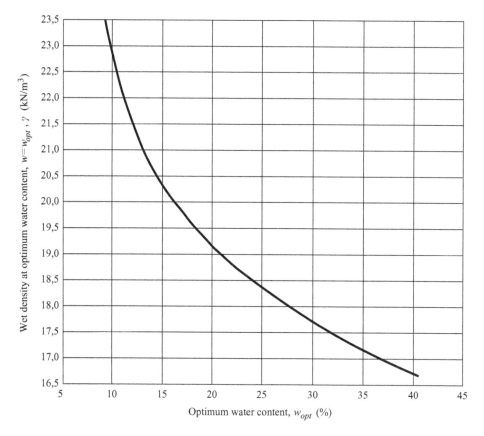

Figure 10.26 Optimum water content *versus* unit weight calculated at a water content equal to the optimum (curve based on 1,300 experimental results; adapted from Hilf, 1959).

---

[1] Note that this unit weight is equal to $f(Z_{max})$ in Figure 10.25.

as seen, the values of the dry unit weight and water content – can be included, at a later stage, to obtain the absolute values of these parameters for record purposes. For this, the water content of the fill should be determined by oven drying.

10.3.7.3 Frequency of the compaction control operations and statistical analysis of results

The control parameters should be determined more often, by any of the described methodologies, in the initial phase of the construction works, and may become less frequent when the procedures become more accurate and when staff has gained a better understanding of these same procedures. For this phase, entities, such as the U.S. Bureau of Reclamation (USBR) and other American public bodies, establish, as a general guidance, that one control test should be carried out for every 1,500 m³ of fill, with a minimum of one test per lift or layer.

A greater frequency of control operations is typically adopted in certain areas that are more sensitive, in which the work conditions can provide less quality of compaction, namely in: i) contact zones of the fill with the abutments and the foundation; ii) fills for drains and cut-off trenches; iii) places where compaction rollers are usually reversed; and iv) areas where there have likely been, for any given reason, misapplications of the material and/or of the specified compaction procedure.

In large earthworks, the number of control operations is very high, with the results having to be statistically analyzed. Therefore, it is essential that the specifications regarding the physical characteristics are appropriately adjusted for this statistical treatment of data. Experience demonstrates that the results of dry unit weight and water content determined in the field follow a normal distribution function. Table 10.5 shows an example of these specifications for an earthfill dam (Hilf, 1991).

The lifts that do not comply with such specifications must be adjusted, which may involve their removal, reworking or a combination of both. Reworking may consist of a number of distinct operations: i) to sprinkle water on the layer, when water content is below the level desired; ii) to rotavate the soil with disc harrows, for aeration, in cases of excess water content; or iii) to carry out additional passes of the compaction rollers, in case relative compaction is insufficient.

As a rule, lifts that have been previously rejected are more strictly monitored, following the appropriate correction procedures.

10.3.8 Control of the compaction through mechanical characteristics

10.3.8.1 Control by means of soil testing

In Section 10.3.5, it was pointed out that, under certain circumstances, field tests should be carried out on the compacted fill itself and/or laboratory tests should be employed

Table 10.5 Example of the specifications for the results of the compaction control for an earthfill dam (Hilf, 1991).

Parameter	Specification	
Water content, w (%)	$w_{opt} - 3.5 \leq w \leq w_{opt} + 1.0$	All the samples
	$w \leq w_{opt} - 3.0$	For not more than 20% of samples
	$w_{opt} + 0.5 \leq w$	For not more than 20% of samples
	$w_{opt} - 1.0 \leq w_{ave} \leq w_{opt} - 0.5$	(w_{ave} - mean value)
Relative compaction, $RC = \gamma_d / \gamma_{d,max}$	$RC \geq 96\%$	All the samples
	$RC \geq 97\%$	For not less than 80% of samples
	$RC_{ave} \geq 100\%$	(RC_{ave} - mean value)

on undisturbed samples, in order to verify the compliance of the fill with the project specifications.

Laboratory tests (triaxial or other) on undisturbed samples of the fill are only performed in certain construction works which involve a high degree of complexity and responsibility, which usually means large dams.

Regarding *in situ* tests, their use has been progressively increased, namely in modern road and railway works (such as platforms for high-speed railway lines). Therefore, technical specifications of these construction works often specify a minimum stiffness modulus for the platforms, which can be measured through the plate load test (PLT) (see Section 1.3.6). This modulus is usually designated EV_2, measured on the second load cycle of this test (AFNOR, 2000; Gomes Correia, 2001).

However, the PLT is relatively costly and takes a considerable amount of time to be performed. These aspects arise as a great limitation to the intensive use of this test in the context of earthworks, in which a great number of lifts and points per lift have to be characterized, almost in real time. With this in mind, several types of more economical load tests of the fill surface have been developed, which are usually quicker than the PLT.

Some of these tests will be briefly described below, namely those that are more frequently used.

The *dynamic plate load test* (also called the light falling weight deflectometer, LFWD), as its name implies, is obviously similar to the plate load test, except for the fact that the plate is loaded by a mass that falls freely from a certain height. Figure 10.27 and Table 10.6 provide

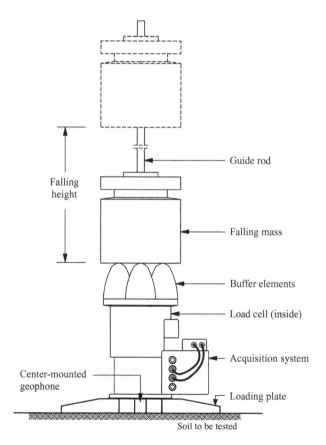

Figure 10.27 Dynamic plate load test.

Table 10.6 Essential characteristics of the system that performs the dynamic plate load test.

Diameter of the loading plate	10 cm (diameters of 20–30 cm are also used)
Falling mass	10 kg (a mass of 20 kg is also used with larger diameter plates)
Falling height	80 cm
Load cell (capacity)	20 kN
Load amplitude	1–10 kN (maximum of 15 kN)
Duration of the impulse	15–20 ms
Strain measurements	Geophones
Total weight of the system	26 kgf (base arrangement)

further explanations of this test. Its main advantages regard the portability of the equipment and the rapid execution of the test.

This equipment essentially consists of a circular rigid load plate with a central hole. Over this plate are successively assembled, from bottom to top, a hollow cylinder, a load cell and a set of buffer elements. A vertical guide rod is fixed to this system, being used to conduct the falling mass. In the upper part of the guide rod, there is a device used to fix the mass in its appropriate position before falling.

The results of this test are registered using geophones, which measure the velocity of the vertical deflection at the point of the surface of the lift to be tested, corresponding to the center of the plate (as a reminder, this plate has a central hole) and, optionally, at two other points on the sides of the plate, along the same radius, up to a maximum distance of 60 cm from the center. By integrating the velocity with respect to time, it is possible to calculate the vertical deflection at each point.

The load cell and the geophones are connected to a computer that, with suitable software, automatically registers and records the applied force, as well as the velocities and resultant deflections, when the mass falls from the reference position.

Similar to the PLT, a full and uniform contact between the plate and surface to be tested is required.

The interpretation of the results of the test is analogous to the static plate load test (see Section 1.3.6), with a dynamic stiffness modulus being obtained in this case. Nevertheless, a calibration of the results, by means of a correlation with the PLT results, is highly recommended (COST 337, 2001).

The *falling weight deflectometer test* (FWD) is very similar to the one that has just been described. However, it involves greater loads, plates of greater diameter, as well as more points along the surface where velocities and deflections are registered. The test was initially developed in order to test road pavements, though it has been increasingly applied to characterize compacted fills (COST 337, 2001; Fontul et al., 2011).

The *portancemeter test* consists of towing a narrow and smooth vibrating roller at constant speed. It is, thus, somehow similar to the smooth wheel vibrating roller, although it is not a compaction roller. Figure 10.28 and Table 10.7 complement the following explanation (Quibel et al., 2001; LCPC, 2008).

The vibrating wheel is equipped with devices able to measure the intensity of the force applied, F_v, and the value of soil deflection, d_v, in real time. As Figure 10.28 shows, the stiffness of the platform is calculated from these measurements, through regression, in the area of the load diagram lying between 30% and 90% of the maximum value of F_v:

$$k = F_v/d_v \tag{10.25}$$

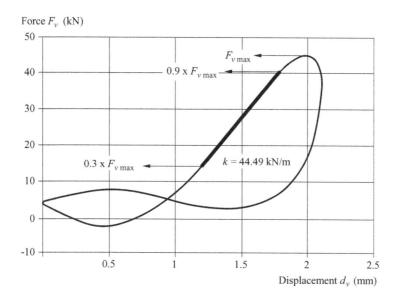

Figure 10.28 Graph of F_v versus d_v, corresponding to the load cycle from Portancemeter test (LCPC, 2008).

Table 10.7 Essential characteristic of a Portancemeter.

Dimensions of the vibrating roller	diameter: 100 cm
	width: 20 cm
Mass of the roller (vibrating mass)	600 kg
Total mass applied to the soil	1000 kg
Vibration	amplitude: 0.5 mm
	frequency: 35 Hz
Travel speed	3.6 km/h
Spacement of the stiffness records	$\approx$1.0 m (average of 30 consecutive cycles)

The software associated with the test then calculates the average values of k after 30 consecutive cycles, obtaining a value for this parameter for every meter in the longitudinal direction. The number of parallel runs made by the equipment over a given layer in order to enable an appropriate characterization naturally depends on the width of this layer. This value is typically equal to two for railway platforms and up to four for road platforms.

The systematic use of the Portancemeter test as a compaction control method requires, at an initial stage, the definition of a correlation between k and the EV_2 modulus (or another stiffness modulus taken as reference), by carrying out PLT tests (or other reference tests) on the same site and using this device (Fortunato et al., 2012).

The *light dynamic probing test* has already been discussed in Chapter 1 (see Section 1.3.4). Its use for compaction control is less frequent, because, being a penetration test, it is less suitable for the characterization of soil stiffness when compared to the tests previously described. Nevertheless, its low cost, simplicity and swift execution are considered to be advantageous.

Fortunato et al. (2009) report the use of the DPL in the renovation and reinforcement works on a platform of an old railway line. The use of the DPL was considered particularly useful as the railway line, with intense traffic, remained operational during the renovation work. It is also important to point out that a calibration of the results from the DPL with other conventional PLT tests was carried out.

10.3.8.2 Control using a compaction plant: intelligent compaction

Similar to what was described in Section 10.3.8.1 for the Portancemeter test, there are on the market compaction rollers (typically smooth wheel vibrating rollers) capable of assessing, in real time, the stiffness of the layers being compacted. This system is known as a *continuous compaction control* (CCC) (Thurner and Landstrom, 2000; Mooney and Adam, 2007).

This function was later combined with other functions by manufacturers of compaction equipment to, automatically and in real time, adjust the factors that control the performance of the equipment, depending on the stiffness assessment, in order to optimize the process. These factors are essentially as follows: i) the vibration amplitude; ii) the vibration frequency; and iii) the travel speed. This methodology is termed *intelligent compaction*.

The system can be pre-conditioned to attain values of the stiffness modulus within a certain range. Therefore, the number of passes of the equipment becomes, naturally, the number strictly necessary to reach these target values, avoiding both under- and over-compaction.

Depending on the results obtained and on the number of passes necessary to reach these target values on a given lift, other adjustments may be required in the subsequent lifts, for instance, to optimize the thickness of the lifts and the water content.

The optimization of the compaction effort, the homogeneity of stiffness in the entire compacted area, and the ability to incorporate a larger (and deeper) volume of compacted fill in the same compaction analysis (when compared to other conventional methods) are seen as the main advantages of using these methods (Brandl, 2001; Briaud and Leo, 2003).

10.3.9 Final remarks

In the previous sections, a discussion was presented on the control of compaction by means of the physical and mechanical characteristics of the fill, highlighting the respective difficulties, advantages, and limitations. It now seems appropriate to make a final remark.

The performance of the earthfill structure depends on its mechanical behavior. Therefore, it may seem reasonable to consider that the control of the compaction is more efficient when using *in situ* measurements of the mechanical parameters, compared with the compaction control through physical characteristics.

This perspective would be entirely correct if the conditions of the fill at the end of compaction remained practically unchanged through the lifetime of the structure. This is not at all true, both in earthfill dams and in earthfills for transport infrastructures! In dams, in particular, the filling of the reservoir takes place immediately after compaction, resulting in the saturation and submersion of a great part of the body of the fill.

A soil with significant fines content can present considerable stiffness at the end of compaction and exhibit relatively low density. This given stiffness can exist as a result of the high values of suction, considering that the soil had been compacted mostly on the dry side of the optimum. In these soils, submersion will result in a significant reduction in stiffness, with large deformations due to collapse.

Therefore, it can be concluded that a compaction control, performed exclusively with reference to the mechanical characteristics, would be insufficient (Briaud and Leo, 2003).

On the other hand, a compaction control essentially reliant on dry unit weight measurements, and, therefore, on relative compaction, would also be insufficient, and would not present a reliable correlation with the stiffness and strength parameters of the compacted soil. In fact, in a soil with significant fines content, the same dry unit weight corresponds to two different water contents, above or below the optimum. Thus, it also corresponds to two very different values of suction, given the sensitivity of suction to water content and to the degree of saturation. Accordingly, the same density would correspond to two different values of stiffness and strength.

It can also be concluded that a control of compaction exclusively through relative compaction would not be enough in many cases (Briaud and Leo, 2003). Thus, information on the compaction water content – with the exception of clean granular materials – is of great importance in the control of compaction.

If the control based on mechanical characteristics is clearly insufficient, in most cases, it plays an important complementary role. This can be explained by the fact that construction works in the field (excavation in the borrowing pit, transportation, spreading, compaction, among others) are radically different from the procedures carried out in the laboratory. In some cases, those works can lead to great structural changes (in terms of particle size distribution, and others), with consequences for the mechanical behavior of the compacted material, which a mere comparison of values of water content and dry unit weight cannot reflect.

In summary, a combination of the compaction control through both forms – based on both physical and mechanical characteristics – is mostly advisable, and, in some cases, is even crucial.

References

AASHTO (2012). *LRFD Bridge Design Specifications*, 6th edition. American Association of State Highway and Transportation Officials.

AASHTO T 224-86 (2018a). *T180, Standard Method of Test for Moisture-Density Relations of Soils Using 2–5-kg Rammer and a 305-mm Drop*. American Association of State Highway and Transportation Officials.

AASHTO T 224-86 (2018b). *T180, Standard Method of Test for Moisture-Density Relations of Soils Using 4–54-kg Rammer and a 457-mm Drop*. American Association of State Highway and Transportation Officials.

Aboshi, H., Ichimoto, E., Harada, K. and Emoki, M. (1979). The composer: A method to improve characteristics of soft clays by inclusion of large diameter sand columns. *Colloque International sur Renforcement des Sols*, ENPC-LCPC, vol. 1, pp. 211–216.

Absi, E. (1962). Équilibre limite des massifs. *Annales de l'Institut Technique du Bâtiment et des Travaux Publics*, September.

AFNOR (2000). *Formation Level Bearing Capacity. Part 1: Plate Test Static Deformation Module (EV2). NF P 94-117-1 – Soils: Investigation and Testing.* Association Française de Normalisation.

AFNOR (2012) *NF P 94-262 – Justification des Ouvrages Géotechniques – Normes d'Application Nationale de l'Eurocode 7 – Fondations Profondes*. AFNOR.

Aggour, M. S. and Brown, C. B. (1974). The prediction of earth pressure on retaining walls due to compaction. *Géotechnique*, 24(4): 489–502.

AGI (1991a). The Leaning Tower of Pisa. Present situation. *10th European Conference on Soil Mechanics and Foundation Engineering*, Associazione Geotecnica Italiana.

AGI (1991b). The contribution of geotechnical engineers to the preservation of Italian historical sites. *10th European Conference on Soil Mechanics and Foundation Engineering*, Associazione Geotecnica Italiana.

Almeida, M. S. S. and Marques, M. E. S. (2018). *Design and Performance of Embankments on Very Soft Soils*. CRC Press.

Almeida, M. S., Futai, M. M., Lacerda, W. A. and Marques, M. E. (2008). Laboratory behaviour of Rio de Janeiro soft clays. Part 1: Index and compression properties. *Soils and Rocks*, 31(2): 69–75.

Alonso, E. (1976). Risk analysis of slopes and its application to slopes in Canadian sensitive clays. *Géotechnique*, 26(3): 453–472.

Alonso, E. (2014). Fracture mechanics and rockfill dams. *Soils and Rocks*, 37(1): 3–35.

Alonso, E. (2019). Personal communication.

Alonso, E., Gens, A. and Hight, D. W. (1987). Special problem soils. General report. *Proceedings of the 9th European Conference on Soil Mechanics and Foundation Engineering*, Dublin, vol. 3, pp. 1087–1146.

Alonso, E., Gens, A. and Josa, A. (1990). A constitutive model for partially saturated soils. *Géotechnique*, 40(3): 405–430.

Alonso, E., Gens, A. and Lloret, A. (1993). The landslide of Cortes de Pallas, Spain. *Géotechnique*, 43(4): 507–521.

Alonso, E., Gens, A. and Lloret, A. (2000). Precompression design for secondary settlement reduction. *Géotechnique*, 50(6): 645–656.

Alonso, E., Pinyol, N. M. and Puzrin, A. M. (2010). *Geomechanics of Failures. Advanced Topics.* Springer.

Ambraseys, N. N. (1988). Engineering seismology: Part I. *Earthquake Engineering and Structural Dynamics*, 17(1): 1–105.

Andrus, R. D. and Stoke, K. H. (2000). Liquefaction resistance of soils from shear-wave velocity. *Journal of Geotechnical and Geoenvironmental Engineering, ASCE*, 126(11): 1015–1025.

Antão, A. N. and Guerra, N. C. (2009). Personal communication.

Aoki, N. and Velloso, D. A. (1975). An approximate method to estimate the bearing capacity of piles. *Proceedings of the 5th Pan American Conference on Soil Mechanics and Foundation Engineering*, Buenos Aires, pp. 367–376.

API (2005). *Recommended Practice for Planning, Designing and Constructing Fixed Offshore Platforms—Working Stress Design, API RP-2A.* American Petroleum Institute.

Asaoka, A. (1978). Observational procedure of settlement prediction. *Soils and Foundations*, 18(4): 87–101.

ASTM D 420 (2018). *Standard Guide for Site Characterization for Engineering Design and Construction Purposes.* ASTM International.

ASTM D 698 (2002). *Standard Test Methods for Laboratory Compaction Characteristics Using Standard Effort.* ASTM International.

ASTM D 1195 / D 1195M – 09 (2015). *Standard Test Method for Repetitive Static Plate Load Tests of Soils and Flexible Pavement Components, for Use in Evaluation and Design of Airport and Highway Pavements.* ASTM International.

ASTM D 1196 / D 1196M – 12 (2016). *Standard Test Method for Nonrepetitive Static Plate Load Tests of Soils and Flexible Pavement Components, for Use in Evaluation and Design of Airport and Highway Pavements.* ASTM International.

ASTM D 1452 / D 1452M (2016). *Standard Practice for Soil Exploration and Sampling by Auger Borings.* ASTM International.

ASTM D 1557 (2002). *Standard Test Methods for Laboratory Compaction Characteristics Using Modified Effort.* ASTM International.

ASTM D 1586 / D 1586M (2018). *Standard Test Method for Standard Penetration Test (SPT) and Split-Barrel Sampling of Soils.* ASTM International.

ASTM D 2488 (2017). *Standard Practice for Description and Identification of Soils (Visual-Manual Procedures).* ASTM International.

ASTM D 2573 / D 2573M (2018). *Standard Test Method for Field Vane Shear Test in Saturated Fine-Grained Soils.* ASTM International.

ASTM D 3441 (2016). *Standard Test Method for Mechanical Cone Penetration Testing of Soils.* ASTM International.

ASTM D 4050 (2014). *Standard Test Method for (Field Procedure) for Withdrawal and Injection Well Tests for Determining Hydraulic Properties of Aquifer Systems.* ASTM International.

ASTM D 4428 / D 4428M (2014). *Standard Test Methods for Crosshole Seismic Testing.* ASTM International.

ASTM D 4643 (2017). *Standard Test Method for Determination of Water Content of Soil and Rock by Microwave Oven Heating.* ASTM International.

ASTM D 4700 (2015). *Standard Guide for Soil Sampling from the Vadose Zone.* ASTM International.

ASTM D 4718 (2007). *Standard Practice for Correction of Unit Weight and Water Content for Soils Containing Oversize Particles.* ASTM International.

ASTM D 5030 (2013). *Standard Test Method for Density of Soil and Rock in Place by the Water Replacement Method in a Test Pit.* ASTM International.

ASTM D 5778 (2012). *Standard Test Method for Electronic Friction Cone and Piezocone Penetration Testing of Soils.* ASTM International.

ASTM D 6635 (2015). *Standard Test Method for Performing the Flat Plate Dilatometer.* ASTM International.

Atkinson, J. (1993). *An Introduction to the Mechanics of Soils and Foundations.* McGraw-Hill.

Atkinson, J. H. (2000). Non-linear soil stiffness in routine design. *Géotechnique*, 50(5): 487–508.

Azzouz, A. S., Baligh, M. M. and Ladd, C. C. (1983). Corrected field vane strength for embankment design. *Journal of Geotechnical Engineering, ASCE*, 109(5): 730–734.

Azzouz, A. S., Krizek, R. J. and Corotis, R. B. (1976). Regression analysis of soil compressibility. *Soils and Foundations*, 16(2): 19–29.

Baligh, M. M., Azzouz, A. S. and Chin, C. T. (1987). Disturbances due to ideal tube sampling. *Journal of Geotechnical Engineering, ASCE*, 113(7): 739–757.

Baligh, M. M. and Levadoux, J. N. (1986). Consolidation after undrained piezocone penetration. II: Interpretation. *Journal of Geotechnical Engineering, ASCE*, 112(7): 727–745.

Barghouthi, A. F. (1990). Active earth pressure on walls with base projection. *Journal of the Geotechnical Engineering Division, ASCE*, 116(10): 1570–1575.

Barksdale, R. D. and Bachus, R. C. (1983). *Design and Construction of Stone Columns. FHWA/ RD-83/026.* Federal Highway Administration.

Barradas, J. (1999). *Long-Term Behaviour of a Natural Slope That Suffer an Excavation at the Toe. Short Course on Slope Stabilization.* University of Porto (in Portuguese).

Barron, R. A. (1944). *The Influence of Drain Wells on the Consolidation of Fine-Grained Soils.* Diss. Providence US Eng. Office.

Barron, R. A. (1947). Consolidation of fine-grained soils by drain wells. *Proceedings of the ASCE*, 73(6): 811–835.

Becker, D. E. (1996a). Eighteenth Canadian Geotechnical Colloquium: Limit states design for foundations. Part I. An overview of the foundation design process. *Canadian Geotechnical Journal*, 33(6): 956–983.

Becker, D. E. (1996b). Eighteenth Canadian Geotechnical Colloquium: Limit states design for foundations. Part II. Development for the National Building Code of Canada. *Canadian Geotechnical Journal*, 33(6): 984–1007.

Beresantev, V. G., Khritoforov, V. S. and Golubkov, V. N. (1961). Load bearing capacity and deformation of piled foundations. *Proceedings of the 5th International Conference on Soil Mechanics and Foundation Engineering*, Paris, Vol. 2, pp. 11–15.

Bergado, D. T., Akasami, H., Alfaro, M. C. and Balasubramaniam, A. S. (1992). Smear effects of vertical drains on soft Bangkok clay. *Journal of Geotechnical Engineering, ASCE*, 117(10): 1509–1530.

Bergdahl, U., Ottosson, E. and Malmborg, B. S. (1993). See Eurocode 7 – Geotechnical design – Part 2: Ground investigation and testing. CEN, March 2007.

Bermingham, P. and Janes, M. (1989). An innovative approach to load testing of high capacity piles. *Proceedings of the International Conference on Piling and Deep Foundations*, J. B. Burland and J. M. Mitchell (eds.), Balkema, Vol. 1, pp. 409–413.

Biot, M. A. (1935). Les problèmes de la consolidation des matières argileuses sous une charge. *Annales de la Société Scientifique de Bruxelles, Séries B* 55: 110–113.

Biot, M. A. (1941a). General theory of three-dimensional consolidation. *Journal of Applied Physics*, 12(2): 155–164.

Biot, M. A. (1941b). Consolidation settlement under a rectangular load distribution. *Journal of Applied Physics*, 12(5): 426–430.

Bishop, A. W. (1955). The use of slip circle in the stability analysis of earth slopes. *Géotechnique*, 5(1): 7–17.

Bishop, A. W. and Bjerrum, L. (1960). The relevance of the triaxial test to the solution of stability problems. *Proceedings of the Conference on Shear Strength of Cohesive Soils*, ASCE, Boulder, CO, pp. 437–501.

Bjerrum, L. (1963). Allowable settlement of structures. *Proceedings of the 3rd European Conference on Soil Mechanics and Foundation Engineering*, Wiesbaden, vol. 2, pp. 135–137.

Bjerrum, L. (1972). Embankments on soft ground. *Proceedings of the ASCE Conference on Performance of Earth and Earth-Supported Structures*, Lafayette, vol. 2, pp. 1–54.

Bond, A. and Harris, A. (2008). *Decoding Eurocode 7.* Taylor and Francis.

Borges, J. F. and Castanheta, M. (1968). *Structural Safety.* LNEC.

Borges, J. L. and Cardoso, A. S. (2002). Overall stability of geosynthetic-reinforced embankments on soft soils. *Geotextiles and Geomembranes*, 20(6): 395–421.

Boulanger, R. W. and Idriss, I. M. (2012). Probabilistic standard penetration test-based liquefaction-triggering procedure. *Journal of Geotechnical and Geoenvironmental Engineering, ASCE,* 138(10): 1185–1195.

Boulanger, R. W. and Idriss, I. M. (2014). CPT and SPT based liquefaction triggering procedures. Report UCD/CGM-14/01. Center for Geotechnical Modelling, University of California, Davis, CA.

Boulanger, R. W. and Idriss, I. M. (2015). Magnitude scaling factors in liquefaction triggering procedures. *Soil Dynamics and Earthquake Engineering,* 79-B(December): 296–303.

Boulanger, R. W. and Idriss, I. M. (2016). CPT-based liquefaction triggering procedure. *Journal of Geotechnical and Geoenvironmental Engineering, ASCE,* 142(2).

Boussinesq, J. (1885). *Application des Potentiels à l'Étude de l'Équilibre et du Mouvement des Solides Élastiques.* Gauthier-Villars.

Bowles, J. E. (1997). *Foundation Analysis and Design,* 5th edition. McGraw-Hill.

Bowles, J. E. (1988). *Foundation Analysis and Design,* 4th edition. McGraw-Hill.

Brand, E. W. (1985). Predicting the performance of residual soil slopes. *Proceedings of the 11th International Conference on Soil Mechanics and Foundation Engineering,* San Francisco, CA, vol. 5, pp. 2541–2578.

Brandl, H. (2001). The importance of optimum compaction of soil and other granular materials. *Geotechnics for Roads, Rail Tracks and Earth Structures. Outcome of European Technical Committee No. 11 (ECT 11).* International Society for Soil Mechanics and Geotechnical Engineering, Balkema, pp. 3–11.

Briaud, J. L. and Leo, J. (2003). Intelligent compaction: Overview and research needs. Texas A & M University Report.

Brinch Hansen, J. (1951). Simple statistical computation of permissible pile loads. *Christiani and Nielsen Post:* No. 13, pp. 14–15.

Brinch Hansen, J. (1953). *Earth Pressure Calculation.* The Danish Technical Press.

Brinch Hansen, J. (1956). Limit state and safety factors in soil mechanics. Danish Geotechnical Institute, Bulletin No. 1.

Brinch Hansen, J. (1961a). A general formula for bearing capacity. Danish Geotechnical Institute, Bulletin No. 11, pp. 38–46.

Brinch Hansen, J. (1961b). The ultimate resistance of rigid piles against transversal forces. Danish Geotechnical Institute. Bulletin No. 12, pp. 5–9.

Brinch Hansen, J. (1970). A revised and extended formula for the bearing capacity. Danish Geotechnical Institute, Bulletin No. 28, pp. 5–11.

Brinch Hansen, J. and Christensen, N. H. (1969). Discussion on theoretical bearing capacity of very shallow footings. *Journal of the Soil Mechanics and Foundation Division,* 95(6): 1568–1572.

Broms, B. B. (1968). Swedish tie-back systems for sheet pile walls. *Proceedings of the 3rd Budapest Conference on Soil Mechanics and Foundation Engineering,* pp. 391–403.

Brown, D. A. and Bollman, H. T. (1993). Pile-supported bridge foundations designed for impact loading. *Transportation Research Record,* 1331: 87–91.

Brown, D. A., Morrison, C. and Reese, L. C. (1988). Lateral load behavior of pile group in sand. *Journal of Geotechnical Engineering, ASCE,* 114(11): 1261–1276.

BS 1377 (1990). *British Standard Methods of Tests for Soils for Civil Engineering Purposes. Part 4. Compaction-Related Tests.* British Standard Institution.

Budin, A. Y. and Demina, G. A. (1979). *Quays Handbook.* Moscow (in Russian).

Buisman, A. S. (1940). *Grondmechanica.* Waltman.

Burland, J. B. (1989). Small is beautiful: The stiffness of soils at small strains. *Canadian Geotechnical Journal,* 26(4): 499–516.

Burland, J. B., Broms, B. B. and De Mello, V. F. (1977). Behaviour of foundations and structures. State-of-the-art report. *Proceedings of the 9th International Conference on Soil Mechanics and Foundation Engineering,* Tokyo, vol. 2, pp. 495–546.

Burland, J.B. and Burbidge, M.C. (1985). Settlement of foundations on sand and gravel. *Proceedings of the Institution of Civil Engineers, Part 1,* 78(6): 1325–1381.

Burland, J. B. and Wroth, C. P. (1974). Allowable and differential settlement of structures including damage and soil-structure interaction. *Proceedings of the Conference on Settlement of Structures*. Pentech Press, Cambridge, pp. 611–654.

Burland, J. B., Jamiolkowski, M., Lancellotta, R., Leonards, G. A. and Viggiani, C. (1993). Leaning Tower of Pisa: What is going on. *ISSMFE News, International Society for Soil Mechanics and Foundation Engineering*, 20(2): 1–3.

Burland, J. B., Jamiolkowski, M., Lancellotta, R., Leonards, G. A. and Viggiani, C. (1994). Pisa update – Behaviour during counterweight application. *ISSMFE News, International Society for Soil Mechanics and Foundation Engineering*, 21(2): 1–2.

Burland, J. B., Jamiolkowski, M. and Viggiani, C. (2009). Leaning Tower of Pisa: Behaviour after stabilization operations. *International Journal of Geoengineering Case Histories*, 1(3): 156–169.

Burland, J. B., Potts, D. M. and Walsh, N. M. (1981). The overall stability of free and propped embedded cantilever retaining walls. *Ground Engineering*, 14(5): 28–38.

Bustamante, M. and Gianeselli, L. (1981). Prévision de la capacité portante des pieux isolés sous charge verticale. Règles pressiométriques et pénétrométriques. *Bulletin de Liaison Laboratoire Ponts et Chaussés*, 113: 83–108.

Canadian Geotechnical Society (1985). *Canadian Foundation Engineering Manual*, 2nd Edition. Bitech Publishers, Ltd.

Caquot, A. (1934). *Équilibre des Massifs à Frottement Interne*. Gauthier-Villars.

Caquot, A. and Kérisel, J. (1949). *Traité de Mécanique des Sols*. Gauthier-Villars.

Caquot, A. and Kérisel, J. (1953). Sur le terme de surface dans le calcul des fondations em milieu pulvérulent. *Proceedings of the 3rd International Conference on Soil Mechanics and Foundation Engineering*, Zurich, vol. 1, pp. 336–337.

Caquot, A. and Kerisel, J. (1956). *Traité de Mécanique des Sols*, 2nd edition. Gauthier Villars.

Cardoso, A. S., Guerra, N. C., Antão, A. N. and Matos Fernandes, M. (2007). Limit analysis of anchored concrete soldier-pile walls in clay under vertical loading. *Canadian Geotechnical Journal*, 43(5): 516–530.

Cardoso, A. S. and Matos Fernandes, M. (2001). Characteristic values of ground parameters and probability of failure in the design according to Eurocode 7. *Géotechnique*, 51(6): 519–531.

Carrillo, N. (1942). Simple two and three dimensional cases in the theory of consolidation of soils. *Journal of Mathematics and Physics*, 21(1): 1–5.

Carter, J. P. and Kulhawy, F. H. (1988). Analysis and design of foundations socketed into rock, Report No. EL-5918. Empire State Electric Engineering Research Corporation and Electric Power Research Institute.

Casagrande, A. (1936). The determination of the preconsolidation load and its practical significance. *Proceedings of the 1st International Conference on Soil Mechanics and Foundation Engineering*, Cambridge, vol. 3, pp. 60–64.

Casagrande, A. (1950). Notes on the design of earth dams. *Journal of the Boston Society of Civil Engineers*, 37: 405–429.

Casagrande, A. and Fadum, R. E. (1940). *Notes on Soil Testing for Engineering Purposes*. Harvard University Graduate School Engineering Publication No. 8.

Chester, D. K. (2001). The 1755 Lisbon earthquake. *Progress in Physical Geography*, 25(3): 363–383.

Christian, J. T. (2004). Geotechnical Engineering reliability: How well do we know what we are doing? *Journal of Geotechnical and Geoenvironmental Engineering*, 130(10): 985–1003.

Chu, S. C. (1991). Rankine's analysis of active and passive pressures in dry sands. *Soils and Foundations*, 31(4): 115–120.

Clayton, C. R. I., Matthews, M. C. and Simons, N. E. (1995). *Site Investigation*, 2nd edition. Blackwell Science.

Clayton, C. R. I., Siddique, A. and Hopper, R. J. (1998). Effects of sampler design on tube sampling disturbance – Numerical and analytical solutions. *Géotechnique*, 48(6): 847–867.

Clough, G. W. and Tsui, Y. (1974). Performance of tied-back walls in clay. *Journal of Geotechnical Engineering, ASCE*, 100(12): 1259–1273.

Coduto, D. P. (2001). *Foundation Design: Principles and Practice*, 2nd edition. Prentice Hall.

Coelho, P. L. F. (2000). Geotechnical characteristics of soft soils. Study of the experimental site of Quinta do Foja (Lower Mondego). MSc thesis, University of Coimbra.

Collotta, T., Manassero, V. and Moretti, P. C. (1988). An advanced technology in deep drainage of slopes. Proceedings of the 5th International Symposium on Landslides, Lausanne, vol. 2, pp. 887–892.

Correia, R. (1982). Application of mathematical models to the study of embankments on soft clayey soils. Specialist thesis, LNEC, Lisbon.

COST (2001). Unbound granular materials for road pavements. Final Report of the Action. COST 337, Directorate-General for Energy and Transport, European Commission.

Coulomb, C. R. (1773). Essai sur une application des règles des maximis et minimis à quelques problèmes de statique rélatifs à l'architecture. Mémoires de Mathématique et de Physique, présentées à l'Académie Royale des Sciences par divers Savants et lus dans ses Assemblées, Paris, 7: 343–382.

Cruz, P. T. (1996). 100 Brazilian Dams. Case Studies. Construction Materials. Design. Oficina de Textos (in Portuguese).

Cruz, P. T. (2018). Personal communication.

Cryer, C. W. (1963). A comparison of the three-dimensional consolidation theories of Biot and Terzaghi. Quarterly Journal of Mechanics and Applied Mathematics, 16(4): 401–412.

Culmann, C. (1875). Die Graphische Statik. Meyer and Zeller.

CUR (1996). Building on Soft Soils. Centre for Civil Engineering Research and Codes, Balkema.

Das, B. M. (2004). Principles of Foundation Engineering. Thompson Brooks/Cole.

Davies, T. G., Richards Jr., R. and Chen, K.-H. (1986). Passive pressure during seismic loading. Journal of Geotechnical Engineering, ASCE, 112(4): 479–483.

Davis, E. H. and Poulos, H. G. (1972). Rate of settlement under three-dimensional conditions. Géotechnique, 22(1): 95–114.

Davisson, M. T. (1972). High capacity piles. Proceedings of the Soil Mechanics Lecture Series on Innovations in Foundation Construction, ASCE. Illinois Section, Chicago, IL, pp. 81–112.

Day, R. A. (1999). Net pressure analysis of cantilever sheet pile walls. Géotechnique, 49(2): 231–245.

De Beer, E. E. (1945). Étude des fondations sur pilotis et des fondations directes. Annales des Travaux Publics de Belgique, 46: 1–78.

De Beer, E. E. (1970). Experimental determination of the shape factors and the bearing capacity factors of sand. Géotechnique, 20(4): 387–411.

De Mello, V. F. B. (1971). Standard penetration test. Proceedings of the 4th Pan-American Conference on Soil Mechanics and Foundation Engineering, San Juan, Puerto Rico, vol. 1, pp. 1–86.

Décourt, L. (1989). SPT, CPT, pressuremeter testing and recent developments in in-situ testing – Part 2: The standard penetration test. State-of-the-art report. Proceedings of the 12th International Conference on Soil Mechanics and Foundation Engineering, Rio de Janeiro, vol. 4, pp. 2405–2416.

Décourt, L. (1998). Analysis and Design of Deep Foundations, in Foundations – Theory and Practice, 2nd edition. Ed. Pini – ABMS/ABEF (in Portuguese).

Décourt, L. and Quaresma, A. R. (1978). Pile capacity based on SPT values. Proceedings of the 6th Brazilian Conference on Soil Mechanics and Foundation Engineering, Rio de Janeiro (in Portuguese), pp. 45–53.

DiBiagio, E. and Bjerrum, L. (1957). Earth pressure measurements in a trench excavated in stiff marine clay. Proceedings of the 4th International Conference on Soil Mechanics and Foundation Engineering, London, vol. 2, pp. 196–202.

Donaghe, R. T. and Townsend, F. C. (1976). Scalping and Replacement Effects on Compaction Characteristics of Earth-Rock Mixtures. ASTM Special Publication, vol. STP 599, pp. 248–277.

Drucker, D. C. (1953). Limit analysis of two and three-dimensional soil mechanics problems. Journal of Mechanics and Physics of the Solids, 1(4): 217–226.

Duncan, J. (2000). Factors of safety and reliability in geotechnical engineering. Journal of Geotechnical and Geoenvironmental Engineering, 126(4): 307–316.

Duncan, J. and Buchignani, A. L. (1976). *An Engineering Manual for Settlement Studies*. Department of Civil Engineering, Institute of Transportation and Traffic Engineering, University of California.

Dupuit, J. (1863). *Études Théoriques et Pratiques sur le Mouvement des Eaux dans les Canaux Découverts et à Travers les Terrains Perméables*. Dunod.

Elms, D. G. and Richards Jr., R. (1990). Seismic design of retaining walls. *Proceedings of the ASCE Conference on Design and Performance of Earth Retaining Structures*. Cornell University, New York, pp. 854–871.

EN 1537:2013. *Execution of Special Geotechnical Works. Ground Anchors*. CEN.

EN 1990-1:2002. *Eurocode 0: Basis of Structural Design*. CEN.

EN 1991-1-1:2002. *Eurocode 1: Actions on Structures – Part 1-1: General Actions – Densities, Self-Weight, Imposed Loads for Buildings*. CEN.

EN 1992-1-1:2004. *Eurocode 2: Design of Concrete Structures – Part 1-1: General Rules and Rules for Buildings*. CEN.

EN 1997-1:2004. *Eurocode 7 – Geotechnical Design – Part 1: General Rules*. CEN.

EN 1997-1:2004/A1:2013. *Eurocode 7 – Geotechnical Design – Part 1: General Rules*. CEN.

EN 1997-2:2007. *Eurocode 7 – Geotechnical design – Part 2: Ground Investigation and Testing*. CEN.

EN 1998-1:2004. *Eurocode 8 – Design of Structures for Earthquake Resistance. Part 1: General Rules, Seismic Action and Rules for Buildings*. CEN.

EN 1998-5:2004. *Eurocode 8 – Design of Structures for Earthquake Resistance. Part 5: Foundations, Retaining Structures and Geotechnical Aspects*. CEN.

ENV 1997-1:1994. *Eurocode 7 – Geotechnical design - Part 1: General Rules*. CEN.

ENV 1997-3:1999. *Eurocode 7 – Geotechnical design – Part 3: Design Assisted by Field Testing*. CEN, Brussels.

Esrig, M. E. and Kirby, R. C. (1979). Advances in general effective stress method for the prediction of axial capacity for driven piles in clay. *Proceedings of the 11th Annual Offshore Technology Conference*, pp. 437–449.

Esteves, E. (2013). Soft silty-clayey alluvium of Portugal. Parametrization for the design of geotechnical structures. PhD thesis, University of Porto.

Fang, H.-Y. (1991). *Foundation Engineering Handbook*. Springer.

Fellenius, B. H. (1972). Downdrag on piles in clay due to negative skin friction. *Canadian Geotechnical Journal*, 9(4): 323–337.

Fellenius, B. H. (2019). *Basics of Foundation Design*. Pile Buck International, Inc., Electronic Edition, www.Fellenius.net.

Fellenius, W. (1927). *Erdstatische Berechnungen mit Reibung und Kohaesion*. Ernst.

Fellenius, W. (1936). Calculation of the stability of earth dams. *Transactions of the 2nd Congress on Large Dams*, Washington, D.C., vol. 4, p. 445.

Ferreira, C. F. (2009). The use of seismic wave velocities in the measurement of stiffness of a residual soil. PhD thesis, University of Porto.

Ferreira, C., Martins, J. P. and Gomes Correia, A. (2014). Determination of the small-strain stiffness of hard soils by means of bender elements and accelerometers. *Geotechnical and Geological Engineering*, 32(6): 1369–1375.

Ferreira, C., Viana da Fonseca, A. and Nash, D. (2011). Shear wave velocities for sample quality assessment on a residual soil. *Soils and Foundations*, 51(4): 683–692.

Finn, W. D. L. (2001). *Earthquake Engineering. Geotechnical and Geoenvironmental Engineering Handbook*, R. K. Rowe (ed.). Kluwer Academic Publishers, Chapter 21, pp. 615–659.

Fishman, K. L., Richards Jr., R. and Yao, D. (2003). Inclination factors for seismic bearing capacity. *Journal of Geotechnical and Geoenvironmental Engineering*, ASCE, 129(9): 861–865.

Flamant, A. (1892). Sur la répartition des pressions dans un solide rectangulaire chargé transversalement. *Comptes Rendues Académie des Sciences de Paris*, 114: 1465–1468.

Fleming, K., Weltman, A., Randolph, M. and Elson, K. (2009). *Piling Engineering*, 3rd edition. Taylor & Francis.

Fontul, S., Fortunato, E. and De Chiara, F. (2011). Non-destructive tests for railway infrastructure stiffness evaluation. *Paper 16, Proceedings of the 13th International Conference on Civil, Structural and Environmental Engineering Computing*, B. H. V. Topping and Y. Tsompanakis (eds.). Civil-Comp Press.

Fortunato, E., Paixão, A. and Fontul, S. (2012). Improving the use of unbound granular materials in railway sub-ballast layer. *Advances in Transportation Geotechnics II*, Miura et al. (eds.). Taylor & Francis Group, ISBN 978-0-415-62135-9, pp. 522–527.

Fortunato, E., Pinelo, A. and Matos Fernandes, M. (2009). In situ characterization of an old railway platform with DCP. *Proceedings of the 17th International Conference on Soil Mechanics and Geotechnical Engineering*, Alexandria, vol. 2, pp. 919–952.

Foti, S., Lancellotta, R., Marchetti, D., Monaco, P. and Totani, G. (2006). Interpretations of SDMT tests in a transversely isotropic medium. *Proceedings of the 2nd International Conference on the Flat Dilatometer*, Washington D.C., pp. 295–305.

Foti, S., Lai, C. and Lancellotta, R. (2002). Porosity of fluid-saturated porous media from measured seismic wave velocities. *Géotechnique*, 52(5): 359–373.

Frank, R. (1991). Quelques développements récents sur le comportement des fondations superficielles. *Proceedings of the 10th European Conference on Soil Mechanics and Foundation Engineering*, Florence, vol. 3, pp. 1003–1030.

Frank, R. (2003). *Calcul des Foundations Superficielles et Profondes*. Presses de l'École Nationale des Ponts et Chaussées.

Frank, R., Bauduin, C., Driscoll, R., Kavvadas, M., Krebs Ovesen, N. and Schuppener, B. (2004). *Designers' Guide to EN 1997-1, Eurocode 7: Geotechnical Design – General Rules*. Thomas Telford.

Futai, M. M. (1999). Theoretical and practical concepts on behaviour analysis of some Rio de Janeiro clays. *Doctoral Seminar COPPE/UFRJ*, Rio de Janeiro, Brazil (in Portuguese).

GEO (1992). *Reassessment of the Po Shan Landslide of 18 June 1992, SPR 16/92*. Geotechnical Engineering Office, Government of Hong Kong.

Giroud, J. P. (1970). Stress under linearly loaded rectangular area. *Journal of the Soil Mechanics and Foundations Division, ASCE*, 98(1): 263–268.

Goble, G. G. and Rausche, F. (1970). *Pile Load Test by Impact Driving*. Highway Research Record, Highway Research Board, No. 333, pp.123–129.

Gomes Correia, A. (2001). Soil mechanics in routine and advanced pavement and rail track rational design. *Geotechnics for Road, Rail Tracks and Earth Structures*, A. Gomes Correia and H. Brandl (eds.), Balkema, pp. 165–187.

González de Vallejo, L. I. and Ferrer, M. (2011). Geological Engineering. CRC Press.

Greco, V. R. (1992). Discussion to "Active earth pressure on walls with base projection". *Journal of Geotechnical Engineering, ASCE*, 118(5): 825–827.

Greco, V. R. (1999). Active earth thrusts on cantilever walls in general conditions. *Soils and Foundations*, 39(6): 65–78.

Greco, V. R. (2001). Active earth thrust on cantilever walls with short heel. *Canadian Geotechnical Journal*, 38(2): 401–409.

Han, J. and Ye, S. L. (2000). Simplified method for consolidation rate of stone column reinforced foundations. *Journal of Geotechnical and Geoenvironmental Engineering, ASCE*, 127(7): 597–603.

Han, J. and Ye, S. L. (2002). A theoretical solution for consolidation rates for stone columns reinforced foundations accounting for smear and well resistance effects. *International Journal on Geomechanics, ASCE*, 2(2): 135–151.

Hanna, T. H. (1983). *Foundations in Tension*. McGraw-Hill.

Hannigan, P. J., Goble, G. G., Thendean, G., Likins, G. E. and Rausche, F. (2006). Design and construction of driven pile foundations, FHWA-NHI-05-042 and NHI-05-043, vols. I and II, Federal Highway Administration, U.S. Department of Transportation.

Hannigan, P. J., Rausche, F., Likins, G. E., Robinson, B. and Becker, M. (2016). *Design and Construction of Driven Pile Foundations*. Publication No. FHWA-NHI-16-009, vol. I. Federal Highway Administration, U.S. Department of Transportation.

Hansbo, S. (1981). Consolidation of fine-grained soils by prefabricated drains. *Proceedings of the 10th International Conference on Soil Mechanics and Foundation Engineering*, Stockholm, vol. 3, pp. 677–682.

Hansbo, S. (1994). *Foundation Engineering: Developments in Geotechnical Engineering*. Elsevier, Vol. 75.

Hansbo, S. (2004). Band drains. *Ground Improvement*, M. P. Moseley and K. Kirsch (eds.). Taylor & Francis, pp. 4–56.

Hatanaka, M. and Uchida, A. (1996). Empirical correlation between penetration resistance and effective friction of sandy soil. *Soils and Foundations*, 36(4): 1–9.

Hight, D. W. (2000). Sampling methods: Evaluation of disturbance and new practical techniques for high quality sampling in soils. *Proceedings of the 7th Portuguese Geotechnical Congress*, Porto, vol. 3, pp. 1275–1309.

Hilf, J. W. (1959). *A Rapid Method of Construction Control for Embankments of Cohesive Soil*. Bureau of Reclamation, Engineering Monographs No. 26.

Hilf, J. W. (1991). Compacted fill. *Foundation Engineering Handbook*, H.-Y. Fang (ed.). Kluwer Academic Publishers, Chapter 8, pp. 249–316.

Hird, C. C. and Moseley, V. J. (2000). Model study of seepage in smear zone around vertical drains in layered soil. *Géotechnique*, 50(1): 89–97.

Hjiaj, M., Lyamin, A. V. and Sloan, S. W. (2005). Numerical limit analysis solutions for the bearing capacity factor N_γ. *International Journal of Solids and Structures*, 42(5–6): 1681–1704.

Hoek, E. and Bray, J. W. (1981). *Rock Slope Engineering*, Revised 3rd edition. Published for the Institution of Mining and Metallurgy, by E & FN Spon.

Honjo, Y. and Kusakabe, O. (2002). Proposal of a comprehensive foundation design code: Geocode ver.2. *Proceedings of the International Workshop on Foundation Design Codes and Soil Investigation in View of International Harmonization and Performance Based Design, IWS Kamakura 2002*, Balkema, pp. 95–103.

Horvath, R. G., and T. C. Kenney (1979). Shaft resistance of rock socketed drilled piers. *Proceedings of the Symposium on Deep Foundations*, ASCE, Atlanta, GA, pp. 182–214.

Hutchinson, J. N. (1977). Assessment of effectiveness of corrective measures in relation to geological conditions on types of slope movements. *Bulletin of the IAEG*, 16: 131–155.

Hvorslev, M. J. (1948). *Subsurface Exploration and Sampling of Soils for Civil Engineering Purposes*. U.S. Army Corps of Engineers, Waterways Experiment Station.

Hvorslev, M. J. (1951). Time lag and soil permeability in ground water observations. U.S. Waterways Experiment Station, Bulletin No. 36. Vicksburg, MS.

Idriss, I. M. and Boulanger, R. W. (2006). Semi-empirical procedures for evaluating liquefation potencial during earthquakes. *Soil Dynamics and Earthquake Engineering*, 26(2–4): 115–130.

Indraratna, B., Geng, X. and Rujikiatkamjorn, C. (2010). Review of methods of analysis for the use of vacuum preloading and vertical drains for soft clay improvement. *International Journal on Geomechanics and Geoengineering*, 5(4): 223–236.

Indraratna, B., Rujikiatkamjorn, C. and Sathananthan, I. (2005). Radial consolidation of clay using compressibility indices and varying horizontal permeability. *Canadian Geotechnical Journal*, 42(5): 1330–1341.

Ishihara, K. (1985). Stability of natural deposits during earthquakes. *Proceedings of the 11th International Conference on Soil Mechanics and Foundation Engineering*, San Francisco, CA, Vol. 1, pp. 321–376.

Ishihara, K. (1993). Liquefaction and flow failure during earthquakes. *Géotechnique*, 43(3): 351–415.

Ishihara, K. (1996). *Soil Behaviour in Earthquake Geotechnics*. Oxford Engineering Science Series 46. Clarendon Press.

ISO 14688-1:2017. *Geotechnical Investigation and Testing - Identification and Classification of Soil - Part 1: Identification and Description*.

ISO 22282-1: 2012. *Geotechnical Investigation and Testing – Geohydraulic Testing – Part 1: General Rules*.

ISO 22282-2: 2012. *Geotechnical Investigation and Testing – Geohydraulic Testing – Part 2: Water Permeability Tests in a Borehole Using Open Systems*.

ISO 22282-4: 2012. *Geotechnical Investigation and Testing – Geohydraulic Testing – Part 4: Pumping Tests.*

ISO 22476-1: 2012. *Geotechnical Investigation and Testing - Field Testing - Part 1: Electrical Cone and Piezocone Penetration Test.*

ISO 22476-2: 2005. *Geotechnical Investigation and Testing – Field Testing – Part 2: Dynamic Probing.*

ISO 22476-3: 2005. *Geotechnical Investigation and Testing – Field Testing – Part 3: Standard Penetration Test.*

ISO 22476-4: 2012. *Geotechnical Investigation and Testing - Field Testing - Part 4: Ménard Pressuremeter Test.*

ISO 22476-6: 2018. *Geotechnical Investigation and Testing - Field Testing - Part 6: Self-Boring Pressuremeter Test.*

ISO 22476-9: 2014. *Geotechnical Investigation and Testing - Field Testing - Part 9: Field Vane Test.*

ISO 22476-11: 2017. *Geotechnical Investigation and Testing – Field Testing – Part 11: Flat Dilatometer Test.*

ISO 22477-5: 2018. *Geotechnical Investigation and Testing – Testing of Geotechnical Structures - Part 5: Testing of Ground Anchors.*

ISSMFE (1988). Standard penetration test (SPT): International Reference Test Procedure. ISSMFE Technical Committee in Penetration Testing, SPT Working Party. *Proceedings of the ISOPT-1*, Orlando, USA, vol. 1, pp. 3–26.

ISSMFE (1989). *Report of the ISSMFE Technical Committee on Penetration Testing of Soils – TC 16 with Reference Test Procedures, CPT-SPT-DP-WST.* Swedish Geotechnical Society, Information 7.

Jaky, J. (1944). The coefficient of earth pressure at rest. *Journal of Hungarian Architects and Engineers*, October: 355–358.

Jamiolkowski, M. (1999). The leaning Tower of Pisa. XIVth Manuel Rocha Lecture. *Geotecnia*, 85: 7–42.

Jamiolkowski, M., Ladd, C. C., Germaine, J. T. and Lancellotta, R. (1985). New developments in field and laboratory testing of soils. *Proceedings of the 11th International Conference on Soil Mechanics and Foundation Engineering*, San Francisco, CA, vol. 1, pp. 57–153.

Janbu, N. (1954). *Stability Analysis of Slopes with Dimensionless Parameters. Harvard Soil Mechanics Series.* Harvard University, vol. 46.

Janbu, N. (1996). Slope stability evaluation in engineering practice. *Proceedings of the 7th International Symposium on Landslides*, Trondheim, Norway, vol. 1, pp. 17–34.

Jefferies, M. and Been, K. (2016). *Soil Liquefaction: A Static Approach.* CRC Press, Taylor & Francis.

Jefferies, M. and Davies, M. P. (1993). Use of CPTU to estimate equivalent SPT N_{60}. *Geotechnical Testing Journal, ASTM*, 16(4): 458–468.

Jelinek, R. and Ostermayer, H. (1967a). On the design of anchored cofferdams and retaining walls. *Bautechnik* (in German), 5: 167–171.

Jelinek, R. and Ostermayer, H. (1967b). On the design of anchored cofferdams and retaining walls. *Bautechnik* (in German), 6: 203–207.

JGS (1998). *Remedial Measures against Liquefaction: From Investigation and Design to Implementation*, Japanese Geotechnical Society (ed.). Balkema.

Jiménez Salas, J. A. (1980). *Geotecnia y Cimientos III*. Editorial Rueda.

Jiménez Salas, J. A., Justo Alpañes, J. L. and Serrano Gonzalez, A. A. (1976). *Geotecnia y Cimientos II, Mecanica del Suelo y de las Rocas.* Editorial Rueda.

Keaveny, J. and Mitchell, J. K. (1986). Strength of fine-grained soils using the piezocone. *Use of In-Situ Tests in Geotechnical Engineering, GSP 6.* ASCE, S. P. Clemence (ed.), pp. 668–685.

Kézdi, A. (1975). Pile foundations. *Foundation Engineering Handbook*, H. F. Winterkorn and H.-Y. Fang (eds.). Van Nostrand Reinhold Company, Chapter 19, pp. 556–599.

King, J. W. (1995). Analysis of cantilever sheet-pile walls in cohesionless soil. *Journal of Geotechnical Engineering, ASCE*, 121(9): 629–635.

Kjellman, W. (1952). Consolidation of clay soil by means of atmospheric pressure. *Proceedings of the Conference on Soil Stabilization*, MIT, Cambridge, pp. 258–263.

Koppula, S. D. (1981). Statistical estimation of compression index. *Geotechnical Testing Journal*, 4(2): 68–73.

Krahn, J. (2003). The 2001 R. M. Hardy Lecture: The limits of the limit equilibrium analyses. *Canadian Geotechnical Journal*, 40(3): 643–660.

Kramer, S. L. (1996). *Geotechnical Earthquake Engineering*. Prentice Hall.

Kranz, E. (1953). *Uber die Verankerung von Spundwanden*. Verlag von Wilhelm Ernst & Sohn.

Kulhawy, F. H. and Duncan, J. M. (1972). Stresses and movements of Oroville dam. *Journal of the Soil Mechanics and Foundation Division, ASCE*, 98(7): 683–685.

Kulhawy, F. H., Trautmann, C. H., Beech, J. F., O'Rourke, T. D., McGuire, W., Wood, W. A. and Campano, C. (1983). Transmission line structure foundations for uplift-compression loading. Report EL-2870. Electric Power research Institute, Palo Alto, CA.

L'Herminier, R. (1967). *Mécanique des Sols et des Chaussées*. Société de Diffusion des Techniques du Bâtiment et des Travaux Publics.

Lacerda, W. A. (2004). The behaviour of colluvial slopes in a tropical environment. *Proceedings of the 9th International Symposium on Landslides*, Rio de Janeiro, vol. 2, pp. 1315–1342.

Ladd, C. C. (1991). Stability evaluation during staged construction. *Journal of Geotechnical Engineering, ASCE*, 117(4): 540–615.

Lambe, T. W. and Whitman, R. V. (1979). *Soil Mechanics, SI Version*. John Wiley & Sons.

Lancellotta, R. (2002). Analytical solution of passive earth pressure. *Géotechnique*, 52(8): 617–619.

Lancellotta, R. (2007). Lower-bound approach for seismic passive earth resistance. *Géotechnique*, 57(3): 319–321.

Lancellotta, R. (2008). *Geotechnical Engineering*, 2nd edition. Taylor & Francis.

Law, K. T. (1985). Use of field vane test under earth structures. *Proceedings of the 11th International Conference on Soil Mechanics and Foundation Engineering*, San Francisco, CA, vol. 2, pp. 893–898.

LCPC (2008). Portance des plates-formes Mesure du module en continu par le Portancemètre. *Guide Technique*, Laboratoire Central des Ponts et Chaussées.

Lee, W. F., Ishihara, K. and Chen, C. C. (2012). Liquefaction of silty sand. Preliminary studies from recent Taiwan, New Zeeland and Japan earthquakes. *Proceedings of the International Symposium on Engineering Lessons Learned from the 2011 Great East Japan Earthquake*, Tokyo, Japan.

Lefebvre, G., Pare, J. J. and Dascal, O. (1987). Undrained shear strength in the superficial weathered crust. *Canadian Geotechnical Journal*, 24(1): 23–34.

Lefranc, E. (1936). Procédé de mesure de la perméabilité des sols. *Génie Civil*: 306–307.

Leonards, G. A. (1968). *Les Fondations*. Dunod.

Leonards, G. A. (1987). Overview and personal commentary. *Engineering Geology*, 24(1–4): 577–612.

Leroueil, S. (2001). Natural slopes and cuts: Movement and failure mechanisms. *Géotechnique*, 51(3): 197–243.

Leroueil, S., Locat, J., Vaunat, J., Picarelli, L., Lee, H. and Faure, R. (1996). Geotechnical characterization of slope movements. *Proceedings of the 7th International Symposium on Landslides*, Trondheim, Kaare Senneset (ed.), Balkema, Vol. 1, pp. 53–74.

Leroueil, S., Magnan, J. P. and Tavenas, F. (1990). *Embankments on Soft Clay*. Ellis Horwood Limited.

Leroueil, S. Locat, J., Sève, G., Picarelli, L. and Faure, R. M. (2001). Slopes and mass movements. *Geotechnical and Geoenvironmental Engineering Handbook*, R. K. Rowe (ed.). Kluwer Academic Publishers, Chapter 14, pp. 397–428.

Leroueil, S. and Rowe, R. K. (2001). Embankments over soft soil and peat. *Geotechnical and Geoenvironmental Engineering Handbook*, R. K. Rowe (ed.). Kluwer Academic Publishers, Chapter 16, pp. 463–499.

Letourneur, J. and Michel, R. (1971). *La Géologie du Génie Civil*. Armand Colin.

Liao, S. S. and Whitman, R. V. (1986). Overburden correction factors for SPT in sand. *Journal of Geotechnical Engineering, ASCE*, 112(3): 373–377.

Littlejohn, G. S. (1972). Anchored diaphragm walls in sand. *Ground Engineering*, 5(1): 12–17.

Littlejohn, G. S. (1976). Contribution to discussion, session IV. *Proceedings of the Conference on A Review of Diaphragm Walls*. ICE, London, pp. 62–63.

Littlejohn, S. (1990). Ground anchorage practice. *Proceedings of the ASCE Conference on Design and Performance of Earth Retaining Structures*. Cornell University, Ithaca, NY, pp. 692–733.

Lo, M. B. (1967). Discussion to paper by B. O. Beredugo. *Canadian Geotechnical Journal*, 4(3): 353–356.

Loehr, J. E., Bowders, J. J., Ge, L., Likos, W., Luna, R., Maerz, N., Rosenblad, B. L. and Stephenson, R. W. (2011). *Engineering Policy Guidelines for Design of Drilled Shafts*. Missouri Department of Transportation (MoDOT).

Lundgren, H. and Mortensen, K. (1953). Determination by the theory of plasticity of the bearing capacity of continuous footings on sand. *Proceedings of the 3rd International Conference on Soil Mechanics and Foundation Engineering*, Zurich, vol. 1, pp. 409–412.

Lunne, T., Robertson, P. K. and Powell, J. J. M. (1997). *Cone Penetration Testing in Geotechnical Practice*. Blackie Academic, EF Spon/Routledge Publ.

Lupini, J. F., Skinner, A. E. and Vaughan, P. R. (1981). The drained residual strength of cohesive soils. *Géotechnique*, 31(2): 181–213.

Lyamin, A. V. and Sloan, S. W. (2002). Lower bound limit analysis using nonlinear programming. *International Journal for Numerical Methods in Engineering*, 55(5): 573–611.

Maccaferri (1987). Technical specifications.

Magnan, J. P., Mieussen, C. and Queyroi, D. (1983). Étude d'un remblai sur sols compressibles: Remblai B du site expérimental de cubzac-les-ponts. Raport de Recherche LPC, No. 127, Laboratoire Central des Ponts et Chaussées.

Mandel, J. and Salençon, J. (1969). Force portante dun sol sur une assisse rigide. *Proceedings of the 7th International Conference on Soil Mechanics and Foundation Engineering*, México, vol. 2, pp. 157–164.

Mandel, J. and Salençon, J. (1972). Force portante d'un sol sur une assisse rigide. Étude théorique. *Géotechnique*, 22(1): 79–93.

Maranha das Neves, E. and Veiga Pinto, A. (1989). Collapse of rockfill. *Procedddings of the 12th International Conference on Soil Mechanics and Foundation Engineering*, Rio de Janeiro, vol. 11, pp. 735–738.

Marchetti, S. (1975). A new in-situ test for the measurement of horizontal soil deformability. *Proceedings of the ASCE Conference on In-Situ Measurement of Soil Properties*, Raleigh, New York, vol. 2, pp.225–259.

Marchetti, S. (1980). In-situ tests by flat dilatometer. *Journal of Geotechnical Engineering, ASCE*, 106(3): 299–321.

Marchetti, S., Monaco, P., Totani, G. and Calabrese, M. (2001). The flat dilatometer test (DMT) in soil investigations. *ISSMGE TC16 Report Proceedings of the International Conference on In-Situ Measurements of Soil Properties*, Bali, Indonesia, 41 p.

Marques, J. C. and Magalhães, T. (2010). Elastic settlement of shallow foundations: A numerical study. *Proceedings of the 12nd Portuguese Geotechnical Congress*, Guimarães, CD-ROM, Abstract vol., pp. 145–146.

Martins, I. S., Santa Maria, P. E. and Lacerda, W. A. (1997) A brief review about the most significant results of COPPE research on rheological behaviour of saturated clays subjected to one-dimensional strain. *Recent Developments in Soil and Pavement Mechanics*, M. Almeida (ed.). Balkema, pp. 255–264.

Martins, I. S. M. (2014). The secondary 1D consolidation: A mechanical concept. *Lecture presented at the XVIIth Brazilian Conference on Soil Mechanics and Geotechnical Engineering*, Goiania (unpublished).

Massad, F. (2010). *Earthworks*. Oficina de Textos (in Portuguese).

Matos Fernandes, M. (2006). *Soil Mechanics, Fundamental Concepts and Principles*. FEUP Ed. (in Portuguese).

Matos Fernandes, M. (2011). *Introduction to Geotechnical Engineering*. FEUP Ed. (in Portuguese).

Matos Fernandes, M. and Falcão, J. C. B. (1988). The nonlinear behaviour of ground anchors and its consideration in finite elements analysis of tied-back walls. *Proceedings of the 6th International Conference Numerical Methods in Geomechanics*, Innsbruck. Balkema, vol. 2, pp. 1243–1248.

Matos Fernandes, M., Vieira, C. F. S. and Almeida e Sousa, J. (2002). Flexible cantilever retaining walls: Design according to Eurocode 7 and classical methods. *Proceedings of the Integrability Workshop on Foundation Design Codes and Soil Investigation in View of International Harmonization and Performance Based Design, IWS Kamakura 2002*. Balkema, pp. 117–125.

Matsuzawa, H., Ishibashi, I. and Kawamura, M. (1985). Dynamic soil and water pressures of submerged soils. *Journal of Geotechnical Engineering, ASCE*, 111(10): 1161–1176.

Mayne, P. and Kulhawy, F. (1982). K_0-OCR relationships in soil. *Journal of Geotechnical Engineering, ASCE*, 108(6): 851–872.

Mayne, P. W., Christopher, B. R. and Dejong, J. (2001). *Manual of Subsurface Investigation*. National Highway Institute, Publication No. FHWA NHI-01-031, Federal Highway Administration.

Mayne, P. W., Jones, J. S. and Dumas, J. C. (1984). Ground response to dynamic compaction. *Journal of Geotechnical Engineering, ASCE*, 110(56): 757–774.

Mayne, P. W., Schneider, J. A. and Martin, G. K. (1999). Small and large strain soil properties from seismic flat dilatometer tests. *Proceedings of the 2nd International Symposium on Pre-Failure Deformation Characteristics of Geomaterials*, Torino, Italy, vol. 1, pp. 419–427.

McCabe, B. A., Sheil, B. B., Long, M. M., Buggy, F. J. and Farrell, E. R. (2014). Empirical correlations for the compression index of Irish soft soils. *Geotechnical Engineering*, 167(6): 510–517.

MELT (1993). *Règles Techniques de Conception et de Calcul des Fondations des Ouvrages de Génie Civil*. Ministère de l'Equipement du Logement et des Transports, France, CCTG, Fascicule No. 62, Titre V.

Ménard, L. (1956). An apparatus for measuring the strength of soils in place. MSc thesis, University of Illinois.

Mesri, G. (2001). Primary compression and secondary compression. *Soil Behavior and Soft Ground Construction*. Geotechnical Special Publication No. 119, J. T. Germaine et al. (eds.), Cambridge, MA.

Meyerhof, G. G. (1948). An investigation of the bearing capacity of shallow footings on dry sand. *Proceedings of the 2nd International Conference on Soil Mechanics and Foundation Engineering*, Rotterdam, vol. 1, pp. 237–243.

Meyerhof, G. G. (1953). The bearing capacity of foundations under eccentric and inclined loads. *Proceedings of the 3rd International Conference on Soil Mechanics and Foundation Engineering*, Zurich, vol. 1, pp. 440–445.

Meyerhof, G. G. (1957). The ultimate bearing capacity of foundations on slopes. *Proceedings of the 4th International Conference on Soil Mechanics and Foundation Engineering*, London, vol. 1, pp. 384–386.

Meyerhof, G. G. (1976). Bearing capacity and settlements of pile foundation. *Journal of the Soil Mechanics and Foundations Division, ASCE*, 102(3): 195–228.

Meyerhof, G. G. (1993). Development of geotechnical limit state design. *Proceedings of the International Symposium on Limit State Design in Geotechnical Engineering*, Copenhagen, vol. 1, pp. 1–12.

Meyerhof, G. G. (1995). Development of geotechnical limit state design. *Canadian Geotechnical Journal*, 32(2): 128–136.

Milititsky, J., Consoli, N. C. and Schnaid, F. (2005). *Pathology of Foundations*. Oficina de Textos (in Portuguese).

Milović, D. (1992). *Stresses and Displacements for Shallow Foundations*. Elsevier Science Publishers.

Mindlin, R. D. (1936). Discussion on pressure distributions on retaining walls. *Proceedings of the 1st International Conference on Soil Mechanics and Foundation Engineering*, Cambridge, MA, vol. 3, pp. 155–157.

Mitchell, J. K. (1981). Soil improvement – State-of-the-art report. *Proceedings of the 10th International Conference on Soil Mechanics and Foundation Engineering*, Stockholm, vol. 4, pp. 509–565.

Mononobe, N. and Matsuo, H. (1929). On the determination of earth pressures during earthquakes. *Proceedings of the World Engineering Conference*, vol. 9, pp. 177–185.

Mooney, M. and Adam, D. (2007). Vibratory roller integrated measurement of earthwork compaction: An overview. *Proceedings of the 7th International Symposium on Field Measurements in Geomechanics (GSP 175)*, Boston, MA.

Morgenstern, N. R. and Price, V. E. (1965). The analysis of the stability of general slip surfaces. *Géotechnique*, 15(1): 79–93.

MTO (1991). *Ontario Highway Bridge Design Code*, 3rd edition. Ontario Ministry of Transportation.

Muller-Salzburg, L. (1987). The Vajont catastrophe – A personal review. *Engineering Geology*, 24(1–4): 423–444.

Munfakh, G., Arman, A., Collin, J. G., Hung, J. C.-J. and Brouillette, R. P. (2001). Shallow Foundations Reference Manual, FHWA-NHI-01-023. Federal Highway Administration, U.S. Department of Transportation.

Newmark, N. M. (1965). Effects of earthquakes on dams and embankments. *Géotechnique*, 15(2): 139–160.

Ng, C. W., Chiu, C. F. and Shen, C. K. (1998). Effects of wetting history on the volumetric deformations of unsaturated loose fill. *Proceedings of the 13th Southeast Asian Geotechnical Conference*, Taipei, Taiwan, vol. 1, pp. 141–146.

Ng, C. W. and Menzies, B. (2007). *Advanced Unsaturated Soil Mechanics and Engineering*. Taylor & Francis.

Nilsen, T. H. and Brabb, E. E. (1975). Landslides. Studies for seismic zonation of the San Fernando Bay Region. Geological Survey Professional Paper 941-A. U. S. Government Printing Office.

NP EN 1998-5 (2010). *Eurocode 8: Design of Structures for Earthquake Resistance – Part 5: General Rules, Seismic Actions and Rules for Buildings*. National Annex, Portuguese Institute for Quality.

NRC (2000). *Seeing into the Earth: Committee for Non-invasive Characterization of the Shallow Subsurface for Environmental and Engineering Applications*. National Academy Press.

Nunes, A. L., Sayão, A. S., Rios Filho, M. G. and Dias, P. H. (2013). Disasters and actions in the slopes of Rio de Janeiro. *Proceedings of the 6th COBRAE – Brazilian Conference on Natural Slopes*, ABMS, Angra dos Reis (in Portuguese), 30 p.

Okabe, S. (1926). General theory of earth pressure. *Journal of the Japonese Society of Civil Engineers*, 12(1).

Oliphant, J. (1993). Managing safety in geotechnical calculations. *Proceedings of the International Symposium on Limit State Design in Geotechnical Engineering*, Copenhagen. Published in Bulletin No. 10, Vol 1/3. Dansk Geoteknisk Forening, vol. 1, pp. 51–60.

O'Neill, M. W. and Reese, L. C. (1999). *Drilled Shafts: Construction Procedures and Design Methods*. Publication No. FHWA-IF-99-025. Federal Highway Administration.

O'Neill, M. W., Townsend, F. C., Hassan, K. H., Buller, A. and Chan, P. S. (1996). Load transfer for drilled shafts in intermediate geomaterials. Publication No. FHWA-RD-95-171. Federal Highway Administration, Mc Clean, VA.

Ortigão, J. A. R. and Sayão, A. S. F. J. (2004). *Handbook of Slope Stabilization*. Springer.

Osterberg, J. O. (1973). An improved hydraulic piston sampler. *Proceedings of the 8th International Conference on Soil Mechanics and Foundation Engineering*, Moscow, vol. 1, pp. 317–321.

Ostermayer, H. (1974). Construction, carrying behaviour and creep characteristics of ground anchors. *Proceedings of the Conference on Diaphragm Walls and Anchorages*. ICE, London, Session, V.: Paper 18.

Ostermayer, H. (1976). Practice in the detail design applications of anchorages. *Proceedings of the Conference on a Review of Diaphragm Walls*, ICE, London, pp. 55–61.

Ostermayer, H. and Scheele, F. (1977). Research on ground anchors in non-cohesive soils. *Proceedings of the 9th International Conference on Soil Mechanics and Foundation Engineering*, Tokyo, Specialty Session N°. 4.

Ou, C. Y. (2006). *Deep Excavations. Theory and Practice*. Taylor & Francis.

Ovesen, N. K. and Orr, T. (1991). Limit states design – The European perspective. *Proceedings of the Geotechnical Engineering Congress*, ASCE Special Publication, vol. 27(2), pp. 1341–1352.

Padfield, C. J. and Mair, R. J. (1984). The design of propped cantilever walls embedded in stiff clays. CIRIA Rep. 104. Construction Industry Research and Information Association.

Pastor, J., Thai, T. H. and Francescato, P. (2000). New bounds for the height limit of a vertical slope. *International Journal for Numerical and Analytical Methods in Geomechanics*, 24(1): 165–182.

Peck, R. B. (1969). Deep excavations and tunnelling in soft ground. *Proceedings of the 7th International Conference on Soil Mechanics and Foundation Engineering*, Mexico City, General Report, State-of-the-Art, vol., pp. 225–290.

Peck, R. B. (1972). Soil-structure interaction. *Proceedings of the Conference on Performance of Earth and Earth Retaining Structures*, ASCE, vol. 2, pp. 145–154 and vol. 3, pp. 249–250.

Penman, A. D. (1986). On the embankment dam. 26th Rankine Lecture. *Géotechnique*, 36(3): 301–348.

Perloff, W. H. (1975). Pressure distribution and settlement. *Foundation Engineering Handbook*, H. F. Winterkorn and H.-Y. Fang (eds.). Van Nostrand Reinhold, Chapter 4, pp. 148–196.

Pile Dynamics (2004). *Pile Driving Analyzer Manual*. Model PAK.

Pinto, C. S. (2002). *Basic Course on Soil Mechanics*. Oficina de Textos (in Portuguese).

Polshin, D. E. and Tokar, R. A. (1957). Maximum allowable non-uniform settlement of structures. *Proceedings of the 4th International Conference on Soil Mechanics and Foundation Engineering*, London, vol. 1, pp. 402–405.

Poulos, H. G. (1971). Behaviour of laterally loaded piles: II-pile groups. *Journal of the Soil Mechanics and Foundations Division*, ASCE, 97(5): 733–751.

Poulos, H. G. and Davis, E. H. (1974). *Elastic Solutions for Soil and Rock Mechanics*. John Wiley & Sons.

Poulos, H. G. and Davis, E. (1980). *Pile Foundation Analysis and Design*. John Wiley & Sons.

Powrie, W. (1996). Limit equilibrium analysis of embedded retaining walls. *Géotechnique*, 46(4): 709–723.

Prandtl, L. (1921). Uber die Eindringungsfestigkeit plastischer Baustoffe und die Festigkeit von Schneiden. *Zeitschrift für Angewandte Mathematik und Mechanik*, 1(1): 15–20.

Priebe, J. (1995). The design of vibro-replacement. *Ground Engineering*, 28(10): 31–37.

Proctor, R. R. (1933). The design and construction of rolled earth dams. *Engineering News Record* III(August 31, September 7): 21 and 28.

Puller, M. (1996). *Deep Excavations: A Practical Manual*. Thomas Telford.

Puzrin, A. M., Alonso, E. E. and Pinyol, N. M. (2010). *Geomechanics of Failures*. Springer.

Pyke, R., Seed, H. B. and Chan, C. K. (1975). Settlement of sands under multi-directional loading. *Journal of Geotechnical Engineering*, ASCE, 101(4): 379–398.

Quibel, A., Havard, H. and Bisson, D. (2001). Control of treated platforms with a new continuous method to assess the modulus. *Proceedings of the 1st International Symposium on Subgrade Stabilization and In Situ Pavement Recycling Using Cement*, Salamanca.

Randolph, M. F. (1981). The response of flexible piles to lateral loading. *Géotechnique*, 31(2): 247–259.

Randolph, M. F. and Murphy, B. S. (1985). Shaft capacity of driven piles in clay. *Proceedings of the Annual Offshore Technical Conference*, Houston, TX, OTC 4883.

Rankine, W. J. (1857). On the stability of loose earth. *Transactions Royal Society, London*, 147(1): 9–27.

Raposo, N., Topa Gomes, A. and Matos Fernandes, M. (2017). Anchored retaining walls in granite residual soils. I – Parametric study. *Soils and Rocks*, 40(3): 229–242.

Rausche, F., Goble, G. G. and Likins, G. E. (1985). Dynamic determination of pile capacity. *Journal of the Geotechnical Engineering Division*, ASCE, 111(3): 367–383.

Reese, L. C. (1984). *Handbook on Design of Piles and Drilled Shafts Under Lateral Load*. U.S. Department of Transportation, Federal Highway Administration, Office of Implementation.

Reese, L. C., Isenhower, W. and Wang, S. (2005). *Analysis and Design of Shallow and Deep Foundations*. John Wiley & Sons.

Reiding, F. J., Middendorp, P., Schoenmaker, R. P., Middendorp, F. M. and Bielefeld, M. W. (1988). FPDS-2. A new generation of foundation diagnostic equipment. *Proceeding of the 3rd International Conference on the Application of Stress Wave Theory to Piles*, B. H. Fellenius Ottawa (ed.). BiTech Publishers, pp. 123–134.

Reissner, H. (1924). Zum Erddruckproblem. *Proceedings of the 1st International Conference on Applied Mechanics*, Delft, pp. 295–311.

Résal, J. (1903). *Poussée des Terres, Première Partie: Stabilité des Murs de Soutènement*. Béranger.

Résal, J. (1910). *Poussée des Terres, Deuxième Partie: Théorie des Terres Cohérentes*. Béranger.

Richards Jr., R. and Elms, D. (1979). Seismic behaviour of gravity retaining walls. *Journal of the Geotechnical Engineering Division, ASCE*, 105(4): 449–464.

Robertson, P. K. (1990). Soil classification using the cone penetration test. *Canadian Geotechnical Journal*, 27(1): 151–158.

Robertson, P. K. (2009). Interpretation of cone penetration tests – A unified approach. *Canadian Geotechnical Journal*, 46(11): 1337–1355.

Robertson, P. K. (2013). The James K. Mitchell Lecture: Interpretation of in-situ tests–some insights. *Geotechnical and Geophysical Site Characterization 4*. R. Q. Coutinho and P. W. Mayne (eds.). Taylor & Francis, pp. 3–23.

Robertson, P. K. and Campanella, R. G. (1983). Interpretation of cone penetration tests. Part I: Sand. *Canadian Geotechnical Journal*, 20(4): 718–733.

Robertson, P. K. and Campanella, R. G. (1988). Guidelines for geotechnical design using CPT and CPTU data. Report FAWA-PA-87-014-84-24. Federal Highway Administration.

Robertson, P. K., Campanella, R. G., Gillespie, D. and Rice, A. (1986). Seismic CPT to measure in situ shear wave velocity. *Journal of Geotechnical Engineering, ASCE*, 112(8): 791–803.

Robertson, P. K., Davies, M. and Campanella, R. (1989). Design of laterally loaded driven piles using the flat dilatometer. *Geotechnical Testing Journal, ASTM*, 12(1): 30–38.

Robertson, P. K., Sully, J. P., Woeller, D. J., Lunne, T., Powell, J. J. and Gillespie, D. G. (1992). Estimating coefficient of consolidation from piezocone tests. *Canadian Geotechnical Journal*, 29(4): 539–550.

Robertson, P. K. and Wride, C. E. (1998). Evaluating cyclic liquefaction potential using the CPT. *Canadian Geotechnical Journal*, 35(3): 442–459.

Rocscience Inc. (2018). Slide version 8.0 – 2D limit equilibrium slope stability analysis, Rocscience Inc., Toronto, Ontario, Canada.

Rollins, K. M., Peterson, K. T. and Weaver, T. J. (1998). Lateral load behavior of full-scale pile group in clay. *Journal of Geotechnical and Geoenvironmental Engineering, ASCE*, 124(6): 468–478.

Rowe, P. W. (1952). Anchored sheet-pile walls. *Proceedings of the ICE*, London, 1(1): 705–710.

Rowe, P. W. and Barden, L. (1966). A new consolidation cell. *Géotechnique*, 16(2): 162–170.

Rowe, R. K. and Armitage, H. H. (1987). Theoretical solutions for the axial deformation of drilled shafts in rock. *Canadian Geotechnical Journal*, 24(1): 114–125 and 126–142.

Rowe, R. K. and Mylleville, B. L. (1993). The stability of embankments reinforced with steel. *Canadian Geotechnical Journal*, 30(5): 768–780.

Salençon, J. (1974). *Théorie de la Plasticité Pour les Applications à la Mécanique des Sols*. Eyrolles.

Salençon, J. (2001). *Handbook of Continuum Mechanics. General Concepts. Thermoelasticy*. Springer, École Polytechnique.

Salgado, R. (2008). *The Engineering of Foundations*. McGraw-Hill.

Sayão, A. S., Sandroni, S. S., Fontoura, S. A. and Ribeiro, R. C. (2012). Considerations on the probability of failure of mine slopes. *Soils and Rocks*, 35(1): 31–38.

Schiffman, R. L., Chen, A. and Jordan, J. C. (1967). The consolidation of a half plane. MATE Report 67-3. University of Illinois, Chicago Circle.

Schmertmann, J. M. (1955). The undisturbed consolidation of clay. *Transactions of the ASCE*, 120: 1201–1233.

Schmertmann, J. M. (1970). Static cone to compute settlement over sand. *Journal of the Soil Mechanics Engineering Division, ASCE*, 96(3): 1011–1043.

Schmertmann, J. M., Hartman, J. P. and Brown, P. R. (1978). Improved strain influence factor diagrams. *Journal of Geotechnical Engineering, ASCE*, 104(8): 1131–1135.

Schmertmann, J. M., Nottingham, L. C. and Renfro, R. H. (1970). Guidelines for use in the soils investigation and design of foundations for bridge structures in the state of Florida. Research Bulletin No. RB121. Florida Department of Transportation, USA.

Schneider, H. (1997). Definition and determination of characteristic soil properties. *Proceedings of the 14th International Conference on Soil Mechanics and Foundation Engineering*, Hamburg, vol. 4, pp. 2271–2274.

Scott, R. F. (1963). *Principles of Soil Mechanics*. Addison-Wesley Publishing Comp..

Seed, H. B. and Idriss, I. M. (1971). Simplified procedures for evaluation of soil liquefaction potential. *Journal of the Soil Mechanics and Foundations Division*. ASCE, 107(9): 1249–1274.

Seed, H. B., Idriss, I. M. and Arango, I. (1983). Evaluation of liquefaction potential using field performance data. *Journal of Geotechnical Engineering, ASCE*, 109(3): 458–482.

Seed, H. B., Idriss, I. M., Makdisi, F. and Banerjee, N. (1975). Representation of irregular stress time histories by equivalent uniform stress series in liquefaction analyses. EERC 75-29. Earthquake Engineering Research Center, University of California, Berkeley, CA.

Seed, H. B., Tokimatsu, L. F., Harder, M. and Riley, M. C. (1985). Influence of SPT procedures in soil liquefaction resistance evaluations. *Journal of Geotechnical Engineering, ASCE*, 111(12): 1425–1445.

Seed, H. B. and Whitman, R. V. (1970). Design of earth retaining structures for dynamic loads. *Proceedings of the ASCE Conference on Lateral Stresses in the Ground and Design of Earth Retaining Structures*, Cornell University, New York, pp. 103–147.

Semenza, E. (2001). *La Storia del Vaiont Raccontata del Geologo che ha Scoperto la Frana*. Tecomproject. Editore Multimediale.

Serajuddin, M. (1987). Universal compression index equation and Bangladesh soil. *Proceedings of the 9th Southeast Asian Geotechnical Conference*, Bangkok, pp. 5.61–5.72.

Sharma, P. V. (1997). *Environmental and Engineering Geophysics*. Cambridge University Press.

Sherif, M. A., Ishibashi, I. and Lee, C. D. (1982). Earth pressures against rigid retaining walls. *Journal of the Geotechnical Engineering, ASCE*, 108(5): 679–695.

Sherif, M. M. and Mackey, R. D. (1977). Pressures on retaining walls with repeated loading. *Journal of Geotechnical Engineering, ASCE*, 103(11): 1341–1345.

Silver, M. L. and Seed, H. B. (1971). Volume changes in sands during cyclic loading. *Journal of the Soil Mechanics and Foundations Division, ASCE*, 97(9): 1171–1182.

Simpson, B. and Driscoll, R. (1998). *Eurocode 7, a Commentary*. Building Research Establishment.

Skempton, A. W. (1951). The bearing capacity of clays. *Proceedings of the Building Research Congress*, London, Div. 1: 180–189.

Skempton, A. W. (1954). The pore-pressure coefficients A and B. *Géotechnique*, 4(4): 143–147.

Skempton, A. W. (1964). Long term stability of clay slopes. *Géotechnique*, 14(2): 77–101.

Skempton, A. W. (1970). First-time slides in over-consolidated clays. *Géotechnique*, 20(3): 320–324.

Skempton, A. W. (1985). Residual strength of clays in landslides, folded strata and the laboratory. *Géotechnique*, 35(1): 3–18.

Skempton, A. W. (1986). Standard penetration test procedures and the effects in sands of overburden pressure, relative density, particle size, ageing and overconsolidation. *Géotechnique*, 36(3): 425–447.

Skempton, A. W. and Bjerrum, L. (1957). A contribution to the settlement analysis of foundation on clay. *Géotechnique*, 7(4): 168–178.

Skempton, A. W. and Mc Donald, D. H. (1956). Allowable settlement of buildings. *Proceedings of the of the Institution of Civil Engineers*, 3(5): 727–768.

Skempton, A. W., Yassin, A. S. and Gibson, R. E. (1953). Théorie de la force portante des pieux. *Annales de l'Institut Technique du Batiment et des Travaux Publics*, 6(63–64): 285–290.

Sliwinski, Z. J. and Fleming, W. G. K. (1983). The integrity and performance of bored piles. *Piling and Ground Treatment for Foundations*, Thomas Telford, pp. 153–165.

Smith, E. A. L. (1960). Pile driving analysis by the wave equation. *Journal of the Soil Mechanics and Foundations Division*. ASCE, 86(4): 35–61.

Sowers, G. F. (1968). Fondations superficielles. *Les Foundations*, G. A. Leonards (ed.). Dunod, Chapter 6, pp. 529–641.

Spangler, M. G. (1936). The distribution of normal pressures on a retaining wall due to a concentrated surface load. *Proceedings of the 1st International Conference on Soil Mechanics and Foundation Engineering*, Cambridge, MA, vol. 1, pp. 200–207.

Spencer, E. (1967). A method of analysis of the stability of embankments assuming parallel inter-slice forces. *Géotechnique*, 17(1): 11–26.

Steedman, R. S. and Zeng, X. (1990). The influence of phase on the calculation of pseudo-static earth pressure on a retaining wall. *Géotechnique*, 40(1): 103–112.

Stokoe II, K. H. and Santamarina, J. C. (2000). Seismic-wave based testing in geotechnical engineering. Plenary Paper, *International Conference on Geotechnical and Geological Engineering, GeoEng 2000*, Melbourne, Australia, pp. 1490–1536.

Stokoe II, K. H., Sung-Ho, J. and Woods, R. D. (2004). Some contributions of in situ geophysical measurements to solving geotechnical engineering problems. *Proceedings of the ISC-2 on Geotechnical and Geophysical Site Characterization*, A. Viana da Fonseca and Paul Mayne (eds.), Millpress, vol. 1, pp. 97–132.

Tatsuoka, F. and Shibuya, S. (1992). Deformation characteristics of soils and rocks from field and laboratory tests. *Report of the Institute of Industrial Science, University of Tokyo*, 37(1): 1–136.

Tavenas, F. and Leroueil, S. (1980). The behavior of embankments on clay foundations. *Canadian Geotechnical Journal*, 17(2): 236–260.

Tavenas, F., Mieussens, C. and Bourges, F. (1979). Lateral displacements in clay foundations under embankments. *Canadian Geotechnical Journal*, 16(3): 532–550.

Taylor, D. W. (1937). Stability of earth slopes. *Journal of Boston Society of Civil Engineers*, 24(3): 197.

Taylor, D. W. (1948). *Fundamentals of Soils Mechanics*. John Wiley & Sons.

Taylor, P. W. (1967). Design of spread footings for earthquake loadings. *Proceedings of the 5th Australia-New Zealand Conference on Soil Mechanics and Foundation Engineering*, pp. 221–229.

Teh, C. I. and Houlsby, G. T. (1991). An analytical study of the cone penetration test in clay. *Géotechnique*, 41(1): 17–34.

Terzaghi, K. (1920). Old earth-pressure theories and new test results. *Engineering News-Record*, 85(14): 632–637.

Terzaghi, K. (1923a). Die berechnung der durchlassigkeitsziffer des tones aus dem verlauf der hydrodynamischen spannungserscheinungen. *Akademie der Wissenchaften in Wein*, 132: 125–138.

Terzaghi, K. (1923b). Die beziehungen zwischen elastizitat und innendruck. *Akademie der Wisserchaften in Wein*, 132: 105–124.

Terzaghi, K. (1925). *Erdbaumechanik*. Franz Deuticke.

Terzaghi, K. (1934). Large retaining wall tests. I. Pressure of dry sand. *Engineering News-Record*, 112(5): 136–140.

Terzaghi, K. (1943). *Theoretical Soil Mechanics*. John Wiley & Sons.

Terzaghi, K. (1953). Anchored bulkheads. *Proceedings of the ASCE, Soil Mechanics and Foundations Division*, vol. 79, Separate n° 262.

Terzaghi, K. and Peck, R. B. (1948). *Soil Mechanics in Engineering Practice*. John Wiley & Sons.

Terzaghi, K. and Peck, R. B. (1967). *Soil Mechanics in Engineering Practice*. 2nd edition, John Wiley & Sons.

Thurner, H. and Landstrom, A. (2000). Continuous compaction control, C.C.C. *Proceedings of the European Workshop on Compaction of Soils and Granular Materials*, Presses Ponts et Chaussées, pp. 237–246.

Timoshenko, S. and Goodier, J. N. (1951). *Theory of Elasticity*. McGraw-Hill.

Tokimatsu, K. and Seed, H. B. (1987). Evaluation of settlements in sands due to earthquake shaking. *Journal of Geotechnical Engineering, ASCE*, 113(8): 861–878.

Tomlinson, M. J. (1987). *Pile Design and Construction Practice*. Viewpoint Publication.

Tomlinson, M. and Woodward, J. (2007). *Pile Design and Construction Practice*. Taylor & Francis.

Topa Gomes, A. M. (2008). Elliptical wells by the sequential excavation method: The case of metro of Porto. PhD thesis, University of Porto (in Portuguese).

Torrey, V. H. and Donaghe, R. T. (1994). Compaction control of earth-rock mixtures: A new approach. *Geotechnical Testing Journal*, 17(3): 371–386.

Torstensson, B. A. (1977). The pore pressure probe. *Nordiske Mote, bergemekanikk*, Oslo, Paper 34, pp. 48–54.

Tschebotarioff, G. P. (1951). *Soil Mechanics, Foundations and Earth Structures*. McGraw-Hill.

Tschebotarioff, G. P. (1973). *Foundations, Retaining and Earth Structures*. McGraw-Hill.

US Army Corps of Engineers (1993). *Technical Engineering and Design Guides as Adapted from the US Army Corps of Engineers, No. 7. Bearing Capacity of Soils*. ASCE Press.

USBR (1998). *Earth Manual*. Bureau of Reclamation, US Department of the Interior, vol. 1, 3rd edition.

Vaughan, P. and Walbancke, J. (1973). Pore pressure changes and the delayed failure of cutting slopes in overconsolidated clay. *Géotechnique*, 23(4): 531–539.

Veiga Pinto, A. (1983). Prediction of the structural behaviour of rockfill dams. Specialist thesis, LNEC (in Portuguese).

Venda Oliveira, P. J., Lemos, L. L. and Coelho, P. A. (2010). Behaviour of an atypical embankment on soft soil: Field observations and numerical simulation. *Journal of Geotechnical and Geoenvironmental Engineering, ASCE*, 136(1): 35–47.

Vesić, A. S. (1961). Bending of beams resting on isotropic elastic solid. *Journal of the Mechanical Engineering Division, ASCE*, 87(2): 35–53.

Vesić, A. (1963). *Bearing Capacity of Deep Foundations in Sand*. National Academy of Sciences, National Research Council, Highway Research Record, vol. 39, pp. 112–153.

Vesić, A. S. (1969). Experiments with instrumented pile groups in sand. *Performance of Deep Foundations*, R. Lundgren and E. D'Appolonia (eds.), pp. 177–222.

Vesić, A. (1975). Bearing capacity of shallow foundations. *Foundation Engineering Handbook*, H. F. Winterkorn and H.-Y. Fang (eds.). Van Nostrand Reinhold, Chapter 3, pp. 121–147.

Vesić, A. S. (1977). *Design of Pile Foundations. NCHRP Synthesis of Practice No. 42*. Transportation Research Board.

Viana da Fonseca, A., Ferreira, C., Ramos, C. and Molina-Gómez, F. (2019). The geotechnical test site in the greater Lisbon area for liquefaction characterisation and sample quality control of cohesionless soils. *AIMS Geosciences*, 5(2): 325–343.

Viana da Fonseca, A., Matos Fernandes, M. and Cardoso, A. S. (1997). Interpretation of a footing load test on a saprolitic soil from granite. *Géotechnique*, 47(3): 633–651.

Viana da Fonseca, A., Molina-Gómez, F., Ferreira, C. and Ramos, C. (2020). The Portuguese experience on the collection of high-quality samples for liquefaction susceptibility assessment in the laboratory. Final report of the LIQ2PROEARTH project, FEUP, University of Porto.

Vijayvergiya, V. N. and Focht, J. A. (1972). A new way to predict the capacity of piles in clay. *Proceedings of the 4th Annual Offshore Technology Conference*, vol. 2, pp. 865–874.

Vucetić, M. and Dobry, R. (1991). Effect of soil plasticity on cyclic response. *Journal of Geotechnical Engineering, ASCE*, 117(1): 89–107.

Walsh, J. M. (2005). *Full Scale Lateral Load Test of a 3x5 Pile Group in Sand*. Department of Civil and Environmental Engineering, Brigham Young University.

Westergaard, H. M. (1933). Water pressures on dams during earthquakes. *Transactions of the American Society of Civil Engineers*, 98(2): 418–433.

Whitman, R. V. (1984). Evaluating calculated risk in Geotechnical Engineering. *Journal of Geotechnical Engineering, ASCE*, 110(2): 145–188.

Whitman, R. V. and Bailey, W. A. (1967). Use of computers for slope stability analysis. *Journal of the Soil Mechanics and Foundation Division, ASCE*, 93(4): 475–498.

Windle, D. and Wroth, C. P. (1977). The use of a self-boring pressuremeter to determine the undrained properties of clays. *Ground Engineering*, 10(6): 37–46.

Winkler, E. (1867). *Die Lehre von Elasticität und Festigkeit*. H. Dominicus.

Winter, M. G., Hólmgeirsdóttir, Th. and Suhardi (2016). The effect of large particles on acceptability determination for earthworks compaction. *Quarterly Journal of Engineering Geology*, 31(3): 247–268.

Wood, J. H. (1973). Earthquake-induced soil pressures on structures. PhD thesis, California Institute of Technology, Pasadena, CA.

Wroth, C. P. and Hughes, J. (1973). An instrument for the in situ measurement of the properties of soft clay. *Proceedings of the 9th International Conference on Soil Mechanics and Foundation Engineering*, Tokyo, vol. 1, pp. 487–494.

Yoon, G. L., Kim, B. T. and Jeon, S. S. (2004). Empirical correlations of compression index for marine clay from regression analysis. *Canadian Geotechnical Journal*, 41(6): 1213–1221.

Yoshimi, Y., Tokimatsu, K. and Ohara, J. (1994). In situ liquefaction resistance of clean sands over a wide density range. *Géotechnique*, 44(3): 479–494.

Youd, T. L. and Idriss, I. M. (2001). Liquefaction resistance of soils: Summary report from the 1996 NCEER and 1998 NCEER/NSF Workshops on evaluation of liquefaction resistance of soils. *Journal of Geotechnical and Geoenvironmental Engineering, ASCE*, 127(4): 297–313.

Yu, H. S. (2004). In situ soil testing: From mechanics to interpretation. *Proceedings of the ISC-2 on Geotechnical and Geophysical Site Characterization*, A. Viana da Fonseca and Paul Mayne (eds.), Millpress, vol. 1, pp. 3–38.

Ziegler, E. J. (1948). Effect of material retained on the No. 4 sieve on the compaction test of soils. *Proceedings of the Highways Research Board*, 28: 409–414.

Final annex

Resolution of the exercises

CHAPTER 2

ER2.1

Table ER2.1a contains the calculation of the safety factor by the Fellenius method, $F = 1.10$. This value was used as a guide for the first iteration for solving the problem by the simplified method of Bishop. This method provided $F = 1.22$ at the third iteration, as shown in Table ER2.1b.

Table ER2.1a Calculation of the safety factor of the cut described in E2.1 by the Fellenius method

Slice	α_i (°)	Δx_i (m)	Δl_i (m)	W_i (kN/m)	u_i (kPa)	$c'_i \Delta l_i + (W_i \cos \alpha_i - u_i \Delta l_i) \tan\phi'_i$ (kN/m)	$W_i \sin\alpha_i$ (kN/m)
1	47.5	7.5	11.1	670	30.4	288.5	494.0
2	36.2	7.5	9.3	1760	84.7	551.2	1039.5
3	26.6	7.5	8.4	2010	112.2	661.5	900.0
4	17.5	7.5	7.9	1910	115.8	681.5	574.3
5	8.7	7.5	7.6	1600	98.1	634.7	242.0
6	0.1	7.5	7.5	1090	77.3	444.6	1.9
7	-8.6	7.5	7.6	410	32.4	243.9	-61.3
						$\Sigma = 3505.9$	$\Sigma = 3190.4$
							$F = 1.099 \approx 1.10$

Table ER2.1b Calculation of the safety factor of the cut described in E2.1 by the simplified Bishop method

	$F = 1.10$		$F = 1.199$		$F = 1.215$	
Slice	$1/M_i(\alpha_i)$	$[c'_i \Delta x_i + (W_i - u_i \Delta x_i) \tan\phi'] [1/M_i(\alpha_i)]$ (kN/m)	$1/M_i(\alpha_i)$	$[c'_i \Delta x_i + (W_i - u_i \Delta x_i) \tan\phi'] [1/M_i(\alpha_i)]$ (kN/m)	$1/M_i(\alpha_i)$	$[c'_i \Delta x_i + (W_i - u_i \Delta x_i) \tan\phi'] [1/M_i(\alpha_i)]$ (kN/m)
1	0.941	381.3	0.970	393.0	0.975	395.1
2	0.895	715.4	0.916	732.2	0.919	734.6
3	0.886	730.7	0.901	743.0	0.903	744.7
4	0.900	676.1	0.910	683.7	0.912	685.2
5	0.936	607.5	0.942	611.3	0.943	612.0
6	0.999	444.1	0.999	444.1	0.999	444.1
7	1.099	270.8	1.091	268.8	1.090	268.6
		$\Sigma = 3825.9$		$\Sigma = 3876.1$		$\Sigma = 3884.3$
		$F = 1.199$		$F = 1.215$		$F = 1.217 \approx 1.22$

ER2.2

a) The stability number is given by Equation 2.49. So, in this case:

$$N_s = \frac{20 \times 9.5}{50} = 3.8$$

This value is very close to the estimations for the critical stability number (see Equation 2.50 and 2.47, as well). Therefore, the safety factor against a sliding failure passing at the base of the cut should be very close to 1.0.

b) Figure ER2.2 shows the output of the application of code SLIDE (Rocscience Inc., 2018). It can be seen that the sliding surface with the minimum safety factor passes at the base of the cut and that the safety factor is practically equal to 1.0.

ER2.3

a) Adopting a safety factor for the undrained shear strength $\gamma_{cu} = 1.40$, one obtains:

$c_{u,d} = 55/1.40 = 39$ kPa

For this design value of the undrained shear strength, the failure sliding of the cut should be imminent. So, this value may be introduced in the slope stability chart of Janbu (Figure 2.19), to obtain the maximum inclination of the slope.

In this case,

$$N_{s,cr} = 20 \times 12/39 = 6.15$$

and

$$d = D/h = 0.5$$

In the chart, the point of the curve representative of $d = 0.5$ and with an ordinate equal to 6.15, corresponds approximately to an abscissa $b = \cotan \beta = 2$. Then, the inclination angle of the slope should be 1(V):2(H), which corresponds to $\beta = 26.6°$.

b) Figure ER2.3 shows the output of the application of code SLIDE (Rocscience Inc., 2018), assuming the factored value of c_u (value of the undrained shear resistance) and the inclination obtained from the chart. It can be seen that the sliding surface with the minimum factor is practically equal to 1.0 and the respective sliding surface passes under the slope base, as predicted in the chart.

ER2.4

Figures ER2.4a to ER2.4c show the output of the application of code SLIDE (Rocscience Inc., 2018) for three situations:

a) the embankment, as defined in Figure E2.4 and Table E2.4; this corresponds to a safety factor $F = 1.23$;

b) the embankment together with lateral stabilizing berms (width = 14.0 m and height = 2.1 m); this corresponds to a safety factor $F = 1.52$;

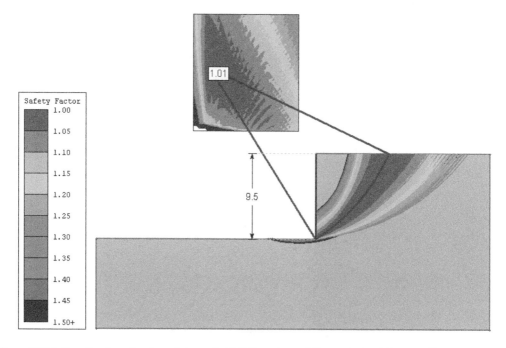

Figure ER2.2 Results of application of the code SLIDE to the stability analysis of the cut of Figure E2.2.

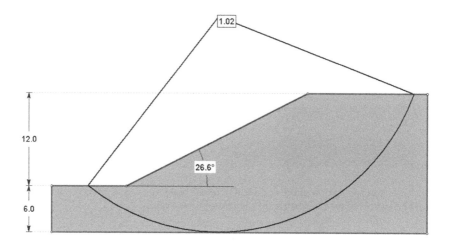

Figure ER2.3 Results of application of the code SLIDE to the stability analysis of the cut of Figure E2.3, designed applying the slope stability chart of Janbu.

c) the embankment with its lateral sides with an inclination of 1(V):7(H); this corresponds to a safety factor $F = 1.48$.

It can be observed that the solution with berms is more economical in terms of fill volume and space occupied than the solution with reduced inclination of the embankment sides.

d) The following, are the steps for designing of the base reinforcement of the embankment applying the method of Leroueil and Rowe (see Annex A2.3).

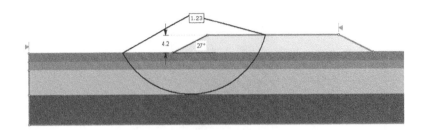

Figure ER2.4a Results of application of the code SLIDE to the stability analysis of the embankment of Figure E2.4.

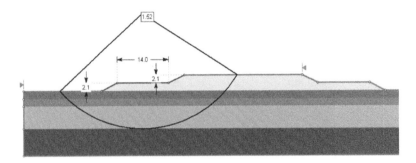

Figure ER2.4b Results of application of the code SLIDE to the stability analysis of the embankment of Figure E2.4, with lateral stabilizing berms.

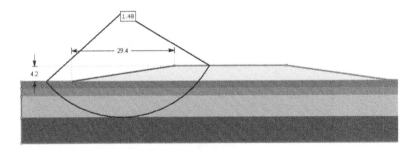

Figure ER2.4c Results of application of the code SLIDE to the stability analysis of the embankment of Figure E2.4, with reduced inclination of the embankment sides.

1. *Establish the evolution with depth of the design values of c_u by application of a safety factor considered to be adequate, $c_{u,d} = c_u/F_d$.*
   The value 1.4 was taken for F_d, giving the results shown in Table ER2.4.
2. *Take for the fill material the maximum expected unit volume and the critical value for the angle of shearing resistance.*
   It was assumed that $\gamma = 22$ kN/m³ and $\phi' = 35°$.
3. *Consider a potential sliding surface of circular shape, so that the vertical line through its intersection with the base reinforcement crosses the upper horizontal boundary of the embankment; it should be noted that the sliding surface does not cross the embankment.*
   Figure ER2.4d shows the result found for the critical surface with code SLIDE, considering the effect of the embankment to be a surcharge on the surface. With

Table ER2.4 Undrained shear strength assumed in the stability analysis applying the method of Leroueil and Rowe.

Material	z, depth (m)	γ (kN/m³)	c_u (kPa)	$c_{u,d}$ (kPa)
Clay - layer 1	z = 0.0 to −2.0	17	24	17.1
Clay - layer 2	z = −2.0 to −4.0	17	22	15.7
Clay - layer 3	z = −4.0 to −10.0	17	20	14.3
Clay - layer 4	z = −10.0 to −17.0	17	28	20.0

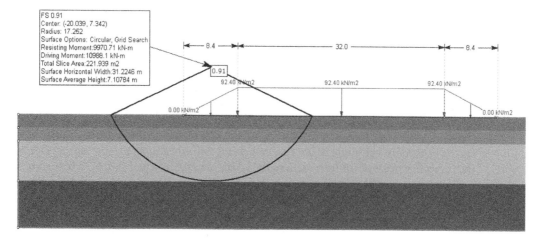

Figure ER2.4d Results of application of the code SLIDE to the stability analysis considering the effect of the embankment to be a surcharge on the surface.

this option the shear strength of the material of the embankment does not influence the result of the stability analysis, according to the proposal of the cited authors.

4. *Calculate the stabilizing moment supplied by the shear strength of the foundation soil.*

    The stabilizing or *resisting moment*, as given by SLIDE (see Figure ER2.4d) is equal to: $M_R = 9970$ kN.m/m.

5. The contribution of the embankment to the destabilizing moment is due to the combined effect of two forces: i) the weight, W, of the fill mass to the left of the vertical line mentioned in step 3; and ii) the horizontal force P_a, applied at one third of the embankment height, measured from its base, and whose value is the Rankine active thrust.

    The value of the first moment was calculated by the code SLIDE (see *driving moment* in Figure ER2.4d) and is equal to 10988 kN.m/m.

    The Rankine active thrust is equal to:

$$P_a = \frac{1}{2} \cdot K_a \cdot \gamma \cdot h^2 = \frac{1}{2} \times 0.271 \times 22 \times 4.2^2 = 52.6 \text{ kN/m}$$

which is applied at a vertical distance from the center of the sliding surface equal to $y_P = 5.94$ m.

6. *Calculate, for the adopted sliding surface, the force T_g in the geosynthetic necessary for moment equilibrium.*

    According to equation A2.3.3, and considering the values calculated above:

$$T_g = \frac{10988.1 + 52.6 \times 5.94 - 9970.7}{7.34} = 181.18 \text{ kN/m}$$

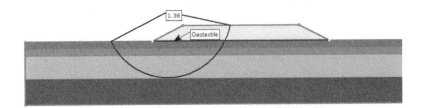

Figure ER2.4e Results of application of the code SLIDE to the stability analysis of the embankment of Figure E2.4, with the base reinforced with a geosynthetic, designed according the method of Leroueil and Rowe.

7. A stability analysis was performed with the code SLIDE introducing the unfactored undrained shear resistance of the foundation soil, considering the shear strength of the embankment and introducing as base reinforcement a geosynthetic whose design value of the tensile resistance is $T_{g,f}=200$ kN/m.

    Figure ER2.4e shows the result obtained, assuming that T_g is horizontal (parallel to the plane of the reinforcement), which is on the side of safety. If it is assumed that T_g is on the bisector of the angle of the reinforcement, with the tangent to the slip surface at the respective point of the intersection, the factor of safety is equal to 1.41. These results are in agreement with the safety factor introduced at step 1 for the strength of the foundation soil. The pull-out resistance of the geotextile, $T_{g,p}$, is comfortably verified.

CHAPTER 4

ER4.1

a) The clay layer can be divided into five sublayers of equal thickness and the at-rest vertical stress is then calculated at the mid-point of each sublayer, as shown in Table ER4.1a.

The permanent lowering of the water table will generate an effective stress increase at the end of consolidation of:

$$\Delta\sigma_v' = 10 \times 9.8 = 98 \text{ kPa}$$

The consolidation settlement of each sublayer is calculated by applying Equation 4.11:

$$s_c = \frac{h_0}{1+e_0} C_c \log\frac{\sigma_{v0}' + \Delta\sigma_v'}{\sigma_{v0}'}$$

The calculations are summarized in Table ER4.1b.

Table ER4.1a Sublayers considered for the calculation of the consolidation settlement

Sublayer	Point	z (m)	σ_{v0} (kPa)	u_0 (kPa)	σ_{v0}' (kPa)
1	A	−11.0	217	107.8	109.2
2	B	−13.0	251	127.4	123.6
3	C	−15.0	285	147.0	138.0
4	D	−17.0	319	166.6	152.4
5	E	−19.0	353	186.2	166.8

Table ER4.1b Calculation of the consolidation settlement due to the permanent water table lowering

Sublayer	h_0 (m)	σ'_{v0} (kPa)	$\Delta\sigma'_v$ (kPa)	s_c (m)
1	2.0	109.2	98	0.12
2	2.0	123.6	98	0.11
3	2.0	138.0	98	0.10
4	2.0	152.4	98	0.10
5	2.0	166.8	98	0.09
			Σ=	0.52

Table ER4.1c Sublayers considered for the calculation of the upward displacement due to the excavation

Sublayer	Point	z (m)	σ_{vi} (kPa)	u_i (kPa)	σ'_{vi} (kPa)
1	A	−11.0	217	9.8	207.2
2	B	−13.0	251	29.4	221.6
3	C	−15.0	285	49.0	236.0
4	D	−17.0	319	68.6	250.4
5	E	−19.0	353	88.2	264.8

The consolidation settlement due to permanent lowering of the water table is $s_c = 0.52$ m.

b) The initial situation is now the end of consolidation associated with lowering of the water table, as shown in Table ER4.1c.

Due to the general excavation to elevation −4.0, the effective stress variation at the end of consolidation is:

$$\Delta\sigma'_v = -4 \times 20 = -80 \text{ kPa}$$

As the final effective stresses are lower than the initial ones, the void ratio variation will follow the unloading branch of the curve. The consolidation settlement is negative (upward displacement) due to soil swelling, and can be calculated by Equation 4.12 (note that the ratio $h/(1 + e)$ is constant, see Section 4.2.3):

$$s_c = \frac{h_0}{1 + e_0} C_r \log \frac{\sigma'_{vi} + \Delta\sigma'_v}{\sigma'_{vi}}$$

The calculations are summarized in Table ER4.1d.

The ground surface displacement due to excavation is $s_c = -0.05$ m.

c) The construction of the building corresponds to a vertical stress increment of 150 kPa. The initial stress state is now the final stress state from the previous situation, as shown in Table ER4.1e.

Due to building construction, the clay layer will be reloaded from the effective stresses calculated in b) to those calculated in a) and loaded in the virgin branch beyond these stresses. The consolidation settlement is calculated by applying Equation 4.10:

$$s_c = \frac{h_0}{1 + e_0} \times \left(C_r \log \frac{\sigma'_p}{\sigma'_{vi}} + C_c \log \frac{\sigma'_{vi} + \Delta\sigma'_v}{\sigma'_p} \right)$$

Table ER4.1d Calculation of the upward displacement due to the excavation

Sublayer	h_0 (m)	σ'_{vi} (kPa)	$\Delta\sigma'_v$ (kPa)	s_c (m)
1	2.0	207.2	−80	−0.01
2	2.0	221.6	−80	−0.01
3	2.0	236.0	−80	−0.01
4	2.0	250.4	−80	−0.01
5	2.0	264.8	−80	−0.01
			$\Sigma=$	−0.05

Table ER4.1e Sublayers considered for the calculation of the consolidation settlement due to the building construction

Sublayer	Point	z (m)	σ_{vi} (kPa)	u_i (kPa)	σ'_{vi} (kPa)
1	A	−11.0	137	9.8	127.2
2	B	−13.0	171	29.4	141.6
3	C	−15.0	205	49.0	156.0
4	D	−17.0	239	68.6	170.4
5	E	−19.0	273	88.2	184.8

Table ER4.1f Calculation of the consolidation settlement due to the building construction

Sublayer	h_0 (m)	σ'_{vi} (kPa)	σ'_p (kPa)	$\Delta\sigma'_v$ (kPa)	s_c (m)
1	2.0	127.2	207.2	150	0.06
2	2.0	141.6	221.6	150	0.06
3	2.0	156.0	236.0	150	0.06
4	2.0	170.4	250.4	150	0.05
5	2.0	184.8	264.8	150	0.05
				$\Sigma=$	0.28

The calculations are summarized in Table ER4.1f.

The consolidation settlement due to building construction is $s_c = 0.28$ m.

d)

0) At-rest $\sigma'_{v0} = 138$ kPa $\log\sigma'_{v0} = 2.140$

$e_0 = 1.3$

a) Lowering of the water table $\sigma'_v = 236$ kPa $\log\sigma'_v = 2.373$

$$\Delta e = -C_c \log\frac{\sigma'_{v0} + \Delta\sigma'_v}{\sigma'_{v0}}$$ $e = e_0 + \Delta e$

$\Delta e = -0.117$ $e = 1.3 - 0.117$
$e = 1.183$

b) General excavation until elevation −4.0 $\sigma'_v = 156$ kPa $\log\sigma'_v = 2.193$

$$\Delta e = -C_r \log\frac{\sigma'_{vi} + \Delta\sigma'_v}{\sigma'_{vi}}$$ $e = e_i + \Delta e$

$\Delta e = 0.009$ $e = 1.183 + 0.009$
$e = 1.192$

c) Construction of the building $\sigma'_v = 306$ kPa $\log\sigma'_v = 2.486$

$$\Delta e = -C_r \log\frac{\sigma'_p}{\sigma'_{vi}} - C_c \log\frac{\sigma'_{vi} + \Delta\sigma'_v}{\sigma'_p}$$ $e = e_i + \Delta e$

$\Delta e = -0.0654$ $e = 1.192 - 0.0654$
$e = 1.127$

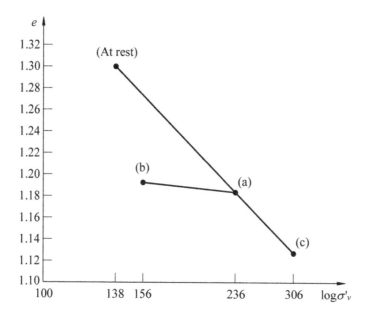

Figure ER4.1 Graph with the at-rest situation and the changes induced by the analyzed construction operations.

Figure ER4.1 presents in a (log σ'_v; e) graph the situation corresponding to the state of stress at-rest and its evolution induced by the various construction operations.

ER4.2

a) Taking the pressure head values given in Table E4.2, the excess pore pressure at the piezometers P_1, P_2 and P_3 can be determined for the situations from t_0 to t_4.

   Tables ER4.2a to ER4.2e summarize the calculations. Note that the filling of the reservoir (t_3) will cause the water table to rise 2.0 m (from elevation 120 to elevation 122), which will change the equilibrium pore pressure, that will become $u_{eq} = u_0 + 2 \times 10$ (kPa).

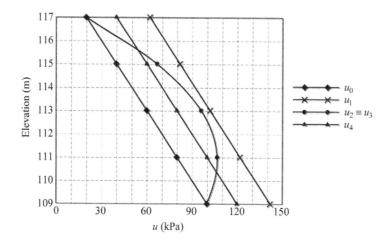

Figure ER4.2a Pore pressure distribution in the clay layer for times t_0 to t_4.

Table ER4.2a Pore pressure at-rest

z (m)	h_w (m)	u_0 (kPa)
115	4	$4 \times 10 = 40$
113	6	$6 \times 10 = 60$
111	8	$8 \times 10 = 80$

Table ER4.2b Pore pressure and excess pore pressure for t_1

z (m)	h_w (m)	u (kPa)	$u_e = u - u_0$ (kPa)
115	8.20	$8.20 \times 10 = 82$	42
113	10.20	$10.20 \times 10 = 102$	42
111	12.20	$12.20 \times 10 = 122$	42

Table ER4.2c Pore pressure and excess pore pressure for t_2

z (m)	h_w (m)	u (kPa)	$u_e = u - u_0$ (kPa)
115	6.68	$6.68 \times 10 = 66.8$	26.8
113	9.61	$9.61 \times 10 = 96.1$	36.1
111	10.68	$10.68 \times 10 = 106.8$	26.8

Table ER4.2d Pore pressure and excess pore pressure for t_3

z (m)	h_w (m)	u (kPa)	$u_e = u - u_{eq}$ (kPa)
115	6.68	$6.68 \times 10 = 66.8$	6.8
113	9.61	$9.61 \times 10 = 96.1$	16.1
111	10.68	$10.68 \times 10 = 106.8$	6.8

Table ER4.2e Pore pressure and excess pore pressure for t_4

z (m)	h_w (m)	u (kPa)	$u_e = u - u_{eq}$ (kPa)
115	6	$6 \times 10 = 60$	0
113	8	$8 \times 10 = 80$	0
111	10	$10 \times 10 = 100$	0

Figure ER4.2a displays the pore pressure distribution in the clay layer for times t_0 to t_4.

On the basis of this figure, Figure ER4.2b was developed, displaying the excess pore pressure distribution. The equilibrium pore pressures increase of 20 kPa at time t_3, gave rise to negative excess pore pressures in two horizons of the clay layer adjacent to the draining boundaries.

b) The layer underneath the clay provides a free-draining boundary, as the excess pore pressure distributions are symmetrical in relation to the center of the clay layer.

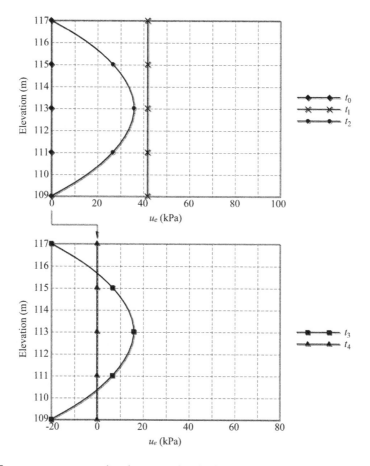

Figure ER4.2b Excess pore pressure distribution in the clay layer for times t_0 to t_4.

c) The placement of the fill layer implies a total vertical stress variation given by:

$$\Delta\sigma_v = 2 \times 21 = 42 \text{ kPa}$$

This vertical stress increment corresponds to an identical excess pore pressure:

$$u_e(0) = 42 \text{ kPa}$$

One year after the placement of the fill, and for the mid-point of the clay layer:

$$u_e(1\,\text{year}) = 36.1 \text{ kPa}$$

This excess pore pressure corresponds to:

$$\frac{u_e(1\,\text{year})}{u_e(0)} = \frac{36.1}{42} = 0.859$$

In the middle of the clay layer, one year after placement of the fill, the consolidation ratio is:

$$U_z = 1 - 0.859 = 0.140$$

Considering that the layer underneath the clay provides a free-draining boundary, $H = h/2$. So, in the middle of the clay layer:

$$Z = \frac{z}{H} = \frac{4}{4} = 1$$

According to Figure 4.10, for $U_z = 0.140$ and $Z = 1$, the time factor is $T \cong 0.15$. So, from:

$$T = \frac{c_v \times t}{H^2} \cong 0.15$$

the coefficient of consolidation is:

$$c_v = \frac{0.15 \times 4^2}{365 \times 24 \times 3600} = 7.61 \times 10^{-8} \text{ m}^2/\text{s}$$

d) The effective stress increment at the end of consolidation due to the placement of the fill is:

$$\Delta\sigma'_v = 2 \times 21 = 42 \text{ kPa}$$

The consolidation settlement of the clay layer can be calculated by Equation 4.11. The calculations are summarized in Table ER4.2f.
The consolidation settlement is $s_c = 0.35$ m.
One year after the placement of the fill (t_2), $T = 0.15$.

Table ER4.2f Calculation of the consolidation settlement

Sublayer	σ_{v0} (kPa)	u_0 (kPa)	σ'_{v0} (kPa)	$\Delta\sigma'_v$ (kPa)	s_c (m)
1	73.8	30	43.8	42	0.11
2	107.4	50	57.4	42	0.09
3	141	70	71.0	42	0.08
4	174.6	90	84.6	42	0.07
				$\Sigma =$	0.35

For this time factor, according to Table 4.4, the average consolidation ratio is:

$$\bar{U}_z = 43.7\%$$

Taking into account Equation 4.30, the consolidation settlement of the clay layer at t_2 is equal to:

$$s_c(t) = 0.437 \times 0.35 = 0.15 \text{ m}$$

ER4.3

a) The construction of the embankment generates an effective vertical stress increment at the end of consolidation given by:

$$\Delta\sigma'_v = 3 \times 21 = 63 \text{ kPa}$$

The consolidation settlement of the clay layer is calculated using Equation 4.11. The calculations are summarized in Table ER4.3a.

The consolidation settlement is $s_c = 0.45$ m.

Paving will be placed when the consolidation ratio is $\bar{U}_z = 90\%$, which corresponds to a time factor $T = 0.848$, according to Table 4.4. Thus, the time required for achieving such consolidation ratio is:

$$t = \frac{T \times H^2}{c_v} = \frac{0.848 \times 3^2}{9 \times 10^{-8}} = 84800000 \text{ s} = 981 \text{ days}$$

b) In order to have the road paved after 12 months, the consolidation settlement must be:

$$s_c(t = 12 \text{ months}) = 0.9 \times 0.45 = 0.405 \text{ m}$$

For $t = 12$ months the time factor is:

$$T = \frac{c_v \times t}{H^2} = \frac{9 \times 10^{-8} \times 365 \times 24 \times 3600}{3^2} = 0.315$$

According to Table 4.4, this time factor corresponds to an average consolidation ratio of

$$\bar{U}_z = 62.7\%$$

Given that $s_c(t) = \bar{U}_z \times s_c$, then

$$0.405 = 0.627 \times s_c$$

$$s_c = 0.646 \text{ m}$$

Replacing in Equation 4.11, and considering a point in the middle of the clay layer:

$$0.646 = \frac{6}{1 + 2.2} \times 0.82 \times \log\frac{66.5 + 63 + \Delta\sigma'_v}{66.5}$$

Table ER4.3a Calculation of the consolidation settlement

Sublayer	σ_{v0} (kPa)	u_0 (kPa)	σ'_{v0} (kPa)	$\Delta\sigma'_v$ (kPa)	s_c (m)
1	100.4	44.1	56.3	63	0.17
2	130.2	63.7	66.5	63	0.15
3	160	83.3	76.7	63	0.13
				$\Sigma=$	0.45

$\Delta\sigma'_v = 45.5 \text{ kPa}$

This stress increment corresponds to an additional fill height of:

$$\Delta b_{fill} = \frac{45.5}{21} = 2.2 \text{ m}$$

c) To calculate the effective stresses and pore water pressures, the clay layer will be divided into 4 sublayers, 1.5 m thick. Tables ER4.3b to ER4.3f summarize the calculations:
   i) Table ER4.3b concerns the at-rest situation;
   ii) Table ER4.3c represents the situation immediately after the placement of the fill and the temporary load:

$\Delta\sigma_v = (3+2.2)\times 21 = 109.2 \text{ kPa}$ and $u_e = \Delta\sigma_v = 109.2 \text{ kPa}$;

   iii) Table ER4.3d represents the situation immediately before the removal of the temporary load (12 months later):

$T = c_v t/H^2 = 0.315$; the values of the column $u_e(t)/u_e(0)$ were obtained from Figure 4.10; the values of the column $u_e(t)$ were calculated taking into account that $u_e(0) = 109.2 \text{ kPa}$;

   iv) Table ER4.3e represents the situation immediately after the removal of the temporary load:

$\Delta\sigma_v = -2.2\times 21 = -46.2 \text{ kPa}$ and $u_e = \Delta\sigma_v = -46.2 \text{ kPa}$

Table ER4.3b At-rest total and effective vertical stress and pore pressure

Point	Elevation (m)	σ_{v0} (kPa)	u_0 (kPa)	σ'_{v0} (kPa)
A	−4.5	85.5	34.3	51.2
B	−6.0	107.9	49.0	58.9
C	−7.5	130.2	63.7	66.5
D	−9.0	152.6	78.4	74.2
E	−10.5	174.9	93.1	81.8

Table ER4.3c Total and effective vertical stress and pore pressure immediately after the placement of the fill and the temporary load

Point	Elevation (m)	$\sigma_v = \sigma_{v0} + \Delta\sigma_v$ (kPa)	$u = u_0 + u_e$ (kPa)	σ'_v (kPa)
A	−4.5	194.7	143.5	51.2
B	−6	217.1	158.2	58.9
C	−7.5	239.4	172.9	66.5
D	−9	261.8	187.6	74.2
E	−10.5	284.1	202.3	81.8

Table ER4.3d Total and effective vertical stress, pore pressure and excess pressure immediately before the removal of the temporary load

Point	Elevation (m)	z (m)	$Z = \dfrac{z}{H}$	σ_v (kPa)	$u_e(t)/u_e(0)$	$u_e(t)$ (kPa)	$u = u_0 + u_e(t)$ (kPa)	σ'_v (kPa)
A	−4.5	0.0	0.0	194.7	0.0	0.0	34.3	160.4
B	−6.0	1.5	0.5	217.1	0.41	44.8	93.8	123.3
C	−7.5	3.0	1.0	239.4	0.58	63.3	127.0	112.4
D	−9.0	4.5	1.5	261.8	0.41	44.8	123.2	138.6
E	−10.5	6.0	2.0	284.1	0.0	0.0	93.1	191.0

v) Table ER4.3f represents the situation at the end of consolidation ($u_e = 0$).
    Figure ER4.3 presents the pore pressure distribution in the clay layer for the various phases.
d) The consolidation settlement 25 years after pavement construction will be 10% of the primary consolidation settlement plus the secondary consolidation settlement given by Equation 4.34:

Table ER4.3e Total and effective vertical stress and pore pressure immediately after the removal of the temporary load

Point	Elevation (m)	$\sigma_v = \sigma_{vi} + \Delta\sigma_v$ (kPa)	$u = u_i + u_e$ (kPa)	σ_v' (kPa)
A	−4.5	148.5	−11.9	160.4
B	−6.0	170.9	47.6	123.3
C	−7.5	193.2	80.8	112.4
D	−9.0	215.6	77.0	138.6
E	−10.5	237.9	46.9	191.0

Table ER4.3f Total and effective vertical stress and pore pressure at the end of consolidation

Point	Elevation (m)	σ_v (kPa)	$u = u_0$ (kPa)	σ_v' (kPa)
A	−4.5	148.5	34.3	114.2
B	−6.0	170.9	49.0	121.9
C	−7.5	193.2	63.7	129.5
D	−9.0	215.6	78.4	137.2
E	−10.5	237.9	93.1	144.8

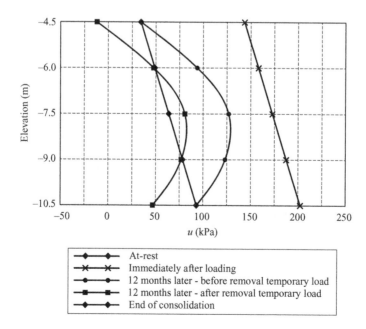

Figure ER4.3 Pore pressure distribution in the clay layer for the various phases.

$$s_d = \frac{h_0}{1+e_0} \times C_\alpha \times \log\frac{t_2}{t_1} = \frac{6}{1+2.2} \times 0.03 \times \log\frac{1+25}{1} = 0.08 \text{ m}$$

The total settlement 25 years after pavement construction is

$$s_{\text{total}} = 0.1 \times s_c + s_d = 0.1 \times 0.45 + 0.08 = 0.125 \text{ m}$$

ER4.4

a) The placement of a 5.0 m thick landfill generates a vertical effective stress increment at the end consolidation of:

$$\Delta\sigma_v' = 5 \times 22 = 110 \text{ kPa}$$

Dividing the two clay layers into 2 m thick (h_0) sublayers, the consolidation settlement can be calculated using Equation 4.11. The calculations are summarized in Table ER4.4a. The total consolidation settlement will be $s_c = 0.79 + 0.40 = 1.19$ m.

As a first approximation, in order to ensure elevation + 5.0 at the end of the consolidation, an additional fill height $\Delta h_{\text{fill}} = 1.19$ m should be used. However, by placing a fill with total height 6.19 m instead of 5.0 m, the consolidation settlement will be greater than that calculated above. This is shown in Table ER4.4b.

Hence, an iterative procedure must be performed in order to find a consolidation settlement equal to the additional fill height. Table ER4.4c corresponds to the last iteration.

In conclusion: the total height of the landfill that is necessary to ensure a +5.0 m surface elevation after primary consolidation is:

$$h_{\text{fill}} = 5.0 + 1.4 = 6.4 \text{ m}$$

b) For an average consolidation ratio $\bar{U}_z = 50\%$, the time factor is $T = 0.197$.
   For the upper clay layer, this time factor corresponds to:

$$T = 0.197 = \frac{c_v \times t}{H^2} = \frac{10^{-7} \times t}{4^2}$$

Table ER4.4a Calculation of the consolidation settlement for a 5.0 m thick landfill

Sublayer	σ_{v0} (kPa)	u_0 (kPa)	σ_{v0}' (kPa)	$\Delta\sigma_v'$ (kPa)	s_c (m)
Upper clay layer					
1	56.7	9.8	46.9	110	0.24
2	90.1	29.4	60.7	110	0.21
3	123.5	49.0	74.5	110	0.18
4	156.9	68.6	88.3	110	0.16
				$\Sigma=$	0.79
Lower clay layer					
1	227.3	107.8	119.5	110	0.11
2	262.7	127.4	135.3	110	0.10
3	298.1	147.0	151.1	110	0.10
4	333.5	166.6	166.9	110	0.09
				$\Sigma=$	0.40

Table ER4.4b Calculation of the consolidation settlement for a 6.19 m thick landfill

Sublayer		σ'_{v0} (kPa)	h_{fill} (m)	$\Delta\sigma'_v$ (kPa)	s_c (m)
1	Upper	46.9	5.00 + 1.19	136.2	0.27
2	clay	60.7	5.00 + 1.19	136.2	0.23
3		74.5	5.00 + 1.19	136.2	0.21
4		88.3	5.00 + 1.19	136.2	0.19
				$\Sigma =$	0.90
1	Lower	119.5	5.00 + 1.19	136.2	0.13
2	clay	135.3	5.00 + 1.19	136.2	0.12
3		151.1	5.00 + 1.19	136.2	0.11
4		166.9	5.00 + 1.19	136.2	0.10
				$\Sigma =$	0.46

$$s_c = 0.90 + 0.46 = 1.36 \text{ m}$$

Table ER4.4c Calculation of the consolidation settlement for a 6.40 m thick landfill

Sublayer		σ'_{v0} (kPa)	h_{fill} (m)	$\Delta\sigma'_v$ (kPa)	s_c (m)
1	Upper	46.9	5.00 + 1.40	140.8	0.28
2	clay	60.7	5.00 + 1.40	140.8	0.24
3		74.5	5.00 + 1.40	140.8	0.21
4		88.3	5.00 + 1.40	140.8	0.19
				$\Sigma =$	0.92
1	Lower	119.5	5.00 + 1.40	140.8	0.14
2	clay	135.3	5.00 + 1.40	140.8	0.12
3		151.1	5.00 + 1.40	140.8	0.11
4		166.9	5.00 + 1.40	140.8	0.11
				$\Sigma =$	0.48

$$s_c = 0.92 + 0.48 = 1.40 \text{ m}$$

$t = 365$ days $= 1$ year

For the lower clay layer, this time factor corresponds to:

$$T = 0.197 = \frac{c_v \times t}{H^2} = \frac{10^{-7} \times t}{8^2}$$

$t = 4$ years

Therefore, it may be concluded that the answer is between 1 and 4 years. Since the upper layer exhibits a higher consolidation settlement, the correct time is closer to the lower limit of the interval. That time must be found by a trial and error procedure.

After some trials, time $t = 1.5$ years is considered.
For the upper clay layer,

$$T = \frac{10^{-7} \times 1.5 \times 365 \times 24 \times 3600}{4^2} = 0.296$$

$\bar{U}_z = 60.87\%$

$s_c(t) = 0.6087 \times 0.92 = 0.56$ m

For the lower clay layer

$$T = \frac{10^{-7} \times 1.5 \times 365 \times 24 \times 3600}{8^2} = 0.074$$

$\bar{U}_z = 30.67\%$

$s_c(t) = 0.3067 \times 0.48 = 0.15$ m

The total settlement at $t = 1.5$ years is

$s_c(t) = 0.56 + 0.15 = 0.71$ m

As a percentage of the settlement by primary consolidation, this value represents

$$\frac{0.71}{1.40} = 0.51 = 51\%$$

So, in order to reach 50% of the total consolidation settlement, it will be necessary to wait approximately 1.5 years.

ER4.5

a) The effective vertical stress increment due to placement of the landfill is, at the end of consolidation:

$\Delta\sigma'_v = 2.5 \times 21 = 52.5$ kPa

The consolidation settlement can be calculated by Equation 4.11. The calculations are summarized in Table ER4.5.
The consolidation settlement is $s_c = 0.92$ m.
b) The system of vertical drains with $r_w = 0.10$ m in a square layout with spacing $l = 2.65$ m corresponds to

Table ER4.5 Calculation of the consolidation settlement

Sublayer	Elev. (m)	σ_{v0} (kPa)	u_0 (kPa)	σ'_{v0} (kPa)	$\Delta\sigma'_v$ (kPa)	s_c (m)
1	−2	35.2	19.6	15.6	52.5	0.28
2	−4	65.6	39.2	26.4	52.5	0.21
3	−6	96.0	58.8	37.2	52.5	0.17
4	−8	126.4	78.4	48.0	52.5	0.14
5	−10	156.8	98.0	58.8	52.5	0.12
					$\Sigma =$	0.92

$R = 0.564 \times l = 1.495$ m

$$n = \frac{R}{r_w} = \frac{1.495}{0.1} = 14.95 \approx 15$$

The time factor for radial consolidation corresponding to 6 months (180 days) is given by Equation 4.45.

$$T_r = \frac{1.13 \times 10^{-7} \times 180 \times 24 \times 3600}{1.4946^2} = 0.787$$

According to Table 4.6, for $n = 15$ and $T_r = 0.787$, the average radial consolidation ratio is $\bar{U}_r = 55\%$.

c) After 6 months, the surface settlement is 0.66 m, which corresponds to an average consolidation ratio of:

$$\bar{U}(t) = \frac{0.66}{0.91} = 0.725 = 72.5\%$$

Considering that for $t = 6$ months, $\bar{U}_r = 55\%$, according to Equation 4.43

$$1 - \bar{U}(t) = \left[1 - \bar{U}_z(t)\right] \cdot \left[1 - \bar{U}_r(t)\right]$$

$$1 - 0.725 = \left[1 - \bar{U}_z(t)\right] \cdot \left[1 - 0.55\right]$$

$$\bar{U}_z = 0.39 = 39\%$$

Assuming that the thin sand layer exists as a draining boundary, the time factor for $t = 6$ months is:

$$T = \frac{c_v t}{H^2} = \frac{5 \times 10^{-8} \times 180 \times 24 \times 3600}{2.5^2}$$

$$T = 0.124$$

According to Table 4.4, the average consolidation ratio for this value of T is

$$\bar{U}_z = 39.7\%$$

This value matches the one previously found, so the existence of a draining layer can be confirmed.

ER4.6

a) The placement of the landfill generates a vertical effective stress increment at the end of consolidation of:

$$\Delta\sigma'_v = 2 \times 21 = 42 \text{ kPa}$$

The consolidation settlement of the clay layer can be calculated by Equation 4.11. The calculations are summarized in Table ER4.6a.

The total consolidation settlement due to construction of the landfill is $s_c = 0.51$ m.

b) Filling the tank with 5 m of water corresponds to the application, on a circle of radius $a = 10$ m, of a pressure:

$$\Delta q_s = 5 \times 9.8 = 49.0 \text{ kPa}$$

The incremental total stresses may be calculated according to Equations A6.2.16 and A6.2.17 (assuming $\nu = 0.5$, since the clay will exhibit undrained behavior). The results are shown in Table ER4.6b.

The excess pore water pressure can be determined by (see Section 4.4.2):

$$\Delta u = \Delta \sigma_3 + A \times (\Delta \sigma_1 - \Delta \sigma_3)$$

and it will be equal to the effective vertical stress increment during the subsequent consolidation process. According to Skempton and Bjerrum (see Section 4.4.2), the consolidation settlement can be calculated introducing these increments in equation 4.11. The calculations are summarized in Table ER4.6c.

The total consolidation settlement due to the filling of the tank is $s_c = 0.19$ m.

Table ER4.6a Calculation of the consolidation settlement

Sublayer	σ_{v0} (kPa)	u_0 (kPa)	σ'_{v0} (kPa)	$\Delta\sigma'_v$ (kPa)	s_c (m)
1	53	29.4	23.6	42	0.18
2	87	49.0	38.0	42	0.13
3	121	68.6	52.4	42	0.11
4	155	88.2	66.8	42	0.09
				$\Sigma =$	0.51

Table ER4.6b Calculation of the incremental total stresses under the center of the tank

z (m)	a/z	$\Delta\sigma_z$ (kPa)	$\Delta\sigma_r$ (kPa)
5	2.00	44.62	18.32
7	1.43	39.76	11.47
9	1.11	34.33	7.17
11	0.91	29.15	4.54

Table ER4.6c Calculation of the consolidation settlement

z (m)	$\Delta\sigma_z$ (kPa)	$\Delta\sigma_x$ (kPa)	Δu (kPa) $= \Delta\sigma'_v$	σ'_{vi} (kPa)	s_c (m)
5	44.62	18.32	34.10	65.60	0.07
7	39.76	11.47	28.44	80.00	0.05
9	34.33	7.17	23.46	94.40	0.04
11	29.15	4.54	19.31	108.80	0.03
				$\Sigma =$	0.19

c) As the lower boundary of the clay layer is silty sand, it configures a draining boundary, hence $H=4$ m. Considering a linear distribution of the excess pore pressure along the clay layer, in one year the time factor is:

$$T = \frac{2 \times 10^{-7} \times 1 \times 365 \times 24 \times 3600}{4^2} = 0.394$$

According to Table 4.4, $\bar{U}_z = 69.3\%$. Then, the settlement will be:

$$s_c(t) = \bar{U}_z(t) \times s_c = 0.693 \times 0.19 = 0.13 \text{ m}$$

The result is an underestimation because in this case we have a three-dimensional consolidation process.

CHAPTER 5

ER5.1

a) The conditions correspond to the Rankine theory. For $\phi' = 30°$, $K_a = 1/3$ and $K_p = 3$. The resultant of the earth pressure diagram on the plate has a minimum value, the active thrust, P_a, and a maximum value, the passive thrust, P_p:

$$P_a = 0.5 K_a \gamma h^2 = 0.5 \times (1/3) \times 20 \times 3^2 = 30 \text{ kN/m}$$

$$P_p = 0.5 K_p \gamma h^2 = 0.5 \times 3 \times 20 \times 3^2 = 270 \text{ kN/m}$$

$$[F_{min}; F_{max}] = [30 \text{ kN/m}; 270 \text{ kN/m}]$$

As the pressure graph is triangular in both cases, $d = (2/3)h = 2$ m.

b) Due to the surcharge, q, the active and passive pressures, both uniformly distributed on the wall face, are, respectively, $K_a q = 5$ kPa e $K_p q = 45$ kPa. The active and passive thrusts due to the surcharge are:

$$P_a^q = K_a qh = 15 \text{ kN/m}; \quad P_p^q = K_p qh = 135 \text{ kN/m}.$$

Thus, the force F interval becomes:

$$[F_{min}; F_{max}] = [45 \text{ kN/m}; 405 \text{ kN/m}]$$

In order to obtain the new value for d, let us take moments relative to the top of the wall face, taking into account that the pressure graphs associated with the soil weight are triangular, while those for the surcharge are uniform.
Active case:

$$P_a \times (2/3)h + P_a^q \times 0.5 \times h = F_{min}d; \quad 30 \times (2/3)3 + 15 \times 0.5 \times 3 = 45d; \quad d = 1.83 \text{ m}$$

Passive case:

$$P_p \times (2/3)h + P_p^q \times 0.5 \times h = F_{max}d; \quad 270 \times (2/3)3 + 135 \times 0.5 \times 3 = 405d; \quad d = 1.83\,\text{m}.$$

ER5.2

a) Imagine, to start with, that each soil mass interacts with the respective plate without any strut connection between plates: the interaction force magnitude is contained in the interval $[P_a; P_p]$. Therefore, given that the conditions correspond to the Rankine theory:
Soil 1:

$$K_{a1} = 0.22; \quad K_{p1} = 4.60; \quad P_{a1} = 0.5K_{a1}\gamma_1 h^2 = 19.6 \text{ kN/m};$$

$$P_{p1} = 0.5K_{p1}\gamma_1 h^2 = 413.9 \text{ kN/m};$$

$$\left[P_{a1}; P_{p1} \right] = \left[19.6 \text{ kN/m}; 413.9 \text{ kN/m} \right]$$

Soil 2:

$$K_{a2} = 0.33; \quad K_{p2} = 3.00; \quad P_{a2} = 0.5K_{a2}\gamma_2 h^2 = 22.5 \text{ kN/m};$$

$$P_{p2} = 0.5K_{p2}\gamma_2 h^2 = 202.5 \text{ kN/m};$$

$$\left[P_{a2}; P_{p2} \right] = \left[22.5 \text{ kN/m}; 202.5 \text{ kN/m} \right]$$

As the two soil masses interact through the struts, the admissible range for the strut forces corresponds to the intersection of the two intervals:

$$\left[F_{min} = 22.5 \text{ kN}; F_{max} = 202.5 \text{ kN} \right].$$

It can be seen that force F is conditioned by soil 2, which provides the largest active thrust and the smallest passive thrust.

b) It is necessary to recalculate the interval for the soil on the left. As that soil is now submerged, the minimum interaction force corresponds to the sum of the active and hydrostatic thrusts, whereas the maximum interaction force is given by the sum of the passive and hydrostatic thrusts.
The hydrostatic thrust amounts to:

$$P_w = 0.5\gamma_w h^2 = 44.1 \text{ kN/m}$$

The active and passive thrusts must now be calculated taking the submerged unit weight: $\gamma'_1 = \gamma_{1,sat} - \gamma_w = 22.3 - 9.8 = 12.5 \text{ kN/m}^3$. So:

$$P_{a1} = 0.5K_{a1}\gamma'_1 h^2 = 12.4 \text{ kN/m}; \quad P_{p1} = 0.5K_{p1}\gamma'_1 h^2 = 258.7 \text{ kN/m}.$$

The new admissible range for the interaction forces between soil 1 (now submerged) and the plate (without any connection with soil 2) is then:

$$\left[P_{a1} + P_w; P_{p1} + P_w \right] = \left[56.5 \text{ kN/m}; 302.8 \text{ kN/m} \right].$$

Taking into account that soils 1 and 2 interact through the struts, the range of the strut forces becomes:

$$[F_{min} = 56.5 \text{ kN}; F_{max} = 202.5 \text{ kN}].$$

ER5.3

a) If the strut force does not vary, this means that the limits of the interval referred in the previous exercise coincide, i.e., the largest active thrust and the smallest passive thrust are equal. Given the height difference between the two plates, soil 1 will probably be in active state and soil 2 in passive state.

Let us then calculate the total thrust acting on the left plate admitting soil 1 in the active state:

$$P_{a1} + P_{w1} = 0.5 \times 0.33 \times (19.8 - 9.8) \times 3^2 + 0.5 \times 9.8 \times 3^2 = 59 \text{ kN/m}.$$

This confirms that soil 1 is in the active state.

Now let us determine what angle of shearing resistance would enable soil 2 to supply a total passive thrust that added to the hydrostatic thrust would provide a force equal to 59 kN/m.

$$P_{p2} + P_{w2} = 0.5 \times K_{p2} \times (20.4 - 9.8)1.5^2 + 0.5 \times 9.8 \times 1.5^2 = 59 \text{ kN/m}.$$

This leads to $K_{p2} = 4.02$, which corresponds to an angle of shearing resistance $\phi' = 37°$ for the soil on the right.

b) Figure ER5.3 illustrates the total and effective stress states at a point in the soil adjacent to each plate at 1.0 m depth.

ER5.4

a) If the plates are kept fixed during the filling operation, the at-rest thrust, P_0, will be acting on them. The tendon force F_i must be equal to this thrust, so that strut removal does not induce plate horizontal displacement. So:

$$F_i = P_0 \times 1 = 0.5 \times 0.5 \times 16 \times 3^2 \times 1 = 36 \text{ kN}.$$

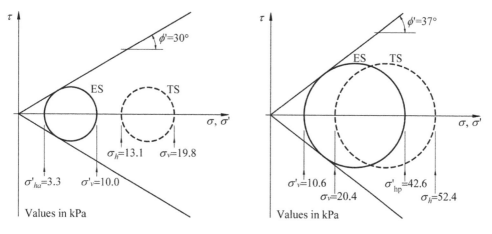

Figure ER5.3 Mohr circles representing the total and effective stress states at a point in the soil adjacent to each plate at 1.0 m depth.

b) The tendon force can vary from a minimum value equal to the active thrust to a maximum equal to the passive thrust. Applying the Rankine theory:

$$P_a = 0.5 \times 0.33 \times 16 \times 3^2 = 24 \text{ kN/m.}$$

$$P_p = 0.5 \times 3.0 \times 16 \times 3^2 = 216 \text{ kN/m.}$$

Thus:

$$\left[F_{min} = 24 \text{ kN}; F_{max} = 216 \text{ kN} \right].$$

c) If the dry sand in active state becomes submerged, the total thrust on the plates will increase, and so will the tensile tendon force. The tendon length will therefore increase and the plates will move outwards. Thus, it can be concluded that the soil remains in the active state.

The tendon force F has to equilibrate the hydrostatic thrust plus the active thrust of the submerged soil:

$$F = \left(P_a + P_w \right) \times 1 = \left[0.5 \times 0.33 \times (19.8 - 9.8) 3^2 + 0.5 \times 9.8 \times 3^2 \right] \times 1$$

$$= 15 + 44.1 = 59.1 \text{ kN.}$$

ER5.5

a) If the plate moved to the left and the soil on the right reached a limit equilibrium state, this is necessarily the active state. As the soil is submerged, the soil pressure is calculated considering the submerged unit weight, $\gamma' = \gamma - \gamma_w = 8.8 \text{ kN/m}^3$. There is a hydrostatic thrust applied on both faces of the plate, so is cancelled out. Figure ER5.5 shows the pressure diagrams and force F.

From the horizontal equilibrium equation:

$$0.5 \times K_a \times 8.8 \times 3^2 + 26.5 = 0.5 \times 8.8 \times 3^2$$

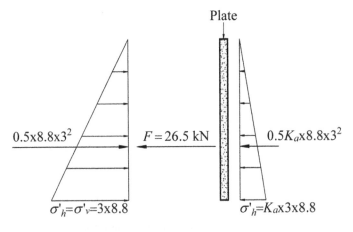

Figure ER5.5 Force F= 26.5 kN and earth pressures on both sides of the plate.

one gets:

$K_a = 0.33$, so $\phi' = 30°$.

b) The maximum force is the one that mobilizes the passive limit state on the left side of the plate. As $\phi' = 30°$, $K_p = 3.00$. So:

$$F_{max} = (P_p - P_a) \times 1 = \left[0.5 \times 8.8 \times 3^2 \times (3.00 - 0.33)\right] \times 1 = 105.7 \text{ kN}.$$

ER5.6

a) The plate displacement to the left, which mobilizes 65% of the soil passive resistance, is certainly much larger than that required to mobilize the active state on the opposite side (see, for example, Figure 5.7). For the specified angle of shearing resistance, $K_a = 0.27$ and $K_p = 3.69$. The horizontal equilibrium equation:

$$0.65P_p = F/1 + P_a$$

permits the calculation of F:

$$F = \left(0.65 \times 0.5 \times 3.69 \times 16 \times 3^2 - 0.5 \times 0.27 \times 16 \times 3^2\right) \times 1 = 153.3 \text{ kN}$$

b) As the ground is now submerged, the pressures and thrusts are calculated taking the submerged unit weight, $\gamma' = \gamma_{sat} - \gamma_w = 10.0 \text{ kN/m}^3$. The hydrostatic thrust is applied on both faces of the plate and, therefore, is cancelled out.
   Let us start by assuming that the soil on the right side remains in the active state. The new value of the active thrust is:

$$P_a = P_{right} = 0.5 \times 0.27 \times 10.0 \times 3^2 = 12.2 \text{ kN/m}$$

The thrust on the left side of the plate will have to balance the sum of the new value of the active thrust and force F, which is assumed to remain constant:

$$P_{left} = (153.3/1) + 12.2 = 165.5 \text{ kN/m}$$

This value of P_{left} corresponds to a given coefficient of earth pressure, K_{left}:

$$165.5 = 0.5 \times K_{left} \times 10 \times 3^2$$

From this equation, one gets $K_{left} = 3.68$, which is almost equal to the value of K_p (3.69). Therefore, the mobilized thrust is very close to the passive thrust, which is now calculated, for confirmation:

$$P_p = 0.5 \times 3.69 \times 10.0 \times 3^2 = 166.1 \text{ kN/m}.$$

The rising of the water level to the ground surface induces an increase of the mobilization of the passive resistance at the left side from 65% to almost 100%. So, the plate moved towards the left side. This conclusion also validates the assumption made at the start of the calculations admitting that the soil at the right remains in the active state.

ER5.7

a) For the problem conditions ($\phi' = 35°$; $\delta = (2/3)\ \phi'$; $\beta = 0$; $\lambda = -10°$), Table A5.1.2 supplies $K_a = 0.18$.

At the base of the wall back face, the active stress due to ground weight is:

$$p_a = K_a \gamma l = 0.18 \times 20 \times 8 / \cos 10° = 29.2 \text{ kPa}$$

so, the ground active thrust is given by:

$$P_a = 0.5 \times 29.2 \times 8 / \cos 10° = 118.6 \text{ kN/m}$$

The active stress due to the surface applied surcharge is:

$$p_a^q = K_a q = 0.18 \times 10 = 1.8 \text{ kPa}$$

so, the active thrust due to the surcharge is:

$$P_a^q = 1.8 \times 8 / \cos 10° = 14.6 \text{ kN/m}.$$

Figure ER5.7a represents the calculated stresses and thrusts.
The horizontal and vertical components of the total active thrust are, respectively:

$$P_{ah} = \left(P_a + P_a^q \right) \cos \left(\delta - |\lambda| \right) = 129.6 \text{ kN/m}$$

$$P_{av} = \left(P_a + P_a^q \right) \sin \left(\delta - |\lambda| \right) = 30.7 \text{ kN/m}$$

b) Taking into account the effective cohesion of the supported soil, two additional stress graphs have to be considered, in accordance to the corresponding state theorem (see 5.5.3). One graph represents the active stresses due to a uniform surcharge normal to the soil surface, the value of which is $c'\cot\an\phi' = 14.3$ kPa. The other graph represents the stresses normal to the wall back face, and directed towards the supported soil, with value $c'\cot\an\phi' = 14.3$ kPa.

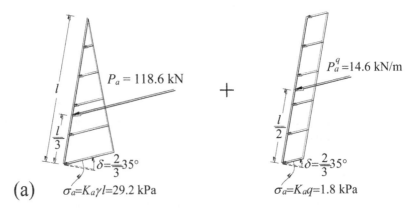

Figure ER5.7a Soil and surcharge active stresses and thrusts on the back of the wall.

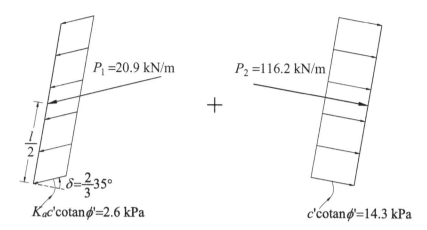

Figure ER5.7b Stresses and forces, in addition to the ones shown in Figure ER5.7a, in order to consider the effective cohesion c'= 10 kPa of the backfill, by applying the corresponding state theorem.

Figure ER5.7b shows the two graphs, whose resultants are, respectively:

$$P_1 = 0.18 \times 14.3 \times 8 / \cos 10° = 20.9 \text{ kN/m}$$

$$P_2 = 14.3 \times 8 / \cos 10° = 116.2 \text{ kN/m}$$

Therefore, the horizontal and vertical components of the total active thrust are, respectively:

$$P_{ah} = \left(P_a + P_a^q + P_1\right)\cos\left(\delta - |\lambda|\right) - P_2 \cos\lambda = 35.5 \text{ kN/m}$$

$$P_{av} = \left(P_a + P_a^q + P_1\right)\sin\left(\delta - |\lambda|\right) + P_2 \sin\lambda = 55.7 \text{ kN/m}$$

ER5.8

In this case, it is not possible to apply the analytical solution due to the complex shape of the surface and due to the surcharge, which is applied on only part of the surface.

The "graphical" solution, considering distinct soil wedges in order to identify the one corresponding to the maximum interaction force, must be used. The application of this solution was performed according to the Culmann construction (see Section 5.6.2).

Figure ER5.8 and Table ER5.8 show the results. It can be concluded that the active thrust, P_a, is about 232.5 kN/m.

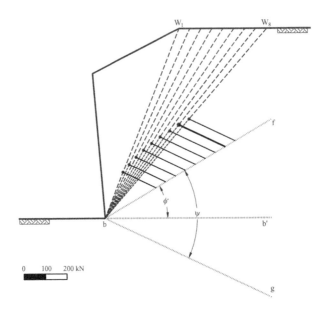

Figure ER5.8 Culmann construction

Table ER5.8 Active thrust calculation.

Wedge	Area (m²)	Weight (kN/m)	Surcharge (kN/m)	$\Sigma (W+Q)$ (kN/m)	Soil thrust (kN/m)
1	13.52	270.35	0	270.35	166.46
2	15.64	312.72	10	322.72	185.58
3	17.75	355.09	20	375.09	201.11
4	19.87	397.47	30	427.47	213.30
5	21.99	439.86	40	479.86	222.34
6	24.11	482.22	50	532.22	228.34
7	26.23	524.60	60	584.60	231.76
8	28.35	566.98	70	636.98	**232.49** $\approx Pa$
9	30.47	609.35	80	689.35	230.77

CHAPTER 6

ER6.1

The structures are isostatic, therefore the foundation load distribution does not depend on the settlement. The pressure applied to the ground is identical for all footings, and equal to Q/B^2.

Structure I is founded on ground with a constant E value. Therefore Equation 6.39 can be applied. Thus, it may be concluded that $s_1/s_2 = 2$.

In Structure II, the settlement of each footing is going to be larger than that of the corresponding footing of Structure I, due to the presence of the more deformable lower layer; this layer influences the behavior of the larger footing more.

In Structure III, the settlement of each footing is going to be smaller than that of the corresponding footing of Structure I, due to the presence of the less deformable lower layer; this layer influences the behavior of the larger footing more.

Thus, it may be concluded that:

$$1 < s_5/s_6 < s_1/s_2 = 2 < s_3/s_4$$

and also:

$$s_6 < s_2 < s_4 < s_5 < s_1 < s_3$$

ER6.2

a) The structure represented is hyperstatic (i.e., statically indeterminate), therefore the load distribution on the foundations depends on the settlement distribution in footings 5 to 8. In the case of footings 1 to 4, Equation 6.39 applies. The ground pressure transmitted by footings 1 to 4 is identical and its value is Q/B^2. So, it can be concluded that:

$$s_2 = s_3 = 2s_1 = 2s_4$$

b) Taking into account that the foundation loads of Figure E6.2b, assuming rigid supports, coincide with those of Figure E6.2a, the footings 6 and 7 will tend to have larger settlements than those of footings 5 and 8. As these four footings are structurally connected, an internal force redistribution will occur that reduces the load in footings 6 and 7, while increasing it in footings 5 and 8. Due to the differences in span, the load redistribution from footing 7 to footing 8 will be larger. Therefore, one can say that:

$$s_1 = s_4 < s_5 < s_8 < s_7 < s_6 < s_2 = s_3$$

c) It can be also concluded that:

$$s_7/s_3 < s_6/s_2 < 1 < s_5/s_1 < s_8/s_4 < 2$$

ER6.3

a) In the isostatic structure, the foundation load distribution does not depend on the settlement. For the loads presented, the pressure applied to the ground is equal to Q/B^2 for the three footings. As E is constant, Equation 6.39 applies. So, it can be concluded that:

$$s_2/s_1 = 2$$

In hyperstatic structures, foundation load distribution depends on the settlement distribution. As the central footings tend to settle more than the lateral ones, an internal force redistribution occurs that provides a load reduction on the central ones and a load increase on the lateral ones. The higher the structural stiffness, the more pronounced this redistribution will be. So, one can conclude that:

$$1 < s_6/s_5 < s_4/s_3 < s_2/s_1 = 2$$

Looking now at Structure IV, the increase of E with depth reduces the influence of the footing dimension on the settlement (see Section 6.3.2.5). Therefore, given that Structures IV and III are equal, the previous inequality can be extended as follows:

$$1 < s_8/s_7 < s_6/s_5 < s_4/s_3 < s_2/s_1 = 2$$

b) The more plausible scenario is clearly the last one, since the deformation modulus increases with the effective stress.

ER6.4

a) For the loads presented, the pressure applied to the foundation soil is equal for the four footings to Q/B^2. If the pressure, the shape and the foundation soil are common to all the footings, the settlement will be greater for footing 3, due to its larger width. Note that, in this case, it is not possible to establish a quantitative relation with the settlement of that footing and that of the remainder, since the deformation modulus is not constant, so Equation 6.39 does not apply. It can be concluded that:

$$s_1 = s_2 = s_4 < s_3$$

b) Due to the reason mentioned above, footing 7 will tend to settle more than the remainder. Therefore, an internal force redistribution will occur that reduces the load in that footing, while increasing it in the adjacent footings 6 and 8. Due to this redistribution, footing 6 will tend to settle more than footing 5; this will provoke some redistribution of the overload received by footing 6 to footing 5. Then, the settlements are in the following order:

$$s_5 < s_6 < s_8 < s_7$$

c) Bearing in mind the considerations presented to answer the previous questions, the global order of the settlements of the eight foundations must be as follows:

$$s_1 = s_2 = s_4 < s_5 < s_6 < s_8 < s_7 < s_3$$

d) Bearing in mind the considerations present to answer the previous questions, the global order of the ratios of the settlements must be as follows:

$$s_7/s_3 < 1 < s_5/s_1 < s_6/s_2 < s_8/s_4$$

e) Clearly, the firm layer will positively influence the behavior of footing 7 more due to its larger width, so the differential settlements will be smaller.

ER6.5

a) Immediately after the first tank filling operation, i.e., in the short-term, the clay layer is under undrained loading conditions. So, according to Equations 6.16 and 6.26:

$$q_{ult} = c_u 5.14 s_c f_c + 0 \times N_q s_q f_q + 0 = c_u 5.14 s_c f_c$$

Considering Figures 6.3 and 6.4, it can be concluded that the potential bearing capacity failure mechanism will involve the whole thickness of the clay layer. Therefore, the average value of the undrained shear strength of the layer will be assumed in the calculations. Additionally, the top of the dense sand layer will be assumed to be a rigid boundary.

Taking as reference the mid-point of the clay layer at $z = -6$ m:

$$\sigma'_{v0,\,\text{average}} = z \times \gamma' = z \times (\gamma - \gamma_w) = 6 \times (17.8 - 9.8) = 48 \text{ kPa}$$

$$c_{u,\,\text{average}} = 17 + 0.30 \times \sigma'_{v0,\,\text{average}} = 17 + 0.30 \times 48 = 31.4 \text{ kPa}$$

The corrective factors are given by ($\phi = 0$):

$$s_c = 1 + 0.2 B/L = 1.2$$

$$\begin{cases} D_f = 12\text{m} \\ B/D_f = 24/12 = 2 \rightarrow f_\gamma = 1.02 \\ \phi = 0 \end{cases}$$

So,

$$q_{\text{ult}} = 31.4 \times 5.14 \times 1.2 \times 1.02 = 197.5 \text{ kPa}$$

and the global safety factor immediately after the first tank filling operation is

$$F = \frac{197.5}{100} = 1.98 \approx 2$$

b) Under long-term conditions, drained loading is considered for the clay layer. So, according to Equations 6.15 and 6.26:

$$q_{\text{ult}} = 0 + 0 + \frac{1}{2} \gamma' B N_\gamma s_\gamma f_\gamma$$

$$\phi' = 25° \rightarrow N_\gamma = 6.46$$

The corrective factors are now given by:

$$s_\gamma = 1 - 0.3 \frac{B}{L} = 0.70$$

and

$$\begin{cases} D_f = 12 \text{ m} \\ B/D_f = 24/12 = 2 \rightarrow f_\gamma = 1.10 \\ \phi' = 25° \end{cases}$$

This leads to:

$$q_{ult} = \frac{1}{2}(17.8 - 9.8) \times 24 \times 6.46 \times 0.70 \times 1.10 = 477.5 \text{ kPa}$$

The global safety factor under long-term conditions is:

$$F = \frac{477.5}{100} = 4.78 \approx 4.8$$

c) According to Equation 6.49 and Figure 6.16 (see Section 6.3.4):

$$s_i = \sum_{j=1}^{n} \frac{1}{E_{uj}} \left[\Delta\sigma_{zj} - v_{uj} \left(\Delta\sigma_{xj} + \Delta\sigma_{yj} \right) \right] h_j$$

$$E_u = Mc_u = 250c_u$$

$$v_u = 0.5$$

$$\Delta q_s = 100 \text{ kPa}$$

The total stress increments can be computed on the basis of Equations A6.2.16 and A6.2.17. Table ER6.5 summarizes the calculations.

In conclusion, the immediate settlement of the center of the foundation is $s_i = 0.066$ m.

ER6.6

a) Foundation load bearing capacity will be computed according to Eurocode 7 (Design Approach 1). In this case, Combination 2 governs the design, so only the calculations concerning that combination are presented below.

$$\tan\phi_d' = \frac{\tan\phi_k'}{\gamma_{\phi k}} = \frac{\tan 40}{1.25} = 0.67 \equiv \phi_d' = 33.87°$$

$$B' = B - 2e = B - 2\frac{M_d}{V_d} = 12 - 2 \times \frac{60000}{15000} = 12 - 2 \times 4 = 4.00 \text{ m}$$

$$L' = L = 12.00 \text{ m}$$

Table ER6.5 Calculation of the immediate settlement of the foundation

z (m)	σ_{v0}' (kPa)	c_u (kPa)	E_u (kPa)	z/b	$\Delta\sigma_z$ (kPa)	$\Delta\sigma_r$ (kPa)	s_i (m)
1.5	12.0	20.6	5150	0.13	99.8	81.5	0.0107
4.5	36.0	27.8	6950	0.38	95.7	49.5	0.0199
7.5	60.0	35.0	8750	0.63	85.1	27.9	0.0196
10.5	84.0	42.2	10550	0.88	71.4	15.5	0.0159
						$\sum s_i =$	0.066

$$V_{Rd} = q_{\text{ult},d}\, B'L'$$

$$q_{\text{ult},d} = qN_q s_q i_q f_q + \frac{1}{2}\gamma' B' N_\gamma s_\gamma i_\gamma f_\gamma$$

$$q = 3 \times \gamma = 63 \text{ kPa}$$

$$\gamma' = 21 - 9.8 = 11.2 \text{ kN/m}^3$$

$$\phi'_d = 33.87° \rightarrow \begin{cases} N_q = 28.99 \\ N_\gamma = 28.09 \end{cases}$$

$$\begin{cases} s_q = 1 + \dfrac{B'}{L'}\sin\phi'_d = 1.19 \\ s_\gamma = 1 - 0.3\dfrac{B'}{L'} = 0.90 \end{cases}$$

$$m = m_B = \frac{2 + B'/L'}{1 + B'/L'} = 1.75 \rightarrow \begin{cases} i_q = \left(1 - \dfrac{H_d}{V_d}\right)^{m_B} = 0.83 \\ i_\gamma = \left(1 - \dfrac{H_d}{V_d}\right)^{m_B+1} = 0.75 \end{cases}$$

$$\begin{cases} D_f = 24 \text{ m} \\ B'/H = 4/24 = 0.17 \\ \phi'_d = 33.87° \end{cases} \begin{cases} f_q = 1 \\ f_\gamma = 1 \end{cases}$$

$$q_{\text{ult},d} = 63 \times 28.99 \times 1.19 \times 0.83 + \frac{1}{2}11.2 \times 4 \times 28.09 \times 0.90 \times 0.75$$
$$= 2229 \text{ kPa}$$

$$V_{Rd} = q_{\text{ult},d}\, B'L' = 2229 \times 4 \times 12 = 106974 \text{ kN}$$

$$106974 \text{ kN} \geq 15000 \text{ kN}$$

b) Foundation settlement

Assuming that E is constant in depth, Equation 6.39 is applicable:

$$s_i = \Delta q_s B \frac{1 - v^2}{E} I_s$$

The factor I_s can be obtained from Table 6.7:

$$L/B = 1; \quad D_f/B = 24/12 = 2 \rightarrow I_s = 0.63$$

Assuming a Poisson ratio of 0.3:

$$s_i = \frac{15000}{12 \times 12} \times 12 \times \frac{1-0.3^2}{60000} \times 0.63 = 0.012 \text{ m}$$

Foundation rotation
According to Equations 6.41 and 6.43:

$$\tan \omega_y = \frac{M_d}{B^2 L} \frac{1-v^2}{E} I_{\omega y}$$

$$I_{\omega y} = \frac{16}{\pi \left(1 + \dfrac{0.22B}{L}\right)} = \frac{16}{\pi \left(1 + 0.22\right)} = 4.175$$

$$\tan \omega_y = \frac{30000}{12^2 \times 12} \times \frac{1-0.3^2}{60000} \times 4.175 = 1.099 \times 10^{-3} \Rightarrow \omega_y = 0.063° < 0.1°$$

ER6.7

a) The bearing capacity equations allow the consideration of only a homogeneous soil, limited or not by a rigid lower boundary. Therefore, for this case, some simplified assumptions must be adopted in order to bound the actual bearing capacity of the foundation. If these assumptions are "pessimistic", they will lead to values below the real bearing capacity. Conversely, "optimistic" assumptions provide values above the real bearing capacity.

"Pessimistic" assumption - Consider the foundation soil to be fine sand down to the bedrock.

$$q_{ult} = qN_q s_q f_q + \frac{1}{2} \gamma' B N_\gamma s_\gamma f_\gamma$$

$$\phi' = 30° \;\rightarrow\; \begin{cases} N_q = 18.40 \\ N_\gamma = 14.62 \end{cases}$$

$$\begin{cases} s_q = 1 + \dfrac{B}{L} \sin \phi' = 1.50 \\ s_\gamma = 1 - 0.3 \dfrac{B}{L} = 0.70 \end{cases}$$

$$\begin{cases} D_f = 9 \text{ m} \\ B/D_f = 0.67 \\ \phi' = 30° \end{cases} \rightarrow \begin{cases} f_q = 1.0 \\ f_\gamma = 1.0 \end{cases}$$

$$q_{ult,min} = 18 \times 18.4 \times 1.50 \times 1.0 + \frac{1}{2}(19.0 - 9.8) \times 6.0 \times 14.62 \times 0.70 \times 1.0$$

$$= 779 \text{ kPa}$$

$$F_{min} = \frac{779}{300} = 2.60$$

"Optimistic" assumption 1 - Consider the foundation soil to be fine sand and the bedrock to be located at the top of the medium sand layer.

$$\begin{cases} D_f = 3 \text{ m} \\ B / D_f = 2.0 \\ \phi' = 30° \end{cases} \rightarrow \begin{cases} f_q = 2.42 \\ f_\gamma = 1.20 \end{cases}$$

$$\begin{aligned} q_{ult,max(i)} &= 18 \times 18.4 \times 1.50 \times 2.42 + \frac{1}{2}(19.0 - 9.8) \times 6.0 \\ &\quad \times 14.62 \times 0.70 \times 1.20 \\ &= 1541 \text{ kPa} \end{aligned}$$

$$F_{max(i)} = \frac{1541}{300} = 5.14$$

"Optimistic" assumption 2 - Consider the foundation soil to be the medium sand layer down to the schist bedrock.

$$\phi' = 35° \rightarrow \begin{cases} N_q = 33.30 \\ N_\gamma = 34.16 \end{cases}$$

$$\begin{cases} s_q = 1 + \frac{B}{L}\sin\phi' = 1.57 \\ s_\gamma = 1 - 0.3\frac{B}{L} = 0.70 \end{cases}$$

$$\begin{cases} D_f = 9 \text{ m} \\ B / D_f = 0.67 \\ \phi' = 35° \end{cases} \rightarrow \begin{cases} f_q = 1.10 \\ f_\gamma = 1.00 \end{cases}$$

$$\begin{aligned} q_{ult,max(ii)} &= 18 \times 33.30 \times 1.57 \times 1.10 + \frac{1}{2}(20.0 - 9.8) \\ &\quad \times 6.0 \times 34.16 \times 0.70 \times 1.00 \\ &= 1767 \text{ kPa} \end{aligned}$$

$$F_{max(ii)} = \frac{1767}{300} = 5.89$$

From the two "optimistic" estimations, the lower value (i.e., the less "optimistic" one) should naturally be selected. In conclusion:

$$2.60 < F < 5.14.$$

b) The immediate settlement of a shallow foundation on a soil with E constant in depth is given by Equation 6.39:

$$s_i = \Delta q_s B \frac{1 - v^2}{E} I_s$$

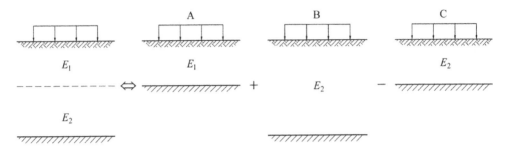

Figure ER6.7 Application of the principle of superposition to the calculation of the settlement of the footing.

The parameter I_s depends on the shape of the foundation and on the ratio D_f/B, with D_f representing the distance between the base of the foundation and a rigid boundary (see Tables 6.6 and 7.7).

The estimation of the settlement in this case can be performed applying the principle of superposition, as explained in Figure ER6.7. In the figure, E_1 and E_2 represent the deformation modulus of the fine sand and the medium sand, respectively. So, the actual problem involving two layers (with distinct moduli) is split into three cases, each one with homogeneous soil limited by a rigid boundary.

Case A (foundation soil is the fine sand limited by a rigid boundary located at the top of medium sand)

$$L/B = 1 \quad D_f/B = 0.5 \quad \rightarrow \quad I_s = 0.32$$

$$s_{i,1} = 300 \times 6 \frac{1 - 0.3^2}{30000} \times 0.32 = 0.0175 \text{ m} = 1.75 \text{ cm}$$

Case B (foundation soil is the medium sand limited by a rigid boundary at the top of schist bedrock)

$$L/B = 1 \quad D_f/B = 1.5 \quad \rightarrow \quad I_s = 0.57$$

$$s_{i,2} = 300 \times 6 \frac{1 - 0.3^2}{90000} \times 0.57 = 0.0104 \text{ m} = 1.04 \text{ cm}$$

Case C (foundation soil is the medium sand limited by a rigid boundary at the top of medium sand)

$$L/B = 1 \quad D_f/B = 0.5 \quad \rightarrow \quad I_s = 0.32$$

$$s_{i,3} = 300 \times 6 \frac{1 - 0.3^2}{90000} \times 0.32 = 0.0058 \text{ m} = 0.58 \text{ cm}$$

Therefore, applying the principle of superposition:

$$s_i = 1.75 + 1.04 - 0.58 = 2.21 \text{ cm}$$

c) According to Schmertmann's method, the immediate settlement is given by:

$$s_i = C_s \Delta q_s \sum_{j=1}^{n} \frac{I_{\varepsilon j}}{E_j} h_j$$

with:

$$\Delta q_s = 300 - \sigma'_{vb} = 300 - 18 = 282 \text{ kPa}$$

$$C_s = 1 - 0.5\left(\frac{\sigma'_{vb}}{\Delta q_s}\right) = 1 - 0.5\left(\frac{18}{282}\right) = 0.968$$

For a square footing:

$$0 \le z_f \le B/2 \qquad I_\varepsilon = 0.1 + (z_f/B)(2I_{\varepsilon p} - 0.2)$$
$$B/2 \le z_f \le 2B \qquad I_\varepsilon = 2/3 I_{\varepsilon p}(2 - z_f/B)$$

$$I_{\varepsilon p} = 0.5 + 0.1\sqrt{\frac{\Delta q_s}{\sigma'_{vp}}}$$

$$\sigma'_{vp} = 18 + 3 \times 9.2 = 45.6 \text{ kPa} \qquad I_{\varepsilon p} = 0.5 + 0.1\sqrt{\frac{282}{45.6}} = 0.749$$

Table ER6.7 summarizes the calculations.

The estimated settlement is $s_i = 2.07$ cm, which is close to the estimation obtained above.

ER6.8

a) According to the correlation between $(N_1)_{60}$ and ϕ' proposed by Décourt (see Figure 1.15):

$$(N_1)_{60} = 15 \rightarrow \phi' \cong 35°$$

$$(N_1)_{60} = 30 \rightarrow \phi' \cong 40°$$

b) According to Figure 6.4, the depth reached by the plastic zone can be estimated:

$$\phi' \cong 35° \;\rightarrow\; d/B \cong 1.4 \;\rightarrow\; d = 1.4 \times 4.0 = 5.6 \text{ m} < 6 \text{ m}$$

So, the plastic zone will be fully within the fine sand layer, and the medium sand layer will not affect the bearing capacity of the foundation.

Table ER6.7 Calculation of the settlement applying the Schmertmann's method

j	z_{fj}/B	$I_{\varepsilon j}$	E_j (kPa)	h_j (m)	s_{ij} (m)
1	0.250	0.425	30000	3	0.0116
2	0.750	0.624	90000	3	0.0057
3	1.250	0.375	90000	3	0.0034
				$s_i =$	0.0207

Given that for ULS at the central footing $V = 6300$ kN and $M_y = 2520$ kN·m, the effective width B' is:

$$B' = B - 2e = B - 2 \times \frac{M_y}{V} = 4 - 2 \times \frac{2520}{6300} = 3.2 \text{ m}$$

and

$$L' = L$$

The safety factor is given by:

$$F = \frac{Q_{ult}}{V} = \frac{q_{ult} \times B' \times L'}{V}$$

So $q_{ult} \times L'$ must be, at least,

$$q_{ult} \times L' = \frac{F \times V}{B'} = \frac{4 \times 6300}{3.2} = 7875 \text{ kN/m}$$

Given the granular nature of the soil, the footing shape and the position of water level, q_{ult}, can be expressed as:

$$q_{ult} = q N_q s_q i_q + \frac{1}{2} \gamma' B' N_\gamma s_\gamma i_\gamma$$

with:

$$q = \gamma_{fill} \times 2.0 = 40 \text{ kPa}; \quad \gamma' = 20 - 9.8 = 10.2 \text{ kN/m}^3$$

$$\phi' \cong 35° \rightarrow N_q = 33.30; \quad N_\gamma = 34.16$$

$$s_q = 1 + \frac{B'}{L'} \sin\phi' = 1 + \frac{1.835}{L'}; \quad s_\gamma = 1 - 0.3 \times \frac{3.2}{L'} = 1 - \frac{0.96}{L'}$$

$$m = \frac{2 + \dfrac{3.2}{L'}}{1 + \dfrac{3.2}{L'}}; \quad i_q = \left(1 - \frac{630}{6300}\right)^m = 0.9^m; \quad i_\gamma = 0.9^{m+1}$$

So,

$$\left[40 \times 33.3 \times \left(1 + \frac{1.835}{L'}\right) \times 0.9^m + \frac{1}{2} \times 10.2 \times 3.2 \times 34.16 \times \left(1 - \frac{0.96}{L'}\right) \times 0.9^{m+1}\right] \times L' = 7875$$

With the help of a spreadsheet, one gets:

$$L' = L = 3.99 \text{ m} \rightarrow L = 4.0 \text{ m}$$

c) The immediate settlement of a shallow foundation on a soil with E constant in depth is given by Equation 6.39:

$$s_i = \Delta q_s B \frac{1 - v^2}{E} I_s$$

- Fine sand: $E = 30,000$ kPa
- Medium sand: $E = 60,000$ kPa

"Pessimistic" assumption: consider the foundation soil to be fine sand down to the bedrock.

Taking into consideration Table 6.7:

$$L/B = 4/4 = 1 \quad D_f/B = 12/4 = 3 \quad \rightarrow \quad I_s = 0.69$$

$$s_{i,max} = \frac{4000}{4 \times 4} \times 4 \times \frac{1 - 0.3^2}{30000} \times 0.69 = 0.021 \text{ m}$$

"Optimistic" assumption 1: consider the foundation soil to be fine sand and the bedrock at the top of the medium sand layer.

$$L/B = 4/4 = 1 \quad D_f/B = 6/4 = 1.5 \quad \rightarrow \quad I_s = 0.57$$

$$s_{i,min(i)} = \frac{4000}{4 \times 4} \times 4 \times \frac{1 - 0.3^2}{30000} \times 0.57 = 0.017 \text{ m}$$

"Optimistic" assumption 2: consider the foundation soil to be medium sand layer to the bedrock.

$$L/B = 4/4 = 1 \quad D_f/B = 12/4 = 3 \quad \rightarrow \quad I_s = 0.69$$

$$s_{i,min(ii)} = \frac{4000}{4 \times 4} \times 4 \times \frac{1 - 0.3^2}{60000} \times 0.69 = 0.010 \text{ m}$$

From the two "optimistic" estimations, the higher value (i.e., the less "optimistic") should naturally be selected. In conclusion:

$$0.017 \text{ m} < s_i < 0.021 \text{ m}$$

It may be interesting to observe that the application of the superposition method, introduced in ER6.7, provides a result on the centre of this interval.

d) According to Schmertmann's method, the immediate settlement is given by:

$$s_i = C_s \Delta q_s \sum_{j=1}^{n} \frac{I_{\varepsilon j}}{E_j} b_j$$

with:

$$\Delta q_s = \frac{V}{BL} - \sigma'_{vb} = \frac{4000}{4^2} - 40 = 210 \text{ kPa}$$

$$C_s = 1 - 0.5 \left(\frac{\sigma'_{vb}}{\Delta q_s} \right) = 1 - 0.5 \left(\frac{40}{210} \right) = 0.905$$

For a square footing:

$$0 \le z_f \le B/2 \qquad I_\varepsilon = 0.1 + (z_f/B)(2I_{\varepsilon p} - 0.2)$$
$$B/2 \le z_f \le 2B \qquad I_\varepsilon = 2/3 I_{\varepsilon p}(2 - z_f/B)$$

$$I_{\varepsilon p} = 0.5 + 0.1\sqrt{\frac{\Delta q_s}{\sigma'_{vp}}}$$

$$\sigma'_{vp} = 40 + 2 \times 10.2 = 60.4 \text{ kPa} \qquad I_{\varepsilon p} = 0.5 + 0.1\sqrt{\frac{210}{60.4}} = 0.686$$

Table ER6.8 summarizes the calculations.

So, $s_i = 0.0173$ m. This result is practically coincident with the lower limit of the interval previously determined.

e) The best criterion would be to adopt the same settlement. The other two criteria would lead to the existence of differential settlements, which would induce redistribution of internal forces in the structure.

ER6.9

The pressure q applied by the silo under service conditions is given by:

$$q = \frac{V}{BL} = \frac{146000}{10 \times 60} = 243 \text{ kN/m}^2$$

The calculation of the differential settlement between the two foundations of the pavilion, requires:

i) computation of the incremental stresses induced by the silo construction along the vertical lines passing through the foundation centres;
ii) use of Equation 6.33 to compute the settlement of each foundation:

$$s_i = \sum_{j=1}^{n} \frac{1}{E_j}\left[\Delta\sigma_{zj} - v_j\left(\Delta\sigma_{xj} + \Delta\sigma_{yj}\right)\right]h_j$$

iii) calculation of the difference between these settlements.

The incremental stresses may be evaluated, with a reasonable approximation, by using Equations A6.2.9 to A6.2.11. It is easy to conclude, that the silo construction will induce

Table ER6.8 Calculation of the settlement applying the Schmertmann's method

j	z_{fj}/B	$I_{\varepsilon j}$	E_j (kPa)	h_j (m)	s_{ij} (m)
1	0.250	0.393	30000	2	0.0050
2	1.000	0.458	30000	4	0.0116
3	1.750	0.114	60000	2	0.0007
				$s_i =$	0.0173

Table ER6.9 Calculation of the settlement of the nearest foundation of the pavilion induced by the silo construction

j	z/b	$\Delta\sigma_z$ (kPa)	$\Delta\sigma_x$ (kPa)	$\Delta\sigma_y$ (kPa)	q_c (MPa)	E (MPa)	s_{ij} (m)
1	0.3	4.84	49.18	18.91	4.5	13.5	−0.0042
2	0.9	38.50	62.10	35.21	4.5	13.5	0.0010
3	1.5	58.47	44.85	36.16	6.5	19.5	0.0046
4	2.1	64.90	30.08	33.24	6.5	19.5	0.0066
5	2.7	64.73	20.15	29.71	6.5	19.5	0.0073
						$s_i =$	0.0153

negligible incremental stresses underneath the most distant foundation, so that the same can be said of the induced settlement. On the contrary, the induced stresses underneath the nearest foundation are significant and provide the data required for the settlement calculation. This calculation is summarized in Table ER6.9, for $x/b = (5+3)/5 = 1.6$ and considering five sublayers 3 m thick. The soil deformation modulus was assumed as $E = 3.q_c$.

The differential settlement is: $s_D = s_{\max} - s_{\min} = 0.015 - 0.000 = 0.015$ m .

CHAPTER 7

ER7.1

Base resistance:

$$q_b = q'N'_q$$

Effective vertical stress at the pile base:

$$q' = 15 \times (19 - 10) + 2 \times (21 - 10) = 157 \text{ kPa}$$

Rigidity index:

$$I_R = \frac{35000}{2(1+0.3)(157 \times \tan 40°)} = 102$$

Considering $\varepsilon_v = 0$, then $I_{RR} = I_R = 102$.

The critical rigidity index is equal to:

$$I_{RC} = 0.5 \exp\left[2.85 \cot\left(45° - \frac{40°}{2}\right)\right] = 226$$

The failure mode is a punching failure beneath the pile base and the corrective factor is equal to:

$$\xi_{qr} = \exp\left[-3.8 \tan 40° + \frac{(3.07 \sin 40°)(\log(2 \times 102))}{1 + \sin 40°}\right] = 0.60$$

The bearing capacity factor is equal to:

$$N_q = e^{\pi \tan 40°} \tan^2\left(45° + \frac{40°}{2}\right) = 64.20$$

The shape and depth corrective factors are equal to:

$$s_q = 1 + \tan 40° = 1.84$$

$$d_q = 1 + 2\tan 40°(1 - \sin 40°)^2 \tan^{-1}\left(\frac{17}{0.4}\right) = 1.33$$

Therefore

$$N'_q = 64.20 \times 0.60 \times 1.84 \times 1.33 = 94.27$$

(Note that if Figure 7.9 is used the factor is equal to $N'_q = 100$) and

$$q_b = 157 \times 94.27 = 14840 \text{ kPa}$$

Finally:

$$R_b = q_b A_b = 14840 \times 0.40^2 = 2374 \text{ kN}$$

Shaft resistance:
For granular materials:

$$q_s = K\sigma'_v \tan\delta$$

Layer 1 – Loose sand
The pile imposes large displacement on the ground and considering the average of the interval in Table 7.3:

$$K = 1.5K_0 = 1.5 \times (1 - \sin(30°)) = 0.75$$

$$\sigma'_v = 7.5 \times (19 - 10) = 67.5 \text{ kPa}$$

$$q_{s;1} = 0.75 \times 67.5 \times \tan(0.9 \times 30°) = 26 \text{ kN/m}^2$$

Layer 2 – Dense sand

$$K = 1.5K_0 = 1.5 \times (1 - \sin(40°)) = 0.54$$

$$\sigma'_v = 15 \times (19 - 10) + 1 \times (21 - 10) = 146 \text{ kPa}$$

$$q_{s;2} = 0.54 \times 146 \times \tan(0.9 \times 40°) = 57 \text{ kN/m}^2$$

Thus, the shaft resistance can be computed as:

$$R_s = 4 \times 0.40 \times (26 \times 15 + 57 \times 2) = 806 \text{ kN}$$

Bearing capacity:

$$R_c = R_b + R_s = 2374 + 806 = 3180 \text{ kN}$$

ER7.2

a) Soil classification:

    According to Table A7.1.1, the upper layer is classified as loose sand and the bottom layer is classified as dense sand.

    Pile Type:

    According to Table A7.1.2, driven concrete piles are classified as Pile type 9, and belong to Group G1, Class 4.

    Unit base resistance:

    Penetrometer base resistance coefficient, k_c, is a function of the effective penetration depth, D_{ef}, which is given by Equation 7.32. For the pile cross-section geometry and CPT test results, $D_{ef} = 2.58$ m.

    Since $D_{ef}/B > 5$, k_c is equal to $k_{c,max}$, the value of which is given in Table A7.1.4 as a function of the pile classification. For concrete driven piles, $k_{c,max} = 0.4$.

    From Equation 7.27:

$$q_b = 0.4 \times 12 = 4.8 \text{ MPa}$$

Unit shaft resistance:

    The $\alpha_{pile\text{-}soil}$ parameter is given in Table A7.1.5. For pile type 9 installed in sandy soils, this parameter is equal to 1.

    The value of f_{soil} depends on CPT cone resistance. The values of the constants needed for definition of this function are given in Table A7.1.6. For sandy soils:

$$a = 0.0012; \quad b = 0.1; \quad c = 0.15.$$

    The unit shaft resistance for each of the layers is included in Table ER7.2a.

Bearing capacity:

    The base resistance is given by Equation 7.3:

$$R_b = 4800 \times 0.4^2 = 768 \text{ kN}$$

Table ER7.2a Unit shaft resistance

Soil Layer	q_c (MPa)	$\alpha_{pile\text{-}soil}$	f_{soil} (kPa)	$q_{s,max}$ (kPa) (1)	q_s (kPa)
Top layer	3.5	1	43	130	43
Bottom layer	12	1	95	130	95

1 - For driven piles in sandy soils $q_{s,max} = 130$ kPa (see Table A7.1.7).

The shaft resistance is given by Equation 7.4:

$$R_s = (0.4 \times 4)(10 \times 43 + 2 \times 95) = 992 \text{ kN}$$

Thus, pile bearing capacity is equal to:

$$R_c = R_b + R_s = 768 + 992 = 1760 \text{ kN}$$

b) Pile Type:
   Continuous flight auger piles are defined as pile type 6, and belong to Group G1, Class 2.
   Unit base resistance:
      Penetrometer base resistance coefficient, k_c, is a function of the effective penetration depth, D_{ef}, which is given by Equation 7.32. Accordingly, given the pile cross-sectional geometry and CPT test results, $D_{ef} = 2.58$ m.
      Since $D_{ef}/B > 5$, k_c is equal to $k_{c,max}$, the value of which is given in Table A7.1.4 as function of the pile classification. For CFA piles, $k_{c,max} = 0.25$.
      From Equation 7.27:

$$q_b = 0.25 \times 12 = 3.0 \text{ MPa}$$

   Unit shaft resistance:
      The $\alpha_{pile-soil}$ parameter is given in Table A7.1.5. For pile type 6 installed in sandy soils, this parameter is equal to 1.25.
      The value of f_{soil} depends on CPT cone resistance. The values of the constants needed for definition of this function are the same as in question a).
      The unit shaft resistance for each of the layers is included in Table ER7.2b.
Bearing capacity:
The base resistance is given by Equation 7.3:

$$R_b = 3000 \times \pi \times \frac{0.5^2}{4} = 589 \text{ kN}$$

The shaft resistance is given by Equation 7.4:

$$R_s = (\pi \times 0.5)(10 \times 53 + 2 \times 119) = 1206 \text{ kN}$$

Thus, the pile bearing capacity is equal to:

$$R_c = R_b + R_s = 589 + 1206 = 1795 \text{ kN}$$

Table ER7.2b Unit shaft resistance

Soil layer	q_c (MPa)	$\alpha_{pile-soil}$	f_{soil} (kPa)	$q_{s,max}$ (kPa) (1)	q_s (kPa)
Top layer	3.5	1.25	43	170	53
Bottom layer	12	1.25	95	170	119

1 - For CFA piles in sandy soils $q_{s,max} = 170$ kPa (see Table A7.1.7).

ER7.3

a) The lateral load is uniformly divided by both piles. Therefore, each pile is subjected to the following loads at the head:

$V = 800$ kN; $H_x = 75$ kN

Assuming that the influence region of the pile is developed only in the upper soil layer, the analytical solution based on the Winkler foundation can be followed.
Taking into account Equation 7.79, the Winkler coefficient is:

$k' \approx E = 15$ MPa

For assessing the characteristic elastic length, the pile bending stiffness computation is required:

$$E_p I_p = 30 \times 10^6 \times \pi \, \frac{0.5^4}{64} = 92039 \text{ kN} \cdot \text{m}^2$$

Taking into account Equation 7.76: $\lambda = 0.45$ m^{-1} and $l_e = 2.23$ m.
Since the thickness of the upper soil layer is more than five times the characteristic elastic length, the contribution of the deeper soil layer is negligible, confirming the previous assumption.
Since the lateral load is applied in the same direction as the pile alignment, head rotation is restricted. Thus, shear stress and bending moment evolution along the pile length are given by equations presented in Table 7.10. For fixed pile head rotation, these equations are:

$$V(z) = -75e^{-0.45z}\left(\cos(0.45z)\right)$$

$$M(z) = \frac{75}{2 \times 0.45} e^{-0.45z}\left(\cos(0.45z) - \sin(0.45z)\right)$$

Figure ER7.3 shows the graphs of shear forces and bending moments over pile length.

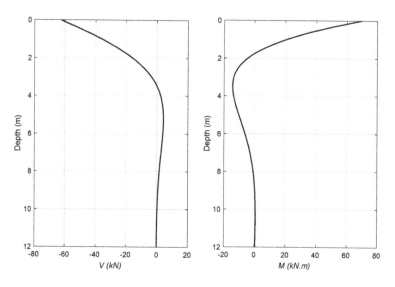

Figure ER7.3 Shear forces and bending moments over pile length.

b) In such conditions, each of the head of the piles is subjected to the following loads:

$$V = 800 \text{ kN}; \quad H_y = 50 \text{ kN}$$

The characteristic elastic length is not dependent on the loading direction or on the boundary conditions at pile head. Therefore: $\lambda = 0.45 \text{ m}^{-1}$ and $l_e = 2.23 \text{ m}$.

Since the load is applied in the direction orthogonal to the pile alignment, the pile head is free to rotate. From Table 7.10:

$$y(z=0) = \frac{H}{2\lambda^3 E_p I_p} = \frac{50}{2 \times 0.45^3 \times 92039} = 2.9 \text{ mm}$$

The maximum moment occurs when shear stress is null. Therefore, the location of the maximum bending moment is given by:

$$V(z) = -He^{-\lambda z}\left(\cos(\lambda z) - \sin(\lambda z)\right) = 0$$

$$z = \frac{\pi}{4\lambda} \approx 1.75 \text{ m}$$

and the value of the maximum moment is given by (see Table 7.10):

$$M(1.75) = -\frac{50}{0.45} e^{-0.45 \times 1.75} \sin(0.45 \times 1.75) = -38.82 \text{ kN.m}$$

ER7.4

The pile perimeter and cross sectional area are equal to:

$$C = \pi \times 0.5 = 1.57 \text{ m}$$

$$A_b = \pi \times \frac{0.5^2}{4} = 0.196 \text{ m}^2$$

The pile cap and pile self-weight are equal to:

$$W_{pc} = 25 \times 1 \times 4 \times 4 = 400 \text{ kN}$$

$$W_p = 25 \times 0.196 \times (9 + h) = 44.2 + 4.9h \, (\text{kN}, \text{m})$$

Two modes of failure must be considered: individual pile failure and block failure.
   Individual pile failure mode:
   The axial compressive force supported by each pile is equal to:

$$F_c = \frac{1000 + 400}{9} = 156 \text{ kN}$$

Since the cap is in contact with the soil and $s = 3B$, the group efficiency is $E_g = 1$. Therefore, no reduction of the bearing capacity is considered due to the group effect.

The unit base resistance is equal to:

$$q_b = c_u N'_c + q = 9 \times 35 + (9 \times 18 + h \times 20) = 477 + 20h$$

Then:

$$R_b = 0.196 \times (477 + 20h) = 93.66 + 3.93h$$

The unit shaft resistance is equal to:

$$q_s = \alpha c_u$$

For layer 1:

$$q_{s1} = 0.55 \times 25 = 13.8 \text{ kPa}$$

$$R_{s1} = 13.8 \times 1.57 \times 9 = 195 \text{ kN}$$

For layer 2:

$$q_{s2} = 0.55 \times 35 = 19.3 \text{ kPa}$$

$$R_{s2} = 19.3 \times 1.57 \times h = 30.3h \; (\text{kN}, \text{m})$$

and

$$R_s = 195 + 30.3h$$

Thus, the ultimate resistance can be computed:

$$R_c = 93.66 + 3.93h + 195 + 30.3h = 288.7 + 34.2h$$

Safety is achieved by imposing:

$$F_c \leq \frac{R_c - W_p}{FS}$$

resulting in:

$$156 \leq (288.7 + 34.2h - 44.2 - 4.9h)/3$$

$$h \geq 7.6 \text{ m}$$

The pile length will then be:

$$D = 9 + 7.6 = 16.6 \text{ m}$$

Block failure mode

Equation 7.84 provides the resistance of the pile group as a block:

$$R_{ug,block} = 2D(L_B + B_B)c_{u_1} + B_B L_B c_{u_2} N_c'$$

Taking $D = 16.6$ m, $B_B = 3.5$ m and $L_B = 3.5$ m, the bearing capacity factor N_c' is equal to:

$$N_c' = 5\left(1 + \frac{D}{5B_B}\right)\left(1 + \frac{B_B}{5L_B}\right) \le 9$$

Lateral resistance:

$$R_{sg,block} = 2 \times 9 \times (3.5 + 3.5) \times 25 + 2 \times 7.6 \times (3.5 + 3.5) \times 35 = 6874 \text{ kN}$$

Base resistance:

$$N_c' = 5\left(1 + \frac{16.6}{5 \times 3.5}\right)\left(1 + \frac{3.5}{5 \times 3.5}\right) \le 9$$

$$N_c' = 9$$

$$R_{bg,block} = 3.5 \times 3.5 \times 35 \times 9 = 3859 \text{ kN}$$

Note that this expression does not take in account the contribution of the total vertical stress at the pile base, and therefore the self-weight of the block does not need to be added to the load from the structure.

Total resistance:

$$R_{ug,block} = 6874 + 3859 = 10732 \text{ kN}$$

Allowable load:

$$R_{ug,block,all} = \frac{10732}{3} = 3577 \text{ KN} \ge 1400 \text{ kN}$$

ER7.5

a.1) The load distribution by the piles is obtained by solving the system of Equations 7.89, which reflects the displacement compatibility and the equilibrium requirements.

Since the displacement of one pile is influenced by the presence of other piles in the vicinity, as shown in Equation 7.85, it is needed to compute the influence coefficients α_{ij} (see Equation 7.87).

Assuming that the pile deflection develops only to a depth lower than the location of the boundary between the two ground layers, the problem is governed by the properties of the upper layer. Hence, it is assumed to be a homogeneous ground with the properties of the shallow layer.

As an example, the influence coefficient α_{12} is presented here:
The ζ_{12} is a dimensionless coefficient given by:

$$\zeta_{12} = \left[\frac{E_p}{G_s\left(1+\frac{3v_s}{4}\right)}\right]^{\frac{1}{7}} \frac{B}{2s_{12}} = \left[\frac{30\times10^6}{7692\left(1+\frac{3\times0.3}{4}\right)}\right]^{\frac{1}{7}} \frac{0.6}{2\times2.4} = 0.3957$$

where

$$G_s = \frac{E}{2(1+v_s)} = \frac{20\times10^3}{2(1+0.3)} = 7692 \text{ kPa}$$

Hence:

$$\alpha_{12} = 0.6\zeta_{12}\left(1+\cos^2\psi_{12}\right) = 0.6\times0.3957\times\left(1+\cos^2(0)\right) = 0.4748$$

The remaining coefficients are given in Table ER7.5a.

The group horizontal stiffness is given by Equation 7.86, being necessary to compute k' and λ:

i)   the Winkler coefficient, k', can be computed from Equation 7.80:

$$k' = 0.65\sqrt[12]{\frac{EB^4}{E_pI_P}}\frac{E}{1-v^2} = 0.65\sqrt[12]{\frac{20\times10^3\times0.6^4}{30\times10^6\times\frac{\pi\times0.6^4}{64}}}\times\frac{20\times10^3}{1-0.3^2} = 9984 \text{ kN/m}^2$$

ii)   the λ parameter is given by Equation 7.76:

$$\lambda = \sqrt[4]{\frac{k'}{4EI}} = \sqrt[4]{\frac{9984}{4\times30\times10^6\times\frac{\pi\times0.6^4}{64}}} = 0.338 \text{ m}^{-1}$$

iii)   thus, K_t is given by:

$$K_t = \frac{k'}{\lambda} = 29523 \text{ kN/m}$$

Table ER7.5a Influence coefficients, α_{ij}

Pile	1	2	3	4	5	6	7	8
1	1.0000	0.4748	0.2374	0.2374	0.1911	0.1187	0.1274	0.1259
2	0.4748	1.0000	0.4748	0.2518	0.2518	0.1274	0.1187	0.1274
3	0.2374	0.4748	1.0000	0.1903	0.2374	0.1259	0.1274	0.1187
4	0.2374	0.2518	0.1903	1.0000	0.2374	0.2374	0.2518	0.1903
5	0.1903	0.2518	0.2374	0.2374	1.0000	0.1903	0.2518	0.2374
6	0.1187	0.1274	0.1259	0.2374	0.1903	1.0000	0.4748	0.2374
7	0.1274	0.1187	0.1274	0.2518	0.2518	0.4748	1.0000	0.4748
8	0.1259	0.1274	0.1187	0.2374	0.2374	0.2374	0.4748	1.0000

Table ER7.5b Lateral load distribution

Pile	1	2	3	4	5	6	7	8
Load (kN)	123.4	51.6	123.4	101.6	101.6	123.4	51.6	123.4

The distribution of load by the different piles is given by solving the system of Equations 7.89, and is included in Table ER7.5b.

As can be seen, the lateral load on the piles located at group corners 1,3, 6 and 8 are subjected to a lateral load that is more than twice the lateral load on the piles located in the intermediate row (2 and 7).

The pile group lateral displacement is computed from Equation 7.85:

$$u_1 = \frac{1}{K_t} \sum_{j=1}^{n} \alpha_{1j} H_j = 8.72 \text{ mm}$$

a.2) Piles 1, 3, 6 and 8 are subjected to the largest lateral load. Taking pile 1, for instance, the lateral load is equal to 123.4 kN and the pile head displacement is equal to 8.72 mm. The Winkler coefficient for pile 1, k'_1, is computed from equation 7.92:

$$k'_1 = \left(\frac{H_1^4}{4 E_p I_p u_1^4} \right)^{\frac{1}{3}} = \left(\frac{123.4^4}{4 \times 30 \times 10^6 \dfrac{\pi \times 0.6^4}{64} \times 0.00872^4} \right)^{\frac{1}{3}} = 3745.24 \text{ kN/m}$$

and

$$\lambda_1 = \sqrt[4]{\frac{k'_1}{4EI}} = \sqrt[4]{\frac{3745.24}{4 \times 30 \times 10^6 \times \dfrac{\pi \times 0.6^4}{64}}} = 0.2647 \text{ m}^{-1}$$

Since the pile head rotation is restricted, the maximum bending moment is located at the pile head, i.e., for $z = 0$. Therefore, taking into account the equations presented in Table 7.10:

$$M_1(z = 0) = \frac{H_1}{2\lambda_1} = \frac{123.4}{2 \times 0.2647} = 233.1 \text{ kN} \cdot \text{m}$$

b) If the pile–soil–pile interaction effects are neglected, the load is uniformly distributed by the eight piles and each one is subjected to a lateral load of 100 kN applied on its head.

As in question a.2), the pile head rotation is restricted and the maximum moment is located at the pile head. Thus:

$$M_1(z = 0) = \frac{H_1}{2\lambda} = \frac{100}{2 \times 0.338} = 148 \text{ kN} \cdot \text{m}$$

Note that the value of λ was computed in question a.1).

Comparing the maximum bending moment, a large difference is observed. In conclusion, pile bending moments are clearly underestimated if pile–soil–pile interaction is neglected.

ER7.6

a) According to Eurocode 7 – Design Approach 1, two load combinations must be considered with different values for the partial factors of safety.
Combination 1

$$E_d = 1.35 \times G_k + 1.5 \times Q_k$$

Thus:

$$V_{Ed} = 1.35 \times 9300 + 1.5 \times 2500 = 16305 \text{ kN}$$

$$M_{x,Ed} = 1.35 \times 100 = 135 \text{ kN}$$

$$M_{y,Ed} = 1.35 \times 1000 + 1.5 \times 2000 = 4350 \text{ kN}$$

Taking into account Equation 7.81, the axial load in the most stressed pile is given by:

$$N_{Ed}^{max} = \frac{16305}{8} + \frac{135 \times 0.9}{8 \times 0.9^2} + \frac{4350 \times 2.7}{4 \times 0.9^2 + 4 \times 2.7^2} = 2419 \text{ kN}$$

Combination 2

$$E_d = 1.0 \times G_k + 1.3 \times Q_k$$

Thus:

$$V_{Ed} = 1.0 \times 9300 + 1.3 \times 2500 = 12550 \text{ kN}$$

$$M_{x,Ed} = 1.0 \times 100 = 100 \text{ kN}$$

$$M_{y,Ed} = 1.0 \times 1000 + 1.3 \times 2000 = 3600 \text{ kN}$$

Taking into account Equation 7.81, the axial load in the most stressed pile is given by:

$$N_{Ed}^{max} = \frac{12550}{8} + \frac{100 \times 0.9}{8 \times 0.9^2} + \frac{3600 \times 2.7}{4 \times 0.9^2 + 4 \times 2.7^2} = 1883 \text{ kN}$$

If Eurocode 7 - Design Approach 2 is used, the load combination is equal to:

$$E_d = 1.35 \times G_k + 1.5 \times Q_k$$

Thus:

$$V_{Ed} = 1.35 \times 9300 + 1.5 \times 2500 = 16305 \text{ kN}$$

$$M_{x,Ed} = 1.35 \times 100 = 135 \text{ kN}$$

$$M_{y,Ed} = 1.35 \times 1000 + 1.5 \times 2000 = 4350 \text{ kN}$$

Taking into account Equation 7.81, the axial load in the most stressed pile is given by:

$$N_{Ed}^{max} = \frac{16305}{8} + \frac{135 \times 0.9}{8 \times 0.9^2} + \frac{4350 \times 2.7}{4 \times 0.9^2 + 4 \times 2.7^2} = 2419 \text{ kN}$$

b) Soil classification:

According to Table A7.1.1, the upper layer is classified as loose sand, the intermediate layer as stiff clay and the bottom layer as dense gravel.

Pile Type:

CFA piles are defined as pile type 6, and belong to Group G1, Class 2.

Unit base resistance:

Penetrometer base resistance coefficient, k_c, is a function of the effective penetration depth, D_{ef}, which is given by Equation 7.32. Since the pile length is unknown, it is assumed, in a first iteration, that the pile toe penetrates the gravel layer and $D_{ef}/B > 5$. Thus, for a first trial, k_c is equal $k_{c,max}$, the value of which is given in Table A7.1.4 as a function of the pile classification. For CFA piles, $k_{c,max} = 0.25$.

From Equation 7.27:

$$q_b = 0.25 \times 18 = 4.5 \text{ MPa}$$

Unit shaft resistance:

The unit shaft resistance is given by equation 7.33.

The $\alpha_{pile-soil}$ parameter is given in Table A7.1.5. For pile type 4 installed in sandy soils and gravels, this parameter is equal to 1.25, while for clayey soils its value is 0.75.

The value of f_{soil} depends on CPT cone resistance and on the soil type, as described by Equation 7.34. The parameters a, b and c are given in Table A7.1.6. For sand and gravel the constants are: $a = 0.0012$, $b = 0.1$, $c = 0.15$. For clayey soils, these constants assume the following values: $a = 0.0018$, $b = 0.1$, $c = 0.4$.

Thus, the unit shaft resistance for each of the layers can be found in Table ER7.6a.

Group effects

For non-displacement piles installed in granular soils with a center-to-center spacing of $3B$, the group efficiency is equal to 0.77 (see Section 7.7.2), so the pile axial resistance is given by:

$$R_c = E_g \times (R_b + R_s)$$

Table ER7.6a Unit shaft resistance

Soil Layer	q_c (MPa)	$\alpha_{pile-soil}$	f_{soil} (kPa)	$q_{s,max}$ (kPa) (1)	q_s (kPa)
Loose sand	3.5	1.25	43	170	53
Stiff Clay	8	0.75	110	90	82
Dense gravel	18	1.25	113	170	142

Bearing capacity:
The base resistance is given by Equation 7.3:

$$R_b = 4500 \times \pi \times \frac{0.6^2}{4} = 1272 \text{ kN}$$

The shaft resistance is given by Equation 7.4:

$$R_s = (\pi \times 0.6)(8 \times 53 + 4 \times 82 + 142 \times L)$$

where L is the pile embedment length in the gravel layer.
Assessment of the pile length according to Eurocode 7
The characteristic pile resistance is given by the introduction of the correlation factors to the computed resistances. Only one test profile was developed, $\xi_3 = \xi_4 = 1.4$ (see Table 7.12), so the characteristic resistance of the pile is given by:

$$R_{ck} = E_g \times \left(\frac{R_b}{1.4} + \frac{R_s}{1.4} \right)$$

Combination 1

$$F_{c,d} \leq R_{c,d}$$

with $F_{c,d} = 2419$ kN, and:

$$R_{ck} = E_g \times \left(\frac{R_b}{1.4 \times \gamma_b} + \frac{R_s}{1.4 \times \gamma_s} \right)$$

where γ_b and γ_s are equal to 1.1 and 1.0, respectively.
Thus:

$$2419 \leq 0.77 \times \left(\frac{1272}{1.4 \times 1.1} + \frac{(\pi \times 0.6)(8 \times 53 + 4 \times 82 + 142 \times L)}{1.4 \times 1.0} \right)$$

$$L \geq 6.8 \text{ m}$$

Combination 2

$$F_{c,d} \leq R_{c,d}$$

with $F_{c,d} = 1883$ kN, and:

$$R_{ck} = E_g \times \left(\frac{R_b}{1.4 \times \gamma_b} + \frac{R_s}{1.4 \times \gamma_s} \right)$$

where γ_b and γ_s are equal to 1.45 and 1.3, respectively.
Thus:

$$1883 \leq 0.77 \times \left(\frac{1272}{1.4 \times 1.45} + \frac{(\pi \times 0.6)(8 \times 53 + 4 \times 82 + 142 \times L)}{1.4 \times 1.3} \right)$$

$$L \geq 7.0 \text{ m}$$

The pile must penetrate 7.0 m into the gravel layer. For this penetration length, $D_{eff}/B > 5$, which means that the assumption of $k_c = k_{c,max}$ is correct.

Taking in consideration Eurocode 7 - Design approach 2:

$$F_{c,d} \leq R_{c,d}$$

with $F_{c,d} = 2419$ kN and:

$$R_{ck} = E_g \times \left(\frac{R_b}{1.4 \times \gamma_b} + \frac{R_s}{1.4 \times \gamma_s} \right)$$

where γ_b and γ_s are equal to 1.1 and 1.1, respectively (R2).
Thus:

$$2419 \leq 0.77 \times \left(\frac{1272}{1.4 \times 1.1} + \frac{(\pi \times 0.6)(8 \times 53 + 4 \times 82 + 142 \times L)}{1.4 \times 1.1} \right)$$

$$L \geq 8.0 \text{ m}$$

For Design Approach 2, the required embedment in the gravel layer is equal to 8.0 m.

c) For the design of the longitudinal steel rebars in the pile, it is necessary to compute the bending moments. The structural design of the piles is governed by Combination 1, thus the design lateral loads acting in the pile group are equal to:

$$H_{Ed,x} = 1.35 \times 100 + 1.5 \times 500 = 885 \text{ kN}$$

$$H_{Ed,y} = 1.35 \times 70 + 1.5 \times 30 = 139.5 \text{ kN}$$

Assuming that pile deflection only develops to a depth lower than the thickness of the upper soil layer, the ground profile can be assumed to be homogeneous with the properties of the loose sand.

As an example, the influence coefficient α_{16} is presented here.
The ζ_{16} is a dimensionless coefficient given by:

$$\zeta_{16} = \left[\frac{E_p}{G_s \left(1 + \frac{3v_s}{4} \right)} \right]^{\frac{1}{7}} \frac{B}{2s_{16}} = \left[\frac{30 \times 10^6}{4700 \left(1 + \frac{3 \times 0.3}{4} \right)} \right]^{\frac{1}{7}} \frac{0.6}{2 \times \sqrt{1.8^2 + 1.8^2}} = 0.4$$

where

$$G_s = \frac{E}{2(1 + v_s)}$$

The deformation modulus can be estimated by correlation with q_c, where $E = 3.5 q_c = 12.25$ MPa.
Thus:

Table ER7.6b Lateral load distribution

Pile	1	2	3	4	5	6	7	8
Load (kN)	190.2	34.0	53.0	170.7	170.72	53.02	34.0	190.2

Table ER7.6c Maximum bending moments

Pile	1	2	3	4	5	6	7	8
M_{ED} (kN.m)	408	129.5	174.1	379.6	379.6	174.1	129.5	408

$$G_s = \frac{12.25}{2(1+0.3)} = 4.7 \text{ MPa}$$

Hence:

$$\alpha_{12} = 0.6\zeta_{16}\left(1+\cos^2\psi_{16}\right) = 0.6 \times 0.3957 \times \left(1+\cos^2(53.96)\right) = 0.323$$

Solving the system of Equations 7.89, the load distribution by the piles is obtained, as shown in Table ER7.6b.

Using Equation 7.85, the displacement of the pile cap is equal to 1.97 cm.

Taking into account Equation 7.92 and the expressions shown in Table 7.10, the maximum moments occur at the pile head location and are presented in Table ER7.6c.

The most unfavorable loading occurs in pile 1, which has the lower compression axial load and the maximum moment. Therefore, the stresses in this pile will be assumed for the design of all piles:

$$M_{ED} = 408 \text{ kN} \cdot \text{m}; \quad N_{ED} = 1656 \text{ kN}$$

Thus, taking into account Eurocode 2 design rules, the longitudinal steel reinforcement should be constituted by 9 rebars with 20 mm diameter.

CHAPTER 8

ER8.1

As the retaining wall is practically rigid, the movement of its back face is controlled by the deformation of the foundation ground. This deformation decreases progressively from case 1 to case 4: i) in case 2 the deformation is smaller than in case 1 because the effective stresses are larger; ii) in cases 3 and 4 the reason is more obvious: the stiffness and the strength of the foundation ground are superior (case 3) or far superior (case 4). So, this leads to:

$$P_1 < P_2 < P_3 < P_4$$

If the deformation is practically null in case 4, so will be the back face movement; therefore, the thrust will be very close to, or practically coincident with, the at-rest thrust, P_0. So, an upper limit can be added to the previous relation:

$$P_1 < P_2 < P_3 < P_4 \leq P_0$$

The retaining wall design has taken as reference the conditions of case 1. The introduction of a safety factor of 2.0 means that the wall resistance is 100% higher than the active thrust, in deterministic terms. Thus, even in case 1 the wall movement will be smaller than required to mobilize the active thrust in the backfill (see Section 8.2.7). So a lower bound can be inserted in the previous relation:

$$P_a < P_1 < P_2 < P_3 < P_4 \leq P_0$$

ER8.2

If the retaining walls have different cross sections (while sharing all the other conditions), this means that they were designed on the basis of distinct safety factors, being larger the one adopted for wall 2. The inclination and eccentricity of the foundation load for wall 2 are smaller, so the ground deformation is smaller. Therefore, the back face displacement for wall 2 is smaller. So, the following relation can be written, to start with:

$$P_1 < P_2$$

As the retaining walls are founded on soil, their base has experienced some displacement, and so:

$$P_1 < P_2 < P_0$$

Additionally, if we assume that even wall 1 has been designed with a reasonable safety margin relative to the active thrust, P_a, then it follows that:

$$P_a < P_1 < P_2 < P_0$$

ER8.3

	Agree	Disagree
1 – It would have been more prudent to design the wall for the passive thrust, which is the largest, rather than the active, which is the smallest.		x
2 – The fact that thrust is, in both cases, larger than the active thrust is reasonable, because safety factors have been introduced in the design process.	x	
3 – The wall is in danger of collapsing because the thrust due to backfill is larger than that considered in the design, the active thrust.		x
4 – The resistance and stiffness of the foundation soil in section B are lower, providing larger wall displacement, so earth pressures are lower.	x	
5 – The thrust being larger in section A means that in A the wall displacement was larger than in section B.		x
6 – The soil thrust in both sections is lower than the at-rest thrust.	x	

1 - Disagree. The statement is absurd. The passive thrust is to be accounted for when the structure (or a structural element) is supported by the ground, receiving pressures from it of reactive nature.
2 - Agree. Due to the introduction of (partial or global) safety factors to account for the inherent-design uncertainties, the wall resistance ends up being larger than the active thrust (in deterministic terms). Therefore, the displacement of the wall back face will be smaller than required for mobilizing the active state.
3 - Disagree. See comment to statement 2.
4 - Agree. The thrust is smaller in section B because the displacement of the back face was larger in that section. As the wall is practically rigid, this happened because in section B the foundation ground experienced larger deformation.
5 - Disagree. See comment to statement 4.
6 - Agree. It seems reasonable that the wall suffered displacement in both sections, so the thrust will have to be lower than the at-rest value.

ER8.4

a) For this example, Eurocode 7 – Design Approach 1 was followed.

For this type of retaining wall and for external design, the active Rankine thrust should be considered at a vertical plane coinciding with the inner limit of the base (see Section 8.2.5.3), as shown in Figure ER8.4a.

Figure ER8.4b summarizes the actions and the respective partial factors for the two combinations. For Combination 1 two situations are considered: i) the effects of the weight of the wall, of the soil block above the base and of the surcharge on this block are favorable; and ii) the effects of the weight of the wall, of the soil block above the base and of the surcharge on this block are unfavorable (see, for instance, Figure 3.14).

For the ULS of base sliding and foundation failure, Combination 2 governs the design. So, the calculations will be presented only for this combination.

Table ER8.4 summarizes the magnitude of the forces above the wall foundation, the respective lever arms in relation to the barycenter of the wall base, and the respective moments in relation to this point.

Base sliding

With regard to base sliding (see Section 8.2.6) the safety factor is assessed by Equation 8.3. Taking into consideration the values of the table above:

$$T_{Sd} = P_a + P_a^q = 179.6 \text{ kN/m}$$

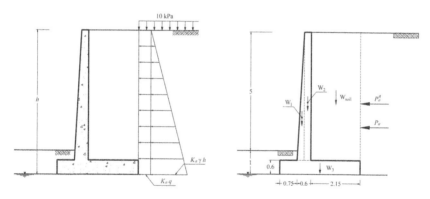

Figure ER8.4a   Earth pressures and gravity forces acting on the retaining wall and on the adjacent soil block.

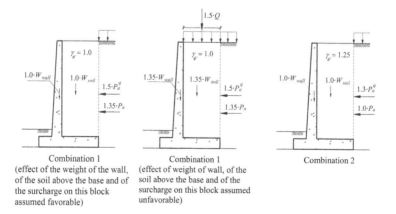

Combination 1
(effect of the weight of the wall, of the soil above the base and of the surcharge on this block assumed favorable)

Combination 1
(effect of weight of wall, of the soil above the base and of the surcharge on this block assumed unfavorable)

Combination 2

Figure ER8.4b   Actions and respective partial safety factors for the Design Approach I of Eurocode 7.

Table ER8.4   Summary of the forces and moments above the wall fondation.

Force	W_1	W_2	W_3	W_{soil}	ΣW_i (2)	P_a	P_a^q	$P_a + P_{aq}$ (3)
Magnitude (kN/m)	19.4	38.9	50.4	232.2	340.9	147.6	32.0	179.6
Lever arm (m) (1)	0.80	0.55	0	0.675	–	2.0	3.0	
Moment (kN.m/m) (1)	15.5	21.4	0	–156.7	–	295.2	96.0	$\Sigma M_i = 271.4$

Notes:

1 – Moments are calculated in relation to the center of the base of the footing.

2 – This force is N_s for the limit state of base sliding and V_s for the limit state of bearing capacity.

3 – This force is T_s for the limit state of base sliding and H_s for the limit state of bearing capacity.

$$T_{Rd} = \Sigma W_i \cdot \tan \delta_{b,d} = 340.9 \times \tan 29.3° = 191.3 \text{ kN/m}$$

Then, it can be concluded that:

$$T_{Sd} \leq T_{Rd}$$

Foundation soil failure

The system of forces applied at the barycenter of the foundation are:

$$(V, H, M) = (340.9 \text{ kN/m}; 179.6 \text{ kN/m}; 271.4 \text{ kN} \cdot \text{m/m})$$

The eccentricity of the vertical force can be calculated:

$$e = M/V = 0.80 \text{ m}$$

Therefore, for computation of the bearing capacity of the foundation, the effective width of the footing is equal to:

$$B' = B - 2e = 1.9 \text{ m}$$

In the present case, among the corrective factors of the resistance, only the ones for load inclination, i_q and i_γ, need to be considered. The following expression allows the determination of the bearing capacity of the foundation, q_{ult}, according to Equation 6.26:

$$q_{ult} = q \cdot N_q \cdot i_q + \frac{1}{2} \cdot B' \cdot \gamma' \cdot N_\gamma \cdot i_\gamma$$

Considering Tables 6.1 and 6.3,

$$\text{for } \phi_d' = 35.8°: \begin{cases} N_q = 36.86; & i_q = 0.224 \\ N_\gamma = 39.43; & i_\gamma = 0.106 \end{cases}$$

The bearing capacity can be computed as follows:

$$q_{ult} = 20 \times 36.86 \times 0.224 + \frac{1}{2} \times 1.9 \times (20 - 9.8) \times 39.43 \times 0.106 = 205.6 \text{ kPa}$$

The value of the resistance to vertical loading can then be computed as:

$$V_{Rd} = q_{ult} \times B' = 390.6 \text{ kN/m}$$

With regard to foundation failure (see Section 8.2.6) the safety factor is assessed by Equation 8.4. Taking into consideration the value above and the value from Table ER8.4:

$$V_{Sd} = \Sigma W_i = 340.9 \text{ kN/m}$$

It can be concluded that:

$$V_{Sd} \leq V_{Rd}$$

b) For internal design, it was found that Combination 1 (assuming the effect of the weight of the wall and of the soil above the base as unfavorable, see middle column of Figure ER8.4b) is the critical one. So, the calculations will be presented only for this combination.

For internal design, it must be taken into account that the supported backfill is not really in the active state (see Section 8.2.7). For this reason, the pressures acting on the wall will be considered on the basis of a pressure coefficient equal to:

$$K = \frac{K_a + K_0}{2} = \frac{\dfrac{1}{3} + \dfrac{1}{2}}{2} = 0.42$$

being

$$K_0 = 1 - \sin\phi' = 0.50$$

For this new earth pressure coefficient, the system of forces applied at the barycenter of the foundation is:

$$(V, H, M) = (492.5 \text{ kN/m}; \ 241.9 \text{ kN/m}; \ 338.1 \text{ kN} \cdot \text{m/m})$$

and the eccentricity of the vertical force becomes:

$$e = M / V = 0.69 \text{ m}$$

Figure ER8.4c shows the distribution of soil pressures under the footing for this situation. Note that since $e > B/6$, part of the footing base is not active.

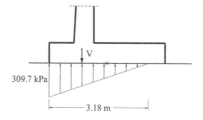

Figure ER8.4c Distribution of the normal pressures under the footing.

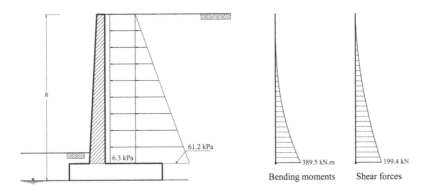

Figure ER8.4d Pressures applied on the stem back face, bending moment and shear forces distributions.

Stem design

The following concrete and steel parameters will be assumed: C30/37 and S500.

Figure ER8.4d shows the pressure graphs applied by the soil at the stem back face, as well as the respective bending moment and shear stress distributions.

Bending reinforcement

The steel reinforcement will be calculated only for the section with maximum bending moment. It will be assumed that:

$$d = h - c = 0.60 - 0.07 = 0.53 \text{ m}$$

where d and h are, respectively, the effective and overall depth of the cross section and c is the reinforcement cover.

Following Eurocode 2 (EN 1992-1-1:2004), to calculate the reinforcement it is necessary to determine the active height of the concrete section:

$$Fc \cdot z = M_{Ed}$$

$$(\chi_1 \cdot b \cdot f_{cd} \cdot x) \cdot (d - \chi_2 \cdot x) = M_{Ed}$$

$$(0.81 \times 1 \times 20000 \cdot x) \times (0.53 - 0.416 \cdot x) = 389.5$$

$$x = 0.047 \text{ m}$$

where:
- F_c is the resultant of the concrete compressive normal stresses at the section;
- z is the lever arm of the internal forces;
- M_{Ed} is the design value of the applied internal bending moment;
- χ_1 is the area coefficient;
- b is the width of the cross section;
- f_{cd} is the design value of concrete compressive strength;
- x represents the height of the concrete section which acts as a resistance to the bending moment;
- χ_2 is the positioning coefficient.

In the following, an equation is formulated to find the required reinforcement to ensure the cross section equilibrium:

$$F_c = F_s$$

$$\chi_1 \cdot b \cdot f_{cd} \cdot x = f_{yd} \cdot A_s$$

$$0.81 \times 1 \times 20000 \times 0.047 = 434783 \cdot A_s$$

$$A_s = 17.5 \text{ cm}^2/\text{m}$$

where:
- F_s is the resultant of the steel tensile stresses at the section;
- f_{yd} is the design yield strength of reinforcement.

Shear reinforcement

According to Eurocode 2, the following equation can be used to determine the maximum shear stress mobilized by the transverse section:

$$V_{Rd,c} = \max\left\{\left[C_{Rd,c} \cdot k \cdot (100 \cdot \rho \cdot f_{ck})^{1/3}\right] \cdot b \cdot d; v_{min} \cdot b \cdot d\right\}$$

$$V_{Rd,c} = \max\left\{\left[0.12 \times 1.614 \times (100 \times 0.002 \times 30)^{1/3}\right] \times 1 \times 530; 0.3932 \times 1 \times 530\right\}$$

$$V_{Rd,c} = \max\{186.5; 208.4\} = 208.4 \text{ kN/m}$$

where:
- $V_{Rd,c}$ is the design shear resistance of the member without shear reinforcement;
- $C_{Rd,c} = \dfrac{0.18}{\gamma_c}$, with γ_c the partial factor for concrete;
- $k = 1 + \sqrt{\dfrac{200}{d}} \leq 2.0$, with d expressed in mm;
- $\rho = \dfrac{A_{sl}}{b \cdot d} < 0.02$, with A_{sl} the area of the tensile reinforcement;
- f_{ck} is the characteristic compressive cylinder strength of concrete at 28 days; and
- $v_{min} = 0.035 \cdot k^{3/2} \cdot f_{ck}^{1/2}$.

As can be seen, the cross section can produce a shear resistance of 208.4 kN/m, which is enough to equilibrate the applied shear stress (199.4 kN/m). Therefore, the minimum reinforcement established in the code must be adopted:

$$\frac{A_{sw}}{s} = \frac{0.08 \cdot z \cdot \sqrt{f_{ck}}}{f_{yk}}$$

$$\frac{A_{sw}}{s} = \frac{0.08 \times 0.9 \times 0.53 \times \sqrt{30}}{500}$$

$$\frac{A_{sw}}{s} = 4.18 \text{ cm}^2/\text{m}$$

where:
- A_{sw} is the cross-sectional area of the shear reinforcement;
- s is the spacing of the stirrups;
- z is the lever arm of the internal forces (the approximate value $z=0.9d$ may normally be used);
- f_{yk} is the characteristic yield strength of reinforcement.

Base design

Figure ER8.4e shows the loading diagram on the base of the wall, as well as the bending moment and shear stress distributions.

Bending reinforcement

Following the same methodology as used for stem design:

$$F_c \cdot z = M_{Ed}$$

$$(0.81 \times 1 \times 20000 \cdot x) \cdot (0.53 - 0.416 \cdot x) = 317.1$$

$$x = 0.038 \text{ m}$$

Imposing the equilibrium of the cross section:

$$F_c = F_s$$

$$0.81 \times 1 \times 20000 \times 0.038 = 434783 \cdot A_s$$

$$A_s = 14.2 \text{ cm}^2/\text{m}$$

Shear reinforcement

$$V_{Rd,c} = 208.4 \text{ kN/m}$$

Since the maximum shear resistance provided by the cross section is not enough to equilibrate the maximum shear stress, the required reinforcement must be determined:

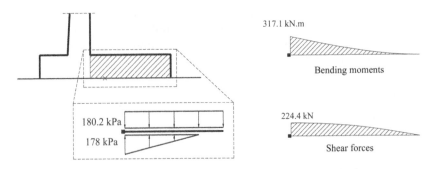

Figure ER8.4e Loading applied on the base, bending moment and shear stress distributions.

$$\frac{A_s}{s} = \frac{V_{Ed}}{z \cdot fyd \cdot \cotan\theta}$$

$$\frac{A_s}{s} = \frac{224.4}{0.9 \times 0.53 \times 434783 \times 2.5}$$

$$\frac{A_s}{s} = 4.33 \text{ cm}^2/\text{m}$$

Toe design

Figure ER8.4f shows the loading diagrams applied on the toe.

The internal stresses are in this case very small. Therefore, the minimum reinforcement is applied.

Bending reinforcement

$$A_s = 0.26 \cdot \frac{f_{ctm}}{f_{yk}} \cdot b \cdot d$$

$$A_s = 0.26 \cdot \frac{2.9}{500} \cdot 1 \cdot 0.53$$

$$A_s = 8 \text{ cm}^2/\text{m}$$

Shear reinforcement

$$\frac{A_{sw}}{s} = \frac{0.08 \cdot z \cdot \sqrt{f_{ck}}}{f_{yk}}$$

$$\frac{A_{sw}}{s} = \frac{0.08 \times 0.9 \times 0.53 \times \sqrt{30}}{500}$$

$$\frac{A_{sw}}{s} = 4.18 \text{ cm}^2/\text{m}$$

Finally, Figure ER8.4g illustrates the simplified reinforcement layout.

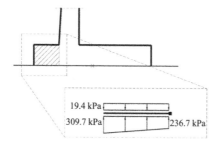

Figure ER8.4f Loading on the toe.

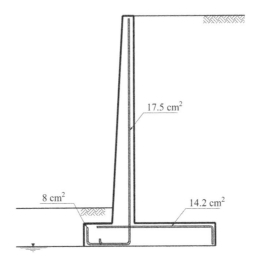

Figure ER8.4g Reinforcement layout

ER8.5

a) In static conditions, the wall is subjected to the loading system indicated in Figure ER8.5a. The passive thrust in front of the wall is neglected.

The wall back face was defined by the dashed line indicated in Figure ER8.5a (see 8.2.5.2). The active thrust coefficient, calculated from Equation 5.69, considering $\phi' = 30°$, $\delta = 30°$, $\beta = 0$ and $\lambda = 17°$, is equal to $K_a = 0.462$. The wall height is:

$$h = \frac{8\cos 17}{\cos(17 + 6)} = 8.31 \text{ m}$$

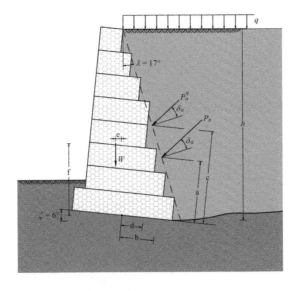

a = 2.67 m
b = 1.51 m
c = 4.00 m
d = 0.92 m
e = 0.04 m

Figure ER8.5a Load system acting on the wall.

The active thrust applied by the backfill is equal to (see Equation 5.68):

$$P_a = \frac{1}{2} K_a \gamma h^2 = \frac{1}{2} \times 0.462 \times 20 \times 8.31^2 = 319.04 \text{ kN/m}$$

and the thrust originated by the surface surcharge is equal to:

$$P_a^q = K_a q h = 0.462 \times 10 \times 8.31 = 38.39 \text{ kN/m}$$

Taking into account the wall cross section represented in Figure ER8.5a and the gabion unit weight ($\gamma_g = 16.25$ kN/m³), the wall weight is:

$W = 333.12$ kN/m.

The tangential force applied at the wall base, T_S, is given by:

$$T_S = P_a \cos(\delta_a + \lambda + \zeta) + P_a^q \cos(\delta_a + \lambda + \zeta) - W \sin \zeta$$

$$= 319.04 \cos(30 + 17 + 6) + 38.48 \cos(30 + 17 + 6) - 333.12 \sin 6$$

$$= 180.29 \text{ kN/m}$$

On the other hand, the tangential resisting force in the plane of the wall base, T_R, is equal to:

$$T_R = N_S \tan \delta_b$$

where N_S is the load acting in the direction normal to the wall base and δ_b is the angle of shearing resistance of the base-soil interface.

Taking into account the load system illustrated in Figure ER8.5a:

$$N_S = W \cos \zeta + P_a \sin(\delta_a + \lambda + \zeta) + P_a^q \sin(\delta_a + \lambda + \zeta)$$

$$= 333.12 \cos 6 + 319.04 \sin(30 + 17 + 6) + 38.39 \sin(30 + 17 + 6)$$

$$= 616.75 \text{ kN/m}$$

Thus,

$$T_R = N_S \tan \delta_b = 616.75 \times \tan 30 = 356.08 \text{ kN/m}$$

The global safety factor against sliding is given by:

$$F_{\text{sliding}} = \frac{T_R}{T_S} = \frac{356.08}{180.29} = 1.98$$

For the calculation of the safety factor, regarding foundation bearing capacity, the load system must be converted in an equivalent system applied at the barycenter of the foundation. The system is composed by a load normal to the wall base, V_S, a load parallel to the wall base, H_S, and a moment, M_S. Loads V_S and H_S have already been computed, and are equal to 616.75 kN/m and 180.29 kN/m, respectively.

The moment transmitted to the wall base barycenter is given by (see the lever arm values in Figure ER8.5a):

$$M_S = P_a \cos(\delta_a + \lambda + \zeta) \times a + P_a^q \cos(\delta_a + \lambda + \zeta) \times c + W \times e$$

$$- P_a \sin(\delta_a + \lambda + \zeta) \times b - P_a^q \sin(\delta_a + \lambda + \zeta) \times d$$

$$= 205.44 \text{ kN} \cdot \text{m/m}$$

The bearing capacity, q_{ult}, is given by Equation 6.26. For $\phi' = 33°$, the bearing capacity factors are:

$$N_c = 38.64; \quad N_q = 26.09; \quad N_g = 24.19.$$

The effective footing width, B', is given by (see Equations 6.17 and 6.19):

$$B' = B - 2 \times \frac{M_S}{V_S} = 4.5 - 2 \times \frac{205.44}{616.75} = 3.83 \text{ m}$$

Regarding the corrective factors, only the load inclination (i factors) and foundation base inclination (b factors) are applicable (see Tables 6.1 to 6.3):

$$i_c = 0.53; \quad i_q = 0.55; \quad i_g = 0.41; \quad b_c = 0.86; \quad b_q = 0.87; \quad b_g = 0.87$$

Regarding the effective vertical stress at the footing base level in front of the wall, q', it is equal to:

$$q' = 1.3 \times 20 = 26 \text{ kPa.}$$

Therefore, applying Equation 6.26:

$$q_{ult} = 808.85 \text{ kPa.}$$

The bearing capacity safety factor is equal to:

$$F_{bearing} = \frac{V_R}{V_S} = \frac{q_{ult} \times B'}{V_S} = 5.02$$

b) Under seismic conditions, the wall is subjected to the loading system illustrated in Figure ER8.5b.

   The incremental lateral earth pressure induced by the earthquake is given by (see Equation 5.105):

$$\Delta P_{ae} = \frac{1}{2} \Delta K_{ae} \gamma h^2$$

where (see Equation 5.104):

$$\Delta K_{ae} = K_{ae}(1 \pm k_v) - K_a$$

The seismic active thrust coefficient, K_{ae}, is computed from the Mononobe-Okabe method (see Equation 5.87).

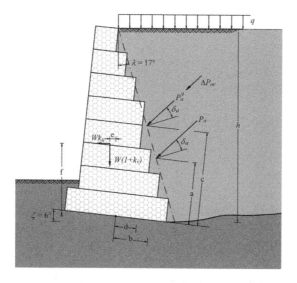

a = 2.67 m
b = 1.51 m
c = 4.00 m
d = 0.92 m
e = 0.04 m
f = 3.10 m

Figure ER8.5b Load system acting on the wall for seismic conditions.

The loading system illustrated in Figure ER8.5b can be converted to an equivalent loading system applied at the barycenter of the wall base and composed by:

$$V_S = W(1 \pm k_v)\cos\zeta + P_a \sin(\delta_a + \lambda + \zeta) + P_a^q \sin(\delta_a + \lambda + \zeta)$$
$$+ \Delta P_{ae} \sin(\delta_a + \lambda + \zeta) + W k_b \sin\zeta$$

$$H_S = P_a \cos(\delta_a + \lambda + \zeta) + P_a^q \cos(\delta_a + \lambda + \zeta) + \Delta P_{ae} \cos(\delta_a + \lambda + \zeta)$$
$$+ W k_b \cos\zeta - W(1 \pm k_v)\sin\zeta$$

$$M_S = P_a \cos(\delta_a + \lambda + \zeta) \times a + P_a^q \cos(\delta_a + \lambda + \zeta) \times c$$
$$+ \Delta P_{ae} \cos(\delta_a + \lambda + \zeta) \times c + W(1 \pm k_v) \times e + W \times k_b \times f$$
$$- P_a \sin(\delta_a + \lambda + \zeta) \times b - P_a^q \sin(\delta_a + \lambda + \zeta) \times d$$
$$- \Delta P_a^q \sin(\delta_a + \lambda + \zeta) \times d$$

Calculation of the safety factors for $k_b = 0.11$ and $k_v = +0.056$
Under such conditions $\theta = 5.95°$ (see Equation 5.80) and $K_{ae} = 0.566$. Thus:

$$\Delta K_{ae} = 0.566 \times (1 + 0.056) - 0.462 = 0.136$$

and

$$\Delta P_{ae} = \frac{1}{2} \Delta K_{ae} \gamma b^2 = \frac{1}{2} \times 0.136 \times 20 \times 8.31^2 = 93.92 \text{ kN/m}$$

Taking into account the static and seismic earth pressures, as well as the wall weight:

$$V_S = 714.14 \text{ kN/m}; \quad H_S = 271.30 \text{ kN/m and } M_S = 476.86 \text{ kN} \cdot \text{m/m}$$

The sliding safety verification is given by Equation 8.2, where $T_S = H_S = 271.30$ kN/m and the resistance, T_R, is given by:

$$T_R = N_S \tan \delta_b$$

where $N_S = V_S = 714.14$ kN/m. Hence,

$$T_R = N_S \tan \delta_b = 714.14 \times \tan 30 = 412.31 \text{ kN/m}$$

$$F_{\text{sliding}} = \frac{T_R}{T_S} = \frac{412.31}{271.30} = 1.52$$

Regarding foundation bearing capacity, the effective footing width, B', is given by:

$$B' = B - 2 \times \frac{M_s}{V_s} = 4.5 - 2 \times \frac{476.86}{714.14} = 3.16 \text{ m}$$

The bearing capacity factors and the corrective factors for the inclination of the foundation base are the same as for static conditions. The corrective factors due to the load inclination have the following new values:

$$i_c = 0.41; \quad i_q = 0.43; \quad i_g = 0.28$$

Therefore, applying Equation 6.26:

$$q_{\text{ult}} = 576.22 \text{ kPa}.$$

and the bearing capacity safety factor is equal to:

$$F_{\text{bearing}} = \frac{V_R}{V_S} = \frac{q_{\text{ult}} \times B'}{V_S} = 2.55$$

Calculation of the safety factors for $k_h = 0.11$ and $k_v = -0.056$
Under such conditions $\theta = 6.64°$ (see Equation 5.80) and $K_{ae} = 0.58$.
Since the procedure is similar to the one explained above (with the sole difference being the sign of k_v), only the main numerical results are presented in what follows:

$$\Delta K_{ae} = 0.58 \times (1 - 0.056) - 0.462 = 0.086$$

$$\Delta P_{ae} = \frac{1}{2} \Delta K_{ae} \gamma h^2 = \frac{1}{2} \times 0.086 \times 20 \times 8.31^2 = 59.39 \text{ kN/m}$$

$$V_S = 649.46 \text{ kN/m}; \quad H_S = 254.4 \text{ kN/m} \quad \text{and} \quad M_S = 417.60 \text{ kN} \cdot \text{m/m}$$

$$T_R = N_S \tan \delta_b = 649.46 \times \tan 30 = 374.97 \text{ kN/m}$$

$$F_{\text{sliding}} = \frac{T_R}{T_S} = \frac{374.97}{254.4} = 1.47$$

$$B' = B - 2 \times \frac{M_S}{V_S} = 4.5 - 2 \times \frac{417.60}{649.46} = 3.21 \text{ m}$$

$$i_c = 0.40; \quad i_q = 0.42; \quad i_\gamma = 0.27$$

$$q_{ult} = 563.19 \text{ kPa}$$

$$F_{bearing} = \frac{V_R}{V_S} = \frac{q_{ult} \times B'}{V_S} = 2.78$$

c) The safety regarding the internal stresses in the wall comprises two main verifications: i) sliding in the block interfaces; ii) excessive compressive stresses in the gabions.

Regarding sliding between blocks, the global factor of safety is given by Equation 8.9. The sliding resistance force is given by Equation 8.10, where ϕ^* is the angle of shear resistance along the interface, which can be estimated by (see Equation 8.11):

$$\phi^* = 25 \times \frac{\gamma_g}{\gamma_w} - 10° = 25 \times \frac{16.25}{9.8} - 10° = 31.5°$$

Table ER8.5a summarizes the results for each level of interface.

Regarding the safety against excessive compression of the gabion blocks, the following procedure is followed:
i)   replace the load system above the interface under analysis to an equivalent system applied at the barycenter of the top face of the lower block (this system is composed of a normal force N, a tangential force, T, and a moment M);
ii) compute the load eccentricity, e, by dividing M by N;
iii) using the criteria summarized in Figure 8.7, compute the value of σ_{max};
iv) compare the value of σ_{max} with the allowable stress, σ_a, given by Equation 8.8:

$$\sigma_a = \left(5 \times \frac{\gamma_g}{\gamma_w} - 3\right) \times 98 = \left(5 \times \frac{16.25}{9.81} - 3\right) \times 98 = 518 \text{ kPa}$$

The results are summarized in Table ER8.5b. As can be seen, the maximum stress values are lower than the allowable stress computed below.

d) Figures ER8.5c and ER8.5d show the stability analysis results, for static and seismic conditions, of a ground mass containing the wall analyzed above and a parallel similar wall. The analysis was performed using the Bishop simplified method.

Table ER8.5a Safety factor against sliding in the distinct gabion interfaces.

Level	B (m)	T_S (kN/m)	T_R (kN/m)	F
1	1.0	4.19	14.69	3.51
2	1.5	13.53	39.21	2.90
3	2	28.02	73.57	2.63
4	2.5	47.66	117.75	2.47
5	2.5	73.31	166.82	2.28
6	3	104.10	225.71	2.17
7	3.5	140.04	294.44	2.10

Table ER8.5b Verification against excessive compression of the gabions.

Level	B (m)	B/6 (m)	M (kN.m/m)	N (kN/m)	e (m)	σ_{max} (kPa)
1	1.0	0.17	−4.24	23.98	−0.18	49.42
2	1.5	0.25	−4.11	63.99	−0.06	53.64
3	2	0.33	6.82	120.05	0.06	70.26
4	2.5	0.42	37.07	192.16	0.19	112.45
5	2.5	0.42	28.95	272.22	0.11	136.68
6	3	0.50	99.07	368.33	0.27	188.23
7	3.5	0.58	200.26	480.48	0.42	235.37

ER8.6

a) The British Method will be adopted (see Section 8.3.2). The Coulomb solution was used to determine the active pressure coefficients, whereas the Caquot-Kérisel tables have been used to select the passive one.

Figure ER8.6a illustrates the horizontal active and passive stresses due to the soil weight and the surface surcharge on both wall faces.

a.1) A safety factor $F_r = 2.0$ will be applied to the net passive resistance below the excavation bottom (see Section 8.3.2.1).

Table ER8.6a presents the basic data and the horizontal thrusts. The first wall analysis will be performed assuming $\delta_a = \delta_p = (2/3) \phi'$.

Figure ER8.6b shows the application points of the horizontal forces.

Horizontal equilibrium

The analysis of the wall horizontal equilibrium involves the formulation of the two following equations, according to Figure 8.21:

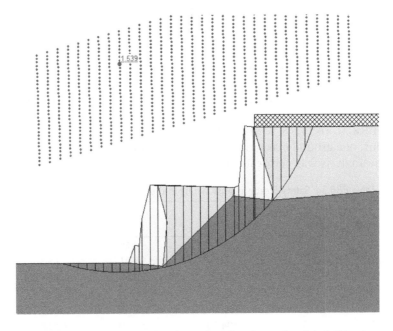

Figure ER8.5c Results of a stability analysis to evaluate the safety factor for global sliding in static conditions by the simplified Bishop method.

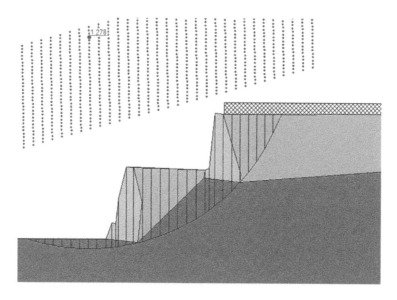

Figure ER8.5d Results of a stability analysis to evaluate the safety factor for global sliding in seismic conditions by the simplified Bishop method.

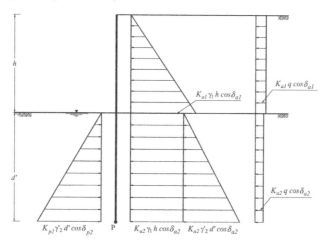

Figure ER8.6a Active and passive horizontal stresses due to the soil weight and the surcharge.

$$\sum M_P = 0$$

$$P_{a1}^q(h) \cdot (d' + 2.5) + P_{a1}(h) \cdot \left(d' + \frac{5}{3}\right) + P_{a2}^1(h) \cdot \frac{d'}{2} + P_{a2}^q(h) \cdot \frac{d'}{2}$$

$$= \frac{\left(P_{p2}(h) - P_{a2}(h)\right)}{F_r} \cdot \frac{d'}{3}$$

$$17.2 \cdot (d' + 2.5) + 81.9 \cdot \left(d' + \frac{5}{3}\right) + 19.86d' \cdot \frac{d'}{2} + 2.09d' \cdot \frac{d'}{2}$$

$$= \frac{\left(43.10d'^2 - 1.07d'^2\right)}{2} \cdot \frac{d'}{3}$$

Table ER8.6a Mechanical properties and acting horizontal forces.

Soil	Limit state	ϕ'	δ	K	Horizontal thrust (kN/m)
Clayey silt (1)	Active	25	$\frac{2}{3}\phi'$	0.36	• $P_{a1}(h) = \frac{1}{2}\cdot\gamma_1\cdot h^2\cdot K_{a1}\cdot\cos\delta_{a1} = 81.9$ • $P_{a1}^q(h) = q\cdot h\cdot K_{a1}\cdot\cos\delta_{a1} = 17.2$
Silty sand (2)	Active	37	$\frac{2}{3}\phi'$	0.23	• $P_{a2}(h) = \frac{1}{2}\cdot\gamma_2'\cdot d'^2\cdot K_{a2}\cdot\cos\delta_{a2} = 1.07d'^2$ • $P_{a2}^1(h) = (\gamma_1\cdot h)\cdot d'\cdot K_{a2}\cdot\cos\delta_{a2} = 19.86d'$ • $P_{a2}^q(h) = q\cdot d'\cdot K_{a2}\cdot\cos\delta_{a2} = 2.09d'$
	Passive	37	$\frac{2}{3}\phi'$	9.3	• $P_{p2}(h) = \frac{1}{2}\cdot\gamma_2'\cdot d'^2\cdot K_{p2}\cdot\cos\delta_{p2} = 43.10d'^2$

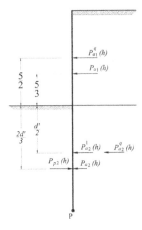

Figure ER8.6b Horizontal thrust application points.

$d' \geq 5.22$ m; $d = 1.2\times 5.22 = 6.26$ m

$\sum F_x = 0$

$$R_d + P_{a1}(h) + P_{a1}^q(h) + P_{a2}^1(h) + P_{a2}^q(h) = \frac{\left(P_{p2}(h) - P_{a2}(h)\right)}{F_r}$$

$$R_d + 81.9 + 17.2 + 103.67 + 10.91 = \frac{(1174.41 - 29.16)}{2}$$

$R_d = 359.0$ kN/m

The force R_d represents the resultant of the soil pressures developed at the embedded wall length between depths d' and $1.2d'$. Figure ER8.6c shows the computation of these pressures. It can be verified that their resultant is greater than the value found for R_d in the analysis above:

$-P_1 + P_2 + P_3 + P_4 \geq R_d$

1430.42 kN $\geq$ 359 kN

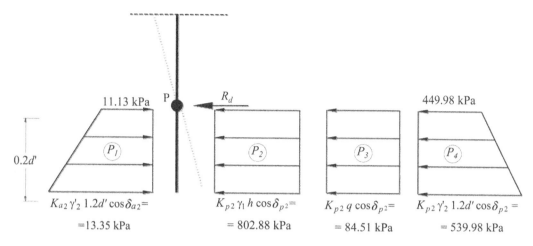

Figure ER8.6c Pressures on the wall between depths d' and $1.2d'$.

Vertical equilibrium

Figure ER8.6d shows the different soil thrusts, which act on the wall with a given inclination. As treated in Section 8.3.2.3, it is necessary to check if the upward tangential component of the passive thrust in front of the wall can be mobilized.

For this check, the wall weight must be determined:

$$W_{w,1} = h \cdot t \cdot \gamma_{conc} + d \cdot t \cdot (\gamma_{conc} - \gamma_w)$$

$$W_{w,1} = 132.13 \text{ kN/m}$$

where:
- t is the wall thickness (0.6 m);
- γ_{conc} is the concrete unit weight;
- the subscript 1 denotes that this calculation corresponds to the 1st iteration (if an iterative process is in fact required).

The check mentioned corresponds to the following inequality:

$$\frac{P_{p2}}{F_r} \cdot \sin\delta_{p2} \leq W_{w,1} + \left(P_{a1} + P_{a1}^q\right) \cdot \sin\delta_{a1} + \left(P_{a2}^1 + P_{a2}^q + \frac{P_{a2}}{F_r}\right) \cdot \sin\delta_{a2}$$

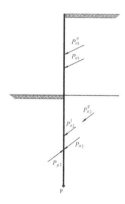

Figure ER8.6d Resultant acting forces.

$$646.2 \cdot \sin\left(\frac{2}{3} \times 37\right) \leq 132.13 + \left(85.5 + 18\right) \cdot \sin\left(\frac{2}{3} \times 25\right)$$

$$+ \left(114.1 + 12.0 + 15.98\right) \cdot \sin\left(\frac{2}{3} \times 37\right)$$

$269.68 \not\leq 221.1$

It can be concluded that the upward force is greater than the available downward forces.

As seen in Section 8.3.2.3, this requires a new wall analysis (a new iteration), by adopting a lower value of the angle δ_p. For the new iteration it was assumed that $\delta_{p,2} = (1/2)\,\phi'$, whereas the value of $\delta_{a,2}$ was maintained to be equal to $(2/3)\,\phi'$.

Table ER8.6b summarizes the data and the results of the 1st and the 2nd iterations, the latter having provided results ensuring vertical equilibrium.

It can be concluded that the embedded wall length is equal to:

$d = 1.2d' = 6.77$ m

which corresponds to a cantilever wall with a total height equal to:

$h + d = 11.77$ m

a.2) The analysis of the wall will now be performed following the Design Approach 1 of Eurocode 7. Tables ER8.6c, ER8.6d, ER8.6e and ER8.7f summarize the data assumed and the results. As indicated in the first column of Table ER8.6c, for Combination 1 it was necessary to perform three iterations, reducing δ_p to $\phi'/3$, to ensure the vertical equilibrium (the results of the 2nd iteration are not shown).

Table ER8.6b Summary of horizontal and vertical equilibrium analyses ($F_r = 2.0$ applied to net passive resistance).

Iteration	Limit state	Soil	ϕ'_k (°)	δ (°)	K	d' (m)	R_d (kN/m)	$d+h$ (m)	W_w (kN/m)	$\sum P_a(v) + W$ (kN/m)	$P_p(v)$ (kN/m)
1	Active	1	25	$\frac{2}{3}\phi'$	0.36	5.22	359.0	11.26	132.13	221.1	269.68
		2	37	$\frac{2}{3}\phi'$	0.23						
	Passive	2	37	$\frac{2}{3}\phi'$	9.3						
2	Active	1	25	$\frac{2}{3}\phi'$	0.36	5.64	355.5	11.77	136.73	231.07	199.3
		2	37	$\frac{2}{3}\phi'$	0.23						
	Passive	2	37	$\frac{1}{2}\phi'$	7.74						

Table ER8.6c Safety factors and data for Design Approach 1 – Combination 1.

Iteration	Limit state	Soil	γ_G	γ_Q	$\gamma_{\phi'}$	ϕ'_k	ϕ'_d	δ	K
1	Active	1	1.35	1.5	1.0	25	25	$\frac{2}{3}\phi'_d$	0.36
		2				37	37	$\frac{2}{3}\phi'_d$	0.23
	Passive	2	1.0			37	37	$\frac{2}{3}\phi'_d$	9.3
3	Active	1	1.35	1.5	1.0	25	25	$\frac{2}{3}\phi'_d$	0.36
		2				37	37	$\frac{2}{3}\phi'_d$	0.23
	Passive	2	1.0			37	37	$\frac{1}{3}\phi'_d$	6.28

It can be concluded from Tables ER8.6d and ER8.6f that Combination 2 governs the wall height (see result in bold):

$$d = 1.2d' = 6.77 \text{ m}$$

and

$$h + d = 11.77 \text{ m}.$$

It is interesting to conclude, as well, that the application of the Design Approach 1 of Eurocode 7 and the application of safety factor $F_r = 2.0$ to the net passive resistance lead to the same result. This corroborates the conclusions extracted from the charts of Figure 8.23.

Table ER8.6d Horizontal thrusts and results of the horizontal and vertical equilibrium, applying Design Approach 1 – Combination 1 of Eurocode 7.

	Horizontal thrust					
Iteration	$P_{a1}(h)$ (kN/m)	$P_{a1}^q(h)$ (kN/m)	$P_{a2}(h)$ (kN/m)	$P_{a2}^1(h)$ (kN/m)	$P_{a2}^q(h)$ (kN/m)	$P_{p2}(h)$ (kN/m)
1	110.6	25.9	$1.44d'^2$	$26.8d'$	$3.14d'$	$43.1d'^2$
2	110.6	25.9	$1.44d'^2$	$26.8d$	$3.14d'$	$31.3d'^2$

	Results					
Iteration	d' (m)	R_d (kN/m)	$d+h$ (m)	W_w (kN/m)	$\sum P_a(v) + W$ (kN/m)	$P_p(v)$ (kN/m)
1	4.32	511.74	10.18	122.3	234.86	369.41
2	5.13	495.5	11.16	131.14	259.9	180.04

Table ER8.6e Safety factors and data for Design Approach 1 – Combination 2.

Iteration	Limit state	Soil	γ_G	γ_Q	$\gamma\phi'$	ϕ'_k	ϕ'_d	δ	K
1	Active	1	1.0	1.3	1.25	25	20.5	$\frac{2}{3}\phi'_d$	0.43
		2				37	31.1	$\frac{2}{3}\phi'_d$	0.28
	Passive	2				37	31.1	$\frac{2}{3}\phi'_d$	5.8
2	Active	1	1.0	1.3	1.25	25	20.5	$\frac{2}{3}\phi'_d$	0.43
		2				37	31.1	$\frac{2}{3}\phi'_d$	0.28
	Passive	2				37	31.1	$\frac{1}{2}\phi'_d$	5.02

Table ER8.6f Horizontal thrusts and results of the horizontal and vertical equilibrium, applying Design Approach 1 – Combination 2 of Eurocode 7.

	Horizontal thrust					
Iteration	$P_{a1}(h)$ (kN/m)	$P_{a1}^q(h)$ (kN/m)	$P_{a2}(h)$ (kN/m)	$P_{a2}^l(h)$ (kN/m)	$P_{a2}^q(h)$ (kN/m)	$P_{p2}(h)$ (kN/m)
1	99.2	27.2	$1.34d'^2$	$24.9d'$	$3.4d'$	$27.7d'^2$
2	99.2	27.2	$1.34d'^2$	$24.9d'$	$3.4d'$	$24.7d'^2$

	Results					
Iteration	d' (m)	R_d (kN/m)	$d+h$ (m)	W_w (kN/m)	$\Sigma P_a(v)+W$ (kN/m)	$P_p(v)$ (kN/m)
1	5.29	460.8	11.35	132.89	234.41	293.05
2	**5.64**	456.2	**11.77**	136.7	243.92	218.3

b) The internal design of the wall will follow the results of the external design applying Design Approach 1 of Eurocode 7, that is, the results presented above. In this case, the results obtained from Combination 1 are the critical ones, and are presented in Figure ER8.6e.

The following concrete and steel parameters will be assumed: C30/37 and S500.

Eurocode 2 was applied to determine the amount of steel needed to satisfy the maximum bending moment, which is equal to 530.9 kN.m/m, and the maximum shear force, which is equal to 495.7 kN/m. Since a detailed explanation of the calculation was provided in exercise ER8.4, only the equations will be included in the present case.

Bending reinforcement

$$F_c \cdot z = M_{Ed}$$

$$(0.81 \times 1 \times 20000 \cdot x) \cdot (0.53 - 0.416 \cdot x) = 530.9$$

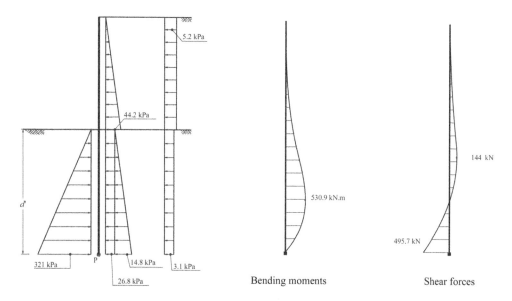

Figure ER8.6e Cantilever wall: loading diagram and bending moment and shear force diagrams from Design Approach I - Combination I.

$x = 0.065$ m

$F_c = F_s$

$0.81 \times 1 \times 20000 \times 0.065 = 434783 \cdot A_s$

$A_s = 24.22$ cm^2/m

Shear reinforcement

$A_s = \dfrac{V_{Ed}}{z \cdot fyd \cdot \mathrm{cotan}\theta}$

$\dfrac{A_s}{s} = \dfrac{495.7}{0.9 \times 0.53 \times 434783 \times 2.5}$

$\dfrac{A_s}{s} = 9.56$ cm^2/m

where 2.5 was assumed to be the maximum value for the cotan θ.

ER8.7

a) The free-earth support method will be adopted (see Section 8.3.3). The Coulomb solution was used to determine the active pressure coefficients, whereas the Caquot-Kérisel tables have been used to select the passive one.

Figure ER8.7a illustrates the active and passive pressures due to the soil weight and the surface surcharge on both wall faces.

a.1)The design will follow Design Approach 1 of Eurocode 7.

Table ER8.7a presents the basic data and the horizontal thrusts for Combination 1.

Figure ER8.7b shows the application points of the horizontal forces.

Horizontal equilibrium

The analysis of the wall horizontal equilibrium involves the formulation of the two following equations, according to Figure 8.25:

$$\sum M_P = 0$$

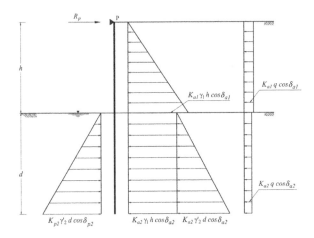

Figure ER8.7a Active and passive horizontal stresses due to the soil weight and the surcharge.

Table ER8.7a Mechanical properties and acting forces for Design Approach 1 – Combination 1 of Eurocode 7.

Soil	Limit state	γ_G	γ_Q	$\gamma_{\phi'}$	ϕ_d'	δ	K	Horizontal thrust (kN/m)
Clayey silt (1)	Active	1.35	1.5	1.0	25	$\frac{2}{3}\phi_d'$	0.36	• $P_{a1}(h) = \dfrac{1}{2}\cdot\gamma_1\cdot h^2\cdot K_{a1}\cdot\cos\delta_{a1} = 110.6$
								• $P_{a1}^q(h) = q\cdot h\cdot K_{a1}\cdot\cos\delta_{a1} = 25.9$
Silty sand (2)	Active	1.35	1.5	1.0	37	$\frac{2}{3}\phi_d'$	0.23	• $P_{a2}(h) = \dfrac{1}{2}\cdot\gamma_2'\cdot d^2\cdot K_{a2}\cdot\cos\delta_{a2} = 1.44d^2$
								• $P_{a2}^l(h) = (\gamma_1\cdot h)\cdot d\cdot K_{a2}\cdot\cos\delta_{a2} = 26.8d$
								• $P_{a2}^q(h) = q\cdot d\ \cdot K_{a2}\cdot\cos\delta_{a2} = 3.14d$
	Passive				37	$\frac{2}{3}\phi_d'$	9.3	• $P_{p2}(h) = \dfrac{1}{2}\cdot\gamma_2'\ \cdot d^2\cdot K_{p2}\cdot\cos\delta_{p2} = 43.10d^2$

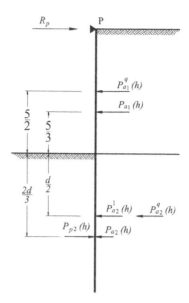

Figure ER8.7b Horizontal thrust application points.

$$P_{a1}^q(h) \cdot 2.5 + P_{a1}(h) \cdot \left(\frac{2}{3} \cdot 5\right) + P_{a2}^q(h) \cdot \left(\frac{d}{2}+5\right) + P_{a2}^1(h) \cdot \left(\frac{d}{2}+5\right)$$

$$+ P_{a2}(h) \cdot \left(\frac{2}{3}d+5\right) = P_{p2}(h) \cdot \left(\frac{2}{3}d+5\right)$$

$$25.9 \cdot (2.5) + 110.6 \cdot \left(\frac{2}{3} \cdot 5\right) + 26.8d \cdot \left(\frac{d}{2}+5\right) + 3.14d \cdot \left(\frac{d}{2}+5\right)$$

$$+ 1.44d^2 \cdot \left(\frac{2}{3}d+5\right) = 43.10d^2 \cdot \left(\frac{2}{3}d+5\right)$$

$d \geq 1.69$ m

$\Sigma F_x = 0$

$$R_p + P_{p2}(h) = P_{a1}^q(h) + P_{a1}(h) + P_{a2}^q(h) + P_{a2}^1(h) + P_{a2}(h)$$

$$R_p + 123.1 = 25.9 + 110.6 + 45.29 + 5.31 + 4.11$$

$$R_p = 68.11 \text{ kN/m}$$

From the previous equations it can be seen that for the horizontal equilibrium to be verified the prop needs to produce a reaction force equal to $R_p = 68.11$ kN/m.

Vertical equilibrium

Figure ER8.7c shows the distinct soil thrusts, which act on the wall with a given inclination. As treated in Section 8.3.3.3, it is necessary to check if the upward tangential component of the passive thrust in front of the wall can be mobilized. For this check, the wall weight must be determined:

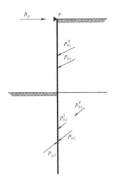

Figure ER8.7c Resultant acting forces.

$$W_{w,1} = h \cdot t \cdot \gamma_{\text{conc}} + d \cdot t \cdot \left(\gamma_{\text{conc}} - \gamma_w \right)$$

$$W_{w,1} = 60.28 \text{ kN/m}$$

where:
- t is the wall thickness (0.4 m);
- γ_{conc} is the concrete unit weight;
- the subscript 1 denotes that this calculation corresponds to the 1st iteration (if an iterative process is in fact required).

The mentioned check corresponds to the following inequality:

$$P_{p2} \cdot \sin\delta_{p2} \leq W_{w,1} + (P_{a1} + P_{a1}^q) \cdot \sin\delta_{a1} + (P_{a2} + P_{a2}^1 + P_{a2}^q) \cdot \sin\delta_{a2}$$

$$135.46 \cdot \sin\left(\frac{2}{3} \times 37 \right) \leq 60.28 + \left(115.5 + 27 \right) \cdot \sin\left(\frac{2}{3} \times 25 \right)$$

$$+ \left(4.52 + 49.84 + 5.84 \right) \cdot \sin\left(\frac{2}{3} \times 37 \right)$$

$$56.53 \text{ kN} \leq 126.27 \text{ kN}$$

It can be concluded that the check of vertical equilibrium is satisfied, so the external design can be considered to be concluded for this combination.

Table ER8.7b Safety factors and data for Design Approach I – Combination I.

Iteration	Limit state	Soil	γ_G	γ_Q	$\gamma_{\phi'}$	ϕ'_k	ϕ'_d	δ	K
I	Active	I	1.35	1.5	1.0	25	25	$\frac{2}{3}\phi'_d$	0.36
		2				37	37	$\frac{2}{3}\phi'_d$	0.23
	Passive	2	1.0			37	37	$\frac{2}{3}\phi'_d$	9.3

Table ER8.7c Horizontal thrusts and results of the horizontal and vertical equilibrium, applying Design Approach 1 – Combination 1 of Eurocode 7.

	Horizontal thrust					
Iteration	$P_{a1}(h)$ (kN/m)	$P_{a1}^q(h)$ (kN/m)	$P_{a2}(h)$ (kN/m)	$P_{a2}^1(h)$ (kN/m)	$P_{a2}^q(h)$ (kN/m)	$P_{a2}(h)$ (kN/m)
1	110.6	25.9	$1.44d^2$	26.8d	3.14d	$43.1d^2$

	Results					
Iteration	d (m)	R_p (kN/m)	$d+h$ (m)	W_w (kN/m)	$\Sigma P_a(v)+W$ (kN/m)	$P_p(v)$ (kN/m)
1	1.69	68.11	6.69	60.26	126.27	56.53

Table ER8.7d Safety factors and data for Design Approach 1 – Combination 2.

Iteration	Limit state	Soil	γ_G	γ_Q	$\gamma_{\phi'}$	ϕ_k'	ϕ_d'	δ	K
1	Active	1	1.0	1.3	1.25	25	20.5	$\frac{2}{3}\phi_d'$	0.43
		2				37	31.1	$\frac{2}{3}\phi_d'$	0.28
	Passive	2	1.0			37	31.1	$\frac{2}{3}\phi_d'$	5.8

Tables ER8.7b to ER8.7e and Figure ER8.7d summarize the calculations according to both combinations of Design Approach 1.

It can be concluded that Combination 2 governs the wall height (see result in bold):

$d = 2.1$ m

and

$h + d = 7.1$ m.

Table ER8.7e Horizontal thrusts and results of the horizontal and vertical equilibrium, applying Design Approach 1 – Combination 2 of Eurocode 7.

	Horizontal thrust					
Iteration	$P_{a1}(h)$ (kN/m)	$P_{a1}^q(h)$ (kN/m)	$P_{a2}(h)$ (kN/m)	$P_{a2}^1(h)$ (kN/m)	$P_{a2}^q(h)$ (kN/m)	$P_{p2}(h)$ (kN/m)
1	99.2	27.2	$1.34d^2$	24.9d	3.4d	$27.7d^2$

	Results					
Iteration	d (m)	R_p (kN/m)	$d+h$ (m)	W_w (kN/m)	$\Sigma P_a(v)+W$ (kN/m)	$P_p(v)$ (kN/m)
1	**2.1**	67.8	**7.1**	62.8	118.3	46.9

a.2) The design will follow now the Design Approach 2 of Eurocode 7.
Tables ER8.7f and ER8.7g and Figure ER8.7e summarize the calculations.
It can be concluded that for Design Approach 2 (see result in bold):

$d = 2.1$ m

and

$h + d = 7.1$ m.

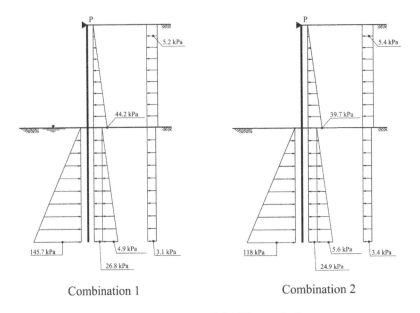

Combination 1   Combination 2

Figure ER8.7d Pressures on the wall - Design Approach I of Eurocode 7.

The comparison of the results from Design Approaches 1 and 2 corroborates the conclusions extracted from the charts of Figure 8.26.

b) Figure ER8.7f summarizes the results of internal stresses obtained from the two design approaches applied in a.1) and a.2). For Design Approach 1 it can be observed that the maximum bending moment is provided by Combination 2, whereas the maximum shear force is obtained from Combination 1.

The following concrete and steel parameters will be assumed: C30/37 and S500. The steel reinforcement calculation is presented in the last column. It can be seen that it is very similar for the two Design Approaches.

Table ER8.7f Safety factors and data for Design Approach 2.

Iteration	Limit state	Soil	γ_G	γ_Q	$\gamma_{\phi'}$	γ_R	ϕ'_k	ϕ'_d	δ	K
I	Active	I	1.35	1.5	1.0	1.4	25	25	$\frac{2}{3}\phi'_d$	0.36
		2					37	37	$\frac{2}{3}\phi'_d$	0.23
	Passive	2	1.0				37	37	$\frac{2}{3}\phi'_d$	9.3

Table ER8.7g Horizontal thrusts and results of the horizontal and vertical equilibrium, applying Design Approach 2 of Eurocode 7.

	Horizontal thrust					
Iteration	$P_{a1}(h)$ (kN/m)	$P^q_{a1}(h)$ (kN/m)	$P_{a2}(h)$ (kN/m)	$P^I_{a2}(h)$ (kN/m)	$P^q_{a2}(h)$ (kN/m)	$P_{p2}(h)$ (kN/m)
I	110.6	25.9	$1.44d^2$	$26.8d$	$3.14d$	$30.8d^2$

	Results					
Iteration	d (m)	R_p (kN/m)	$d+h$ (m)	W_w (kN/m)	$\Sigma P_a(v)+W$ (kN/m)	$P_p(v)$ (kN/m)
I	2.1	69.89	7.1	62.77	135.4	62.35

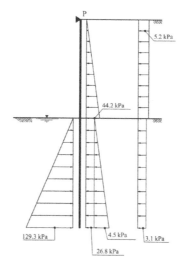

Figure ER8.7e Pressures on the wall - Design Approach 2.

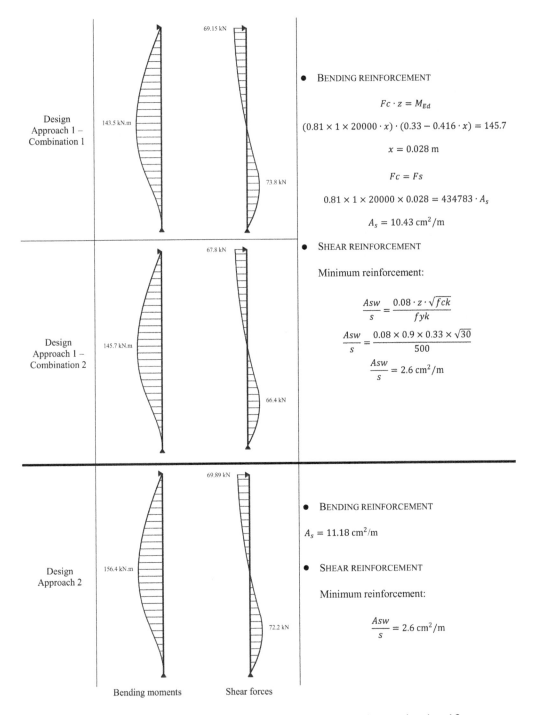

Figure ER8.7f Bending moment and shear force distribution from Design Approaches I and 2.

CHAPTER 9

ER9.1

After excavation, the thickness of the residual soil becomes $9-d$. Consider an elemental prism bounded by vertical faces, by the ground surface, and by a plane parallel to the ground surface at a depth of the contact with the underlying rock mass (see Section 9.2). The height of the prism is $9-d$, with unit transversal and longitudinal base dimensions, which results in the forces and stresses coinciding at the prism base.

The volume of such a prism is equal to:

$$V = (9-d)\cos\beta$$

As the water level coincides with the ground surface, the submerged weight of the block is given by:

$$W' = \gamma'(9-d)\cos\beta$$

For seepage parallel to the ground surface, the hydraulic gradient is equal to (see Figure 9.9):

$$i = \sin\beta$$

and the seepage force (per unit volume) is equal to:

$$j = i\gamma_w = \gamma_w \sin\beta$$

The resultant of the seepage forces acting on the block is expressed by:

$$J = Vj = \gamma_w(9-d)\sin\beta\cos\beta$$

This force has the same direction of the flow, so it is parallel to the ground surface. It is now possible to calculate the stresses at the base of the block:

$$\sigma' = W'\cos\beta = \gamma'(9-d)\cos^2\beta$$

$$\tau = W'\sin\beta + J = \gamma'(9-d)\sin\beta\cos\beta + \gamma_w(9-d)\sin\beta\cos\beta$$

$$= \gamma(9-d)\sin\beta\cos\beta$$

The safety factor is defined as:

$$F = \frac{c' + \sigma'\tan\phi'}{\tau}$$

Therefore:

$$F = \frac{c' + \left[\gamma'(9-d)\cos^2\beta\right]\tan\phi'}{\gamma(9-d)\sin\beta\cos\beta}$$

Introducing $F \geq 1.5$ leads to $d \geq 7$ m.

ER9.2

a) The strength of the weaker material (i.e., marl) will be the determining factor for sliding. There are two hypotheses for the location of the sliding surface: i) at $z = 3.0$ m, i.e., along the top surface of the marl layer; or ii) at $z = 3.3$ m, i.e., along its bottom surface.

   The second hypothesis is the most unfavourable because, in this type of situation, the safety factor decreases with depth (see Section 9.2.3).

   Since the case can be treated as an infinite slope, a square prism will again be considered (as in ER9.1) the base of unit area of which is at a depth of 3.3 m.

   The weight of the prism can be calculated as follows (where 1 and 2 refer to limestone and marl, respectively):

$$W = \left(\gamma_1 z_1 + \gamma_2 z_2\right)\cos\beta = \left(23 \times 3 + 21.5 \times 0.30\right)\cos 35 = 61.8 \text{ kN/m}$$

   The stresses at the base are equal to:

$$\sigma' = W \cos\beta = 50.6 \text{ kPa}$$

$$\tau = W \sin\beta = 35.4 \text{ kPa}$$

   Now, the available shear strength at the base of the block can be calculated:

$$\tau_f = c_d' + \sigma' \tan\phi_d' = 10 + 50.6 \times \tan 20 = 28.4 \text{ kPa}$$

   So, it may be concluded that the soil mass will very probably experience sliding as $\tau > \tau_f$.

b) The stresses at the base of the prism maintain their values.

   Applying Eurocode 7, the design values of the shear strength parameters are (see Section 3.6.5):

$$c_d' = \frac{c_k'}{\gamma_{c'}} = \frac{10}{1.25} = 8 \text{ kPa}$$

$$\tan\phi_d' = \frac{\tan\phi_k'}{\gamma_{\phi k}} = \frac{\tan 20}{1.25} = 0.29 \equiv \phi_d' = 16.2°$$

   The area of influence of each nail is a^2. Therefore, the nails will make a contribution to the shear strength at the sliding surface of $R_d = 75/a^2$ kPa.

   This shear strength is now calculated, depending on the value of a:

$$\tau_f = c_d' + \sigma' \tan\phi_d' + R_d \geq \tau$$

   or

$$\tau_f = 8 + 50.6 \times \tan 16.2 + 75/a^2 \geq 35.4 \text{ kPa}$$

   From this expression, it may be concluded that:

$$a \leq 2.4 \text{ m}$$

ER9.3

a) Since the case can be treated as an infinite slope, this may be represented by a prismatic volume, bounded by vertical faces, by the ground surface and by the base parallel to the ground surface (at a depth, z, of 5.0 m). If the area of the base is assumed with unit value, forces and stresses have the same value.

The weight of the prism may be calculated as follows:

$$W = \gamma z \cos \beta$$

The anchor force will affect both normal and shear stresses at the sliding surface:

$$\sigma' = W \cos \beta + 200 \cos 20/a^2$$

$$\tau = W \sin \beta - 200 \sin 20/a^2$$

It is necessary to satisfy the condition:

$$F = \frac{c' + \sigma' \tan \phi'}{\tau} \geq 1.5$$

or

$$F = \frac{c' + \left(W \cos \beta + 200 \cos 20/a^2 \right) \tan \phi'}{W \sin \beta - 200 \sin 20/a^2} \geq 1.5$$

Note that the strength parameters in the expression above must correspond to the weaker material, i.e., the clayey sandstone.

From this expression, it may be concluded that:

$$a \leq 2.96 \text{ m}$$

b) In seismic conditions, a horizontal inertial force, $k_h W$, is applied to the block with $k_h = 0.12$, as well as a vertical inertial force, $k_v W$, with $k_v = \pm 0.06$. The horizontal force points towards the left and the vertical force should be considered both in the upward and downward directions (see 9.2.5).

The expressions of the stresses at the base of the block now become:

$$\sigma' = \left(1 \pm k_v \right) W \cos \beta - k_h W \sin \beta + 200 \cos 20/a^2$$

$$\tau = \left(1 \pm k_v \right) W \sin \beta + k_h W \cos \beta - 200 \sin 20/a^2$$

It is necessary to satisfy the condition:

$$F = \frac{c' + \sigma' \tan \phi'}{\tau} \geq 1.15$$

Substituting above the expressions of the normal and shear stresses, the following expression is obtained:

$$\frac{c' + \left[(1 \pm k_v)W\cos\beta - k_hW\sin\beta + 200\cos 20/a^2\right]\tan\phi'}{(1 \pm k_v)W\sin\beta + k_hW\cos\beta - 200\sin 20/a^2} \geq 1.15$$

From this expression, it may be concluded that for $k_v = +0.06$

$a \leq 2.82$ m

and for $k_v = -0.06$

$a \leq 2.93$ m.

In conclusion, the design should consider $a = 2.8$ m.

ER9.4

a) Figure ER.9.4 shows the two blocks and the forces applied.
The weight of block 1 is:

$$W_1 = \frac{1}{2} \times 60 \times 50 \times 24 = 36000 \text{ kN/m}$$

At the interface between block 1 and the fault, the normal and tangential forces are related by:

$$T_1 = N_1 \frac{\tan\phi'(\text{Mylonite})}{F} = N_1 \frac{\tan 30}{F}$$

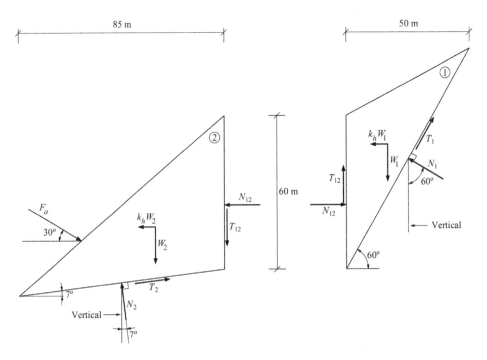

Figure ER9.4 Forces applied to blocks 1 and 2.

At the interface between blocks 1 and 2, the normal and tangential forces are related by:

$$T_{12} = 60 \times \frac{c'(\text{Limestone})}{F} + N_{12}\frac{\tan\phi'(\text{Limestone})}{F} = 60 \times \frac{100}{F} + N_{12}\frac{\tan 40}{F}$$

The equilibrium equations for block 1 are:

$$\begin{cases} N_{12} = N_1\sin 60 - T_1\cos 60 = N_1\sin 60 - N_1\dfrac{\tan 30}{F}\cos 60 \\[3mm] T_{12} = W_1 - N_1\cos 60 - T_1\sin 60 = W_1 - N_1\cos 60 - N_1\dfrac{\tan 30}{F}\sin 60 \end{cases}$$

The weight of block 2 is:

$$W_2 = \tfrac{1}{2} \times 60 \times 85 \times 24 = 61200 \text{ kN/m}$$

At the interface between block 2 and the marl, the normal and tangential forces are related by:

$$T_2 = N_2\frac{\tan\phi'(\text{Marl})}{F} = N_2\frac{\tan 22}{F}$$

The equilibrium equations for block 2 are:

$$\begin{cases} N_{12} = -N_2\sin 7 + T_2\cos 7 = -N_2\sin 7 + N_2\dfrac{\tan 22}{F}\cos 7 \\[3mm] T_{12} = -W_2 + N_2\cos 7 + T_2\sin 7 = -W_2 + N_2\cos 7 + N_2\dfrac{\tan 22}{F}\sin 7 \end{cases}$$

A system of four equilibrium equations has been obtained relating four unknowns: N_1, N_{12}, N_2 and F.

Using, for example, a spreadsheet, the unknown values are as indicated in Table ER9.4 (with the values of tangential forces also provided for verification).

b) The sequence of calculations is similar. However, inertial horizontal forces, $k_h W_1$ and $k_h W_2$, must be included in the equilibrium equations:

$$\begin{cases} N_{12} = k_h W_1 + N_1\sin 60 - T_1\cos 60 = k_h W_1 + N_1\sin 60 - N_1\dfrac{\tan 30}{F}\cos 60 \\[3mm] T_{12} = W_1 - N_1\cos 60 - T_1\sin 60 = W_1 - N_1\cos 60 - N_1\dfrac{\tan 30}{F}\sin 60 \end{cases}$$

Table ER9.4 Results of the equilibrium equations for static conditions.

F	N_1 kN/m	T_1 kN/m	N_{12} kN/m	T_{12} kN/m	N_2 kN/m	T_2 kN/m
1.241	23 412	10 892	14 830	14 861	73 687	23 989

and

$$\begin{cases} N_{12} = -k_h W_2 - N_2 \sin 7 + T_2 \cos 7 = -k_h W_2 - N_2 \sin 7 + N_2 \dfrac{\tan 22}{F} \cos 7 \\[3mm] T_{12} = -W_2 + N_2 \cos 7 + T_2 \sin 7 = -W_2 + N_2 \cos 7 + N_2 \dfrac{\tan 22}{F} \sin 7 \end{cases}$$

The resolution of this new system of equations leads to:

$$F = 1.022$$

c) The sequence of calculations is again similar. In relation to the previous system of equations, the anchor force F_a intervenes in the equilibrium equations for block 2. Now, as the safety factor is imposed ($F = 1.15$), the system of four equilibrium equations relates these four unknowns, N_1, N_{12}, N_2 and F_a.

The new equations for block 1 are:

$$\begin{cases} N_{12} = k_h W_1 + N_1 \sin 60 - T_1 \cos 60 = k_h W_1 + N_1 \sin 60 - N_1 \dfrac{\tan 30}{1.15} \cos 60 \\[3mm] T_{12} = W_1 - N_1 \cos 60 - T_1 \sin 60 = W_1 - N_1 \cos 60 - N_1 \dfrac{\tan 30}{1.15} \sin 60 \end{cases}$$

The new equations for block 2 are:

$$\begin{cases} N_{12} = F_a \cos 30 - k_h W_2 - N_2 \sin 7 + T_2 \cos 7 \\[3mm] \qquad = F_a \cos 30 - k_h W_2 - N_2 \sin 7 + N_2 \dfrac{\tan 22}{1.15} \cos 7 \\[3mm] T_{12} = -F_a \sin 30 - W_2 + N_2 \cos 7 + T_2 \sin 7 \\[3mm] \qquad = -F_a \sin 30 - W_2 + N_2 \cos 7 + N_2 \dfrac{\tan 22}{1.15} \sin 7 \end{cases}$$

The resolution of this system of equations leads to:

$$F_a = 5236 \text{ kN/m}$$

ER9.5

Figures ER9.5a to ER9.5c show the results of the application of the Morgenstern and Price method to the stability analysis of the natural slope represented in Figure E9.5. The results have been obtained with the code SLIDE (Rocscience Inc., 2018).

The results obtained suggest some comments:

i) the thin layer of plastic clay controls the position of the critical slip surfaces;
ii) therefore, in slopes of complex geology, the critical slip surfaces are not circular;

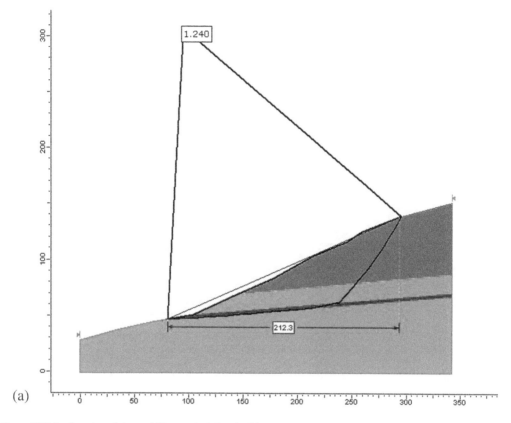

(a)

Figure ER9.5a Results of the stability analysis by the Morgenstern and Price method, without consideration of the water level (dimensions in m).

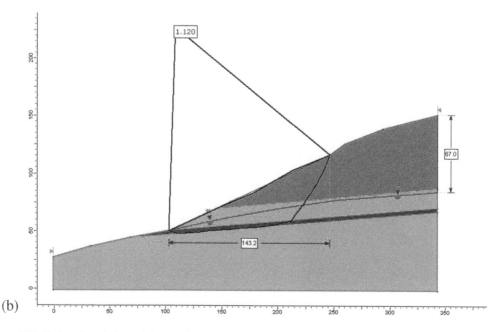

(b)

Figure ER9.5b Results of the stability analysis by the Morgenstern and Price method, assuming the water level is in position 1, as indicated in Figure E9.5 (dimensions in m).

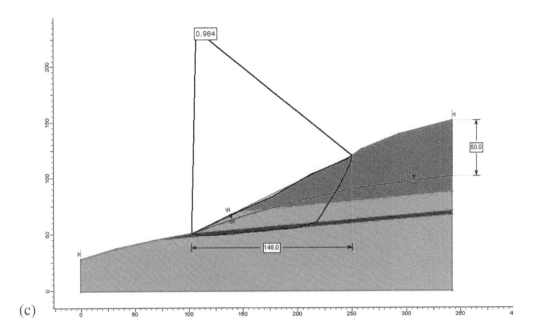

(c)

Figure ER9.5c Results of the stability analysis by the Morgenstern and Price method, assuming the water level is in position 2, as indicated in Figure E9.5 (dimensions in m).

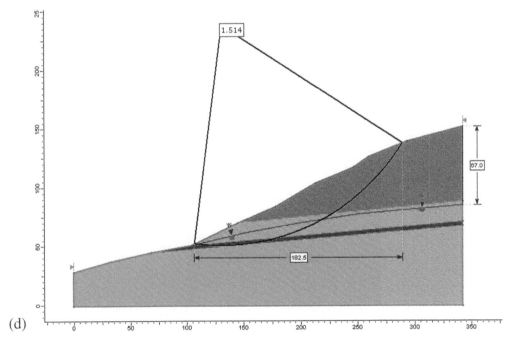

(d)

Figure ER9.5d Results by the simplified Bishop method, assuming the water level is in position 1, as indicated in Figure E9.5 (dimensions in m).

iii) it is of the utmost importance in such cases to use a method like that of Morgenstern and Price, that is prepared to consider complex geometries of the slip surface;

iv) the software used was capable of capturing the main failure mechanism, through the search of a very large number of slip surfaces;

v) the presence of a water table in the slope has a very substantial influence on the value of the safety factor; this explains why the instabilities of natural slopes are very often linked to periods of very intense rainfall, that leads to the water table being closer the ground surface;

vi) in spite of the geometry of the sliding mass limited by the critical slip surface being similar for the three situations analyzed, its actual size is quite different.

Figure ER9.5d shows the results of the stability analysis (with the water table in position 1) by the simplified Bishop method, which assumes circular slip surfaces. The comparison between Figures ER9.5b and ER9.5d is illustrative of the limitations of slice methods assuming circular slip surfaces in the stability analysis of natural slopes with complex geology.

Symbols

ENGLISH

Symbol	Represents	Reference
A	Area	
	pore pressure parameter	Ch. 2
	accidental action	Ch. 3
A_{col}	area of horizontal section of stone column	Eq. 2.37
A_E	seismic action	
A_{ef}	effective area of a shallow foundation	Eq. 6.21
A_N	area of section of CPTu tip inside the porous ring	Fig. 1.16
A_{soil}	area of soil involving a stone column in a square layout	Eq. 2.37, Fig. 2.11
A_T	area of maximum section of CPTu tip	Fig. 1.16
A_t	cross sectional area of anchor tendon	Eq. 8.22
AR	area ratio of a tube sampler	Eq. 1.6
a	geometrical data	Ch. 3
a_g	design ground acceleration on type A ground	Ch. 3
a_{gR}	reference peak ground acceleration on type A ground	Ch. 3
a_{max}	peak surface acceleration	Eq. 6.57
a_s	are replacement ratio applicable to stone columns	Eq. 2.37
a_v	coefficient of compressibility	Eq. 4.3
a_{vg}	design ground acceleration in the vertical direction	Ch. 3
B	width or diameter of loaded area or foundation	
B'	effective width of a shallow foundation	Eq. 6.19
B_B	block width	Eq. 7.83
B_f	footing width	Fig. 1.35
B_p	width of the plate in PLT test	Fig. 1.35
B_q	pore pressure ratio in CPTu test	Eq. 1.19
B_0	reference width of a shallow foundation	Eq. A6.3.3
B	width of wick drain	Eq. 4.49, An. A6.2
	half-width of loaded area	
b_σ, b_q, b_γ	corrective factors of bearing capacity of a shallow foundation to take into account the inclination of the foundation base	Tab. 6.2
C	pile perimeter	Eq. 7.4
	compactability	Eq. 10.3
		(*Continued*)

Symbol	Represents	Reference
C_c	compression index	Fig. 4.7
C_D	coefficient to correct N_{SPT} to take into account borehole diameter	Tab. 1.5
C_E	coefficient to correct N_{SPT} to take into account the energy delivered into the rod stem	Eq. 1.11
C_N	coefficient to correct N_{SPT} to take into account the effective vertical stress	Eq. 1.13
C_R	coefficient to correct N_{SPT} to take into account the rod length	Tab. 1.4
C_r	recompression index	Fig. 4.7
C_s	swelling index	Fig. 4.7
	coefficient to take into account the effect of the foundation embedment in Schmertmann's method	Eq. 6.46
C_α	coefficient of secondary consolidation	Eq. 4.33
CRR	cyclic resistance ratio	Ch. 6
CSR	cyclic stress ratio	Eq. 6.61
c	cohesion in total stresses	
	velocity of the light	Eq. 1.5
	number of compacted layers in a Proctor test	Eq. 10.6
c'	effective cohesion	
c_{fv}	undrained shear strength measured by field vane test	Eq. 1.28
c_h	coefficient of radial or horizontal consolidation	Eq. 4.41
c'_r	residual effective cohesion	Ch. 9
c_u	undrained shear strength	
c_{uc}	undrained shear strength of the foundation soil of an embankment in contact with geosynthetic	An. A2.3
$c_{u,calc}$	undrained shear strength at the surface of the desiccated crust to be taken in the stability calculations	Fig. 2.8
c_{um}	undrained shear strength measured at the base of the desiccated crust	Fig. 2.8
c_{ur}	undrained shear strength of a remolded soil	Eq. 1.30
c_{u0}	undrained shear strength measured at the surface of the desiccated crust	Fig. 2.8
c_{u1}	undrained shear strength along the block lateral surface	Eq. 7.83
c_{u2}	undrained shear strength at the block base	Eq. 7.83
c_v	coefficient of vertical consolidation	Eq. 4.20
D	external diameter of CPT tip	Fig. 1.16
	diameter of the cylindrical failure zone in field vane test	Fig. 1.30
	diameter of borehole in a permeability test	Fig. 1.46
	distance from the base of a cut to a rigid boundary	Fig. 2.19
	depth of the base of a shallow foundation	Ch. 6
	pile length	Ch. 7
D_{aq}	thickness of a confined aquifer in a pumping test	Fig. 1.44
D_B	depth of pile tip	Eq. 7.21
D_e	equivalent depth of foundation	Eq. A6.1.3
D_{ef}	pile effective penetration depth	Eq. 7.32
D_f	depth of bedrock under a shallow foundation	Tab. 6.2
D_r	relative density	
D_{50}	diameter corresponding to 50% of passed material in a grain size distribution curve	
d	displacement	

(Continued)

Symbol	Represents	Reference
	thickness of block or layer	
	distance between source and receiver in refraction method	Fig. 1.2
	diameter of CPTu behind the cone tip inside the porous ring	Fig. 1.16
	diameter of piezometer pipe	Fig. 1.46
	maximum depth reached by shear wedges under a shallow foundation in a bearing capacity failure	Fig. 6.3
	wall embedment depth	Ch. 8
d'	depth of the rotation point below the bottom of excavation in an embedded cantilever wall	Fig. 8.21
d_r	allowable horizontal permanent displacement of a retaining wall under seismic loading	Tab. A8.1.2
d_t	tangential displacement at soil-grout interface	Fig. 8.35
d_v	vertical displacement	Eq. 10.25
E	Young's modulus or deformation modulus	
	energy delivered into the rod stem per blow in SPT test	Ch. 1
	effect of action	Ch. 3
E'	deformation modulus for drained loading	
E_0	deformation modulus corresponding to G_0	Eq. 1.34
E_{50}	deformation modulus for $SL = 50\%$	Fig. 1.47
E_c	value of E_{PMT} from the base of the foundation to a depth equal to $B/2$, for calculating $s_{c,PMT}$	Eq. A6.3.2
	compaction effort per unit of volume	Eq. 10.6
E_d	value of E_{PMT} until a depth equal to $8B$ below the base of the foundation for calculating $s_{d,PMT}$	Eq. A6.3.3
E_g	pile group efficiency factor	Eq. 7.82
E_i	initial tangent deformation modulus	Fig. 1.47
E_l, E_r	normal component of the interaction force at the left and right faces of a soil slice	Fig. 2.4
E'_l, E'_r	resultant of effective normal stresses at the left and right faces of a soil slice	Fig. 2.4
E_{max}	elastic modulus (for small strains)	Fig. 1.47
E_{oed}	oedometer or constrained modulus	Tab. 1.13
E_{PMT}	pressuremeter modulus of PMT	Eq. 1.47
E_p	potential energy of the hammer before falling in SPT	Ch. 1
E_R	rod energy ratio in SPT	Eq. 1.9
E_{sec}	secant deformation modulus	Fig. 1.47
$E_{s,DP}$	effort per blow in DP test	Eq. 1.26
E_t	elastic modulus of an anchor tendon	Eq. 8.22
E_{tan}	tangent deformation modulus	Fig. 1.47
$E_{ULS;d}$	ULS design value of the force to be resisted by an anchor	Eq. 8.24
E_u	deformation modulus for undrained loading	
E_{ur}	deformation modulus for an unloading-reloading cycle	Fig. 1.47
E_{V1}	deformation modulus of the 1st loading cycle in PLT	Ch. 1
E_{V2}	deformation modulus of the 2nd loading cycle in PLT	Ch. 1
$E_{\varepsilon h=0}$	same meaning as E_{oed}	Eq. 4.16
EI	pile bending stiffness	Eq. 7.72
ESP	effective stress path	

(Continued)

Symbol	Represents	Reference
e	void ratio	
	eccentricity of the normal load on the base of a foundation	
e_{max}	void ratio of granular soil in least dense condition	Ch. 10
e_{min}	void ratio of granular soil in densest condition	Ch. 10
e_0	initial void ratio	Ch. 4
F	safety factor	
	force	
	action	
	shape factor in a permeability test	Ch. 1
	function intervening in radial consolidation	Eq. 4.48, 4.52
F_a	anchor force	Ch. 8
F_b	factor of safety applied to pile base resistance	Eq. 7.94
$F_{c;d}$	design pile axial compression force	Eq. 7.96
F_d	safety factor for embedded walls factoring the minimum embedment depth	Ch. 8
F_f	safety factor satisfying force equilibrium in the Spencer method	Fig. 2.6
F_i	inertial force	Eq. 6.57
F_L	safety factor to liquefaction	Eq. 6.63
F_m	safety factor satisfying moment equilibrium in the Spencer method	Fig. 2.6
F_p	safety factor for embedded walls factoring the passive resistance	Ch. 8
F_R	normalized friction ratio in CPT	Eq. 1.18
F_r	safety factor for embedded walls factoring the net passive resistance	Ch. 8
F_{SERV}	value of the anchor force required to prevent a SLS in the supported structure	Eq. 8.24
F_s	safety factor for embedded walls factoring soil shear strength parameters	Ch. 8
	spring force in consolidation hydromechanical analogy	Fig. 4.3, 4.16
	factor of safety applied to pile shaft resistance	Eq. 7.94
F_t	pile axial tensile force	Eq. 7.97
F_{ULS}	value of the anchor force required to prevent ULS in the supported structure	Eq. 8.24
F_v	vertical dynamic force	Eq. 10.25
$F(x)$	cumulative probability density function	Ch. 3
FC	% of fines	Ch. 6
FS_B	safety factor for global stability of an anchored wall, according to the Broms method	Eq. 8.43
FS_K	safety factor for global stability of an anchored wall, according to the Kranz method	Eq. 8.42
f	maximum lateral distance reached by shear wedges in a bearing capacity failure of shallow foundation	Fig. 6.3
$f_\sigma\, f_\varphi\, f_\gamma$	corrective factors of the bearing capacity of a shallow foundation to take into account the effect of a firm stratum underlying the foundation soil	Tab. 6.2
f_s	sleeve friction in CPT	Fig. 1.16
	function for the calculation of the elastic settlement of a shallow foundation	Eq. 6.37
f_{soil}	function used to compute the unit pile shaft resistance	Eq. 7.33, Fig. 7.12

(Continued)

Symbol	Represents	Reference
$f_{t0.1k}, f_{t0.2k}$	characteristic yield stress corresponding to 0.1 and 0.2 % yield strain of anchor tendon	Ch. 8
$f(x)$	probability density function of a random variable	Ch. 3
	shape function in the method of Morgenstern and Price	Ch. 9
$f(z)$	unit weight curve according to the method of Hilf	Fig. 10.25
$f_1(z)$	converted unit weight curve according to the method of Hilf	Fig. 10.25
G	shear modulus	
	permanent action	Ch. 3
G_{max}	elastic shear modulus (for small strains)	Ch. 1
G_{PMT}	shear modulus obtained directly from PMT	Eq. 1.46
G_s	density of the solid particles	
G_0	elastic shear modulus (for small strains)	Ch. 1
g	gravity acceleration	
g_c, g_q, g_γ	corrective factors of the bearing capacity of a shallow foundation to take into account the effect of the inclination of the ground surface	Tab. 6.2
H	height	
	falling height of hammer in DP test	Tab. 1.9
	height of the cylindrical failure zone in field vane test	Fig. 1.30
	height of water above ground water level in a permeability test	Fig. 1.46
	maximum path of water in a layer under vertical consolidation	Ch. 4
	variable intervening in the theorem of corresponding states	Eq. 5.59
h	hydraulic head	
	thickness of a layer or fill	
	excavation depth	
	falling height of the hammer in Proctor test	Eq. 10.6
h_{cr}	critical depth of a cut	Ch. 2
	critical depth of an infinite slope	Ch. 9
h_M	maximum depth of a vertical cut from the upper bound theorem	Ch. 2
h_m	maximum depth of a vertical cut from the lower bound theorem	Ch. 2
h_s	equivalent soil thickness under a shallow foundation for calculating the elastic settlement	Eq. 6.38
	fraction of the height of a specimen corresponding to the solid phase	Fig. A.4.1
h_v	fraction of the height of a specimen corresponding to the voids	Fig. A.4.1
h_w	difference of elevations between the water level behind and in front of a retaining wall face	
	pressure head	
I_c	soil behavior type index in CPT	Eq. 1.20, Fig. 1.23
I_D	density index	Eq. 10.4, 10.5
I_{DMT}	material index in DMT	Eq. 1.48
I_P	plasticity index	
I_R	rigidity index	Eq. 1.24, Eq. 7.9
I_{RC}	critical rigidity index	Eq. 7.13
I_{RR}	reduced rigidity index	Eq. 7.10
I_s	parameter for calculating elastic settlement of a shallow foundation	Eq. 6.38
I_ε	influence factor of the vertical strain for calculating the settlement of a shallow foundation by Schmertmann's method	Eq. 6.45, 6.47
I_ω	parameter for calculating elastic rotation of a shallow foundation	Eq. 6.42, 6.43

(Continued)

Symbol	Represents	Reference
ICA	inside cutting edge angle	Ch. 1, Fig. 1.9
ICR	inside clearance ratio of a tube sampler	Eq. 1.7
IGM	intermediate geomaterial	Tab. A3.4.1
i	hydraulic gradient	
i_c	critical angle of wave refraction in the seismic refraction method	Fig. 1.2
i_o, i_q, i_γ	corrective factors of the bearing capacity of a shallow foundation to take into account the effect of load inclination	Tab. 6.2
J	resultant of seepage forces over a given volume	Ch. 9
J_c	damping coefficient near pile tip	Eq. 7.68
j	seepage force	
K	lateral pressure coefficient	Ch. 5
	coefficient to compute pile base resistance from SPT tests	Eq. 7.41, Tab. 7.5
	discrete spring stiffness	Eq. 7.71
K_a	active pressure coefficient	Ch. 5
K_{ae}	active pressure coefficient under seismic loading	Ch. 5
K'_{ae}	active pressure coefficient under seismic loading for a highly permeable soil below the water level	Ch. 5
K_{DMT}	horizontal stress index in DMT	Eq. 1.50
K^F	bulk deformation modulus of the fluid of the pores	Eq. 1.35
K_f	effective stress failure envelope in a system s', t	
K_p	passive pressure coefficient	Ch. 5
K_{pe}	passive pressure coefficient under seismic loading	Ch. 5
K_{sp}	coefficient to compute pile base resistance on rock	Eq. 7.56
K_t	horizontal stiffness of an isolated pile	Eq. 7.86
K_0	at-rest pressure coefficient	
K_{0e}	pressure coefficient for a fixed vertical wall under seismic loading	Eq. 5.114
K_0 (NC)	at rest pressure coefficient in a normally consolidated soil	Eq. 5.4
K_0 (OC)	at rest pressure coefficient in an over-consolidated soil	Eq. 5.4
k	coefficient of permeability	
	stiffness	
	soil dielectric constant	Eq. 1.5
	subgrade reaction coefficient	Eq. 7.69
k'	reaction modulus or Winkler coefficient	Eq. 7.70
k_c	penetrometer pile base resistance coefficient	Eq. 7.27
k_h	coefficient of permeability in the horizontal direction	Ch. 5
	horizontal seismic coefficient	
k_i	initial stiffness of an anchor	Ch. 8
k_p	factor of resistance of a shallow foundation in relation to vertical loading from results of PMT	Eq. A6.1.4
	pressuremeter pile base resistance coefficient	Fig. 7.35
k_s	coefficient of permeability in the smear zone around a wick drain	Fig. 4.31
k_{theo}	theoretical stiffness of an anchor	Eq. 8.23
k_{ur}	unload-reload stiffness of an anchor	Ch. 8
k_v	coefficient of permeability in the vertical direction	
	vertical seismic coefficient	Ch. 5
L	longitudinal dimension of a loaded area or foundation	
	length	
	penetration depth in DP test	Tab. 1.9

(Continued)

Symbol	Represents	Reference
	intake length of a borehole in a permeability field test	Fig. 1.46
L'	effective length of a shallow foundation	Eq. 6.20
L_{app}	apparent anchor tendon free length	Eq. 8.38
L_B	block length	Eq. 7.83
L_e	external length of tendon measured from the tendon anchorage in the anchor head to the anchorage point in the stressing jack	Eq. 8.39
L_{fixed}	fixed anchor length	Fig. 8.33
L_{free}	free anchor length	Fig. 8.33
L_{tb}	anchor tendon bond length	Fig. 8.33
L_{tf}	anchor tendon free length	Fig. 8.33
L_w	distance from the cutting base beyond which the position of the ground water level remains unaffected	Fig. 2.20
LBT	lower bound theorem	Ch. 2
l	length	
	distance	
l_e	pile characteristic elastic length	Eq. 7.76
M	moment	
	mass of hammer in DP test	Tab. 1.9
	safety margin	Fig. 3.4, Eq. 3.6
	non-dimensional variable	Eq. 4.27
	coefficient relating E_u and c_u	Eq. 6.51
	magnitude of an earthquake	Ch. 6
$M_i(\alpha)$	variable intervening in the safety factor of the simplified Bishop method	Eq. 2.23, 2.24
M_R	moment of the resisting forces in relation to the center of a potential slip surface	
M_{Rg}	moment of the tensile force in a geosynthetic in relation to the center of a potential slip surface	Eq. 2.39
M_S	moment of the unfavorable forces in relation to the center of a potential slip surface	
M_t	torsional moment in field vane test	Fig. 1.30
M_{tf}	maximum value of M_t	Eq. 1.28
MSF	corrective factor for CRR, dependent on the magnitude of the earthquake	Eq. 6.62
m	stone column stress concentration ratio	Eq. A2.2.1
	non-dimensional integer variable	Eq. 4.27
	exponent in the equation relating K_0 (NC) and K_0 (OC)	Eq. 5.4
	exponent in the expressions of i_q and i_γ	Tab. 6.3
m_B	exponent in the expressions of i_q and i_γ	Tab. 6.3
m_L	exponent in the expressions of i_q and i_γ	Tab. 6.3
m_v	coefficient of volumetric compressibility	Eq. 4.4
m_θ	exponent in the expressions of i_q and i_γ	Tab. 6.3
N	normal component of a force on a given section	
	number of blows for driving SPT sampler 30 cm in the 2nd phase of the test	Ch. 1
	number of blows for driving a given length, L, in DP test	Tab. 1.9
N'	resultant of the effective stresses in a given area	
N_c, N_q, N_γ	vertical bearing capacity factors of a foundation	Ch. 6
N_c', N_q'	modified pile bearing capacity factors	Eq. 7.5

(Continued)

Symbol	Represents	Reference
N_{col}	value of N_{SPT} in the axis of a stone column	Eq. 6.79
N_k	parameter of the correlation between c_u and q_c in CPT	Ch. 1
N_{kt}	parameter of the correlation between c_u and q_t in CPTu	Eq. 1.22
N_{qe}, $N_{\gamma e}$	values of the bearing capacity factor of a shallow foundation for seismic loading	Fig. 6.8
N_{res}	value of N_{SPT} of a soil treated with stone columns	Eq. 6.79
N_s	stability base number of a cut in clay	Eq. 2.49
N_{soil}	value of N_{SPT} of a granular soil in the center of the area between stone columns after the construction of these columns	Eq. 6.79
Ns	pile-bearing capacity factor	Eq. 7.8
N_0	value of N_{SPT} of a granular soil before the construction of stone columns	Fig. 6.30
N_1	value of N_{SPT} of a granular soil between stone columns after the construction of the columns	Fig. 6.30
N_{60}	N_{SPT} corrected for $E_R = 60\%$	Eq. 1.10
$(N_1)_{60}$	N_{60} corrected for an effective vertical stress of 1 atmosphere	Eq. 1.12
NC	normally consolidated soil	
n	porosity	
	ratio of the radius of the influence zone of a vertical drain by the radius of the drain	Eq. 4.44
	minimum number of investigation and/or suitability anchor tests	Ch. 8
	number of blows per layer in a Proctor test	Eq. 10.6
OC	over-consolidated soil	
OCA	outside cutting edge angle	Ch. 1, Fig. 1.9
OCR	over-consolidation ratio	
P	soil thrust	Ch. 5
P_a	active thrust	Ch. 5
P_{ae}	active thrust under seismic loading	Ch. 5
P_C	percentage by mass of oversize particles in a compaction test	Eq. 10.8
P_f	probability of failure	Ch. 3
P_p	force intervening in the calculation of the bearing capacity of a shallow foundation	Fig. 6.3
	passive thrust	Ch. 5
P_{pe}	passive thrust under seismic loading	Ch. 5
P_s	force intervening in the pull-out resistance of geosynthetic	Eq. A2.3.7
P_0	at-rest thrust	Ch. 5
P_{0e}	thrust against a fixed wall under seismic loading	Eq. 5.113
p	pressure	
p_a	atmospheric pressure	
	active pressure	
p_f	creep pressure in PMT	Fig. 1.42
p_l	limit pressure in PMT	Fig. 1.42
p_p	passive pressure	
p_l^*	net limit pressure in PMT	Eq. A6.1.1
p_{le}^*	equivalent net limit pressure in PMT	Eq. A6.1.2
p_{we}	water pressure under seismic loading	Eq. 5.107
p_0	ordinate of the initial point of the linear part of the diagram from PMT	Fig. 1.42

(Continued)

Symbol	Represents	Reference
p_1	pressure on the embedded length of a cantilever wall in the American method	Fig. 8.22
p_2	pressure at the tip of an embedded cantilever wall in the American method	Fig. 8.22
Q	rate of flow	
	variable action	Ch. 3
	force applied to the vertical face of a soil element	Ch. 5
	vertical load on a foundation	Ch. 6
Q_{all}	allowable pile axial load	Eq. 7.93
Q_c	vertical force measured by the cell inside the CPT tip	Fig. 1.16
Q_i	resultant of the interaction forces applied to slice i in the method of Spencer	Fig. 2.5
Q_L	line load	Fig. 8.16
Q_P	point load	Fig. 8.16
Q_t	normalized cone resistance in CPT	Eq. 1.17
Q_{ult}	ultimate vertical load of a shallow foundation	Ch. 6
q	surcharge	
q_a	active stress	Ch. 5
q_b	unit pile base resistance	Eq. 7.3
q_c	cone resistance in CPT	Fig. 1.16
q_{cc}	corrected cone resistance for pile base resistance computation	Eq. 7.28
q_{ce}	equivalent cone resistance for pile base resistance computation	Eq. 7.27
q_{cm}	average cone resistance around pile tip	Eq. 7.31
q_{c1N}	q_c corrected for a vertical effective stress of 1 atmosphere	Eq. 1.21
q_n	negative pile skin friction	Ch. 7
q_p	passive stress	Ch. 5
q_s	unit pile shaft resistance	Eq. 7.4
q_t	total cone resistance in CPTu	Fig. 1.16
q_{t1N}	q_t corrected for a vertical effective stress of 1 atmosphere	Eq. 1.21
q_u	uniaxial compressive strength	
q_{ult}	bearing capacity (pressure) to vertical loading of a foundation	Eq. 6.1
R	resistance	
	reactive force at the base of an elemental volume	
	radius	
	radius of the influence zone of vertical drain	Fig. 4.30
R_b	pile base resistance	Ch. 7
	resistance to vertical loading at the base of an embedded wall	Fig. 8.27
R_c	pile bearing capacity for compressive loading	Eq. 7.1
R_{cg}	pile group bearing capacity	Eq. 7.82
R_d	force applied at the center of rotation of an embedded cantilever wall	Fig. 8.21
R_f	friction ratio in CPT	Eq. 1.16
R_M	factor intervening in the correlation between E_{oed} and E_{DMT}	Eq. 1.54
R_m	measured value of the resistance of an anchor or pile	
R_{OC}	over-consolidation ratio	
R_p	prop force in an embedded wall	Fig. 8.25
R_s	pile shaft resistance	Ch. 7

(Continued)

Symbol	Represents	Reference
R_{SLS}	value of the resistance in Serviceability Limit State	
$R_{SLS,m}$	measured value of the resistance of an anchor or pile in SLS	Ch. 7, 8
R_t	resistance of the structural elements of an anchor	Eq. 8.27
	ratio of the creep settlement of a shallow foundation after 3 years per log time cycle by the immediate settlement	Eq. 6.52
	pile bearing capacity for tensile loading	Eq. 7.2
R_{ULS}	value of the resistance in Ultimate Limit state	
$R_{ULS,m}$	measured value of the resistance of an anchor or pile in ULS	Ch. 7, 8
R_3	ratio of the settlement of a shallow foundation after 3 years by the immediate settlement	Eq. 6.52
RC	relative compaction	Eq. 10.2
RTL	pile total dynamic resistance	Eq. 7.67
r	radius	
	coefficient of correlation	Table 4.3
	space variable	Fig. 5.20
	coefficient of reduction of the seismic horizontal acceleration to take into account the capacity of the retaining wall to accommodate permanent horizontal displacements	Eq. A8.1.1
	density correction factor	Eq. 10.11
r_d	reduction factor of shear stress induced by the horizontal seismic acceleration to take into account soil deformation	Eq. 6.59
r_s	radius of the smear zone around a wick drain	Fig. 4.31
r_w	radius of vertical drain	Fig. 4.30
S	area or surface	
	demand, effect of destabilizing forces, effect of actions	Ch. 2, 3
	area of the base of the cone in DP test	Tab. 1.9
	coefficient of amplification of the seismic accelerations between the substratum and the ground surface	Eq. 3.23
S_r	saturation degree	
S_T	factor of topographic amplification of seismic accelerations	Eq. A9.1.1
S_t	sensitivity of clay	Eq. 1.30
SL	stress level	Eq. 1.66
SLS	serviceability limit state	
s	settlement	
	half-sum of maximum and minimum total principal stresses	Fig. 2.10
	ratio of the radius of the smear zone by the radius of the wick drain	Eq. 4.53
	anchor tendon head displacement	Fig. 8.35
s'	half-sum of maximum and minimum effective principal stresses	Fig. 2.10
s_c	consolidation settlement	Ch. 4, 6
s_e	elastic tendon head displacement	Eq. 8.38
s_σ, s_q, s_γ	corrective factors of the bearing capacity of a shallow foundation to take into account the shape of the foundation	Tab. 6.2
$s_{c,PMT}$	settlement of a shallow foundation associated with volumetric strains until a depth of $B/2$ is evaluated from PMT results	Eq. A6.3.2
s_D	differential settlement	Fig. 6.17
$s_{d,PMT}$	settlement of a shallow foundation associated with shear strains until a depth of $8B$ is evaluated from PMT results	Eq. A6.3.3

(Continued)

Symbol	Represents	Reference
s_f	footing settlement	Fig. 1.35
	pile settlement at failure	Eq. 7.62
s_i	immediate settlement	
s_p	settlement of the plate in PLT	Fig. 1.35
s_{PMT}	settlement of a shallow foundation evaluated from PMT results	Eq. A6.3.1
s_t	settlement of a shallow foundation for t greater than three years	Eq. 6.52
T	tangential component of the force in a given section	
	basic time lag in a permeability test	Eq. 1.59
	time factor in a consolidation process	Eq. 4.22
	tensile force on an anchor tendon	Ch. 8
T^*	modified time factor in a dissipation test with CPTu	Eq. 1.23
T_a	datum load of an anchor load test	Fig. 8.35
T_g	tensile force in a geosynthetic	Fig. 2.12, Eq. 2.39
$T_{g,f}$	tensile resistance of the material of a geosynthetic	An. A.2.3
$T_{g,lim}$	limit tensile force of a geosynthetic	An. A.2.3
$T_{g,p}$	pull-out resistance of a geosynthetic	An. A.2.3
$T_{g,req}$	tensile force in a geosynthetic required to ensure stability	An. A.2.3
T_p	proof load, maximum test load to which an anchor is subjected in a particular test	Ch. 8
T_R	tangential resistance force at the base of a foundation	
T_r	time factor in a radial consolidation process	Eq. 4.45
T_S	tangential load at the base of a foundation	
TSP	total stress path	
T_0	lock-off load, load left in the anchor head at the completion of the stressing operation	Fig. 8.38
$T_\%$	time factor corresponding to a given value of the average consolidation ratio (e.g., T_{90}, 90%)	Ch. 4
t	time variable	
	half-difference of the maximum and minimum principal stresses	Fig. 2.10
	thickness of wick drain	Eq. 4.49
$t_\%$	t for a given % of the average consolidation ratio (e.g., t_{90}, 90%)	Ch. 4
U	resultant of water pressures over a given area	
	consolidation ratio	Ch. 4
$\bar{U}$	average consolidation ratio	Ch. 4
$\bar{U}_r$	average radial consolidation ratio	Ch. 4
U_z	consolidation ratio in vertical consolidation	Eq. 4.29
$\bar{U}_z$	average vertical consolidation ratio	Ch. 4
UBT	upper bound theorem	Ch. 2
ULS	ultimate limit state	
u	pore pressure	
	water pressure	
u_a	air pressure in the pores of an unsaturated soil	Fig. 10.17
u_e	excess pore pressure	Ch. 4
u_{ss}	steady state pore pressure	Fig. 4.4
u_0	at-rest pore pressure	
V	volume	
	vertical force	
	velocity	

(Continued)

Symbol	Represents	Reference
	coefficient of variation of a random variable	Ch. 3
V_c	volume of the pressure cell in PMT	Ch. 1
V_P	velocity of P (compression) waves	Ch. 1
V_R	resistance of a foundation to vertical loading	
V_S	vertical load on a foundation	
V_s	velocity of S (shear) waves	Ch. 1
V_{ult}	ultimate vertical load of a pile foundation	Ch. 7
V_0	initial volume of the cavity in PMT	Ch. 1
v	velocity	
v_c	pressure wave propagation velocity	Eq. 7.64
v_f	abscissa corresponding to p_f in the diagram from PMT	Fig. 1.42
v_l	abscissa corresponding to p_l in the diagram from PMT	Fig. 1.42
v_0	abscissa of the initial point of the linear part of the diagram from PMT	Fig. 1.42
W	total weight	
W'	buoyant weight	
W_e	resultant of the weight of a soil wedge with the inertial forces	Fig. 5.32
W_h	weight of the hammer in Proctor test	Eq. 10.6
W_s	weight of solid particles in a soil specimen	Ch. 10
W_w	weight of retaining wall	Ch. 8
W_0	piston weight in the consolidation hydromechanical analogy	Fig. 4.3, 4.16
WL	water level	
w	water content	
w_C	water content of oversize fraction	Eq. 10.10
w_F	water content of finer fraction	Eq. 10.10
w_{fill}	water content of a compacted fill	Ch. 10
w_L	liquid limit	
w_{opt}	optimum water content	Ch. 10
w_P	plasticity limit	
w_T	water content of a sample containing finer and oversize fractions	Eq. 10.10
X	material property	Ch. 3
X_l, X_r	tangential component of the interaction force at the left and right faces of the soil slice	Fig. 2.4
x, y, z	space variables	
$y'(x)$	line of thrust according to the method of Morgenstern and Price	Ch. 9
Z	ratio of the weight of the added water by the weight of a sample of compacted fill according to the method of Hilf	Eq. 10.13
	depth factor	Eq. 4.21
	pile impedance	Ch. 7
z	elevation	
z_f	depth under the footing base in Schmertmann's method	Fig. 6.14
z_w	depth of groundwater level	
z_0	maximum depth of a tensile surface crack	Ch. 2

GREEK

α	angle defining the inclination of a slip surface in relation to the horizontal	
	outside cutting-edge angle of a tube sampler	Fig. 1.9
	fraction of the external load supported by the water for $t = 0$ in the generalized consolidation hydromechanical analogy	Fig. 4.16
	angle of the surcharge with the perpendicular to the surface	Fig. 5.24
	angular distortion	Fig. 6.17
	factor relating E with q_c	Eq. 6.48
	rheological factor to calculate $s_{c,PMT}$	Eq. A6.3.2
	anchor creep coefficient	Ch. 8
	non-dimensional factor	Eq. 6.65, 6.66
	adhesion factor	Eq. 7.24
	coefficient to compute unit pile base resistance from SPT tests	Eq. 7.41, Tab. 7.5
α_a, α_p	angle with the horizontal of the plane limiting the active and passive wedges	Fig. 5.29, Eq. 5.72
α_{ae}, α_{pe}	angle with the horizontal of the plane limiting the active and passive wedges under seismic loading	Eq. 5.99
α_{ij}	influence factor between pile i and pile j	Eq. 7.87
$\alpha_{pile\text{-}soil}$	coefficient to compute unit pile shaft resistance from field tests	Eq. 7.33
β	angle of ground surface with horizontal	
	inside cutting-edge angle of a tube sampler	Fig. 1.9
	reliability index	Ch. 3
	non-dimensional factor	Eq. 6.65, 6.66
	coefficient to compute unit pile shaft resistance from SPT tests	Eq. 7.42, Table 7.7
β_1	angle of Asaoka construction	Fig. 4.36
γ	unit weight	
	shear strain	
γ'	submerged or buoyant unit weight	
γ_a	partial safety factor for anchors	Ch. 3, 8
γ_b	partial safety factor for pile base resistance	Ch. 3, Eq. 7.100
$\gamma_{c'}$	partial safety factor for effective cohesion	Ch. 3
γ_{col}	unit weight of the material of a stone column	Eq. 2.38
γ_{cu}	partial safety factor for undrained shear strength	Ch. 3
γ_{cyc}	cyclic shear strain	Eq. 6.78
$\gamma_{cy,u}$	partial safety factor for cyclic undrained shear strength	Ch. 3
γ_d	dry unit weight	
$\gamma_{d,fill}$	dry unit weight of a compacted soil in the field	Eq. 10.2
γ_d^F	dry unit weight of the finer fraction of a compacted soil	Eq. 10.8
γ_d^T	dry unit weight of the total material (finer and oversize fractions) of a compacted soil	Eq. 10.7
$\gamma_{d,max}$	maximum dry unit weight of a clean sandy soil for $e = e_{min}$	
	dry unit weight of a compacted soil with a significant fraction of fines for $w = w_{opt}$	Ch. 10
$\gamma_{d,max}^F$	dry unit weight of the finer fraction of a compacted soil for $w = w_{opt}$	Ch. 10

$\gamma^T_{d,max}$	dry unit weight of the total material (finer and oversized fractions) of a compacted soil for $w = w_{opt}$	Ch. 10
$\gamma_{d,min}$	minimum dry unit weight of a clean sandy soil for $e = e_{max}$	
γ_E	partial safety factor for an action or for the action effect	Ch. 3
γ_G	partial safety factor for a permanent action	Ch. 3
γ_g	unit weight of a gabion retaining wall	Ch. 8
γ_I	coefficient of importance of a structure	Tab. 3.14
γ_M	partial safety factor for a soil parameter	Ch. 3
γ_P	load factor for permanent action in LRFD	Tab. 3.9
γ_Q	partial safety factor for a variable action	Ch. 3
γ_{qu}	partial safety factor for unconfined strength	Ch. 3
γ_R	partial safety factor for a resistance	Ch. 3
γ_{SERV}	partial safety factor for anchor design	Eq. 8.25
γ_s	unit weight of soil particle	
	partial safety factor for pile shaft resistance	Ch. 3, Eq. 7.100
γ_{sat}	unit weight of saturated soil	
$\gamma_{s,t}$	partial safety factor for tensile pile resistance	Ch. 3
γ_t	partial safety factor for pile total resistance	Ch. 3, Eq. 7.99
γ_w	unit weight of water	
γ_γ	partial safety factor for unit weight	Ch. 3
$\gamma_{\phi'}$	partial safety factor for angle of shearing resistance	Ch. 3
Δ	change, e.g., $\Delta\sigma$	
Δl	length of the base of a soil slice	Fig. 2.4
ΔQ_s	vertical point load at ground surface	Fig. A6.2.1
ΔW	load on the piston in the consolidation hydromechanical analogy	Fig. 4.3, 4.16
Δq_s	surcharge applied at ground surface	
Δu	excess pore pressure	
Δx	horizontal projection of Δl	
δ	displacement vector	
	angle of rotation of a soil mass limited by a circular slip surface	Fig. 2.16
	angle of friction of a soil–structure interface	Ch. 5, 8
δ_a	angle of the active stress with the perpendicular to the face of the wall	Ch. 5, 8
δ_b	angle of friction of the soil-foundation base interface	Ch. 8
δ_p	angle of the passive stress with the perpendicular to the face of the wall	Ch. 5, 8
ε	linear strain	
	angle of anchor with the horizontal	Fig. 8.27
$\varepsilon_a, \varepsilon_r$	axial and radial strains	Table 1.13
$\varepsilon_v, \varepsilon_h$	vertical and horizontal strains	
ε_{vol}	volumetric strain	
ζ	inclination of the base of a shallow foundation	Tab. 6.2
ζ_{ij}	dimensionless coefficient	Eq. 7.87
η	variable whose sine is a shape function of the method of Morgenstern and Price	Fig. 9.14
	angle of the Lancellota solution for the passive thrust	Eq. 5.79
θ	rotation angle in the field vane test	Fig. 1.30
	angle defining the slip surface	Fig. 2.16

	angle with the vertical of the resultant of soil weight and of inertial forces	Fig. 5.32, 9.11
θ'	angle with the vertical of the resultant of soil weight and of inertial forces for a very permeable submerged soil	Eq. 5.110, Fig. 5.37
λ	damping	Fig. 1.51
	angle defining the slip surface	Fig. 2.16
	angle of the wall face with the vertical	Ch. 5
	scale factor according to the method of Morgenstern and Price	Eq. 9.28
λ_c	shape factor in the calculation of $s_{c,PMT}$	Eq. A6.3.2
λ_d	shape factor in the calculation de $s_{d,PMT}$	Eq. A6.3.3
λ_R	Rayleigh wavelength in SASW method	Fig. 1.3
μ	coefficient for correcting vane test results	Eq. 1.29
μ_{col}	vertical stress concentration factor for stone columns	Eq. A2.2.4
μ_{soil}	vertical stress relief factor for soil between stone columns	Eq. A2.2.3
ν	Poisson ratio	
ν_{dyn}	dynamic Poisson ratio	Eq. 1.33
ν^{SK}	Poisson ratio of the solid skeleton	Eq. 1.35
ν_u	undrained Poisson ratio	Ch. 6
ξ	ratio k_h/k_v	Fig. 4.20
ξ_{cr}	rigidity factor for computing N'_c	Eq. 7.18
ξ_{qr}	rigidity factor for computing N'_q	Eq. 7.12
ξ_1, ξ_2	correlation factors for static pile load tests	Eq. 7.101
ξ_3, ξ_4	correlation factors to derive the pile resistance from ground investigation results	Eq. 7.102
ξ_5, ξ_6	correlation factors to derive the pile resistance from dynamic impact tests	Eq. 7.108
ρ	unit mass	
	soil electrical resistivity	Ch. 1
	inclination of the resultant of the interaction forces applied to a soil slice with the horizontal in the Spencer method	Fig. 2.5
ρ^F	unit mass of the pore fluid	Eq. 1.35
ρ_{RS}	correlation between R and S	Ch. 3
ρ^S	unit mass of soil particle	Eq. 1.35
σ	normal total stress	
	standard deviation	Ch. 3
σ'	effective stress	
σ_a, σ_r	axial and radial total stress in the triaxial cell	Tab. 1.13
σ_{ha}, σ_{hp}	horizontal total active and passive stresses	Ch. 5
$\sigma'_{ha}, \sigma'_{hp}$	horizontal effective active and passive stresses	Ch. 5
σ'_p	maximum past vertical consolidation stress, pre-consolidation effective vertical stress	Ch. 4
σ_v, σ_h	vertical and horizontal total stresses	
σ'_v, σ'_h	vertical and horizontal effective stresses	
σ_{v0}, σ_{h0}	at-rest vertical and horizontal total stresses	
$\sigma'_{v0}, \sigma'_{h0}$	at-rest vertical and horizontal effective stresses	
$\sigma_x, \sigma_y, \sigma_z$	stresses in the direction of axes xx, yy, zz	
$\sigma_1, \sigma_2, \sigma_3$	principal stresses	

τ	shear stress	
τ_{col}	shear stress in a stone column	Fig. 2.11
τ_{cyc}	cyclic shear stress	Eq. 6.60
τ_p	peak strength of soil-grout interface in a ground anchor	Fig. 8.35
τ_r	residual strength of soil-grout interface in a ground anchor	Fig. 8.35
Φ	resistance factor in LRFD	Ch. 3
ϕ	angle of friction or of shearing resistance in total stresses	
ϕ'	angle of friction or of shearing resistance in effective stresses	
ϕ'_{col}	angle of friction of a stone column	Eq. 2.38
ϕ'_{cr}	critical angle of friction	
ϕ'_{cv}	angle of friction at a constant volume	
ϕ'_r	residual angle of friction	Ch. 9
$\phi*$	angle of friction between gabion baskets	Eq. 8.10
ψ	internal pressure in SBPT	Ch. 1
	factor for converting the characteristic value to the representative value of an action	Ch. 3
	angle intervening in the Culmann construction	Fig. 5.26
	angle with the horizontal of the planes that limit the active wedge in a bearing capacity failure of a shallow foundation	Fig. 6.3
ψ'	angle with the horizontal of the planes that limit the passive wedges in a bearing capacity failure of a shallow foundation	Fig. 6.3
ψ_{ij}	angle between load direction and the alignment of piles i and j	Eq. 7.87
ψ_0	factor for combination value of a variable action	Ch. 6
ψ_1	factor for frequent value of a variable action	Ch. 6
ψ_2	factor for quasi-permanent value of a variable action	Ch. 6
χ	factor reducing the strength of a soil interacting with a geosynthetic	Eq. A2.3.7
ω	angle of rotation of a shallow foundation	Fig. 6.12
	angle of the inclinometer axis with the vertical	Fig. A9.2.2

Index

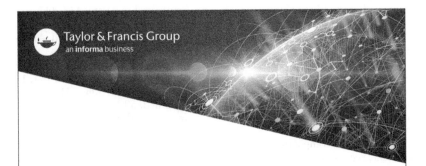

Printed and bound by CPI Group (UK) Ltd, Croydon, CR0 4YY

23/10/2024

01778253-0010